Svalbard (Norway)
RUSSIAN FEDERATION
SWEDEN
FINLAND
ESTONIA
LATVIA
LITHUANIA
BELARUS
POLAND
CZECH REP
SLOVAKIA
UKRAINE
AUSTRIA
MOLDOVA
ROMANIA
BULGARIA
ITALY
GREECE
MALTA
TUNISIA
Black Sea
GEORGIA
ARMENIA
AZERBAIJAN
Caspian Sea
TURKEY
CYPRUS
LEBANON
ISRAEL
SYRIA
IRAQ
JORDAN
IRAN
KUWAIT
Mediterranean Sea
LIBYA
EGYPT
BAHRAIN
QATAR
SAUDI ARABIA
UNITED ARAB EMIRATES
OMAN
YEMEN
Red Sea
KAZAKHSTAN
UZBEKISTAN
TURKMENISTAN
KYRGYZSTAN
TAJIKISTAN
AFGHANISTAN
PAKISTAN
MONGOLIA
CHINA
NORTH KOREA
SOUTH KOREA
JAPAN
Sea of Okhotsk
NEPAL
BHUTAN
BANGLADESH
INDIA
MYANMAR
LAOS
THAILAND
VIETNAM
CAMBODIA
Taiwan (China)
Pacific Ocean
Arabian Sea
Bay of Bengal
South China Sea
PHILIPPINES
PALAU
MICRONESIA
MARSHALL ISLANDS
SRI LANKA
MALDIVES
BRUNEI
MALAYSIA
SINGAPORE
KIRIBATI
NAURU
INDONESIA
PAPUA NEW GUINEA
SOLOMON ISLANDS
TUVALU
EAST TIMOR
VANUATU
FIJI
New Caledonia (France)
AUSTRALIA
NEW ZEALAND
Indian Ocean
NIGER
CHAD
SUDAN
ERITREA
DJIBOUTI
ETHIOPIA
SOMALIA
CAMEROON
CENTRAL AFRICAN REP
SOUTH SUDAN
UGANDA
KENYA
GABON
CONGO
DEMOCRATIC REPUBLIC OF THE CONGO
RWANDA
BURUNDI
TANZANIA
SEYCHELLES
COMOROS
ANGOLA
ZAMBIA
MALAWI
MOZAMBIQUE
MADAGASCAR
MAURITIUS
ZIMBABWE
NAMIBIA
BOTSWANA
SWAZILAND
SOUTH AFRICA
LESOTHO

HUMAN GEOGRAPHY

Ninth Edition

HUMAN GEOGRAPHY

William Norton with Michael Mercier

Oxford University Press is a department of the University of Oxford.
It furthers the University's objective of excellence in research, scholarship,
and education by publishing worldwide. Oxford is a registered trade mark of
Oxford University Press in the UK and in certain other countries.

Published in Canada by
Oxford University Press
8 Sampson Mews, Suite 204,
Don Mills, Ontario M3C 0H5 Canada

www.oupcanada.com

First Edition published in 1992
Second Edition published in 1995
Third Edition published in 1998
Fourth Edition published in 2002
Fifth Edition published in 2004
Sixth Edition published in 2007
Seventh Edition published in 2010
Eight Edition published in 2013

Library and Archives Canada Cataloguing in Publication

Norton, William, 1944-, author
Human geography / William Norton, Michael
Mercier. -- Ninth edition.

Includes bibliographical references and index.
ISBN 978-0-19-901955-7 (hardback)

1. Human geography--Textbooks. I. Mercier, Michael
E. (Michael Ernest), 1970-, author II. Title.

GF41.N67 2016 304.2 C2016-900340-X

Cover image: Michael Wheatley/Getty Images
Introduction: Paul Horsley/All Canada Photos; Chapter 1: All Canada Photos; Chapter 2: Jon Arnold/Getty Images; Chapter 3: Bernhard Lang/Getty Images; Chapter 4: Photo by Wolfgang Kaehler/LightRocket via Getty Images; Chapter 5: Steve Raymer/Getty Images; Chapter 6: Nadeem Khawar/Getty Images; Chapter 7: Claude Robidoux/Getty Images; Chapter 8: Alan Copson/Getty Images; Chapter 9: Photo by Jeff J Mitchell/Getty Images; Chapter 10: Barrett & Mackay/All Canada Photos; Chapter 11: M Swiet Productions/Getty Images; Chapter 12: Chris Cheadle/Getty Images; Chapter 13: Photo by David Boyer/National Geographic/Getty Images; Conclusion: Michael Marquand/Getty Images

Oxford University Press is committed to our environment.
This book is printed on Forest Stewardship Council® certified paper
and comes from responsible sources.

Printed and bound in the United States of America

2 3 4 — 19 18 17

CONTENTS

Figures, Tables, and Boxes x

Preface xv

Ninth Edition: Special Features xv

Features xix

Acknowledgements xxii

Introduction: The Roads Ahead xxiv

Three Recurring Themes xxiv

Why "Human" Geography? xxvii

The Goal of Human Geography xxvii

About This Book xxviii

The Human Geographer at Work xxix

Conclusion xxxii

Summary xxxii

Questions for Critical Thought xxxiii

Suggested Readings xxxiii

1 What Is Human Geography? 2

Preclassical Geography 4

Classical Geography 4

The Fifth to Fifteenth Centuries: Geography in Europe, China, and the Islamic World 6

The Age of European Overseas Movement 9

Geography Rethought 11

Institutionalization: 1874–1903 13

Prelude to the Present: 1903–1970 15

Human Geography Today: Recent Trends and Subdisciplines 17

Conclusion 20

Summary 21

Links To Other Chapters 22

Questions for Critical Thought 23

Suggested Readings 23

2 Studying Human Geography 24

Philosophical Options 26

Human Geographic Concepts 33

Techniques of Analysis 46

Conclusion 53

Summary 54
Links to Other Chapters 55
Questions for Critical Thought 55
Suggested Readings 56

3 Geographies of Globalization 58

Introducing Globalization 60
Geography as a Discipline in Distance 61
Overcoming Distance: Transportation 62
Overcoming Distance: Trade 66
Overcoming Distance: Transnationals 71
Overcoming Distance: Transmitting Information 72
Interpreting, Conceptualizing, and Measuring Globalization 75
The Global Economic System 78
Cultural Globalization 80
Political Globalization 83
Globalization: Good or Bad? 85
Conclusion 90

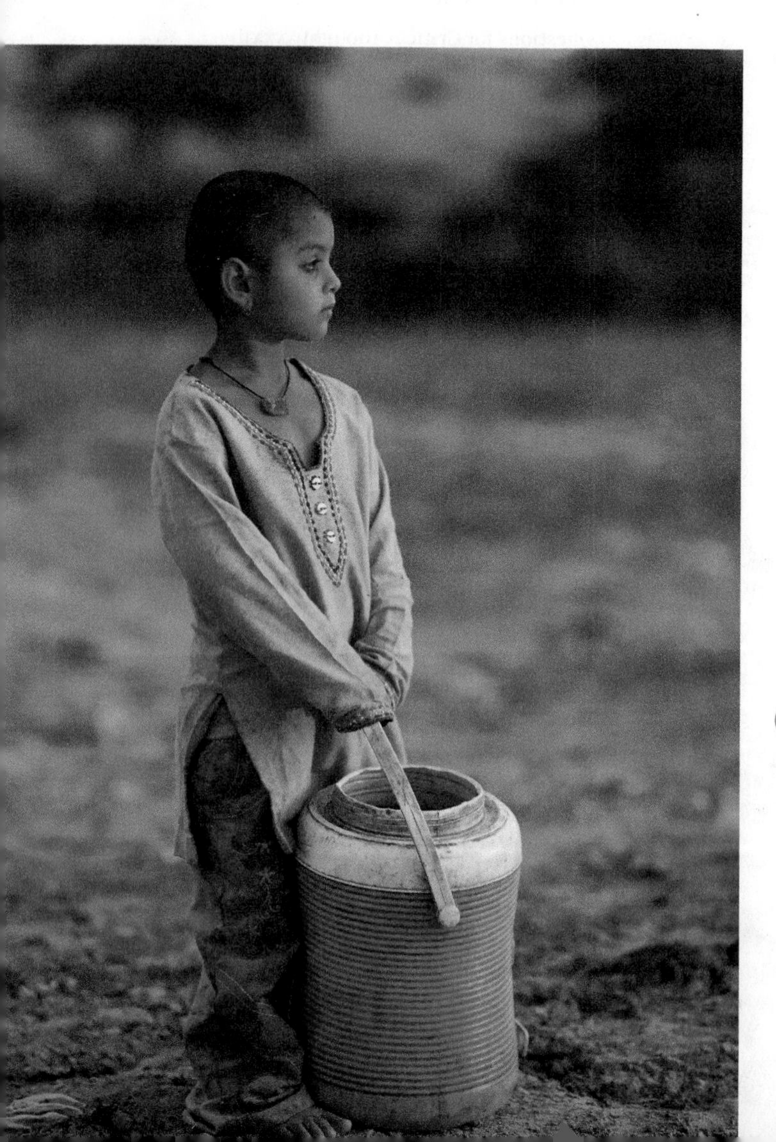

Summary 91
Links to Other Chapters 92
Questions for Critical Thought 92
Suggested Readings 92

4 Humans and Environment 94

A Global Perspective 96
Environmental Ethics 104
Human Impacts on Vegetation 107
Human Impacts on Animals 112
Human Impacts on Land, Soil, Air, and Water 115
Human Impacts on Climate 119
Earth's Vital Signs 124
Sustainability and Sustainable Development 126
Conclusion 129
Summary 130
Links to Other Chapters 131
Questions for Critical Thought 131
Suggested Readings 131

5 Population Geography 132

Fertility 134
Mortality 139
Natural Increase 142
Government Policies 144
The Composition of a Population 146
History of Population Growth 149
Explaining Population Growth 155
Distribution and Density 158
Migration 160
Conclusion 168
Summary 169
Links to Other Chapters 171
Questions for Critical Thought 171
Suggested Readings 171

6 Global Inequalities 172

Explaining Global Inequalities 174
Identifying Global Inequalities 176
Feeding the World 183
Refugees 191
Disasters and Diseases 194
Prospects for Economic Growth 202
Striving for Equality, Fairness, and Social Justice 205
Conclusion 209

Summary 209
Links to Other Chapters 210
Questions for Critical Thought 210
Suggested Readings 211

7 Geographies of Culture and Landscape 212
A World Divided by Culture? 214
Formal Cultural Regions 214
Vernacular Cultural Regions 218
The Making of Cultural Landscapes 222
Language 229
Religion 241
Conclusion 252
Summary 252
Links to Other Chapters 254
Questions for Critical Thought 254
Suggested Readings 254

8 Geographies of Identity and Difference 256
The Cultural Turn 258
The Myth of Race 262
Ethnicity 268
Gender 271
Sexuality 276
Identities and Landscapes 279
Geographies of Well-Being 281
Folk Culture and Popular Culture 285
Tourism 289
Conclusion 297
Summary 297
Links to Other Chapters 299
Questions for Critical Thought 299
Suggested Readings 299

9 Political Geography 300
State Creation 302
Introducing Geopolitics (and *Geopolitik*) 307
Unstable States 310
Groupings of States 321
The Role of the State 323
Elections: Geography Matters 325
The Geography of Peace and War 328
Our Geopolitical Future? 334
Conclusion 338
Summary 339
Links to Other Chapters 340
Questions for Critical Thought 340
Suggested Readings 341

10 Agricultural Geography 342
The Agricultural Location Problem 344
Distance, Land Value, and Land Use 350
Domesticating Plants and Animals 355
The Evolution of World Agricultural Landscapes 357
World Agriculture Today: Types and Regions 366
Global Agricultural Restructuring 370
Food Production, Food Consumption, and Identity 378
Conclusion 382

Summary 382
Links to Other Chapters 384
Questions for Critical Thought 385
Suggested Readings 385

11 An Urban World 386
An Urbanizing World 388
The Origins and Growth of Cities 398
The Location of Cities 405
Urban Systems and Hierarchies 406
Global Cities 412
Conclusion 416
Summary 417
Links to Other Chapters 419
Questions for Critical Thought 419
Suggested Readings 419

12 The City and Urban Form 420
Explaining Urban Form 422
Housing and Neighbourhoods 430
Suburbs and Sprawl 437
Inequality and Poverty 440
Cities as Centres of Production and Consumption 446
Transportation and Communication 447
Planning the City 449
Cities of the Less Developed World 455
Conclusion 460
Summary 461
Links to Other Chapters 463
Questions for Critical Thought 464
Suggested Readings 464

13 Geographies of Energy and Industry 466
The Industrial Location Problem 468
The Industrial Revolution 473
Fossil Fuel Sources of Energy 476
World Industrial Geography 483
Globalization and Industrial Geographies 492
Uneven Development in More Developed Countries 501
Conclusion 503
Summary 504
Links to Other Chapters 506
Questions for Critical Thought 506
Suggested Readings 506

Conclusion: Where Next? 508
Changing Human Geographies 508
A Changing Discipline 512
Being a Human Geographer: Where We Began 515
Summary 516
Suggested Readings 516

Appendix 1: On the Web 517

Online Appendix 2: Global Physical Geography

A Habitable Planet

Global Environments

Suggested Readings

On the Web

Online Appendix 3: The Evolution of Life

Life on Earth

Human Origins

Suggested Readings

On the Web

Glossary 521

References 529

Index 536

FIGURES, TABLES, AND BOXES

FIGURES

1.1 The world according to Eratosthenes 5
1.2 The world according to Ptolemy 6
1.3 An example of a T-O map 7
1.4 An example of a Portolano chart 7
2.1 The scientific method 27
2.2 The site of Winnipeg 35
2.3 The situation of Winnipeg within North America 36
2.4 Clustered, random, and uniform point patterns 37
2.5 A typical distance decay curve 38
2.6 Shortest distance route 39
2.7 A one-way system 39
2.8 Time distance in Edmonton 39
2.9 Toronto in physical space and time space 40
2.10 The impact of spatial scale 41
2.11 Urban centres in Manitoba 41
2.12 A typical S-shaped growth curve 42
2.13 Images of North America in 1763 43
2.14 Mapping at a scale of 1:250,000 47
2.15 Mapping at a scale of 1:50,000 47
2.16 Schematic representation of a dot map 48
2.17 Schematic representation of a choropleth map 48
2.18 Schematic representation of an isopleth map 48
2.19 Two topographic maps of the Love Canal area, Niagara Falls, New York 50
3.1 Diffusion of transport innovations in Britain, 1650–1930 64
3.2 The impact of the Suez and Panama canals on ocean travel distances 65
3.3 Selected economic groupings of countries 69
3.4 Global ICT Developments, 2000-2010 73
3.5 Internet users, 2000-2010 73
3.6 KOF Index of Globalization worldwide 78
3.7 Depth Index of Globalization, world, 2005–2012 78
3.8 World distribution of Depth Index of Globalization, 2013 79
3.9 Territorial interpenetration: The "incorporation" of parts of a state's territory into a transnational 79
3.10 The contemporary geo-economy 81
3.11 World merchandise exports by region, 2012 85
4.1 Chemical cycling and energy flows 97
4.2 Spread of radiation from Chernobyl across Europe, 3 May 1986 102
4.3 Some consequences of human-induced vegetation change 108
4.4 Past and present location of tropical rain forests 110
4.5 Australia's rabbit and dog fences 113
4.6 The global water cycle 117
4.7 The impact of sea-level change on Bangladesh 120
4.8 Global distribution of some major environmental problems 125
5.1 World distribution of crude birth rates, 2014 138
5.2 Death rates and age 140
5.3 World distribution of crude death rates, 2014 141
5.4 World distribution of life expectancy, 2014 141
5.5 World distribution of rates of natural increase, 2014 143
5.6 Age and sex structure in China in the late twentieth century 147
5.7 Age structure of populations 148
5.8 Age and sex structure in Brazil: 1975, 2000, 2025 149
5.9 Proportion of the population aged 60 years or over: World and development regions 1950-2050 150
5.10 Age and sex structure in Canada: 1861, 1921, 1981, 2036 151
5.11 World population growth 155
5.12 The demographic transition model 157
5.13 World population distribution and density 160
5.14 Major world migrations, 1500–1900 166
6.1 Civilizations of the ancient world 174
6.2 The shape of continents 175
6.3 Factors underlying the broadest patterns of history 176
6.4 The world system: Core, semi-periphery, and periphery 177
6.5 More, less, and least developed countries 178
6.6 National economies: income per person, 2013 181
6.7 Global distribution of human development, 2014 183
6.8 Quality of governance among different types of countries 187
6.9 Refugee numbers, 1960–2013 192
6.10 Major source countries of refugees, end of 2013 193
6.11 The Horn of Africa 195
6.12 Countries affected by the Asian tsunami, December 2004 197
6.13 The path followed by Hurricane Katrina, August 2005 198
6.14 Malaria in Africa 199

7.1	Cultural regions of the world	216
7.2	Regions of North America	217
7.3	Europe defined	218
7.4	Cultural regions of the United States	218
7.5	Regions of Canada	219
7.6	American vernacular regions	220
7.7	The nine nations of North America: a journalist's perception	221
7.8	Core, domain, and sphere	225
7.9	Diffusion of cholera, North America, 1832 and 1866	228
7.10	Effects of pre-emption on the adoption curve	230
7.11	World distribution of language families before European expansion	233
7.12	Linguistic densities	234
7.13	Initial diffusion of Indo-European languages	235
7.14	Diffusion of Indo-European languages into England	235
7.15	Four official languages in Switzerland	236
7.16	Flemish, French, and German in Belgium	236
7.17	French and English in North America	238
7.18	Hearth areas and diffusion of four major religions	242
7.19	World distribution of major religions	242
8.1	South African "homelands"	265
8.2	Mental maps	269
8.3	Global distribution of the GII	277
8.4	World distribution of happiness	285
8.5	The tourism system of place construction	294
9.1	The British Empire in the late nineteenth century	304
9.2	Principal elements in the process of exploration	305
9.3	Territorial expansion of the United States	305
9.4	European imperial coverage of the globe	306
9.5	Mackinder's heartland theory	308
9.6	African ethnic regions	313
9.7	African political areas in the sixteenth, eighteenth, and nineteenth centuries	313
9.8	The former Yugoslavia	316
9.9	The former USSR	317
9.10	Some areas of conflict in South Asia	319
9.11	European ethnic regions	322
9.12	The original "gerrymander"	326
9.13	Gerrymandering in Mississippi	327
9.14	Voing and place in South Carolina	328
9.15	Global distribution of freedom, 2014	335
9.16	World civilizations	337
10.1	Relationship between mean annual rainfall and wheat yield in the US, 1909	344
10.2	Crop and livestock combinations along the US–Canada border	346
10.3	Supply and demand curves	347
10.4	Rent-paying abilities of selected land uses	347
10.5	Economic rent lines for three crops and related zones of land use	348
10.6	Agricultural land use in the isolated state	350
10.7	Relaxing a von Thünen assumption	351
10.8	Relaxing two von Thünen assumptions	351
10.9	Scatter graphs, best-fit lines, and *r* values	353
10.10	Agricultural land use in Uruguay	354
10.11	Areas of agricultural domestication and early diffusion	356
10.12	Tillage system trends on the Canadian prairies	366
10.13	World agricultural regions	367
10.14	Change in area devoted to wheat and specialty crops, Canadian prairies	370
10.15	Percentage of labour force in agriculture by country, 2012	373
10.16	The food supply system	374
10.17	Global dietary patterns	378
11.1	World rural and urban populations, 1950–2050	392
11.2	Rural and urban populations, more developed and less developed regions, 1950–2050	392
11.3	Percentage urban population by country, 2014	394
11.4	Rural and urban populations, Canada, 1950–2050	396
11.5	Canada: Census metropolitan areas, 2011	396
11.6	Mega-cities, 2014 and 2030	399
11.7	The six urban hearths	399
11.8	Theoretical hinterlands (or market areas) for central places	407
11.9	A simplified (two-order) central place system	408
11.10	Alpha global cities, 2012	415
11.11	The Global Cities Index, 2014	415
11.12	Global cities and spheres of influence	417
12.1	Urban land values	423
12.2	Three classic models of the internal structure of urban areas	424
12.3	The West European city	426
12.4	Modelling the Latin American city	426
12.5	Modelling the Asian colonial city	428
12.6	Modelling the Southeast Asian city	429
12.7	White's model of the twenty-first-century city	430
12.8	Locating Eight Mile Road	444
12.9	Space–time prism for Ellie, March 1998 to March 2000	445
12.10	Street layout in Charlottetown	450
12.11	Ebenezer Howard's garden city and its agricultural belt	451
12.12	The layout of Wildwood	452
12.13	Economic geographies of Accra and Mumbai during the colonial phase	457
12.14	Economic geographies of Accra and Mumbai during the global phase	458

12.15 Incidence of urban slums as a percentage of urban population, 2009 459
13.1 A locational triangle 470
13.2 A simple isotim map 471
13.3 An isodapane map 471
13.4 Transport cost and distance 472
13.5 Stepped transport costs 472
13.6 Major oil trade movements (million tonnes), 2014 478
13.7 Major world industrial regions 486
13.8 Export-processing zones 489
13.9 Special economic zones in China 490
13.10 Economic growth and employment distribution 495
13.11 The changing structure of world employment 495
13.12 Percentage of labour force in industry by country, 2010 496
13.13 Percentage of labour force in services by country, 2010 497
A2.1 The revolution of the earth around the sun 3
A2.2 Moving continents 4
A2.3 The major ocean currents 5
A2.4 Global cordilleran belts 6
A2.5 Global distribution of soil types 7
A2.6 Global distribution of natural vegetation 8
A2.7 Global distribution of climate 9
A2.8 Generalized global environments 10
A3.1 Probable directions of movement out of Africa by early humans 6

TABLES

I.1 Subdisciplines of human geography and text coverage xxix
I.2 Selected demographic data xxxi
3.1 Revenue data for the top 10 transnationals, 2014, and gross national income for selected countries, 2013 71
3.2 Foreign direct investment inflows by major regions, 2013 (US$ billions) 72
3.3 Three theses about globalization 77
3.4 KOF Index of Globalization: Top 10 and bottom 10 countries, 2014 78
3.5 Globalization theses and economic geography 80
3.6 Globalization theses and cultural geography 83
3.7 Globalization theses and political geography 83
3.8 Globalization theses and development 86
3.9 Globalization theses and environmental issues 87
4.1 Global deforestation: Estimated areas cleared (thousands km^2) 108
5.1 Contraceptive use by region, 2014 137
5.2 Population data, Canada, 2014 139
5.3 Countries with the highest rates of natural increase (3.0 and above), 2014 142
5.4 Countries with the lowest rates of natural increase (less than 0.0), 2014 142
5.5 Projected population growth, 2014–2050 143
5.6 Global aging, 1950–2050 150
5.7 Major epidemics, 1500–1700 153
5.8 Adding the billions: Actual and projected 153
5.9 Estimating how many people have ever lived on earth 154
5.10 World population distribution by major area (percentage): Current and projected 161
5.11 The 10 most populous countries: Current and projected 161
5.12 Population densities of the 10 most populous countries, 2014 161
5.13 Some push–pull factors 163
5.14 Some typical moorings 165
6.1 Extremes of human development, 2014 184
6.2 Population densities, selected countries, 2014 185
6.3 Main origins of refugees, 2013 192
6.4 Main countries of asylum, 2013 194
6.5 Countries with highest levels of adult (ages 15–49) HIV/AIDS prevalence, 2014 200
7.1 Languages with more than 100 million native speakers, 2014 231
7.2 Language families 234
7.3 Major world religions: Number of adherents (thousands) by continental region, 2010 243
8.1 Examples of scales of difference 261
8.2 Examples of scales of inclusions and exclusions 262
8.3 Gender Inequality Index 276
8.4 Characteristic tendencies: Conventional mass tourism vs alternative tourism 293
9.1 Ethnic groups in the former Yugoslavia 316
9.2 Ethnic groups in the former USSR 318
9.3 Global trends in the spread of democracy, 1977–2014 335
10.1 Average distances from London to regions of import derivation (miles) 352
10.2 Comparing tillage strategies 365
10.3 Transnationals and crops in the less developed world 377
11.1 Some definitions of urban centres 389
11.2 Population growth rates (total, urban, rural): 1950–2000, 2000–2015, and 2015–2030 393
11.3 Canada: Total, urban, and rural population (thousands), 1950–2050 395

11.4 Canada: Census metropolitan area populations (thousands), 2011 395
11.5 Cities with more than 10 million people, 2014 and 2030 397
11.6 Change in population in rust belt vs sun belt cities in the United States, 1950–2030 406
11.7 The hierarchical Canadian urban system, 2011 410
11.8 Canada's urban system and the rank-size distribution, 2014 411
11.9 Urban systems of the more and less developed worlds with primate cities, 2014 412
12.1 Selected life-cycle events that can cause residential relocation (chronological) 433
12.2 Factors underlying neighbourhood decline or revitalization 433
13.1 Principal oil-consuming countries, 2013 479
13.2 Top 12 countries in oil production and oil reserves, 2013 478
13.3 Natural gas: Proven reserves, 2013 481
13.4 Natural gas production, 2013 481
13.5 Natural gas consumption, 2013 482
13.6 Coal: Proven reserves, production, and consumption, 2013 482
13.7 Employment by sector, 1990 and 2001, selected transition economies 497
13.8 Information technology and business process outsourcing employment in India (thousands of jobs) 498
13.9 Labour markets: From Fordism to Post-Fordism 501
A3.1 Basic chronology of life on earth 3
A3.2 Basic chronology of human evolution 4

BOXES

Focus on Geographers

Chapter 2 Geographic Literacy (Walter Peace and Michael Mercier) 34
Chapter 2 Participatory Mapping (Jon Corbett) 44
Chapter 3 Regional Integration, Political Independence, and the Canada–US Relationship (Emily Gilbert) 71
Chapter 4 Jobs in the Environment and Sustainability Sector (Dan Shrubsole) 126
Chapter 7 Family Geographies (Bonnie C. Hallman) 223
Chapter 8 Urbanization, Gender, and Everyday Life in Botswana (Alice J. Hovorka) 273
Chapter 8 The Paradox of Polar Bear Ecotourism in Churchill, Manitoba (Charles Greenberg) 291
Chapter 9 Human Geography and Postwar Recovery of Land and Property Rights (John D. Unruh) 329
Chapter 10 Seeking Equity through the Study of Agricultural Geography (Raju J. Das) 375
Chapter 12 Human Geography and the Housing Experiences of Immigrants (Carlos Teixeira) 438
Chapter 13 Historical Industrial Geography: Population, Health, and the Cape Breton Sydney Coalfield (Natalie C. Ludlow) 499

Examining the Issues

1.1 Telling a Story of Geography 5
1.4 Environmental Determinism 16
1.5 Why Learn about the History of Geography? 20
2.1 Positivistic Human Geography 28
2.2 Humanistic Human Geography 28
2.3 Marxist Human Geography 31
2.5 Some Qualitative Resources 52
3.1 The Tyranny of Distance 62
3.2 Explaining Commodity Flows 68
3.6 Capitalism Is Good and Bad 88
4.3 The Tragedy of the Commons or Collective Responsibility? 106
4.6 Climate Change Front and Centre 121
5.6 How Many People Have Ever Lived on Earth? 154
5.7 World Population Density 159
5.8 The Ravenstein Laws 164
6.2 The Idea of Growth through Developmental Stages 182
6.3 A Food Crisis in the Early Twenty-First Century 188
6.4 Emerging Africa: The Hopeful Continent 190
6.5 War, Famine, and Refugees in the Horn of Africa 195
8.2 Constructing Geographic Knowledge 260
8.3 Species and Races 263
8.4 A History of Racism 263
8.6 The Geography of Fear 283
8.8 Consigned to the Shadows 286
8.9 Geography, Consumption, and Identity Formation 288
9.3 Regional Identities and Political Aspirations: The Example of Canada 311
9.6 Creating Electoral Bias 326
10.2 Calculating Economic Rent 348

10.3 Correlation and Regression Analysis 353
10.5 Organic Farming 360
10.8 Canadian Farmers: Fewer and Older 371
10.9 Marxist Political Economy 372
10.10 You Are What, and Where, You Eat
11.3 Application of Central Places to Hockey Associations and Teams in Canada 409
12.1 Social Trends and the Social Geography of the City 431
13.1 Factors Related to Industrial Location 469
13.2 Testing Weberian Theory 472
13.3 The Period of the Industrial Revolution 473
13.5 The World's Biggest Coalmine 483
C.1 Moving Beyond the Introductory Level: Be Prepared to Think Again 513

In the News

1.2 Did China "Discover the World" in 1421? 8
2.4 The Power of Maps 49
3.4 Social Media, Revolutions, and Riots 74
4.2 The Chernobyl Nuclear Accident 102
5.5 Causes and Consequences of Population Aging 152
6.7 Defeating AIDS? 201
7.6 Islamic and Christian Identities 247
8.7 Less Developed Canada? 283
9.2 The Jewish State 309
10.7 For and Against Genetically Modified Crops 363
11.1 Listing the "20 Largest Cities" 390
12.2 Making the Most of Space in Modern Cities 441
12.7 Slum Areas as Gateways to Prosperity 461
13.6 Mongolia—A New Resource Frontier 484

Around the Globe

1.3 The Southern Continent 10
3.3 The European Union 69
3.5 The World Trade Organization 81
4.1 Lessons from Easter Island 98
4.4 Unwanted Guests 114
4.5 A Tale of Three Water Bodies 118
4.7 Clayoquot Sound: The Case for Sustainability 127
5.1 Declining Fertility in the Less Developed World 136
5.2 Declining Fertility in the More Developed World 139
5.3 Fertility in Romania, 1966–1989 144
5.4 Population in China 147
6.1 Less Developed World Case Studies: Ethiopia, Sri Lanka, and Haiti 179
6.6 Flooding in Bangladesh 196
6.8 The Grameen Bank, Bangladesh 204
7.1 Europe as a Cultural Region 217
7.2 The Mormon Landscape 226
7.3 Cholera Diffusion 228
7.4 Linguistic Territorialization in Belgium and Canada 237
7.5 The Celtic Languages 239
7.7 Religious Landscapes: Hutterites and Doukhobors in the Canadian West 251
8.1 Iconographic Analyses of Canadian Landscapes 259
8.5 Immigration and Ethnic Diversity in Canada 270
8.10 Tourism in Sri Lanka 296
9.1 Remnants of Empire 304
9.4 The Plight of the Kurds 312
9.5 Conflicts in the Former Yugoslavia 315
10.1 Government and the Agricultural Landscape 346
10.4 Agricultural Core Areas, *c.* 500 BCE 357
10.6 The Green Revolution in India 362
11.2 Changing Settlement in Canada 394
12.3 Eight Mile Road 444
12.4 The Homeless Experience 445
12.5 Wildwood Park Community in Winnipeg 452
12.6 Globalization and the Less Developed World City 457
13.4 Oil Reserves and Oil Production 478
13.7 Industry in Canada 487
13.8 China: From Struggling Peasant Economy to Industrial Giant 490
13.9 The Deindustrialization Revolution in the United Kingdom 494

PREFACE

The need for an education in geography—knowing where things are, why they are there, and why this knowledge matters—has always been paramount in all societies. It is no different today. Both geographic knowledge and an appreciation of the value of a geographic perspective are essential to help individuals and groups make sense of the changing worlds in which we live.

This book attempts to capture both the spirit and the practical merit of our contemporary human geography. Like the discipline, it encompasses an extraordinarily broad range of subject matter and no single approach or methodology dominates. Loosely organized around three recurring themes—relations between humans and land, regional studies, and spatial analysis—this text emphasizes how human geography has developed in response to society's needs and continues to change accordingly. It also stresses the links between human geography and other disciplines, not only to clarify the various philosophies behind different types of human geographic work but also to encourage students to apply their human geographic knowledge and understanding in other academic contexts.

NINTH EDITION: SPECIAL FEATURES

Because the practice and subject matter of human geography are always changing, this edition has been revised in three ways: the text's flow, clarity, and readability have been improved; new content has been included; and, most significant, the coverage of urban geography has been streamlined into two chapters, providing a more focused overview of this subdiscipline. Michael Mercier undertook this restructuring and rewriting—his willingness to contribute to the revision has resulted in a marked improvement to both the urban content and the text more generally.

This edition's most important additions, deletions, and revisions include the following:

- **Introduction: The Roads Ahead** Now more focused, the goal of the introduction is to enable students to grasp the essence of what is to come without providing unnecessary detail.
- **Chapter 1: What Is Human Geography?** Some detailed content was deleted, making the chapter a more readable overview of human geography's long history.
- **Chapter 2: Studying Human Geography** Again, some detailed content was removed to make the key facts, ideas, and arguments more evident and accessible.

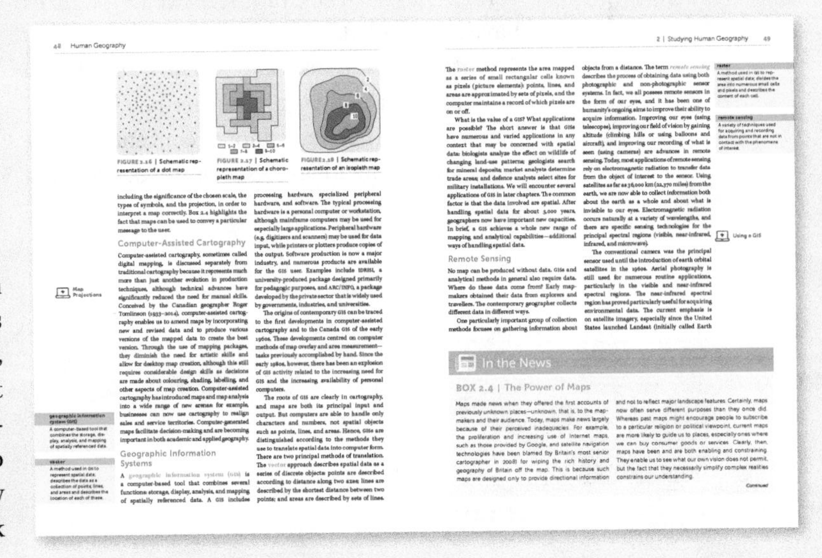

48 Human Geography

FIGURE 2.16 | Schematic representation of a dot map

FIGURE 2.17 | Schematic representation of a choropleth map

FIGURE 2.18 | Schematic representation of an isopleth map

including the significance of the chosen scale, the types of symbols, and the projection, in order to interpret a map correctly. Box 2.4 highlights the fact that maps can be used to convey a particular message to the user.

Computer-Assisted Cartography

Computer-assisted cartography, sometimes called digital mapping, is discussed separately from traditional cartography because it represents much more than just another evolution in production techniques, although technical advances have significantly reduced the need for manual skills. Conceived by the Canadian geographer Roger Tomlinson (1933–2014), computer-assisted cartography enables us to amend maps by incorporating new and revised data and to produce various versions of the mapped data to create the best version. Through the use of mapping packages, they diminish the need for artistic skills and allow for desktop map creation, although this still requires considerable design skills as decisions are made about colouring, shading, labelling, and other aspects of map creation. Computer-assisted cartography has introduced maps and map analysis into a wide range of new arenas: for example, businesses can now use cartography to realign sales and service territories. Computer-generated maps facilitate decision-making and are becoming important in both academic and applied geography.

Map Projections

Geographic Information Systems

A geographic information system (GIS) is a computer-based tool that combines several functions: storage, display, analysis, and mapping of spatially referenced data. A GIS includes processing hardware, specialized peripheral hardware, and software. The typical processing hardware is a personal computer or workstation, although mainframe computers may be used for especially large applications. Peripheral hardware (e.g. digitizers and scanners) may be used for data input, while printers or plotters produce copies of the output. Software production is now a major industry, and numerous products are available for the GIS user. Examples include IDRISI, a university-produced package designed primarily for pedagogic purposes, and ARC/INFO, a package developed by the private sector that is widely used by governments, industries, and universities.

The origins of contemporary GIS can be traced to the first developments in computer-assisted cartography and to the Canada GIS of the early 1960s. These developments centred on computer methods of map overlay and area measurement—tasks previously accomplished by hand. Since the early 1980s, however, there has been an explosion of GIS activity related to the increasing need for GIS and the increasing availability of personal computers.

The roots of GIS are clearly in cartography, and maps are both its principal input and output. But computers are able to handle only characters and numbers, not spatial objects such as points, lines, and areas. Hence, GISs are distinguished according to the methods they use to translate spatial data into computer form. There are two principal methods of translation. The vector approach describes spatial data as a series of discrete objects: points are described according to distance along two axes; lines are described by the shortest distance between two points; and areas are described by sets of lines.

geographic information system (GIS)
A computer-based tool that combines the storage, display, analysis, and mapping of spatially referenced data.

vector
A method used in GIS to represent spatial data; describes the data as a collection of points, lines, and areas and describes the location of each of these.

2 | Studying Human Geography 49

The raster method represents the area mapped as a series of small rectangular cells known as pixels (picture elements): points, lines, and areas are approximated by sets of pixels, and the computer maintains a record of which pixels are on or off.

What is the value of a GIS? What applications are possible? The short answer is that GISs have numerous and varied applications in any context that may be concerned with spatial data: biologists analyze the effect on wildlife of changing land-use patterns; geologists search for mineral deposits; market analysts determine trade areas; and defence analysts select sites for military installations. We will encounter several applications of GIS in later chapters. The common factor is that the data involved are spatial. After handling spatial data for about 3,000 years, geographers now have important new capacities. In brief, a GIS achieves a whole new range of mapping and analytical capabilities—additional ways of handling spatial data.

Remote Sensing

No map can be produced without data. GISs and analytical methods in general also require data. Where do these data come from? Early mapmakers obtained their data from explorers and travellers. The contemporary geographer collects different data in different ways.

One particularly important group of collection methods focuses on gathering information about objects from a distance. The term remote sensing describes the process of obtaining data using both photographic and non-photographic sensor systems. In fact, we all possess remote sensors in the form of our eyes, and it has been one of humanity's ongoing aims to improve their ability to acquire information. Improving our eyes (using telescopes), improving our field of vision by gaining altitude (climbing hills or using balloons and aircraft), and improving our recording of what is seen (using cameras) are advances in remote sensing. Today, most applications of remote sensing rely on electromagnetic radiation to transfer data from the object of interest to the sensor. Using satellites as far as 36,000 km (22,370 miles) from the earth, we are now able to collect information both about the earth as a whole and about what is invisible to our eyes. Electromagnetic radiation occurs naturally at a variety of wavelengths, and there are specific sensing technologies for the principal spectral regions (visible, near-infrared, infrared, and microwave).

The conventional camera was the principal sensor used until the introduction of earth orbital satellites in the 1960s. Aerial photography is still used for numerous routine applications, particularly in the visible and near-infrared spectral regions. The near-infrared spectral region has proved particularly useful for acquiring environmental data. The current emphasis is on satellite imagery, especially since the United States launched Landsat (initially called Earth

raster
A method used in GIS to represent spatial data; divides the area into numerous small cells and pixels and describes the content of each cell.

remote sensing
A variety of techniques used for acquiring and recording data from points that are not in contact with the phenomena of interest.

Using a GIS

In the News

BOX 2.4 | The Power of Maps

Maps made news when they offered the first accounts of previously unknown places—unknown, that is, to the mapmakers and their audience. Today, maps make news largely because of their perceived inadequacies. For example, the proliferation and increasing use of Internet maps, such as those provided by Google, and satellite navigation technologies have been blamed (by Britain's most senior cartographer in 2008) for wiping the rich history and geography of Britain off the map. This is because such maps are designed only to provide directional information and not to reflect major landscape features. Certainly, maps now often serve different purposes than they once did. Whereas past maps might encourage people to subscribe to a particular religion or political viewpoint, current maps are more likely to guide us to places, especially ones where we can buy consumer goods or services. Clearly, then, maps have been and are both enabling and constraining. They enable us to see what our own vision does not permit, but the fact that they necessarily simplify complex realities constrains our understanding.

Continued

Issues of spatial and social inequality were introduced in Chapter 3's account of globalization, while the terms *more developed world* and *less developed world*, defined in the introduction, were employed regularly in Chapter 5's discussions of population. In this chapter, we ask why the world is divided into more and less developed countries. To answer this complex question, we consider various factors, including physical geography, agricultural domestication, and, of profound significance, historical and current political and economic relationships between countries. We then discuss problems of defining and measuring development, paying particular attention to the distinction between economic measures and measures that focus on quality of life. An overview of major problems in less developed countries, such as food supply, refugees, disasters, diseases, and national debt, includes several detailed examples. The final sections of the chapter consider some of the strategies designed to further economic growth and improve quality of life in less developed areas, with emphasis on the question of political governance.

Here are three points to consider as you read this chapter.

- All too clearly, and tragically, the world comprises a collection of unequal places. Why is this the case?
- Is the distinction between more and less developed worlds meaningful and helpful? Or are the differences within each category sufficiently great to render it meaningless? Is it useful to identify a least developed world, mostly comprising countries in sub-Saharan Africa?
- How important is good government, especially democracy, to the well-being of a country and its citizens?

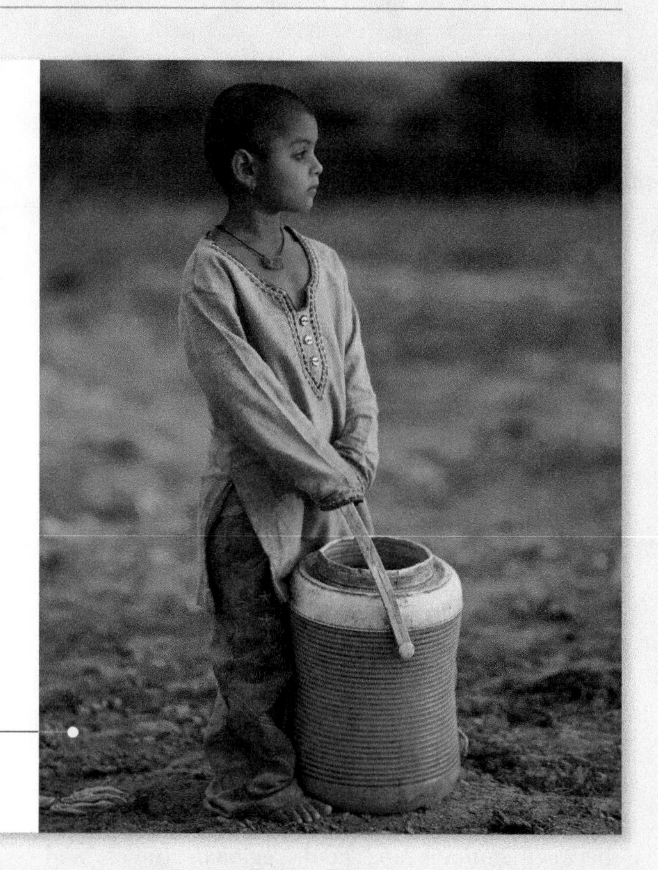

A gypsy girl searches for water in a slum area of Lahore, Pakistan. Access to clean water, proper sanitation, and sufficient food and opportunities are some of the discrepancies between the more developed and less developed worlds.

- **Chapter 3: Geographies of Globalization** A second attempt at objective measurement, one focusing on economic variables, enhances the understanding of globalization. Some chapter content has been streamlined; two new figures have been added.
- **Chapter 4: Humans and Environment** The discussion of the Anthropocene has been enhanced. Much of the chapter content reflects new facts and ideas, especially concerning anthropogenic global warming, with the 2014 IPCC report highlighted.
- **Chapter 5: Population Geography** As with all previous editions, this discussion features numerous detailed updates. The account of AIDS and other major diseases is now in Chapter 6. A new box discusses the thought-provoking question of how many people have ever lived on earth.
- **Chapter 6: Global Inequalities** To heighten emphasis on key points, the opening contextual account of global inequality has been shortened and some content has been moved online. The chapter includes a restructured discussion of global food shortages and a new box on the prospects for growth in Africa.
- **Chapter 7: Geographies of Culture and Landscape** The opening conceptual comments are shorter, with some content moved online. The sections on psychogeography, religions as civilizations, and religion and identity are now online.
- **Chapter 8: Geographies of Identity and Difference** Simplified discussions of key concepts including race and racism help tighten the chapter's focus. Examples of social inequality have been updated as appropriate.

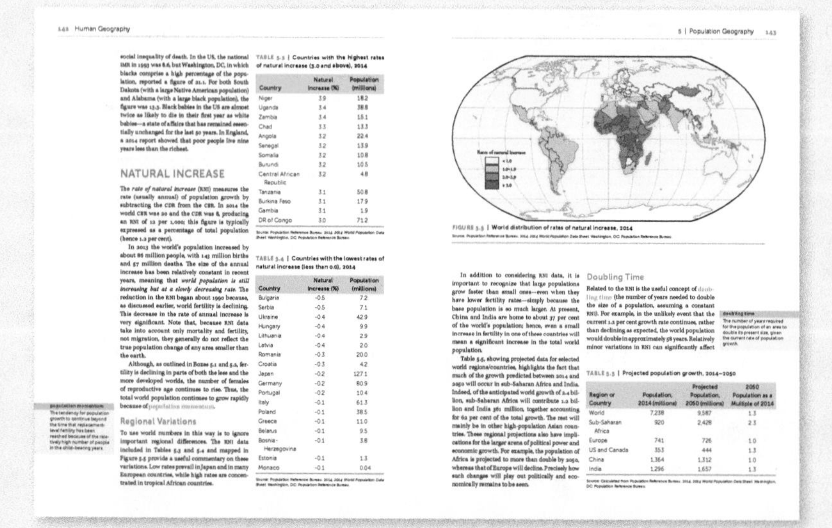

The various vibrant ethnic enclaves contained within Vancouver, such as the Punjabi Market area and Chinatown, are evidence of acculturation.

The political logic for the introduction of multiculturalism was succinctly stated by then prime minister Pierre Elliott Trudeau: "Although there are two official languages, there is no official culture, nor does any ethnic group take precedence over any other" (quoted in Kobayashi, 1993: 205). Box 8.5 discusses immigration and ethnic diversity in Canada. Versions of multiculturalism are in place in many other countries, but the idea and policy are contentious.

Some see multiculturalism as a vision of national identity based on pluralism; others see it as divisive, a form of "have a nice day" racism. This is an extremely challenging circumstance in many European countries, where it is frequently argued that multiculturalism facilitates extremism, specifically Islamic extremism, because it tacitly encourages newcomers or those whose

Around the Globe

BOX 8.5 | Immigration and Ethnic Diversity in Canada

The population of Canada has changed dramatically since the beginning of the twentieth century, both in total numbers and in ethnic composition. In 1901 the country had 5.3 million people, most of British or French origin; in 2014 Canadians numbered 35.5 million and were one of the most ethnically diverse populations in the world. Four relatively distinct waves of migration can be identified during the twentieth century. The years 1901 to 1914 brought dramatic growth—many immigrants came from the new source areas of Eastern and Southeastern Europe and headed for the prairies. For some years during this period, the annual population growth rate was almost 3 per cent, but these rates slowed with the onset of World War I and remained low, at about 1.4 per cent, until the end of World War II.

The next major increase in immigration, between 1951 and 1961, joined with high fertility to produce an average annual growth rate of 2.7 per cent. In this period, during which the Canadian government was anxious to fill labour shortages in both the agricultural and industrial sectors, immigration policy continued to show a strong preference for British and other European settlers. As noted earlier in this chapter, potential immigrants from Asia were subject to quotas.

Since 1961 both immigration and fertility in Canada have declined, resulting in annual growth rates of about 1.3 per

New Canadians swear an oath of citizenship during a citizenship ceremony in Edmonton, Alberta.

cent. In 1962 Canada's immigration rules dropped all references to supposed race and nationality, and in 1967 a points system was introduced that reflected economic needs and gave particular weight to education, employment qualifications, English- or French-language competence, and family reunification. Thus, policies in this most recent phase have been relatively liberal, permitting significant diversification in the ethnic mix of immigrants. Immigrants from Europe and the United States accounted for about 95 per cent of the total in 1960 but presently make up only about 20 per cent. Application of the points system has resulted in an increase in the number of qualified professionals, many from Asia and the Caribbean. More generally, there have been huge increases in immigrant numbers from less developed countries. A new Immigration Act in 1978 had three principal objectives: to facilitate family reunification, to encourage regional economic growth, and to fulfill moral obligations to refugees and persecuted people. Although subsequent legislation has modified policy details concerning refugee claimants and illegal immigration, the overall thrust remained the same until the Conservative election victory in 2006. Today, there is increased emphasis on admitting immigrants who already have offers of employment.

Canada's liberal immigration policies over the last four decades have benefited the country in many ways. Yet increasing ethnic diversity has made the establishment of a coherent cultural identity an elusive goal. Moreover, some Canadians' reactions against that diversity make it clear that we still have a long way to go if we are to free ourselves of racist attitudes.

ethnicity is "different" not to assimilate. Policymakers in several European countries currently are seeking to identify and implement some alternative policy that promotes pluralism without creating divisiveness. One strategy favoured by conservative thinkers involves the promotion of a stronger national identity.

GENDER

As discussed in the Chapter 2 account of feminism, our identities are gendered, a fact that introduces another category of difference. Gender is traditionally understood to involve power relations between dominant males and subordinate females. Indeed, so widespread is this pattern of relations that the few exceptions are well known. For example, the Indian state of Meghalaya has a matrilineal system with all wealth passing from mother to daughter and with routine discrimination against men. The system is so deeply embedded in local culture that many nouns assume a feminine form only if the thing referred to becomes useful; tree is masculine, but wood is feminine.

Human geographic recognition of gendered identities and related differences owes much to the work of mostly female geographers inspired by various bodies of feminist theory. Feminist research has also been inspired by Marxist ideas about inequality and need for social justice and by postmodern ideas about excluded groups.

Feminist Geography

Building on Pratt (2004: 128), the history of feminist geography can be summarized as follows. In the 1970s a critique of the sexism and patriarchy that prevailed, both institutionally and intellectually, in the discipline of geography was initiated. At that time, geography was exceptionally male-dominated, at least partly because of the long-standing links with fieldwork and because any form of outdoor activity typically was presumed to be within the male domain. The new interests in Marxism and humanism did little to correct the traditional gender blindness of the discipline. The feminist geographic critique employed the concept of contextualism to focus on the varied ways that geography ignored the everyday lives of women, including identification of research topics and analytical strategies.

A significant body of feminist geographic research soon emerged with the aim of making space for women, in the everyday world of the university and society more generally and also in the knowledge produced by geographers. This involved acknowledging that people were active agents, living their lives in households and other places.

As this research tradition evolved, it became evident that it was not necessarily appropriate to treat women as a universal category and by the 1990s feminist geographers were considering additional bases for difference besides gender,

- **Chapter 9: Political Geography** Several sections are now online, including those concerned with predecessors of the nation-state, effects of colonialism, conceptual accounts of state creation, African ethnicity and conflict, and substate governments. There is a new box on Canadian regional identities and substantial updating of current political geographic issues.
- **Chapter 10: Agricultural Geography** Sections moved online include discussions of pastoral nomadism in the Sahara and Mongolia, agricultural change in Bhutan, and peasant–herder conflict in Côte d'Ivoire. The account of the agricultural location problem has been shortened without detracting from the key argument.
- **Chapter 11: An Urban World** This condensed chapter focuses on cities; therefore, content on rural settlement patterns and economic base theory has been removed. The chapter includes new material on urban areas in China and Canada; urban hearths; and pre-industrial cities in Greece and the eastern Mediterranean, the Roman Empire, China and eastern Asia, the Islamic World, and Europe.
- **Chapter 12: The City and the Urban Form** This chapter combines chapters 12 and 13 of the previous edition. Content on city models, urban government, and suburbanization was condensed or removed. New or expanded content includes material on urban structure, modelling the North American city, social segregation, cities as production and consumption centres, and informal settlements.

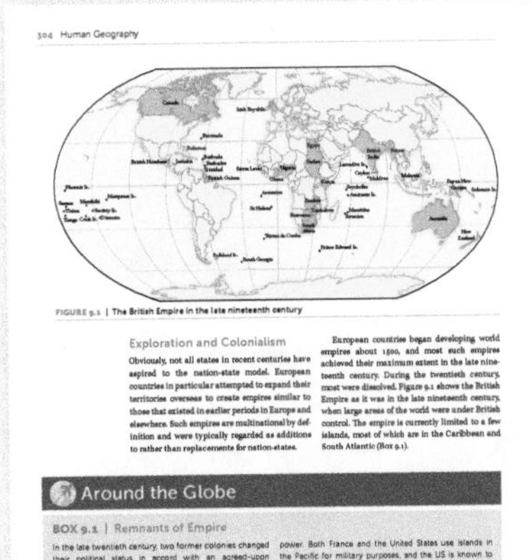

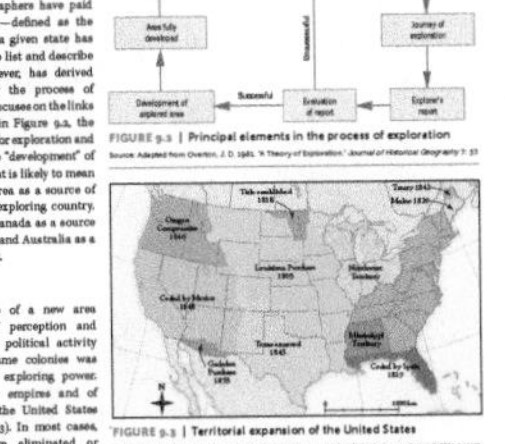

- **Chapter 13: Geographies of Energy and Industry** The discussion of fossil fuel energy sources—reserves, production, movement, and consumption—is significantly enhanced, with much new information on natural gas and coal as well as an updated account of oil. There are new boxes concerned with the North Antelope Rochelle Mine in the Powder River basin of Wyoming and with deindustrialization in the UK.
- **Appendix 2: Global Physical Geography** This content is now online.
- **Appendix 3: The Evolution of Life** This content is now online.

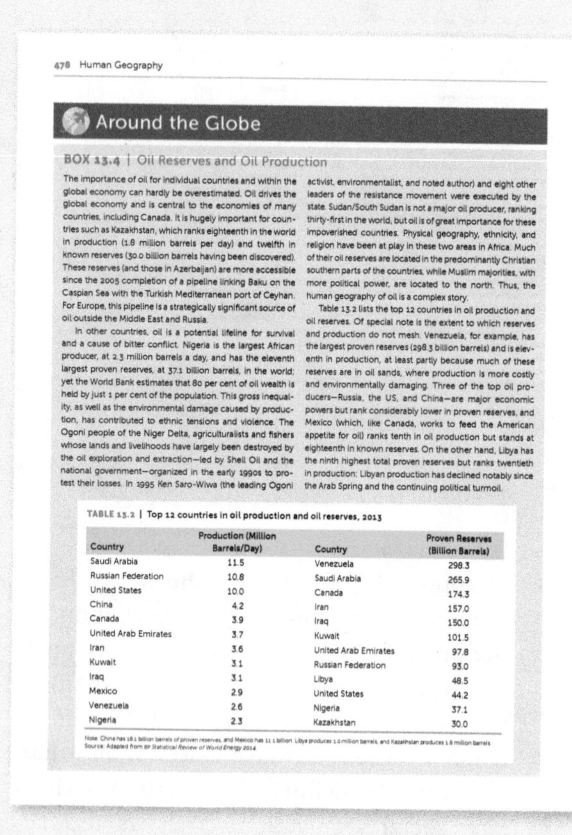

478 Human Geography

Around the Globe

BOX 13.4 | Oil Reserves and Oil Production

The importance of oil for individual countries and within the global economy can hardly be overestimated. Oil drives the global economy and is central to the economies of many countries, including Canada. It is hugely important for countries such as Kazakhstan, which ranks eighteenth in the world in production (1.8 million barrels per day) and twelfth in known reserves (30.0 billion barrels having been discovered). These reserves (and those in Azerbaijan) are more accessible since the 2005 completion of a pipeline linking Baku on the Caspian Sea with the Turkish Mediterranean port of Ceyhan. For Europe, this pipeline is a strategically significant source of oil outside the Middle East and Russia.

In other countries, oil is a potential lifeline for survival and a cause of bitter conflict. Nigeria is the largest African producer, at 2.3 million barrels a day, and has the eleventh largest proven reserves, at 37.1 billion barrels, in the world; yet the World Bank estimates that 80 per cent of oil wealth is held by just 1 per cent of the population. This gross inequality, as well as the environmental damage caused by production, has contributed to ethnic tensions and violence. The Ogoni people of the Niger Delta, agriculturalists and fishers whose lands and livelihoods have largely been destroyed by the oil exploration and extraction—led by Shell Oil and the national government—organized in the early 1990s to protest their losses. In 1995 Ken Saro-Wiwa (the leading Ogoni activist, environmentalist, and noted author) and eight other leaders of the resistance movement were executed by the state. Sudan/South Sudan is not a major oil producer, ranking thirty-first in the world, but oil is of great importance for these impoverished countries. Physical geography, ethnicity, and religion have been at play in these two areas in Africa. Much of their oil reserves are located in the predominantly Christian southern parts of the countries, while Muslim majorities, with more political power, are located to the north. Thus, the human geography of oil is a complex story.

Table 13.2 lists the top 12 countries in oil production and oil reserves. Of special note is the extent to which reserves and production do not mesh. Venezuela, for example, has the largest proven reserves (298.3 billion barrels) and is eleventh in production, at least partly because much of these reserves are in oil sands, where production is more costly and environmentally damaging. Three of the top oil producers—Russia, the US, and China—are major economic powers but rank considerably lower in proven reserves, and Mexico (which, like Canada, works to feed the American appetite for oil) ranks tenth in oil production but stands at eighteenth in known reserves. On the other hand, Libya has the ninth highest total proven reserves but ranks twentieth in production: Libyan production has declined notably since the Arab Spring and the continuing political turmoil.

TABLE 13.2 | Top 12 countries in oil production and oil reserves, 2013

Country	Production (Million Barrels/Day)	Country	Proven Reserves (Billion Barrels)
Saudi Arabia	11.5	Venezuela	298.3
Russian Federation	10.8	Saudi Arabia	265.9
United States	10.0	Canada	174.3
China	4.2	Iran	157.0
Canada	3.9	Iraq	150.0
United Arab Emirates	3.7	Kuwait	101.5
Iran	3.6	United Arab Emirates	97.8
Kuwait	3.1	Russian Federation	93.0
Iraq	3.1	Libya	48.5
Mexico	2.9	United States	44.2
Venezuela	2.6	Nigeria	37.1
Nigeria	2.3	Kazakhstan	30.0

Note: China has 18.1 billion barrels of proven reserves, and Mexico has 11.1 billion. Libya produces 1.0 million barrels, and Kazakhstan produces 1.8 million barrels.
Source: Adapted from *BP Statistical Review of World Energy 2014*

13 | Geographies of Energy and Industry 479

per cent of the global total. As detailed in Box 13.4, a few countries tend to dominate world production.

As shown in Table 13.1, the United States is the world's leading consumer of oil, with about 21.9 per cent of global consumption, followed by China and Japan. In general, most of the principal consuming countries are likely to become increasingly dependent on imports. In short, there is a growing gap between the places where oil is produced and where it is consumed. For this fundamental geographic reason, there is much movement of oil from producing to consuming countries. Figure 13.6 shows the major patterns of trade in oil.

The Future of Oil

One fundamental question is: What will happen when the oil runs out? Note that the data in Box 13.4 cannot answer this question. Proven oil reserves in the Middle East and North Africa increased by more than 80 per cent since 1973, and other reserves continue to be discovered. Such new discoveries and, eventually, exploitation could see a significant upward bump in the coming years as the Arctic Ocean becomes more accessible, a consequence of climate change and ice melt. Nevertheless, our current dependence on oil cannot be sustained. Notwithstanding the recent fall in price, the ever-increasing demand both in the more developed world and in industrializing countries (especially China) will likely

TABLE 13.1 | Principal oil-consuming countries, 2013

Country	Share of Total (%)
United States	19.9
China (including Hong Kong)	12.1
Japan	5.0
India	4.2
Russian Federation	3.7
Saudi Arabia	3.2
Brazil	3.2
Germany	2.7
South Korea	2.6
Canada	2.5
Iran	2.2
Mexico	2.1
France	1.9
Indonesia	1.8
United Kingdom	1.7

Source: *BP Statistical Review of World Energy 2014*

FIGURE 13.6 | Major oil trade movements (million tonnes), 2014

Source: Updated from *BP Statistical Review of World Energy 2011*. www.bp.com/content/dam/bp-country/de_de/PDFs/brochures/statistical_review_of_world_energy_full_report_2011.pdf

FEATURES

LEARNING TOOLS

Along with thorough chapter introductions and summaries, the text offers the following learning tools:

- Links to other chapters underline connections between topics that may not be apparent at first glance.
- Questions for critical thought and a running glossary reinforce understanding and encourage discussion of core concepts.

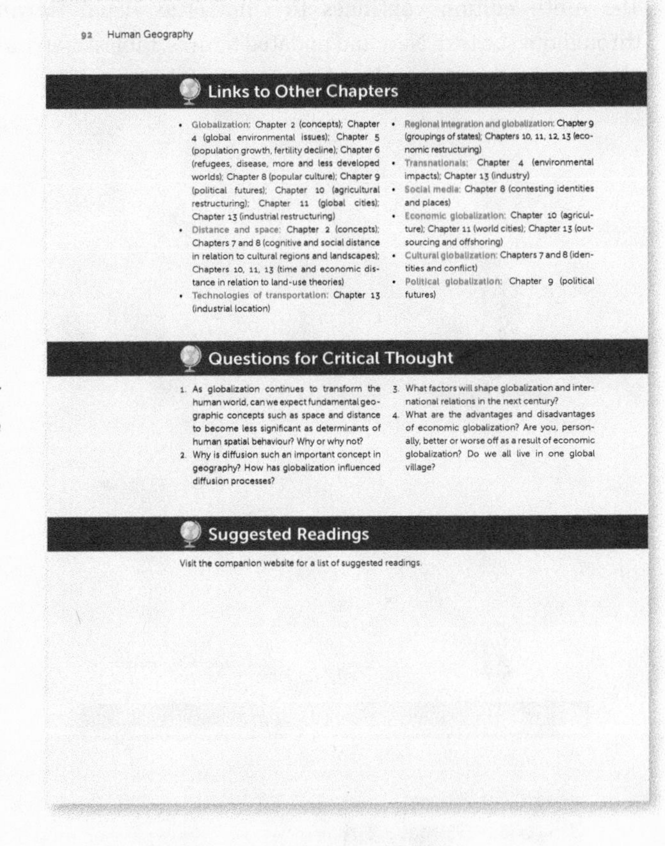

92 Human Geography

Links to Other Chapters

- Globalization: Chapter 2 (concepts); Chapter 4 (global environmental issues); Chapter 5 (population growth, fertility decline); Chapter 6 (refugees, disease, more and less developed worlds); Chapter 8 (popular culture); Chapter 9 (political futures); Chapter 10 (agricultural restructuring); Chapter 11 (global cities); Chapter 13 (industrial restructuring)
- Distance and space: Chapter 2 (concepts); Chapters 7 and 8 (cognitive and social distance in relation to cultural regions and landscapes); Chapters 10, 11, 13 (time and economic distance in relation to land-use theories)
- Technologies of transportation: Chapter 13 (industrial location)
- Regional integration and globalization: Chapter 9 (groupings of states); Chapters 10, 11, 12, 13 (economic restructuring)
- Transnationals: Chapter 4 (environmental impacts); Chapter 13 (industry)
- Social media: Chapter 8 (contesting identities and places)
- Economic globalization: Chapter 10 (agriculture); Chapter 11 (world cities); Chapter 13 (outsourcing and offshoring)
- Cultural globalization: Chapters 7 and 8 (identities and conflict)
- Political globalization: Chapter 9 (political futures)

Questions for Critical Thought

1. As globalization continues to transform the human world, can we expect fundamental geographic concepts such as space and distance to become less significant as determinants of human spatial behaviour? Why or why not?
2. Why is diffusion such an important concept in geography? How has globalization influenced diffusion processes?
3. What factors will shape globalization and international relations in the next century?
4. What are the advantages and disadvantages of economic globalization? Are you, personally, better or worse off as a result of economic globalization? Do we all live in one global village?

Suggested Readings

Visit the companion website for a list of suggested readings.

6 | Global Inequalities 201

In the News

BOX 6.7 | Defeating AIDS?

Talking about the "end of AIDS" seems reasonable. Although many factors can influence the course of the battle against the disease, positive signs are emerging, and the UN AIDS agency is now talking about the epidemic being under control by 2030. Crucially, the number of new infections has fallen from 3.4 million in 2001 to 2.1 million in 2013; also, the number of deaths from AIDS and related illnesses has fallen from 2.4 million in 2005 to 1.5 million in 2013.

The reason for these positive changes appears to be a combination of education, changing sexual behaviour, and increased availability of antiretroviral drugs. Countries such as Thailand, the Philippines, and Brazil have been fighting the disease with much success since the 1990s through prevention programs designed to educate people about such matters as condom use. In sub-Saharan Africa, the principal reasons for these positive changes appear to be that condom use is more widespread, people are engaging in sexual behaviour at a later age, and the more developed world has intensified efforts to combat the disease, including helping to make drugs more readily available.

In Uganda the adult infection rate has dropped from 30 per cent in 1992 to less than 6 per cent. Perhaps surprisingly, this reduction is attributed not only to increased use of condoms but also to a government campaign stressing abstinence and fidelity. In many other countries, including South Africa, people are increasingly well educated concerning the causes of AIDS, and there are clear signs that attitudes towards sexual behaviour are changing. In Botswana, some mining companies that employ large numbers of adult males provide hospital facilities for those already infected and are working to prevent AIDS not only through education but also by making free condoms readily available.

Of course, condom use does meet with opposition. A controversy erupted in 2009 following remarks by Pope Benedict XVI prior to a visit to Africa. Stating that condoms did not help solve the problem but rather exacerbated the effect of HIV/AIDS, he advocated the traditional church teachings of abstinence and fidelity. In a response that was unprecedented in its forcefulness and condemnation, the prestigious medical journal *Lancet* wrote that the pope was publicly distorting well-established scientific facts in order to promote Catholic doctrine.

Another important factor in the worldwide fight against AIDS is the response of the more developed world. Here, too, the signs are positive. A 2001 UN summit agreed to set up a global fund to fight AIDS, along with tuberculosis and malaria. Despite initial funding problems, there is reason to believe that the errors of the 1990s, when the more developed world quite literally looked away, are being corrected. In 2003 the United States committed up to $15 billion over five years to combat AIDS in Africa and the Caribbean, although much of this funding is targeted to programs advocating abstinence.

Changing attitudes in both less and more developed countries and increased use of antiretroviral drugs appear to be both limiting the spread of AIDS and reducing the number of deaths.

Women make a ribbon formation with candles to mark World AIDS Day in Amadabad. The World Health Organization established the day in 1988 to further global awareness and put a focus on prevention.

EXTENSIVE BOX PROGRAM

- Contributed "Focus on Geographers" boxes highlight the research of various human geographers, giving students insight into what it is like to work in the field of human geography.
- "In the News" boxes draw students' attention to current events relating to human geography.
- "Around the Globe" boxes introduce students to issues regarding a particular location.
- "Examining the Issues" boxes help students examine theoretical matters that apply to the broader world.

RICHLY ILLUSTRATED CONTENT

The ninth edition continues to emphasize visual learning throughout the text. New and updated figures, tables, maps, and photos enhance students' understanding of the material.

FIGURE 11.4 | Rural and urban populations, Canada, 1950–2050

Source: Adapted from United Nations Department of Economic and Social Affairs, Population Division. 2015. World Urbanization Prospects: The 2014 Revision. New York: United Nations. http://esa.un.org/unpd/wup/

FIGURE 11.5 | Canada: Census metropolitan areas, 2011

Megacities and Megaregions

megacity Metropolitan areas with populations of more than 10 million.

The urban population is increasing rapidly, both in absolute numbers and in relation to the rural population. This growth is both concentrated in a handful of very large cities and spread across many smaller ones. While the populations of megacities are increasing and new ones appear regularly, most additional urban population growth is actually occurring in smaller cities, some with fewer than 500,000 people. Despite the shock value of considering the consequences of cities with 10 million people or more, the more typical urban experience is living in smaller cities, as approximately half of all urban dwellers currently live in cities with less than 500,000 people.

The growing number of megacities, and their increasing size, is highlighted by the following facts:

- In 1950 there were just two megacities: New York with 12.3 million people and Tokyo with 11.3.
- By 1975 there were four megacities: Tokyo with 26.6 million people, New York with 15.9, Shanghai with 11.4, and Mexico City with 10.7.
- By 2014 there were 28 megacities; that number is expected to rise to 40 by 2030, with most of the new megacities located in the less developed world.

Table 11.5 provides data on the number and size of megacities in 2014 and projections for 2030, and Figure 11.6 maps the two patterns (but recall the cautionary comments about city size made in Box 11.1). The vast majority of megacities are located in less developed countries. In 2014 only 6 of the 28—Tokyo, New York, London,

People walk across the Viaduto Santa Ifigenia in São Paulo, Brazil, one of the world's megacities.

Around the Globe

BOX 6.1 | Less Developed World Case Studies: Ethiopia, Sri Lanka, and Haiti

A recent photograph of the Merkato marketplace in Addis Ababa, Ethiopia's capital. Located at the approximate geographical centre of the country, Addis Ababa is home to over 2.7 million of the nation's inhabitants.

Ethiopia is one of several countries in the Horn of Africa suffering from a tragic combination of human and environmental factors. It is categorized as a least developed country.

Ethiopia was settled by Hamitic peoples of North African origin. Following an in-movement of Semitic peoples from southern Arabia in the first millennium BCE, a Semitic empire was founded at Axsum (it became Christian in the fourth century CE). The rise of Islam displaced that empire southwards and established Islam as the dominant religion of the larger area; the empire was overthrown in the twelfth century CE. Ethiopia escaped European colonial rule, but its borders were determined by Europeans occupying the surrounding areas, and the Eritrea region was colonized by Italy from the late nineteenth century until 1945. Attempts by the central government in Addis Ababa to gain control over both the Eritreans and the Somali group in the southeast generated much conflict. In 1974–5 a revolution displaced the long-serving ruler, Haile Selassie, and established a socialist state.

Ethiopia is slightly larger than Ontario—1.2 million km² (463,400 square miles). Much of the country is tropical highlands, typically densely populated because such areas have good soils and are free of many diseases. In 2014 the population was estimated at 95.9 million; the CBR was 28, the CDR 8, and the RNI 2.3. Although it is a leading coffee producer, Ethiopia is one of the poorest states in Africa, with many people reliant on overseas food aid. Population growth is highly uneven, and urban growth has actually declined since 1975 because of socialist land reform policies and the low quality of life in urban areas.

Most Ethiopians have little or no formal education, especially among the (predominant) rural population. The key social and economic unit is the family, in which women are subordinate, first to their fathers, then to their husbands and, if widowed, to their adult sons. Health care varies substantially between urban and rural areas.

The country also has a particular regional problem. People in the Lower Omo River Valley live in difficult circumstances. The Omo tribes are agricultural peoples who depend on the annual cycle of river flooding. Each year, communities on the riverbanks move to higher ground prior to the flood, then return to plant their crops (mostly sorghum) in the replenished soil when the waters recede. But this resource and their cattle are often insufficient. Furthermore, disputes between groups about water rights are common, with there being evidence that many of the males carry an automatic weapon.

The situation may deteriorate further. The Ethiopian government is constructing a huge dam (Gibe III) upstream that, the Omo people claim, will mean the end of the annual flooding that forms the basis of their livelihood. The government argues otherwise, maintaining that it will be possible to regulate flooding as needed and that the Omo people will not be disadvantaged. Many scientists consider the Omo position legitimate, and European and British politicians have expressed concern. It is notable that the Omo were not involved in any consultations about construction.

Sri Lanka

Sri Lanka (Ceylon until 1972) is an island located in the Indian Ocean, south of India. Mostly low-lying, it has a tropical climate and a limited resource base. Minerals are in short supply, as are sources of power. The dominant economic activity is agriculture; about 36 per cent of the country is under cultivation.

Continued

NATIONAL AND GLOBAL PERSPECTIVES

Current Canadian and international examples are used throughout the text. Students will gain a global perspective and a greater understanding of issues such as climate change, world population density, and food shortages.

STUDENT AND INSTRUCTOR SUPPLEMENTS TO THE TEXT

The ninth edition of *Human Geography* is accompanied by a wide range of supplementary online items for students and instructors alike, all designed to enhance and complete the learning and teaching experiences. These resources are available at **www.oupcanada.com/Human9e.**

For Students

Online Resources
Available at www.oupcanada.com/Human9e

A comprehensive student study guide of review material—including research questions, links to human geography websites, interactive practice quizzes and study flash cards, YouTube videos, and Google Earth exercises—is designed to reinforce understanding of the material and provide directions for further research. Two appendices and a substantial portion of material from the previous edition, on topics ranging from trade theories to religious identity, have been placed online; an icon has been added to the text to indicate where these sections relate to chapter content. The website also features a Google Maps guide to key areas discussed in the book, a Google Earth tutorial session, and a streaming video. A mobile study unit provides a screen-ready, condensed version of the chapter pedagogy and elements of the study guide.

For Instructors

The following instructor's resources are available to qualifying adopters. Please contact your OUP sales representative for more information.

- An instructor's manual simplifies class planning by providing learning objectives, expanded key concepts, teaching aids, and discussion topics for each chapter.
- A test generator offers an array of true/false, multiple-choice, short-answer, and essay questions, making test formulating a snap.
- PowerPoint slides provide valuable visual aids for classroom use.
- An image bank containing hundreds of full-colour figures, photographs, and tables makes classroom discussion more engaging and relevant.

William Norton and Michael Mercier

***Human Geography*, Ninth Edition**
ISBN 13: 9780199019557

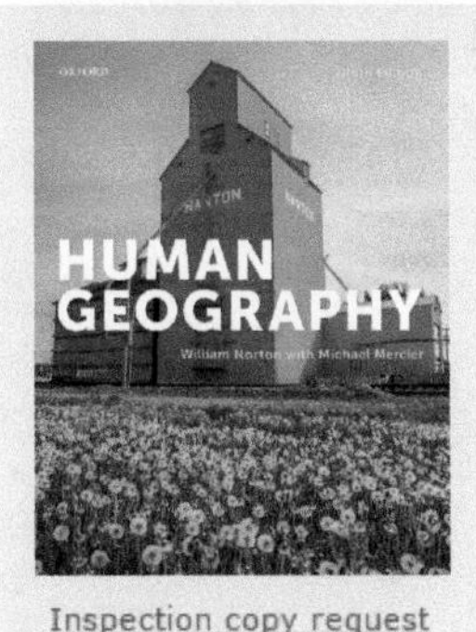

Inspection copy request

Ordering information

Contact & Comments

About the Book

A bestseller for over twenty years and now in its ninth edition, *Human Geography* continues to offer an authoritative and comprehensive examination of human behaviour's effects on the earth's surface. Focusing on three themes-relations between humans and land, regional studies, and spatial analysis-this introduction engages with such areas as philosophy, culture, identity, politics, and economics. It includes the most current statistics and data, as well as revamped and streamlined chapters on urban issues, providing students with a thorough discussion of the discipline's fundamental concepts.

Instructor Resources

You need a password to access these resources. Please contact your local Sales and Editorial Representative for more information.

Student Resources

ACKNOWLEDGEMENTS

WILLIAM NORTON

I am pleased to welcome Michael Mercier as a co-author and acknowledge two colleagues for their significant contributions to this new edition: Doug Fast has drawn all the new figures and revised many others; Barry Kaye has continued to draw my attention to numerous new books and relevant news items. I am most grateful for their ongoing interest and involvement in this work. In addition, many other geographers and many of my students have commented on earlier editions, and most of their suggestions are incorporated in this edition. Finally, many thanks to my wife, Pauline, for her unceasing support of my academic activities.

MICHAEL MERCIER

I would like to express my gratitude to Bill Norton and the Oxford team for providing me with the opportunity to contribute to such an outstanding book. Several of my colleagues (Walter Peace, Richard Harris, Rob Wilton, and Bruce Newbold) provided insights and have helped guide my teaching and writing throughout my career. Additionally, I'd like to thank the thousands of students in my introductory human geography classes, who inspired me to find the most relevant and illustrative examples of human geographic patterns and processes. I would also like to thank my wife, Patty, and our two boys, Graham and Bennett, for encouraging me to take on this endeavour and supporting me along the way.

Both authors acknowledge Peter Chambers, whose editorial suggestions were of tremendous help in shaping this ninth edition. Peter has a thorough understanding of the subject, and we thank him for his professionalism, friendship, and tremendous support of us and this project. Similarly, Janna Green's efficient and conscientious editing of the text resulted in numerous and significant improvements. She has a meticulous attention to detail, always appreciated the larger intent of the authors, and made many helpful suggestions to improve flow and readability.

As always, staff at Oxford University Press supported this book in every way possible. We are most appreciative for the enthusiastic support of Lisa Ball, Katherine Skene, Caroline Starr, and Phyllis Wilson.

Oxford University Press and the authors would also like to recognize the many reviewers whose comments have proved invaluable over the years. In addition to those who provided anonymous feedback on this ninth edition, the authors and the publisher thank the following reviewers, whose thoughtful comments and suggestions have helped to shape this text:

Martin A. Andresen
Simon Fraser University
Godwin Arku
Western University
Scott Bell
University of Alberta
Anna Pojadas Botey
University of Alberta
Leith Deacon
University of Alberta
Brent Doberstein
University of Waterloo
Sean Doherty
Wilfrid Laurier University
Matthew Evenden
University of British Columbia
Michael Fox
Mount Allison University
Jason Grek-Martin
Saint Mary's University
Edgar L. Jackson
University of Alberta
Marilyn Lewry
University of Regina
Ian MacLachlan
University of Lethbridge
Bernard Momer
University of British Columbia Okanagan
Raj Navaratnam
Red Deer College
Bob Patrick
University of Saskatchewan
Walter Peace
McMaster University
Marc Vachon
University of Winnipeg
Susan Wurtele
Trent University

INTRODUCTION

THE ROADS AHEAD

You know that this book is an introduction to human geography—but what does that term mean? Answer this question in a few key words. It should be interesting to compare your initial perception with your understanding after you have read this introductory chapter.

Consider the word *geography*. Its roots are Greek: *geo* means "the world"; *graphei* means "to write." Literally, then, geographers write about the world. The sheer breadth of this task has demanded that it be divided into two relatively distinct disciplines. Physical geography is concerned with the physical world (e.g. climates and landforms), and human geography is concerned with the human world (e.g. agricultural activities and settlement patterns).

So human geographers write about the human world. But what questions do they ask? How do they approach their work? What methods do they employ? Not surprisingly, human geographers ask various questions and have diverse approaches and methods from which to choose. Nevertheless, three themes are central to any study of the human world: relations between humans and land, regionalization, and spatial analysis. Keeping these themes in mind from the start will help you to follow and understand connections among the many aspects of human geography.

THREE RECURRING THEMES

Humans and Land

The human world is not in any sense preordained. It is not the result of any single cause, such as climate, physiography, religion, or culture. It is the ever-changing product of the activities of human beings, as individuals and as group members, working within human and institutional frameworks to modify pre-existing physical conditions. Thus, human geographers often focus on the evolution of the human world with reference to people, their cultures, and their physical environments.

Tourists in Alberta's Banff National Park follow the Johnston Canyon trail. The interactions between humans and landscapes, and their effects, are a central theme in human geography.

landscape
A major concern of geographic study; the characteristics of a particular area especially as created through human activity.

location
A specific part of the earth's surface; an area where something is situated.

globalization
A complex combination of economic, political, and cultural changes that have long been evident but that have accelerated markedly since about 1980, bringing about a seemingly ever-increasing connectedness of both people and places.

region
A part of the earth's surface that displays internal homogeneity and is relatively distinct from surrounding areas according to some criterion or criteria; regions are intellectual creations.

We usually describe the human world as a **landscape**. There are two closely related aspects to the term in this sense:

1. Landscape is what exists as a result of human modifications to physical geography. This aspect of landscape includes crops, buildings, lines of communication, and other visible, material features.
2. Landscape has significant symbolic content, or meaning. It is hard to view a church, statue, or skyscraper without appreciating that it is something more than a visible, material human addition to the physical geography of a place: such features are expressions of the cultures that produced them.

As human geographers, we are interested in landscape both for what it is and for what it means to live in it: we interpret landscape as the outcome of particular relationships between humans and land. For some geographers, both physical and human, the study of humans and land together is geography. This view, which relates human and physical variables and identifies links, is sometimes called ecological analysis.

Regional Studies

To facilitate their task of writing about the world, human geographers often divide large areas into smaller ones that have one or more features in common. These smaller areas are **regions**. To regionalize is to classify on the basis of one or more variables. Our ability to regionalize tells us that human landscapes make sense; they are not random assemblages of features. Groups of people occupying particular areas over a period of time create regions, human landscapes that reflect their occupancy and that differ from other landscapes. Much contemporary regional study focuses on this social organization of space, on the impact of creating regions on social and economic life, on interactions between regions, and on regional boundaries as barriers.

Building on a long tradition, contemporary human geography considers regions at a wide range of scales—from the local (a suburban neighbourhood, for example) all the way through to the global (the world as a single region). In acknowledging the relevance of different spatial scales of analysis, human geography reveals the importance of place in all aspects of our lives.

Spatial Analysis

Understanding the human world requires that we explain **location**, or why things are where they are. Typically, a geographer using a spatial analysis approach tackles this question through theory construction, models, and hypothesis testing (using quantitative methods). For example, we may find that towns in an area are spaced at relatively equal distances apart or that industrial plants are located close to one another. The primary goal of spatial analysis is to explain such locational regularities. Thus, we have already identified two types of spatial organization: the creation of regions and the identification of distinct patterns of locations. A secondary goal is often identifying alternative locational patterns that might be more efficient or more equitable.

Central to this spatial analysis is the realization that all things are related. For geographers, to say that things are related is usually to say that they interact. Perhaps the best examples of interaction in the contemporary world involve what is often described as **globalization**—a complex set of processes with economic, political, and cultural dimensions.

A Common Thread

Our recurring themes reflect three separate but overlapping traditions (to be outlined in Chapter 1). One common thread among them, however, is the fact that the human world is always changing. Thus, human geography, regardless of the specific focus, typically incorporates a time dimension. Change can be a response to either internal or external factors. Some changes are rapid, some gradual. Landscapes change in content and meaning; regions gain or lose their distinctive features; and locations adjust to changing circumstances.

One compelling example of changing geographies can be found in the terrorist attacks of 11 September 2001. The consequences of those attacks, some immediate, some longer-term and still uncertain, have served to reshape both local and global geographies. Locally, the skyline of New York City will never be the same. Nor will the pattern of global politics, since the United States responded to the attacks by invading first Afghanistan, where the ruling Taliban had supported terrorist groups, and then Iraq. Many commentators suggested that global politics would henceforth be characterized by a "clash of civilizations," and there was widespread acknowledgement that fighting terrorism would require in-depth understanding of local geographies, both physical and human.

In identifying our three principal themes, we explicitly acknowledge the legitimacy of multiple approaches to our subject matter: the human world. It is also important to note how the application of these approaches changes over time. Such change is equally legitimate—and necessary—in any dynamic discipline.

WHY "HUMAN" GEOGRAPHY?

Although many geographers have striven to make their field a truly integrated one, combining physical and human components (as in the humans-and-land theme), human geography and physical geography are typically acknowledged to be relatively distinct disciplines. Contemporary human geographers do not deny the relevance of physical geography to their work, but they do not insist that their studies always include a physical component. Because human geography studies human beings, it has close ties with social sciences such as history, economics, anthropology, sociology, psychology, and political studies. One of the particular strengths of human geography is that it considers the human world in multivariate terms, incorporating physical and human factors as necessary. Where appropriate, then, this book introduces aspects of physical geography. For example, to understand world population distributions, we obviously need to know about global climates. However, we also need to know about human perceptions of those climates. Thus, human geography is related to but separate from physical geography in much the same way as it is related to but separate from the other disciplines that study human beings.

The central subject matter of human geography is human behaviour as it affects the earth's surface. Expressed in this way, the subject matter of human geography is very similar to that of the various other social sciences, all of which focus on human behaviour in some specific context.

Defining Human Geography

Providing a meaningful one-sentence definition of any academic discipline is a real challenge. In the case of human geography, however, the American geographer Charles Gritzner has suggested a useful definition in the form of three closely related questions:

> What is where, why there, and why care? (Gritzner, 2002)

Every exercise in human geography, regardless of which theme it highlights, begins with the spatial question: "Where?" Once the basic environmental, regional, and spatial facts are known, the geographer focuses on understanding, or explaining, "Why there?" Finally, the third part of the question—"Why care?"—draws attention to the pragmatic nature of human geography: geographic facts matter because they reflect and affect human life. This is true of any geographic fact, whether it is something as seemingly mundane as the distance between your home and the nearest convenience store or something as far-reaching or serious as a drought or a civil war that causes peasant farmers in Ethiopia to lose all their crops.

THE GOAL OF HUMAN GEOGRAPHY

Writing about the human world to increase our understanding of it has been the goal of human geography at least since the time of the ancient Greeks and the rise of Chinese civilization. Two statements of this continuing goal help to explain why it remains so important:

> The function of geography is to train future citizens to imagine accurately the conditions of the great world stage and so help them to think sanely about political and social problems in the world around. (Fairgrieve, 1926: 18)

> Geography is the only subject that asks you to look at the world and try to make sense of it. The field never stops being exciting, because that's what geography [is] all about—trying to make sense of the world. (Lewis, 2002: 4)

Although written some 80 years apart, these two statements express essentially the same view. Human geography is a practical and socially relevant discipline that has a great deal to teach us about the world we live in and how we live in the world. The goal of this book is to provide a basis for comprehending the human world as it is today and as it has evolved. If it succeeds, it will have made a contribution to the more general goal of advancing a just global society. Students of human geography are in an enviable position to accomplish this more general goal. A holistic discipline, human geography is not restricted to any narrowly defined subject or single theme that could limit our ability to appreciate what we might call the interrelatedness of things. Traditional links with physical geography clearly play a key role, but human geographers do not hesitate to draw on material and ideas from many other disciplines. This book overtly recognizes and reflects the controlled diversity of human geography.

society
The interrelationships that connect individuals as members of a culture.

ABOUT THIS BOOK

This diversity means that we have much subject matter and many methods to introduce. Chapters 1 and 2 address two basic issues. Chapter 1 investigates the nature of human geography and its history as an endeavour of research and exploration. Our discipline has a long and fascinating history, involving many different groups of people and many noteworthy individuals working in different places at different times. The close ties between human and physical geography are evident. By the time you reach the end of this chapter, the origins of our three recurring themes should be clear. In Chapter 2, we look at the many ways in which human geography is now studied. This subject naturally includes many technical matters. Before turning to them, however, we will look at the philosophies behind each of our three recurring themes. Some of these ideas may seem somewhat abstract on your first reading, but as you read further you will find that they are central to the issues discussed in later chapters.

less developed world
All countries not classified as more developed (see more developed world); countries characterized by a low standard of living.

more developed world
(According to a United Nations classification) Europe, North America, Australia, Japan, and New Zealand; countries characterized by a high standard of living.

The increasing necessity for human geographers to discuss many topics at a global (as opposed to a local, regional, or national) scale is a feature of the contemporary world. To point out this trend is one way of saying that our world is becoming more and more integrated and interconnected. It is not only where things are located that matters but also the connections between locations. Chapter 3, "Geographies of Globalization," details these globalization processes and thus provides an essential context for understanding contemporary human geography.

Chapter 4 offers an overview of humans' global impact on and adjustment to the physical environment. Surveying humans' occupation of the earth through time, we consider the use and abuse of resources and related environmental issues, which are of crucial importance to contemporary life. The theme of the relationship between humans and land is particularly evident in this chapter.

The next two chapters deal with population topics, both globally and in major world regions. Chapter 5 describes and suggests explanations for the exponential growth in global population, especially since about 1650, and identifies some consequences of that growth. The use of population data as indicators of regional inequalities is examined in Chapter 6, including a discussion of the idea that the world's countries can be divided into two general groups, less developed and more developed. All three of our recurring themes come into play in these chapters.

Much human geography is best studied on a group scale, employing the concepts of culture and society as they relate to human landscapes. Chapter 7 focuses on the evolution and regionalization of landscapes and includes discussions of language and religion as these involve the formation of groups and the creation of landscapes. Chapter 8 continues the discussions of group identities and related landscapes with reference to the myth of race, ethnicity, gender, and sexuality. Like Chapter 6, this chapter encourages us to think about various examples of inequality in the human experience from place to place. Two of our three recurring themes, humans and land and regional studies, are central to these discussions.

Political groupings of people in space are the subject of Chapter 9, which looks at the reasons behind the current partitioning of the world into territories. Because division into political units is the most fundamental way in which humans divide the world, this chapter includes discussions of the significance of these divisions for our way of life. Here again, the themes of humans and land and regional studies are crucial.

The next four chapters include a good deal of spatial analysis—our third recurring theme—while also reflecting our other two themes. Chapter 10, on agriculture, asks why specific agricultural activities are located where they are; describes how the various major agricultural regions around the world have evolved; discusses contemporary agricultural landscapes in both the less and the more developed worlds, using a political economy perspective and explicit reference to restructuring processes; and considers issues related to food consumption.

Chapter 11, the first of two chapters on the urban world, discusses the origins and growth of urban centres; world urbanization; and pre-industrial and world cities. Chapter 12 looks at how cities are structured and governed; at how people move around cities; at the spread of cities into previously rural areas; at cities as homes, with an emphasis on neighbourhoods and segregation; at reasons for poverty and deprivation inside cities; at cities as centres of production and consumption; and at city life in the less developed world.

Chapter 13, on industrial activity, looks at the reasons behind industrial location decisions, the period known as the Industrial Revolution, sources of energy, contemporary industrial circumstances in a global context, and the complex processes of industrial restructuring that are possibly related to a new post-industrial phase. Finally, the conclusion highlights some of the changing ways that human geographers study our changing world. For a summary of the structure of this book, indicating the chapters of most relevance for the various subdisciplines of human geography, see Table I.1.

You are probably beginning to realize that human geographers seek to understand both the world and people's lives within it. As previously noted, this book works to achieve these aims by incorporating not only a wide variety of facts about humans and their world but also a number of different approaches to those facts.

TABLE I.1 | Subdisciplines of human geography and text coverage

Subdiscipline	Chapters
Transportation geography	3
Geography of trade	3
Geography of natural resources	4
Environmental geography	4
Geography of energy	4 and 14
Population geography	5 and 6
Geography of migration	5
Geography of the less developed world	6
Regional development and planning	6 and 13
Cultural geography	7 and 8
Social geography	8 and 12
Geography of tourism	8
Feminist geography	8
Political geography	9
Economic geography	3, 10, 11, 12, and 13
Agricultural and food geography	10
Settlement geography	11 and 12
Rural geography	10 and 11
Urban geography	11 and 12
Industrial geography	13

THE HUMAN GEOGRAPHER AT WORK

What do human geographers actually do? What problems do they strive to solve? The following two vignettes offer some preliminary insights into the concerns and activities of human geographers that will be further explored in later chapters. The first looks at some of the ways our world is divided; the second looks at some of the ongoing transformations of global life.

A World Divided

Political Units

The map inside this book's front cover shows how the earth's land surface is divided into countries—one example of the diversity of human life. Are the divisions changing today or even between the time when we are writing and the time when you read these words? Is the total number of countries increasing or decreasing? What do these political divisions reflect: population distribution, cultures, economies? Is it possible to aggregate countries? Are there meaningful groupings? Do basic aggregations such as North versus South or East versus West have any value? These are some of the questions that human geographers ask.

We will answer the first of these questions now. The political world changes substantially. Consider that the map shows more than 190 countries; as recently as the 1930s, there were only about 70. Several major changes have increased the number of countries, including the decolonization of Africa and Asia, the political consequences of World War II, and the post-1989

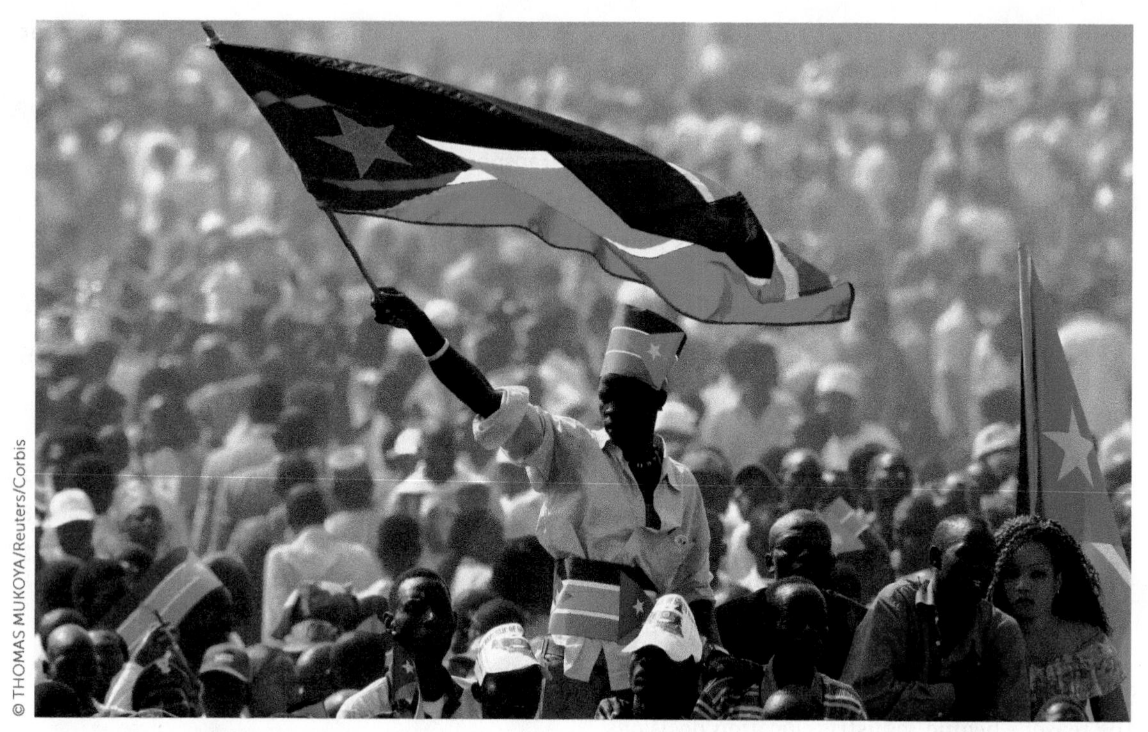
© THOMAS MUKOYA/Reuters/Corbis

After a long conflict with the north, South Sudan declared its independence on 9 July 2011. Independence day celebrations attracted tens of thousands of people, many displaying the country's flag.

political transformation of Europe and the former USSR. As of 2015, the most recent addition to the world map is South Sudan. More changes seem likely because the contemporary political world is characterized by two tendencies in particular: a tendency for regions to begin a process of integration, as in the case of the European Union, and a tendency for portions of some countries to seek a separate political identity, as in the case of Quebec within Canada.

Quality of Life

The world is divided into political units and, as previously mentioned, into larger areas often referred to as more developed or less developed. Our human world is full of diversity: languages, religions, ethnic identities, and standards of living all vary from place to place. There are differences among individuals and also more general differences between groups of individuals. Similarly, landscapes vary. Each particular location is unique, but it is often more useful to note the more general differences that exist between regions. The photographs to the right suggest a basic distinction in our contemporary world. In the first, we see middle-class people in a prosperous area of a modern Western urban centre; we can probably assume that most of them are employed, own or

Jeff Whyte/Shutterstock

Stephen Avenue, a popular pedestrian mall, in Calgary, Alberta.

© Dennis Cox/Alamy

A farm labourer with his family in front of their house in rural Rivas Department, west of Lake Nicaragua.

TABLE I.2 | Selected demographic data

World Population Totals (millions)		Birth Rates and Death Rates[1] (per 1,000), Finland			Life Expectancy at Birth (number of years), France[2]	
Year	Total	Year	Birth Rate	Death Rate	Year	Life Expectancy
1700	680	1755	45.3	28.6	1750	27
1800	954	1805	38.4	24.7	1825	41
1900	1,600	1909	31.0	17.7	1905	49
2014	7,200	2014	11.0	9.0	2014	81

Notes:
1. Birth rate refers to the total number of live births in one year for every 1,000 people living. Death rate refers to the total number of deaths in one year for every 1,000 people living.
2. Finland and France are selected because both countries have better than average records.

rent comfortable homes, and have surplus ("disposable") income for purposes such as recreation. In the second image, we see a family in a rural area of what is often called the less developed world; the people in the photo are small farmers with little more than what they need for subsistence. Pictures like these suggest that the human world is, in reality, at least two worlds.

Of course, differences exist between major world regions and between other spatial entities, such as regions inside countries. Even within a single city, the quality of life can differ enormously between the outer suburbs and the inner-city skid row. For more on this theme, see especially Chapters 6, 8, and 12.

A World Transforming

Modernization

Table I.2 provides basic sample demographic data for various years from 1700 to 2014. Population totals and life expectancy have increased while the birth and death rates have declined. In sum, these demographic changes reflect a process that we might label "modernization." Birth rates decline primarily because of changing social and economic conditions; death rates decline and life expectancy increases primarily because of technological advances related to sanitation and health. Population totals increase because of differences between birth and death rates. Clearly, however, this process of modernization has not had the same effects in all areas of the world.

Modernization involves much more than these demographic changes; cultural and economic aspects of landscape and life have also been transformed over the past three centuries, and human geographers employ their various perspectives—those suggested in our three recurring themes—to describe and explain these changes. Questions of population growth and change are discussed in Chapter 5.

Globalization

One set of processes evident today (especially in advanced economies) that is causing changes in both landscape and way of life is globalization and the rise of post-industrial society. Among the changes associated with this process are an increased importance of information technologies, an economic transformation from a goods-producing to a service economy, the rise of professional and technical workers as a new middle class, and the related decline of the working class. These ongoing changes could have consequences for our social and economic identities that will be just as dramatic as those of the Industrial Revolution, which initiated the demographic changes noted in Table I.2. In North American and European cities, for example, traditional industries are declining, the professional workforce is growing, and the process of decentralization is accelerating. Such post-industrial cities include large numbers of people who are disadvantaged, unemployed, or the "working poor" (many of the last group are part-time workers).

There is a new geography of employment, of housing, and of neighbourhoods. Here again, the human geographer uses her or his particular approach (any one of our three recurring themes) to identify these ongoing trends and the reasons for them and their character and to analyze their impacts on landscapes and ways of life. These and related topics are discussed in detail in Chapters 11–13.

CONCLUSION

Human geography is worth studying not only for the importance of the subject matter but also for the career training that it offers. Many graduating students find that a background in human geography, often in conjunction with physical geography, leads to diverse employment opportunities in such areas as education, business, and government. Geography students are employed as cartographers, geographic information specialists, demographers, land-use and environmental consultants, social service advisers, and industrial and transportation planners, to mention only a few possibilities.

Three distinctive features of an education in geography make graduates particularly valuable to employers:

- subject matter, physical and human, that provides the geographer with a specific perspective and knowledge base
- technical experience with data collection and analysis
- recognition that any attempt to understand our contemporary world must take into account the crucial roles played by space and place

Summary

Geography
Literally, writing about the world.

Three Recurring Themes
1. Humans and land: The human world, a landscape, is continuously changing because of human actions within institutional and physical frameworks.
2. Regional studies: Describing the earth and dividing the whole into parts—regions—in which one or more variables exhibit some uniformity.
3. Spatial analysis: Explaining the locations of geographic phenomena using abstract arguments and quantitative procedures.

Because the human world is constantly changing, human geography typically includes a time dimension.

Human Geography/Physical Geography
Related disciplines, both traditionally and for logical reasons. Best introduced separately at the introductory level because physical geography is a physical science, while human geography is a human science.

Our Subject Matter
Human behaviour as it affects the earth's surface.

Goal of the Book
To provide a basis for comprehending the human world as it is today and as it has evolved. As a human geographer, you will have an understanding of our world and a sound training for a variety of careers.

Questions for Critical Thought

1. What distinguishes human geography from other disciplines in the social sciences?
2. What are the similarities and differences between a human geography of the world in the early twentieth century compared to today?
3. Why does the geographical perspective matter?

Suggested Readings

Visit the companion website for a list of suggested readings.

1 WHAT IS HUMAN GEOGRAPHY?

This chapter provides an overview of the origins and evolution of human geography, describing advances in geographic knowledge through time, noting how this knowledge was gradually organized to form a new academic discipline, and outlining the changing character of contemporary human geography. Our approach is chronological, from the major contributions of the Greek, Chinese, and Islamic civilizations to European overseas exploration, which began in the fifteenth century and led to the development of new mapping techniques and improved descriptions of the world. Beginning with the attempts by seventeenth- and eighteenth-century writers such as Varenius and Kant to organize geographic knowledge and inspired by the ambitious writings of the great nineteenth-century geographers such as Humboldt and Ritter, the formal discipline of geography emerged in European and North American universities during the late nineteenth and early twentieth centuries. Since 1900, there have been several notable changes in the practice of human geography.

Here are three points to consider as you read this chapter.

- Is a chronological account misleading, in that it carries an implicit understanding that human geography's present state is an inevitable consequence of past circumstances?
- Maps are used to describe places, but do they say as much about the map-makers' motives and circumstances as they do about the area being mapped?
- Despite a long shared history, human and physical geography are now typically seen as related but separate disciplines. Is this a regrettable situation given our obvious need to address a host of environmental issues?

The study of landscape has been a constant part of human geography throughout the discipline's history. Individual locations also have their own histories. For example, Gran Paradiso—Italy's oldest national park—was declared a royal hunting reserve by King Vittorio Emanuele II in 1856 and was donated to the Italian state in 1920. The park was established two years later.

The GA.SUR, or Nuzi, map. This early map was engraved on a clay tablet *c.* 2200 BCE and discovered in 1930–1 by archaeologists in present-day northeastern Iraq. (The original tablet has since been lost.) Measuring just 7.6 × 6.5 centimetres, the map was oriented with the east at the top and showed an area bounded by two ranges of hills and bisected by a stream. It seems to have been intended to indicate the location of a parcel of land west of the stream, perhaps for purposes of defining ownership or assessing taxes.

Full of drama and intriguing individuals, the history of geography is a fascinating subject in its own right. Yet it also provides the background required for a full understanding of contemporary geography. In other words, to comprehend contemporary human geography, it is essential to grasp how it has changed through time. Indeed, there would be cause for great concern if such change did not occur. Geography, like most academic disciplines, functions to serve society. As society and societal requirements change, so does geography. Geographers have always had a consistent purpose: to describe and explain the world. Only the manner in which they approach this task changes. For example, for many years geographers were principally involved in discovering, describing, and explaining an increasingly better-known world. As unknown areas became known, geographers worked feverishly either to fit new facts into established knowledge or to propose radically new knowledge bases. Their most important tool was the map. Since the nineteenth century, geographers have reoriented their activities. Once basic global descriptions were in place, the emphasis shifted to developing better explanations and clearer understandings of geographic facts.

PRECLASSICAL GEOGRAPHY

It seems likely that the earliest geographic descriptions took the form of maps—a simple but effective means of communicating spatial information. No doubt maps have been created, temporarily at least, by all human groups. Roughly sketched with a stick in sand or soil or carefully scratched into rock or wood, maps could be used to show the location of water, game, or a hostile group. To the extent that geography is about maps, humans have always been geographers. Lack of hard evidence need not prevent us from recognizing geography's centrality to our human existence. Our ancestors could not function without maps any more than we can.

The world's first civilization—for a discussion of this often-contentious term, see Chapter 6—emerged in Mesopotamia (now southern Iraq) some time after 4000 BCE. Drawn on clay tablets, surviving Mesopotamian maps typically show local areas, reflecting limited knowledge of them beyond the immediate environment. Geographic knowledge probably was similarly limited for all civilizations before that of classical Greece, namely, those in the Nile Valley, the Indus Valley, China, Crete, the Greek mainland (Minoan and Mycenaean), and southeast Mexico (Olmec).

CLASSICAL GEOGRAPHY

With the emergence of classical Greece, shortly after 1000 BCE, geographic understanding—and hence maps—began to cover more than small local areas for the first time. The reason was that the Greeks were the first civilization to become geographically mobile and to establish colonies. Greek scholars initiated two major geographic traditions. The first is literary, involving written descriptions of the known world. Many Greek scholars contributed to this tradition. Herodotus

Examining the Issues

BOX 1.1 | Telling a Story of Geography

This chapter aims to tell the story of geography as a tradition of scholarly inquiry, as one way of looking at and seeking understanding of the world in which we live. This goal sounds straightforward enough at first blush, but "the deceptively simple word 'geography' embraces a deeply contested intellectual project of great antiquity and extraordinary complexity" (Heffernan, 2003: 3). There are two closely related challenges faced by anyone who tries to write the story of geography.

First, there is more than one story of geography. In any recounting of the past, some facts are included and others are excluded. The conventional story focuses on the achievements of influential scholars, on ever-more sophisticated key ideas, on advances in factual knowledge, on influential books, and on technical advances in mapping.

Another story that might be told is that of changing geographical ideas, both true and false. This way of thinking about the past of geography was advocated some years ago and labelled geosophy (Wright, 1926, 1947). Following this argument, the important subject matter to be included is what people believed at the time. After all, what people believe is a principal cause of their behaviour and therefore of the changing world. Wright favoured a story about geographical ideas rather than the more usual story of geographical realities.

Second, there is more than one way of telling the chosen story of geography. Inevitably, facts are interpreted and presented in accord with a specific way of thinking. In the English-speaking world the conventional version emphasizes achievement and progress from humble beginnings to present-day successes. Histories are written this way, at least partly, because we judge the past from the perspective of the present and we make the assumption that progress is an integral part of change over time. Indeed, Nisbet (1980: 4) claimed that "no single idea has been more important than, perhaps as important as, the idea of progress in Western civilization for nearly three thousand years."

A different version might focus on a theme other than progress. Both the choice of facts to be included and the interpretation of them reflect the prevailing **discourse** as this validates a particular version of a story and a particular way of telling it. In other words, who is telling the story and whose story is it? For example, was the period of European exploration and discovery not also a time of invasion and conquest?

discourse
A system of ideas or knowledge that serves as the context through which new facts and ideas are understood.

In summary, it is always important while reading to wonder, why *this* story, and why *this version* of the story. With these cautionary comments in mind, this chapter tells its version of the story of geography.

(484–*c.* 425 BCE), for example, provided descriptions of lands and peoples, based on observations made during his extensive travels to such areas as Egypt, Ukraine, and Italy. Possible relationships between latitude, climate, and population density were noted by Aristotle (384–322 BCE), who also speculated about the ideal locations for cities and the conflicts between rich and poor groups.

Eratosthenes (*c.* 273–*c.* 192 BCE)—often considered the father of geography because he coined the word—contributed to the literary tradition by writing a book describing the known world, though unfortunately no copy has survived. He also mapped the known world at the time: the Mediterranean region and adjacent areas in Europe, Africa, and Asia (Figure 1.1). We know

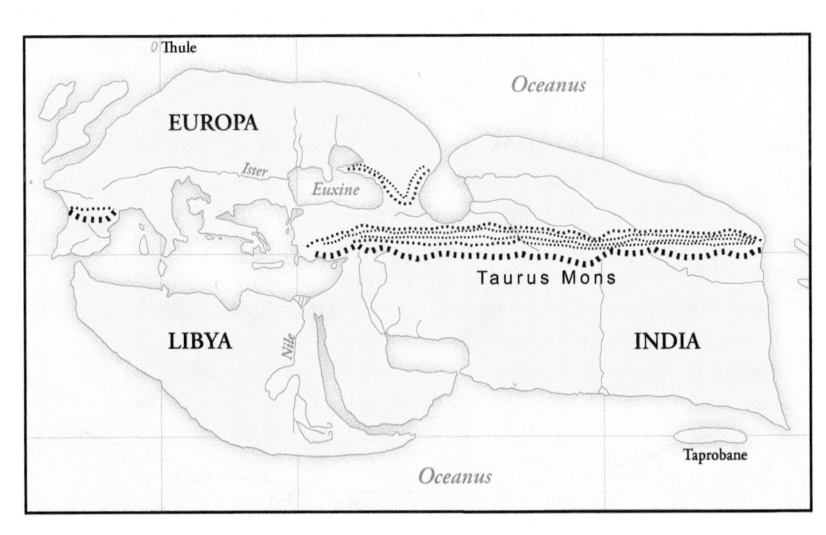

FIGURE 1.1 | The world according to Eratosthenes

longitude
Angular distance on the surface of the earth, measured in degrees, minutes, and seconds, east and west of the prime meridian (the line of 0° longitude that runs through Greenwich, England); lines of constant longitude are called meridians.

latitude
Angular distance on the surface of the earth, measured in degrees, minutes, and seconds, north and south of the equator (which is the line of 0° latitude); lines of constant latitude are called parallels.

teleology
The doctrine that everything in the world has been designed by God; also refers to the study of purposiveness in the world and to a recurring theme in history, such as progress or class conflict.

about the contributions of Eratosthenes and others because the literary tradition they began was summarized by Strabo (64 BCE–20 CE) in *Geographia*, a multi-volume work that has survived. An encyclopedic description of the entire world known to the Greeks, it consists of two introductory books, eight books on Europe, six on Asia, and one on Africa.

The second tradition is mathematical. By the fifth century BCE the Greeks knew that the earth was a sphere, and in the second century BCE Eratosthenes calculated the circumference of the earth. Shortly thereafter, Hipparchus devised a grid system of imaginary lines on the earth's surface, based on the poles and the equator. These lines, **longitude** and **latitude**, made mapping considerably more accurate. Determining the exact position of any location, however, remained difficult. Latitude was relatively easy to calculate by using an early version of a sundial to observe the angle of the sun's shadow, but longitude continued to require estimation because there was no way to measure time precisely, especially at sea. In addition, Hipparchus was the first person to tackle a problem that remains with us: how to map the curved surface of the earth on a flat surface.

Much of this mathematical tradition was summarized by the Alexandrian Ptolemy (fl. second century CE) in his eight-volume *Guide to Geography*. Using the grid system devised by Hipparchus, Ptolemy produced the first index of places, or gazetteer, of the world with coordinates provided for about 8,000 locations. Although his gazetteer is sometimes fanciful (for instance, he believed that a great continent must exist in the Southern Ocean) and includes numerous errors because latitude was not carefully measured and longitude was necessarily estimated, he also produced a world map (Figure 1.2) that includes a grid system and that, in its details, is generally a clear improvement over Eratosthenes's map (Figure 1.1). The mapping procedures devised by Ptolemy have persisted to the present.

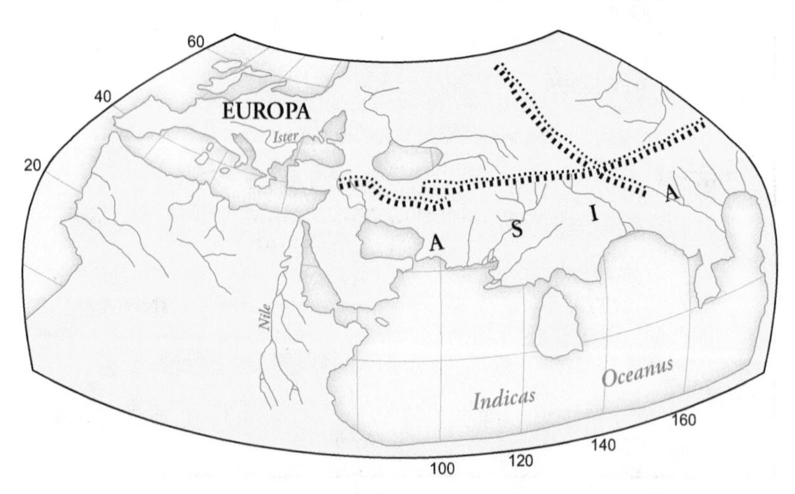

FIGURE 1.2 | **The world according to Ptolemy**

Together, the written works of Strabo and Ptolemy and the latter's world map provide a useful indication of Greek civilization's achievements in geography, for they make it clear that geography emerged and evolved in response to larger societal requirements. The civilization of ancient Rome added little to geographic knowledge, despite the expansion of the Roman Empire. The classical geography of the Greeks is the true beginning of our contemporary discipline.

THE FIFTH TO FIFTEENTH CENTURIES: GEOGRAPHY IN EUROPE, CHINA, AND THE ISLAMIC WORLD

The period from the fifth to the fifteenth centuries was one of only sporadic and limited geographic work in Europe. Elsewhere, especially in China and the Islamic world, geography flowered. Again, we find a close relationship between expanding societies and a thirst for geographic knowledge.

The European Decline

The word *geography* did not enter the English language until the sixteenth century. Medieval Europeans knew little beyond their immediate environment. Over the centuries, much of the knowledge gained by the Greeks was lost, and the only place where any kind of geographic work continued was in the monasteries. The general assumption, inside and outside the monasteries, was that God had designed the earth for humans (this doctrine is called **teleology**). In effect, geography as such no longer existed during the Middle Ages.

The clearest evidence of decline can be seen in maps from the period. The ancient Greek maps had been drawn by scholars with expertise in astronomy, geometry, and mathematics. The medieval European map-makers, by contrast, were more interested in symbolism (particularly scriptural

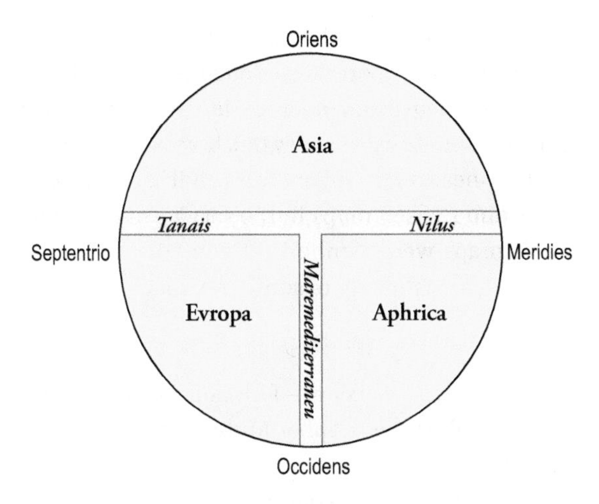

FIGURE 1.3 | An example of a T-O map

dogma—what Christians are expected to believe) than in scientific facts. Stylizing geographic reality in order to arrive at a predetermined structure, they produced less detailed and accurate maps than those produced 1,500 years earlier by the Greeks. The best examples are the T-O maps produced between the twelfth and the fifteenth centuries. Consisting of a T drawn within an O (Figure 1.3), they show the world as a circle divided by a T-shaped body of water. East is at the top of the map; above the T is Asia; below left is Europe; and below right is Africa. The cross of the T is the Danube–Nile axis; the perpendicular part is the Mediterranean; and the map is centred on Jerusalem.

Perhaps the only medieval maps that served a practical purpose were the ones known as Portolano (meaning "handy") charts (Figure 1.4). Dating from about 1300, these maps depicted a series of radiating lines from several centres on the map. Usually there are 8 or 16 such lines from each centre, and these correspond to the points of the compass. Sailors were able to lay out compass courses using these lines.

Although Norsemen sailed to Greenland and North America and Christian Europeans embarked on a series of crusades and military invasions to the Holy Land during this period, the results were minimal as far as geographic knowledge is concerned. The most significant exploratory journey was that of Marco Polo (1254–1323), a Venetian who visited China and wrote a description of the places he saw. He was unable to add to Greek knowledge because, not being a scholar, he was largely unaware of it. The distinction is not always an easy one to grasp, but Marco Polo was an explorer, not a geographer.

Geography in China

During the period after the Greeks and before the fifteenth century, the major geographic advances took place in China and the Islamic world. A great civilization—clearly the ancestor of contemporary China—developed in the former before 2000 BCE. Writings describing the known world of the Chinese date back to at least the fifth century BCE.

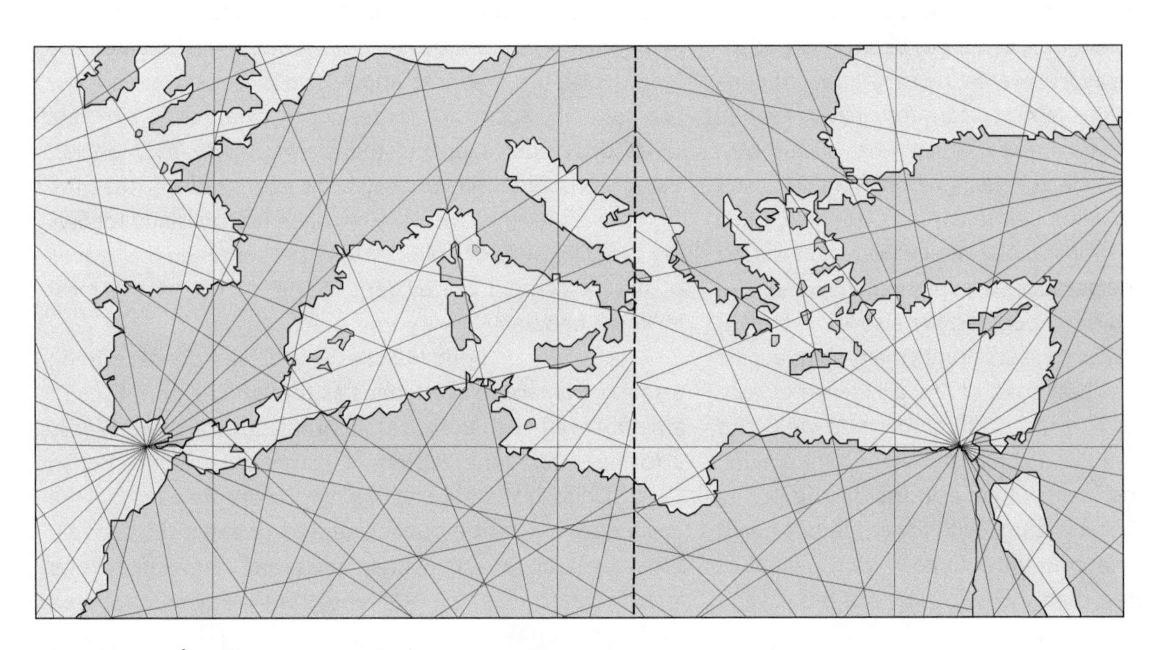

FIGURE 1.4 | An example of a Portolano chart

Later, the Chinese also explored and described areas beyond their borders; in 128 BCE, for example, Chang Chi'en discovered the Mediterranean region, described his travels, and initiated a trade route. Other Chinese geographers reached India, central Asia, Rome, and Paris. Indeed, Chinese travellers reached Europe before Marco Polo reached China (Box 1.2).

Early Chinese geography differed from the European equivalent in one important respect. It is a difference of geographic perspective—a different way of looking at the world. Traditionally, Chinese culture has viewed the individual as *a part of* nature, whereas the Greeks and subsequent European culture have typically viewed the individual as *apart from* nature. Given their view of humans and land as one, it was natural for Chinese descriptive geographers to integrate human and physical description.

Maps were central to geography in China, as elsewhere, and there is evidence that a grid system was in use during the Han dynasty (third century BCE–third century CE). It appears that the first Chinese map-makers were civil servants who drew and revised maps in the service of the state. Their maps were symbolic statements, asserting the state's ownership of some territory.

Geography in the Islamic World

As the religion of Islam—founded in the seventh century CE by the prophet Muhammad (d. 632)—spread, it served as a unifying force, bringing together previously disparate tribes. Consequently, at the time when Europe was immersed in the Dark Ages, civilization flourished in Arabia. As Islamic conquests spread beyond the Arab region, the geographic knowledge base expanded to include North Africa, the Iberian Peninsula, and India.

In the News

BOX 1.2 | Did China "Discover the World" in 1421?

Our understanding of the past is far from complete. Necessarily, numerous details remain unknown, but more fundamentally the validity of some of the most significant achievements are questioned. Consider the well-known "fact" that the first circumnavigation of the globe was accomplished by Ferdinand Magellan in the years 1519–22. In recent years an alternative "fact" concerning China has been proposed.

China was a great naval power in the early fifteenth century—so great that Chinese navigators may have reached North America 70 years before Christopher Columbus, circumnavigated the globe 100 years before Magellan, and reached Australia 350 years before James Cook. We know that Emperor Zhu Di sponsored several voyages during his reign (1403–24) for the purposes of exploring, mapping, and collecting tribute from other peoples. We also know that, in 1421, the great admiral Zheng set sail with many ships across the Indian Ocean to the east coast of Africa. However, Gavin Menzies's (2002) *1421: The Year China Discovered the World* suggests that the expedition continued around the southern capes of Africa and South America and across the Pacific before returning to China in 1423. With the emperor's death the following year, the country entered a long period of isolation, which is presumably why the story of the voyage remained unknown in Europe.

Menzies's theory—based in part on the existence of European maps from as early as 1428 depicting regions that Europeans themselves had not yet seen—is fascinating, although cartographers are not likely to rewrite their histories without some additional evidence. A 1763 Chinese map, discovered and unveiled in 2006, purports to be the needed additional evidence but has not resolved this debate. The map shows the Americas and Australia and is claimed (but not generally accepted) to be a copy of a 1419 map. Paul Chiasson (2006) supports the claims made by Menzies. Using aerial and site photographs, he suggests that the Chinese established a colony on Cape Breton Island following the discovery of gold. Further, he contends that traces of this settlement are evident in the culture of the local Mi'kmaq people.

The works of both Menzies and Chiasson have been dismissed by some academics as the fanciful narratives of amateurs rather than those of careful scholars (Menzies is a former submarine lieutenant commander in the British Navy; Chiasson is a Canadian architect). Regardless, the ideas continue to be publicized and debated. In Canada, Chiasson's book was widely reviewed, the author was interviewed on *Canada AM*, and the topic was discussed on *The National*.

By the ninth century Islamic geographers were recalculating the circumference of the earth. From then until the fifteenth century, they and their successors produced a wealth of geographic writings and maps based on earlier Greek work as well as Islamic travels. Among the most notable contributions were those of al-Idrisi (1099–1180), whose book on world geography corrected many of Ptolemy's errors. Perhaps the best-known traveller was ibn-Battuta (1304–*c.* 1368), who journeyed extensively in Europe, Asia, and Africa. A third major addition to geography came from ibn-Khaldun (1332–1406), a historian who wrote at length about the relations between humans and the environment. Maps produced by Islamic geographers centred on Arabia.

An important eleventh-century Arabic atlas, previously unknown to modern geographers, was discovered in a private collection in 2002 and is now housed in a library in Oxford, England. Widely regarded as a missing link in the history of cartography, this two-volume, 96-page manuscript includes 17 maps, 2 of them depicting the world as it was known at the beginning of the second millennium. The fact that some of the maps show travel routes suggests that, unlike earlier Greek and later European maps, they were intended not to represent actual landscapes but to serve a practical purpose as memory aids for travellers.

Chinese and Islamic geographies prior to the fifteenth century were roughly comparable to Greek geography. Two traditions (mathematical and literary) developed, and map-making was central to most geographic work. In all cases, the geographers' work reflected the knowledge and needs of particular societies.

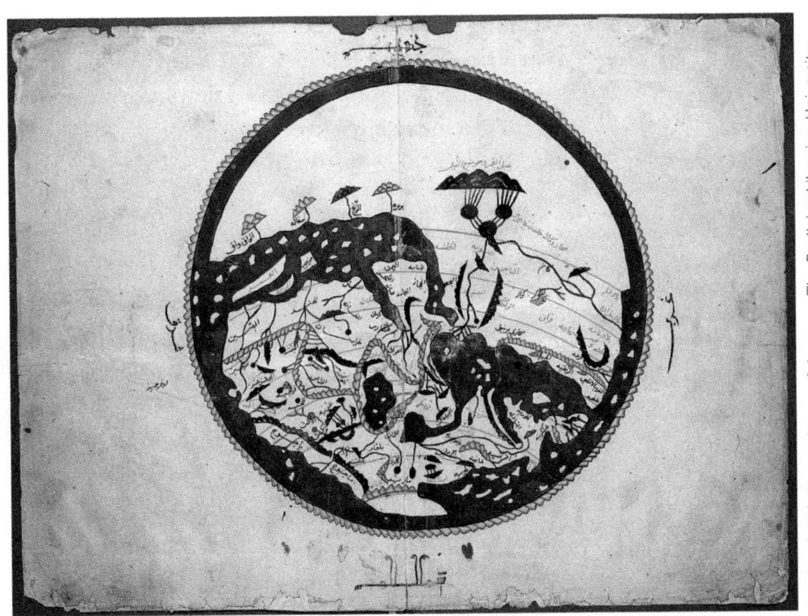

Shelfmark: MS. Popcocke 375, fols. 3v-4r. The Bodleian Libraries, University of Oxford.

A copy (dated 1456) of al-Idrisi's twelfth-century circular world map. As early as the ninth century, Muslim traders had travelled east to China by land and by sea, south along the Indian Ocean coast of Africa, north into Russia, and west as far as the Atlantic. Al-Idrisi combined the knowledge acquired through these journeys with that available from the works of Greek and Persian geographers to produce the geographic description of which this map was a part. Depicting Europe and Africa as well as Asia, it is oriented with south at the top.

THE AGE OF EUROPEAN OVERSEAS MOVEMENT

By 1400, then, geographic knowledge had grown considerably, but the "known world" on which it was based was still limited to a small portion of the earth. In the fifteenth century European scholars began to recognize this fact, and the impact on geography was dramatic and significant. All three components discussed so far—mathematical, literary, and cartographic—underwent rapid change. Between the fifteenth and seventeenth centuries, Europeans embarked on a period of unprecedented exploratory activity that happened to coincide with a decline in Chinese and Islamic explorations.

Many factors, among them the desire to spread Christianity and to establish trade routes, contributed to the surge in European exploration, but there were two interesting additional factors. Printing technology, which was first applied to maps by the Chinese in 1155, was first used for map production in Europe in 1472. In 1410, editions of Ptolemy's work were printed; editions including maps (redrawn using new projections) appeared in 1477. (A copy of Ptolemy's atlas printed at this time sold at a 2006 auction for almost US$4 million.) Printing allowed information, including that contained on maps, to diffuse rapidly.

The second additional factor, which reflected the period's thirst for knowledge, was the establishment of what might be called centres of geographic analysis. The first was initiated by Prince Henry of Portugal in 1418 and focused on key geographic questions such as the size of the earth and the suitability of tropical environments for habitation. In addition, techniques of navigation were taught there, and Prince Henry set in motion a series of explorations along the western coast of Africa. Interestingly, the maps produced following these voyages did not always reflect new discoveries. It seems that a 1459 map by Fra Mauro deliberately concealed new information in order to maintain secrecy.

Exploration

Exploration is not, of course, geography, but it furnished new facts and provided the basis for new maps, books, and descriptive geographies, which encouraged further exploration. The major explorations were led by Bartolomeu Dias around southern Africa (1486–7), Columbus to North America (1492–1504), Vasco da Gama to India (1497–9), and Magellan, who reached Asia by sailing west (1519–22). The last in this line of great explorers was James Cook, who made three voyages into the Pacific (1769–80). It was Cook who corrected one of Ptolemy's greatest errors, revealing that the supposed southern continent did not exist, at least not as it had been envisioned (see Box 1.3). By 1780 the basic outlines of the world map were established, adding considerably to Europeans' knowledge of the world and thus to the geographer's task.

Exploratory activities became easier as several basic hurdles were overcome. Most important was the absence, recognized over a thousand years earlier by the Greeks, of an accurate method of determining location. As we have seen, establishing latitude was no problem, but an instrument for establishing longitude at sea was not available until 1761 and was not used on a major voyage until Cook's second in 1772–5.

Mapping

During the early phase of European overseas movement, science in general changed from being a practice controlled by the church to one concerned with the acquisition of knowledge. In response to the demands of sea travellers, maps in

Around the Globe

BOX 1.3 | The Southern Continent

The Greeks were the first to suggest that a large land mass should exist in the southern hemisphere to balance the large known areas of land in the North. Along with so many other provocative ideas, this notion was lost during the medieval period; the medieval belief in a flat world did not require any such symmetry. The first knowledge of the basic distribution of land and sea in the southern hemisphere was gained with the beginnings of European overseas exploration in the fifteenth century. Early voyages often were prompted by incorrect information and sometimes garnered fiction as well as fact. For example, in the early sixteenth century Binot Paulmier Gonneville claimed to have discovered a tropical paradise in the South, inhabited by an easygoing, contented people. All records of this voyage were lost, but the claim clearly influenced French exploration for the next 200 years. Subsequent French and other explorers sought to rediscover this idyllic "Gonneville land."

When Magellan, on his voyage of 1518, discovered the Straits of Magellan, separating the island of Tierra del Fuego from the mainland of South America, it was thought that the island was the edge of the huge southern continent. Cartographers such as Orontius (in 1531) responded with maps that included the continent. By 1578, however, Francis Drake had shown that Tierra del Fuego was not a large continent. Exploration then turned east with the discovery of the islands of South Georgia by Anthony de la Roché in 1675. In 1772 Breton-French explorer Yves Joseph de Kerguelen reported that he had discovered a place promising all the resources the mother country desired—wood, minerals, diamonds, and rubies. Many thought that this area, which Kerguelen called South France, was the long-sought Gonneville land. Find Kerguelen Island in your atlas and you will see just how wrong they were: the island is in an isolated, far from tropical location in the south Indian Ocean. Clearly Kerguelen discovered what he wanted to discover, just as Gonneville had.

The distribution of land and sea in the southern hemisphere became clearer in the late eighteenth century, even as Dalrymple was unequivocally asserting that almost all the area between the equator and 50°S was land. In 1773 Cook spent much of a long voyage searching for the southern continent, sailing farther south than any previous explorer but finding nothing but pack ice. Although he discovered numerous small islands on his 1775 voyage, all were inhospitable environments. The coastline of Antarctica was finally mapped during the nineteenth century.

this period returned to the model developed by the Greeks—a model in which facts triumphed over imagination. Unlike medieval maps, with their mixture of fantasy and dogma, the typical map from the fifteenth century onward was functional. Maps showing the grid of latitude and longitude replaced T-O maps centred on Jerusalem.

The demand for geographic knowledge, and hence for maps, coincided with the rise of printing to encourage new developments. Gerardus Mercator (1512–94) was undoubtedly the most influential of the new map-makers. He tackled the crucial problem of representing a sphere on a flat surface. His answer was the famous 1569 Mercator projection, which is still used extensively. This projection, which showed the earth as a flat rectangle with a grid of latitude and longitude lines, was enormously useful to sea travellers because a straight line on the map was a course of constant compass bearing. By the early seventeenth century, Mercator's map had replaced all earlier charts used at sea, including the Portolanos. Another map-maker, Abraham Ortelius (1527–98), produced the first modern atlas in 1570; such was the demand for geographic knowledge that it ran into 41 editions by 1612.

Library and Archives Canada, e004414662

Mercator's map of the North Pole, 1595.

Geographic Description

The awakening of Europe and the burst of exploratory activity had an impact not only on map-making but also on geographic description. European scholars had to make sense of the "new world." Once again, society turned to geography to provide answers to a multitude of questions concerning the shape of the earth, the location of places, physical processes, and human lifestyles. Just as earlier Greek, Chinese, and Islamic geographers had described the worlds known to them, the European geographers of the sixteenth century onward needed to describe the world they knew—one that was constantly expanding and changing. Now geographers faced an enormous task: writing about all aspects of the entire world.

The resulting flurry of geographic writing was of mixed quality. Some works of fiction were regarded as factual. Imaginative works by an author of uncertain identity, Sir John Mandeville, were reprinted three times in 1530 alone. Other geographers, however, largely reflected available knowledge and excluded content that was essentially surmise. Their works, along with developments in map-making, provide a clear indication of the progress of geography and geographic understanding. Among them was Peter Apian (1495–1552), a map-maker and writer who in 1524 published a book that divided the earth into five zones (one torrid, two temperate, and two frigid), provided notes on each continent, and listed major towns. Sebastian Münster (1488–1552), a contemporary of Apian, produced *Cosmography* in 1544, the first major work following the initial burst of European expansion activities that included descriptions of the earth's major regions.

GEOGRAPHY RETHOUGHT

Varenius

Contemporary geography continues to be concerned with map-making and description, but it also addresses many other questions. The first indication of these new questions appeared in 1650 with the publication of *Geographia Generalis* by Bernhardus Varenius (1622–50), which remained the standard geographic text for at least a century; the last English-language edition appeared in 1765. What made this work so distinctive and

 Ritter

Humboldt

important was that Varenius provided an explicit definition of geography as the study of the state of the earth, both physical and human, and also emphasized the need for both detailed description (what he called special or particular geography) and generalizations (what he called general or universal geography).

The one and a half centuries following the publication of Varenius's major work witnessed major advances in a wide range of geographic issues. Exploratory activity and map-making continued apace. In addition, geographic questions were asked about such fundamental issues as the physical environment's role as a cause of the growth of civilization, the unity of the human race, and the relationship between population density and productivity.

Kant

Immanuel Kant (1724–1804) taught at the University of Königsberg and is best known for his work in logic and metaphysics. But he also lectured in geography for a period of 40 years. He has been described as "the outstanding example in western thought of a professional philosopher concerned with geography" (May, 1970: 3).

Alexander von Humboldt

© Archive Pics/Alamy

To introduce his lectures on geography, Kant emphasized that the subject involved the description or classification of facts in their spatial context. As a result, he has been interpreted by some later geographers, especially Alfred Hettner, as an explicit advocate of geography as a regional study. This interpretation is also based on Kant's argument that geography is description according to space and that history is description according to time. In *Physische Geografie* (1802), Kant asserted that geography and history together comprise all knowledge.

Universal Geography: 1800–1874

Carl Ritter

© Historical image collection by Bildagentur-online/Alamy

The first major work of this period was published by a Danish geographer, Conrad Malte-Brun (1775–1826), between 1810 and 1829. Following in the established tradition, this all-embracing study includes both general and special geography as defined by Varenius. Mathematical, physical, and political principles are discussed along with physical phenomena, including animals and plants, and human matters, including race, language, beliefs, and law. Describing all areas of the known world, Malte-Brun succeeded in producing a complete geography.

However, geography in the first half of the nineteenth century was dominated by two German scholars, Alexander von Humboldt and Carl Ritter. Humboldt's greatest work was *Cosmos*, a five-volume study published between 1845 and 1862; the title is the Greek term for an orderly universe (as opposed to chaos). Like earlier geographers, Humboldt wanted to describe the universe, but that was not all. As he wrote, "my true purpose is to investigate the interaction of all the forces of nature." For him, humans were part of nature—a major shift of emphasis in the European world. In addition to conventional regional descriptions, he strove to offer a complete account of the way all things are related. General concepts were carefully blended with precise observation.

Beginning in 1820, long before the first departments of geography were established, Ritter held a chair in geography at the University of Berlin. He was concerned with relationships and argued for coherence in describing the way things are located on the earth's surface. Like Humboldt, he expressed interest in moving from description alone to description and laws. This, of course, was exactly where Varenius had directed the attention of scholars. These interests are paramount in *Die Erdkunde*, an only partially complete world geography comprising 19 volumes published between 1817 and 1859.

All three of our recurring themes are evident in the work of Humboldt and Ritter. The study of humans and land is central to their conception of geography; their interest in the formulation of concepts as general statements that aid in the understanding of specific facts anticipate a focus on regions and the concerns of spatial analysis. Together, Humboldt and Ritter represent a fitting conclusion and an appropriate beginning. Although much of the world was still unknown to Europeans—Africa was largely a mystery, as was Asia—these two men produced two great works of geography. Both recognized the need for complete geographic descriptions and yet neither could truly fulfill that aim. By the year of their deaths, 1859, there was simply too much geography. A complete, encyclopedic description of the world was now beyond the reach of a single scholar.

Humboldt and Ritter were the first geographers to pay full attention to concept formulation, that is, to the derivation of general statements from the detailed factual information available. Their interest in the relations between things and their inclusion of humans as part of nature established the basis for much subsequent geography. Thus, they represent both an end to the long road begun by the Greeks and a beginning to a geography combining regional description and concept formulation.

Map-Making, Exploration, and Geographical Societies

Despite considerable new thinking about geography, there was also a steady continuing interest in some of the traditional geographic concerns. Map-making was considered so important that governments began to assume responsibility for the task (you will recall that civil servants were responsible for map-making in China, probably as early as the third century CE) and in England the Ordnance Survey was founded in 1791. Large-scale topographic maps, showing small areas in considerable detail, became possible with the development of exact survey techniques in eighteenth-century France. Atlases also became much more sophisticated. Exploration continued and was greatly assisted by the founding of geographical societies in Paris (1821), Berlin (1828), London (1830), St Petersburg (1845), and New York (1851). The ever-growing interest in overseas areas encouraged prosperous individuals as well as governments to support these organizations.

The home of the Royal Geographical Society (RGS) in London. The RGS is the largest geographical society in Europe and one of the largest in the world. In the nineteenth century, such societies helped geographers to organize their field of study and promoted recognition of their work. Most also encouraged exploration and implicitly—if not explicitly—supported colonial activity. In 2009 the RGS hosted a spirited, even contentious, debate regarding its required role in the contemporary world. Some members argued for continued funding of major exploratory activity, while others argued for more and less expensive academic projects. The latter viewpoint was successful, heralding a symbolic end to the geography-as-exploration era. (Royal Geographical Society)

INSTITUTIONALIZATION: 1874–1903

The year 1874 marks the formal beginning of geography as an institutionalized academic discipline. In 1903, the first North American university department of geography was established. This 30-year period was one of dramatic developments in geography, and scholars had to consider the appropriate weight to give the many aspects of established geographic traditions, namely, mapping, description, general and special geography, regions, and human–land relations. They also had to take into account ongoing developments in all the physical and social sciences, as well as the hugely controversial issues raised by Darwinian thought and the moral questions associated with expansion overseas. In brief, from 1874 onward, geographers had a great deal to accommodate. How did they respond?

Germany

In 1874 the Prussian government established geography departments in all Prussian universities. Why? Possibly because the time was ripe; possibly because the value of geography was especially evident following the Franco-Prussian War of 1870–1; possibly because of a belief that geographic knowledge would facilitate political expansion. Whatever the reasons, it became the responsibility of these new departments to clarify what geography entailed. Several leading scholars attempted this task. For Ferdinand von Richthofen (1833–1905), geography was—as it had been earlier—the science of the earth's surface. Richthofen maintained the distinction between special and general geography and further argued that the two could be combined to form a chorological (regional) approach. The subject matter of this regional geography (or **chorology**) included human activities but only in relation to the physical environment.

Friedrich Ratzel

Friedrich Ratzel focused on human geography, or what he termed anthropogeography. His major work, *Anthropogeographie*, was published in two volumes (1882 and 1891). The overriding theme of the first volume is the influence of physical geography on humans, while the second volume focuses on humans using the earth. Interestingly, the first volume had the greatest immediate impact, possibly

chorology
A Greek term revived by nineteenth-century German geographers as a synonym for *regional geography*.

Ratzel

because its inherently simple cause-and-effect logic was in accord with mechanistic views and helped to assert the importance of physical geography. It is appropriate to regard Ratzel as a founder of human geography because he probably was the first to focus on human-made landscape.

Alfred Hettner (1859–1941) was the most influential follower of Richthofen. In his methodological and descriptive work, geography was unequivocally regarded as the chorological science of the earth's surface, a view that had been clearly enunciated first by Kant. A persuasive advocate of regional geography, Hettner was highly influential in the United States. However, his work excluded studies of human–land relations, as well as studies involving time.

Thus, German geography following 1874 consisted of three quite different interpretations: geography as chorology (Richthofen and Hettner), geography as the influence of physical geography on humans (Ratzel, volume l), and geography as the study of the human landscape (Ratzel, volume 2). These differences continued well into the twentieth century.

France

Vidal

Vidal

Geography in France followed a route independent of German developments, although it reflected many of the same ideas expounded by Ratzel in the second volume of *Anthropogeographie*. Elisée Réclus (1830–1905) paved the way for later French geography. A geographer and anarchist who was barred from France and imprisoned at various times, he published a descriptive systematic geography of the world and a 19-volume universal geography. His work was very much in the Ritter tradition.

The most important French geographer, Paul Vidal de la Blache, laid down ideas that, once accepted and amplified by others, established a dominant French geographic tradition known as ***géographie Vidalienne***, or ***la tradition Vidalienne***. His overriding concerns were the relations between humans and land, the evolution of human landscapes, and the description of distinctive local regions. For Vidal, geography should consider both physical geographic impacts on humans and human modification of physical geography. The parallels with Humboldt, Ritter, and Ratzel (volume 2) are clear. Vidal and his many followers continue to exert enormous influence on human geography.

***géographie Vidalienne* (or *la tradition Vidalienne*)**
French school of geography initiated by Paul Vidal de la Blache at the end of the nineteenth century and still influential today, focusing on the study of human-made (cultural) landscapes.

Britain

Britain's first professorship of geography was established at University College London in 1833. In 1887 geography lectureships were created at Oxford and Cambridge universities with funding provided by the Royal Geographical Society (RGS), but it was not until 1900 that the first British geography department was set up, at Oxford, again largely as a result of RGS efforts. Halford J. Mackinder (1861–1947) was the first dominant influence. Interestingly, he determined to become the first European to climb Mount Kenya, in Africa, to ensure that his geographic views would be favourably received; he believed that one had to be a successful explorer to be respected as a geographer! For Mackinder, geography and history were closely related: a global geographic perspective was essential, and, following Richthofen, physical geography was a prerequisite for human studies.

United States

The establishment of the first North American department of geography, at the University of Chicago in 1903, came at a time when American geography was influenced largely by German scholars. Physical geography (physiography) was dominant, possibly because of the powerful personality of William Morris Davis (1850–1934), a geologist who promulgated the German view that physical geography influenced human landscapes and that geography was essentially a regional science. Ellen Churchill Semple (1863–1932) followed the early Ratzel to become a leading proponent of the "physical influences" school of thought, while many American geographers quickly adopted a regional focus. At the time there was no evidence of a distinctive school analyzing the relations between humans and physical geography. Neither the general legacies of Humboldt and Ritter nor the views of Vidal were in evidence, although George Perkins Marsh (1801–82), an American geographer and congressman, had earlier focused on the same issues.

Geography in 1903

We have once again reached an important turning point. By 1903 geography was a full-fledged academic discipline in many European countries and in the US. Elsewhere geography was institutionalized somewhat later, largely as a consequence of academic contacts with these pioneering countries. In Canada, for example, a partial department of geography was created at the University of British Columbia in 1923, 12 years before the first full department was established in 1935 at Toronto. Canada currently has more than 40 geography departments. An interesting recent trend in Canada and some other countries, notably Australia, involves the explicit linking of geography and environmental programs in a single university department.

In 1903 the general subject matter was clear—there had been no real change since Greek times—and a number of different approaches were advocated. Many people studied physical geography in its own right; some focused on physical geography as the cause of human landscapes; others concentrated on regional description; still others centred their attention on humans and nature combined. These general threads continued through much of the last century. Geography also continued its association with mapping and, to a lesser extent, exploration.

PRELUDE TO THE PRESENT: 1903–1970

The period from 1903 to 1970 was characterized by several different approaches to geographic subject matter. As in the past, academic geography continued to emphasize the three principal areas of study—physical geography as cause, humans and land, and regional studies—and added a fourth, relatively novel, approach: spatial analysis. The physical-geography-as-cause approach proved relatively short-lived; the other three, of course, are our recurring themes.

Physical Geography as Cause

Much of the attraction of the emphasis on physical geography as cause (most clearly stated in Ratzel's volume 1) lay in its relative simplicity. Arguing that human landscapes and cultures resulted primarily from physical geography reduced the need to think about economics, politics, societies, and so forth. Among the well-known geographers who favoured this approach were Semple, Ellsworth Huntington (1876–1947), and Griffith Taylor. Nevertheless, the principal significance of the physical-geography-as-cause perspective (also known as **environmental determinism**) lies in the fact that it simply was taken for granted as a self-evident truth. Much twentieth-century geography that did not explicitly take this perspective did so implicitly.

There are important and often intimate relations between physical and human geography, but these are never so straightforward as to allow us to assume that the former is the cause and the latter the effect. Examples of these intimate relations include much evidence that movements of early humans and technological changes such as the development of agriculture often were linked to changes in climate (noted in Chapters 4 and 10). It is also clear that the global distribution of humans corresponds closely to global environmental regions (discussed in Chapter 5; see Figure 5.13).

Fortunately for geography, environmental determinism as an explicit identification of physical cause and human effect, although very popular in the first half of the twentieth century and occasionally taken to extremes, never became the focus of a formal school of geography and is now discredited (Box 1.4).

Humans and Land

The view of geography as the study of all things physical and human, specifically as the study of relationships between physical and human facts, needed a better definition of its subject matter and a more nuanced formal methodology than that provided by environmental determinism. These were provided by three scholars.

Vidal, as we have seen, developed *géographie Vidalienne*, beginning about 1899. Otto Schlüter (1872–1952) founded a German school of **Landschaftskunde**—"landscape science" or "landscape geography"—in 1906, which provided a clear definition of subject matter. The third scholar was Carl Sauer, an American who effectively introduced the various European ideas to North America in 1925 and elaborated on them. The **landscape school** that grew out of his work focuses on human cultural groups' transformation of the physical geographic landscape over time.

Together, the ideas of Vidal, Schlüter, and Sauer provide us with a clearly defined approach to our subject matter. Landscape geography explicitly rejects any suggestion of physical-geography-as-cause by rejecting environmental determinism in favour of **possibilism**, the view that human activity is determined not by physical environments but by choices that humans make.

Taylor

University of Toronto Image Bank

Griffith Taylor (1880–1963), born in Britain, was responsible for introducing geography as a university discipline in Australia and Canada, and was a leading environmental determinist.

environmental determinism
The view that human activities are controlled by the physical environment.

Sauer

Landschaftskunde
A German term, introduced in the late nineteenth century and best translated as "landscape science"; refers to geography as the study of the landscapes of particular regions.

landscape school
American school of geography initiated by Carl Sauer in the 1920s and still influential today; an alternative to environmental determinism, focusing on human-made (cultural) landscapes.

possibilism
The view that the environment does not determine either human history or present conditions; rather, humans pursue a course of action that they select from among a number of possibilities.

Examining the Issues

BOX 1.4 | Environmental Determinism

"In the early part of this century, a major stumbling block to meaningful objective research in American geography was the assumption that the physical environment largely determined the cultural landscape" (Dohrs and Sommers, 1967: 121).

There is no denying the long scholarly pedigree of this view in Western thought. Both Plato and Aristotle regarded Greece as having an ideal climate for government. Montesquieu contended that areas of high winds and frequent storms were particularly conducive to human progress. As the French philosopher Victor Cousin put it:

> Yes, gentlemen, give me the map of a country, its configuration, its climate, its waters, its winds and all its physical geography; give me its natural productions, its flora, its zoology, and I pledge myself to tell you, a priori, what the man of this country will be, and what part this country will play in history, not by accident but of necessity, not at one epoch but at all epochs. (quoted in Febvre, 1925: 10)

In addition to this scholarly background, environmental determinism proved attractive to geography for four reasons.

1. According to the prevailing late nineteenth-century view, it was the job of science to explain phenomena in terms of cause and effect. Thus, a geography centred on environmental determinism was a geography with scientific credibility. Specifically, its emphasis on cause and effect linked it philosophically to Charles Darwin's landmark work, *On the Origin of Species*, published in 1859.
2. When geography was in its institutional infancy, any approach that explicitly asserted the importance of geography to human affairs would enhance the new discipline's status.
3. The apparent logic of the environmental determinist argument was attractive in itself. As Taylor once wrote, "as young people we were thrilled with the idea that there was a pattern anywhere, so we were enthusiasts for determinism" (cited in Spate, 1952: 425).
4. Environmental determinism was seen by some as offering a distinct and necessary basis for the new discipline of geography.

Nevertheless, in retrospect it still is surprising that so many geographers accepted environmental determinism in a relatively uncritical manner. Davis, for instance, considered any statement to be geographic if it linked physical factors to some human response, while Semple (1911: 1) wrote, "Man is a product of the earth's surface." Most significantly, until the 1950s environmental determinism was implicitly accepted in much geography, including much of the regional geography that was the dominant approach at the time.

Taken out of context, many of the apparently extreme statements of this view, such as those quoted from Cousin and Semple, are easily misinterpreted. However, careful reading suggests that such dramatic statements are not substantiated in the larger body of their work—see, for example, G.R. Lewthwaite (1966: 9). Indeed, Huntington (1927: vi), a human geographer usually regarded as an archetypal determinist, wrote, "Physical environment never compels man to do anything: the compulsion lies in his own nature. But the environment does say that some courses of conduct are permissible and others impossible."

Debates about the relative merits of determinism and alternative interpretations became common in the 1950s. Environmental determinists posited that it was essential to talk in terms of cause and effect, and therefore it was necessary to be a determinist. The basic counter to this assertion was that environmental determinist statements were incapable of being tested; they were merely working assumptions, useful only for suggesting research directions. According to this logic, the fundamental error of much environmental determinism is that it simply assumes a certain cause and effect to be true, when in actuality there is no basis for any such assumption.

Many other social scientists regarded geography with some dismay because of what they perceived as an overemphasis on the role of physical factors in discussions of human behaviour. Émile Durkheim even objected to the inclusion of physical factors in the French school of geography initiated by Vidal, a school that itself opposed environmental determinism.

Regardless of where one stands on the relative importance of physical and human factors, it is always an oversimplification to treat physical geography as a cause of human behaviour.

Possibilism emerged as a component of the newly institutionalized discipline of geography at much the same time as environmental determinism for three reasons:

1. A substantial literature, centred on humans and land, already presumed that humans could make decisions not explicitly caused by the physical environment. (Humboldt and Ritter were its two most notable authors.)
2. Geographers could point to many instances in which different human landscapes were evident in essentially similar physical landscapes.
3. Possibilism corresponded to the historically popular view that every event is the result of individual human decision-making.

Regional Studies

Regional geography, or chorology, proved to be the most popular focus during the first half of the nineteenth century. Pioneering work by German geographers such as Richthofen and Hettner was carried into English-language geography, and there was a prevailing understanding that the ultimate task of geography was to delimit regions. In America this view was attractive for physical geographers in particular; for example, W.M. Davis produced a map of regions as early as 1899. In 1905 the British geographer A.J. Herbertson (1865–1915) proposed an outline of the world's natural regions. These developments culminated with the 1939 publication of *The Nature of Geography* by the American Richard Hartshorne. This substantial contribution to geographic scholarship argued forcefully for geography as the study of regions—what Hartshorne often called **areal differentiation**—a view that was very much in accord with prevailing American opinion and that continued to dominate geography until the mid-1950s.

Geography in the North American world by 1953 was thus characterized by two related but different emphases: analysis of the relationship between humans and land and regional studies. Both were the products of long geographic traditions, and both have continued to change. In 1953, however, a new focus was forcefully introduced.

Spatial Analysis

A 1953 paper by F.W. Schaefer (see Chapter 2) is usually seen as the beginning of spatial analysis. This approach, as we saw in the introduction, focuses on explaining the location of geographic facts. Schaefer argued that geographers should move away from simple description in regional studies to a more explanatory framework based on scientific methods such as the construction of theory and the use of quantitative methods. Here again we can recognize the special/general distinction identified by Varenius. Schaefer objected to what he saw as the overly special focus of regional geography. Regardless of the specific merits of Schaefer's arguments, his views struck home with many. What followed might be called a revolution, in which the emphasis shifted to quantitative studies. The 1960s were characterized by phenomenal growth in detailed analyses based on the proposal and testing of hypotheses by means of quantitative procedures.

By 1970, regional geography was receding in popularity and spatial analysis had found a niche alongside somewhat modified versions of both the landscape and regional approaches.

HUMAN GEOGRAPHY TODAY: RECENT TRENDS AND SUBDISCIPLINES

As discussed in the introduction and throughout this chapter, human geography is a remarkably wide-ranging discipline that employs a variety of approaches and is divided into a number of subdisciplines (see Table I.1)—hardly surprising given the initial goal of geography as providing a description of the earth and our definition of human geography as asking "Where?" "Why there?" and "Why care?" The different approaches reflect different ways of thinking about human geography, while the presence of subdisciplines reflects the fact that the subject matter of human geography is so diverse. Packaging the content of human geography into subdisciplines is essentially a way of imposing some order on what we do.

areal differentiation
From Hartshorne, a synonym for *regional geography*.

Since about 1970, human geography has seen a number of important revisions and additions both to subdiscipline identification and to traditional interests and approaches. These are discussed in the following sections.

Physical and Human Geography

We have seen that geography from the Greeks onward has had both physical and human components. Until the nineteenth century, however, the clear tendency was to treat them separately. Humboldt and Ritter, by contrast, focused closely on integrating the two. Subsequently, some geographers saw physical geography as paramount (the environmental determinist school); some saw the human landscape as the relevant subject matter (the landscape school); and some saw the two as separate (the regional school). It was not unusual for geographers to assert a unity between physical and human geography, but there was not much evidence of such unity. Very few geographers deny the relevance of physical to human geography, but equally few see the one as necessary to the other. Therefore, we tend to teach and research the two separately, an approach that applies even in most of the newly formed departments that include both geographers and environmental scholars.

Of course, some regional accounts and some specific issues require consideration of both dimensions. Chapter 4 of this book, on environmental geography and natural resources, demonstrates the value of a geography that is both physical and human. And a basic understanding of global physical geography is obviously relevant to much work in human geography (see Appendix 1). Overall, though, the two are no longer seen as so closely related that we need to discuss them as one. Contemporary human geography is a social science discipline but one with special and valuable ties to physical sciences, especially physical geography.

Contemporary Landscape Geography

At least since the writings of Vidal, Schlüter, and Sauer, the study of landscape has been a consistent presence in human geography. It typically involves studies of human ways of life, cultural regions and related landscapes, and relationships between human and physical landscapes. But the landscape approach has recently been enhanced by the inclusion of new conceptual concerns. Since about 1970, ideas associated with humanism, Marxism, and feminism, along with a generally increased awareness of advances in other social science disciplines, specifically those relating to postmodernism, have enriched landscape studies.

Social and cultural geographic studies of landscape are now concerned both with visible features (such as fields, fences, and buildings) and with symbolic features (such as meaning and values). Thus, many contemporary landscape geographers focus on the human experience of being in landscape; here landscapes are regarded as relevant not simply because of what they are but also because of what we think they are. Further, there is explicit acknowledgement that landscapes, like regions, both reflect and affect cultural, social, political, and economic processes.

Contemporary Regional Geography

The rise of spatial analysis came largely at the expense of the areal differentiation articulated by Hartshorne. Since 1970, however, regional geography, like landscape geography, has resurfaced in a somewhat different guise and has once again become a central perspective. Most human geographers accept that the earlier regional geography is no longer appropriate; there is a clear need to move beyond regional classification and description. Yet the regional approach is obviously a valuable and distinctive geographic device. One geographer described regional geography as "the highest form of the geographer's art" (Hart, 1982: 1).

Regional geography currently emphasizes the understanding and description of a particular region and what it means for different people to live there. This emphasis reflects at least three general concerns: regions as settings or locales for human activity; uneven economic and social development between regions, including a focus on the changing division of labour; and the ways in which regions reflect the characteristics of the occupying society and in turn affect that society. Looking at regional geography in this light, we can see it as increasingly satisfying the requirement that human geography serve society by addressing a range of economic and social problems.

Contemporary Spatial Analysis

The spatial analytic approach that first became a prime interest of human geographers in the mid-1950s was very influential until about 1970. It has continued to be an integral part of human geography, but it is generally regarded as only one among several widely accepted approaches. A central concern is that the theoretical constructs it uses to explain locations are somewhat limited and hence tend to emphasize generalizations at the expense of specifics. The topics studied by spatial analysts are mostly in the subdisciplines of economic, agricultural, settlement, urban, and industrial geography (for example, the locations of industrial plants), as opposed to political, cultural, or social. This tendency reflects the influence of various economic location theories.

There is a clear distinction between the spatial analytic approach and the contemporary regional and landscape approaches. Indeed, some geographers believe that spatial analysis is overly concerned with spatial issues, perhaps to the point of seeing space as a cause of human landscapes, and as a result ignores the full range of human variables as causes of landscapes. Contemporary regional and landscape geographers explicitly argue that space is important only when it is analyzed in terms of the use that humans make of it. This important philosophical question will be discussed in the next chapter.

Global Issues

Geography has always been a global discipline. Much of the work of the early Greek, Chinese, and Islamic geographers, for example, was concerned with describing the limits of the known world and identifying links between places. European geography, as it developed from about 1450 onward, focused on understanding where places were, mapping routes, and describing what was previously unknown.

But there is more behind our current concern with global issues. The fact is that people and places in the contemporary world increasingly are connected and interconnected. Today we are interested in the global movements of people, products, ideas, and capital. We are interested in how the many and diverse peoples and places scattered throughout the world are increasingly associated with one another. Human impacts on the earth, population growth, the spread of diseases (such as AIDS, SARS, H1N1, and Ebola), international migration and refugee problems, food shortages, vanishing languages, the spread of democracy, agricultural change, urban growth, and industrial restructuring are just a few of the major issues in the world that require thinking on the global scale. More than ever before, human geographers are finding it necessary to look at the big picture—the world—to understand the detail at the local level.

Applied Geography

Because geography is an academic discipline that serves society, it has always had an applied component in which geographic skills are used to solve problems. Geography as exploration is a prime example of such applied work. Since about 1930, geographers have played a major role in land-use studies. For example, the studies of the British Land Utilization Survey directed by L.D. Stamp in the 1930s proved enormously valuable during World War II, and American geographers made important contributions in both land-use and military matters.

All subdisciplines of contemporary human geography respond to the social and environmental issues that confront us with regular contributions on such topics as peace, energy supply and use, food availability, population control, and social inequalities. These issues are covered in this book, for human geographers recognize that geography has responsibilities to society and the world.

Lunenburg, Nova Scotia, from the harbour. Geographers often suggest that landscapes can speak to us, communicating information about the people who live there, what they do, and the values they hold. Established by the British in 1753 as their second outpost in Nova Scotia (the former French colony of Acadia), Lunenburg was first settled by Protestants specifically recruited from Germany, Switzerland, and France. Farming, fishing, shipbuilding, and ocean-based commerce provided the foundations of a vibrant economy, and today tourism is a major source of income. This photograph reflects elements of the town's economic history and its character. For example, there is no evidence of the ongoing physical change—whether construction or demolition—so typical of most urban areas. Rather, the presence of many carefully maintained old buildings suggests permanence and stability.

Technical Advances

Human geography is now greatly aided by a variety of technical advances in areas such as navigation-assisted exploration. Among the technical advances that have facilitated data acquisition are aerial photography and both infrared and satellite imagery. Similarly, advances in computer technologies and geographic information systems have facilitated mapping and data analysis. These important geographic techniques will be discussed in the next chapter.

CONCLUSION

As befits geographers, we have travelled a long way in this chapter (Box 1.5). From the earliest map-makers to the present, we have seen the evolution of geography as an academic discipline. We know that human geography has a rich heritage and exciting contemporary developments.

Here are three important conclusions:

- Human geography is currently a responsible social science with the basic aim of advancing knowledge and serving society. In this respect it is no different from Greek, Chinese, Islamic, or later European geographies.
- Our subject matter is clear and our approaches necessarily various.
- Our key goal—to provide a basis for comprehending the human world as it is today and as it has evolved—is of crucial importance to contemporary world society.

But we have not yet travelled far enough. The next chapter presents fuller discussions of the various contemporary developments that have received only brief mention so far.

Examining the Issues

BOX 1.5 | Why Learn about the History of Geography?

Geography matters, but why does the history of geography matter? The simple answer is that "what's past is prologue" (Shakespeare, *The Tempest*, Act 2, Scene 1). Necessarily, then, understanding the history of geography enhances our understanding of contemporary geography and even enables us to think about what might happen in the future. This point is demonstrated by two examples.

First, the story told in this chapter shows that, for many years, acquiring factual information was a principal concern of geography. Facts were required to enable geographers to accomplish the basic task of description, including mapping. Strabo's *Geographia*, Ptolemy's *Guide to Geography*, Varenius's *Geographia Generalis*, and many more books attempted to reflect the existing geographic knowledge base. As each new book was written, the body of geographic knowledge became more comprehensive and better established, with earlier knowledge verified or corrected.

It is no different today. All geographic research relies on a body of factual knowledge, and much geographic writing reports on new factual knowledge or on reinterpretations of established knowledge. Much of our new knowledge is acquired by different means from those employed in the past (e.g. by using satellites), but the basic goal is no different. Also, much new knowledge may refer to recently defined geographic problems, such as homeless living, but again the basic goal is the same. Geographic research, like any other, requires careful observation as a first step.

Second, looking at the history of geography also clarifies our understanding of how geographers interpret facts. For most of the past 3,000 years, new facts resulted from people travelling to lands new to them. These facts were then understood in the context of the existing knowledge base and recorded accordingly. Consider, for example, the maps shown in Figures 1.1–1.4. Each map reflects not only the geographic knowledge base but also its creator's larger understanding of the world. Consider also the knowledge that Europeans acquired about other lands and peoples as they moved overseas. How did they accommodate this new knowledge? The answer is that the new facts were understood in the context of the prevailing discourse, or what can be more simply called the larger social and intellectual context. Specifically, as the process of European overseas expansion proceeded, new lands and people were reinterpreted, especially through colonial and racist lenses.

Geographic facts are still interpreted and recorded in ways that reflect prevailing discourses. For example, in Chapter 4 we consider human impacts on the earth; in Chapter 5 we consider whether or not there are too many of us living on earth; and in Chapter 6 we ask why the world is such an unequal place. In all three cases, the answers are necessarily framed in the context of our larger intellectual framework, that is, the prevailing discourse. But our study of the history of geography has taught us that these discourses are always changing—overt colonialism and racism no longer play the important role they once did. Therefore, this text is always careful to acknowledge that our explanations are necessarily somewhat uncertain and that alternative understandings of facts may be appropriate.

The story told in this chapter has made it clear that all knowledge is contextual and thus needs to be understood as it relates to other knowledge. This is central to understanding how we, as humans, interact with and shape the natural environment, and it is an important lesson to learn.

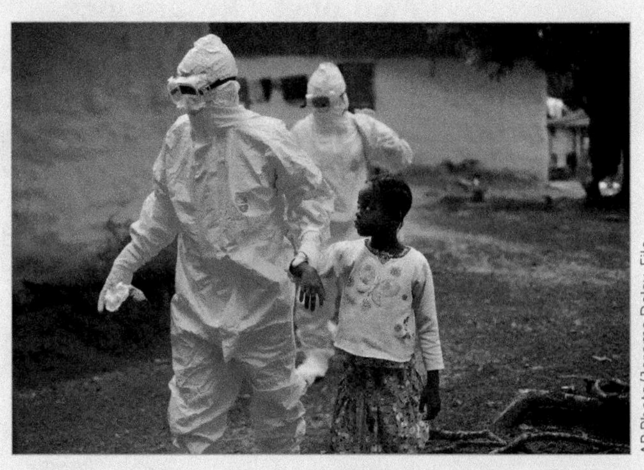
AP Photo/Jerome Delay, File

Health workers in the village of Freeman Reserve, Liberia, lead nine-year-old Nowa Paye to an ambulance after she showed signs of the Ebola virus. Beginning in March 2014, the epidemic in West Africa was the largest and longest Ebola outbreak in history. It was declared over in May 2015; as of October 2015, there have been over 28,000 reported cases and over 11,000 deaths.

Summary

A Raison D'être for Geography

To serve society.

The Story of Geography

There is more than one story about geography and more than one way of telling the chosen story. In common with other intellectual projects, geography is deeply contested.

Preclassical Geography

The earliest geographic descriptions were maps. The earliest surviving maps, from ancient Mesopotamia, showed only local areas.

Classical Geography

Because the Greeks were geographically mobile, they began a descriptive literary tradition that included the mapping of the entire world as they knew it. Major figures in this tradition include Eratosthenes and Strabo. The Greeks also initiated a mathematical tradition; Eratosthenes measured the circumference of the earth and Hipparchus devised a grid system for maps. Ptolemy summarized the mathematical tradition. Roman civilization added little to geographic knowledge.

The Fifth to Fifteenth Centuries

From the fifth century to the fifteenth, only sporadic and limited geographic work was accomplished in Europe. Christian dogma determined the contents of T-O maps. Only Portolano charts represented a development in mapping. Thus, geographic study largely disappeared. In China from *c.* 2000 BCE onward, and in the Islamic world from the seventh century CE on, geography flourished; significant progress was made in exploration, mapping, description, and mathematical work.

A European Resurgence

Beginning in the fifteenth century, Europe expanded overseas; cartography was transformed with new map projections, especially that of Mercator; and geographic description reappeared with major works by Apian and Münster. By the mid-seventeenth century, geographers had added significantly to our knowledge of ourselves and our world; geography itself continued to be concerned with mapping and describing the earth.

Rethinking Geography

Varenius was the first scholar to formally define geography and distinguish between general and special geography (in 1650). Kant contributed further to the understanding of geography by identifying it as the science that describes or classifies things in terms of area (*c.* 1765). Between roughly 1650 and 1800, geography broadened to include discussion of such issues as physical-geography-as-cause, the growth of civilizations, the unity of the human race, and population density and productivity relationships. Descriptive geography and mapping continued.

Universal Geography

The mid-nineteenth century was the high point for geography as mapping and description. By this time considerable geographic knowledge had been accumulated, and specialist sciences were emerging. Major nineteenth-century geographies include those by Humboldt and Ritter, both of whom made novel contributions by discussing human–land relations and providing comprehensive descriptions.

Institutionalization

Geography was firmly established as a university discipline in Prussia in 1874. Other countries rapidly followed suit. Programmatic statements about geography were plentiful: Richthofen and Hettner formulated geography as chorology; Ratzel introduced anthropogeography and Vidal followed his lead; Mackinder focused on a world view; and Davis pioneered physiography. The first North American department of geography was established at the University of Chicago in 1903.

Prelude to the Present

Four principal emphases were evident between 1903 and 1970. Physical-geography-as-cause flowered initially. Sophisticated human and land views developed under Vidal, Schlüter, and Sauer and have persisted to the present. Regional geography dominated North American work until the 1950s. Spatial analysis appeared in the 1950s and flourished in the 1960s but is now regarded as just one of several valid approaches.

Human Geography Today

Contemporary human geography has close ties to the long and illustrious history of geography and is affected by several relatively new influences. These days human geography is easily distinguished from physical geography, comprises several different but related subdisciplines, includes revitalized regional and landscape approaches, takes an ongoing interest in spatial analysis, is increasingly concerned with applied issues in the global context, and has a continually expanding technical component.

Conclusion

Knowledge of geography is essential to the education of all humans. Its fundamental goal is still to serve the society of which it is a part.

Links To Other Chapters

Because Chapter 1 traces the roots and development of human geography over time, it has links with all later chapters.

- **Philosophies, concepts, and techniques:** Chapter 2
- **Humboldt, Ritter, Vidal, and Sauer:** Chapter 4 (human impacts on land); Chapter 7 (approaches to landscape study)
- **The history of geography:** Chapter 9 (European colonial expansion); Chapter 13 (Industrial Revolution)

Questions for Critical Thought

1. What are the major developments in human geography's evolution as an academic discipline since the time of classical Greece?
2. Why are maps so important to the study of geography? Can we do geography without them?
3. How does the perspective of geography differ from that of other academic disciplines?
4. Why is it important to study the history of one's discipline? What do we gain by knowing the "history of geography"?

Suggested Readings

Visit the companion website for a list of suggested readings.

2 STUDYING HUMAN GEOGRAPHY

To make sense of the diverse approaches to the remarkably broad subject matter of human geography, this chapter is divided into three closely related parts: philosophical, conceptual, and analytical. First, we consider the philosophical component: why geographers ask the questions they do. The answer may seem self-evident, but it is not. Different geographers ask different questions, and we need to know why. Second, we consider the conceptual element: what concepts human geographers use to assist their research activities. Once again, we find several core concepts that are central to almost any piece of geographic research and several other concepts that tend to be associated with particular types of questions. Third, we look at the many analytical techniques available to help human geographers answer the questions they ask.

Here are three **points to consider** as you read this chapter.

- A basic assumption of a positivist philosophy is that it is possible for a human geographer to conduct research objectively, without being affected by her or his personal beliefs about the world and the way it ought to be. Does this sound reasonable?
- A basic assumption of a Marxist philosophy is that, rather than strive for objectivity, it is essential that a human geographer conduct research with a specific ideological agenda in mind. Does this sound reasonable?
- Human geographers who subscribe to one of the several versions of a humanist philosophy acknowledge that the research they conduct is not easily verified by others. Does this matter?

Running along the coast of Havana from the Havana Harbor to the Vedado neighbourhood, the Malecón is one of the city's most popular spots. As political and social change continues in Cuba, the importance of such sites will be considered from various perspectives.

Philosophy is important to the beginning geographer because our choice of philosophical perspective explains our specific content, concepts, and analytical techniques. In human geography, as in other disciplines, facts, concepts, and techniques are logically interrelated. But it is not enough simply to accept that these are related: we need to know *why*. This means we must understand the philosophical viewpoints that serve as the "glue." Much geographic work is guided by a specific philosophical viewpoint. Before we embark on such work, we need to learn about our philosophical options—to understand why we ask the questions we do, why we conceptualize as we do, and why we use particular techniques. Philosophy is at the heart not only of this chapter but also of much of this book. We begin this chapter with an overview of our philosophical options.

empiricism
A philosophy of science based on the belief that all knowledge results from experience and therefore gives priority to factual observations over theoretical statements.

positivism
A philosophy that contends that science is able to deal with only empirical questions (those with factual content), that scientific observations are repeatable, and that science progresses through the construction of theories and derivation of laws.

scientific method
The various steps taken in a science to obtain knowledge; a phrase most commonly associated with a positivist philosophy.

theory
In positivist philosophy, an interconnected set of statements, often called assumptions or axioms, that deductively generates testable hypotheses.

hypothesis
In positivist philosophy, a general statement deduced from theory but not yet verified.

law
In positivist philosophy, a hypothesis that has been proven correct and is taken to be universally true; once formulated, laws can be used to construct theories.

PHILOSOPHICAL OPTIONS

The philosophical diversity of contemporary human geography reflects the diversity of our subject matter, namely, human behaviour in a spatial context. It also means that we draw on a wide variety of concepts and methods. Some aspects of the human geography presented in this book are clearly aphilosophical or empiricist; others are positivist, humanist, Marxist, feminist, or postmodernist.

Empiricism

Some human geography, even today, may appear to be aphilosophical. Certainly most human geographers before the 1950s ignored philosophical issues and simply conducted research that was considered appropriate in the light of the discipline's historical development. Such work, particularly in regional and cultural studies, made no claim to have a philosophical base; human geographers were just doing what human geographers did. Nevertheless, it can be argued that there was an implicit philosophy in such regional and cultural work. This is the philosophy of **empiricism**, which contends that we know through experience and that we experience only those things that actually exist. Empiricism typically sees knowledge acquisition as an ongoing process of verifying and, as necessary, correcting factual statements. In practical terms, an empiricist approach allows the facts to speak for themselves. By definition, empiricism rejects any philosophy that purports to be an all-embracing system.

Empiricism, in turn, is rejected by most other philosophies. It is, however, a fundamental assumption of positivism, a philosophy that builds upon the basic empiricist foundation to include such strategies as theory construction and hypothesis testing.

Positivism

For some human geographers **positivism** is a very attractive philosophy because it is rigorous and formal. In principle a clear and straightforward philosophy for human geography, positivism makes the following arguments:

1. Human geography needs to be objective; the personal beliefs of the geographer should not influence research activity. Do you agree with this principle? If so, do you believe that it is possible to research human geography without being affected by your personal beliefs? According to humanism and Marxism (outlined in the two following sections), objectivity is not only undesirable but also, in fact, impossible.
2. Human geography can be studied in much the same way as any other science. For the positivist, there is really no such thing as a separate geographic method; all sciences rely on the same method. Specifically, positivism first found favour in the physical sciences, and its applications in human geography reflect the belief that humans and physical objects can be treated in a similar fashion. Once again, humanism and Marxism reject this assumption, believing that it dehumanizes human geography.
3. The specific method that positivism sees as appropriate for all sciences, physical and human, is known as the **scientific method** (Figure 2.1): reflecting the empiricist character of the philosophy, research begins with facts; a **theory** is derived from those facts, together with any available laws or appropriate assumptions; a **hypothesis** (or a set of hypotheses) is derived from the theory; and that hypothesis becomes a **law** when verified by the real world of facts. Thus, the scientific method consists of the study of facts, the construction of theory, the derivation of hypotheses, and the related

recognition of laws (Box 2.1). For the positivist, any science rests on the twin pillars of facts and theory, and a disciplinary focus on one at the expense of the other is wrong.

Positivist philosophy was introduced into human geography relatively late (1953) and was closely associated with quantitative methods and theory development during the 1960s. Positivism is an integral part of what we described earlier as the spatial analysis approach. Its introduction to geography was controversial because it directly challenged the regional approach dominant at the time.

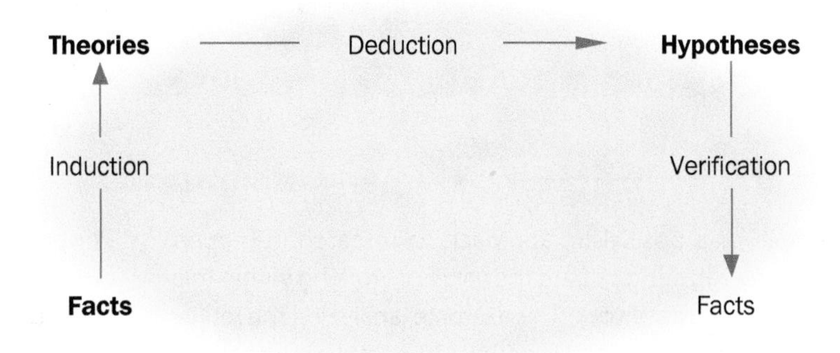

FIGURE 2.1 | The scientific method
Science stands on the twin legs of facts and theory; theoretical work depends on and stimulates factual work.

Source: Adapted from Kemeny, J. G. 1959. *A Philosopher Looks at Science*. Englewood Cliffs, NJ: Prentice-Hall, 81.

Humanism

Compared to positivism, **humanism** is a much more loosely structured set of ideas. Humanistic geography developed from about 1970 onward, initially in strong opposition to positivism. It focuses on humans as individual decision-makers, on the way humans perceive their world, and emphasizes subjectivity in general. Discussions of humanism easily become complicated because there are several humanistic philosophies. **Pragmatism** is one example; for human geographers the most important is **phenomenology**.

Phenomenology has many variations, all of them based on the idea that knowledge is subjective. To understand human behaviour, therefore, it is necessary to reconstruct the worlds of individuals. A central component of phenomenology is the idea that researchers need to demonstrate ***verstehen***, or sympathetic understanding, of the issue being researched. The distinction with respect to positivism is clear. Phenomenology seeks an empathetic understanding of the lived worlds of individual human subjects, whereas positivism seeks objective causal explanations of human behaviour, without reference to individual human differences. One especially thoughtful geographer, Yi-Fu Tuan, has written a series of books and articles that are phenomenological in focus (Box 2.2).

Several other humanistic philosophies, such as individuals' personal **existentialism** and **idealism**, have been advocated by geographers, but they have not proved to be highly influential.

Our brief discussion of humanism raises two general issues. First, the basic distinction between positivism and humanism is one of objectivism versus subjectivism. Positivism contends that the study of human phenomena can be objective; humanism says that it cannot. This argument is an acknowledged, long-established, and unresolved issue in the social sciences. Three questions are relevant here:

1. Is there an interaction between the researcher and the research subject that invalidates the information collected?
2. Does the researcher have a personal background that effectively influences her or his choice of problem, methods, and interpretation of results?
3. If we view humans objectively, does this mean we see them as objects? If so, is this approach dehumanizing?

Answering "no" to these three questions means that you have positivistic tendencies; "yes" means that you have humanistic tendencies. Answering "don't know" is perfectly understandable at this stage in your studies.

The second issue involves social scale. Do we as human geographers study individuals or groups of people and, if the latter, what size of group? Traditionally, we have focused on groups, but you will have noted that the classic humanistic philosophies place some emphasis on individuals. For many human geographers, this emphasis on individuals is inappropriate, and much humanistic work in geography has been done on a group scale. Typically, as discussed in Chapters 7, 8, and 9, groups are defined by culture, religion, language, ethnicity, gender, or sexuality.

The Schaefer–Hartshorne Debate

humanism
A philosophy centred on such aspects of human life as value, quality, meaning, and significance.

pragmatism
A humanistic philosophy that focuses on the construction of meaning through the practical activities of humans.

phenomenology
A humanistic philosophy based on the ways in which humans experience everyday life and imbue activities with meaning.

verstehen
A research method, associated primarily with phenomenology, in which the researcher adopts the perspective of the individual or group under investigation; a German term best translated as "empathetic understanding."

existentialism
A philosophy that sees humans as responsible for making their own natures; it stresses personal freedom, decision-making, and commitment in a world without absolute values outside individuals' personal preferences.

idealism
A humanistic philosophy according to which human actions can be understood only by reference to the thought behind them.

Examining the Issues

BOX 2.1 | Positivistic Human Geography

A sample of a positivistic approach to research in human geography will demonstrate the value of working within this philosophical framework. This example, analyzing the retailing and wholesaling activities in urban centres, proceeds through five conventional stages (Yeates, 1968: 99–107).

1. *Statement of problem:* The purpose of the study is to determine the relationships between the numbers of retailing and wholesaling establishments and the size of urban centres in a part of southern Ontario. This problem is suggested by earlier work, both factual and theoretical, indicating that urban land uses and activities may be explained in terms of variables such as size.
2. *Hypotheses:* Three specific hypotheses are identified on the basis of available human geographic theory (the theory in question is known as central place theory and is discussed in Chapter 11). These are (1) the number of retail establishments and the population size of an urban centre are directly related; (2) the number of retail and service functions in an urban centre increases with the size of that centre, but at a decreasing rate; and (3) the number of wholesale establishments and the population size of an urban centre are directly related.
3. *Data collection:* Population, retail, and wholesale data are collected for all urban centres of 4,000 or more population. In addition to fieldwork, the data sources used include city directories, telephone directories, and censuses. Note that the use of a minimum population of 4,000 is an example of an operational definition (a way of describing how an individual object can be identified as belonging to a general set).
4. *Data analysis:* At this stage, information is subjected to appropriate analysis to formally test the stated hypotheses (the verification stage). In this study, the data are graphed (for example, the number of establishments against the population of urban centres), and a statistical procedure known as correlation analysis is used to calculate the extent of the relationship between the two sets of data.
5. *Stating conclusions:* In standard positivistic research, the formal testing of a hypothesis results in a statement concerning the validity of the hypothesis. In this study, all three of the stated hypotheses are confirmed by the statistical analysis; such analysis increases our confidence in the theory that generated the hypotheses and also provides insights into the specific landscape investigated.

As this example shows, it is no exaggeration to describe positivism as a rigorous and formal philosophy.

Examining the Issues

BOX 2.2 | Humanistic Human Geography

What questions does humanistic geography ask? What types of research issues does it address? What kinds of answers does it look for? Although the "humanistic" name covers many different approaches, one way to think about these questions is to consider the work of Yi-Fu Tuan.

Tuan was born in China in 1930 and was educated at London and Oxford universities in England and at Berkeley in California. An initial interest in physical geography soon expanded into a broad focus on questions of humans and nature. Tuan, the leading humanistic geographer during the 1970s and 1980s, was the first to make explicit reference to philosophy, although he never claimed to subscribe to any single approach: "My point of departure is a simple one, namely, that the quality of human experience in an environment (physical and human) is given by people's capacity—mediated through culture—to feel, think and act" (Tuan, 1984: ix).

With that aim in mind, Tuan was critical of much geography for ignoring individuals and the relationships they have with one another. For him, there was a need to be comprehensive, like a novelist: "comprehensiveness—this attempt at complex, realistic description—is itself of high intellectual value; places and people do exist, and we need to see them as they are even if the effort to do so requires the sacrifice of logical rigour and coherence" (Tuan, 1983: 72).

Three examples of Tuan's work, all well received within and beyond our discipline, illustrate the broad range of his interests: the concept of landscapes of fear, which proposes that fear is a component part of some environments and, in turn, helps create environments (Tuan, 1979); the idea of gardens as an example of our attempts to dominate environments (Tuan, 1984); and his discussion of the links between territory and self (Tuan, 1982). In the last, Tuan contends that the presence of segmented spaces—for instance, the many discrete spaces in any urban area—leads to segmented societies. Humans retreat into segmented social worlds in response to the difficulties of coping in a complex society. There is a clear distinction between a small settlement in the less developed world, which has few human-constructed barriers and few spaces assigned for exclusive activities, and a modern city with numerous physical and social barriers. The human experiences in these two environments are very different, with significant implications for individual and group behaviour.

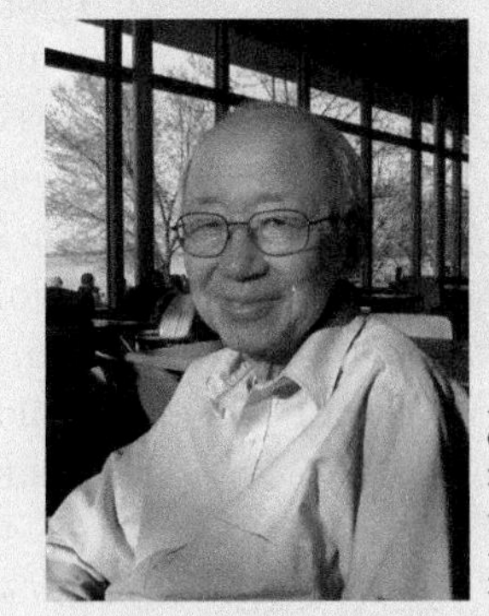
Melanie McCalmont

Yi-Fu Tuan, Madison, Wisconsin on 15 April 2005.

Marxism

Whether or not one espouses a class-based materialist analysis of society, the ideas of Karl Marx (1818–83) are foundational to an understanding of the human–land interface and must be considered. Unfortunately, Marx and his collaborator, Friedrich Engels, are difficult writers to summarize. Nowhere are their ideas presented in a clear, unequivocal fashion, and any interpretation of **Marxism** is bound to give rise to disagreement. In broad terms, Marx was a political, social, economic, and philosophical theorist who worked to construct a body of social theory that would explain how society actually worked. He was also a revolutionary who aimed to facilitate a change in the economic and political structure of society, from **capitalism** to communism.

The Marxist perspective is often described as **historical materialism**. This term refers to Marx's concern with the material basis of society and his effort to understand society and social change by referring to historical changes in social relations. Marx summarized the character of society as follows. First, there are forces of production: the raw materials, implements, and workers that actually produce goods. Second, there are relations of production: the economic structures of a society, that is, the ways in which the production process is organized. The most important relations are those of ownership and control. In a capitalist society, for example, those relations are such that workers are able to sell their labour on the open market. Thus, although workers—the labour force—produce goods, the relations of production determine how the production process is organized. Together, these forces and relations make up what is called the **mode of production**. This concept is the key to understanding the composition of society. Examples of modes of production include **slavery**, **feudalism**, capitalism, and **socialism**. It is helpful to think of the mode of production as a stage in a process.

Marx used these ideas to sketch a history of economic change that traces the transitions from one mode of production to another, for example, from feudalism to capitalism to socialism. Marx was especially critical of the capitalism that flourished in his lifetime because of his belief that one **class** (owners) exploited the other (workers) in order to maximize its own profits. Exploitation itself was not new; both slavery and serfdom were obvious forms of exploitation. What Marx recognized was that the same kind of exploitation continued under capitalism, except it was disguised. For this fundamental reason—the exploitation of workers by owners—Marx called for a socialist revolution in the form of a class struggle to overturn one mode of production and replace it with another under the control of the workers themselves.

Marxism differs profoundly from both positivism and humanism in that it sees human behaviour as being constrained by economic processes. Marx held human societies alone responsible for the conditions of life within them. In other words, he believed that social institutions were created by humans and that, when those institutions no longer served a society's needs, they could and should be changed.

Marxist Thought and Practice

Marxism
The body of social and political theory developed by Karl Marx, in which mode of production is the key to understanding society and class struggle is the key to historical change.

capitalism
A social and economic system for the production of goods and services based on private enterprise.

historical materialism
An approach associated with Marxism that explains social change by reference to historical changes in social and material relations.

mode of production
The organized social relations through which a human society organizes productive activity.

slavery
Labour that is controlled through compulsion and is not remunerated; in Marxist terminology, one particular mode of production.

feudalism
A social and economic system prevalent in Europe prior to the Industrial Revolution; land was owned by the monarch, controlled by lords, and worked by peasants who were bound to the land and subject to the lords' authority.

© GL Archive/Alamy

Karl Marx

socialism

A social and economic system that involves common ownership of the means of production and distribution.

class

A large group of individuals of similar social status, income, and culture.

superstructure

A Marxist concept that refers to the political, legal, and social systems of a society.

feminism

The movement for and advocacy of equal rights for women and men and a commitment to improve the position of women in society.

Thus, Marxism has a significant revolutionary dimension. Working within Marxism usually implies that one is striving both to understand the human world and to change it. Box 2.3 summarizes one example of the Marxist concern with the global spread of capitalist culture. The long-term aim of such work is to facilitate a social and economic transformation from capitalism to socialism. In the shorter term, Marxism is an attractive philosophy for many human geographers who feel strongly that their work should focus explicitly on social and environmental ills and contribute to solutions to those ills. Other philosophies, such as positivism, are seen as merely describing the **superstructure**, which is implicitly accepted as appropriate. It is worth noting that Marxist geography is not the only kind of geography that strives to contribute solutions to environmental and social ills; however, it is fair to say that this focus is most explicit in Marxist work.

Feminist Thought

It is difficult to exaggerate the importance of feminist thought for contemporary human geography, as we will see on many occasions in this book. There is no single feminist philosophy or body of feminist theory; rather, various schools of feminist thought are associated with larger bodies of theory such as liberalism, Marxism, socialism, or postmodernism. Nevertheless, all schools of **feminism** are united in their commitment to improving the social status of women and securing equal rights with men. This

© Fox Photos/Getty Images

Manchester cotton mills, *c.* 1940. Similar landscapes developed in many parts of northern England during the Industrial Revolution. Friedrich Engels wrote of Manchester a century before this photograph was taken: "If any one wishes to see in how little space a human being can move, how little air—and such air!—he can breathe, how little of civilisation he may share and yet live, it is only necessary to travel hither" (*The Condition of the Working-Class in England in 1844* [London, 1892, p. 53]). In "dark Satanic mills" like these (to use William Blake's famous phrase), workers manufactured cotton goods, spinning the yarn and weaving it on huge looms. Some factories were attached to orphanages that offered children room and board in return for their labour. Small children—employed to clear out cotton fluff from the looms—were sometimes crushed and killed by the fast-moving machines. In the nineteenth century, mill owners who used child labour were often praised for their philanthropy in keeping the children off the streets and providing them with food and shelter. Some mills remain as museum pieces, cleaned up and Disneyfied for tourist consumption.

Examining the Issues

BOX 2.3 | Marxist Human Geography

Much contemporary human geography employs—implicitly or explicitly—a Marxist philosophical framework. The following discussion of the global spread of a capitalist culture aims to clarify what doing Marxist research really means (Peet, 1989).

What is culture? How is it formed? Why has capitalist culture been able to spread across the globe and supplant previously existing cultures? A Marxist response to these questions begins with the idea of a culture's consciousness—that is, the general way that humans think about things. Put simply, different human experiences in different regions of the world result in differences in regional consciousness and culture. Before the introduction of capitalism as a new economic and cultural system (a new mode of production), the world comprised a variety of regional cultures, each with its own distinct kind of consciousness and specific means and modes of production. One way to explain these regional variations is to recognize that human culture was closely related to physical geography (this is not an environmental determinist assertion but a simple recognition that humans with limited technology are especially closely related to their environment).

To understand cultural change as a result of the rise of capitalism, a Marxist acknowledges that human history is a story of domination by forces beyond human control. Initially, as we have seen, the dominating force was nature—physical geography—but over time capitalism came to replace nature as the dominating force.

Like nature, capitalism has the ability to cause human wealth or poverty, success or failure. It is a powerful and all-pervading economic and cultural system that can overwhelm and replace earlier regionally based economies and cultures. Capitalist culture is now a world culture, while regional non-capitalist cultures in the less developed world either have been or are being dramatically transformed. In explicitly Marxist terminology, this process can be seen as reflecting change in the mode of production.

A human geographer with a Marxist perspective regards these developments as undesirable because the regional cultures that are disappearing had adapted to deal with a wide range of environmental problems, while the new culture, with its new ways of life and thought, is unsuited to handling those issues.

emphasis on identifying social problems and seeking to change society for the better is central to all critical theories, especially Marxist, feminist, and post-colonial/anti-racist theories.

Although there are significant differences from country to country, women are usually systematically disadvantaged in most areas of contemporary life. The fundamental reason for this inequality is patriarchy, a social system in which men dominate women. Under the traditional division of labour, women were economically dependent on male breadwinners, and the few women who did work outside the home were paid less than what men received for equivalent work. Culture is seen as a key factor in the construction of gender differences through various socialization processes. Sexuality and violence are seen as forms of social control over women. Finally, the state is seen as typically reinforcing traditional households and failing to intervene in cases of violence against women.

Of the several traditions of feminist thought and action, the oldest, dating back to the late eighteenth century, is *liberal feminism*, aimed at securing equal rights and opportunities for women. Two of the most important recent traditions argue that the oppression of women cannot be corrected by superficial change because it is embedded in deep psychic and cultural processes that need to be fundamentally changed. *Radical feminism* contends that gender differentiation results from gender inequality and that the subordination of women is separate from other forms of social inequality, such as those based on class. *Socialist feminism* similarly emphasizes gender inequality but links that inequality to class; it argues that both men and capital benefit from the subordination of women.

For feminist geographers, the key category of analysis is gender. Gender, like class, implies a distinction between power groups—in this case, dominant males and subordinate females—and

patriarchy
A social system in which men dominate, oppress, and exploit women.

gender
The social aspect of the relations between the sexes.

constructionism
The school of thought according to which all our conceptual underpinnings (e.g. ideas about identity) are socially constructed and therefore contingent and dynamic, not given or absolute.

essentialism
Belief in the existence of fixed unchanging properties; attribution of "essential" characteristics to groups.

postmodernism
A movement in philosophy, social science, and the arts; based on the idea that reality cannot be studied objectively and that multiple interpretations are possible.

modernism
A view that assumes the existence of a reality characterized by structure, order, pattern, and causality.

involves specific roles for males and females. That is, our identities are gendered. Feminist geographers have usually accepted that gender is a social construction deriving largely from the natural category of biological sex; thus, discussions of gender are premised on the logic of **constructionism**, not **essentialism**. Gender is formed initially through the differential treatment of girls and boys and continues to be reinforced throughout the life cycle. This differential treatment is accompanied by different societal expectations of the values, attitudes, and behaviours of boys and girls.

Human geographers did not begin paying serious attention to gender differences until around 1980, which meant, in effect, that the former insistence on discussing humans in general resulted in ignoring all the ways in which the lives and experiences of women and men are different around the world. For example, women still typically perform domestic work, which is unpaid, repetitious, and often boring, and the jobs available to women outside the home are often low-paid and low-skilled. In short, women and men tend to do different work in different places and to lead different lives, which include different visions of the world and of themselves.

But it is not sufficient to note that girls and boys are raised differently and expected to be different throughout their lives. They also are raised unequally, with boys socialized to be aggressive and to assume leadership roles while girls are socialized to be passive and to be compliant followers. Such characteristics possibly have some initiating biological cause, but even if this is so, the socialization process clearly emphasizes and increases any natural differences and minimizes movement across the categories of male and female. These differences are most evident in a country such as Saudi Arabia, which uses religious arguments to rigorously enforce gender distinctions.

Overall, this process works to the advantage of men. In the context of work, for example, women are relegated to the private domain or to limited opportunities in the public domain. The differing employment experiences of women and men even in the more developed world were highlighted by recent data for the European Union countries showing that women achieve a better education overall but earn less and occupy fewer top jobs. Part of the explanation for this apparent contradiction is that women continue to perform more domestic work and more women hold part-time jobs. Data such as these are published regularly, often on the occasion of International Women's Day, 8 March. As noted, these differences are not explained in terms of biological differences (sex) but in terms of cultural differences (gender).

Fully veiled women work in a military uniforms factory in Riyadh, Saudi Arabia. Saudi law, which is based on a conservative interpretation of Islam, requires the separation of men and women in public spaces, limiting women's ability to participate fully in the workplace as well as engage in entrepreneurial activities.

Postmodern Thought

As the name implies, **postmodernism** emerged as a reaction to **modernism**, a general term used to refer to any number of movements beginning in the mid-nineteenth century that broke with earlier traditions. Modernism developed most fully in art and architecture, but social science methodology arose via the positivism that first appeared in the nineteenth century. It assumes that reality can be studied objectively and can be validly represented by theories (as in positivism) and that scientific knowledge is practical and desirable. Modernism is also closely linked to the Industrial Revolution and the rise of capitalism, emphasizing such classic liberal themes as the rationality of humans, the privileged position of science, human control over the physical environment, the inevitability of human progress, and a search for universal truths.

What was the postmodern reaction? Most contemporary human geographers would agree that postmodernism is an especially difficult body of ideas to understand. At least one reason is that postmodernism is, by its reactive nature, unstructured and ambiguous; anarchic concepts may be embraced. Another is that there are many versions of postmodernist theory, most of them developed in disciplines far removed from human geography—most notably architecture, literature, and other expressions of culture. Despite these difficulties, postmodernism is playing an increasingly important role in contemporary human geographic research.

The Postmodern Alternative

Postmodernism rejects all the assumptions of modernism. Reality cannot be studied objectively because it is based on language. Rather, reality should be thought of as a **text** in which all aspects are related (it is "intertextual"). Therefore, reality cannot be accurately represented. Taken to the extreme, this means that truth is relative and, for practical purposes, non-existent. Causality does not exist, and theory construction has no meaning.

So how do postmodernists approach an understanding of the world we live in? They emphasize the **deconstruction** of texts and the construction of narratives that do not make claims about truthfulness; such narratives tend to focus on differences, uniqueness, irrationality, and marginal populations. Deconstruction questions the established readings of a text and highlights alternative readings. Overall, postmodernism considers modernist claims to be arrogant, even authoritarian. "Postmodernism and deconstruction question the implicit or explicit rationality of all academic discourse" (Dear, 1988: 271).

Among the principal attractions of postmodernism for contemporary human geography is its emphasis on cultural otherness, its openness to previously repressed experiences such as those of women, the lesbian, gay, bisexual, transgender, and queer/questioning (LGBTQ) community, and, in general, those lacking power and authority. Later sections of this chapter will reflect this emphasis on the diversity of human experience.

The Diversity of Postmodernism

Not all postmodernism is quite as described above. In fact, the concept varies considerably between disciplines and even within them. Some who embrace postmodernism nevertheless continue in the progressive directions suggested by modernism, becoming involved in social movements or working to break down the barriers between researchers and subjects so that people are allowed to speak for themselves. These versions are relatively close to some earlier concepts of culture, such as symbolic interactionism, and some other social philosophies, such as humanism. The lack of a single unequivocal version of postmodernism in human geography is not surprising, given that the central message of postmodernism is the importance of diversity. While human geographers often acknowledge the strengths of the postmodern perspective, such as its emphasis on cultural otherness, some express concern about the postmodern tendency to focus on topics that could be seen as trivial. For example, Hamnett (2003: 1) worries that human geography has become a "theoretical playground where its practitioners stimulate or entertain themselves and a handful of readers, but have in the process become increasingly detached from contemporary social issues and concerns."

Positivist and Marxist Critiques of Postmodernism

text
A term that originally referred to the written or printed page but that has broadened to include such products of culture as maps and landscape; postmodernists recognize that there may be any number of realities, depending on how a text is read.

deconstruction
A method of critical interpretation applied to texts, including landscapes, that aims to show how an author's or reader's multiple positioning (in terms of class, gender, and so on) affects the creation or reading of the text.

HUMAN GEOGRAPHIC CONCEPTS

Like any other academic discipline, human geography involves two basic endeavours. First is the need to establish facts. A vital starting point, certainly, is to know where places are and to be aware of their fundamental characteristics. This is what we might call geographic literacy. Second is the need to understand and explain the facts—to know why the facts are the way they are. This is what we might call geographic knowledge. Understanding and explaining require that we ask intelligent questions; this, in turn, requires the conscious adoption of an appropriate philosophical stance.

But much more is involved in the acquisition of geographic knowledge. Philosophy guides us, but it does not necessarily provide all the tools we need. Our task now is to identify those tools, first in the form of concepts and then, in the next section, in the form of techniques. Some of these concepts and techniques are relevant regardless of philosophical bent, while others are philosophy-specific. Where appropriate in this account, concepts are linked to the parent philosophy. Concepts and techniques that are not tied to particular philosophies focus on fac-

Focus on Geographers

Geographic Literacy | WALTER PEACE and MICHAEL MERCIER

Having taught introductory human geography for many years, we recognize first-hand the relevance and value of what Norton and Mercier refer to as geographic literacy. In simple terms, geography has profound and far-reaching influences on the ways we experience our world. How we understand and make sense of the world is, in part, a product of the way basic geographic principles and relationships affect our daily lives. The following two examples illustrate geography's relevance and importance as a conditioner of our everyday experiences.

Consider the effects of the recent and ongoing drought conditions in California. Droughts are among the most costly weather-related events in terms of economics. Perhaps the most far-reaching consequence is the increase in food prices in the US, Canada, and many other parts of the world that depend on the fruits and vegetables grown in California. How might a geographic perspective help us make sense of this event? The geographically literate person might take into account the fact that environmental conditions (e.g. weather and climate) and human practices (e.g. how agriculture is carried out) vary from place to place, making some areas more (or less) susceptible to adverse environmental conditions such as a drought. Furthermore, while specific agricultural practices and environmental events are located in specific geographic spaces (regions), the outcomes are felt both locally and in other parts of the world. One reason for this is the high degree of interdependence or interconnectivity of regions due to globalization. This is particularly true in the case of California.

Now consider how the world map has changed in political and economic terms since the beginning of the twentieth century. In 1900 the British Empire's political and economic dominance was beginning to decline as the United States stood poised to become the next superpower. Today's world map reveals enormous (and ongoing) change since this time. Of particular note are the following three trends: the independence of many former colonies, especially in Africa, Asia, and South America; the emergence of regional political and economic associations, such as the European Union (EU), the South American Common Market (Mercosur), and the North American Free Trade zone (NAFTA); and the rise of China as a challenger to the century-long economic and political dominance of the United States. Consider, for example, that nearly one-fifth of the world's seven billion people live in China alone. Furthermore, with its rapid economic growth over the past three decades, China now has the world's second largest economy (after the US) and is poised to become the largest by 2020. It is the world's largest exporter and the second largest importer of goods. China's military is now the second largest in the world, and only the United States spends more annually on its military. The combined effects of these geopolitical changes from decolonization, globalization, and the rise and fall of global political and economic superpowers are felt throughout the world at all scales.

In summary, space and place are fundamental concepts in geography. Our awareness of and familiarity with such concepts as geographic literacy enhance our ability to understand our world and our place in it. By expanding our geographic literacy, we can become better global citizens.

WALTER PEACE recently retired from the School of Geography and Earth Sciences at McMaster University.

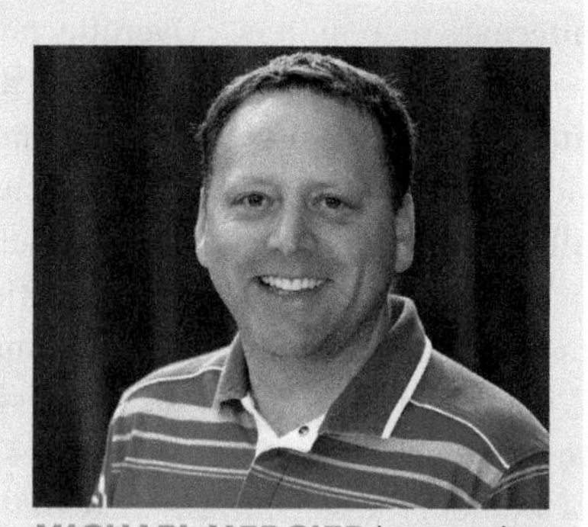

MICHAEL MERCIER is an assistant professor in the School of Geography and Earth Sciences at McMaster University.

tual matters, such as determining where things are on the surface of the earth; those tied to a philosophy focus on understanding or explaining why things are where they are. A useful way to think about concepts is to understand them as specifically human geographic vocabulary or, indeed, as part of the prevailing discourse of human geography.

An important distinction needs to be drawn here between idiographic and nomothetic methods. An **idiographic** approach is concerned with individual phenomena; traditional empiricist regional geography—Hartshorne's areal differentiation—was idiographic in focus, as is much humanistic and postmodern geography. A **nomothetic** approach, on the other hand, formulates generalizations or laws. Research in a positivist tradition is nomothetic, as is much Marxist human geography and some feminist geography.

Space

This account of concepts begins with a quartet of closely related terms: *space*, *location*, *place*, and *region*. We have already had occasion to use the word ***space*** several times in this book. In fact, it is not uncommon to describe human geography as a spatial discipline, one that deals primarily with space. (Note that the term is not being used in the context of outer space but in the context of the earth's surface.) To understand the geographic meaning of this term, it is helpful to distinguish between absolute and relative space. Absolute space is objective: it exists in the areal relations among phenomena on the earth's surface. This conception of space is at the heart of map-making, the study of regions, and spatial analysis, and it is central to the ideas of Kant, who saw the geographer as ordering phenomena in space. Relative space, by contrast, is perceptual: it is socially produced and therefore, unlike absolute space, is subject to continuous change.

Location

Regardless of philosophical persuasion, every human geographer begins with one central question: Where? Location (defined in the introduction) is the basic concept; it refers to a particular position within space, usually (but not necessarily) a position on the earth's surface. As we have just noted with regard to space, there are two ways to describe location. Absolute location identifies position by reference to an arbitrary mathematical grid system such as latitude and longitude; for example, Winnipeg is located at 49° 53'N latitude, 97° 09'W longitude. Such mathematically precise statements often are essential. Sometimes, though, they may not be as meaningful as a statement of relative location, the location of one place relative to that of one or more other places. Absolute locations are unchanging, whereas some relative locations do change. The location of Winnipeg, relative to lines of communication such as roads and air routes, changes over time.

There are other means by which the location of a geographic fact can be identified. A location can be described simply by reference to its place name or toponym, for instance (with increasing degrees of exactness), Canada, Manitoba, Winnipeg, Portage Avenue.

Geographers also distinguish between site and situation. **Site** refers to the local characteristics of a location, whereas **situation** refers to a location relative to others. Figures 2.2 and 2.3 illustrate the site and the situation of Winnipeg, respectively.

Place

The term ***place*** has developed a special meaning in human geography. It refers not only to a location but also and more specifically to the values that we associate with that location. We all recognize that some locations are in some way distinctive to us or to others. A place is a location that has a particular identity, such as our home, our place of worship,

idiographic
Concerned with the unique and particular.

nomothetic
Concerned with the universal and the general.

space
A real extent; used in both absolute (objective) and relative (perceptual) forms.

site
The location of a geographic fact with reference to the immediate local environment.

situation
The location of a geographic fact with reference to the broad spatial system of which it is a part.

place
Location; in humanistic geography, place has acquired a particular meaning as a context for human action that is rich in human significance and meaning.

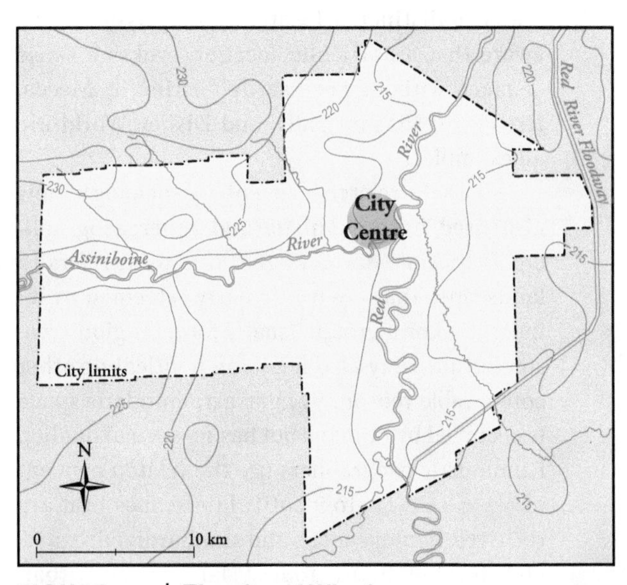

FIGURE 2.2 | The site of Winnipeg

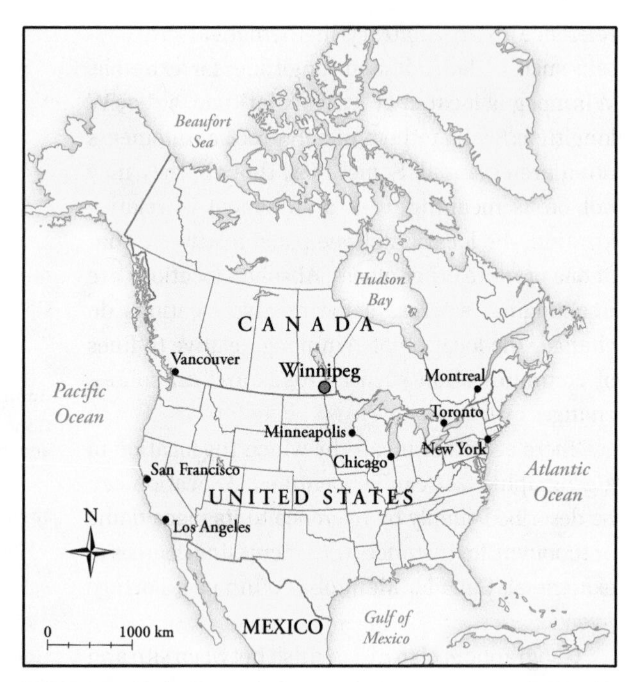

FIGURE 2.3 | The situation of Winnipeg within North America

and the shops we frequent. Clearly, for one person or another, many of earth's locations qualify as places. To distinguish place from concepts such as space and location, we might note that place is not about where we live but rather about how we live where we live.

The term *sense of place* was popularized in the 1970s by humanistic geographers with philosophical roots in phenomenology. It refers to our attachments to locations with personal significance, such as our home, and to especially memorable or distinctive locations. It is possible to be aware that a particular location evokes a sense of place without necessarily visiting it; Mecca, Jerusalem, Niagara Falls, and Disney World are all examples.

A closely related concept in sociology, now also used by humanistic and other geographers, is that of **sacred space**. This term refers to landscapes that are particularly esteemed by an individual or a group, usually for a religious reason but possibly also for some political or other comparable reason. By contrast, mundane space is occupied by humans but has no special quality. Humanistic geographers use the related concept of **placelessness** to identify landscapes that are relatively homogeneous and standardized; examples include many tourist landscapes, urban commercial strips, and suburbs (Relph, 1976). There is an interesting parallel here with the Marxist perspectives outlined in Box 2.3: places may be thought of as especially characteristic of pre-capitalist cultures, whereas placelessness is more evident in the industrial and post-industrial world.

Tuan, the humanistic geographer whose work is discussed in Box 2.2, introduced the concept of **topophilia**, literally "love of place." This term refers to the positive feelings that link humans to particular environments. By contrast, the less frequently used term ***topophobia*** refers to dislike of a landscape that may prompt feelings of anxiety, fear, or suffering.

These concepts add valuable refinement not only to the more general concept of location but also to any human geographic work that aspires to understand people and their relationships with land. Indeed, we can think of places as emotional anchors for much human activity.

Region

Region is one of the most useful yet most confusing of geographic concepts:

> So much geography is written on a regional basis that the idea of the region and the regional method is as familiar and as accepted as is Mercator's map in an atlas. Yet as with so many other familiar ideas which we use every day and take for granted, the concept of the region floats away when one tries to grasp it, and disappears when one looks directly at it and tries to focus. (Minshull, 1967: 13)

Chapter 1 identified regional geography as a traditional focus. By the early twentieth century, however, regional geographers had come to a formal definition of region as "a device for selecting and studying areal groupings of the complex phenomena found on the earth. Any segment or portion of the earth's surface is a region if it is homogeneous in terms of such an areal grouping" (Whittlesey, 1954: 30). Such a definition can accommodate many types of region, both physical and human.

Dividing a large area into regions, or **regionalization**, is a process of classification in which each specific location is assigned to a region. Human geographers recognize various types of regions, most notably the **formal (or uniform) region**, an area with one or more traits in com-

sense of place
The deep attachments that humans have to specific locations, such as home, and to particularly distinctive locations.

sacred space
A landscape particularly esteemed by an individual or a group, usually (but not necessarily) for religious reasons.

placelessness
Homogeneous and standardized landscapes that lack local variety and character.

topophilia
The affective ties that humans have with particular places; literally, love of place.

topophobia
The feelings of dislike, anxiety, fear, or suffering associated with a particular landscape.

regionalization
A special kind of classification in which locations on the earth's surface are assigned to various regions, which must be contiguous spatial units.

mon, and the functional (or nodal) region, an area with locations related either to each other or to a specific location. Thus, an area of German-speaking people qualifies as a formal region, while the sales distribution area of a city newspaper qualifies as a functional region. Delimiting formal regions is notoriously troublesome. Consider, for example, how you might delimit a wheat-growing area in the Canadian prairies. Would you delimit it according to the percentage of wheat farmers per township, the percentage of acreage under wheat, or the percentage of farm income derived from wheat? These are only some of the possibilities. Clearly, the choice of measure involves considerable subjectivity. Another problem inherent in region delimitation is the implication that regions are geographically meaningful: in fact, a prairie wheat-growing region may or may not be a significant portion of geographic space.

With the rise of spatial analysis in the 1960s, the traditional concept of formal regions declined and the concept of functional regions increased in popularity. By about 1970, humanistic geographers began arguing for a revitalized regional geography involving vernacular regions (regions perceived to exist by people either within or outside them). The "Bible belt" of the American South and Midwest is one example. This region is readily identifiable to most North Americans as an area with a particularly strong commitment to various conservative Protestant religions. For many living in the region, the name is a source of pride; Oklahoma City declares itself to be the "buckle on the Bible belt."

Distance

Since our first four concepts—space, location, place, and region—all refer to portions of the earth's surface, logically we are also interested in the distance between locations, between places, and between or within regions. Indeed, in Chapter 3 we consider the claim that geography can be understood as a discipline in distance.

The distribution of geographic facts—such as churches or towns—can often be explained by reference to the distance between them and other geographic facts. Geographers often talk about distributions, patterns, or forms in reference to the mapped appearance of spatial facts (see Figure 2.4). It is often the case that such maps reveal some sort of order. Specialty retail stores (jewellers and shoe stores, for example) tend to locate near one another. Similarly, financial institutions prefer to be close to one another. It is useful to characterize such distributions as resulting from a clustering process. Urban centres, on the other hand, typically develop at some distance

formal (or uniform) region
A region identified as such because of the presence of some particular characteristic(s).

functional (or nodal) region
A region that comprises a series of linked locations.

vernacular region
A region identified on the basis of the perceptions held by people inside and outside the region.

distance
The spatial dimension of separation; a fundamental concept in spatial analysis.

distribution
The pattern of geographic facts (for example, people) within an area.

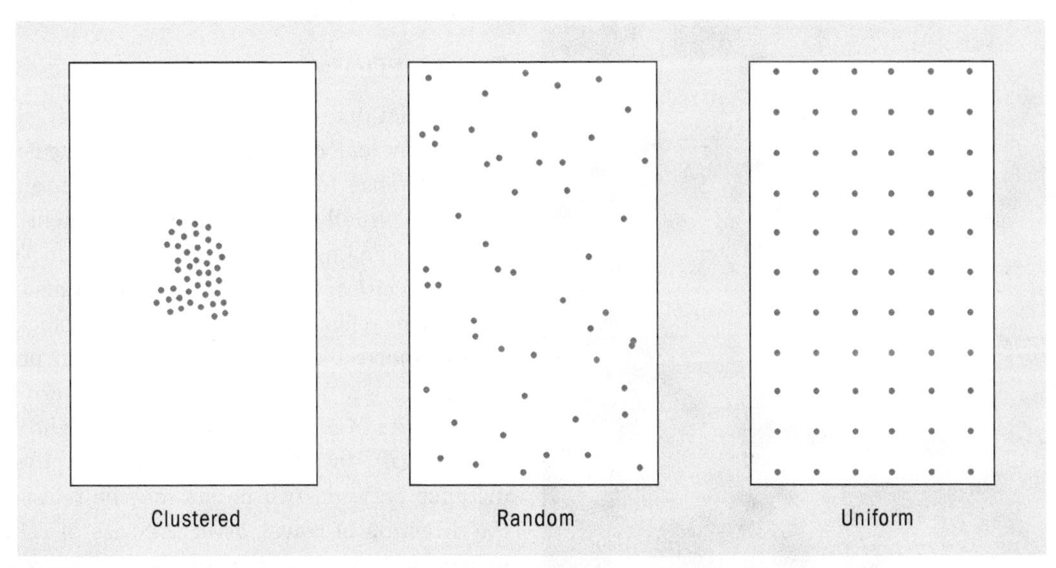

FIGURE 2.4 | Clustered, random, and uniform point patterns

A clustered pattern is typically produced when the geographic facts benefit from close proximity, for example, specialty retail outlets such as jewellery stores. A random pattern may be the result of a random process, may reflect the influence of more than one process, or may represent a temporary stage during changing circumstances. A uniform, or dispersed, pattern is produced when geographic facts benefit from separation because they compete for surrounding space, for example, urban centres in a region.

distance decay
The declining intensity of any pattern or process with increasing distance from a given location.

effect (or friction) of distance
A measure of the restraining effect of distance on human movement.

accessibility
A variable quality of a location, expressing the ease with which it may be reached from other locations.

interaction
The relationship or linkage between locations.

agglomeration
The spatial grouping of humans or human activities to minimize the distances between them.

deglomeration
The spatial separation of humans or human activities so as to maximize the distances between them.

from one another, since one of the principal reasons for a particular location is to provide services to surrounding rural populations. Urban centres also locate at least partly in response to competition. Other geographic facts locate apart because they involve the provision of services without involving competition; hospitals, community centres, and recreation areas are examples.

Much of what has been said about distance so far is evident in what Waldo R. Tobler (1970) referred to as the first law of geography: everything is related to everything else, but near things are more related than distant things. This is the notion of **distance decay**, graphically depicted in Figure 2.5 and sometimes referred to as the **effect**, or **friction**, **of distance**; typically both time and cost are involved in overcoming distance. This concept lies at the heart of much spatial analysis.

The distance concept is related to several other ideas. **Accessibility** refers to the relative ease with which a given location can be reached from other locations and therefore indicates the relative opportunities for contact and interaction; it is a key concept in the agricultural, settlement, and industrial location theories that we will encounter later in this text. **Interaction** refers to the act of movement, trading, or any other form of communication between locations. **Agglomeration** describes situations in which locations (usually of activities related to production or consumption) are in close proximity to one another. Conversely, **deglomeration** refers to situations in which those locations are characterized by separation from one another.

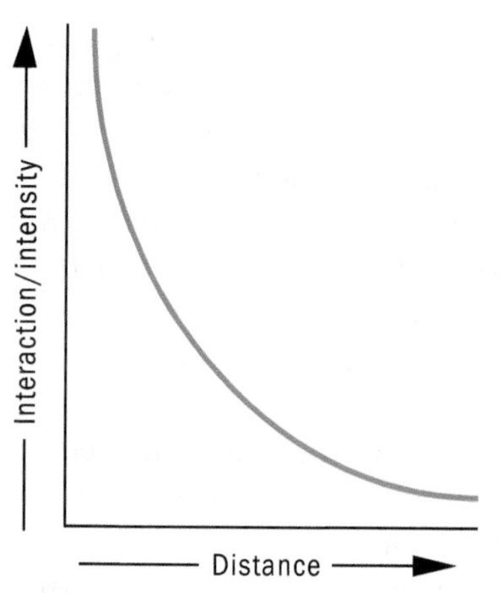

FIGURE 2.5 | A typical distance decay curve
The specific slope is a function of the particular variable being plotted against distance. The slope usually becomes less steep over time, reflecting decreasing distance friction as a result of constantly improving technology.

Distances may be measured in many ways, but most measuring systems typically involve some standard unit such as the kilometre. However, human geographers also measure distances in terms of time or cost.

© Scoast/iStockphoto

"Honeymoon capital of the world." Many North Americans would easily recognize Niagara Falls from this label alone. One of the most readily identifiable places in the world, Niagara Falls is a distinctive physical landscape with a distinctive cultural image. The reason for the association with honeymoons is debated, but there is no doubt that the Falls has traditionally represented a "trip of a lifetime" destination.

Physical Distance

The spatial interval between points in space is the physical distance. It is often measured with reference to some standard system and is therefore usually a precise measurement. An example of an imprecise measurement is in North American cities, where it is common to measure distance by reference to numbers of city blocks.

The shortest travel distance between points is often not a straight line. In a grid-pattern city, it is a series of differently oriented straight lines (Figure 2.6). In other instances, the physical distance between two points may be related to the direction of travel, as in the case of vehicle movement in a one-way street system (Figure 2.7).

Time Distance

For some movements, especially those of people as opposed to materials or products, time is important; hence, the preferred route may be the quickest rather than the shortest. Time distance is

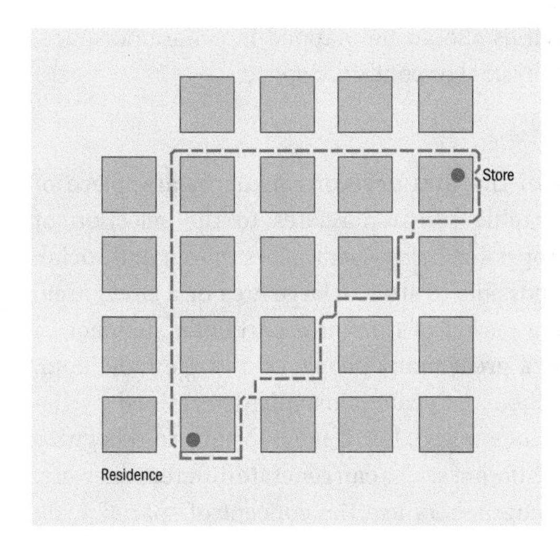

FIGURE 2.6 | Shortest distance route
A non-straight line, shortest distance route.

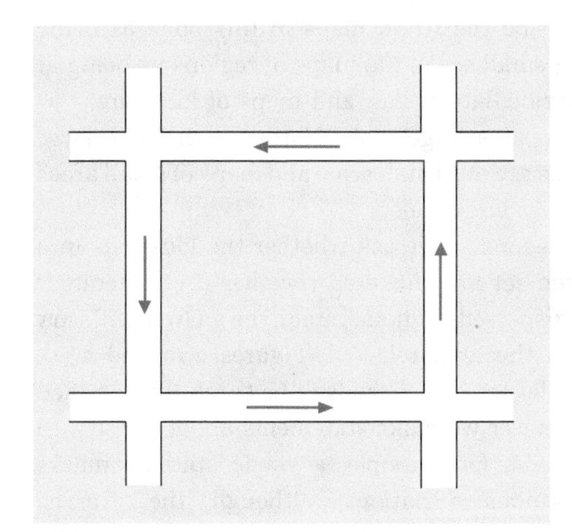

FIGURE 2.7 | A one-way system
To travel a block, the distance covered may be one block or three, depending on direction.

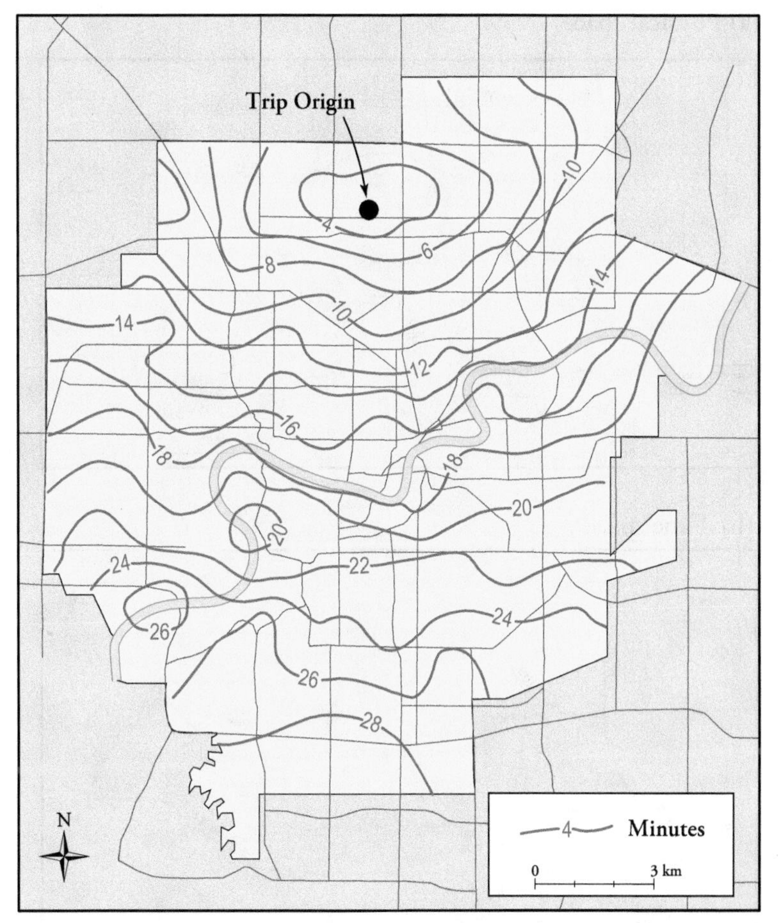

FIGURE 2.8 | Time distance in Edmonton
Source: Adapted from Muller, J. C. 1978. "The Mapping of Travel Time in Alberta, Canada." *Canadian Geographer* 22: 197–8.

related to the mode of movement, traffic densities, and various regulations regarding movement. Figure 2.8 depicts isochrones—lines joining points of equal time distance from a single location—for the city of Edmonton. Clearly, travel time is not directly proportional to physical distance; if it were, the isochrones would be equally spaced concentric circles. Figure 2.9 presents an imaginative extension of these ideas depicting Toronto in (a) physical space and (b) time space. The time-space map shows space stretching in the congested central area and shrinking in the outlying areas—a direct consequence of the greater time needed to travel a given distance on congested as opposed to freely flowing routes. This pattern is not fixed: the extent of stretching and shrinking varies according to time of day and day of the week (rush hour on a business day generates the most stretching).

isochrones
Lines on a map of equal travel time from a given starting point. One example of an isoline, which generally allows map readers to infer change with distance and to estimate specific values at any location on the map.

Economic Distance

Movement from one location in space to another usually entails an economic cost of one kind or another. Thus, economic distance can be defined as the cost incurred to overcome physical distance. As with time distance, there is not necessarily a direct relationship between physical distance and other measures. If we consider the movement of commodities, for example, costs frequently increase in a step-like fashion and the cost curve is convex (see Figure 13.5). Similarly, in some cities taxi fares are determined not by physical distance but by the number of zones crossed.

Social and Cognitive Distance

scale
The resolution level(s) used in any human geographic research; most characteristically refers to the size of the area studied but also to the time period covered and the number of people investigated.

For many industries and other businesses, cost distance is of paramount importance. There is considerable logic to the notion that economic

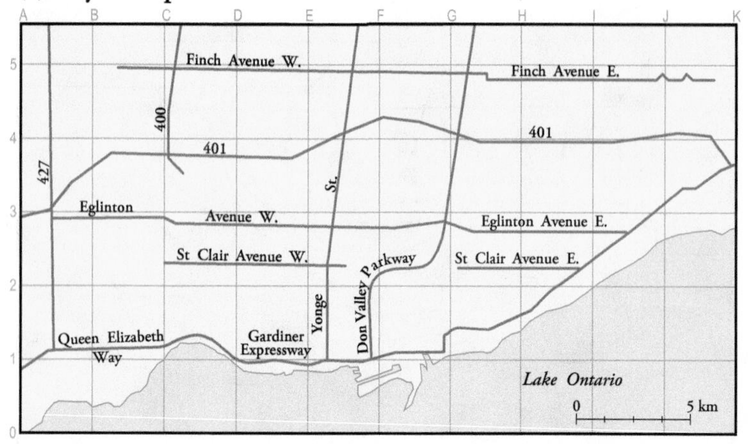

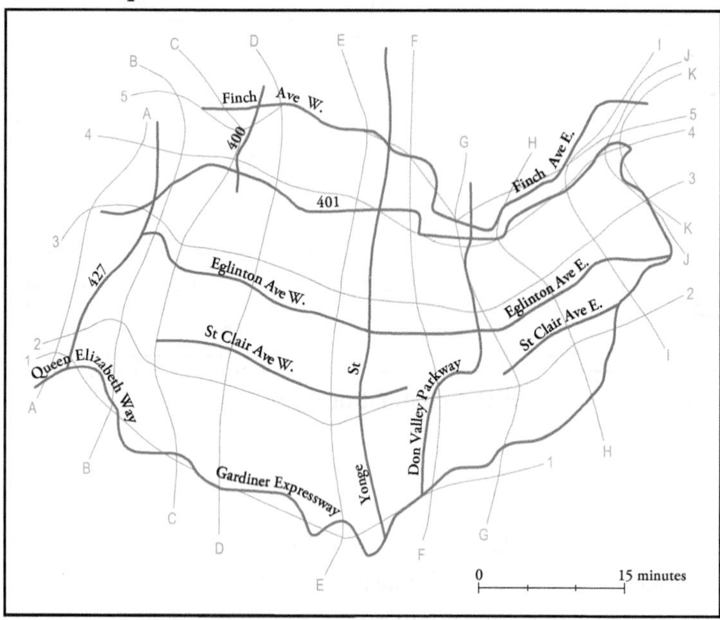

FIGURE 2.9 | Toronto in physical space and time space

Source: Ewing, G. O., and R. Wolfe. 1977. "Surface Feature Interpolation on Two-Dimensional Time–Space Maps." *Environment and Planning A* 9: 430, 435. Pion Limited, London. www.pion.co.uk, www.envplan.com.

Library and Archives Canada

Landing the telegraph cable at Heart's Content Bay, Newfoundland, from the *London Illustrated News*, 8 September 1866.

activities should be mapped in economic space, not physical or container space.

Scale

One of the first decisions made in any piece of geographic research relates to the selection of appropriate scales—spatial, temporal, and social. It is possible to study a large area or a small area, a long period of time or a particular moment in time, a great many people or a single individual. The choice of scale is usually determined by the questions posed, but it is important to recognize that different scales can generate different answers.

Geographers use the concept of spatial scale in three distinct ways. First, in accordance with a technical meaning associated with the use of maps, scale is the ratio of distance on a map to distance on the ground. In this sense we might describe the world maps in this book as being at a small scale, the maps of regions as being at intermediate scales, and maps of local areas as being at a large scale. Note that maps of large areas are at a small scale and maps of small areas are at a large scale.

Second, if we ask whether the locations in a given set are clustered together (agglomerated) or dispersed (deglomerated), the answer will vary with the area selected (Figures 2.10 and 2.11). Spatial scale also needs to be carefully identified whenever we make statements about density. In Canada, for example, a single statistic masks enormous variations; although the average population density for Canada as a whole is 3.25 persons per km^2, for an urban centre the density may be 10,000 per km^2.

Third, spatial scale refers more generally to the specific identification of the area being studied. Although there is no direct link between philosophical emphasis and scale employed, the choice of scale does relate to the purpose of the research.

Some research is concerned with a local area. Studies may focus on a shopping area, business district, small ethnically distinct residential area, neighbourhood, village, local agricultural area, or an agglomeration of industrial plants. Much humanistic research, concerned as it often is with individuals, favours this scale of analysis. Other geographic studies are more concerned with a larger area, what is often called a regional scale. Thus, there are studies of the geography of the Canadian North, the Amazon Basin, or the Australian outback. Some of these studies

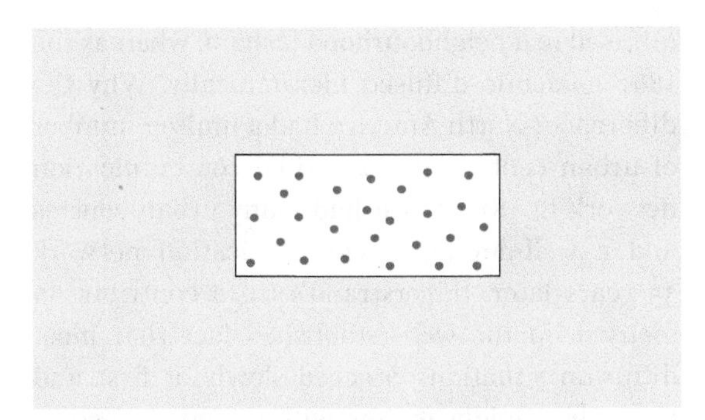

FIGURE 2.10 | The impact of spatial scale
Whether we describe this point pattern as clustered or dispersed depends on the area within which it is contained. Using the inner boundary, the pattern is dispersed; using the outer boundary, it is clustered.

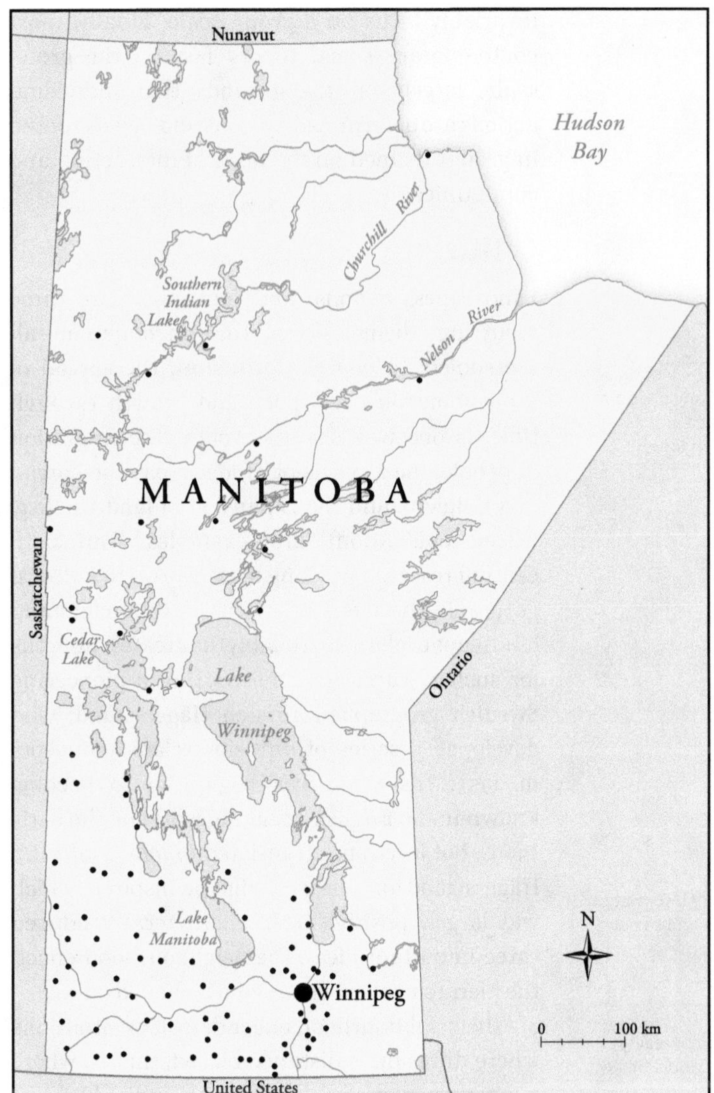

FIGURE 2.11 | Urban centres in Manitoba
This real-world pattern encompasses two relatively distinct regions: a sparsely settled centre and north and, by comparison, a densely settled south.

are empiricist in character, belonging to the genre we call regional geography. Of course, a national scale (a single country, large or small) is often the basis for analysis. The concern may be with the industrial geography of Canada, for example. Still other studies look at a group of countries, such as those that comprise Central America, Southeast Asia, or sub-Saharan Africa. Groupings of countries are usually identified because they are understood to share some important characteristics. Finally, the entire world is analyzed in many instances, as in studies of world population or the global economy. Positivist and Marxist analyses are likely to be conducted at all these scales.

The choice of temporal scale is also important. If questions are asked concerning the evolution of landscape, a temporal perspective is essential; but if questions are asked concerning the manner in which a given area functions, a temporal perspective may be quite irrelevant. Historical and cultural geography usually emphasize time, while chorological work and spatial analysis tend to emphasize the present. Contemporary human geography uses whatever temporal scale is most appropriate.

The identification of social scale is a major concern. Selecting the scale most appropriate to the question posed, and that will therefore lead most directly to the correct answer, is not simple. Scales must be selected with care and be properly justified if we are to be confident of our results. As students of human populations, we may in principle choose to work at any scale, from the individual to the world population, through the intermediate scales of nuclear families, extended families, friendship circles, voluntary associations, involuntary associations, institutions, and nations. Selecting the appropriate scale is a function of the particular type of study being conducted and thus reflects philosophical preference.

Transforming Distance and Gravity and Potential Model

Those with a humanistic focus recognize the need to study the intentions and actions of people both as individuals and as members of groups. Those with a Marxist perspective focus on groups because they believe that individuals cannot be understood without reference to the appropriate larger cultural context and, more specifically, to the overarching social and economic mode of production. Most traditional cultural geography has favoured a group scale because it is best suited to the typical geographic interest in the world or regions of the world. Given the concern with theories, laws, and hypothesis testing, positivists invariably focus on a group scale. Finally, most contemporary social theory favours the group scale, largely on the grounds that individual actions are determined by ideas and beliefs rooted in groups defined on the basis of interaction and communication.

Diffusion

diffusion
The spread of any phenomenon over space and its growth through time.

Landscapes, regions, and locations—our three recurring themes from the introduction—all are subject to change. **Diffusion**, the spread of a phenomenon over space and growth through time, is one way change occurs. The migration of people, the movement of ideas (e.g. the spread of a religion), and the expansion of land use (e.g. wheat cultivation) are examples. Diffusion-centred research has long been central to cultural geography because of the need to understand landscape evolution. Probably the greatest impetus for such research was the work of a pioneering Swedish geographer, Torsten Hägerstrand, who developed a series of diffusion-related concepts in 1953. His work was beginning to become known in the English-speaking world by the early 1960s, but it was not translated *in toto* until 1967. Hägerstrand and the work that he inspired, which was largely positivistic in character, introduced three important ideas: the neighbourhood effect, the hierarchical effect, and the S-shaped curve.

The neighbourhood effect describes situations where diffusion is distance-biased, that is, where a phenomenon spreads first to individuals or groups nearest its place of origin. Other situations, however, involve a hierarchical effect. In this case, the phenomenon first diffuses to large centres, then to centres of decreasing size. One geographer noted that the 1832 cholera epidemic in North America diffused in a neighbourhood fashion, whereas the 1867 epidemic diffused hierarchically. Why the difference? North America had a limited number of urban centres and a limited communication network in 1832, but it had many urban centres and a well-developed communication network 35 years later. Hägerstrand's third contribution focused on the well-established fact that most diffusion situations proceed slowly at first and then very rapidly, ending with a final slow stage to produce an S-shaped curve (Figure 2.12). The diffusion concept is perhaps best described as a process that prompts changes in landscapes, regions, and locations.

100
% of population who have Accepted innovation
0
Time

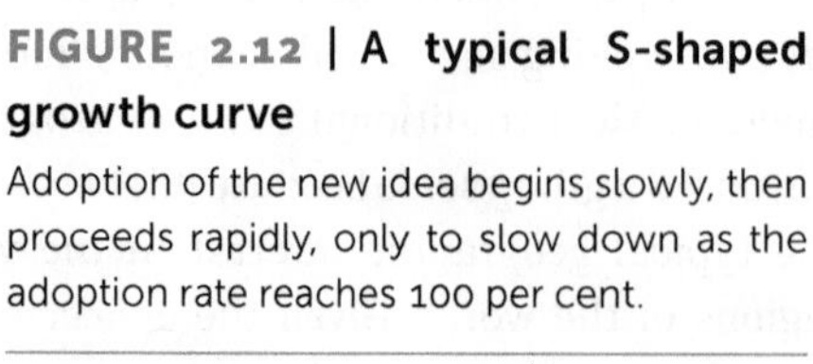

FIGURE 2.12 | A typical S-shaped growth curve

Adoption of the new idea begins slowly, then proceeds rapidly, only to slow down as the adoption rate reaches 100 per cent.

Perception

perception
The process by which humans acquire information about physical and social environments.

image
The perception of reality held by an individual or group.

mental map
The individual psychological representation of space.

In 1850, Humboldt noted that, "in order to comprehend nature in all its vast sublimity, it would be necessary to present it under a twofold aspect, first objectively, as an actual phenomenon, and next subjectively as it is reflected in the feelings of mankind" (quoted in Saarinen, 1974: 255–6). Despite this pioneering statement, geographers paid relatively little attention to subjective matters, especially the perceived environment, until the late 1960s. We now recognize that all humans relate not to some real physical or social environment but rather to their **perception** of that environment—a perception that varies with knowledge and is closely related to cultural and social considerations. Humanistic geographers in particular discuss the mental **images** of places and other people and seek to describe and understand the images, or **mental maps**, that we carry in our heads. Mental maps of relatively unknown areas are especially subject to error. Figure 2.13 is an interesting composite of the image of North America held by Europeans in the mid-eighteenth century.

Clearly, we acquire our mental maps from a variety of sources, but one is of particular relevance to us: human geography teaches us about the world, about where things are located, why they are there, and what they really are. By the time you have completed this course, you may notice a significant improvement in your own mental maps and images!

Development

In their studies of the landscapes created by humans, human geographers recognize that any one area changes through time and that different

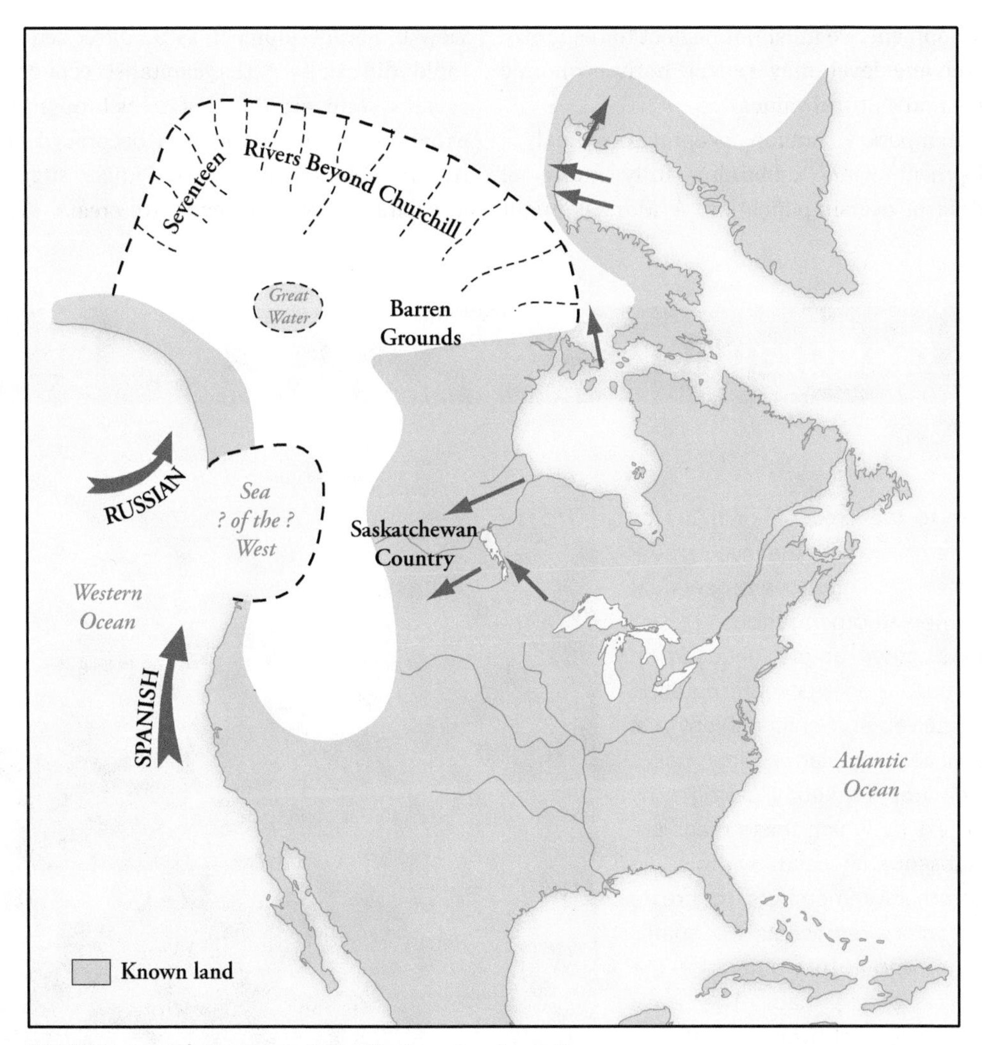

FIGURE 2.13 | Images of North America in 1763

Limited knowledge resulted in many inappropriate decisions. As reality became better understood, spatial behaviour became increasingly appropriate.

Source: Ruggles, R. I. 1971. "The West of Canada: Imagination and Reality." *Canadian Geographer* 15: 237. Published by Blackwell Publishing Ltd.

areas have different landscapes. Such conditions are interpreted in terms of **development**. The general meaning of this term involves measures of economic growth, social welfare, and modernization. It is possible to define, for any given area at any given time, the level or state of development. Following this logic, certain areas are qualified as more developed and others as less developed. (This distinction is discussed further in Chapters 3, 4, 5, and 6.)

development
A term that should be handled with caution because it has often been used in an ethnocentric fashion; typically understood to refer to a process of becoming larger, more mature, and better organized; often measured by economic criteria.

Such basic identifications can be of real value, but we need to use caution in applying them. It is important that human geographers highlight spatial disparities in economic well-being, but it is also important that we interpret variations with reference to cultural and social considerations. If, for example, we use income levels as a measure

RyersonClark/iStockphoto

A musher and his team participate in a dog-sled race during Yellowknife's Caribou Carnival. Many people who have never visited Canada might consider this image to represent typical Canadian life, especially if they perceive the country as a place of constant snow and cold weather where all citizens play hockey and travel by snowmobile and/or dog sled.

of development, we must not neglect to recognize that income level may reflect both economic success and cultural values.

Contemporary human geographers analyze development while remaining fully aware of the risks of oversimplification. A Marxist might view underdevelopment as a consequence of the rapid diffusion of the capitalist economic and social system, arguing that areas brought into the expanding capitalist system become dependent. Another Marxist perspective might suggest that a capitalist system tends to create depressed

Focus on Geographers

Participatory Mapping | JON CORBETT

Participatory mapping refers to the creation of maps by local communities. This is often done with the involvement of supporting organizations, including governments (at various levels), non-governmental organizations (NGOs), and universities, as well as other actors engaged in planning. Participatory maps provide a visual representation of what the community perceives as significant physical and socio-cultural features, issues, and places. What makes participatory mapping different from traditional cartography and map-making is the process by which these maps are created and how they are subsequently used. By providing skills and expertise for community members to create the maps themselves, the content represents the spatial knowledge of these members, who in turn determine the ownership and communication of the maps. The levels of community involvement and control over the mapping process can vary considerably.

JON CORBETT is an associate professor in the Community, Culture and Global Studies Unit at the University of British Columbia, Okanagan, and the co-director of the Centre for Social, Spatial and Economic Justice.

Participatory mapping can have an important influence on the process of social change. For example, it can transform a community's internal dynamic by contributing to building community cohesion, help stimulate community members to engage in land-related decision-making, raise awareness about pressing land-related issues, and ultimately contribute to empowering local communities and their members. The general aims and specific objectives of participatory mapping initiatives vary depending on the map's end use, which in turn is influenced by the audience that will view and make decisions related to the content of these maps. Participatory mapping projects can take on an advocacy role and actively seek recognition for community spaces through identifying traditional lands and resources and demarcating ancestral domain. In this way, participatory maps play an important role in helping marginalized groups (including Indigenous communities) work towards legal recognition of customary land rights.

Participatory mapping uses a range of spatial tools, including mental maps, ephemeral mapping, participatory sketch mapping, transect mapping, and participatory three-dimensional modelling. In recent years initiatives have begun to use more technically advanced geographic information technologies including global positioning systems (GPS), aerial photos and remote-sensed images (from satellites), geographic information systems (GIS), and other digital computer-based technologies. The breadth of tools available in the mapper's toolbox makes participatory mapping highly flexible and valuable for use in spatial planning initiatives that demand the input of multiple stakeholders. Yet these mapping initiatives can also be ineffective and generate confusion and conflict if implemented without a working knowledge of cartography and participatory processes, as well as community facilitation and organization skills.

In the last 10 years the increase in Internet capability has changed the shape and direction of participatory mapping, in particular, the development of what is now referred to as the Geospatial Web, or Geoweb. The Geoweb is the geographic platform for Web 2.0 digital social networking applications; it includes applications such as Google Earth, Google Maps, Bing Maps, and other location-based Internet technologies. The Geoweb has achieved broad acceptance thanks to its widespread availability on the Internet, its platform independence, and its being superficially free to use. (There are associated costs, for example, some services claim ownership over data collected through their systems and preserve the right to reuse it.) Another reason for the Geoweb's popularity is its ability to aggregate user-contributed digital information from multiple web-based applications and to capture and present user-generated digital content, referred to as crowdsourcing. Within the discipline of geography this spatial crowdsourced information is referred to as volunteered geographic information (VGI).

areas within any given country, thus prompting uneven development. Such variations are evident throughout what we generally label the more developed world. In Canada, for example, an economically advantaged core area, the St Lawrence lowlands of Quebec and Ontario, is surrounded by a number of relatively disadvantaged peripheries, such as the Maritime area and the North.

Discourse

The root meaning of the word *discourse* (defined in Chapter 1) is "speech." In the social sciences, however, the term also refers to a way of communicating, in speech or in writing, that serves to identify the person communicating as a member of a particular group. (Note that language, dialect, or accent can serve the same function.) For example, the technical vocabulary introduced above—*space*, *location*, and so on—is part of the discourse of human geography and serves to identify those who use that vocabulary as members of the group of human geographers.

However, as suggested in Box 1.1, discourse also has a more profound meaning, derived from the work of the French social theorist Michel Foucault. Foucauldian theory was introduced into the literature of human geography in the 1980s as one aspect of social theory. According to Foucault, the history of ideas is a history of changing discourses in which (a) there is a fundamental connection between power and knowledge, and (b) truth is not absolute but relative, dependent on the power relations within the societies that construct it. This more complex meaning is pursued in Chapter 8, in connection with feminism and postmodernism. The latter are bodies of social theory that challenge established discourses because they are seen as products of people in positions of academic power who are able to define the truth in their terms, to the exclusion of the concerns of other, usually marginalized, groups.

Globalization

Our final concept, globalization (defined in the introduction), integrates—some might even say replaces—several of the concepts we have introduced, namely, space, location, place, region, and distance. For this reason, it is the subject of the following chapter. The most fundamental consequence of globalization is that our complex and varied human worlds are becoming, however unevenly, more and more like a single world. Accordingly, we now identify globalization as an overriding metaconcept, providing human geographers with a body of ideas that may facilitate analysis of environmental, cultural, political, and economic topics. At the same time, as discussed in Chapter 3, we acknowledge the many uncertainties concerning what globalization is and how it affects the world.

The term *globalization* came into widespread use only in the 1980s. In general, it refers to the idea that the world is becoming increasingly homogenized economically, politically, and culturally. Globalization is both a result and a cause of an ever-increasing connectedness of places and peoples as economic, political, and cultural institutions, flows, and networks all combine to bring previously separated peoples and places together. Among the most obvious components of globalization are advances in communications technologies and the increasing dominance of **transnationals**. Among the most obvious consequences is the fact that, because we are generally able to overcome distances much more easily today than we were in the past, distance no longer plays the critical role it once did in promoting the development of separate human geographic worlds.

transnationals
Large business organizations that operate in two or more countries.

cartography
The conception, production, dissemination, and study of maps.

Concepts: A Concluding Comment

This discussion of human geographic concepts has introduced the terms that are central to understanding how human behaviour affects the earth's surface, along with various other concepts that can be conveniently subsumed under those headings. In most cases, as we have seen, a concept can be understood and used in multiple ways, depending on philosophical approach; think especially about region, distance, scale, and development. All the terms introduced here will reappear in subsequent accounts of the substantive work of human geographers and will be put into practice.

TECHNIQUES OF ANALYSIS

Now that we have a basic understanding of why human geographers ask particular research questions, what their philosophical options are, and what concepts they use to structure their research activity, we are ready to look at the techniques they use to collect, display, and analyze data.

The first three techniques discussed are cartography, computer-assisted cartography, and geographic information systems. Each of these closely related techniques is inherently geographic, having been largely developed within geography, and each involves inputting, storing, analyzing, and outputting spatial data. A group of techniques that relate primarily to the collection of data—remote sensing—is discussed next. Finally, we consider both qualitative and quantitative methods of collecting and analyzing data.

choropleth map
A thematic map using colour (or shading) to indicate density of a particular phenomenon in a given area.

isopleth map
A map using lines to connect locations of equal data value.

Cartography

Chapter 1 highlighted the centrality of maps to the geographic enterprise. The science of map-making is known as **cartography**. Until the 1960s, cartography was essentially limited to map production, following data collection by surveyors and preceding analysis by geographers. There was much emphasis on manual skills. The primary purpose of such maps was to communicate information. Quite simply, maps are an efficient means of portraying and communicating spatial data. Today, however, cartography is less dependent on manual skills and is closely integrated with analysis. Regardless, maps are no less important to us today than they were 500 years ago; we carry them on our smartphones and other devices.

As communication tools, maps describe the location of geographic facts. As analytical tools, they can be used to clarify questions and suggest research directions. In the production of maps, cartographers need to decide on questions of scale, type, and projection, which can significantly affect map appearance and quality.

As we have already noted, map scale relates to the area covered and the detail presented. Large-scale maps portray small areas in considerable detail, while small-scale maps portray large areas with little detail. Scale is always indicated on a map, whether as a fraction, a ratio, a written statement, or a graphic scale. Canada has a National Topographic System (NTS), with maps at a scale of 1:250,000 covering 15,539 km² (6,000 square miles) and maps at a scale of 1:50,000 covering 1,036 km² (400 square miles) in greater detail. On a 1:50,000 map, farm buildings are located, and contours (lines of equal elevation) are shown at 7.5 m (25 ft) intervals. Figures 2.14 and 2.15 provide examples of each type.

The type of map constructed depends on the information being presented. Dot maps are useful for data such as towns, wheat farming, cemeteries, incidence of diseases, and so forth. Typically, each dot represents one occurrence of the phenomenon being mapped (Figure 2.16). A **choropleth map** displays data by using tonal shadings that are proportional to the density of the phenomena in each of the defined areal units (Figure 2.17). Choropleth maps sacrifice detail for improved appearance. An **isopleth map** consists of a series of lines—isopleths or isolines—that link points having the same value (Figure 2.18). Examples include contour maps, isochrome maps showing

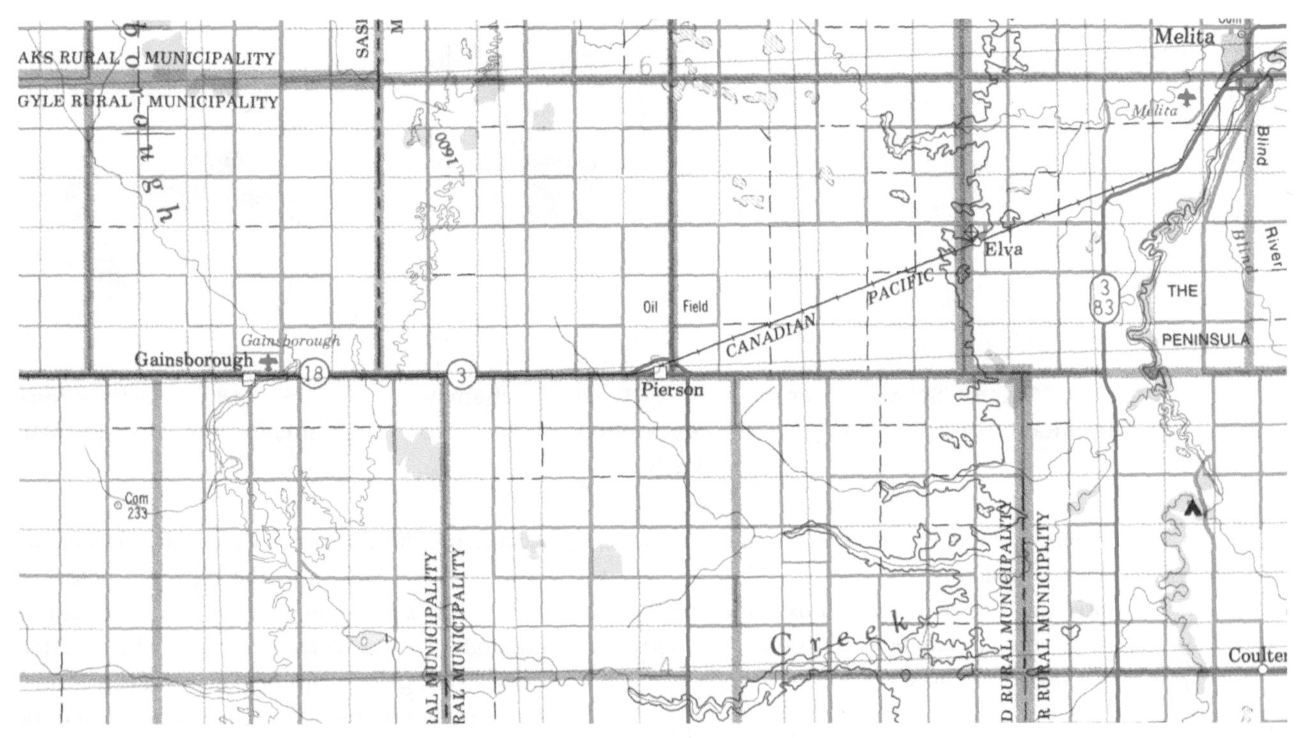

FIGURE 2.14 | Mapping at a scale of 1:250,000

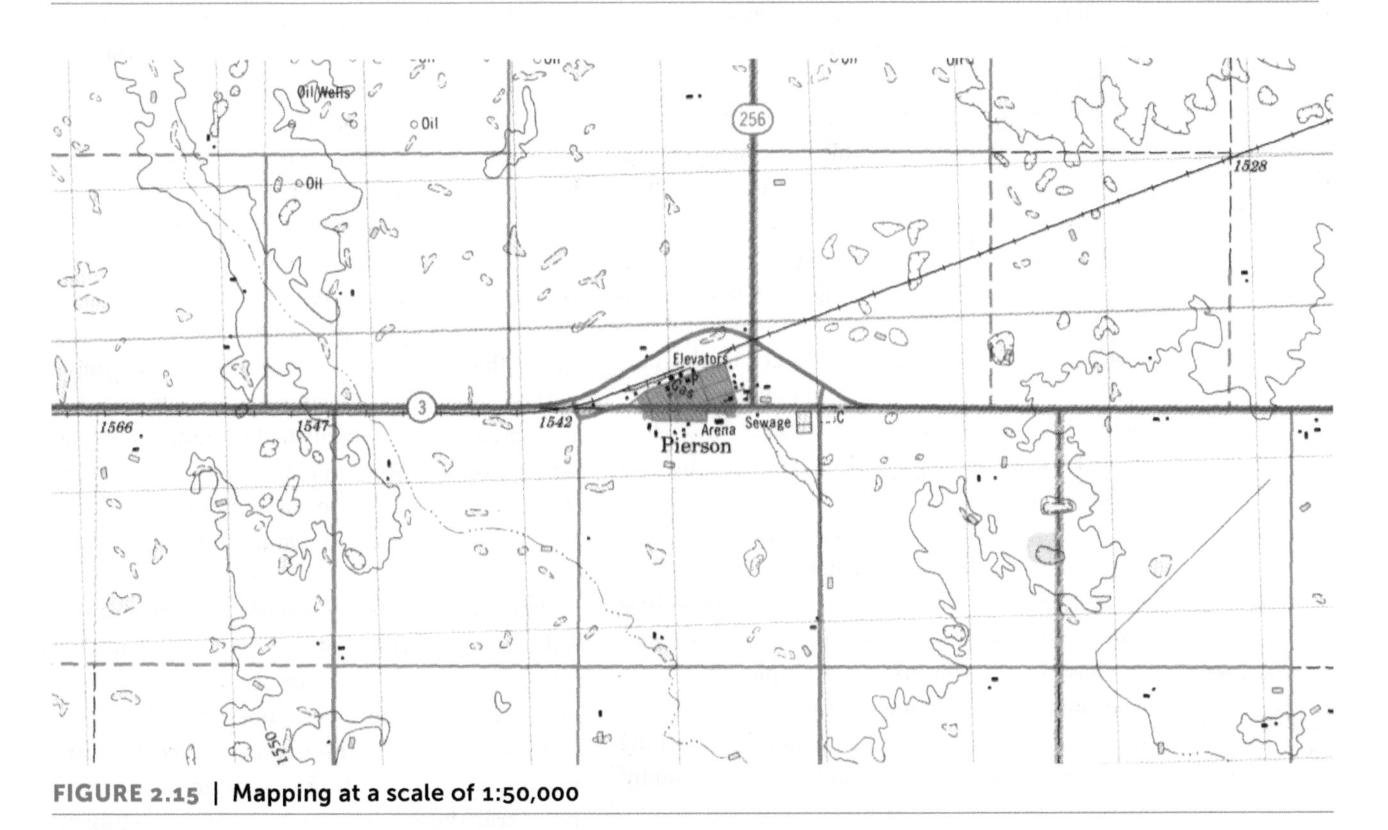

FIGURE 2.15 | Mapping at a scale of 1:50,000

lines of equal time, and isotim maps showing lines of equal transport cost.

Small-scale maps (of large areas such as the earth) raise a fundamental question: How can we best represent a nearly spherical earth on a flat surface (a process of conversion known as a **projection**)? No satisfactory answer to this question has yet been found, although hundreds of projections have been developed. Map users need to understand the cartographic design process,

projection
Any procedure employed to represent positions of all or a part of the earth's spherical (three-dimensional) surface onto a flat (two-dimensional) surface.

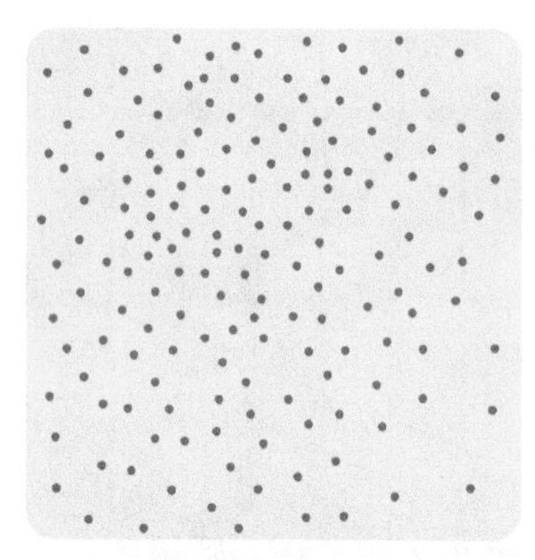

FIGURE 2.16 | Schematic representation of a dot map

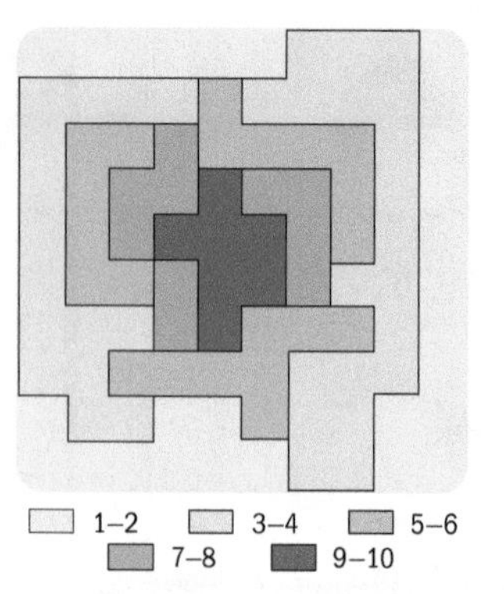

FIGURE 2.17 | Schematic representation of a choropleth map

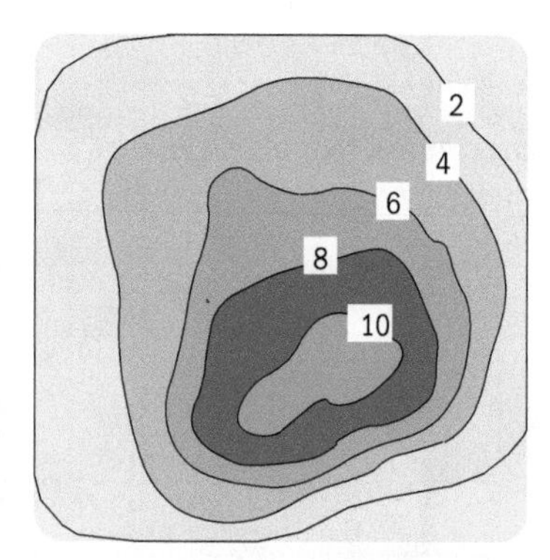

FIGURE 2.18 | Schematic representation of an isopleth map

including the significance of the chosen scale, the types of symbols, and the projection, in order to interpret a map correctly. Box 2.4 highlights the fact that maps can be used to convey a particular message to the user.

Computer-Assisted Cartography

Map Projections

Computer-assisted cartography, sometimes called digital mapping, is discussed separately from traditional cartography because it represents much more than just another evolution in production techniques, although technical advances have significantly reduced the need for manual skills. Conceived by the Canadian geographer Roger Tomlinson (1933–2014), computer-assisted cartography enables us to amend maps by incorporating new and revised data and to produce various versions of the mapped data to create the best version. Through the use of mapping packages, they diminish the need for artistic skills and allow for desktop map creation, although this still requires considerable design skills as decisions are made about colouring, shading, labelling, and other aspects of map creation. Computer-assisted cartography has introduced maps and map analysis into a wide range of new arenas: for example, businesses can now use cartography to realign sales and service territories. Computer-generated maps facilitate decision-making and are becoming important in both academic and applied geography.

geographic information system (GIS)
A computer-based tool that combines the storage, display, analysis, and mapping of spatially referenced data.

vector
A method used in GIS to represent spatial data; describes the data as a collection of points, lines, and areas and describes the location of each of these.

Geographic Information Systems

A **geographic information system (GIS)** is a computer-based tool that combines several functions: storage, display, analysis, and mapping of spatially referenced data. A GIS includes processing hardware, specialized peripheral hardware, and software. The typical processing hardware is a personal computer or workstation, although mainframe computers may be used for especially large applications. Peripheral hardware (e.g. digitizers and scanners) may be used for data input, while printers or plotters produce copies of the output. Software production is now a major industry, and numerous products are available for the GIS user. Examples include IDRISI, a university-produced package designed primarily for pedagogic purposes, and ARC/INFO, a package developed by the private sector that is widely used by governments, industries, and universities.

The origins of contemporary GIS can be traced to the first developments in computer-assisted cartography and to the Canada GIS of the early 1960s. These developments centred on computer methods of map overlay and area measurement—tasks previously accomplished by hand. Since the early 1980s, however, there has been an explosion of GIS activity related to the increasing need for GIS and the increasing availability of personal computers.

The roots of GIS are clearly in cartography, and maps are both its principal input and output. But computers are able to handle only characters and numbers, not spatial objects such as points, lines, and areas. Hence, GISs are distinguished according to the methods they use to translate spatial data into computer form. There are two principal methods of translation. The **vector** approach describes spatial data as a series of discrete objects: points are described according to distance along two axes; lines are described by the shortest distance between two points; and areas are described by sets of lines.

The raster method represents the area mapped as a series of small rectangular cells known as pixels (picture elements): points, lines, and areas are approximated by sets of pixels, and the computer maintains a record of which pixels are on or off.

What is the value of a GIS? What applications are possible? The short answer is that GISs have numerous and varied applications in any context that may be concerned with spatial data: biologists analyze the effect on wildlife of changing land-use patterns; geologists search for mineral deposits; market analysts determine trade areas; and defence analysts select sites for military installations. We will encounter several applications of GIS in later chapters. The common factor is that the data involved are spatial. After handling spatial data for about 3,000 years, geographers now have important new capacities. In brief, a GIS achieves a whole new range of mapping and analytical capabilities—additional ways of handling spatial data.

Remote Sensing

No map can be produced without data. GISs and analytical methods in general also require data. Where do these data come from? Early map-makers obtained their data from explorers and travellers. The contemporary geographer collects different data in different ways.

One particularly important group of collection methods focuses on gathering information about objects from a distance. The term *remote sensing* describes the process of obtaining data using both photographic and non-photographic sensor systems. In fact, we all possess remote sensors in the form of our eyes, and it has been one of humanity's ongoing aims to improve their ability to acquire information. Improving our eyes (using telescopes), improving our field of vision by gaining altitude (climbing hills or using balloons and aircraft), and improving our recording of what is seen (using cameras) are advances in remote sensing. Today, most applications of remote sensing rely on electromagnetic radiation to transfer data from the object of interest to the sensor. Using satellites as far as 36,000 km (22,370 miles) from the earth, we are now able to collect information both about the earth as a whole and about what is invisible to our eyes. Electromagnetic radiation occurs naturally at a variety of wavelengths, and there are specific sensing technologies for the principal spectral regions (visible, near-infrared, infrared, and microwave).

The conventional camera was the principal sensor used until the introduction of earth orbital satellites in the 1960s. Aerial photography is still used for numerous routine applications, particularly in the visible and near-infrared spectral regions. The near-infrared spectral region has proved particularly useful for acquiring environmental data. The current emphasis is on satellite imagery, especially since the United States launched Landsat (initially called Earth

raster
A method used in GIS to represent spatial data; divides the area into numerous small cells and pixels and describes the content of each cell.

remote sensing
A variety of techniques used for acquiring and recording data from points that are not in contact with the phenomena of interest.

In the News

BOX 2.4 | The Power of Maps

Maps made news when they offered the first accounts of previously unknown places—unknown, that is, to the map-makers and their audience. Today, maps make news largely because of their perceived inadequacies. For example, the proliferation and increasing use of Internet maps, such as those provided by Google, and satellite navigation technologies have been blamed (by Britain's most senior cartographer in 2008) for wiping the rich history and geography of Britain off the map. This is because such maps are designed only to provide directional information and not to reflect major landscape features. Certainly, maps now often serve different purposes than they once did. Whereas past maps might encourage people to subscribe to a particular religion or political viewpoint, current maps are more likely to guide us to places, especially ones where we can buy consumer goods or services. Clearly, then, maps have been and are both enabling and constraining. They enable us to see what our own vision does not permit, but the fact that they necessarily simplify complex realities constrains our understanding.

Continued

Recall the T-O maps introduced in Chapter 1; these European maps clearly reflect medieval Christian values, with Jerusalem placed at the centre of the world. Other maps from that place and time show *terra incognita* to the south and make extensive use of decorative pictures. Nineteenth-century European maps reflect the political and economic concerns of their creators: colonial possessions, for example, are prominently displayed. These maps, building on Ptolemaic traditions, established the conventions that most of us take for granted today: north at the top, 0° longitude running through Greenwich, England, and the map centred on either North America or Western Europe. These conventions are totally arbitrary and yet their influence is enormous because, in effect, they tell us that certain countries are world centres and others are outliers.

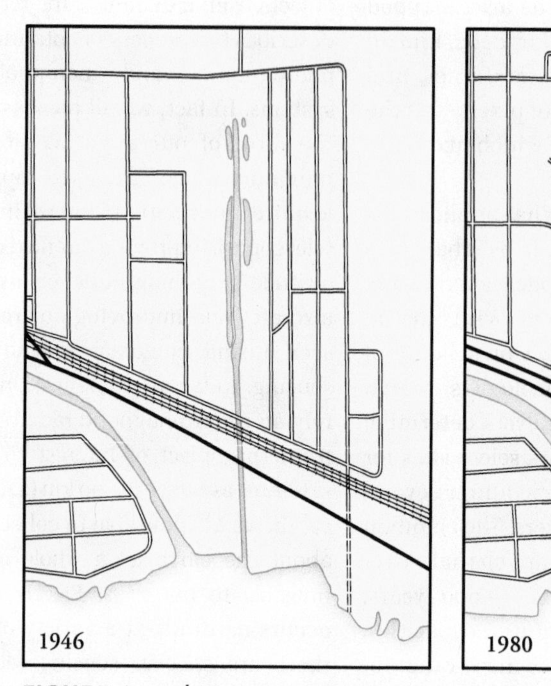

FIGURE 2.19 | Two topographic maps of the Love Canal area, Niagara Falls, New York

The 1946 map shows the canal (the long vertical feature in the centre of the map), but does not show the use made of it, beginning in 1942, as a dump for chemical waste. The 1980 map does not show the filled-in canal, nor does it indicate that dumping continued until 1954, even though the environmental damage led the New York health commissioner to declare a state of emergency and relocate 239 families in 1978.

Source: Monmonier, M. 1991. *How to Lie with Maps*. Chicago: University of Chicago Press, 121–2 (1946: US Army Map Service, 1946, Tonawanda West, NY, 7.5-minute quadrangle map. 1980: US Geological Survey, 1980, Tonawanda West, NY, 7.5-minute quadrangle map).

Other maps may convey a biased message either deliberately or because the map-maker has some ulterior motive. As Monmonier (1991) suggests in his book *How to Lie with Maps*, there are many ways to deceive. Political propaganda maps represent an extreme example. Less extreme but still significant are maps that simply ignore important features, for example, recent maps of the Love Canal area in Niagara Falls, New York (Figure 2.19), which simply fail to include extremely important information. The message here is that "a single map is but one of an indefinitely large number of maps that might be produced for the same situation or from the same data" (Monmonier, 1991: 2).

This simple example highlights what has remained all too commonplace, namely, that maps embody and project the map-makers' intentions and concerns. Maps are useful indicators not only of what we know about the world but also of how we see ourselves in the world. It might be said that, while we create maps, maps in turn recreate us. They can be, and often are, used to convey messages and are thus powerful symbolic representations.

Resources Technology Satellite [ERTS]) in 1972, the first unmanned satellite with the express purpose of providing detailed geophysical data. Satellite scanners numerically record radiation and transmit numbers to a receiving station; these numbers are used to computer-generate pixel-based images. In other words, satellite images are not photographs.

There are several principal advantages to satellite remote sensing. First, repeated coverage of an area facilitates analysis of land-use change. The most recent Landsat satellite covers most of the earth's surface every 16 days; these data are homogeneous and comprehensive. Second, because the data collected are in digital format, rapid data transmission and image manipulation are possible. Third, for many parts of the globe, these are the only useful data available. Finally, remote sensing allows the collection of entirely new sets of data: satellite data first alerted us to the changing patterns of atmospheric ozone in high-latitude areas. For the human geographer, remote sensing is especially valuable in aiding understanding of human use of the earth; much of the discussion in Chapter 4 relies on remotely sensed data. Remote sensing is less useful if we are concerned with underlying economic, cultural, or political processes.

Most satellites have been launched by either the US or the European Space Agency. Canada launched a satellite, Radarsat, in 1996. A recent substantial achievement—indeed, a breakthrough in the science of remote sensing—was the remarkably detailed mapping of the earth's surface in 2000 by a manned NASA space shuttle. This mission involved a partnership among the military, intelligence-gathering, and environmental communities and resulted in a topographic map of the earth's land mass between 60°N and 56°S that is about 30 times as precise as the best maps available before the mission.

Besides remote sensing, some data can be collected by another means. In the early 1990s geographers began to make use of another new digital geographic technology, the global positioning system (GPS). A GPS is an instrument (either hand-held or installed in a personal computer) that uses signals emitted by satellites to calculate location and elevation. Along with remotely sensed data, GPS data can be integrated into a GIS.

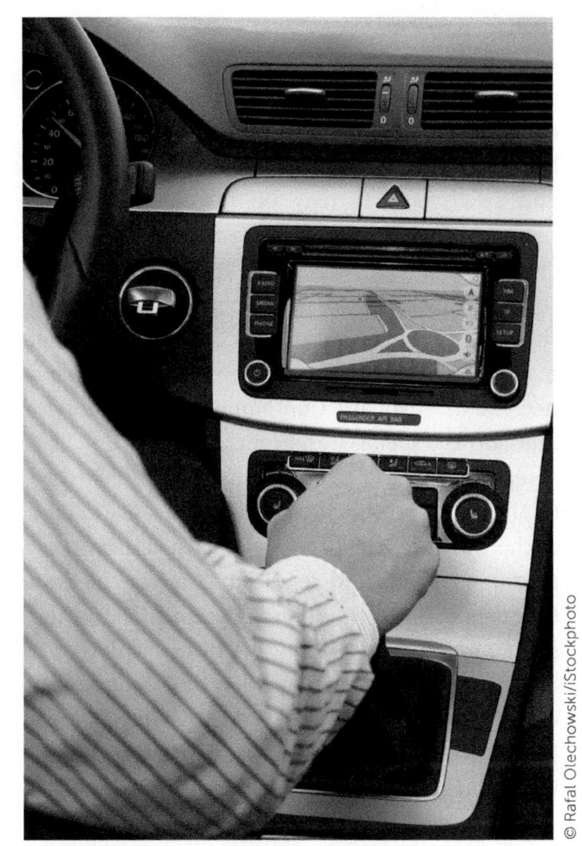

© Rafal Olechowski/iStockphoto

When geographers collect data in the field, they need to specify the geographic locations of their measurements. These locations are mapped to facilitate identification in a corresponding remote-sensing image. Today, precise field locations can be established using a GPS. Originally developed by the US military, a GPS involves a group of satellites circling the earth in precisely known orbits and transmitting signals that are recorded by ground-based receivers built into a variety of devices such as cellphones and in-car navigation systems.

Qualitative Methods

Human geographers collect and analyze data using a broad range of **qualitative methods**—a term widely used in other social sciences that refers to research with a focus on the attitudes, behaviour, and personal observation of human subjects. Qualitative methods are a part of **ethnography**, a general approach that requires researcher involvement in the subject studied.

Much **fieldwork**—a traditional term for the methods that geographers use to obtain primary data (for example, observation and interviews with informants)—is qualitative in character. Observation of landscapes was central to much early regional and cultural geography, but such observation decreased in importance around 1960 as geographers began to focus more on secondary data. Recently, however, new types of fieldwork have appeared in response to humanistic concerns, and human geographers now use a range of qualitative methods for collecting and analyzing data (see Box 2.5). Early fieldwork was not philosophically motivated, although it was implicitly empiricist because it assumed that reality was present in appearance. By contrast, contemporary qualitative fieldwork is by nature humanistic as a response to the humanistic requirement that human geography strive to understand the nature of the social world.

For the humanist, qualitative methods that involve a researcher's observation of and involvement in everyday life are central to understanding humans and human landscapes. **Participant observation**, a standard method in anthropology and sociology, is now a popular geographic approach. The principal advantage of this method is its explicit recognition that people and their lives do matter.

To conduct research using qualitative methods requires considerable skill. A subjective procedure

qualitative methods
A set of tools used to collect and analyze data in order to subjectively understand the phenomena being studied; the methods include passive observation, participation, and active intervention.

ethnography
The study and description of social groups based on researcher involvement and first-hand observation in the field; a qualitative rather than quantitative approach.

fieldwork
A means of data collection; includes both qualitative (for example, observation) and quantitative (for example, questionnaire) methods.

participant observation
A qualitative method in which the researcher is directly involved with the subjects in question.

ethnocentrism
A form of prejudice or stereotyping that presumes that one's own culture is normal and natural and that all other cultures are inferior.

questionnaire
A structured and ordered set of questions designed to collect unambiguous and unbiased data.

sampling
The selection of a subset from a defined population of individuals to acquire data representative of that larger population.

random sampling
Sampling such that every part of the study area or every item in the data set has an equal chance of being selected and is independent of all other parts or items.

model
An idealized and structured representation of the real world.

such as participant observation does not provide any means for the researcher to objectively control the relationship between observer and observed. You will recall that this was one of the key issues in the differences between humanism and positivism. It may be the case that the researcher, who is often of a higher social status, is ethnocentric (**ethnocentrism** is the presumption that one's own culture is normal and natural and that other cultures are inferior), or the researcher may identify with the group being observed and become its advocate. Contemporary geographers and other social researchers pursuing field research, however, seek to bring a "reflexivity" to their fieldwork, which includes an awareness of their own real or potential biases, of how their presumed status and gender may affect the data they collect from human subjects, and of how their simple presence inevitably will alter the dynamic of that which they seek to observe and understand. Other disadvantages of qualitative procedures include the risk that the researcher will begin with a biased or otherwise inappropriate idea about the data to be collected or that the subjects of the study may not be sufficiently representative to provide an accurate picture. You will have an opportunity to assess the quality of some of this work, especially in Chapters 8 and 12.

Quantitative Methods

Some fieldwork is explicitly quantitative in character, notably, the use of a **questionnaire** to survey people. Like the traditional qualitative fieldwork procedure of observation, a questionnaire is part of an empiricist research activity. Unlike qualitative fieldwork, however, it asks all individuals the same questions in the same way. The value of questionnaire results depends on the response rate achieved and the way potential respondents are selected, namely, the **sampling** method. Proper sampling methods, based on statistical sampling theory, allow the sample results to be treated as representative of the population, within certain error limits. The most common technique used for selecting respondents is **random sampling**.

During the 1960s, quantitative methods in general developed extensively in association with the spatial analysis school and the general acceptance of a positivist philosophy. The principal methods used were statistical, and the purposes were to describe data (by calculating a mean and a standard deviation, for example) and to test hypotheses generated by theory (using correlation tests, e.g. as in Box 2.1, point 4).

The spatial analysis school recognized early that models could play a much greater role in analyzing data. A **model** is an idealized, simplified representation of the real world that highlights its key properties and eliminates incidental information. Many of the earliest spatial models employed by human geographers were based on generalizations about the relationships between the distribution of geographic facts and distance.

Geographers use quantitative techniques for a wide variety of purposes but especially for analyzing relationships between spatial patterns and for classifying data. Describing relationships is fundamental in producing explanations and is usually broached by proposing a functional relationship such that one variable is dependent

Examining the Issues

BOX 2.5 | Some Qualitative Resources

The term *qualitative resources* is used to refer to both methods and data sources. Holland et al. (1991) have identified nine of the many such resources available. Here is their list, accompanied by some comments that point to the breadth of humanistic research today:

1. *Talking and listening:* The typical quantitative interview uses a carefully structured questionnaire to ensure that multiple respondents answer a specific set of questions. By contrast, a qualitative interview seeks to gain insight into the human worlds of individuals. It is particularly important to talk with and to listen to respondents when the human world under investigation is different from the interviewer's own. In planning such research, you should always try to take into account the respondents' likely perspectives and be ready to change your questions, depending on what they say.

2. *A feminist perspective:* Feminist research focuses on the social significance of gender and especially the unequal distribution of power between women and men. A feminist perspective will influence the questions posed as well as the techniques used to carry out research.
3. *Literature:* Why do humans change landscapes as we do? Some geographers have turned to literature to answer this question. Creative writing often reflects cultural attitudes.
4. *Cartoons:* Although, like works of literature, cartoons are usually the work of an individual, they often reflect a broad social consensus on political, environmental, or economic issues.
5. *Photographs:* Human geographers frequently turn to photographs for information about the past. Photos can be invaluable not only as records of historical landscapes but as evidence of cultural attitudes. Choice of subject matter (including what the photographer omits or ignores) and manner of presentation can be extremely revealing.
6. *Art:* Many works of art have been interpreted as reflecting the distinctive character of a particular time and place.
7. *Popular music:* Music is often closely associated with place—just think of blues and the American South. Today much popular music is a direct reflection of inner-city urban experiences, highlighting social and environmental ills.
8. *Buildings:* Buildings are among the most important features of human landscapes, especially in urban areas. Different styles and functions reflect larger geographies, past and present.
9. *Monuments, memorials, and cemeteries:* Careful analysis of these landscape features can tell us a lot about the cultural identities of their creators.

These are only a few examples of the many qualitative possibilities open to the skilled researcher.

Cartoon by Chic *Damn! Somebody's pinched our spot!*

on one or more other variables. The relationship specified is, ideally, derived from appropriate theory in accordance with the scientific method outlined earlier. Classifying imposes order on data, and a number of techniques facilitate that activity.

CONCLUSION

Together, Chapters 1 and 2 provide a solid foundation for understanding human geography, both past and present, for appreciating the discourse of the discipline and its diverse subject matter, and for recognizing that there are several different but equally legitimate approaches to researching that subject matter. We conclude this chapter with a provocative, if somewhat crude, distinction.

- Positivism and the quantitative procedures associated with it have a tendency to exclude the individual human element from research, preferring to focus on aggregate data; critics consider such work to be dehumanized human geography. We might call it a geography of people.
- Humanism, postmodernism, and the qualitative procedures associated with

these, on the other hand, emphasize the integration of researcher and researched, thus generating a geography with people.
- Marxism, along with feminism, aspires to solve problems associated with inequality and lack of social justice, thus aiming to be a geography for people.

We do not, however, need to debate the relative merits of differing research strategies; we simply need to acknowledge that contemporary human geography is an exciting and diverse discipline incorporating a variety of useful philosophies, concepts, and techniques that complement each other and combine to offer an enviable range of procedures. As we will see in the chapters that follow, all our philosophies have much to contribute.

Summary

The Importance of Philosophy

The subject matter of human geography, the questions asked, and the aids used in obtaining answers are best understood by reference to various underlying philosophies. Most human geography is explicitly or implicitly guided by one of several particular philosophical viewpoints.

Philosophical Options

Contemporary human geography draws on several principal philosophies. Empiricism gives priority to facts over generalizations; traditional regional geography, although presumed to be aphilosophical, was empiricist in character. Positivism accepts and moves beyond the basic empiricist assumption to argue for use of the scientific method, as in the physical sciences, and purports to be objective. The approach known as spatial analysis follows a positivistic logic. Humanism is a set of subjective philosophies, including pragmatism and phenomenology, with a focus on humans. Early cultural geography had some humanistic content, but humanist philosophies did not begin to play a major role in human geography until about 1970. Marxism, the oldest and most fully developed of the critical theories, is a loosely structured philosophy that aims to understand and change the human world; work inspired by Marxist thought tends to centre on social and environmental problems. Feminism is both pragmatic and theoretical, stressing the need for equality and acknowledging the role played by gendered identities. Postmodernism has contributed several key ideas, such as that of social construction, and currently plays a major role in human geography today.

Human Geographic Concepts

A great many concepts are used in human geography. We constantly refer to space (we even think of geography as a spatial discipline), locations (where a geographic fact is present), place (the quality of a given location), region (a grouping of similar locations or places), and distance (the interval, however measured, between locations or places). Geographic research is conducted at specific spatial, temporal, and social scales. Explaining and understanding landscape change is often enhanced by considering both diffusion—the spread and growth of geographic facts—and perception—how landscapes are viewed by individuals and groups. Analyses of landscapes frequently distinguish different degrees of development. The technical vocabulary of human geography—for example, conceptual terms such as *place* and *location*—is an example of discourse. Most notably, human geographers identify globalization as an overriding metaconcept, providing a body of ideas that may facilitate analysis of environmental, cultural, political, and economic topics.

Techniques of Analysis

Geographic data are collected and handled by a variety of techniques. Maps are used to store, display, and analyze data and can be produced manually or by computer. Geographic information systems are computer-based tools that analyze and map data and have a wide range of applications. An important method of collecting data involves remote-sensing techniques, such as aerial photography and satellite imagery. Fieldwork, both observation and interviewing, is an essentially qualitative procedure for the collection and analysis of data and is favoured by humanists. The use of models and quantification is associated with positivistic spatial analysis and involves abstraction and empirical testing of ideas.

Links to Other Chapters

- **Philosophies, concepts, and techniques:** Chapter 1 (history of geography)
- **Environmental determinism and possibilism:** Chapter 4 (natural resources and environmental ethics); Chapter 6 (population density and distribution); Chapter 7 (cultural regions and landscapes); Chapter 10 (agricultural regions)
- **Positivism:** Chapters 10, 11, and 13 (spatial analysis in economic geography and location theories of von Thünen, Christaller, and Weber, respectively)
- **Humanism:** Chapters 7 and, especially, 8 (landscapes); Chapter 12 (contemporary urban experience)
- **Marxism:** Chapter 5 (Marxist theory of population growth); Chapter 6 (world systems approach); Chapter 8 (inequalities); Chapter 10 (Marxist economic geography); Chapter 12 (contemporary urban experience)
- **Feminism:** Chapter 5 (fertility decline); Chapter 8 (gender and sexuality); Chapter 13 (gendered industrial landscapes)
- **Postmodernism:** Chapter 8 (identity and inequality); Chapter 12 (urban theory)
- **Concepts of space, location, and distance:** Chapters 3, 10, 11, 12, and 13 (interaction and location theories)
- **Concept of place:** Chapters 7, 8, and 12 (understandings of landscape)
- **Concept of region:** Chapters 5 and 6 (demographic maps); Chapter 7 (cultural regions); Chapter 10 (agricultural regions); Chapter 13 (industrial regions)
- **Spatial scale:** Tables and maps generally, especially in Chapters 5 and 6 (population)
- **Concept of diffusion:** Chapter 3 (transport systems); Chapter 5 (migration)
- **"Neighbourhood effect":** Chapter 10 (von Thünen theory)
- **Diffusion and political economy:** Chapter 10 (political ecology and agriculture)
- **Concept of perception:** Chapter 5 (migration); Chapters 7 and 8 (group identities); Chapter 12 (urban areas)
- **Concept of development:** Chapter 6 (more and less developed worlds); Chapter 13 (uneven spatial development)
- **Concept of discourse:** Chapter 8 (the cultural turn)
- **Globalization:** Chapter 3 (entire chapter); Chapter 4 (global environmental issues); Chapter 5 (population growth, fertility decline); Chapter 6 (refugees, disease, more and less developed worlds); Chapter 8 (popular culture); Chapter 9 (political futures); Chapter 10 (agricultural restructuring); Chapter 11 (global cities); Chapter 13 (industrial restructuring)
- **Map projections:** World maps (all chapters)
- **GIS and remote sensing:** Data collection and analysis (all chapters)
- **Qualitative methods:** Chapters 8 and 12
- **Quantitative methods:** Chapters 3, 10, 11, 12, and 13 (spatial analysis)

Questions for Critical Thought

1. Why is it important to be familiar with the various philosophical underpinnings of contemporary geography?
2. Why are geographic literacy and geographic knowledge important tools or skills?

3. Why are maps arguably the most important tools of geographers? Could we study geography without maps?
4. In what ways are qualitative and quantitative methods in geography complementary?

Suggested Readings

Visit the companion website for a list of suggested readings.

3 GEOGRAPHIES OF GLOBALIZATION

This chapter begins by suggesting that distance is now less important as a human geographic concept than it was in the past. This is because a collection of globalization processes, both technological and organizational, make it easier to move ourselves, goods, and information from place to place.

A useful long-term way of thinking about globalization that helps locate current processes in an appropriate larger context is outlined here, followed by a series of accounts considering the various reasons—improved transportation, increased trade and related regional integration, rise of transnational companies, and information technologies—that distance is becoming less important. The chapter also looks at accounts of different understandings of globalization and of two ways to measure the specifics of recent and ongoing change, and it discusses contemporary geo-economy, cultural and political globalization, and the pros and cons of globalization.

Here are three **points to consider** as you read this chapter.

- If geography is, as was often suggested in the past, a discipline in distance and if distance is becoming less significant, are we approaching the "end of geography"?
- Because the world is increasingly being economically dominated by three large regions—North America, Europe, and East Asia—is it becoming characterized by greater spatial inequalities?
- Opinions about whether globalization contributes to a better or a worse world vary but are closely related to individuals' ideological perspective, especially their political—left or right wing—inclinations. Is it desirable, or even possible, to think about globalization in a relatively objective way?

Shipping containers await transport at Bremerhaven, Germany. One of the largest ports in the world, Bremerhaven shipped 5.9 million units in 2011 and more than 2 million automobiles in peak years. Transporting goods around the world is a key part of globalization.

INTRODUCING GLOBALIZATION

An understanding of human geography today—of the world we live in and how we live in it—is incomplete without awareness of past physical and human geographies. But these cannot provide a full understanding. Many observers consider that a new set of processes, called globalization, is the most important change or (more accurately) set of changes, currently affecting human geography. One broad conclusion is that we are now able to talk intelligibly on a global scale as well as in local and regional terms, an idea that lies at the heart of many accounts of our contemporary world.

Put simply, our long history of living distinctive lives and creating distinctive landscapes in essentially separate parts of the world may be ending. More and more, we are seeing groups of people and the places they occupy changing in accord with larger global (as opposed to national, regional, or local) forces. There is no single reason for these changes, but human geographers are always quick to point out that a key factor is humans' increasing ability to overcome the friction of distance (formally defined in Chapter 2 as the restraining effect of distance on human movement). We can move our goods and ourselves from one place to another much more rapidly than ever before, while ideas and capital can travel anywhere in the world almost instantaneously.

Globalization increases both the quantities of goods, information, and people moving across national boundaries and the speed with which they do so. Most readers of this book are accustomed to having significant world happenings, such as sports competitions, concerts, and political events, beamed directly to their TVs and other devices; to receiving information and goods from faraway places; and to seeing local decisions affected by larger global circumstances. For many who regularly travel to other countries, the sense of place-to-place difference is vanishing as a few dominant brands—CNN, the BBC, McDonald's, Coca-Cola, Benetton, Nike—become ever more ubiquitous. Many of us have also had direct experience of financial homogenization: we can purchase goods or local currency almost anywhere in the world using only one major credit card. Virtually all these developments have come about since 1980.

Some observers go so far as to predict the emergence of a global village, a global culture. Although such a development may still be far in the future, we do appear to be entering a new era involving the creation of global landscapes, in which places separated by great distances are nevertheless very similar in character. Think of shopping malls, major hotels, fast-food restaurants, and airports. In these instances, similar settings may be patronized by people with similar lifestyles and similar aspirations. At the centre of all this increasing interaction and homogenization is the capitalist mode of production. We will see evidence of the trend towards globalization throughout this book. Interestingly, the 2008–09 global economic recession and subsequent economic downturns threatened economic globalization, as the decline in trade prompted national governments to introduce various measures of protectionism.

In this chapter, globalization is considered from cultural, political, and economic perspectives. Culturally, as we see also in Chapters 4–8, a certain set of attitudes, beliefs, and behaviours is spreading. These elements reflect a more ecocentric environmental ethic, an increasing insistence on gender equality, and a better understanding of the linguistic, religious, and ethnic bases for differences between groups of people. More generally, we are seeing the extension of some American characteristics to other parts of the world—what has been described as "Coca colonization." Politically, as discussed in Chapter 9, many countries are working together for strategic reasons (the UN is the prime example of such a grouping). Economically, as discussed further in Chapters 10–13, the connections between different parts of the world are constantly increasing in terms of product movement and capital investment. For example, corporate headquarters in major cities control economic activities throughout the world.

The Local Still Matters

These preliminary observations about globalization are necessarily partial and therefore misleading. It may be tempting to jump on the globalization bandwagon and begin interpreting all human geographic change in these terms. But even a brief reflection on our personal lives or a cursory glance at newspaper headlines make it clear that the local still matters to us, with distance continuing to play a central role in everyday life. Think of your own life, of the places you frequent and the people who matter to you. Your home, local community, family, friends, and neighbours likely give shape and meaning to your life. Think about travelling between home and school, visiting friends, choosing which restaurant or movie theatre to go to, and it is clear that the friction of distance remains an important consideration in your decision-making.

Distance has not become irrelevant, and the local has not been submerged under a tidal wave of globalization. Many groups of people want to see their specific culture recognized and their particular region attain some autonomous status; in the Canadian context alone, we can point to several separatist movements, the political aspirations of First Nations, and the tribal urge sometimes evident in the behaviours of some cultural groups. The forces associated with globalization are definitely powerful, and their impacts are undoubted, but they are not the only forces at work.

Dick Hemingway Photographs

In Toronto's Kensington Market neighbourhood, shops open onto the streets, forming an old-fashioned market. Local markets offer a pleasant alternative to the more common urban experience of indoor malls and major grocery chains. Kensington's appeal is enhanced by its ethnic diversity. In earlier years many Jewish and Italian immigrants settled here, but eventually they moved on to more affluent neighbourhoods and were succeeded by immigrants from Africa, the Caribbean, Southeast Asia, and Latin America. As land prices rise and the neighbourhood becomes increasingly fashionable, the market's residents and business owners find themselves fighting against commercial chain stores attempting to establish themselves in the area.

GEOGRAPHY AS A DISCIPLINE IN DISTANCE

A standard assumption in human geography has long been that location—whether of people, farms, towns, industries, or anything else—is not random but follows a pattern. Sometimes the phenomena of human geography are close to one another and sometimes they are separated by great distances, but spatial regularity enables us to make sense of those locations. With this idea of spatial regularity in mind, one way to think about human geography is as a discipline in distance.

Why is distance important in this way? Because overcoming it requires effort and, as the **principle of least effort** suggests, all our behaviour aims to minimize effort. Thus, location decisions are made to minimize the effort required to overcome the friction of distance. Human geographers agree that many of our current distribution patterns reflect this friction (Box 3.1). This is evident in the first law of geography noted in Chapter 2, namely that near things are more related than distant things. Of course, distance can be measured in several different ways, including physical, time, cost, cognitive, and social.

Converging Locations

This friction of distance seems to be decreasing in many instances, perhaps even to the point where at least some distances will become frictionless. The idea that travel times typically decrease with improvements in transportation technology has been labelled **time–space convergence** (Janelle, 1969). We can conceive of locations converging on each other; these are locational changes in relative space (recall that, in Chapter 2, we described relative space as subject to continuous change).

We can calculate a convergence rate as the average rate at which the time required to move from one location to another decreases over time. For example, Janelle (1968) calculated the convergence rate between London, England, and Edinburgh, Scotland, from 1776 to 1966 as 29.3 minutes per year. More generally, it took Magellan three years (1519–22) to circumnavigate the globe. Today we can fly around the world in less than two days. With the laying of the first telegraph cable across the North Atlantic seabed in 1858, the "distance" between Europe and North America was reduced from weeks to minutes, and now information can be transmitted around the world almost instantaneously.

If it is true that some distances, at least, will become frictionless, is human geography still a discipline in distance? As already suggested, the

principle of least effort
Considered a guiding principle in human activities; for human geographers, refers to minimizing distances and related movements.

time–space convergence
A decrease in the friction of distance between locations as a result of improvements in transportation and communication technologies.

Examining the Issues

BOX 3.1 | The Tyranny of Distance

The Tyranny of Distance is the title of the eminent Australian historian Geoffrey Blainey's history of his country. Blainey (1968: 2) begins: "In the eighteenth century the world was becoming one world but Australia was still a world of its own. It was untouched by Europe's customs and commerce. It was more isolated than the Himalayas or the heart of Siberia." The twin ideas of distance and isolation frame the processes of change in Australia during the European era. The four principal "distances" are

1. between Australia and Europe;
2. between Australia and nearer lands;
3. between Australian ports and the interior; and
4. along the Australian coast.

The distance concept proposes new ways of explaining why Britain sent settlers to Australia in 1788 and why the colony was at first weak but gradually succeeded. Similarly, it aids understanding of Chinese immigration in the 1850s and later immigrations from Southern Europe. Within Australia, the distances from coast to interior and along the coast are closely linked to economic and other change. The first Australian staple, wool, was effectively able to overcome the distance from pasture land to port and the distance from port to Europe. Wool opened up much of southeastern Australia and was followed briefly by gold and then wheat. These products, along with technological **innovations**, combined to overcome the problems of distance.

innovations
Introduction of new inventions or ideas, especially ones that lead to change in human behaviour or production processes.

Blainey is acutely aware that distance was not the only relevant variable—climate, resources, European ideas, and events were also important. Similarly, he continually acknowledges that distance is not a simple concept: it is fluid, not static, and can be measured in many ways. In later writings, Blainey noted that distance is also a key to understanding the history of Australian Aboriginals. Sensibly used, the distance concept can help us to understand the unfolding of life in many parts of the world. Blainey's work is eminently geographical.

Although the distance concept has not been explicitly applied by Canadian historians, it is central to a leading thesis in Canadian history. This "Laurentian" thesis, proposed by Harold Innis, places the Canadian historical experience in the context of staple exports and communications that—as in Australia—effectively overcame the tyranny of distance.

answer is both yes and no. On the one hand, even a cursory glance at the world shows that distance still matters, since in most cases overcoming it still takes some effort. On the other hand, that same glance is likely to show that, in many instances, distance matters somewhat less than it did in the past. As noted, the reason is that friction decreases as communication technologies improve: what is increasingly important is how locations are integrated with high-speed communication networks. For most of us, the most potent examples of an essentially frictionless distance are e-mail and the Internet—technologies that make almost instantaneous contact with faraway places a routine matter. Recent estimates suggest that over 80 per cent of the world's population use a mobile phone and about 30 per cent use the Internet. There is, of course, a digital divide evident between rich and poor countries. This divide is discussed later in this chapter.

OVERCOMING DISTANCE: TRANSPORTATION

Humans have continually striven to facilitate movement across the earth's surface by reducing distance friction and the costs of interaction. One way to achieve this is to construct transport systems. Complex components of the human landscape, transport systems use specific modes (such as road, rail, water, or air) to move people and materials, to allow for spatial interaction, and to link centres of supply and demand. Global movement of goods relies primarily on water transportation, although road and rail are needed for movement to and from ports. The fact that transport systems in much of the less developed world are largely deficient has a direct economic effect, limiting activities such as commercial agriculture and mineral production.

In some countries, a transport system has been constructed to encourage national unity and open up new settlement regions. It is not a coincidence that this has occurred in Canada and Russia, the two largest countries in the world and both with large areas of inhospitable environment. Canada's transcontinental railway, completed in 1886, served as a crucial centripetal factor and encouraged settlement of the prairie region. The trans-Siberian railway between Moscow and Vladivostok has been operating since the beginning of the twentieth century.

Nineteenth-century geographers such as Ratzel and Hettner viewed transport routes as landscape features and as factors related to more general landscape changes, and the early twentieth-century French school of human geography regarded transport as a critical component of the geography of circulation. Nevertheless, transport geography remained largely unexplored until the 1950s, when several geographers published studies of particular modes of transport. The most significant developments came with the rise of spatial analysis; they involved quantitative and modelling studies, many of which had a planning orientation. Even today, transport geography has a notably positivistic flavour and has been relatively unaffected by humanist, Marxist, and other types of social theory.

Evolution of Transport Systems

Three processes characterize the evolution of transport systems: intensification, or the filling of space; diffusion, or spread across space; and articulation, or the development of more efficient spatial structures. However, changes in transport systems do not occur at a steady pace but at times of revolutionary (as opposed to evolutionary) change. Of course, the changes that have occurred in transportation reflect many of the same basic forces that have prompted revolutionary change in other aspects of the human geographic landscape. In short, transport systems change in response to (1) advances in technology and (2) various social and political factors. A review of the evolution of transport networks in Britain will clarify these generalizations.

Evolution of Transport Systems in Britain

The first organized transport system in Britain was created by the Romans between 100 and 400 CE for political and military reasons. Unlike the earlier system of local routes intended simply to permit social interaction between settlement centres, the Roman system focused specifically on facilitating rapid movement between London and the key Roman centres. Once the Romans withdrew from Britain, their system fell into disuse because it did not serve the needs of the local population. Only in the seventeenth century did a new national system reflecting the national interest emerge, in the form of carriers and coach services. By the early eighteenth century, two innovations diffused rapidly across Britain: turnpike roads (an organizational innovation) and navigable waterways, particularly canals (a technological innovation). In the nineteenth century the technological innovation of railways appeared, and in the twentieth century the road system responded to the new technology of automobiles. Finally, air transport added a new dimension to the overall transport system.

Turnpike roads charged tolls, thereby transferring the cost of road maintenance from local residents to actual road users. No new technology was involved; thus, in principle, a turnpike system could have been set up at any time. Beginning about 1700, turnpikes diffused rapidly. During the early industrial period, *canals* played an important role in serving mines and ironworks. After 1790 they became the dominant transport mode, and until about 1830 they played an important role in industrial location decisions and related population distributions. Canals were tied to industry because they were best suited for moving heavy raw materials, whereas turnpikes were better suited to the movement of people and information. Thus, the two new systems were largely complementary rather than competitive.

Theories of Transport System Evolution, Networks, and Graph Theory

Neither of these eighteenth-century developments compared to railways in long-term impact. The first British railway was completed in 1825. The early railway network was relatively local and related to industry, but after 1850 railways linked all urban centres and dominated British transport until about 1920. Railway construction in Britain was organized and financed by small groups of entrepreneurs with minimal government intervention. Elsewhere in Europe and overseas, governments typically played a larger role in railway construction.

As shown in Figure 3.1, each of these three innovations displays the characteristic S-shaped growth curve (Figure 2.12): a slow start, rapid

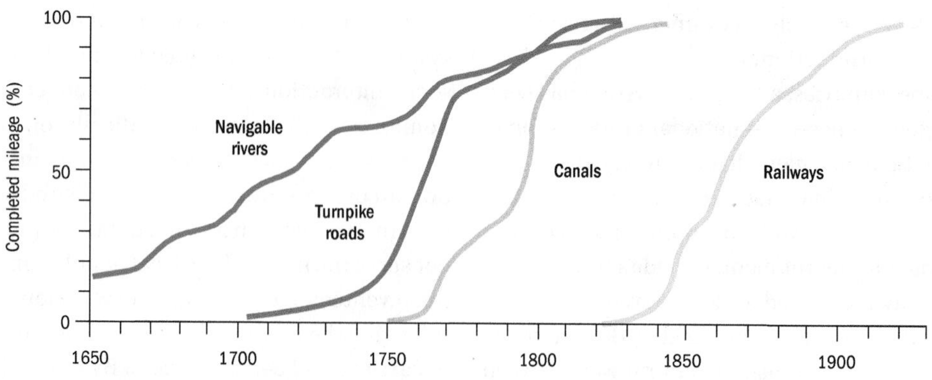

FIGURE 3.1 | Diffusion of transport innovations in Britain, 1650–1930

Source: Pawson, E. 1977. *Transport and Economy: The Turnpike Roads of Eighteenth Century Britain*. London and New York: Academic Press, 13. Used with permission of the author.

expansion, and a slow completion period. The pattern by which transport systems evolved in Britain is fairly typical. Most parts of the developed world have transport systems that include navigable waterways, railways, roads, and air traffic and are continually changing in response to technology and demand. Most notably, transport innovations within and between countries serve to decrease the friction of distance, thus facilitating globalization.

Modes of Transport

As the British example makes clear, some modes of transport evolved at different times and under very different human geographic circumstances, eventually combining to create an integrated system. In other cases, route duplication and a general lack of coordination may combine to create a system that is somewhat incoherent as a whole, although each specific mode may be a perfectly logical system in itself. Sometimes dramatic change takes place very rapidly. This is the case in contemporary China where, between 2001 and 2005, more was spent on transport than in the previous 50 years and, notwithstanding the 2008–09 economic slowdown, massive expenditures continue apace.

Water Transportation

Each transport mode has advantages and disadvantages. Water is a particularly inexpensive method of moving people and goods over long distances because it offers minimal resistance to movement and because waterways are often available for use at no charge. But it is slow, may be circuitous, and not all locations are accessible by water. In the case of inland waterways, topographic features such as variations in relief can be a major obstacle. In some areas, weather conditions—especially freezing and storms—may cause certain problems for water transport. In northern Manitoba, for example, the port of Churchill on Hudson Bay is open only from mid-July to mid-November.

The opening of the Suez (1869) and Panama (1914) Canals resulted in dramatic changes to travel distances. The Suez Canal reduced the distance between the Indian and Atlantic Oceans (Asia and Europe), while the Panama Canal reduced the distance between the Pacific and Atlantic Oceans (Figure 3.2). Following failed negotiations with Colombia to build an alternative to the Panama Canal, in the form of a 791 km railway from Buenaventura on the Pacific coast to the Atlantic coast, China has reached an agreement with Nicaragua to build a canal that runs through Lake Nicaragua. Construction began in late 2014 and, if completed, this canal will be three times longer than the Panama Canal. More significantly it will be able to handle much larger vessels, being 28 m deep and 55 m wide (the Panama Canal is 16 m deep and 55 m wide). In the same year Egypt announced plans to upgrade the Suez Canal for the first time since it opened; the upgrade was completed in 2015.

Canals continue to be a critical component of many continental transportation systems. In Europe, the Rhine–Main–Danube Canal was completed in 1992, allowing for a continuous water link between the North Sea and the Black Sea.

Railway Transportation

Land transportation includes railways and roads. Railways are the second least expensive form of

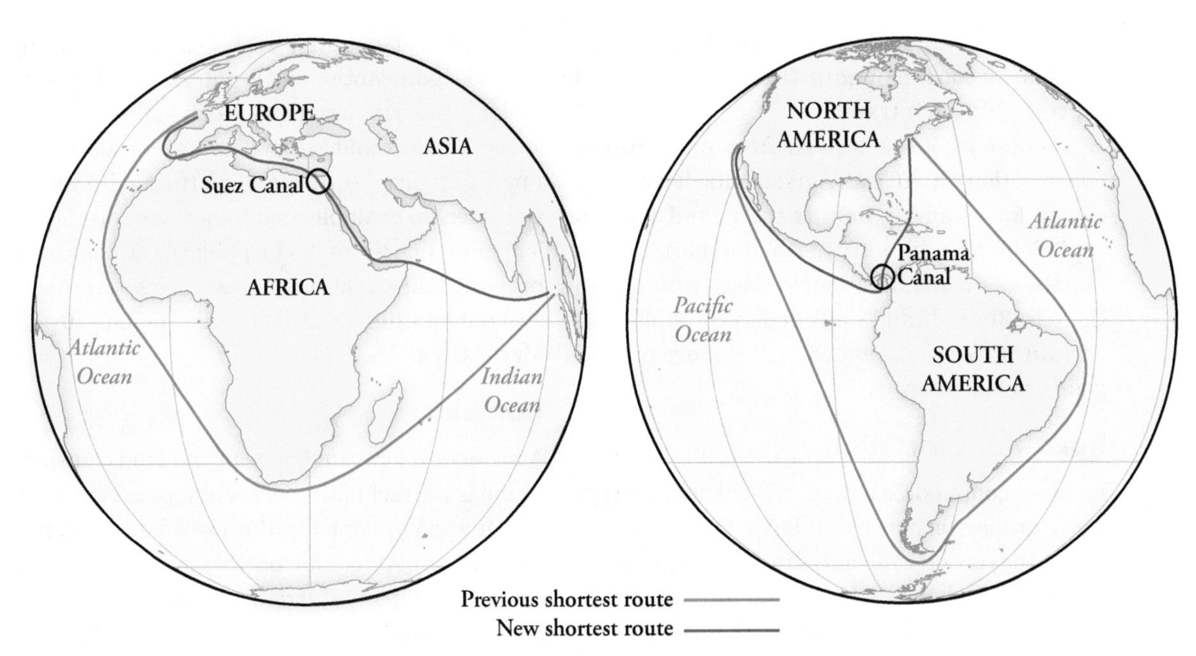

FIGURE 3.2 | The impact of the Suez and Panama Canals on ocean travel distances

Source: Adapted from Rodrigue, J.-P., C. Comtois, and B. Slack. 2006. *The Geography of Transport Systems*. New York: Routledge, 21, Figure 1.12.

transportation, after water, and are suitable for moving bulk materials when waterways are not available. However, in some areas local topography makes railway construction (and maintenance) expensive, if not impossible. The most significant railway construction in recent years has taken place in Australia, China, and Europe.

In Australia, a railway line between Adelaide and Darwin that was first promised in 1911 was finally completed in 2004; this railway between north and south is a powerful symbolic achievement. Bullet trains in many Chinese cities, including between Beijing and its nearest port, Tianjin, have greatly reduced travel times. The world's longest high-speed route, linking Beijing with the southern commercial centre of Guangzhou, formally opened in late 2012. Also in China, a new railway line including more than 1,000 km of fresh track was completed in 2005 between Golmud in the far west and Lhasa in Tibet. This line is just one example of ongoing efforts to develop an integrated rail network, funded by Chinese capital, for the larger region of East and Southeast Asia.

In Europe, the Channel Tunnel between England and France, opened in 1994, is actually three tunnels—two for trains and one for access by maintenance workers. Scanlink, completed in 2000, is a series of railways and roads (including tunnels and bridges) linking Sweden and Norway through Denmark to Germany. In recent years several train tunnel projects have been undertaken in the Alpine region, on the routes between Italy and Germany. The most notable of these is the Loetschberg rail tunnel, the longest in the Alps at 34 km (21 miles). Also under construction is the 57 km New Gotthard rail tunnel, scheduled to open in late 2016 with a second tunnel being completed in 2020. Much of the recent growth in Europe has been high-speed

topography
A Greek term, revived by nineteenth-century German geographers to refer to regional descriptions of local areas.

The upgraded Suez Canal features a deeper and longer section of the existing canal and the creation of a second shipping lane. For much of the canal's length, traffic can flow in both directions, thereby decreasing wait times and increasing capacity.

rail; in 1990 Europe had 1,000 km of high-speed track and 7,000 km by 2015. On some routes, rail is preferable to air travel.

Chinese capital will likely fund much future construction in Africa. Agreements have been signed for a railway linking the inland Kenyan capital of Nairobi with the major port city of Mombasa, with additional lines linking to Uganda, Burundi, and South Sudan. Most planned African projects are aimed at the movement of goods, not passengers.

Road Transportation

Roads accommodate a range of vehicles and typically are less expensive to build than railways; hence, they are often favoured in regional planning and related schemes. Most of the internal movement within Europe is by road, with inland waterways second and railways third. Construction of bridges often reduces road distances. The 2006 completion of a second major bridge, both road and rail, over the Orinoco River, makes it much quicker for Brazilian goods to reach Venezuela's Caribbean seaports. Not surprisingly, the most dramatic examples of recent and ongoing bridge construction are in China. The world's longest bridge over water (42 km) was completed in 2011, linking the port city of Qingdao with the suburb of Huangdao. Construction of an even longer bridge (48 km), linking Guangdong province in southern China to Hong Kong and Macau, is expected to be completed by 2016.

Air Transportation

Transportation by air is the most expensive method, but it is rapid and is usually favoured for small-bulk, high-value products. Air transport is particularly subject to rapid technological change and to political influence. Even small countries tend to favour maintaining a national airline for reasons of status and tourist development. In many cases airport construction and maintenance are funded by governments.

A new Beijing airport terminal was constructed prior to the city's hosting of the 2008 Olympic Games. Built in only four years by 50,000 workers, the terminal is 3 km long and the floor space is 17 per cent larger than all five of London Heathrow's terminals. But even this is inadequate for existing needs. An additional airport, the world's largest, is scheduled for completion south of Beijing in 2018. The fact that China is a communist country, with secretive planning and no significant public input, means that such major infrastructural change can happen at great speed and without serious consideration of the impact on people's lives and on the environment. Consider, for example, that it took about the same length of time to conduct a public inquiry into the proposed construction of Heathrow's Terminal 5 as it did for China to build the new Beijing airport for the Olympics.

Containers

A major development in ocean and land transportation is the technology of the shipping container, which was developed in the 1950s by Malcolm P. McLean, an American businessman who ran a trucking company. Currently, as many as 300 million containers are used. The widespread use of containers has resulted in dramatically reduced transportation costs, despite the requirement of an entirely new way of handling freight, including new storage facilities, different cranes, changes at ports, and changes to trains and trucks. Theft is reduced, the storage and sorting of breakbulk cargo is eliminated, and time taken to load and unload is halved. Most container ships carry about 4,000 containers, although a few carry up to 18,000. The largest ships are unable to use the Panama Canal and can access only a few major ports (such as Los Angeles, Rotterdam, Hong Kong, and Singapore). The important role played by containers in facilitating economic globalization is difficult to exaggerate, but the subtitle of a book on container shipping, *How the Shipping Container Made the World Smaller and the World Economy Bigger*, is appropriate (Levinson, 2006).

OVERCOMING DISTANCE: TRADE

Transport systems decrease the friction of distance and facilitate movement of people and materials; thus, technological and organizational advances in transport are a key factor promoting and reflecting globalization. One specific consequence of such advances is that trade has become an increasingly important part of the global economy. Indeed, since 1945, trade has increased much more rapidly than production—a clear indication of how much more interconnected the world has become. Simply put, trade can take place if the difference between the cost of production in one area and the market price in another will at least cover the cost of movement. In principle, domestic trade is identical to international trade. The major difference is that it is less likely to be hindered by human-created barriers.

Factors Affecting Trade

The movement of goods from one location to another reflects spatial variations in resources, technology, and culture. The single most relevant variable related to trade is distance. Like all forms of movement, trade is highly vulnerable to the friction of distance. Other relevant variables include

1. the specific resource base of a given area (needed materials are imported and surplus materials exported);
2. the size and quality of the labour force (a country with a small labour force but plentiful resources is likely to produce and export raw materials); and
3. the amount of capital in a country (higher capital prompts export of high-quality, high-value goods).

All these variables are closely linked to level of economic development. Trade between more and less developed countries is often an unequal exchange: the former export goods to the latter at prices above their value, while the latter export goods to the former at prices below their value. Moreover, the less developed countries are inclined to excessive specialization in a small number of primary products or even a single staple. More developed countries may also specialize, of course, but they do so in manufactured goods. This difference reflects the technologically induced division of labour between the two worlds. Less developed countries' dependence on the more developed countries as trading partners is central to world systems theory, discussed in Chapter 6. Today, the majority of world trade moves between developed countries, although the rise of the newly industrializing countries (see Chapter 13) is affecting trade flows. For a simplified explanation of commodity flows, see Box 3.2.

Regional Integration

Because trade occurs across international boundaries, it can be regulated. Most governments actively intervene in the importing and exporting of goods. For example, they may set up trade barriers to protect domestic production against relatively inexpensive imports. The most popular form of regulation involves the imposition of a **tariff**.

© Agencja Fotograficzna Caro/Alamy Stock Photo

The Gare du Nord in Paris is Europe's and (outside Japan) the world's busiest railway station. It connects to several of the city's transportation lines, including the Métro, as well as to northern France, Belgium, Germany, the Netherlands, and the UK.

tariff
A tax or customs duty on imports from other countries.

 **Trade Theories**

Globalization, especially through the activities of the World Trade Organization (WTO), promotes global multilateral free trade, but this does not mean that trade barriers no longer exist. In addition to some barriers set up by individual states, groupings of states (in the form of regional trade blocs) establish additional ones designed to help them compete in the global market and to expand potential markets. Such regional integration has become important since the end of World War II. Until recently, most integration was quite limited, but there is now compelling evidence of a widespread desire among separate states to unite economically, and the world is therefore increasingly divided into trade blocs. In many cases economic integration also involves some degree of political integration. Five stages in the process of integration have been identified (Conkling and Yeates, 1976: 237):

1. The loosest form of integration is the free trade area, consisting of a group of states that have agreed to remove artificial barriers, such as import and export duties, to allow movement and trade among themselves. Each state retains a separate policy on trade with other countries. Two major examples of free trade areas are NAFTA, consisting of

Examining the Issues

BOX 3.2 | Explaining Commodity Flows

The study of commodity flows has long been a topic of concern to geographers. In North America, the most influential writer on this topic has been Edward Ullman. Ullman was especially interested in transportation and even defined geography as the study of spatial interaction (apparently after hearing a sociologist define sociology as social interaction). On the basis of detailed studies, Ullman (1956) concluded that commodity flows could be reasonably explained by reference to three variables:

1. *Complementarity.* Different places have different resources. When a surplus of a certain commodity (resource) in one area is matched by a deficit of the same commodity in a second area, this complementarity permits the commodity to flow from the first area to the second. This logic applies at all spatial scales.
2. *Intervening opportunity.* Complementarity can result in product movement only if no other area between the complementary pair is an area of either surplus or deficit. If such an intervening opportunity exists, it will either serve the deficit area or receive the surplus because it is closer to both areas than they are to each other.
3. *Transferability.* Commodity flows are affected by the ease with which a commodity can be transferred. A product that is difficult or expensive to move will be less mobile than one that is easier or less expensive.

Together, these three factors provide a simplified explanation of commodity flows at various times and at a multitude of spatial scales.

Canada, Mexico, and the United States and established in 1994, and the Association of Southeast Asian Nations (ASEAN), originally comprising Brunei, Indonesia, Malaysia, the Philippines, Singapore, and Thailand and established in 1967.

2. The second stage in international economic integration is the creation of a customs union, in which member states not only remove trade barriers between themselves but also impose a common tariff barrier. In this case there is free trade within the union and a common external tariff on goods from other states.
3. A common market has all the characteristics of a customs union and allows the factors of production, such as capital and labour, to flow freely among member states. Members also adopt a common trade policy towards non-member states.
4. There are two levels of integration beyond the common market that groups of states may achieve. An economic union is a form of international economic integration that includes a common market and harmonization of certain economic policies, such as currency controls and tax policies.
5. At the level of economic integration member states have common social policies and some supranational body is established with authority over all. The extent to which individual states may be prepared to sacrifice national identity and independence to achieve full economic integration is not yet clear.

Figure 3.3 maps the four major trade blocs. At present, the most developed example of regional economic integration is the EU, which established a single European market in 1992 and has 28 member states (Box 3.3). The importance of transportation as a means of encouraging integration is evident in the several ambitious projects recently completed or underway (the Channel Tunnel, Alpine tunnel routes, and Scanlink). As the process of regional integration continues, states left outside the major groupings may suffer economically (Cleary and Bedford, 1993). This is one reason why many countries—including countries in Western Asia and North Africa that would generally be considered non-European—aspire to join the EU.

Formed in 1967, ASEAN currently consists of 10 countries—Brunei, Cambodia, Indonesia, Laos, Malaysia, Burma, the Philippines, Thailand, Singapore, and Vietnam—and encourages free trade between members. It is also moving towards free trade with China and strengthening trading connections with Japan and India.

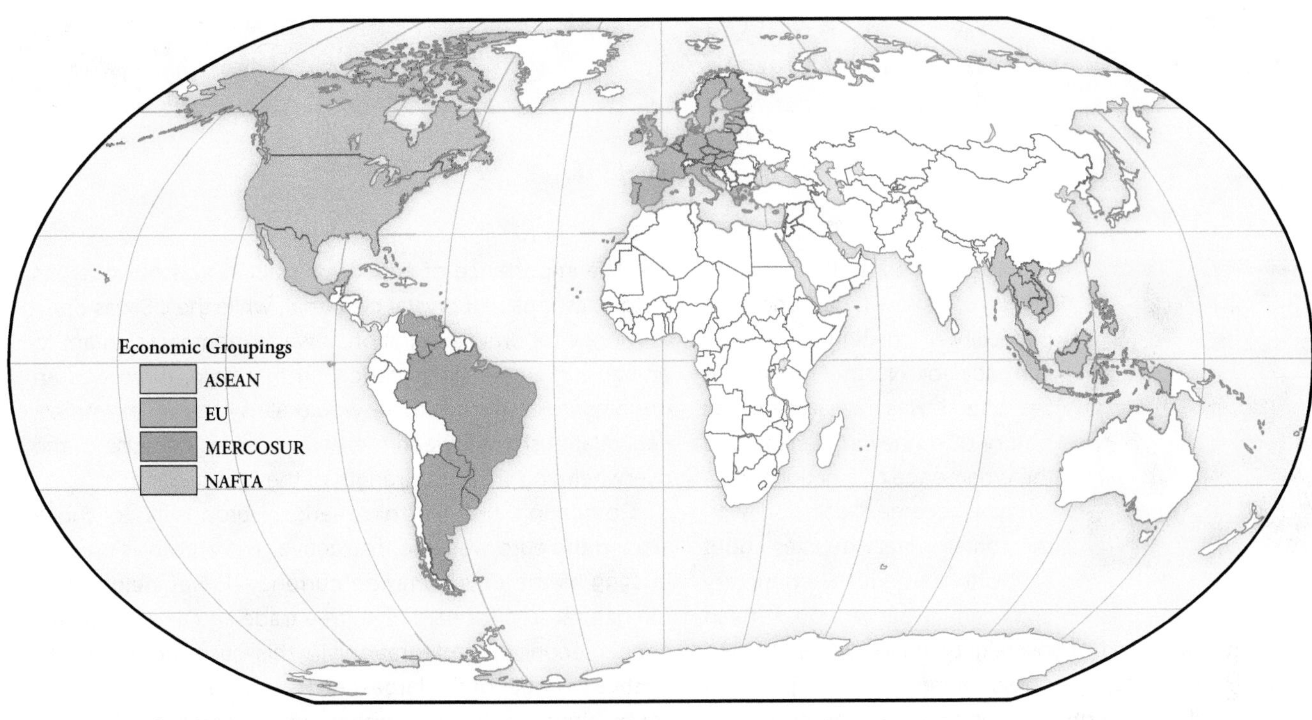

FIGURE 3.3 | Selected economic groupings of countries

Around the Globe

BOX 3.3 | The European Union

The principal example of a voluntary grouping of countries was established following the end of World War II in 1945—the date that marked the end of the old Europe. Various moves towards European union resulted in the 1957 formation of the European Economic Community (EEC) by France, Belgium, the Netherlands, Luxembourg, Italy, and West Germany. These six states had already created a European Coal and Steel Community and a European Atomic Energy Community, both of which merged with the EEC in 1967 to form a single Commission of the European Communities. A common agricultural policy was adopted. Other European countries that did not favour such close integration formed the European Free Trade Association in 1960.

The EEC, now the EU, gradually assumed dominance, with Britain, Denmark, and Ireland joining in 1973, Greece in 1981, Portugal and Spain in 1986, and Austria, Sweden, and Finland in 1995. (Norwegians voted against joining in 1994; indeed, perhaps the most surprising fact about the EU is that neither Norway nor Switzerland, two countries with very high per capita incomes, has opted to join.) The most dramatic expansion of the EU took place in 2004 with the addition of 10 countries. Eight of the new members—Estonia, Latvia, Lithuania, Poland, the Czech Republic, Slovakia, Hungary, Slovenia—are in Eastern Europe. The remaining two are the Mediterranean islands of Malta and Cyprus. The latter poses a political problem because it was partitioned after an attempted Greek coup in 1974, and since then a buffer zone policed by the UN has divided the Greek-controlled Republic of Cyprus from the northern area, which is controlled by Turkey and not internationally recognized.

When these 10 countries were accepted for membership, Romania, Bulgaria, and Turkey applied but were considered not ready to join. Romania and Bulgaria joined in 2007, although with some initial membership restrictions. The case of Turkey is seen as problematic as the country has a record of human rights violations. Croatia joined in 2013, bringing the number of member countries to 28. The remaining Baltic countries and several in Eastern Europe will likely join the EU within the next few years. It is also possible that the EU will expand into North Africa, which would raise complex questions about "European" identity.

Focus on Geographers

Regional Integration, Political Independence, and the Canada–US Relationship | EMILY GILBERT

In the late 1990s, there was a flurry of interest around the possibility of creating a common currency for North America. At the time I was researching the history of money in Canada and the emergence of a national currency. I became fascinated with the contemporary debates about a North American monetary union and what it would mean for our understanding of sovereignty, national identity, and Canada–US relations. I was also interested in the potential social and cultural outcomes of monetary union—issues that weren't getting much public attention at the time.

A better understanding of these issues required a thorough comprehension of the multiple proposals being advanced. Some advocated for a trilateral arrangement that would include Canada, the US, and Mexico, while others argued that only Canada and the US were ready for such an integrative step. I wanted to investigate the underlying logics of these arguments. Or, to put it differently, if monetary union was being presented as the solution, what was the problem? Getting at these questions required a discursive analysis that examined the unstated presumptions of the proposals, the interests being represented, the contradictions or duplicities in the logic, and what was left out or sidelined. Reading across multiple documents such as government studies and briefings, reports in the media, and public polling enabled the perspectives of some of the different stakeholders to emerge.

Most of the proposals for monetary union originated from Canada. The fluctuating value of the Canadian dollar and the uncertainties created for trade were presented as the main impetus for a common currency. At the time, the US economy was thriving, and pinning Canadian money to the American was seen as a way to create some stability. But what, I wanted to know, would this mean for Canadian sovereignty and political relations between the two countries?

Situating the debates within their historical and geographical context was crucial to understanding these potential implications. Since the introduction of free trade in the 1980s, the two economies have become the most interdependent in the world, with Canada especially reliant on its exports to the US. The importance of economic stability was obvious. Yet history also made it crystal clear that, while the US was open to more economic integration, it was extremely reluctant to embark on shared governance. Furthermore, there was an ongoing concern that there would always be an asymmetrical relationship between the two countries because of the overwhelming size and strength of the US economy.

Comparing the North American proposals to those around the euro was also instructive. The euro was created in 1999 as the EU's common currency. It was held out as the natural step forward from free trade and as a model for deeper economic integration. In this case, however, the relative strength of the largest economy, Germany, was offset by almost a dozen member states. Moreover, the euro zone included some mechanisms for both transnational political governance and shared input on monetary policy, issues not really on the table in North America.

A further difference between the EU and North America was mobility. In the EU a shared passport was introduced and national borders were becoming less of an obstacle. By contrast, mobility rights were being extended to only the business classes in North America. Monetary union would not bring about a new vision of citizenship to ensure political representation and accountability.

My research thus concluded that the primary interests reflected in the proposals for North American monetary union were economic and that little attention had been paid to issues of political governance and representation or to citizenship and mobility. These conclusions were informed by a three-pronged analysis that included a discursive analysis of the proposals; contextualization in terms of both history and geography; and a comparative analysis with the euro. It also required an approach that was sensitive to the multiple dimensions of a monetary union: economic, political, social, and cultural. One of the strengths of the discipline of human geography is that it encourages and supports this kind of multidisciplinary and comparative research. It means that geographers are ideally suited both to examine the complexities of territorial restructuring under globalization and to contribute to the public debates as they unfold.

EMILY GILBERT is an associate professor in the Department of Geography & Planning and director of the Canadian Studies Program at the University of Toronto.

MERCOSUR, the South American economic grouping of Argentina, Brazil, Paraguay, Uruguay, and Venezuela was formed in 1991. It is a customs union that has experienced real difficulties because of financial crises in Argentina. With oil-rich Venezuela becoming a full member in 2012; Bolivia likely to gain full-member status in the near future; Chile, Peru, Colombia, and Ecuador as associate members at present; and Mexico expected to move from observer to associate status, the region might follow the route taken by the EU. One potential stumbling block is the risk that further integration might imply some loss of national sovereignty, although a majority of MERCOSUR states have rejected the US proposal of forming a Free Trade Area of the Americas (FTAA) and perhaps view their participation in this Latin American customs union as not only economically important but also ideologically significant.

The fourth major trade bloc is NAFTA, mentioned earlier. As with the other groupings, the impetus for formation was elimination of trade barriers among members, increased free trade, and increased investment opportunities. However, from a Canadian perspective, the Comprehensive Economic and Trade Agreement reached with the EU in 2014 may prove of greater significance.

Many other economic groupings of states are regionally significant but less important globally. Current regional organizations include the Caribbean Community (CARICOM), the Central American Common Market (CACM), the Commonwealth of Independent States (CIS), the Economic Community of West Africa (ECOWAS), the Central African Customs and Economic Union (UDEAC), the Maghreb Union of five North African Muslim countries, the Southern African Development Coordination Conference (SADCC), and the Common Market for Eastern and Southern Africa (COMESA).

It seems likely that African countries will move towards further integration in the near future, particularly with the 2002 creation of the African Union (see Chapter 9). Among the problems that these countries face are a generally low level of economic development by global standards and that agriculture—one of the few areas in which African countries are most competitive—is the one major area still subject to protective tariffs in the West.

TABLE 3.1 | Revenue data for the top 10 transnationals, 2014, and gross national income for selected countries, 2013

Transnational or Country	Revenues or GNI (US$ billions)
Norway	522
Walmart	476
Royal Dutch Shell	460
Sinopec	457
Iran	448
China National Petroleum Corporation	432
Austria	412
EXXON MOBIL	408
BP	396
Venezuela	382
State Grid Corporation of China	333
Philippines	321
Volkswagen	262
Toyota Motor	257
Glencore	233
Algeria	208

Sources: World Bank. 2015. GNI Ranking, Atlas Method. http://data.worldbank.org/data-catalog/GNI-Atlas-method-table; Fortune. 2014. "Fortune Global 500, 2014." http://fortune.com/global500/.

OVERCOMING DISTANCE: TRANSNATIONALS

Transnational corporations (defined in Chapter 2) are distinguished from earlier business organizations by several important characteristics. Most fundamentally, they are able to command and control production and sales at a global scale and usually can relocate production and other facilities with relative ease. They are also functionally integrated and able to benefit from geographic variations in capital, knowledge, labour, resources, national regulations, and taxes.

Table 3.1 provides sales data for the 10 leading transnational corporations, with **gross national income (GNI)** data from selected countries for

gross national income (GNI) or gross national product (GNP)
A monetary measure of the market value of goods and services produced by a country, plus net income from abroad, over a given period (usually one year).

comparative purposes. Remarkably, only 20 countries have a GNI larger than the sales of the top transnational, Walmart.

Because transnationals are able to adjust their activities within and between countries, they can take advantage of variations in factors such as land costs and labour costs at both scales. They undermine national economies because they are able to organize movements of information, technology, and capital between countries and, to a considerable extent, site production and profit in countries with relatively low wages and taxation.

Different transnationals carry out different proportions of their production at home and abroad. For several of the large automobile companies, about 50 per cent of employment is foreign, whereas the figure for Swiss-based Nestlé and UK-based BAT Industries is closer to 90 per cent.

Most of the transnationals are based in the US, Europe (mainly the UK and Germany), or Japan. But others have their home bases in Asia's newly industrializing countries or even in countries (such as Canada) that are more usually regarded as host countries for **foreign direct investment (FDI)**—the investment of capital into manufacturing plants located in countries other than the one where the company is based. In general, countries that are major sources of FDI are also major hosts. Indeed, one way to identify economic globalization is to focus on the amount and location of these capital flows, which have grown even more rapidly than international trade since World War II. In most cases both the investors and the recipients are located in the more developed countries (Table 3.2).

foreign direct investment (FDI)
Direct investment by a government or multinational corporation in another country, often in the form of a manufacturing plant.

international division of labour
The current tendency for high-wage and high-skill employment opportunities, often in the service sector, to be located in the more developed world, while low-wage and low-skill employment opportunities, often in the industrial sector, are located in the less developed world.

TABLE 3.2 | Foreign direct investment inflows by major regions, 2013 (US$ billions)

Region/Country	FDI Inflows
World	1,461
Europe	296
North America	223
Africa	56
Latin America and the Caribbean	294
Developing Asia	406
China	127

Source: United Nations Conference on Trade and Development. 2014 (January). *Global Investment Trends Monitor.* http://unctad.org/en/PublicationsLibrary/webdiaeia2014d1_en.pdf.

Typically, a transnational will have its headquarters in the more developed world but locate its manufacturing activities in the less developed world, where wages and environmental and safety standards are lower. There is, then, a division of production referred to as the **international division of labour** (see Chapter 13, especially Figures 13.12 and 13.13, which show the percentages of national labour forces in industrial and service employment, respectively).

Most transnational products fall into two groups: either they are technologically sophisticated or they involve mass production and mass marketing. In addition, much of the recent growth in trade reflects the increasing role of services—commercial, financial, and business activities—conducted by transnationals.

OVERCOMING DISTANCE: TRANSMITTING INFORMATION

Of all the space-shrinking and time-compressing technologies that have been developed in recent years, the most important to economic globalization are those that facilitate the almost instantaneous transmission of information regardless of distance. These include advances in the mass communications media and in the use of cables, faxes, satellites, and information communication technologies (ICTs). Figure 3.4 summarizes global ICT growth from 2001 to 2015.

The Digital Divide

The perceived importance of being able to access information through ICTs is made evident by the case of Finland, which has declared Internet access a basic human right. Of course, the opportunity to access communications media is not uniformly distributed throughout the world and the availability of ICTs differs significantly from place to place. The idea of the digital divide can be simply understood as the gap between those with the most and those with the least access to computers and the Internet. As with many other differences between people, this divide is closely

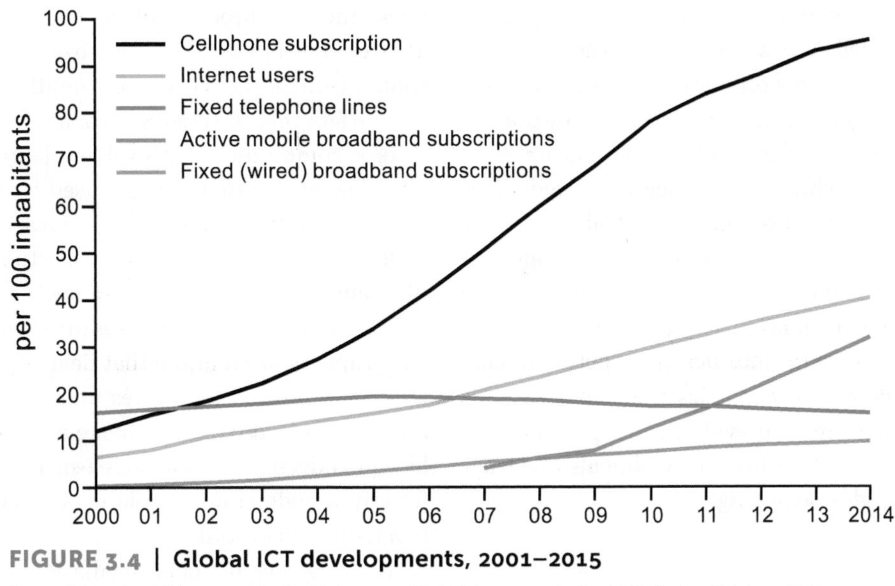

FIGURE 3.4 | Global ICT developments, 2001–2015

Source: International Telecommunication Union. 2015. World Telecommunication/ICT Indicators Database, 2015. www.itu.int/en/ITU-D/Statistics/Pages/publications/wtid.aspx.

related to place, gender, income, and level of education. There is typically a difference between urban and rural areas within a country. A 2010 article in *The Globe and Mail* highlighted the Canadian example:

> This growing digital divide makes rural economic prosperity increasingly elusive. Canadians living in rural areas already have incomes well below their urban counterparts (14 per cent lower than the national average, according to a recent study that used earlier census data), and the earnings gap exists in every province. In areas that have an abundance of oil, potash or other key commodities demanded by the world's economic powers, fast Internet connections might not be so important, but for the rest, they're crucial to pulling in new employers. Communities that cannot plug into the high-speed digital economy cannot attract new businesses that rely on basic services such as electronic invoicing, Internet conferencing and large digital file transfers.

At the global level the divide is most evident when comparing a more developed country such as

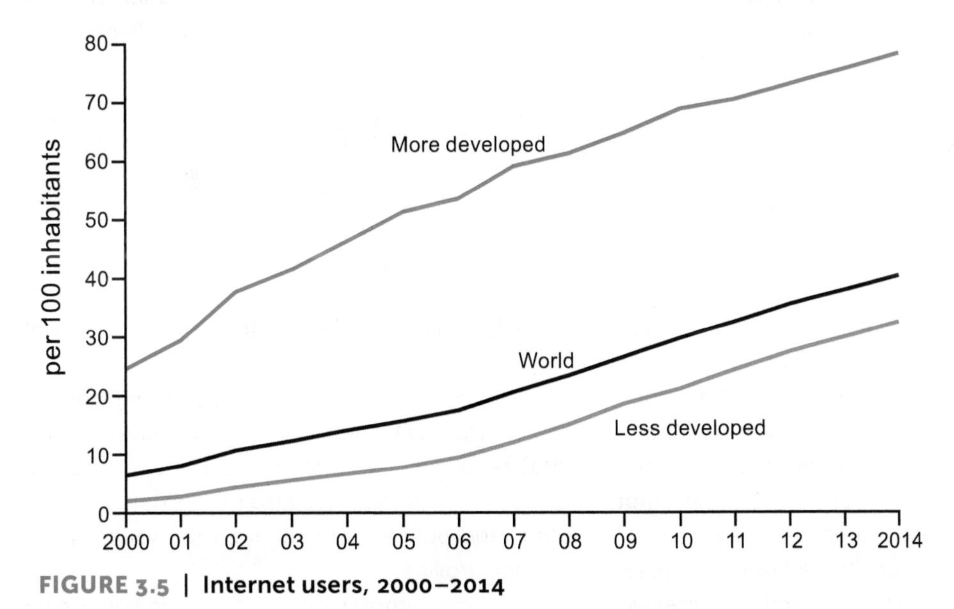

FIGURE 3.5 | Internet users, 2000–2014

Source: International Telecommunication Union. 2014. World Telecommunication/ICT Indicators Database, 2014. www.itu.int/en/ITU-D/Statistics/Pages/publications/wtid.aspx.

Iceland, with more than 90 per cent of the population connected, with a less developed country such as Burma, with only 0.4 per cent connected. As of 2014, the most connected countries—those that are leading the way in the use of information and communications technology—are Iceland, Sweden, the Netherlands, Denmark, and Finland (all over 80 per cent connected). Burma, Bangladesh, Ethiopia, the Democratic Republic (DR) of Congo, and Cambodia are the least connected (0.5 per cent or below). Figure 3.5 compares Internet usage between more and less developed countries from 2001 to 2015. The numbers are clear evidence of a global digital divide—another kind of poverty—but also evidence that the divide is lessening.

The Rise of Social Media

The proliferation of social media in recent years adds a new dimension to the ways technologies are able to overcome distance, effectively transforming communication into interactive dialogue. Blogs were the first form of social media, soon followed by sites that allowed a specific type of content to be shared, such as YouTube for videos and Flickr for images. The emergence of social networking sites, such as Facebook and MySpace, allowed people to input personal information and enable access to selected individuals. Twitter allows unlimited access to content. Some other sites allow users to participate in video conversations as though they were sharing the same physical space. Using these sites has transformed the everyday social lives of many, particularly those of young people in more developed countries. The top three social networks—Facebook, Twitter, and LinkedIn—collectively receive more than several billion visits each month.

The fact that these sites allow a version of face-to-face communication involving large numbers of people means that they are used for many purposes other than socializing. Educational institutions may use sites to connect students from different parts of the world, allowing for better understanding of different lives in different places. Geographers often argue that field trips promote appreciation of other places and other peoples, and virtual field trips may make a similar contribution. Indeed, there is good reason to believe that improved understanding of others may contribute to a reduction of conflict.

Business use of social media is considerable, hardly surprising given that networking has always been key to product development and marketing. Meetings can be held without the need for participants to travel, saving both time and money. Potential customers can be contacted and current customers kept informed of developments.

Social media are also used effectively by groups planning demonstrations, civil unrest, and revolutionary uprisings. In response, some governments attempt to block such opportunities for mass social interaction. As discussed in Box 3.4, social media are now widely used to organize people for peaceful or violent protests.

The long-term impact of social media is undeniable. Nevertheless, for many people and businesses, face-to-face interaction in a shared physical space continues to have real value. Because the nuances

In the News

BOX 3.4 | Social Media, Revolutions, and Riots

During the Arab Spring of 2011, social media were hailed as instrumental tools of the revolution. In Egypt, social media were credited with allowing young people to share their frustrations at repressive governments and with facilitating protest. Both Twitter and Facebook were blocked by the Egyptian government in January 2011 in a failed attempt to prevent them from being used to encourage a popular uprising. Blocking these and other sites did not, of course, prevent people from using cellphones and text messages to organize protests or various 24-hour news channels, especially Al-Jazeera, from providing continuous coverage.

In Vancouver, riots ensued following the Canucks' defeat in the 2011 NHL finals, and Robinson and colleagues (2011) reported that Twitter was used to encourage the rioting. Social media were also used extensively to identify rioters, and police acknowledged the value of the digital evidence provided.

That same summer, riots occurred in London, England. The police and government pointed fingers at social media as a key strategy employed by rioters and looters. These sites enabled individuals to act as members of groups and facilitated mass movement from place to place, avoidance of police lines, and increased opportunities for property damage and theft. Reflecting a widespread concern, Prime Minister David Cameron announced that "everyone watching these horrific actions will be struck by how they were organised via social media. Free flow of information can be used for good. But it can also be used for ill" (Warman, 2011).

One of the principal tools employed in London was the popular Blackberry Messenger (BBM) service. Blackberry handsets were the favoured smartphone for many British teens, partly because BBM allows users to send free and instant one-to-many messages to their network of contacts. Unlike Twitter and Facebook, messages could not be traced. A BBM broadcast sent on the second of four nights of rioting called on people to vandalize shops on Oxford Street:

> Everyone from all sides of London meet up at the heart of London (central) OXFORD CIRCUS!!, Bare SHOPS are gonna get smashed up so come get some (free stuff!!!) fuck the feds we will send them back with OUR riot!>:O Dead the ends and colour war for now so if you see a brother . . . SALUT! if you see a fed . . . SHOOT!" (Halliday, 2011)

It later emerged that, by hacking into the arrested rioters' BBM service, London police thwarted a planned attack on the Olympics site just hours before it was to take place.

In response to this use of social media, the British government suggested that people be prevented from using Twitter, Facebook, and BBM at specific times, but any such proposal raises many technical and legal issues. State control over social media likely cannot be implemented in a democracy. Indeed, many might argue that such proposals are completely inappropriate and align Britain with repressive regimes such as China, which routinely blocks social networking and many other Internet sites. For example, in August 2011 a peaceful protest demanding closure of a polluting chemical plant in Dalian, China, prompted authorities to block specific Twitter searches. Unsurprisingly, immediately following the London riots, Facebook and Twitter were used in a positive way to mobilize volunteers to help clean up the damage.

Perhaps the most dramatic use of technology to help promote specific messages is that associated with the rise of the Islamic State (IS), beginning in 2014. IS successfully recruited thousands of young people from Europe, North America, and elsewhere through professionally created propaganda videos and social media statements. Most notoriously, the group posted videos of beheadings of Western journalists. In response to the release of these brutal videos, the US State Department created a Twitter account intended to discourage people from joining IS.

of body language are typically absent from social media interactions, they may lack the capacity for empathy and understanding. There is much merit in both online and offline social interactions; one does not effectively replace the other.

INTERPRETING, CONCEPTUALIZING, AND MEASURING GLOBALIZATION

Globalization is a highly contested term and a much debated topic. One way to think about it is to consider three of the principal interpretations advocated by leading scholars. A second way identifies three theses concerning what globalization is (and is not) and how it is (or is not) transforming the world. A third way is to focus on how globalization can be measured and thus determine regional variations.

In April 2014, the Nigerian terrorist group Boko Haram kidnapped 276 girls from their school in Chibok. Frustrated by what they perceived as the government's inadequate response to the kidnapping, parents and other citizens used social media to shed light on the situation. Part of this campaign was #BringBackOurGirls, which trended globally on Twitter, helping to spread awareness and to demand action. Over a year later, with most of the girls still missing, other hashtags have kept focus on this tragedy.

Three Interpretations

The first interpretation was put forward by Thomas L. Friedman (2005), the noted *New York Times* columnist. He asserts that the world is now what he describes as "flat," a result of improved communications and the free movement of capital. In this view, the global economy has become a level playing field for the first time, as societies around the world conform to free-market principles and practices. Further, Friedman argues that globalization is intensifying.

Second, the Canadian philosopher and novelist John Ralston Saul sees the world quite differently. He goes so far as to suggest that globalization, seen essentially not as a historical process of innovation and technological change but rather as a neo-conservative ideology, is in retreat. Saul (2005: 270) writes:

> At the most basic level of societal knowledge, we do know that Globalization—as announced, promised and asserted to be inevitable in the 1970s, '80s and much of the '90s—has now petered out. Bits and pieces continue. Other bits have collapsed or are collapsing. Some are blocked. And a flood of other forces have come into play, dragging us in a multitude of directions.

Rather than Friedman's flattening of the world, Saul sees a series of regional dislocations and inequalities resulting from the various successes and failures of globalization, most notably benefitting the more developed world and further disadvantaging the less developed world (Chapter 6 has a detailed account of these global inequalities). Saul's image of the world prompted a book, *The Collapse of Globalism: And the Reinvention of the World,* which often reads like a manifesto designed to encourage the reader to become involved and to care about depopulation in the heartland of North America, about the inexorable spread and growth of Walmart, and (above all) about the plight of the poor in the less developed world, all of which are understood to be evidence of the failings of globalization. Not only is it in retreat, according to Saul, but it also has failed to make the world a better place.

In his book, Saul dates the onset of globalization from 1971 (when the United States went off the gold standard and settled on a floating dollar, to which most other currencies were pegged), claiming that it emerged fully grown. The heyday was in the mid-1990s, by which time trade liberalization had proceeded apace, global markets were dominant, and the WTO was in place to preside over the new global economy. But all was not well, as evidenced by the Asian financial crisis of 1997–8 and by the explosion of anti-globalization activity that began in earnest in Seattle in 1999. Building on these facts, Saul contends that a key defeat for globalization was the 2003 US decision to invade Iraq without the support of other major powers. Thus, it is suggested that globalization might reverse because of an increasingly nationalistic US foreign policy. But this thought-provoking and highly readable book leaves the critical question largely unanswered: What does the future hold if globalization is in retreat? Certainly, a return to a world dominated by nation-states seems unlikely.

The third reading of globalization is provided by Pankaj Ghemawat (2011) in his book, *World 3.0*. His basic thesis is that globalization is exaggerated and that, at best, we live in a world of semi-globalization. Here are some of the figures that support this claim.

- Less than 1 per cent of American companies have foreign operations.
- FDI is only 9 per cent of all fixed investment.
- Sixty years ago half of the world's car production was controlled by just two companies; today, six companies share half the production.
- Trade between countries is 42 per cent greater if those countries speak the same language, and 47 per cent greater if they belong to the same trading bloc.

Ghemawat acknowledges that there is much disagreement as to whether globalization is increasing or lessening, but he recognizes what we noted early in this chapter, namely, that distance—geography—still matters.

Three Globalization Theses

The *hyperglobalist* thesis views globalization as a new age, while *skeptics* question the very existence of globalization. In between these two contrasting viewpoints is the *transformationalist* thesis. Table 3.3 summarizes these three important ideas.

Measuring Globalization

A third and more rigorous way to think about globalization is to attempt to measure it objectively.

TABLE 3.3 | Three theses about globalization

	Hyperglobalist	Skeptic	Transformationalist
What is happening?	The global era.	Increased regionalism.	Unprecedented interconnectedness.
Central features	Global civilization based on global capitalism and governance.	Core-led regionalism makes globe less interconnected than in late nineteenth century.	"Thick globalization." High intensity, extensity, and velocity of globalization.
Driving processes	Technology, capitalism, and human ingenuity.	Nation-states and the market.	"Modern" forces in unison.
Patterns of differentiation	Collapsing of welfare differentials over time as market equalizes.	Core–periphery structure reinforced, leading to greater global inequality.	New networks of inclusion/exclusion that are more complex than old patterns.
Conceptualization of globalization	Borderless world and perfect markets.	Regionalization, internationalization, and imperfect markets.	Time–space compression that increases interaction between places.
Implications for the nation-state	Eroded or made irrelevant.	Strengthened and made more relevant.	Transformed governance patterns and new state imperatives.
Historical path	Global civilization based on new transnational elite and cross-class groups.	Neo-imperialism and civilizational clashes through actions of regional blocs and neo-liberal agenda.	Indeterminate; depends on construction and action of nation-states and civil society.
Core position	Triumph of capitalism and the market over nation-states.	Powerful states create globalization agenda to perpetuate their dominant position.	Transformation of governance at all scales and new networks of power.
Moral message	Pro-globalization. Globalization is real, and is a moral good and force for progress.	Mostly anti-globalization. Globalization is a discourse propagated by powerful interests.	Anti-globalization. Globalization is real and requires regulation to optimize it.

Sources: Adapted from Held, D., A. G. McGrew, D. Goldblatt, and J. Perraton. 1999. *Global Transformations: Politics, Economics and Culture*. Cambridge: Polity Press, 1999, 10; Murray, W. E. 2006. *Geographies of Globalization*. New York: Routledge, 353–4.

Two influential approaches are discussed. First, the KOF Index of Globalization, published annually by the KOF Swiss Economic Institute, conceives of globalization in economic, social or cultural, and political terms and defines it as the

> process of creating networks of connections among actors at multi-continental distances, mediated through a variety of flows including people, information and ideas, capital and goods. Globalization is conceptualized as a process that erodes national boundaries, integrates national economies, cultures, technologies and governance and produces complex relations of mutual interdependence. (Swiss Economic Institute, 2011)

Note that the KOF's selection of specific indicators is intended to be representative of globalization processes, not comprehensive. As is the case with any attempt to measure globalization, it is possible to be critical of the chosen indicators and the weight attached to each and to suggest alternatives.

KOF Index of Globalization: Indices and Variables

Figure 3.6 tracks overall globalization since 1980 and distinguishes the economic, social, and political components. We can see a marked increase at the end of the Cold War (1990), a slight slowdown related to the 9/11 terror attacks in 2001, and a significant setback caused by the 2008 recession. More generally, thinking about globalization in this way shows that it has increased significantly.

Cold War
The period of confrontation without direct military conflict between Western (led by the US) and communist (led by the USSR) powers that began shortly after the end of World War II and lasted until the early 1990s.

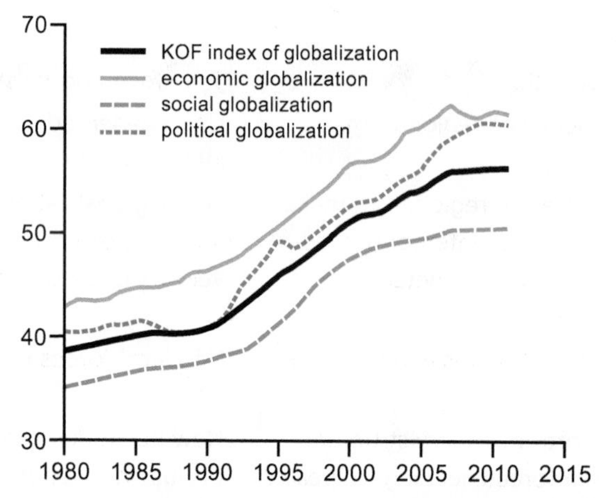

FIGURE 3.6 | KOF Index of Globalization worldwide

Source: KOF Swiss Economic Institute. 2014 (16 April). "KOF Index of Globalization 2014: Switzerland No Longer among the Top Ten." (Press Release). http://globalization.kof.ethz.ch/media/filer_public/2014/04/16/press_release_2014_en.pdf.

TABLE 3.4 | KOF Index of Globalization: Top 10 and bottom 10 countries, 2014

	Top 10			Bottom 10	
	Country	**Index**		**Country**	**Index**
1.	Ireland	92.17	182.	Vanuatu	31.59
2.	Belgium	91.61	183.	Comoros	31.31
3.	Netherlands	91.33	184.	Afghanistan	29.91
4.	Austria	90.48	185.	Bhutan	29.17
5.	Singapore	88.63	186.	Equatorial Guinea	27.70
6.	Denmark	87.43	187.	Lao PDR	27.07
7.	Sweden	87.39	188.	Eritrea	27.03
8.	Portugal	87.01	189.	Kiribata	26.20
9.	Hungary	85.91	190.	Solomon Islands	25.43
10.	Finland	85.87	191.	Somalia	24.03

Source: 2014 KOF Index of Globalization.
http://globalization.kof.ethz.ch/media/filer_public/2014/04/15/rankings_2014.pdf.

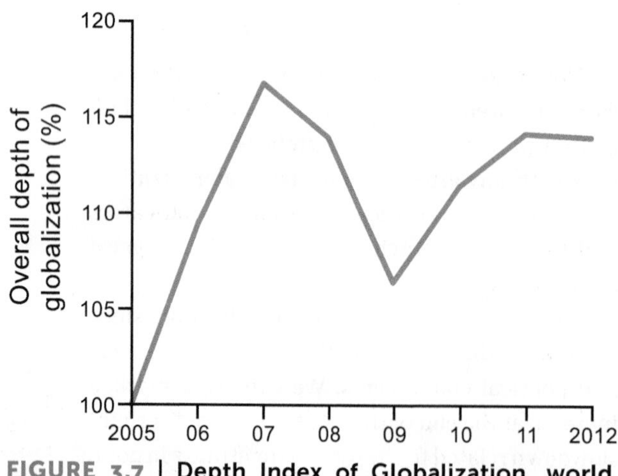

FIGURE 3.7 | Depth Index of Globalization, world, 2005–2012

Source: Ghemawat, P. 2013. Depth Index of Globalization 2013, 13. www.ghemawat.com/Dig/Files/Depth_Index_of_Globalization_2013_(Full Report).pdf.

Table 3.4 details the top 10 and bottom 10 countries on the overall globalization scale (maximum 100, minimum 0). Not surprisingly, nine of the most globalized countries are European; Canada ranks twelfth, with an index of 85.63, and the United States is thirty-second, with an index of 74.94. The bottom 10 are a mix of African, Asian, and very small countries. Note that the difference between particular country indices in both lists is often quite small, meaning that detailed rankings are subject to change from year to year and thus need to be interpreted with some caution.

Ghemawat provides the second influential approach, which focuses on economic variables. A recent report by Ghemawat and Steven A. Altman (2013) notes that globalization has lessened (i.e. the world is less connected) since the economic uncertainties that began in 2008. This conclusion is reached through calculation of a Depth Index of Globalization (DIG). The index measures globalization for 139 countries, which accounts for 99 per cent of global **gross domestic product (GDP)** and 95 per cent of global population. The index is calculated using data on trade, capital movement, the transmission of information, and population movement (Figure 3.7).

In 2013 the top five globalized countries (of the 139 analyzed) were Hong Kong SAR (China), Singapore, Luxembourg, Ireland, and Belgium; the bottom five were Iran, Burma, Bangladesh, Pakistan, and Burundi. Clearly, countries with a high DIG tend to be well developed and relatively small; this is to be expected given that they have small national markets and thus a large share of their economic activity in other countries. Large countries with emerging economies have a lower DIG. For example, Russia is ranked 95, China 122, and India 125. Canada is ranked 45 and the US 91. Figure 3.8 maps global DIG.

The role of distance is significant in the depth of globalization, especially because many interactions (e.g. trade and tourism) depend on transportation and because digital interactions decrease with increasing distance.

THE GLOBAL ECONOMIC SYSTEM

Economic globalization is not the same thing as economic internationalization. The internationalization of economic activity began with the

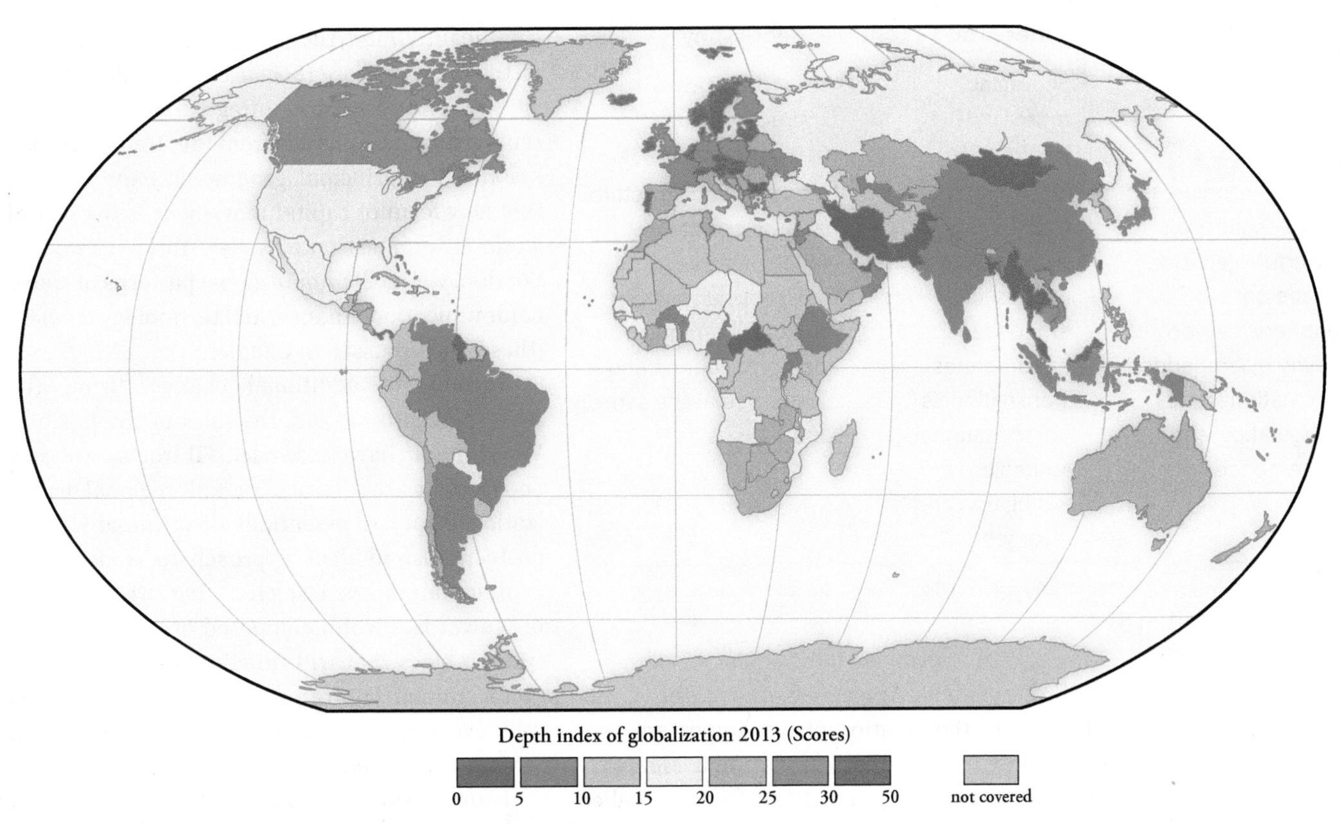

FIGURE 3.8 | World distribution of Depth Index of Globalization, 2013

Source: Ghemawat, P. www.ghemawat.com/dig/default.aspx.

development first of empires and then of economic links—particularly through the movement of capital and goods—between colonial powers and their colonies. But this process did not extend to production, which continued to take place in a single national territory. Trade between countries is not globalization. Only when production and distribution are no longer contained by national boundaries, a change that has been brought about primarily by the rise of transnational corporations, can we properly speak of globalization. As Figure 3.9 shows, the increasing economic importance of transnationals means that the production process is now organized *across* rather than *within* national boundaries, with transnationals effectively penetrating into countries. Table 3.5 provides a useful introduction to this account, highlighting as it does the three globalization theses noted earlier in this chapter and links to economic geography.

To understand globalization processes, and specifically how transnational corporations emerged, it may be helpful to review recent changes in the global economic system. Economic co-operation between states has ebbed and flowed over the past century. In much of the Western world, including Canada, the 1920s and 1930s were a period of trade barriers and limits on foreign investment.

gross domestic product (GDP)
A monetary measure of the market value of goods and services produced by a country over a given time period (usually one year); provides a better indication of domestic production than gnp.

The Bretton Woods Agreement

The beginnings of what we now think of as economic globalization came near the end of World War II, when the United States became the world's principal economic power and was able to initiate

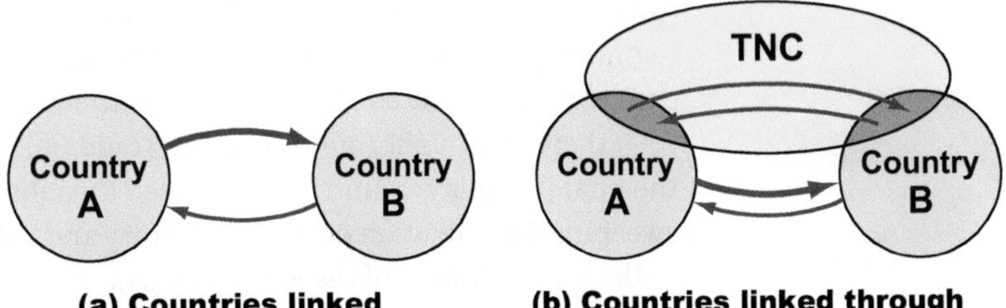

FIGURE 3.9 | Territorial interpenetration: The "incorporation" of parts of a state's territory into a transnational

Source: Dicken, P. 2011. *Global Shift: Mapping the Changing Contours of the World Economy*, 6th edn. New York: Guilford Press, 224.

TABLE 3.5 | Globalization theses and economic geography

Hyperglobalist: The Global Era	Skeptical: Increased Regionalism	Transformationalist: Unprecedented Interconnectedness
Market is omnipresent and transnationals powerful. Networks replace core–periphery division. Activity disembedded from nation-states. Deregulation and convergence.	Trading blocs formed. Core–periphery division increased. Activity embedded in nation-states. Transnationals reflect national strategies. Regulation and divergence.	Networks and structural patterns coexist. Nation-states and transnationals govern markets. Re-regulation and convergence and divergence at the same time.

Source: Adapted from Murray, W. E. 2006. *Geographies of Globalization*. New York: Routledge, 353–4.

From Protectionism to Free Trade: A Chronology

the Bretton Woods Agreement of 1944. This agreement led to the creation of two important economic institutions: the International Monetary Fund (IMF) and the World Bank (initially called the International Bank for Reconstruction and Development). The IMF was to provide short-term assistance to countries whose currencies were tied either to gold or to the US dollar, and the World Bank was to provide development assistance to Europe after the war. An important component of the economic order introduced at Bretton Woods concerned international capital movement, which was to be regulated primarily through national systems of exchange controls, with the US dollar playing the key role.

Another feature of the postwar economic system was free trade. This was to be the focus of a planned third international institution, the International Trade Organization (ITO), but the US Senate refused to ratify the ITO Charter. Instead, the General Agreement on Tariffs and Trade (GATT), a temporary and less formalized trade regime intended to reduce trade barriers between countries, was created in 1947. Between then and 1994 there were eight rounds of GATT trade talks; the final (Uruguay) round accomplished the most sweeping liberalization of trade in history and led to the establishment of the WTO (Box 3.5).

Recent Changes to the Global Economic System

The economic order introduced in the 1940s under the leadership of the United States has changed significantly. First, the movement of capital is now virtually immediate (because it is achieved electronically) and almost unregulated. This represents a dramatic change from the Bretton Woods system. The principal geographic expression of this new form of capital movement is the rise of world cities, most notably New York, Tokyo, and London, which dominate global patterns of trade, communication, finance, and technology transfer (these are discussed in Chapter 11).

Second, in additional changes from the Bretton Woods system, the roles of the IMF and World Bank have expanded. Third, as we saw above, there are several close-knit regional organizations that are essentially discriminatory and protectionist in their approach to trade. Some commentators even suggest that we are moving towards a world composed of three regional trading blocs that will function not only as economic unions (Figure 3.10) but as political ones as well, eventually replacing states. This prediction remains to be seen.

Fourth, political changes in the late 1980s and early 1990s had significant economic repercussions. With the collapse of communism in the former Soviet Union and Eastern Europe and the end of apartheid in South Africa, several new participants entered the world trade arena. And the country with the largest population in the world, China, is adopting increasingly capitalist economic policies.

CULTURAL GLOBALIZATION

In addition to these primarily economic issues, there is evidence of increasing cultural homogenization and related loss of national and local cultural identity. Identity changes seem likely, but it is a matter of opinion as to whether changes are good or bad. It can be argued that national and ethnic identities especially contribute to conflict; this argument is debated in Chapters 8 and 9.

One way to conceive of cultural globalization is to imagine a process that began before the rise of the nation-state, when cultures and identities were essentially local. With the emergence of the nation-state, a second option became available: membership in a national culture. In this sense, nation-states can be seen as cultural integrators, bringing together various local identities in such

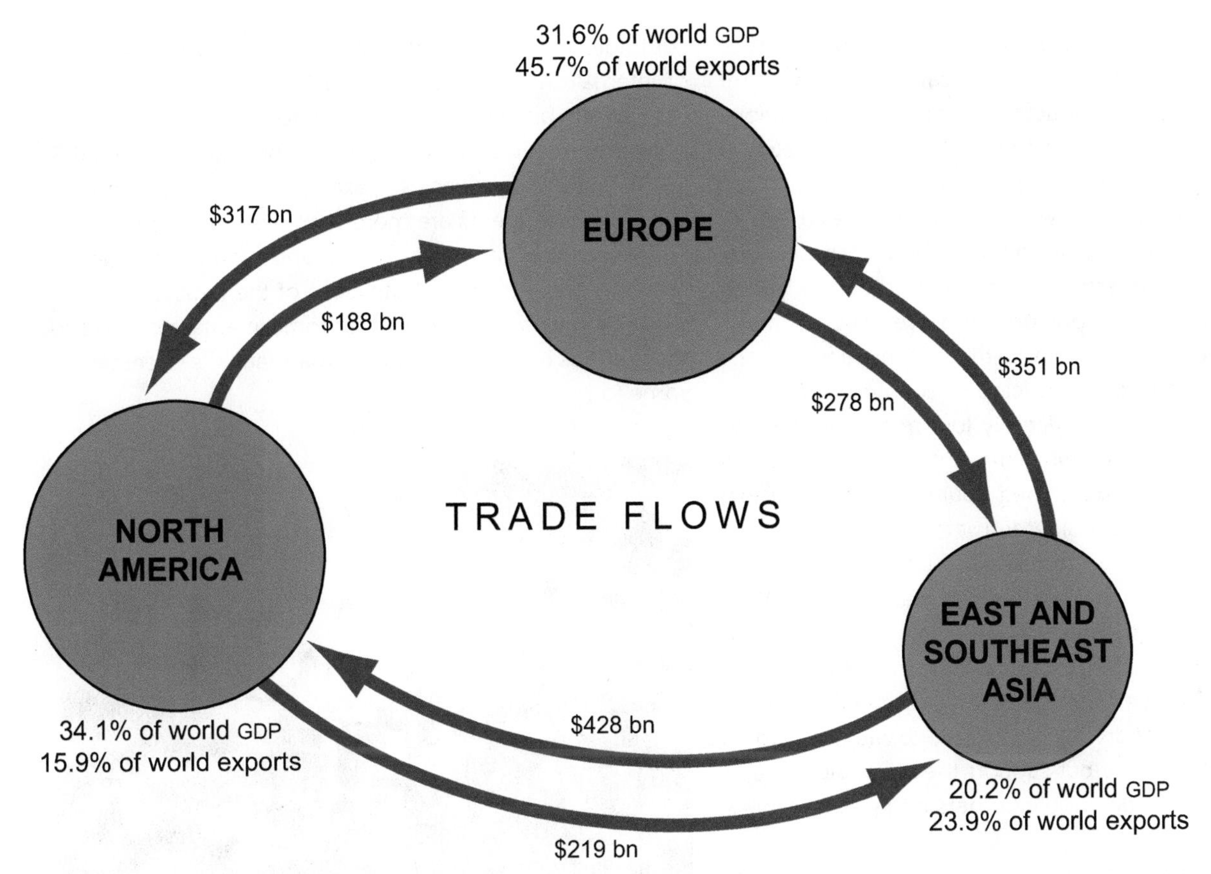

FIGURE 3.10 | The contemporary geo-economy

Dicken emphasizes that the power structure of the contemporary world economy is based on three regions: North America, the EU, and East and Southeast Asia. Each contains one of the three most important and powerful global cities—New York, London, and Tokyo—discussed in more detail in "Global Cities," beginning on page 412. Clearly a key consideration in the creation of these trading blocs is spatial proximity. Dicken writes: "The 'triad' appears to sit astride the global economy like a modern three-legged Colossus, constituting the world's 'mega-markets' and 'sucking in' more and more of the world's production, trade and direct investment." This situation represents a significant change from the mid-twentieth century, when European colonial empires were the crucial regions and the international division of labour was a simple core-periphery structure.

Source: Adapted from Dicken, P. 2007. *Global Shift: Mapping the Changing Contours of the World Economy*, 5th edn. New York: Guilford, 39, Figure 2.5.

Around the Globe

BOX 3.5 | The World Trade Organization

Formed in 1995 to replace the GATT as the guardian of international trade, the WTO has become closely associated with globalization. There are currently 153 member countries (representing more than 97 per cent of the world population), and at least 25 more are hoping to join. The organization's mandate includes facilitating smooth trade flows, resolving disputes between countries, and organizing negotiations. The WTO is the only international body with such authority, and its decisions are absolute and binding on member countries (Hoad, 2002). The GATT and the WTO have been spectacularly successful—world trade is increasing rapidly, and tariffs average about one-tenth what they were in 1947. Today, protectionism remains significant in just one important area: agriculture.

In late November–early December 1999, a WTO meeting of member countries' trade ministers in Seattle was overwhelmed by well-organized demonstrations in which labour organizations, environmental activists, consumer groups, and human rights activists shared their concerns about the circumstances and presumed consequences of economic globalization.

The concerns are many and varied. Some argue that the WTO is too secretive and that its discussions should be held

Continued

in a more public context, involving a much broader range of participants. Others accuse the organization of being a front for corporate interests, particularly influential transnationals. Some see it as benefiting the wealthy, more developed countries and neglecting the poor of the less developed world (certainly the most powerful players are the EU, Japan, and the US). Still others point out that the WTO is inconsistent in promoting unrestricted movement for some goods while permitting restrictions on the movement of others. These critics expressly object to the continuation of agricultural subsidies in more developed countries that allow those nations to export artificially low-priced produce to the less developed world while preventing the importation of produce from less developed countries. Perhaps most fundamentally, some consider it ironic that the wealthiest member states continue to resist the unrestricted movement of labour in the form of international migration. With this contradiction in mind, Massey (2002) points out that many politicians and business leaders in the more developed world take different stances depending on the issue under discussion: they support the idea of a world without borders if the issue is free trade but support the nationalist idea of separate states with defensible borders if the issue is international migration.

The WTO rejects most of these criticisms, arguing that the best way to address global inequality, especially the problems of the less developed world (discussed in Chapter 6), is through trade liberalization so that free trade becomes fair trade. In many respects the WTO is the most democratic of international organizations, having a one member–one vote system and serving as a forum for discussion between governments, the vast majority of which are democratically elected. Rigg (2001), for example, notes that many of the WTO's opponents are more secretive, less accountable, and less democratic than the organization itself. There is probably some truth on both sides of the debate, with the WTO acknowledging the need for some internal reforms and many critics accepting that the organization is necessary in some form.

Trade representatives, including government ministers, speak at the 2015 WTO Public Forum in Geneva, Switzerland. The theme of the forum was "Trade Works!"

a way that it became possible for individuals to understand themselves and their lives in both traditional local and newer national contexts—although (as we will see in Chapter 9) the transition from a singular to a dual cultural identity has not been easily accomplished in many parts of the world. Some suggest that we have reached a stage where a global cultural identity is developing.

Proponents of this argument see a homogeneous global culture replacing the multitude of local cultures that has been characteristic for most of human history (for an essentially Marxist interpretation of this phenomenon, recall Box 2.3). The diffusing culture is Western in character and largely derived from the United States. Among the mechanisms that allow this culture to spread spatially are various aspects of the mass media and consumer culture—newspapers, magazines, the Internet, music, television, films, videos, fast-food franchises, fashions. That these aspects of popular culture are disseminating around the world and influencing aspects of non-material culture such as religion and language is undeniable. However, assessing their significance is not as easy as it might seem.

Landscape features that are local in origin, such as a festival or an indigenous building style, enhance place identity, whereas those that are global in origin, such as chain department stores or an imported building style, contribute to placelessness. Even if all placeless areas do not look alike, they may lack the local characteristics that make a place distinctive, comfortable for insiders, and perhaps incomprehensible to outsiders. Global characteristics are those that make a place similar to other places and therefore relatively placeless, perhaps less comfortable for insiders but more comprehensible for outsiders. The concept of placelessness is an important one because it suggests that the more places become placeless, the more similar human experience becomes.

It appears unlikely that globalization can erase the power of local places and local identities—what Vidal called *pays* and *genres de vie*. Physical environments vary throughout the world, and there is no denying that the connections between physical and human geographies are often intimate. Most places still look different from other places. In some ways, at least, their inhabitants continue to behave differently from other people, and they still hold some attitudes, feelings, and beliefs that differ from those of other people. Despite clear evidence that the number of the world's languages is decreasing, the roughly 6,000 languages that still exist are compelling examples of cultural variation from place to place (see Chapter 7). We are human, but significant differences remain between places and groups of people, resulting primarily from differences in cultural identity. The world we live in is not homogeneous. Thus, a conflict between parochial ethnicity and global commerce may be developing.

These comments once again reflect some of the uncertainties about precisely what globalization is and what impacts it is having. As noted in Table 3.6, the three theses about globalization posit quite different cultural consequences.

Thinking about cultural globalization obliges us to acknowledge that there are no uncontested and unidirectional processes at play in the contemporary world. Globalization, in the sense of an ever-increasing connectedness of places and peoples, is a fact but not the only important one. Some regional economies remain distinctive, and it is possible that some regional differences actually are being enhanced, as suggested by the rise of regional trading blocs. It is hard to believe that the cultural world is becoming uniform at a time when so many ethnic groups are reasserting their identities—at least partly in reaction against the declining importance of national political and cultural identities.

TABLE 3.6 | Globalization theses and cultural geography

Hyperglobalist: The Global Era	Skeptical: Increased Regionalism	Transformationalist: Unprecedented Interconnectedness
New global civilization.	Clashing cultures.	New global and local hybrid cultures.
Homogeneous consumerist culture and global brands dominant.	Civilizational blocs entrenched and cultural identities differentiated and relativized.	Possibility of progressive cultural change, but Western values and tastes currently dominate.
Universalization of cultural identities.		

Source: Adapted from Murray, W. E. 2006. *Geographies of Globalization*. New York: Routledge, 353–4.

POLITICAL GLOBALIZATION

There is much evidence to support the claim that a process of political globalization has been underway for quite some time, although, as suggested in Table 3.7, the significance of this theory is uncertain. As with economic and cultural globalization trends, political globalization results from technology, specifically communications technology, while the spread of increasingly dominant transnational corporations produces an integrated world economy in which, as trade increases, consumer behaviour patterns become more and more similar. The idea of political globalization, however, also incorporates two explicitly political circumstances.

First, following the end of the Cold War, the new world political order meant that, in principle, the UN finally was in a position to play the role for which it was created. Previously, it could be argued, both the US and the Soviet Union actively sought to prevent the organization from fulfilling its mandate. However, it might also be argued that the UN faces many challenges at least partly because it continues to be dominated by the five powers that have permanent seats and vetoes on the Security Council.

TABLE 3.7 | Globalization theses and political geography

Hyperglobalist: The Global Era	Skeptical: Increased Regionalism	Transformationalist: Unprecedented Interconnectedness
Market-led global governance. Nation-state replaced by "natural" region states. Boundaries dissolve and sovereignty surrendered to global market.	Core-led regionalism. Sovereignty surrendered to regional groupings designed by powerful nation-states. Boundaries retrenched.	Multi-layered governance at three scales: global, national, local. Nation-state remains central. Centralization and devolution at same time.

Source: Adapted from Murray, W. E. 2006. *Geographies of Globalization*. New York: Routledge, 353–4.

Second, globalization not only relies on the spread of democracy but also tends to promote it. Because democratic states tend to form closer links with one another than with other political forms, states that want to trade with democratic states will be persuaded to become more democratic. Perhaps more important, globalization may also promote peace if the "perpetual peace" scenario (see Chapter 9) is correct in suggesting that democracies do not wage war with one another.

It can be argued that the process of political globalization has been underway for a very long time: in 1500 BCE the world contained perhaps 600,000 autonomous polities, and the long-term trend has been towards political integration. On the other hand, developments in the twentieth century suggest that, in the short term at least, the trend may be towards more rather than fewer political units. The end of World War I brought the collapse of several empires and the rise of several new states. The end of World War II ushered in a period of previously unheard-of co-operation between European states and a cold war that divided much of the world into two camps, American and Soviet. The decades since 1945 have seen some apparently anti-globalizing trends as colonial areas have achieved independence; approximately 95 new states have come into being this way, and the collapse of the Soviet Union and the disintegration of Yugoslavia were responsible for the creation of about 19 more.

Political States in the Contemporary World

Because political globalization challenges the integrity of the individual state, it entails a reduction in the importance of the role played by states. By the early 1990s, Ohmae (1993: 78) was arguing that the nation-state "has become an unnatural, even dysfunctional unit for organizing human activity and managing economic endeavor in a borderless world" and suggesting that on "the global economic map the lines that now matter are those defining what may be called 'region states.'" A region state is a natural economic area. It may be part of a nation-state, as in the case of northern Italy, or it may cross boundaries, as in the case of Hong Kong and south China before 1997 (when China regained control of Hong Kong). Its principal economic links are global rather than national. In the Canadian context, it might be suggested that British Columbia forms a region state with the American west coast; other parts of Canada also have north–south relationships with areas of the US. In Canada, as elsewhere, the rise of region states has significant implications for national unity.

The role of the state is definitely changing. But perhaps states will be able to adjust to the changing circumstances resulting from political globalization. In the past, many states sought power as measured by territory and population; today, states seek markets. The challenge for the state in the twenty-first century may well be to integrate with other states without sacrificing national identity. On the other hand, the nation-state may prevail after all, since people tend to resist change as long as they are reasonably comfortable. Thus, most Canadians want Quebec to remain part of Canada, if only because of the insecurity that its secession would entail.

Tribalism

What about globalism's opposite number: tribalism, or the tendency for groups to assert their right to a state separate from the one they occupy? Those who see tribalism as gaining ground point to secession movements, civil war, terrorism, and cross-border ethnic conflict fed by the resurgence of local identities. Support for this scenario can be found in two other trends: distinctiveness and multiculturalism.

As we will see in Chapters 7, 8, and 9, groups of people with some common identity are increasingly emphasizing their distinctiveness, often at the expense of the states to which they "belong." During the 1990s a number of new states were created as, for example, the former Czechoslovakia divided peacefully into the Czech Republic and Slovakia and the former Yugoslavia disintegrated in bitter conflict (see Box 9.5). Most recently, the new state of South Sudan was formed in 2011 following prolonged conflict. At least some of the new states created in this way may turn out to be more culturally meaningful than their predecessors. In any event, the potential for the creation of additional new states is considerable.

In some states, such as Canada and Australia, the principle of multiculturalism is widely accepted and cultural pluralism is valued. It is possible that acceptance of multiculturalism will lead to a weakening of central authority as "tribal" identities become stronger.

GLOBALIZATION: GOOD OR BAD?

You may not be surprised to learn that this textbook will not answer this question. Instead, it will provide some food for thought, in the hope that you will use it to clarify your own thinking. As with environmental issues (Chapter 4) and the state of the world generally (Chapter 6), many people express strong opinions on globalization that often reflect ideological positions—personal views of the world and how it ought to be. Let us begin with some of the arguments put forward by those who oppose globalization.

Opposing Globalization

Why is opposition to globalization so widespread? There are several interrelated reasons.

Who Benefits?

In the broadest terms, globalization is seen by some as benefiting the more developed world at the expense of the less developed. More specifically, it is seen as benefiting those few countries and private investors within the more developed world, where transnational corporations are based, at the expense of many (specifically working) people, especially those in the less developed world. More specific criticisms are directed at the WTO in its capacity as guardian of world trade. For example, as discussed in Box 3.5, WTO members tend to insist on closed meetings. By excluding other interested parties, such as labour unions and environmental and anti-poverty groups, the WTO gives the impression that it is undemocratic and working to benefit only its members or, since it now includes less as well as more developed countries, only a select few of those.

The movement towards free trade—or what critics describe as selective free trade—can be interpreted as serving the interests of the major economic powers, such as the US, Japan, and the EU, at the expense of the less developed countries that are becoming industrialized. Figure 3.11 shows that the rich countries of the world continue to dominate trade, with about 15 per cent of the world population accounting for about 70 per cent of world exports. It is worth noting that, when the major powers were industrializing in the nineteenth and early twentieth centuries, they all enjoyed the benefits of protectionist trade policies, government intervention, subsidies designed to encourage domestic industrial growth, and few if any restrictions on environmental degradation.

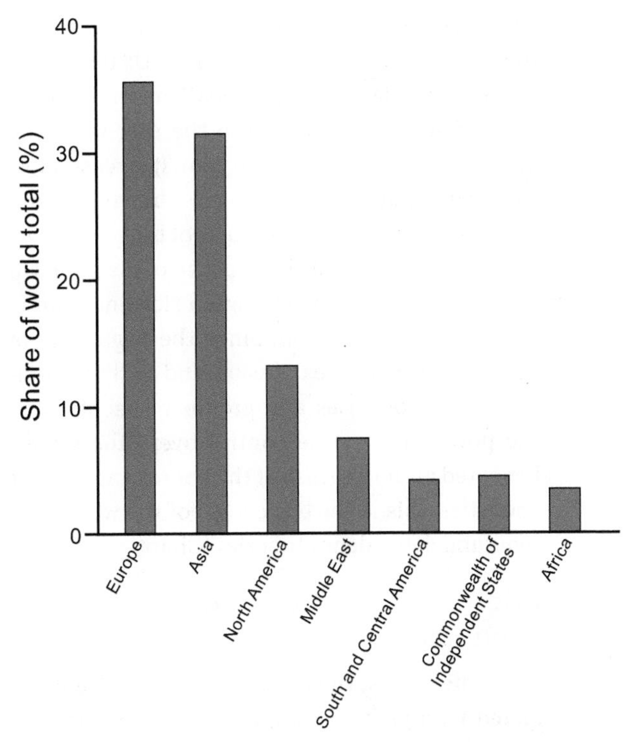

FIGURE 3.11 | World merchandise exports by region, 2012

Source: World Trade Organization. 2013. *Statistics: International Trade Statistics, 2013, World Trade Developments*. Geneva: World Trade Organization, Table 1.5. www.wto.org/english/res_e/statis_e/its2013_e/its13_world_trade_dev_e.htm. © World Trade Organization (WTO) 2015.

The benefits of globalization are unevenly distributed. Most have accrued to countries in East and South Asia (mainly China), while sub-Saharan Africa and, at least for a few years, the countries of the former Soviet bloc have lost rather than gained. Part of the explanation lies in specific circumstances. China, for example, has a long tradition of trading and is embracing a new capitalist approach. The Soviet bloc, in contrast, was sheltered for decades from free-market forces and has not found it easy to embrace the emerging economic order. Numerous problems of conflict, inadequate infrastructure, and some cases of corrupt government in sub-Saharan Africa all work together to limit economic opportunities.

Some recent changes are likely to exacerbate these circumstances. Consider, for example, that all quotas and restrictions on clothing imports were, after a 40-year process, abolished in 2005, making world trade in textiles fully liberalized. As a result, some of the smaller producers, such as the Philippines, Bangladesh, and Mexico, suffered, while China and India significantly increased their exports, especially to the huge US market.

More fundamentally, globalization is often charged with contributing to the ever-widening gap between the rich and the poor. It is reasonable to suspect that, if those who are already rich are in a position to dictate the rules of the game, they may become richer at the expense of the poor. On the other hand, the gap between rich and poor is not a new phenomenon. Since the beginning of European overseas expansion and early capitalism, those countries and groups of people with the power to exercise control over others have benefited at the expense of the poorer and weaker countries; this is the basic logic of the world systems and dependency theories (Chapter 6).

Prioritizing Export-Centred Economies

For some critics, the most serious problem associated with globalization is that it gives priority to export-centred economies, with the result that the merits of locally sustainable economies are downplayed. Exploitation of resources in the less developed world to satisfy demand in the more developed world will likely have a negative impact on local economies. Increasingly, local landscapes are treated as global commodities, and local communities are losing their traditional resource bases. The larger consequences of globalization for the less developed world are debated, as suggested by Table 3.8.

Environmental Issues

There are worries about the environmental consequences of globalization, particularly with respect to resource extraction. Many human impacts on the environment are exacerbated by current trends towards globalization, especially the activities of transnationals. It seems clear that transnationals are among the principal actors in many environmental issues. For example, fast-food corporations have been blamed for the destruction of tropical rain forests because they use meat from animals raised in areas that have been cleared specifically to create pasture. Similarly, oil companies that explore for and produce oil in such environmentally sensitive areas as the Amazon basin, the Canadian Arctic, and the mangrove swamps of the Niger delta are particularly condemned by environmentalists, as are the trawlers of transnational fish-processing companies that have been a major factor in the collapse of local fisheries from New Zealand to Newfoundland.

Amnesty International (2003) suggests that the recently completed oil and gas pipeline from the Caspian Sea port of Baku in Azerbaijan to the Mediterranean Turkish port of Ceyhan has infringed on the human rights of thousands of people and will cause massive environmental damage. The pipeline's route—constructed by a BP-led consortium—was designed to avoid passing through Russian or Iranian territory. As a result, according to Amnesty International, 30,000 villagers were forced to give up their land rights. In addition, the organization claimed that the health and safety precautions are inadequate and that local protesters faced state oppression. BP, for its part, argued that the project followed the highest international standards, that appropriate compensation has been paid, and that there will not be any environmental damage.

Transnational companies have sometimes been charged with actively perpetuating local social injustices. Talisman Oil, the largest independent Canadian oil and gas producer, has operations in Canada, the North Sea, and Indonesia and is conducting explorations in Algeria and Trinidad. Between 1998 and 2003, the company was also active in Sudan. Canadian social justice and church groups criticized these operations, based on claims that the company actively supported the Sudanese government during the country's civil war (which resulted in the independence of South Sudan). Evidence showed that the government, headed by the National Islamic Front and consistently identified as violating basic human rights, forcibly removed people to make southern Sudan safe for Talisman. The

TABLE 3.8 | Globalization theses and development

Hyperglobalist: The Global Era	Skeptical: Increased Regionalism	Transformationalist: Unprecedented Interconnectedness
Development attained by taking part in globalization. Poverty solved by market. Integration unavoidable.	Development threatened by global capitalist expansion. Marginalization increased by spread of market.	Globalization offers threats and opportunities for development. Progress contingent on careful management.

Source: Adapted from Murray, W. E. 2006. *Geographies of Globalization*. New York: Routledge, 353–4.

company denied complicity in any such social injustices and, in 1999, in response to a request from the Canadian government, formally adopted the International Code of Ethics for Canadian Business, which was developed in 1997 by a group of Canadian companies with transnational operations. The difficult political and social environment in Sudan was at least part of the reason that Talisman sold all its interests there in 2003.

Notwithstanding these comments, it is important to appreciate the many uncertainties about globalization. Using the three globalization theses discussed earlier, Table 3.9 highlights varying understandings of the links between globalization and environmental issues.

TABLE 3.9 | Globalization theses and environmental issues

Hyperglobalist: The Global Era	Skeptical: Increased Regionalism	Transformationalist: Unprecedented Interconnectedness
Sustainability solved by the market. Market will price the environment efficiently and technology will solve scarcity.	Sustainability threatened by global capitalism. Spread of modernization and consumerism pushes environment to limits.	Sustainability attainable through political action. Environmental problems and concerns globalize. Regulation from above and below required.

Source: Adapted from Murray, W. E. 2006. *Geographies of Globalization*. New York: Routledge, 353–4.

"Another World Is Possible"

The extent of the opposition to globalization seems undiminished each year. One indicator is the annual World Social Forum, first held in 2001 to counter the World Economic Forum, an exclusive global "think tank" held annually in Davos, Switzerland, and comprised of business and political leaders and academics who largely support economic globalization. The World Social Forum and its various related regional forums provide a platform for anti-globalization activists to talk and plan strategy. Four of the first five annual meetings were held at Porto Alegre (Brazil), with the 2004 meeting in Mumbai. More recent meetings have been held in other places, including three sites in 2006 (Caracas in Venezuela, Bamako in Mali, and Karachi in Pakistan); in Nairobi (2007); in numerous local sites (2008); in Belem, Brazil (2009); Porto Alegre and other locations (2010 and 2012); Dakar, Senegal (2011); Tunis (2013 and 2015); and Ottawa (2014). Participants reject the claim that globalization is in some way inevitable and argue that "another world is possible."

Supporting Globalization

Are the reasons for opposing globalization the only ones worth thinking about? Of course not. Most of those who support globalization do so for business-oriented reasons, pointing to economic growth. However, Johan Norberg defends globalization for reasons that are essentially moral.

A Moral Argument

In his book *In Defence of Global Capitalism* (2001), Norberg argues in favour of capitalistic economic globalization on the grounds that it represents the best hope for eliminating poverty and fulfilling human potential. Using data from many countries, including the two largest (China and India), he argues that lowering trade barriers and investing capital in less developed countries help to reduce global poverty and hence is not only defensible but also morally imperative.

As an example of the way transnational investment in less developed countries increases wages, Norberg points to a Nike shoe manufacturing plant in Vietnam. When it first opened, in the early 1990s, workers walked to the factory; three years later they used bicycles; three years after that they used scooters; and finally they are

© Peter Essick/Aurora Photos

Women sort through electronic waste, or e-waste, in a village outside New Delhi, India. Used and unwanted electronics are often exported to developing countries, where people attempt to extract useable parts and scrap metals. E-waste leaches toxic chemicals into the environment and the methods of extraction are hazardous to workers.

beginning to use cars. Although Norberg differs from many other proponents of globalization in favouring the free movement of people as well as goods, he generally maintains that improvements in food supply, education, equality, human rights, and gender issues show that globalization is making the world a better place for most people.

To the extent that global capitalism increases opportunities and wealth for the world's poor, it is obviously a positive thing. The problem, of course, is that globalization is far too complex for us to be sure that all its consequences will be so benign. As is so often the case, reasonable people may take very different positions and many will come down somewhere in the middle, recognizing the benefits that globalization can bring but also aware of the constant effort required to prevent or mitigate adverse consequences. Box 3.6 is a critical discussion of the capitalist mode of production.

Examining the Issues

BOX 3.6 | Capitalism Is Good and Bad

Of course, the key idea of capitalism—that making money is a good thing—was around long before the rise of capitalism, sometimes viewed favourably and sometimes not. Some early Jewish and Buddhist stories noted the high status of merchants, while Hinduism, Confucianism, and Christianity often looked unfavourably on commerce. Once in place as a social and economic system (a mode of production, using Marxist terminology), it proved highly successful at managing national economies through market forces rather than through state control.

Widely acknowledged as imperfect because of frequent evidence of corporate power abuse and the creation of social and economic inequalities (which became all too apparent with the 2008 global recession), capitalism seems to be both a flawed system and, perhaps, the best system available for managing economies. John Maynard Keynes, a central figure in the creation of the Bretton Woods system and in the development of the Western welfare state, famously observed:

> For my part, I think that Capitalism, wisely managed, can probably be made more efficient for attaining economic ends than any alternative system yet in sight, but that in itself it is in many ways extremely objectionable. Our problem is to work out a social organisation which shall be as efficient as possible without offending our notions of a satisfactory way of life. (Keynes, 1926: 52–3)

Capitalism has undergone several changes since Keynes wrote these words, but there is little evidence that the changes have ameliorated the system's negative social and economic components. After World War II, **competitive capitalism** (as it is now usually called) changed considerably as a result of the rapid growth of major (often transnational) corporations and the state's increased involvement in the economy (often through public ownership). This is known as **organized capitalism**. Most recent evidence suggests that a further transformation is underway: **disorganized capitalism** refers to a new form characterized by a process of disorganization and industrial restructuring. The transition from organized to disorganized capitalism (also referred to as the transition from **Fordism** to **post-Fordism**) is in part a reflection of the malfunctioning of capitalism suggested by economic instability, social injustice, poverty, and unemployment. The significance of these transitions is evident in the accounts of economic restructuring in Chapters 10–13.

competitive capitalism
The first of three phases of capitalism, beginning in the early eighteenth century; characterized by free-market competition and laissez-faire economic development.

organized capitalism
The second phase of capitalism, beginning after World War II; increased growth of major corporations and increased state involvement in the economy.

disorganized capitalism
The most recent form of capitalism, characterized by disorganization and industrial restructuring.

Fordism
A group of industrial and broader social practices introduced by Henry Ford, including the mass-production assembly line, higher wages, and shorter working hours.

post-Fordism
A group of industrial and broader social practices evident in industrial countries since about 1970; involves more flexible production methods than those associated with Fordism.

Marx argued that another of capitalism's major failings is its dehumanizing effect on individuals living within it, an effect that he termed **alienation**. Because they sell their labour and do not control the means of production, workers lack control over their own lives. At the same time, the capitalist state uses democracy and guarantees of individual human rights to legitimize the maldistribution of political and economic power. It has even been suggested that the state evolved to legitimize capitalism and to prevent substantial popular opposition to the circumstances of alienation (Johnston, 1986: 176). As capitalism has spread across the globe, the people of states at all levels—core, semi-periphery, and periphery—have experienced alienation. Although capitalism is most detrimental in the periphery and least in the core (since the core's living standards are highest), the basic effects are the same.

alienation
The circumstance in which a person is indifferent to or estranged from nature or the means of production.

The concept of alienation is central to many discussions of global problems. For example, the alienated individual can no longer interact directly with the natural world; all our relations with that world are in some way organized by forces that are not part of us or of nature. It also is important in everyday life, with people experiencing frustration, powerlessness, and an overriding sense of helplessness, circumstances most fully developed within existentialist philosophy.

So, is capitalism good or is it bad? Many critics hedge their bets, arguing that capitalism is sick but disagreeing as to whether the illness is curable or terminal. More usefully, perhaps, judgments as to whether capitalism is good or bad often conclude that it is the best of a set of unsatisfactory options.

Increasing Participation in Economic Decision-Making

The 2008 and 2009 G20 summits held in Washington and London, respectively, were perhaps another striking sign that globalization is changing the established economic order in a positive way. These global economic summits of the leaders of the world's largest capitalist democracies, which were initiated in the mid-1970s, used to involve the G7 countries (the US, UK, France, Germany, Italy, Japan, and Canada), then the G8 (the G7 and Russia). The fact that a meeting of 20 countries—including China, India, Brazil, and Saudi Arabia—was held indicates that a changing cast of characters is responsible for discussing global economic issues. Indeed, increasing evidence shows that what are often called emerging national economies are playing larger roles in the global economy. In 2008 the *Fortune 500* list of companies included 62 from Brazil, Russia, India, and China (sometimes called the BRIC economies).

However, in some instances, the consequences of change are less clear. In late 2005, the EU agreed to reduce the subsidies paid to Europe's sugar farmers in accord with demands from the WTO after formal complaints from Australia, Brazil, and Thailand (the EU had been paying producers three times the world price). In principle, reduction of subsidies will benefit poor countries that produce sugar, but the situation is more complicated because 18 former colonies of Europe—countries such as Fiji, Barbados, and Mauritius—also had access to the subsidies.

Reducing Poverty?

Many supporters of globalization acknowledge problems but remain convinced that the more general trend is one of trade liberalization leading to rising incomes and longer-term development for the world's poor. Two pieces of evidence in favour of this claim are that, in 2005, the emerging economies of the world (not the rich countries) produced a little more than 50 per cent of world output measured at purchasing-power parity and also accounted for more than 50 per cent of the increase in global GDP. We can interpret these statistics as indicative of the biggest shift in economic strength since the rise of the United States about 100 years ago. In short, more and more countries are industrializing.

Enhancing Mutual Respect?

It can be argued that cultural globalization is emancipating. If Western values are spreading around the world, they include the basic ideals of liberal democracy, including freedom of speech and freedom of cultural expression. If we are free to be what we wish to be, we might not choose either to be the same as others or to separate ourselves from them. Perhaps we will choose to value both our own personal rights and freedoms and those of others, including the inalienable right to improve their economic circumstances. Our future—wherever we are—may well be one of greater pluralism, more choices, and (most important) enhanced mutual respect. These comments, which are in close accord with the 1948 United Nations Universal Declaration of Human Rights, offer a more hopeful counterpart to possible negative consequences of cultural globalization.

CONCLUSION

Space is shrinking, time is being compressed, and national borders are becoming less important than they used to be. These transformations are particularly evident in four recent developments.

- New markets have arisen: global markets in services; global consumer markets with global brands; and deregulated, globally linked financial markets.
- New participants have emerged: most important among these are transnationals, the WTO, and the several regional trading blocs; there have also been numerous corporate mergers and acquisitions.
- Faster and less costly methods of communication have developed: the Internet and smartphones facilitate the transmission of information, and there are also ongoing technological advances in the movement of goods by air, rail, and road.
- New policies and practices are being put into place concerning matters such as human rights and environmental issues.

AP Photo/Kin Cheung

Local and mainland Chinese university student protesters boycott the iPad outside an Apple retailer in Hong Kong. Foxconn, the Taiwanese company that manufactures the Apple iPad, provides jobs to hundreds of thousands of Chinese workers, but the company has also been plagued by accusations of poor working conditions.

But do all these ongoing changes mean that the world we live in, the lives we live, and the landscapes we occupy are changing? Specifically, do we all live in one global village? Marshall McLuhan famously spoke of such a village as early as the 1960s and, as we have seen, geographers have long recognized that our increasing technological ability to overcome distance has resulted in what we might call a "shrinking world," one in which places are closer together in terms of both space and time. Most of us live in one place, but through travel and broadcast and telecommunications media we are increasingly aware of many other places. Consequently, our experiences are a strange mix of near and far. Many of us are also finding that our social relations are being stretched, with face-to-face interactions less important as many of our contacts are with those who are physically distant.

And yet, local places and local times continue to matter. The distinguished human geographer David Harvey (1990: 427) wrote:

> The more global interrelations become, the more internationalized our dinner ingredients and our money flows, and the more spatial barriers disintegrate, so more rather than less of the world's population clings to place and neighborhood or to nation, region, ethnic grouping, or religious belief as specific marks of identity. . . . Who are we and to what space/place do we belong? Am I a citizen of the world, the nation, the locality? Not for the first time in capitalist history . . . the diminution of spatial barriers has provoked an increasing sense of nationalism and localism, and excessive geopolitical rivalries and tensions, precisely because of the reduction in the power of spatial barriers to separate and defend against others.

Reflecting the skeptical thesis, Harvey recognizes that there is a second way of thinking about globalization. Instead of making the world more and more homogeneous, perhaps it is reinforcing the distinctiveness of local places and identities.

Certainly, globalization does not appear to signify the end of diversity or the imposition of a single global culture. We will elaborate on this idea in Chapters 7, 8, and 9.

Globalization is a reality but by no means an uncontested one. We may think of the world as a global village, but the truth is that most of its people are not yet villagers.

Summary

A Discipline in Distance?

The spatial location of geographic facts is not random: humans typically choose to minimize movement in order to minimize the frictional effects of distance. But this long-standing impact of distance is of less importance today because of a collection of globalization processes.

Transport Modes and Systems

Transport systems are both the cause and effect of other economic aspects of landscape. Major changes are associated with political expansion and technological advances. In the case of Britain, the most dramatic changes came with the Industrial Revolution: turnpike roads, navigable waterways, railways, and an improved road network. Ongoing changes in transportation, including improved networks and containerization, are a principal component of globalization.

Trade

Although trade is clearly related to the distance between locations, the three principal branches of trade theory all exclude any reference to distance. Trade is also related to resource base, labour force, and capital. Much world trade is affected by WTO policies and by various agreements between countries concerning regional economic integration.

Transnationals

Production is no longer contained by national boundaries, a change brought about primarily by the rise of transnational corporations that operate in two or more countries. Transnationals are able to organize movements of information, technology, and capital between countries and, to a considerable extent, site production and profit in countries with relatively low wages and taxation.

Transmitting Information

Perhaps the most significant technological changes facilitating economic globalization are those that reduce, or even eliminate, distance through the almost instantaneous transmission of information. ICTs now play a key role in many aspects of our lives, although access to these technologies mirrors the larger distinction between more and less developed worlds, thus leading to a digital divide. Numerous and ever-changing forms of social media have introduced new ways of interacting.

Understanding Globalization

What globalization is, and is not, is regularly debated. Three possible versions are introduced, and one way of measuring the extent of globalization in its economic, social, and political forms is described.

Economic Globalization

Since 1945 trade has increased much more rapidly than production—a clear indication of how much more interconnected the world has become. It is possible that a global geo-economy with three dominant regions (Europe, North America, and East and Southeast Asia) is emerging.

Evaluating Globalization

Although the reality of economic globalization seems clear, the details of its impact are contested. On the one hand, a new global economy is emerging in response to the internationalization of capital, production, and services; on the other hand, there are several major examples of regional integration and associated regional protectionism. According to some observers, trade liberalization is leading to rising incomes and longer-term development for the world's poor. However, one thing is certain: globalization is meeting considerable opposition from a variety of groups.

 Links to Other Chapters

- **Globalization:** Chapter 2 (concepts); Chapter 4 (global environmental issues); Chapter 5 (population growth, fertility decline); Chapter 6 (refugees, disease, more and less developed worlds); Chapter 8 (popular culture); Chapter 9 (political futures); Chapter 10 (agricultural restructuring); Chapter 11 (global cities); Chapter 13 (industrial restructuring)
- **Distance and space:** Chapter 2 (concepts); Chapters 7 and 8 (cognitive and social distance in relation to cultural regions and landscapes); Chapters 10, 11, 13 (time and economic distance in relation to land-use theories)
- **Technologies of transportation:** Chapter 13 (industrial location)
- **Regional integration and globalization:** Chapter 9 (groupings of states); Chapters 10, 11, 12, 13 (economic restructuring)
- **Transnationals:** Chapter 4 (environmental impacts); Chapter 13 (industry)
- **Social media:** Chapter 8 (contesting identities and places)
- **Economic globalization:** Chapter 10 (agriculture); Chapter 11 (world cities); Chapter 13 (outsourcing and offshoring)
- **Cultural globalization:** Chapters 7 and 8 (identities and conflict)
- **Political globalization:** Chapter 9 (political futures)

 Questions for Critical Thought

1. As globalization continues to transform the human world, can we expect fundamental geographic concepts such as space and distance to become less significant as determinants of human spatial behaviour? Why or why not?
2. Why is diffusion such an important concept in geography? How has globalization influenced diffusion processes?
3. What factors will shape globalization and international relations in the next century?
4. What are the advantages and disadvantages of economic globalization? Are you, personally, better or worse off as a result of economic globalization? Do we all live in one global village?

 Suggested Readings

Visit the companion website for a list of suggested readings.

4 HUMANS AND ENVIRONMENT

Most human impacts on the earth are the result of efforts to improve our well-being. Yet these efforts sometimes have the opposite effect. To understand our impacts, we need to consider two facts. First, everything in nature is related. For example, clearing forest to practise agriculture necessarily has many unintentional environmental consequences. Second, human impacts increase through time because of a growing population, technological advances, increased use of resources, and the desire to improve our quality of life.

This chapter discusses natural resources, energy use, technology, and environmental ethics from a global perspective and outlines specific global, regional, and local human impacts on the environment—on vegetation; on animals; on land, soil, and air; on water; and on climate. Finally, differing perspectives concerning the state of the world today are debated.

Here are three **points to consider** as you read this chapter.

- The idea that we are entering a new geological epoch, the Anthropocene, is gaining currency. As you read this chapter about human impacts, are you inclined to agree with this claim?
- In Western culture, the dominant way of thinking about humans and nature has typically involved an anthropocentric (human-centred) world view. What evidence suggests that an ecocentric (environment-centred) world view is assuming prominence?
- Our discussion of the earth's current state raises a provocative question that will resurface in various forms in later chapters: Is a global disaster looming or will a different way of thinking about humans and environment, human ingenuity, and technology solve our problems?

A farmer burns land near Mantasoa, Madagascar. The farming technique known as slash and burn destroys forest or woodland plants so that the land can be used as fields. The practice, primarily used in areas of subsistence farming, has many environmental costs, including desertification and destruction of wildlife habitats.

Human history is a history of impacts on the natural environment; we must have an impact on land, air, and water to survive. Only recently have we realized that some of our impacts actually threaten our continued survival. Here are five fundamental driving forces.

1. Small, often insignificant, changes to the environment can have major impacts if they are repeated enough. Arable and pastoral activities can lead, over time, to major environmental problems.
2. Technological developments related to energy demands continually change the environment.
3. The lifestyles promoted by technological advances also work to change the environment.
4. Increasing human populations are a threat to the environment.
5. Increasing connections between different regions of the globe mean that human activities that used to have merely local or regional consequences are now more likely to be global in their impact. The most obvious example is global warming.

For most of the time that humans and their ancestors have been using the earth, their impacts on the planet have been slight because technologies were limited and populations were small. Technologies such as stone tools and the use of fire and a total human population of about 4 million as recently as 12,000 years ago meant that the few environmental changes brought about by humans were temporary and restricted to a local scale. With a substantial array of agricultural and industrial technologies and a population of over seven billion, today's situation is vastly different. Our interactions with and impacts on the environment are numerous, relatively permanent, and frequently on a global scale.

Our impacts are so significant that many scientists contend that we no longer live in the **Holocene** (recent) geological epoch that began at the end of the most recent Ice Age, about 12,000 years ago, but have entered the Anthropocene (human) epoch. The date of the first nuclear test detonation, 16 July 1945, has been suggested as the outset of this new geological epoch. The idea of the Anthropocene means that we have become a force of nature, transforming the planet on a geological scale. For this contention to be correct, human impacts on earth must show up, in due course, in the geological record. For example, burning fossil fuels causes ocean acidification; at some point, this process will prevent coral reef formation, a circumstance that will register in the geological record. More specifically, in 2011 a leading scientist suggested that coral reefs will be the first ecosystem entirely eliminated by human activity (by 2100). Humans are also causing forests to disappear, deserts to expand, rivers and lakes to dry up, and climate to change. The significance of the Anthropocene is difficult to exaggerate: the future of our planet is at least partly in our hands.

Holocene
Literally, "wholly recent"; the post-glacial period that began 10,000 years ago and was preceded by the Pleistocene.

system
A set of interrelated components or objects linked together to form a unified whole.

Our ability to collect and analyze information about current human impacts is greatly enhanced by technologies such as remote sensing and GIS (introduced in Chapter 2). Different wavebands of the electromagnetic spectrum can be used to gather information on particular land covers, such as areas under crop, pasture land, forest, and wetlands. It is possible to measure such ecosystem properties as soil surface moisture and temperature, to map soils, and to "sense" biogeochemical cycles. Much of the detailed information in this discussion of human impacts is derived from remote sensing by satellite and analyzed in a GIS.

A GLOBAL PERSPECTIVE

A global perspective is employed because everything is related to everything else; one cannot change an aspect of nature without directly or indirectly affecting others. In principle, then, human activity in any single area has the potential to affect all other areas; in practice, the evidence that it does so is overwhelming. One way to approach these issues is to consider the question of how a civilization survives. Smil (1987: 1) answered, "It survives by harnessing enough energy and providing enough food without imperilling the provision of irreplaceable environmental services. Everything else is secondary."

In exploring how harnessing energy and producing food have changed our environment, and whether the changing environment threatens human survival, many human geographers find that the most valuable concepts are those of systems and ecology.

Systems, Ecology, and Ecosystems

In theory, the concept of the **system** (sets of interrelated parts) is simple and widely applicable. It is therefore surprising that it has not been more widely used in studies of the relationship between humans and land. The attraction of the systems concept is its ability to describe a wide range of phenomena and offer a simplified description of what is usually a complex reality. Descriptions of systems typically focus on distinctions and relationships between the parts of the system. Open systems interact with elements outside the system, necessitating the study of inputs and outputs of energy and matter. Closed systems—lacking inputs and outputs—are less common. Finally, relationships between a system's parts or between a system and some external elements often are described as feedbacks. Positive feedback reinforces some change; negative feedback counters some change.

Ecology is the study of organisms in their homes. The concept of an **ecosystem**, integrating systems and ecology, was formally developed by English botanist A.G. Tansley in 1935. Ecosystems can be identified on a wide range of scales. The global ecosystem (also known as the ecosphere or biosphere) is the home of all life on earth. It is a thin (about 14 km/8.5 mile) shell of air, water, and soil. Ecosystems can be identified within the larger ecosphere: any self-sustaining collection of living organisms and their environment is an ecosystem. Thus, the ecosystem concept refers to distinct groupings of things and the relationships between them.

On the global scale, the three basic parts of the ecosystem are the atmosphere (air), the hydrosphere (water), and the **lithosphere** (land). This global ecosystem has one key input: the sun—the source of the energy that warms the earth, which provides energy for photosynthesis and powers the water cycle to provide fresh water. In order to be self-sustaining, our global ecosystem depends on the cycling of matter and the flow of energy. Figure 4.1 outlines the cycling of critical chemicals and the one-way flow of energy. Matter must be cycled; the law of conservation of matter states that matter cannot be created or destroyed, only changed in form. Energy flows through the system because of the second law of thermodynamics: energy quality cannot be recycled. We call the circular pathways of chemicals biogeochemical cycles (*bio* refers to

ecology
The study of relationships between organisms and their environments.

ecosystem
An ecological system; comprises a set of interacting and interdependent organisms and their physical, chemical, and biological environment.

lithosphere
The outer layer of rock on earth; includes crust and upper mantle.

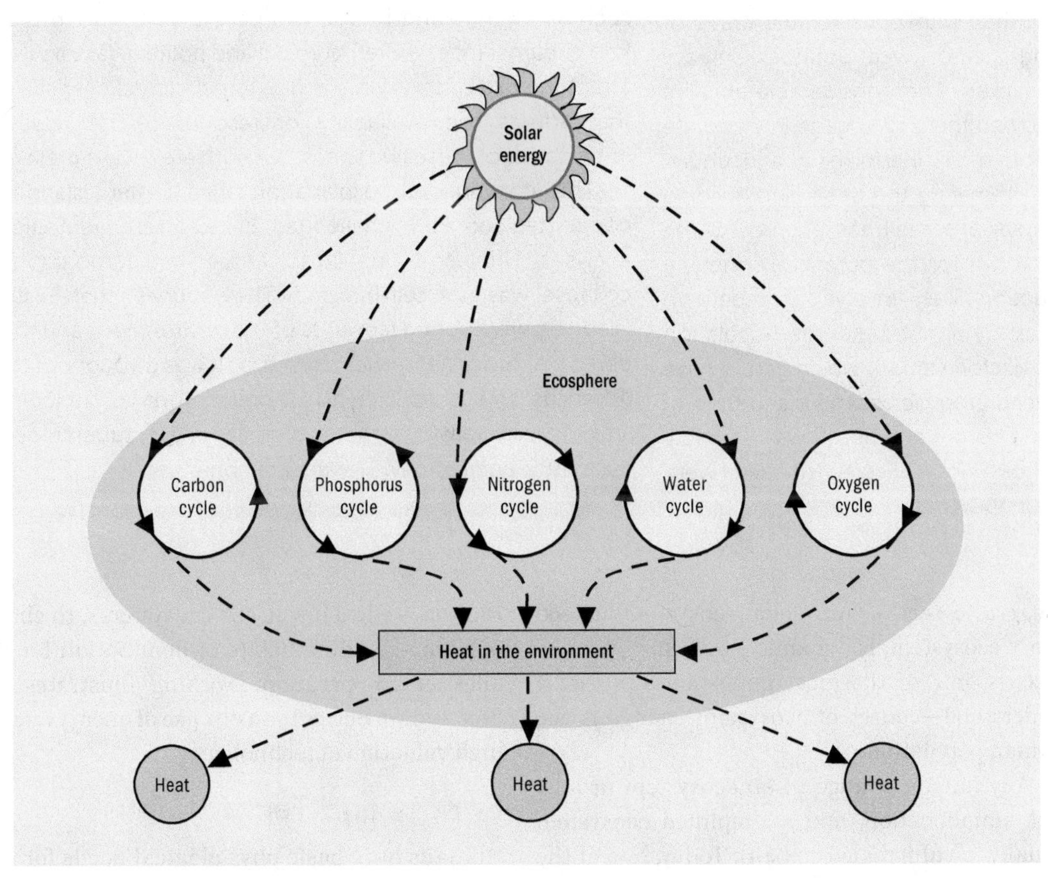

FIGURE 4.1 | Chemical cycling and energy flows

Around the Globe

BOX 4.1 | Lessons from Easter Island

One of the most remote inhabited places on earth, Easter Island is situated in the Pacific Ocean, 3,747 km (2,328 miles) off the coast of South America and 2,250 km (1,400 miles) southeast of Pitcairn Island, the nearest inhabitable land. When Europeans first visited the island on Easter Sunday 1722, they encountered a population of about 3,000 living in a state of warfare, with limited food resources and a treeless landscape (Ponting, 1991: 1–7). They also saw evidence of a once-flourishing earlier culture: between 800 and 1,000 huge stone statues, or moai. Because visiting Europeans believed that these statues could not have been carved, transported, and erected by the impoverished local population, it was long assumed that Easter Island must have been visited by a more culturally and technologically advanced group.

There is a simpler but more disturbing answer to this "mystery" (Diamond, 2005). Radiocarbon dating of evidence suggests that Easter Island was settled by the seventh century, possibly as early as the fifth; the first settlers were likely Polynesians, not South Americans. They arrived at an island with few species of plants and animals and limited opportunities for fishing but with considerable areas of woodland. Their diet consisted of sweet potatoes (an easy crop to cultivate) and chicken. The considerable amount of free time available allowed them to engage in elaborate rituals and construct the moai. Performing agricultural activities, cooking food, cremating the dead, and building canoes required the removal of some trees, but most of the deforestation was carried out for the purpose of moving statues. Statue construction likely involved competition between different groups. By about 1550, the population peaked at roughly 7,000. Deforestation, which would have led to soil erosion, reduced crop yields, and a shortage of building materials for both homes and boats, was probably complete by 1600. Without wood, Easter Islanders were unable even to build canoes to catch porpoises (their principal source of protein) or to escape the island.

It appears that total deforestation initiated the collapse of the Easter Island culture and economy. Bahn and Flenley (1992: 212–13) discuss the significance of this situation:

> We consider that Easter Island was a microcosm that provides a model for the whole planet. Like the Earth, Easter Island was an isolated system. The people there believed that they were the only survivors on Earth, all other land having sunk beneath the sea. They carried out for us the experiment of permitting unrestricted population growth, profligate use of resources, destruction of the environment and boundless confidence in their religion to take care of the future. The result was an ecological disaster leading to a population crash.

Diamond (2005) makes the point that the collapse of many ancient cultures, as well as more recent examples such as Rwanda and Darfur, can be explained in terms of two circumstances—internal social and political factors that affect the use of resources and external climatic variability that affects the availability of resources. The relative importance of these two varies, but both are always present.

The conventional explanation that Easter Islanders committed ecocide is challenged. Based on archaeological analyses, Hunt and Lipo (2011) argue that demographic collapse was not self-inflicted. They suggest that Easter Islanders were sound stewards of the environment and that demographic collapse was caused, as it was throughout the Americas, by susceptibility to European diseases, including smallpox, sexually transmitted diseases (STDs), tuberculosis, dysentery, and leprosy. The debate continues.

life, *geo* to earth). Our global ecosystem, indeed every ecosystem, is dynamic. As we are about to discuss, one of the most important—and least understood—causes of ecosystem change is the human population.

Any human change to an ecosystem usually is a simplification, and a simplified ecosystem usually is vulnerable (Box 4.1). Today, one of the most urgent issues facing us is the need to change the way we live inside our ecosystems, to change our long-standing habits of domination for new ones of co-operation. Nothing illustrates this domination better than our use of energy and our high valuation of technology.

Energy and Technology

Humans have basic physiological needs for food and drink; we also have a host of culturally based wants that appear to have no upper limit. Needs and wants are satisfied largely as a result of humans' using their **energy** (the capacity to do work) to harness other forms of energy. The more successfully we can utilize other energy, the more easily we can fulfill our needs and wants and the more profoundly we can affect the environment. We become aware of other energy sources and acquire the ability to use those sources through the development of new technology.

All forms of **technology** represent ways of converting energy into useful forms. An important technological advance was the human use of fire to convert inedible plants into an edible form for human use. In domesticating plants, humans gained control over a natural energy converter: the plants that convert solar energy into organic material via photosynthesis. Similarly, in domesticating animals, humans took control over another natural converter: the animals that change one form of chemical energy (inedible plants, usually) into another form usable by humans (such as animal protein). In this sense, the **domestication** of plants and animals, sometimes known as the **agricultural revolution**, is an example of how humans have used technology to tap new energy sources.

Between the agricultural revolution (that began approximately 12,000 years ago) and the eighteenth century, many other technological changes permitted increasing human control of energy sources: new plants and animals were domesticated and new tools and techniques were invented. In addition, three new energy converters—the water mill, the windmill, and the sailing craft—were developed to utilize the energy in water and wind. But with the **Industrial Revolution** of the eighteenth century, most particularly the invention of the steam engine, humans began using inanimate converters on a large scale to tap into new energy sources, specifically, coal in the second half of the eighteenth century, oil and electricity in the second half of the nineteenth century, and nuclear power in the middle of the twentieth century.

This brief account raises two key points. First, energy resources such as coal and oil are not literally "new." They are fossil fuels, formed from ancient organic matter and non-renewable. Second, we are using these resources in ways that are often harmful to our ecosystem.

Natural Resources and Human Values

Humans continually evaluate physical environments. As human culture (especially technology) changes, so do those evaluations. Thus, something becomes a resource only if humans perceive it as technologically, politically, economically, or socially useful. Different groups do not necessarily agree on what is a resource. Some people may see an area of wetland in the Prairies as a valuable scientific or recreational landscape; others will see it as potential farm or building land. Different interest groups use different evaluation criteria and hence may hold radically different views.

Traditionally, geographers have divided resources into two types. **Stock resources** include all minerals and land and are essentially fixed, as they take a very long time to form (by human standards). **Renewable resources**, such as air and water, are continually forming. This simple distinction is generally useful, but it blurs some key issues. Numerous resources lie somewhere between the two extremes, and continuing availability depends on how we manage those resources. Obvious examples include game populations for hunting economies and fish populations for much of the contemporary world. In both cases, there may be a perceived need for conservation but also some compelling reasons to continue depleting the resources. In Canada, the east coast fishery has been especially susceptible to controversies over such issues.

Renewable Energy Sources

Globally, fossil fuel sources (oil, coal, and natural gas) comprise 87 per cent of global energy consumption. Chapter 13 includes a discussion of the production, movement, and consumption of these energy sources. Renewable energy sources, including hydroelectric, nuclear, solar, wind, geothermal, and biofuels, account for the remaining 13 per cent. In parts of the less developed world, traditional **biomass** sources such as wood, crop waste, and animal dung are an important energy source.

The International Energy Authority (IEA) predicts an approximate 50 per cent rise in global energy demand by 2030. One of the most asked questions today is "What will happen if the oil, gas, and coal supplies cannot satisfy demand?" Nobody knows when non-renewable energy

energy
The capacity of a physical system for doing work.

technology
The ability to convert energy into forms useful to humans.

domestication
The process of making plants and/or animals more useful to humans through selective breeding.

agricultural revolution
The slow transition, beginning about 12,000 years ago, from foraging to food production through plant and animal domestication.

Industrial Revolution
The process that converted a fundamentally rural society into an industrial society, beginning in England around 1750; primarily a technological revolution associated with new energy sources.

stock resources
Minerals and land that take a long time to form and hence, from a human perspective, are fixed in supply.

renewable resources
Resources that regenerate naturally to provide a new supply within a human lifespan.

biomass
The mass of biological material present in an area, including both living and dead plant material.

sources will run out because, although demand increases all the time (particularly with the rise of China as an industrial power), new reserves are being discovered and new technologies are being developed to make other known reserves, such as the Alberta oil sands, profitable to exploit. Despite the uncertainties, the question is pressing. To put energy use in context, consider that human "progress" has always been tied to increasing energy use, from the use of fire to agricultural domestication, the Industrial Revolution, and the use of fossil fuels.

The Need for Renewable Energy

Decreasing our reliance on non-renewable fossil fuels as energy sources and finding ways to generate more energy from renewable sources are currently major technological challenges. There are many reasons to favour the latter (Goodall, 2008; IPCC, 2011). Environmental reasons include global warming (discussed later in this chapter). Political reasons include the pressure exerted on governments by voters who are increasingly aware of environmental issues; the need for national economies to be self-sufficient in energy; and the need to diversify in order to reduce dependence on particular suppliers of energy sources. Economic reasons centre on the fact that oil, natural gas, and coal will become more expensive as supplies become scarcer.

pollution
The release of substances that degrade air, land, or water into the environment.

Many of these arguments first arose during the energy crisis of the early 1970s, although interest waned when energy costs fell again in the 1980s. They have returned to the fore, especially as a result of the dramatic increases in oil prices that occurred in 2007–08 and notwithstanding the price decreases that occurred in 2014–15. Indeed, even those uncomfortable with environmental arguments acknowledge such concerns, as evidenced by the generally well-received book by the conservative pro-globalization American scholar Thomas Friedman (2008). In *Hot, Flat, and Crowded*, he joins those calling for a transformation of energy systems, specifically in the United States, involving a move away from fossil fuels, less use of electricity, and requirements that power companies buy energy from cleaner sources. Such a transformation would comprise federal government intervention, including a new tax regime. Perhaps the most compelling argument came in a report from the UN's Intergovernmental Panel on Climate Change (IPCC), which stated that there needs to be a massive shift from fossil fuels to renewable energy sources (Connor, 2014). Interestingly, natural gas was seen as a crucial bridge to this change.

At present, renewable energy sources are typically tapped on a relatively local scale, with the main potential sources being water (rivers, waves, and tides), sun, wind, and geothermal energy (the natural heat inside the earth). Nuclear power can also be considered renewable, since uranium reserves are expected to last about 1,000 years if an efficient process is used. (However, nuclear power is not considered a renewable source in an IPCC [2011] report on renewable energy sources and climate change mitigation.) Finally, some crops are being used to produce biofuels as an alternative to gasoline.

Hydropower

Global output of hydroelectric power is increasing annually, especially because of new capacity in China. Many countries, such as Norway, India, New Zealand, and Canada, as well as countries in South America and Africa, rely significantly on hydroelectric power. This resource is generated by extracting energy from moving water. To date, the principal technology involves damming rivers. In addition to being a renewable source, hydroelectric power is very inexpensive (once dams and generating stations are constructed) and, in principle, creates no waste or **pollution**.

Many large dams have been constructed in parts of the less developed world, often with the financial support of the World Bank. Although such projects make effective use of a renewable energy source, they can cause serious damage to the environment. The massive Three Gorges project in China was refused World Bank funding, partly because of the anticipated human and environmental impacts. Nevertheless, the raising of the world's third largest river, the Yangtze, began and the dam, the largest in the world, opened in 2006. Between 1 and 2 million people were displaced as 13 cities, 140 towns, and 1,350 villages were submerged, while evidence of the brutal crushing of protests and of officials pocketing funds intended for resettlement projects revealed further misery. Many ancient fortresses, temples, and tombs were submerged; it is also feared that the reservoir may become a giant cesspool filled with sediment washed down from the surrounding deforested mountains. As early as 2007, these fears appeared to be becoming reality. Four years later, a range of environmental problems, such as landslides, and social issues, such as a lack of jobs for the displaced, were officially acknowledged. Another Chinese megaproject, the principal purpose of which is to divert water from the south to the north, is subject to similar criticisms.

Other major ongoing Asian projects include a dam being constructed in Laos, which is expected to help lift much of the population out of poverty, and a series of five dams planned for the Salween River, along the Burma–Thailand border. In Brazil's Amazon Basin, the huge Belo Monte project is expected to be the third largest in the world (after the Three Gorges and the Itaipu, located on the Brazil–Paraguay border).

Generating power through tidal and wave movement is not yet well developed. Necessarily, tides can be used only in those parts of the world that have a sufficiently high tidal range, making the locations of potential production limited. That said, many projects are in the development or planning stages. The Bay of Fundy, between New Brunswick and Nova Scotia, has an exceptionally high tidal range and is a promising site for development. A tidal and wave energy project that will supply all the electricity needs for the 3,500 inhabitants of the Scottish Isle of Islay is scheduled for completion in 2020.

Nuclear Power

There are more than 400 nuclear reactors in the world, most of which use the basic nuclear fission process, although future reactors are likely to use the more efficient "fast breed" process. Nuclear power per se does not pollute, but the residue from nuclear power plants is extremely polluting and long-lasting. For many observers, this type of power has not been established as a source of safe and inexpensive energy. Nuclear reactors are heavily used in France, Belgium, and South Korea, but cost may be a very real issue for future expansion. As just one example, a power station being built on the island of Olkiluoto in western Finland was scheduled to open in 2008 but was delayed until 2012, at least partly because the cost doubled. In the United States, power companies tend not to favour nuclear precisely because of such concerns.

AP Photo

Ships emerge from the five-stage lock at the Three Gorges Dam on the Yangtze River. More than 2 km wide and 185 metres high, the dam has created a reservoir extending nearly 650 km into the country's interior. Ocean-going vessels have access to agricultural and manufactured products from a vast region, while the dam's hydropower turbines generate as much electricity as 18 nuclear power plants. The dam is the largest construction project in China since the Great Wall.

Increased use of nuclear power is likely, despite several serious accidents. As discussed in Box 4.2, the most devastating accident was the 1986 Chernobyl explosion in northern Ukraine (then part of the USSR). Following the 2011 nuclear power plant failures in Japan, caused by an earthquake and subsequent tsunami, a vibrant debate was initiated in the Western environmental movement. Some leading figures, such as George Monbiot, argued against the conventional environmentalist position that opposed nuclear power primarily because of the possibility of

In the News

BOX 4.2 | The Chernobyl Nuclear Accident

On 26 April 1986, one of four nuclear reactors at the Chernobyl power station exploded after a failed experiment on whether the cooling system could operate effectively without the auxiliary electrical supply. The accident was particularly devastating because the Soviet-built reactor was not housed in a reinforced concrete shell, as was the usual practice in most countries. The building itself was severely damaged; more important, large quantities of radioactive debris—at least a hundred times more than the atom bombs dropped on Nagasaki and Hiroshima in 1945—were released into the atmosphere.

The UN declared Chernobyl's emergency phase ended in 2007 and urged a move to a recovery phase, helping communities to begin reversing the domino effect of poor health, poverty, and fear. Rather surprisingly, there has been a notable revival of local wildlife—wild horses, boars, wolves, and lynx—and birds have successfully nested in the reactor building. Nevertheless, the need to secure the site is compelling. A huge arch, 257 m across and 108 m high, is being constructed to be placed over the ruins. Completion is expected in 2016.

Figure 4.2 maps the spread of the radioactive fallout as measured one week after the accident. Most was deposited in the immediate vicinity of Chernobyl in Ukraine and in nearby areas of Russia and Belarus, but it affected almost all countries in the northern hemisphere. The specific details shown in the figure reflect wind direction and rainfall.

Determining the human consequences of accidents such as Chernobyl is very difficult. About 350,000 people have been resettled away from the worst affected areas, but another 5.5 million remain. For several years, it was estimated that as many as 7,000 people died and up to 3.5 million suffered from diseases related to the release of radioactive material. More recently, a UN report estimated that, as of 2005, the number of deaths was less than 4,000. However, a 2006 report from Greenpeace claimed that, in the coming years, about 100,000 residents would die of cancer, especially thyroid cancer. The twenty-fifth anniversary of the disaster was marked by mass protests in France and Germany, calling for an end to the use of nuclear power.

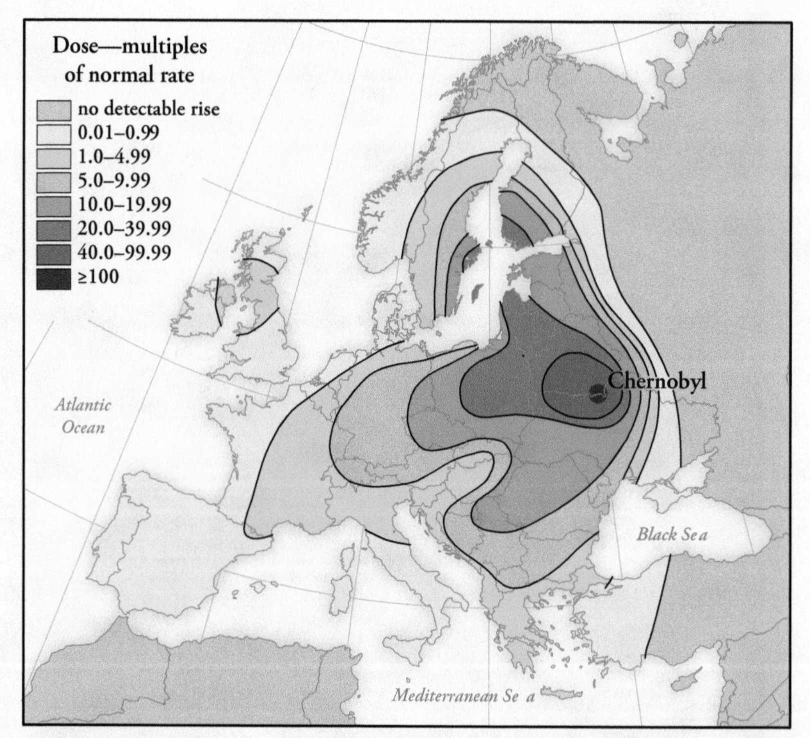

FIGURE 4.2 | Spread of radiation from Chernobyl across Europe, 3 May 1986

Source: Adapted from news.bbc.co.uk/2/shared/spl/hi/guides/456900/456957/html/nn3page1.stm.

major accidents, a view to which they previously subscribed. The argument in favour of nuclear power was twofold: the Japanese experience, it was claimed, demonstrates that even a major accident did not have devastating consequences; and nuclear power is needed as a partial replacement for fossil fuels, which cannot possibly satisfy our energy needs in the next few decades.

Solar Power

Solar power has been widely discussed, and new technologies are being developed that will reduce costs (although it remains expensive). Spanish and German companies are installing large-scale solar power plants in North Africa, a development that requires power transmission over long distances. There is also great potential for the use of solar panels in private homes and businesses. The prospects for solar power production at reasonable costs seem positive.

© Dave Pattinson/Alamy

With 114 turbines, the McBride Lake Wind Farm in southern Alberta generates approximately 235,000 megawatt hours of electricity a year—enough energy to power more than 32,500 homes. Similar projects are underway in many parts of the world, but local responses are not always favourable. In the Lake District of northwest England, for example, opponents claim that turbines are an eyesore in a landscape widely considered to be idyllic.

Wind Power

Several countries are making effective use of wind power and, as with solar power, the prospects of it becoming a significant source are improving. Wind turbines can be used individually but are more typically grouped in wind farms. Denmark is a leader in wind farm technology, and the numbers of such farms in Canada, the United States, and in various European countries are increasing. In Britain, the preferred renewable source is wind.

Although clean and renewable, wind farm technology is not without critics. The sites are often seen as blighting the landscape, especially when located in areas judged to be of outstanding natural beauty, and, when sited in close proximity to human populations, for causing health problems related to the low-frequency sound emitted. Interestingly, critics of this resource include environmentalists who are torn between the need for renewable energy and the need to preserve natural landscapes. More significantly, wind power can be unreliable because wind speeds vary and there is no way to store surplus electricity produced when winds are strong.

Geothermal Energy

Heat from the earth is both clean and renewable. The upper three metres of the earth's surface is typically at about 10 to 16°C, and heat pumps can access this resource to both heat and cool buildings. Some locations contain hot rocks underground that heat water to produce steam, frequently released as geysers. A few kilometres deep, temperatures increase to about 250°C, but the technologies to tap this resource are not yet available. Today, there are geothermal power stations accessing heat sources close to the surface in several countries, including New Zealand, Iceland, Japan, the Philippines, and the United States. The one Canadian location where geothermal energy is being actively pursued is Meager Mountain, a volcanic region in British Columbia.

Biofuels

Recent years have witnessed increasing use of crops to produce biofuels as an alternative to gasoline. Countries such as Brazil and Canada are converting agricultural products to alcohol that can be blended with gasoline; indeed, government regulations in Brazil require that all gasoline sold be mixed with at least 20 per cent ethanol, a by-product of sugar cane. A projection from the IEA suggested that, at best, ethanol will provide 10 per cent of the world's gasoline by 2025.

But biofuels might be a problem rather than part of the solution to climate change. A 2009 report commissioned by the International Council for Science concluded that farming biofuel crops such as corn and canola release enough nitrous oxide (N_2O), a potent greenhouse gas, to negate the benefits of reduced carbon dioxide (CO_2) emissions. Further, it is widely accepted that using crops to produce biofuels instead of using them for human consumption contributed to the 2007 world food price crisis.

The Future for Renewables?

Major oil companies are investing heavily in renewable energy sources. BP, for example, announced its intent to spend US$8 billion over 10 years on the development of such sources. There is, of course, a strong argument for national governments to promote alternatives to fossil fuels. Most notably, Sweden has established a goal of weaning itself off oil by about 2020. This objective is to be accomplished not through constructing any new nuclear power stations but through use of other renewable energy sources. Other countries are less ambitious but heading in the same direction.

How the energy picture will change in the twenty-first century is difficult to predict, although there is optimism that renewable sources (not including nuclear) could, theoretically, supply 80 per cent of global energy needs by 2050. Of course, for this to happen, governments need to actively pursue a full range of renewable technologies. The key point here is that it is not the availability of renewables that is at issue but the public policies that need to be put in place. Certainly, a key problem when discussing future energy sources is the absence of a framework that can inform discussions and assessments of comparative costs and externalities (the costs that occur outside the market, such as pollution).

ENVIRONMENTAL ETHICS

Western Environmental Concern before 1900

Use or abuse? It is not always easy to distinguish the two as they are, after all, relative terms. Yet there is considerable evidence to suggest that we are currently causing damage—perhaps irreparable damage—to our environment. Concerns about the consequences of human activities were raised by the ancient Greeks; Plato noted the detrimental effects of agricultural activities on soil. Despite such early observations, this general question received relatively little attention in the Western world until the eighteenth century. Before that time, geographers were most interested in the earth as a home for humans made by God (teleology) and the land as a cause of human activity (environmental determinism). Significantly, the Europeans' general failure to appreciate the potential dangers of certain human activities appears to have been unique: many other cultures have recognized the necessity of protecting natural "resources." The eighteenth-century origins of Western environmental concern reflected the overseas movement of Europeans, particularly their colonization of tropical areas that, in Europe, had long been regarded as pristine Utopias. It soon became obvious that European activity in those areas was environmentally destructive.

The general question of human impact on the land first received scholarly attention from Buffon (1707–88) in discussions of the contrasts between settled and unsettled areas and of the human domestication of plants and animals. Buffon believed that humans inhabit the earth in order to transform it. Malthus (1760–1834) established the terms of the present debate by focusing attention on the relationship between available resources and numbers of people. On a more specific level, Humboldt, during his travels in South America, explicitly identified lowered water levels in lakes as human impacts and explained that they were caused by deforestation for the purpose of agricultural activities.

Probably the earliest systematic work on human impacts was done by G.P. Marsh (1801–82), an American geographer and congressman. His *Man and Nature, or Physical Geography as Modified by Human Action* (1864, with revised editions in 1874 and 1885) was intended

> to indicate the character and, approximately, the extent of the changes produced by human action in the physical conditions of the globe we inhabit; to point out the dangers of imprudence and the necessity of caution in all operations which, on a large scale, interfere with the spontaneous arrangements of the organic and of the inorganic worlds; to suggest the possibility and the importance of the restoration of disturbed harmonies and the material improvement of wasted and exhausted regions; and, incidentally, to illustrate the doctrine that man is, in both kind and degree,

> a power of a higher order than any of the other forms of animated life, which, like him, are nourished at the table of bounteous nature. (Marsh, 1965 [1864]: 3)

This powerful statement seems more reminiscent of the 1960s than the 1860s. The late nineteenth-century Western world, heavily involved in colonial expansion, became concerned about environmental change only when it had negative impacts on human economic interests. "If a single lesson can be drawn from the early history of conservation, it is that states will act to prevent environmental degradation only when their economic interests are shown to be directly threatened. Philosophical ideas, science, indigenous knowledge and people and species are, unfortunately, not enough to precipitate such decisions" (Grove, 1992: 47).

The Current Debate: Origins

The first sign of a real shift in our appreciation of human impacts on ecosystems came in the 1960s with the publication of Rachel Carson's *Silent Spring* (1962), which reveals the dangers associated with indiscriminate use of pesticides. A few years earlier, a seminal academic work, totalling 1,194 pages and entitled *Man's Role in Changing the Face of the Earth* (Thomas et al., 1956), had come to many of the same conclusions. The decade saw increasing pressure from advocates of wilderness preservation and new scientific evidence about worsening air pollution. Two popular explanations for the environmental "crisis" pointed to the Judeo-Christian belief that humans had been placed on earth to subjugate nature and to the failings of capitalism. The first of these explanations is too simplistic and ignores the complexity of Christian attitudes. The second explanation is one aspect of an ideological approach that stresses the links between different parts of the world: "Clearly there are problems, many of them—and all of them intertwined in the operations of a capitalist world economy, which is hell-bent on annihilating space and place. Those problems are severe now at a global scale, and life-destroying in some places" (Johnston and Taylor, 1986: 9). Our awareness of environmental impacts outside the capitalist world economy, in Eastern Europe and the former Soviet Union prior to the major political changes that began in 1989, should lead us to question such assertions, although the basic idea of interrelatedness is sound.

The Current Debate: Political Overtones

Environmental issues entered the political arena in the early 1970s with the creation in the United States of the Environmental Protection Agency, while the first major international meeting on the subject, the UN Conference on Human Environment, was held in Stockholm in 1972. By 1980 the more developed world was becoming increasingly aware of environmental and related food supply problems in the less developed world, with food shortages in India and droughts in the Sahel region of Africa. A series of disasters and discoveries during the decade ensured that the environment was always in the news. These included the 1984 leak of methyl isocyanate from a pesticide plant in Bhopal, India, that killed perhaps as many as 10,000 and disabled up to 20,000; the 1986 nuclear disaster in Chernobyl; the 1985 discovery of a seasonal ozone hole over Antarctica; recognition of both the rapidity and the consequences of tropical rain forest removal, specifically in Brazil; and, more generally, an increasing concern for numerous local environmental problems.

By the late 1980s the environment was on the national agenda of many countries, as well as the international political agenda. At the national level, green political parties first appeared in West Germany in 1979 and were present in most countries in the more developed world by 1990. Further, many countries have some form of green plan; an encyclopedic survey of the Canadian environment is available as a part of Canada's Green Plan (Environment Canada, 1991).

International agreement is the best way to solve those environmental problems that transcend national boundaries, some of which have global impacts. There have been many calls for the creation of international institutions and policies, most notably by the Brundtland Commission (World Commission on Environment and Development, 1987). Not surprisingly, although most governments agree on the need for international policies, many are unwilling to sacrifice their sovereignty; before they can reach agreements, countries need to work together to resolve their conflicting goals and priorities. Major international developments include the 1987 Montreal Protocol, aimed at the reduction and eventual elimination of chlorofluorocarbons (CFCs), which are one cause of

Examining the Issues

BOX 4.3 | The Tragedy of the Commons or Collective Responsibility?

> Imagine that you and a group of friends are dining at a fine restaurant and that you all have an unspoken agreement to divide the cheque evenly. What do you order? Do you choose the modest chicken entrée or the pricey lamb chops? The house wine or the Cabernet Sauvignon 1983? If you are extravagant, you could enjoy a superlative dinner at a bargain price. But if everyone in the party reasons as you do, the group will end up with a hefty bill. And why should others settle for pasta primavera when someone is having grilled pheasant at their expense? (Glance and Huberman, 1994: 76)

This situation accurately depicts the clash between individual and collective attitudes that, in the context of human use of the environment, was so forcefully put forward by Hardin (1968) as follows:

- A group of graziers use an area of common land.
- They continually add to their herds so long as the marginal return from the additional animals is positive, even though the common resource is being depleted and the average return per animal is falling.
- Indeed, individual graziers are obliged to add to their herds because the average return per animal is falling.
- Clearly, efficient use of the common resource requires restricted herd sizes.
- But individuals will not reduce herd sizes on the common land unless all other group members similarly reduce their herd numbers.
- Hence the metaphor "the tragedy of the commons."

Individual rational behaviour does not result in a collectively prudent outcome when individuals have access to common resources.

Both examples prompt the same question: How do we ensure that individuals act for the common good rather than for personal gain? With reference to the environment, three solutions have been proposed (Johnston, 1992):

1. Resources may be privatized, with the private owners implementing strategies for environmental preservation not available to group owners.
2. The group owners may be able to devise an agreement about the use of the common resource that they are able to implement themselves. According to some recent work in social theory, the success of local and regional recycling programs depends on such group co-operation (Glance and Huberman, 1994: 80).
3. The common resource may be subject to some external control, such as licensing and quota systems for resource harvesters. However difficult it may be to implement, this third proposed solution appears most necessary for ensuring the reduction of deleterious human impacts on the environment.

global warming; the 1992 UN Conference on Environment and Development (the Rio Earth Summit); the 1997 Kyoto Protocol, designed to reduce emissions of greenhouse gases; and the 2002 UN World Sustainable Development Summit, held in Johannesburg. Both of the UN-sponsored meetings were attended by thousands of delegates and various world leaders.

The Current Debate: Three Contentious Issues

Before we address humans' current impacts on the environment, three issues should be identified. The first concerns relationships between the environment and the economy. It is usually argued that market forces are unlikely to solve environmental problems, as market-based decisions rarely consider environmental factors as equal to or above the need to make a profit, even at the national level. More than 20 years ago, a detailed analysis of economic decision-makers in Canada showed that only 6 per cent gave significant consideration to the environment (Gale, 1992). Such findings suggest that the integration of economic and environmental concerns is a major challenge. But these relationships are far from simple. Increasing evidence suggests that economic growth leads to a reduction of environmental problems, as long as growth is accompanied by good governance. The richer a country is and the better it is governed, the more it invests in environmental protection such as cleaning water supplies, reducing pollution, and improving sanitation.

The second contentious issue is that environmental problems are increasingly affecting relationships between countries—not only because of the international implications of many human impacts but also because environmentalists, such as those in North America, are anxious to impose their standards on other countries. Payment is one solution: the 1987 Montreal Protocol included a fund to assist those countries most likely to suffer economically as a result of the agreement. International disapproval is another possible solution: Britain eventually agreed to cease dumping sewage sludge in the North Sea because of the political costs of the dumping. Finally, trade policies are among the few weapons available to one national government to persuade another to amend its environmental behaviour.

The third issue, closely related to the second, concerns the behaviour of individuals as group members (Box 4.3). The ecophilosopher Arne Naess argued that humans need to develop a new **ecocentric** world view recognizing how we are all connected, that we need to work with and not against nature, and that a central goal of human activity is the preservation of ecosystems. This idea is in stark contrast to the **anthropocentric** notion that humans are the source of all value and that land exists for human use, as well as the misconception that energy and other resources are unlimited. *Deep ecology*, introduced by Naess in 1979, is a term sometimes used to describe this viewpoint (see Katz, Light, and Rothenberg, 2000). These concepts have informed many of the ideas of contemporary green movements.

HUMAN IMPACTS ON VEGETATION

When we consider human impacts on environment, it is appropriate to discuss vegetation first, since modification of plant cover results in changing soils, climates, geomorphic processes, and water. "Indeed, the nature of whole landscapes has been transformed by man-induced vegetation change" (Goudie, 1981: 25). Figure 4.3 summarizes some of the consequences of vegetation change. Table 4.1 lists deforestation through time by major world regions, showing that most clearing has occurred since 1850 and that Europe, Asia, and the former USSR were the first regions to experience any considerable impact. It is only relatively recently that the tropical and subtropical areas of Central and South America have been subjected to significant deforestation.

In Europe, large-scale deforestation occurred from about the tenth century. By the eighteenth century, the continent was a largely agricultural region with few remnants of the once dominant forest. For eighteenth-century and later Europeans, cleared land represented progress and the triumph of technology. When Europeans moved to temperate areas overseas, they confronted a very different environment to the one they had left. Eastern North America in particular was densely wooded. Aboriginal populations typically cleared only small areas and then moved elsewhere, allowing the forest to regrow. Among the nineteenth-century British in Ontario, the prevailing attitude towards the forest was antagonistic: "Settlers stripped the trees from their land as quickly as possible, shrinking only from burning them as they stood. They attacked the forest with a savagery greater than that justified by the need to clear the land for cultivation, for the forest smothered, threatened and oppressed them" (Kelly, 1974: 67). Indeed, for the majority of settlers, the forest was a symbol of nature's domination over humans. Deforestation in Ontario proceeded apace as humans established their dominance over the land. By the 1860s, settlers were becoming aware of the disadvantages of deforestation, such as lack of shelter belts, fuel, and building materials. But by then the damage was done.

Today, remote sensing is a versatile and effective tool for monitoring forestry operations. In Canada, Landsat and other imagery allow forest managers to collect data on forest inventory, depletion, and regeneration in areas as small as 2 hectares (5 acres) and with boundary accuracy within 25 metres (82 feet). Such data are invaluable in the accurate mapping of, for example, clear-cut areas.

ecocentric
Emphasizing the value of all parts of an ecosystem rather than, for example, placing humans at the centre, as in an anthropocentric emphasis.

anthropocentric
Regarding humans as the central fact of the world; stressing the centrality of humans to the detriment of the rest of the world.

Introducing Gaia

Shallow and Deep Ecology

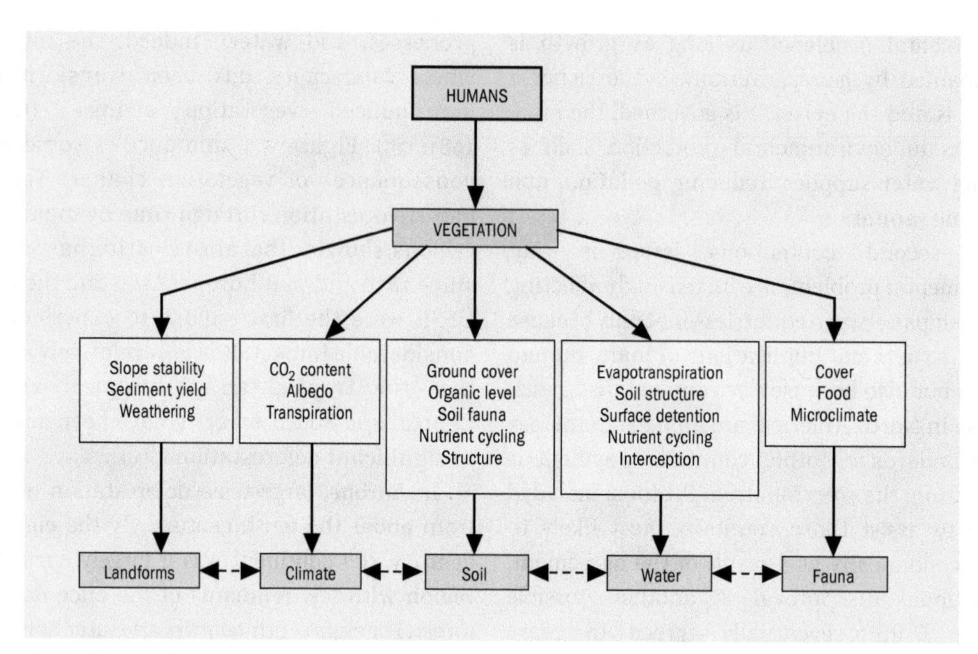

FIGURE 4.3 | Some consequences of human-induced vegetation change

Source: Adapted from Goudie, A. 1981. *The Human Impact: Man's Role in Environmental Change*. Oxford: Blackwell, 25.

Fire

For at least a half-million years and possibly as long ago as 1.5 million years, modern humans and their predecessors, *Homo erectus*, used fire deliberately to modify the environment. Initially, vegetation removal probably resulted in increased animal numbers and greater mobility for human hunters. Fire also offered security and a social setting at night and encouraged movement to colder areas. For later agriculturalists, fire was a key method of clearing land for agriculture and improving grazing areas; it continues to serve these and similar functions. Indeed, deforestation by fire or other means has been prompted largely by the need to clear land for agricultural activities, both pastoral and arable. Together, deliberate burning and the natural fires that are much more common—it has been estimated that lightning strikes some 100,000 times each day (Tuan, 1971: 12)—have drastically modified vegetation cover.

Fire has played a major role in creating some vegetation systems—savannas, mid-latitude grasslands, and Mediterranean scrub lands are prime examples—and has probably affected all such systems except tropical rain forests. Areas

TABLE 4.1 | Global deforestation: Estimated areas cleared (thousands km²)

Region	Pre-1650	1650–1749	1750–1849	1850–1978	Total
North America	6	80	380	641	1,107
Central America	15	30	40	200	285
South America	15	100	170	637	922
Oceania	4	5	6	362	377
Former USSR	56	155	260	575	1,046
Europe	190	60	166	81	497
Asia	807	196	601	1,220	2,824
Africa	161	52	29	469	711
Total	1,254	678	1,652	4,185	7,769

Source: Adapted from Williams, M. 1990. "Forests." In *The Earth as Transformed by Human Action: Global and Regional Changes in the Biosphere over the Past 300 Years*, edited by B. L. Turner, W. C. Clark, R. W. Kates, J. F. Richards, J. T. Mathews, and W. B. Meyer, 180. Cambridge: Cambridge University Press.

significantly affected by fire typically possess considerable species variety.

Plant Domestication

Domestication is a process whereby a plant is modified in order to fulfill a specific human desire; once domesticated, the plant is permanently different from the original. This process is ongoing and is an important part of agricultural research. Associated with plant domestication has been the labelling and removal of plants that are not domesticated as weeds. Once again, such human activity contributes to ecosystem simplification. Early domesticates included wheat, barley, oats (Southwest Asia); sorghum, millet (West Africa); rice (Southeast Asia); yams (tropical areas); potato (Andes); and manioc and sweet potato (lowland South America). Other domesticates include such pulses as peas and beans and such trees/vines as peach and grape.

It is thought that many fruit and nut species, including apples, apricots, plums, cherries, and walnuts, were first domesticated in an extensive forested area in the mountainous landscape of Central Asia—Almaty, the former capital of Kazakhstan, means "Father of Apples." In recent years this area has been subject to extensive deforestation because of overgrazing and other human activities, and there is concern that the loss of original wild species in this biological Eden might threaten the future of these foods, given the uncertainties of climate change.

A Great Reversal?

Without human activity, forests would cover most of the earth's land surface. Large-scale deforestation accompanied the rise of the Chinese, Mediterranean, and Western European civilizations, as well as the nineteenth-century expansion of settlement in North America and Russia, and continues today.

Consider that forests are able to grow both by spreading outwards and by becoming denser. Most research has focused on just the first of

Philip Dearden

Stretching from Prince George in central British Columbia south to central Idaho, the world's largest remaining temperate rain forest includes at least 15 tree species—western red cedar, western hemlock, mountain hemlock, ponderosa pine, Douglas fir, western larch, lodgepole pine, western white pine, subalpine fir, western yew, trembling aspen, paper birch, and three species of spruce—and is a priceless habitat for a wide range of life forms.

In the background of this photo is a hillside that has been clear-cut. Such indiscriminate logging destroys both the trees themselves and the biologically diverse ecosystem that they support. It can also have drastic effects on streams like this one when logging debris accumulates after flooding.

these two measures. But evidence from a major 2011 report suggests that forest density (more and thicker trees) is increasing in many parts of the world after several decades of decline, a change labelled the Great Reversal by the report authors (Rautiainen et al., 2011). Specifically, forest density is thickening in 45 of 68 countries that together account for 72 per cent of global forests. Increasing densities were evident first in Europe, then North America, and then Asia. The increases are uneven but evident in all areas studied. Even in tropical rain forests, discussed below, there is evidence that denser forests are at least partially compensating for a declining areal extent. This dramatic finding has positive implications for carbon capture and therefore for climate change. Also noteworthy is the 2010 Canadian agreement to protect two-thirds of the country's forests (over 72 million hectares) from unsustainable logging. The protected zone extends across Canada from the Pacific to the Atlantic.

Tropical Rain Forest Removal

Notwithstanding the hopeful idea of a great reversal, human removal of tropical rain forest continues to be of major concern. Deforestation is presently concentrated in the tropical areas of the world. Viewed in historical perspective, the current removal of rain forest may not seem excessive, but there are important differences between temperate and tropical deforestation. Tropical forests typically grow on much poorer soils that are unable to sustain the permanent agriculture now practised in temperate areas. Also, tropical rain forests play a major role in the health of our global ecosystem—a fact that has been significantly acknowledged only in recent years. Figure 4.4 shows the past and present distribution of tropical rain forests. These rain forests cover about 8.6 million km² (3.3 million square miles).

The present rate of depletion and the loss to date are both open to dispute. Fortunately, remote sensing using Landsat, together with other satellite data, permits some objective assessment of clearance rates. These data show that annual clearance rates in Brazil and elsewhere were several times greater than the early 1980s estimates made by the UN Food and Agriculture Organization (FAO; Repetto, 1990). On the other hand, the World Bank's 1989 estimate of a 12 per cent loss of Brazilian rain forest was shown to be about twice the actual loss. Although there are still uncertainties about the details of rain

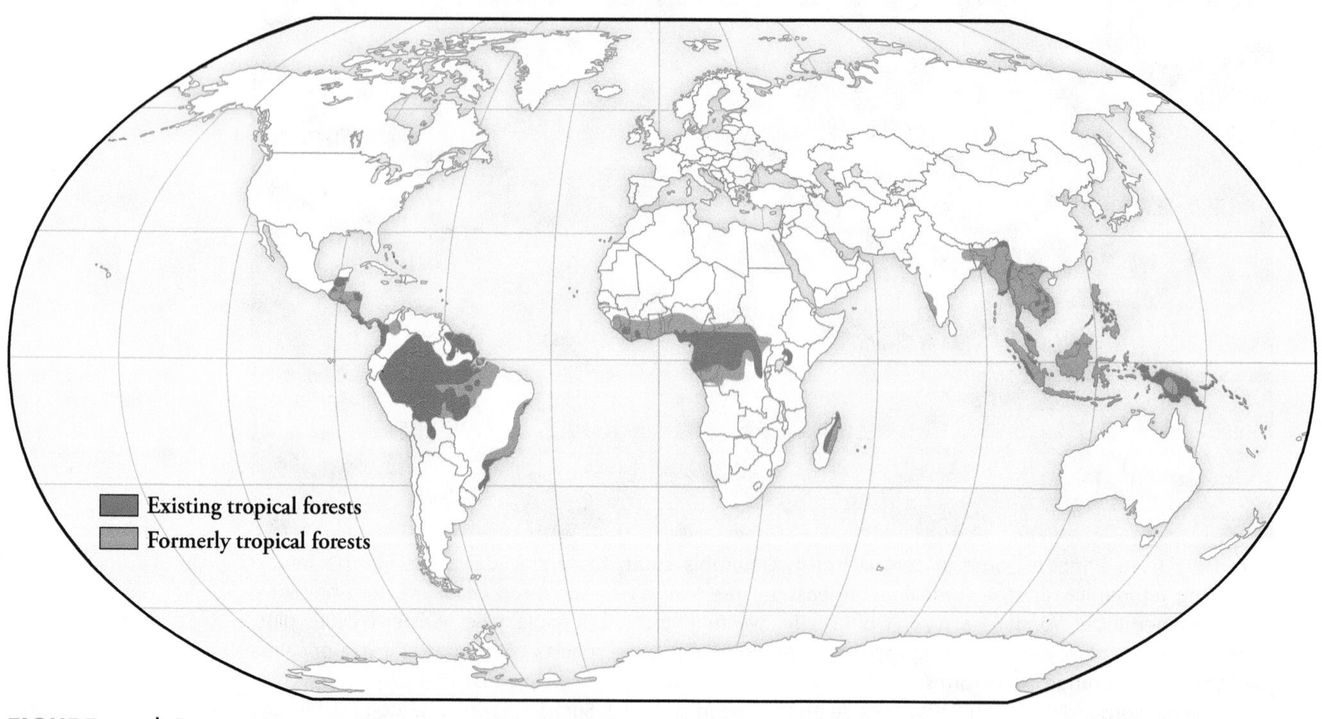

FIGURE 4.4 | Past and present location of tropical rain forests

Source: Murray, W. E. 2006. *Geographies of Globalization*, 328. Copyright © 2006 Taylor & Francis UK. Reproduced by permission of Taylor & Francis Books UK.

forest removal—the FAO recognized in 2008 that it is difficult to demonstrate convincingly that deforestation is occurring—estimates derived from satellite imagery in 2005 suggest annual rates of about 15,500 km^2 (6,000 square miles) due to selective cutting and a similar reduction due to clear-cutting. Indeed, data suggest the rate of depletion is less than 6,000 km^2 per year compared to about 20,000 in the late 1990s. This reduction appears to be mirrored in other countries with tropical rain forest, according to the FAO.

Regardless of the details of tropical rain forest removal, two key questions are "Why are the rain forests being removed?" and "What are the ecological impacts?" As shown in Figure 4.4, rain forests are located primarily in the world's less developed areas, such as Bolivia, Brazil, Colombia, Venezuela, Gabon, DR of Congo, Indonesia, and Malaysia, but the more developed areas are a leading cause of deforestation because of their enormous appetite for tropical timber and the inexpensive beef produced in areas cleared of tropical timber. In addition, poor people in the less developed world use cleared land for some subsistence farming. Often such farming is possible only for a few years because of cultivation techniques that rapidly deplete already poor soils of key nutrients. Cattle ranching also quickly becomes less profitable because rain forest soil supports grazing for only a few short years. Nevertheless, as already noted, the countries experiencing rain forest clearance might view such activity as a means of reducing population pressure elsewhere and as being generally equivalent to the resource and settlement frontier of, say, North America during the past 200 years and Europe during the past 1,000 years.

Rain forest removal has two principal ecological consequences. First, it is a major cause of species extinction because the rain forests are home to at least 50 per cent of all species (the total has been estimated at 8.7 million). From a strictly utilitarian viewpoint, many tropical forest species are (or may prove to be) important to humans as foods, medicines, sources of fibres, and petroleum substitutes.

The second principal consequence involves global warming. Carbon is stored in trees and, when burning occurs, is transferred to the atmosphere as carbon dioxide. Soil is a source of carbon dioxide, methane, and nitrous oxide, all of which are released into the atmosphere as a result of forest removal and farming. Each of these gases contributes to what we call the greenhouse effect—a topic that we will consider shortly.

Desertification

Desertification is land deterioration caused by climatic change and/or by human activities in semi-arid and arid areas: "It is the process of change in these ecosystems that can be measured by reduced productivity of desirable plants, alterations in the biomass and the diversity of the micro and macro fauna and flora, accelerated soil deterioration, and increased hazards for human occupancy" (Dregne, 1977: 324). The significance of desertification lies not simply in the clearing of vegetation but also in the consequences of clearing, which include soil erosion by wind and water and possible alterations of the water cycle. A 2007 UN report estimated that 2 billion people live in drylands susceptible to desertification in sub-Saharan Africa and in Central Asia and that as many as 50 million are in danger of being driven from their homes by about 2017. As is the case with tropical rain forest removal, desertification is a major concern.

desertification
The process by which an area of land becomes a desert; typically involves the impoverishment of an ecosystem because of climate change and/or human impact.

Deserts are natural phenomena, but desertification is the expansion of desert areas. The human causes are complex but typically involve vegetation removal as a result of overgrazing, fuel-gathering, intensive cultivation, waterlogging, and salinization of irrigated lands. The reasons behind these activities are usually population pressure and/or poor land management. Climate change likely also plays a role, but it is much more difficult to measure this impact.

Considerable confusion surrounds the spatial extent of affected areas as there is no clear single definition of desertification. The most publicized area experiencing desertification is the sub-Saharan Sahel zone of West Africa, an area first brought to world attention following the 1968–73 drought. Here, desertification is caused by population pressure, inappropriate human activity, human conflict, and periods of drought. Elsewhere, the process similarly has multiple causes and no simple solution. Many of the human causes may be related to the pressures placed on local people by the introduction of capitalist imperatives into traditional farming systems.

Fortunately, the technology to combat desertification is available; unfortunately, the will to use it is rare. A major international effort to

AP Photo/MaxVision

The desert is encroaching daily onto fertile planting fields about 100 kilometres west of Yin Chuan, capital of Ningxia Hui Autonomous Region in northwest China, and threatening to engulf outlaying railroads and villages.

combat desertification, the 1977 UN Plan of Action, was a failure for two reasons: (1) technical solutions were applied to areas where the key causes were economic, social, and political, and these underlying causes were not addressed; and (2) local populations were not involved in the search for solutions. A potentially important development is discussion concerning an international convention to combat desertification as first proposed at the 1992 Earth Summit in Rio de Janeiro.

But perhaps most significant are those attempts that are initiated at a local level. Led by 2004 Nobel Peace Prize winner Wangari Maathai, the Kenyan Green Belt Movement had its origins in a 1974 Nairobi tree-planting scheme that focused on the value of working on a community level, especially with women (Agnew, 1990). The movement's objectives are many and varied, as befits any effort to solve so difficult a problem. The central goal is to reclaim land lost to desert and to guarantee future fertility. Specific objectives include conserving water, increasing agricultural yields, limiting soil erosion, and increasing wood supplies. The ideas of working on the local level and involving women are crucial. Reclaiming land is important to the local people, and their direct involvement makes the exercise much more meaningful to them, while women (the principal wood-gatherers in Kenya and in most of the Sahel) are increasingly aware of future needs. Planting crops around trees improves yields—an important and direct consequence in a subsistence economy. Tree-planting to combat desertification is far from a panacea, but it is a positive development. Other parts of the world are adopting the strategies of the Kenyan movement, particularly its community focus. The parallels with the Grameen Bank (Box 6.8) are intriguing.

As is the case with the tropical rain forests, areas subject to desertification are home to poor people who often are unable to adopt appropriate remedies and lack the necessary political influence. A proper solution requires that the dryland ecosystems be treated as a whole—land management is needed. Population pressures must be reduced, land needs to be more equitably distributed, greater security of land tenure is required, and global warming must be combatted.

HUMAN IMPACTS ON ANIMALS

Animal domestication serves many purposes, providing foods such as meat and milk (cows, pigs, sheep, goats) and furs, fur fibres, and skins for clothing and accessories, as well as draft animals (horses, donkeys, camels, oxen) and pets (dogs, cats). Once domesticated, animals have frequently been moved from place to place, both deliberately and accidentally. Some deliberate introductions, such as that of the European rabbit to Australia, have had drastic ecological consequences. Whalers and sealers were probably the first to introduce rabbits into the Australian region, but the key arrival was in 1859, when a few pairs were introduced into the southeast area to provide so-called sport for sheep-station owners. Following this introduction, rabbits spread rapidly across the non-tropical parts of the continent, prompting a series of "unrelenting, devastating" rabbit plagues (Powell, 1976: 117). Rabbits consume vegetation needed by sheep and remain a problem despite the introduction of the disease myxomatosis and of rabbit-proof fences (Figure 4.5). The European rabbit in Australia has probably caused more damage than any other introduced animal anywhere in the world—but it is only one of many unwanted guests (Box 4.4).

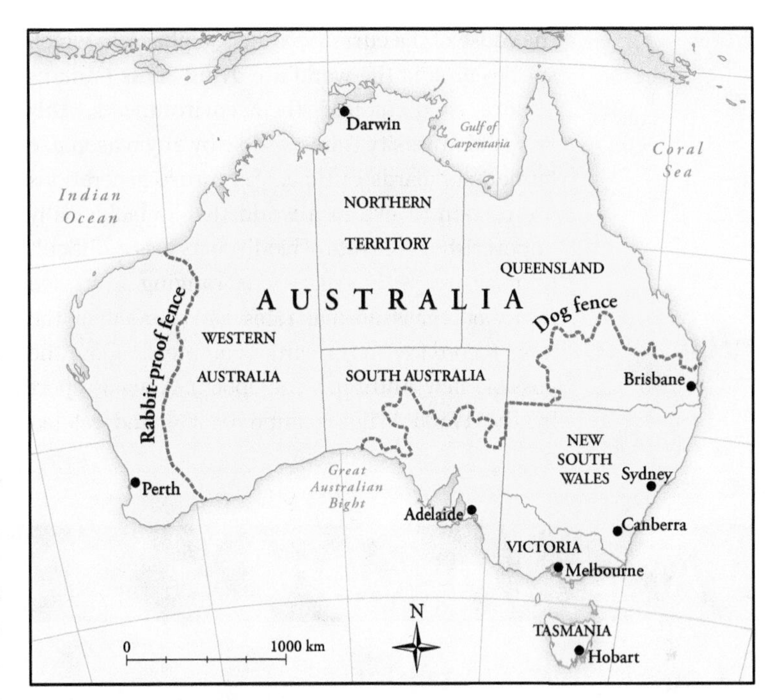

FIGURE 4.5 | Australia's rabbit and dog fences
Fences built to protect Australian agriculture from rabbits and dingoes (wild dogs thought to have been introduced into the continent by Asian seafarers more than 3,500 years ago). Stretching 1,833 km (1,139 miles), the rabbit fence was built in the 1890s in an attempt to confine the introduced species to the central desert, but it was too late; rabbits had already entered western Australia. The dingo fence in eastern Australia stretches 5,321 km (3,307 miles) and was built to protect sheep.

A 2014 report suggested that human activities have caused populations of mammals, birds, reptiles, amphibians, and fish to decline by an average of 52 per cent since 1970 (WWF, 2014). As human numbers and levels of technology have increased, so have animal extinctions. It is possible that humans have caused species extinction since 200,000 BCE, although the evidence is uncertain. What is certain is that hunting populations have caused extinctions. All moa species became extinct after Europeans arrived in New Zealand, and there is much evidence to suggest that animal extinctions in North America coincided with human arrival. The 1859 publication of Darwin's *On the Origin of Species* helped to place the extinction of plant and animal species in context and was followed by protectionist legislation in several of the British colonial areas; an early example was the 1860 Tasmanian law protecting indigenous birds.

Today, our role in animal extinction is clearer. When natural animal habitats are removed, as in the case of tropical rain forest, extinction follows. This is one component of a major threat to biodiversity.

Biodiversity Loss

E.O. Wilson, today's best-known Darwinian thinker, and many other scholars, believe that the earth is entering a new evolutionary era involving the greatest mass extinction since the end of the Mesozoic era, 65 million years ago. Humans are

Swift fox pups at their den near Grasslands National Park, Saskatchewan. During the twentieth century, the swift fox disappeared completely from the wild in Canada. Releases of captive-bred foxes and foxes taken from the wild in the United States led to the reintroduction of the species, but the animal remains listed as "endangered" under the Species at Risk Act.

the cause of the current extinction phase, in which species around the world are dying off as humans remove or destabilize their environments. This loss of biodiversity is irreversible by any reasonable human standards of time, and future generations are certain to live in a world that is biologically impoverished. The rate of biodiversity loss is difficult to measure, with suggestions ranging from 100 to 10,000 times normal rates as recorded in the fossil record (recall the earlier comments about the possible new Anthropocene epoch). A 2008 report by the World Wildlife Fund (WWF) and related organizations concluded that biodiversity had plummeted by one-third between 1970 and 2005.

Most commentators explain biodiversity loss by reference to population growth and increasing consumption of energy and resources. With more and more people around the world aspiring to the lifestyle of the affluent minority, the human ecological footprint is outgrowing the resources needed to support it. Wilson extends this argument in a more controversial direction to suggest that the impulse towards environmental destruction is innate, hard-wired into us—in

Around the Globe

BOX 4.4 | Unwanted Guests

Beginning in 1788, the European colonization of Australia ended a long period of isolation and introduced numerous animals and plants. Many of these species, whether introduced deliberately or accidentally, proliferated in their new environment, where the natural controls on their numbers (such as predators and diseases) were absent.

Rats and mice that arrived in the holds of ships multiplied rapidly after being accidentally introduced. Sheep and cattle were brought over to satisfy British economic needs, while animals brought over to satisfy the "sporting" habits of British settlers included deer, fox, and, of course, rabbits. The interior deserts prompted the introduction of pack animals such as burros, asses, and camels, and soon there were more camels in Australia than in all of Arabia. Many varieties of introduced livestock and pets have reverted to a feral (wild after previously being domesticated) existence, including pigs, horses, camels, water buffalo, and even house cats.

© National Geographic Image Collection/Alamy

Damage to farmland in Australia by feral pigs digging into the soil.

Feral water buffalo damage forest ecosystems by destroying trees and eroding soil, while feral pigs (of which there are an estimated 23 million) damage sugar cane and banana crops, threaten wildlife, and spread disease. More generally, because all feral species require a supply of water, the most precious resource in the interior, they reduce the amount available for native species. Many of Australia's native animals have disappeared because of the presence of the introduced species. Of the smaller marsupials, 17 have become extinct and another 29 are considered endangered.

A recent introduction causing considerable damage is the cane toad, first brought to Queensland from Hawaii in the 1930s to combat the cane grub, which was damaging the sugar cane crop. Unfortunately, cane toads do not eat only cane grubs—they eat almost anything. They are also capable of much more rapid reproduction than native toad species, with females producing up to 40,000 eggs a year; they compete effectively with the native species; and they have no natural predators. Cane toads have moved far beyond the sugar cane area, advancing about 50 km (31 miles) per year, and there are no signs of an end to their movement. In response, authorities announced in 2005 that they are offering a reward to anyone who comes up with an effective way to limit further advance. One initiative begun in 2009 was the declaration of a "Toad Day Out," with everyone in Queensland asked to capture as many toads as possible and hand them to the authorities alive and unharmed, after which they were humanely killed.

short, that clearing forests and killing animals is instinctual for humans. Furthermore, he says, we are unable to see things in the long term. Whether or not Wilson is correct, the evidence of biodiversity loss is compelling.

HUMAN IMPACTS ON LAND, SOIL, AIR, AND WATER

Because we humans live on the land and use soil extensively, we are geomorphic agents who change land and affect that thin and vulnerable resource, soil. The earlier account of desertification, an impact both on vegetation and soil, is a clear example. Also, industrial activities add substances to the atmosphere that can cause harm to people and to environments; these include carbon monoxide, nitrogen oxides, sulphur oxides, hydrocarbons, and particulate matter (solid and liquid). Finally, we use and pollute water, a valuable and threatened resource.

Land

Most human activities create landforms. Excavation of resources such as common rocks (limestone, chalk, sand, gravel), clays (stoneware clay, china clay), minerals (dolomite, quartz, asbestos, alum), precious metals (gold, silver), and fossil fuels (oil, coal, peat) can have major impacts: changing ecosystems, lowering land surfaces, flooding, building waste heaps, creating toxic wastes, and leaving scenic scars. These are not the only human activities that make us geomorphic agents. When we modify river channels, for example, sand dunes are affected; other effects include coastal erosion and coastal deposition. There are many causes and consequences of human impact on land.

Degradation and loss of arable land are occurring throughout the world as a result of population increases, industrialization, and improper agricultural practices. Smil (1993: 67) reported that the average annual loss of farmland in China between 1957 and 1980 was a mammoth 1 million hectares (2.5 million acres). About 40 per cent of the world's agricultural land is seriously degraded, with Latin America the most impoverished at 75 per cent.

One example of a current human impact on land is in Canada, where exploitation of the Alberta

Mining trucks carry loads of oil-laden sand at the Albian Sands oil sands project at Fort McMurray, Alberta.

oil sands began in earnest in the 1990s, prompted by technological advances that reduced costs and resulting in significant environmental damage. The oil sands are a mixture of sand, water, and heavy crude oil. After boreal forest is cleared and peat bog removed, oil-saturated sand remains. Extracting the oil is difficult, is expensive, and requires large amounts of natural gas and water. The water, drawn from the Athabasca River and consequently affecting water levels and the health of communities downstream, cannot be reused and ends up in tailings ponds (a generic term for mining by-products) with other waste materials. In April 2008, about 1,600 migratory waterfowl returning north landed on what would have looked to them like a lake, but they died on contact with an oil slick on top of the tailings pond. This incident was a public relations disaster for the Alberta government and a major embarrassment for the company involved, Syncrude, a consortium that includes Petro-Canada. Syncrude was charged by Environment Canada under the Migratory Birds Convention Act, convicted, and fined $3 million.

Soil

By its very nature, soil is especially susceptible to abuse, and humans use and abuse soils extensively. Agricultural activities have the greatest impact on soil, not only through erosion but also through chemical changes. Humans increase the salinity (salt content) of soil, which has a negative effect on plant growth, largely through irrigation. Humans also increase the laterite content of soil by removing vegetation—laterite is an iron- or aluminum-rich duricrust naturally present in tropical soils and is essentially hostile to agriculture.

Soil erosion is associated with deforestation and agriculture. Forests protect soil from runoff, and roots bind soil. Probably the best-known example of human-induced soil erosion is the "Dust Bowl" in 1930s North American mid-latitude grasslands. Among the causes of this phenomenon were a series of low-rainfall years, overgrazing, and inappropriate cultivation procedures associated with wheat farming. The resulting "black blizzards" led many people to leave the prairies.

Air

Although air pollution is a problem almost everywhere, the pollution related specifically to smoke-producing industries in more developed countries has been significantly reduced, but by no means eliminated, as a result of changes to industrial processes and careful management and control. Many large cities throughout the world—Los Angeles is a particularly notorious example—suffer the problem of chemical smog (smoke + fog), largely as a result of automobile emissions.

An important atmospheric concern is the ozone layer. Ozone (O_3) is a form of oxygen that occurs naturally in the cool upper atmosphere. It serves as a protective sunscreen for the earth by absorbing the ultraviolet solar radiation that can cause skin cancer, cataracts, and weakening of the human immune system; ultraviolet radiation is also damaging to vegetation. Ozone depletion was first recognized in 1985 by scientists with the British Antarctic Survey. An ozone hole over Antarctica appeared to be about half the size of Canada. The principal culprits were CFCs and some greenhouse gases. CFCs exemplify the devastating environmental impact of many efforts to satisfy human demands through the use of technology. CFCS were not even synthesized until the late 1920s, when they represented a remarkable advance. They are ideal as coolants because they vaporize at low temperatures and also serve well as insulators. Most important, they are easy and inexpensive to produce, hence their widespread use since World War II as coolants in refrigerators, as propellant gases in spray cans, and as ingredients in a wide range of plastic foams. As these rise into the atmosphere, chemical reactions occur and ozone is destroyed.

ozone layer
Layer in the atmosphere 16–40 km (10–25 miles) above the earth that absorbs dangerous ultraviolet solar radiation; ozone is a gas composed of molecules consisting of three atoms of oxygen (O_3).

Recognition of CFCs' impact on the atmosphere, especially on ozone, prompted 24 countries to gather in Montreal in 1987 and agree to reduce CFC production by 35 per cent by 1999. Most environmental experts argued that this target was inadequate. Subsequently, a 1990 London agreement set the goal of eliminating CFC production by the year 2000. Neither of these goals was fully achieved. However, UN studies have concluded that depletion of the ozone layer peaked by 2005 and that recovery was evident by 2014.

Water

Water is an essential ingredient of all life. We know this and yet we frequently choose to ignore it. Rather than carefully safeguarding water quantity and quality, we cause shortages and continually contaminate it. The two issues of scarcity and contamination dominate our consideration of the human impact on water.

The Global Water Cycle

How much water is available? Figure 4.6 outlines the global water cycle, identifying three principal paths—precipitation, evaporation, and vapour transport. Total annual global precipitation is estimated at 496,000 km^3 (119,000 cubic miles), most of which (385,000 km^3/92,000 cubic miles) falls over oceans and cannot be easily used. Water returns to the atmosphere via evaporation from the oceans (425,000 km^3/102,000 cubic miles) and from inland waters and land as well as by transpiration from plants (71,000 km^3/17,000 cubic miles combined). In addition, some of the precipitation that falls on land is transported to oceans via surface runoff or groundwater flow (41,000 km^3/9,800 cubic miles) and some water evaporated from the oceans is transported by atmospheric currents and subsequently falls as precipitation over land (again, some 41,000 km^3/9,800 cubic miles). In principle, this much water is available each year globally, but this figure is reduced by 27,000 km^3 (6,500 cubic miles) lost as flood runoff to the oceans and by another 5,000 km^3 (1,200 cubic miles) flowing into the oceans in unpopulated areas. Perhaps 9,000 km^3 (2,200 cubic miles) are readily available for human use.

This amount of water is possibly enough for 20 billion people. However, some states have a plentiful amount and others have an inadequate amount. Where water is abundant, it is treated as though it were virtually free; where it is scarce, it is a precious resource. Thus, the average US citizen

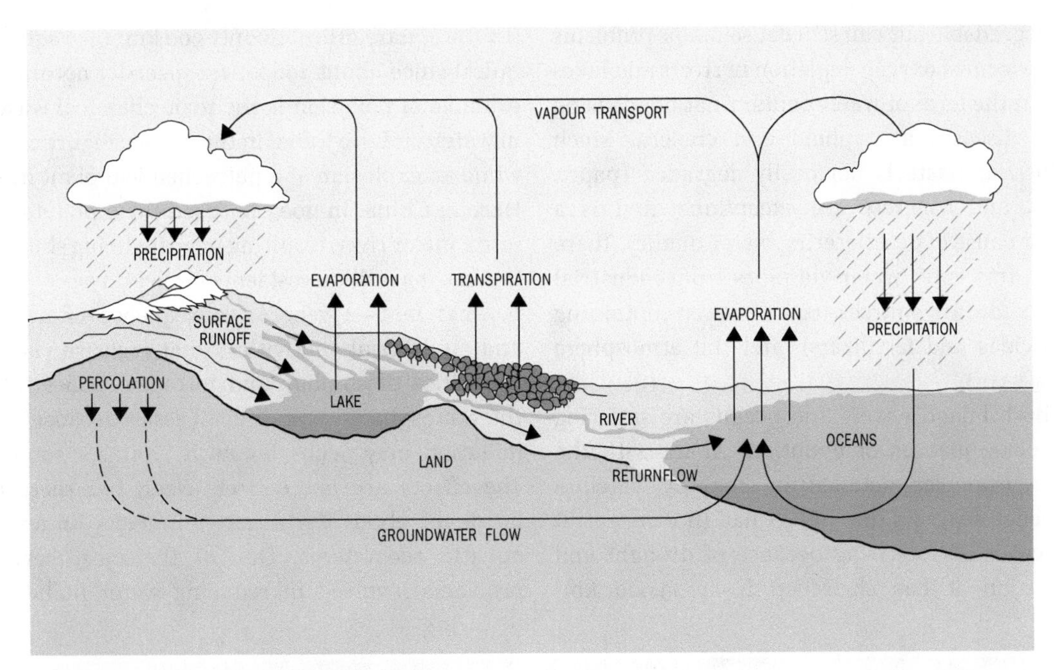

FIGURE 4.6 | The global water cycle

annually consumes 70 times more water—through combined household, industrial, and agricultural uses—than does the average citizen of Ghana.

Using Water

Agriculture consumes 73 per cent of global water supplies, often highly inefficiently. In second place is industry, which consumes perhaps 10 per cent of global supplies. The bulk of what is left goes to supply basic human needs. As each of these uses increases in any given area, whether through agricultural and industrial expansion or population growth, water quantity may become a problem. Shortages are common in many areas, some continuous and some periodic. Bahrain, for example, has virtually no fresh water and relies on the desalinization of sea water. Groundwater depletion is common in the US, China, and India, and water levels have fallen in both Lake Baikal (southern Siberia, the world's deepest lake) and the Aral Sea (Kazakhstan–Uzbekistan) (Box 4.5). In the early twenty-first century, Australia experienced a 10-year drought that devastated agricultural landscapes, while more recent periods of drought in Brazil and South Africa have meant there has been insufficient water to generate hydroelectricity. Consider also that many of the world's great rivers, some of which flow through major grain-growing regions, no longer reach the sea—these include the Huang He River in China, Murray-Darling in Australia, Indus in India and Pakistan, Rio Grande in the United States and Mexico, and Colorado in the United States.

Do these regional problems mean that there is, or soon will be, a global water shortage? The simple, but misleading, answer to this question is no. This is because humans use only about 9 per cent of the water that flows through the global water cycle. However, all other life on earth also uses water. Further, human use of water is growing significantly because of increasing population numbers, improved living standards, and climate change. For example, most of the addition to the world population occurs in the cities of less developed countries, and urban dwellers use more water than rural dwellers do. Improved living standards typically involve a shift in diet from vegetarian to meat-eating. Growing 1 kg of wheat uses about 1,000 litres of water, while producing 1 kg of beef requires 15,000 litres.

Polluting Water

As water passes through the cycle described in Figure 4.6, it is polluted in two ways. Organic waste (from humans, animals, and plants) is biodegradable but can still cause major problems in the form of oxygen depletion in rivers and lakes and in the form of water contamination, causing such diseases as typhoid and cholera. Much industrial waste is not easily degraded (paper, glass, and concrete are exceptions) and is a major cause of deteriorating water quality. These pollutants enter water via pipes from industrial plants, diffuse sources (runoff water containing pesticides and fertilizers), and the atmosphere (acid rain).

acid rain
The deposition on the earth's surface of sulphuric and nitric acids formed in the atmosphere as a result of fossil fuel and biomass burning; causes significant damage to vegetation, lakes, wildlife, and built environments.

Both inland waters and oceans are suffering the consequences of pollution. Along with the three examples detailed in Box 4.5, satellite evidence suggests that Lake Chad (north-central Africa) is disappearing because of drought and irrigation: it has shrivelled from 23,000 km² (14,300 square miles) to only 900 km² (350 square miles) since about 1960. An especially notorious example of pollution is the toxic chemical waste in waters at Love Canal in the US (see Figure 2.19), while an explosion at a petrochemical plant near Harbin, China, in 2005 released toxic pollutants into a major river, resulting in water being shut off to almost 4 million residents for five days.

Acid rain—a general consequence of urban and industrial activities that release large quantities of sulphur and nitrogen oxides into the atmosphere—is a difficult issue because the pollution may occur far away from its source. The effects are not entirely clear, but there is no doubt about the negative impacts on some aquatic ecosystems. One of the most severe problems involved in reducing water pollution

Around the Globe

BOX 4.5 | A Tale of Three Water Bodies

Until the 1960s, the Aral Sea was the fourth largest inland water body in the world. By the early 1990s, however, remote-sensing imagery showed that the area had been reduced by more than 40 per cent. The explanation was that, since the 1950s, much of the sea's water had been diverted to irrigate cotton. Although data for the late 1990s show some signs of recovery—water levels had risen as a result of dam construction and reductions in water use—NASA photographs taken in 2014 indicate that a large area of the remaining sea is completely dry as a result of a prolonged drought and the diversion of a river that flowed into the sea. Tragically, because cotton requires large quantities of defoliants, pesticides, and fertilizers, the Aral Sea has also become dangerously polluted, and the local area has high rates of infectious disease, cancer, miscarriage, and fetal abnormalities. These problems are not likely to be solved any time soon.

The Rhine River, one of the world's most important international waterways, is 1,320 km (820 miles) in length and has a catchment area of 185,000 km² (71,428 square miles) populated by about 50 million people in six European countries. The river receives pollutants from several heavily industrialized areas, in addition to domestic sewage and shipping discharges. By the late 1980s, these human activities had so compromised the water quality that the river had to be treated before it could be used for drinking or crop irrigation in the Netherlands. Local ecosystems were also changed, destroying much animal and plant life. However, considerable success in combatting these problems has been achieved through the activities of the International Commission for the Protection of the Rhine against Pollution, established in 1950. Today, the Rhine is in relatively good health; one indicator was the return of salmon in 2000.

When Europeans first reached Lake Erie, the water was clear and supported a large fish population. By the 1960s, however, it was one of the most seriously polluted large water bodies in the world. With some 13 million people and much heavy industry in its watershed, the lake was eventually polluted by domestic sewage, household detergents, agricultural wastes, and industrial wastes. Clear water became green, fish perished, and algae growth was enhanced by the addition of phosphates and nitrates. These problems were acknowledged, and the Canadian and American governments, urged to take action by newly formed environmental groups, initiated a comprehensive program to solve these problems in the 1970s. Known as the Great Lakes Water Quality Agreement, this program has been remarkably successful and Lake Erie has been rehabilitated.

These three examples demonstrate that, given effective management strategies, seriously polluted water bodies can recover; however, such strategies are more likely to be employed in the more developed than in the less developed world.

is the difficulty of securing the necessary international co-operation. Efforts to combat acid rain, particularly in Europe, require co-operation among many more countries.

International agreement is essential to combat ocean pollution. In a classic example of the tragedy of the commons (see Box 4.3), many states exploit oceans, but no state is prepared to assume responsibility for the effects of human activities. Estimates suggest that only 3 per cent of the world's oceans—remote icy waters near the poles—remain undamaged, with overfishing and climate change the main causes of damage. More than 50 per cent of the world's population live close to the sea, and most of the world's ocean fish are taken from coastal waters. Water quality in the oceans, particularly coastal zones, is seriously endangered, and many ocean ecosystems have been damaged. News reports often refer to dead zones, usually areas close to land that are starved of oxygen because fertilizer flows down rivers into the sea and results in loss of fish and underwater vegetation. Remote sensing enables objective assessments of water pollution; images of the Mediterranean Sea show a marked contrast between the northern shore, heavily polluted by water from major European rivers and coastal towns, and the southern shore. Remote sensing also permits effective monitoring of oil spills, as in the *Exxon Valdez* incident off the Alaskan coast in 1989. Once again, we do not know enough about the consequences of our activities, but we do know that restoring the quality of ocean water is likely to be much more difficult than is the case with surface inland water. One thing seems certain: we need institutions and regulations to enable the long-term interests of all users to take precedence over the short-term interests of individual users.

HUMAN IMPACTS ON CLIMATE

Nowhere is the concept of interrelations better demonstrated than in a consideration of human impacts on climate. Scientists and environmentalists feel that our most damaging impacts of all are on global climate. Unfortunately, as in our discussions of other human impacts, we are confronted with two general areas of uncertainty: the role played by humans (as opposed to natural physical factors) and the extent of any human-induced change. But note that, while these two areas of uncertainty do prevail, what is certain is that humans are causing climate change.

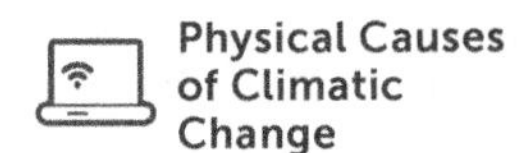

The Natural Greenhouse Effect

Temperatures on the earth's surface result from a balance between incoming solar radiation and loss of energy from earth to space. If the earth had no atmosphere, the average surface temperature would be about −19°C (−3°F), but the presence of an atmosphere results in an actual average surface temperature of about 15°C (60°F). The atmosphere causes this increase because it prevents about half of the outgoing radiation from reaching space; some is absorbed and some bounces back to earth. This natural "greenhouse" effect is not related to human activity but is the result of the presence of water vapour (the most significant greenhouse gas), CO_2, O_3, and other gases in the atmosphere. These greenhouse gases are only a fraction of the atmosphere—nitrogen and oxygen make up 99.9 per cent of it (excluding the widely varying amounts of water vapour)—but their impact is nevertheless considerable. We are increasing the greenhouse effect by adding CO_2 and some other gases that perform a similar function, such as sulphur dioxide (SO_2), N_2O, methane (CH_4), and a variety of CFCs. How are we doing this?

Human-Induced Global Warming

In the most general sense, human additions to the natural greenhouse effect are the product of our increasing population numbers and advancing technology. More specifically, they are the result of fossil fuels use, increased fertilizer use, increased animal husbandry, and deforestation. Until recently, much carbon was stored in the earth in the form of coal, oil, and natural gas. Burning these resources releases CO_2, water vapour, SO_2, and other gases that are then added to the atmosphere. Burning wood also adds CO_2 to the atmosphere. Further, soil contains large quantities of organic carbon in the form of humus, and agricultural activity speeds up the process by which this carbon adds CO_2 to the atmosphere. Estimates suggest that the concentration of CO_2 in the atmosphere has increased from 260 ppm (parts per million) 200 years ago to 400 ppm in 2013 (recorded at a laboratory in Hawaii) and might be up to 550 ppm in 2030. The level is now

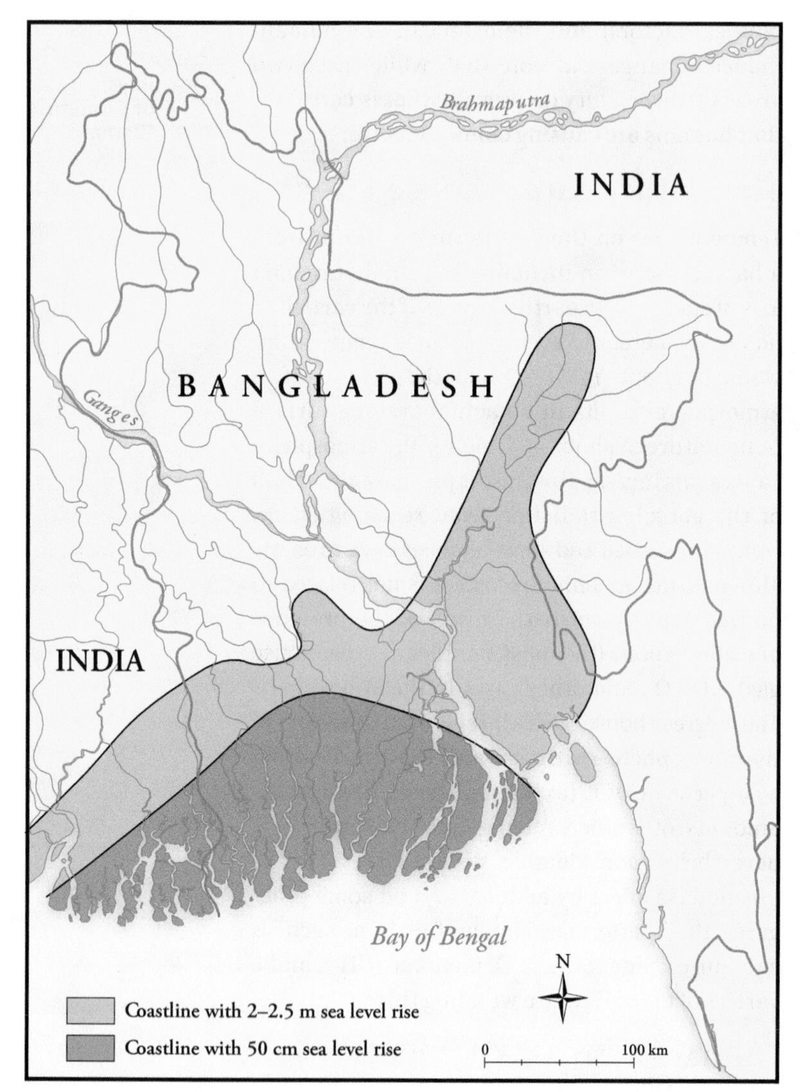

FIGURE 4.7 | The impact of sea-level change on Bangladesh

Source: Based on McKay, G. A., and H. Hengeveld. 1990. "The Changing Atmosphere." In *Planet under Stress*, edited by C. Mungall and D. J. McLaren, 66. Toronto: Oxford University Press.

higher than at any time in the past 650,000 years. Agricultural activity also adds CH_4 and N_2O to the atmosphere. CH_4 is increasing both as a result of paddy rice cultivation and the large number of flatulent farm animals. Recall from the earlier reference to biofuels that growing crops such as corn and canola releases N_2O.

Although most research rightly focuses on what is happening today, it is possible that deforestation resulting from the spread of agriculture through much of Europe and China about 5,000 years ago initiated warming comparable to that evident since the Industrial Revolution. Indeed, Ruddiman (2005) suggests that the spread of early agriculture prevented the onset of a period of much colder temperatures.

It is widely accepted that this human-induced global warming is raising the average temperature of the earth. The fifth report of the UN's authoritative IPCC, published in 2014, noted that global temperatures have increased 0.9°C in the past hundred years and, if CO_2 emissions remain unchecked, predicted a further rise of between 1.4°C and 2.6°C by 2050 and up to 4.8°C by 2100. Some of the key facts about and consequences of warming are noted in Box 4.6.

Human-induced global warming has also caused global sea level to rise by 20 cm in about the last hundred years, and an additional 45–82 cm increase is predicted by 2100. If this occurs, the consequences for many populated areas may be catastrophic. Up to 15 per cent of Egypt's arable land would be at risk, and many coastal cities such as New York and London would be below sea level. The consequences for one of the most densely populated areas in the world—coastal Bangladesh—will be disastrous unless we are able to adapt (Figure 4.7). Indeed, even before sea levels rise substantially, many coastal areas would be in danger because of storm surges.

The Netherlands, a more developed country, has successfully adapted to a situation where much of its current area is already almost 4 metres (13 feet) below sea level. In principle, the Netherlands model—construction of levees and dikes—can be followed. Venice is already pursuing a similar strategy, constructing a flexible seawall to protect the city against Adriatic storms. One estimate places the cost to protect such major areas, not including coastal margins, at approximately US$300 billion.

In theory, there is an alternative to adaptation: moving away from the threatened areas. But this is hardly an option for many people in the less developed world and is culturally and economically unthinkable for many in the more developed world.

Responding to Human-Induced Global Warming

The principal response to the fact of global warming and its probable consequences has been an effort to implement policies that will reduce the emission of greenhouse gases. Most notably, the UN-sponsored Kyoto Protocol established goals that, if met, were expected to slow the rate of global warming. This legally binding agreement to cut greenhouse gas emissions was reached in 1997

Because of its location in the low-lying Ganges Delta and its poor drainage system, Dhaka and nearby areas in Bangladesh are prone to frequent flooding, which kills many people and destroys homes every year. Rising sea levels continue to put lives, food production, and infrastructure at risk.

Examining the Issues

BOX 4.6 | Climate Change Front and Centre

Should we worry about global warming? The response from the scientific community is "Yes, most definitely." The 2014 IPCC report described the evidence as "unequivocal." Although details are necessarily uncertain and specific regional predictions are difficult to make, scientists are expressing real concerns about the possible consequences of human-induced global warming. In the first few years of the twenty-first century, global warming without doubt became *the* critical environmental issue. Few see any realistic likelihood that the major contributors to CO_2 emissions—the United States, the countries of the former Soviet Union, and the emerging industrial economies of China and India—will cut back these emissions sufficiently to prevent further temperature increases.

Consider the following:

- The nine hottest years since effective record-keeping began in the 1890s were, in descending order, 2014, 2010, 2005, 1998, 2002, 2003, 2013, 2004, and 2007.
- The area covered by sea ice in the Arctic Ocean during the summer is shrinking each year; a nearly ice-free September Arctic is possible by 2050.
- The Greenland ice sheet, second in size only to the Antarctic ice sheet, appears to be crumbling at a dramatically increased rate, as shown by recent satellite data.
- The Antarctic ice sheet, which contains 90 per cent of the world's ice, is shrinking because of melting from underneath and surface melting.
- Permafrost is melting rapidly in the Siberian tundra, a process that releases additional methane into the atmosphere.

Continued

These and other facts about the consequences of global warming are reported in the 2014 IPCC report and are discussed regularly in major scientific journals, including *Nature, Scientific American*, and *Science*.

There are real concerns about the negative impacts of global warming, especially because we have made settlement decisions and developed lifestyles with the prevailing climate as background. What are some likely consequences of global warming?

- It is likely that the increased intensity of extreme weather conditions, such as hurricanes, is related to increased sea surface temperatures that destabilize the atmosphere.
- Reports from the World Bank link much illness and death in the less developed world to infectious and respiratory diseases, including malaria and dengue fever, that are spreading as climate warms.
- Increased incidence of disease might lead to conflict, population displacement, and more authoritarian governments.
- The impact on crop yields is unclear. On the one hand, it is well recognized that additional carbon dioxide in the atmosphere is likely to act as fertilizer, potentially increasing yields of all major cereal crops except corn and sorghum. On the other hand, longer droughts and more ground-level ozone may significantly reduce yields.
- The impact on drought conditions in Africa is also unclear. According to many scientists, the region contributing least to global warming (Africa) will be the most adversely affected, with increased incidence and length of droughts exacerbating current problems. But other research claims that rainfall in the semi-arid Sahel region will increase, not decrease.
- There is likely to be a devastating impact on biodiversity. The most vulnerable species may be those in extensive areas with little variation in altitude, as species would have to move long distances to find suitable environments.
- The most significant possible negative consequence of global warming is rising sea levels caused by ice melt.

by about 150 countries but only came into force in early 2005, with most participating countries agreeing to reduce emissions by a specific percentage. It is notable, however, that the United States never ratified the agreement, principally because legislators saw it as unfair that neither China nor India was required to cut emissions during the first phase (through to 2012). This is especially significant because the United States and China are major emitters of greenhouse gases.

How have the participating countries performed? Germany and Britain are generally judged to have responded positively, with about 10 per cent reductions, but most have failed to meet their targets. Canada, for example, agreed to cut emissions by 6 per cent from the 1990 level but, by 2006, emissions had increased by 21.3 per cent. In 2011, the Conservative government formally withdrew from the protocol, arguing instead for an alternative voluntary approach to emission reductions. This decision highlights the close links between environmental action and political ideology. Discussions concerning the details of human-caused global warming and possible responses to that warming continue on a regular basis but tend to be compromised by many national governments' perceived need to focus on economic growth.

Progress on Kyoto has been assessed during UN conferences on climate change, which are held on a regular basis. Most observers judge these meetings as limited successes but with some progress made in five broad areas. First, many remaining details about Kyoto have been finalized, including what is to happen if countries fail to meet targets. Second, the Kyoto Protocol has been extended to 2020. Third, all participants agreed to talk about a possible UN climate pact that will include countries that have not signed up to Kyoto, most importantly the United States. Fourth, participants agreed to promote carbon capture and sequestration technologies. Fifth, a fund has been set up to help less developed countries cope with climate change. However, for many commentators, the perceived failure of many of these meetings has revealed major flaws in existing policy, notably in the inability to agree on plausible policies to reduce emissions after the Kyoto agreement expires.

A separate international agreement was reached in mid-2005. Six countries—the US, China, India, South Korea, Japan, and Australia—agreed to reduce emissions, but this pact has received much criticism because it is non-binding. A feature of this agreement is the emphasis on tackling emissions through new technologies rather than through any reductions in economic growth.

Some argue that these agreements are not the best route to take. Recognizing that increases in atmospheric CO_2 will last a very long time, Victor (2011) noted that it is not sufficient to merely lower emission levels; rather, they need to be close to zero. Further, any reductions resulting from current policies are difficult to predict, which means that setting quantitative goals is not very helpful. Far better than continuing the present approach, which is based on what Victor describes as wishful thinking and has led to global warming gridlock, is for countries to adopt specific policies relating to energy research and development. The needed policies should be based on bottom-up initiatives at national, regional, and global scales.

One other option is to do nothing on the grounds that there might be positive consequences of global warming. For example, the melting of Arctic ice might open the Northwest Passage, significantly reducing the sailing distance between Europe and Asia compared to travelling through the Panama Canal. (Of course, such an eventuality would also bring to the fore acrimonious international conflict regarding sovereignty in the Far North.) The melting of Arctic ice might also mean that drilling for oil in the region will be more economically feasible, a circumstance prompting further interest in issues of Arctic sovereignty, with the five countries with Arctic coasts—Canada, the US, Russia, Denmark, and Norway—having legitimate sectoral claims to portions of the Arctic seabed and many other countries, notably Britain and Japan, claiming an interest in the Arctic waters being "international" for shipping purposes. (This topic is discussed further in Chapter 9.) Another possible advantage is that the growing season might be extended in some areas. The general consensus, however, is that the probable negative consequences of global warming significantly outweigh any possible advantages.

Acknowledging Uncertainty

Although human-induced global warming is a fact, it is important to appreciate that many of the popular accounts of global warming, and some of the scientific ones, make little if any reference to the uncertainties involved in predicting either the magnitude or the consequences of the human-induced greenhouse effect. In particular, there is considerable uncertainty about the regional consequences of global warming, a reflection of the basic ecological fact that all things are related. One of the few specific predictions that scientists feel confident in making is that, as warming occurs, the earth's polar regions will be the most seriously affected. By the late 1990s, it was clear that such warming was already underway, and not all the evidence to this effect came from scientific sources. In northern Canada, Inuit elders and hunters reported that glaciers were receding and coastlines eroding, that the fall freeze was arriving later, and that winters were becoming less severe. Although the evidence of warming in lower latitudes is less clear, it is generally agreed that the next several decades will see a poleward retreat of cold areas, a corresponding expansion of forests and agricultural areas, changes in the distribution of arid areas, and, of course, rising sea levels as the ice caps melt.

Indeed, much detail is not known. For example, scientific evidence suggests that the higher precipitation and related melting of snow and ice in northern areas that are likely to accompany global warming will lead to increased river discharge into the Arctic and North Atlantic Oceans. The resulting decrease in ocean salinity may disturb the balance of water flow (known as the thermohaline circulation) that involves a warm northward flow of surface water and a returning southward flow of deep water. A possible consequence of reduced thermohaline circulation might be to flip the North Atlantic back into a glacial mode and to substantially lower temperatures in Northwest Europe (Anderson, 2000).

On the important issue of impacts on precipitation in Africa, consider the following quote from the 2014 World Development Report:

> Uncertainty is especially deep in "emerging risks" or in areas where scientific uncertainty is the greatest (genetically modified crops, hydropower dams, nuclear energy, climate change). A common example is the uncertainty about future changes in local climates. Different scientific teams develop simulations of climate systems that differ in their technical implementations, but these climate models are based on the same widely accepted laws of physics. And while these models agree on the large patterns of climate change, they can point in opposite directions at the local scale and for some parameters. For example, depending on the model, rainfall in West Africa could

> increase or decrease by 25 percent by the end of this century. . . . Such uncertainty is clearly an obstacle to the design of water infrastructure able to deal with floods and droughts in the region. (World Bank, 2013: 94)

There are other specific uncertainties, as evidenced by two research studies published in major journals (*Nature Geoscience* and *Science*) in 2011. First, contrary to the established scientific consensus, it is claimed that some glaciers in the Himalayas are advancing rather than retreating and that the deciding factor is not climate change but the amount of surface debris. Also challenging established understanding is research by Danish scientists that shows concerns about Arctic ice reaching a tipping point that will prompt a rapid melt of remaining ice are misplaced. The lesson from these two studies is simple but important. Scientific understanding is always subject to evaluation and possible change. At this time, these examples serve to highlight uncertainty, not to overturn conventional scientific understanding.

catastrophists
Those who argue that population increases and continuing environmental deterioration are leading to a nightmarish future of food shortages, disease, and conflict.

cornucopians
Those who argue that advances in science and technology will continue to create resources sufficient to support the growing world population.

The Point of No Return?

Is it possible to arrive at any definitive conclusions about global warming? James Lovelock, originator of the Gaia concept, sees it as inevitable but argues that our intelligence will allow us to cope. James Hansen, a leading US researcher on human-caused climate change, argued: "The Earth's climate is nearing, but has not passed, a tipping point beyond which it will be impossible to avoid climate change with far-ranging undesirable consequences. These include not only the loss of the Arctic as we know it, with all that implies for wildlife and indigenous peoples, but losses on a much vaster scale due to rising seas" (quoted in McKibben, 2006: 18). In a similar vein, the 2014 IPCC report argued that averting catastrophe is possible and affordable. The key is to abandon our dependence on dirty fossil fuels and make much greater use of renewable energy sources. As noted earlier, the report suggested that, to facilitate this transition, natural gas (including that produced through fracking) should play an important role.

The Skeptical Environmentalist

The Politicization of Climate Science

Polluted Landscapes

EARTH'S VITAL SIGNS

As the preceding account demonstrates, our current impacts on ecosystems, from the global to the local, are greater than ever before and are increasing as a general result of the growth of population and technology (Figure 4.8). Debate continues, however, concerning the present condition of the environment and the probable future scenario. As Smil (1993: 35) puts it, "Confident diagnoses of the state of our environment remain elusive," and this observation seems as valid today as it was in the early 1990s.

Apocalypse Now, Deferred, or Never?

In fact, there are many different opinions about the impact of human activities on the environment. At one extreme are **catastrophists**, who view the present situation and future prospects in totally negative terms—an opinion articulated by Kaplan (1994, 1996). At the other extreme are **cornucopians**, who believe that the gravity of current problems has been greatly exaggerated and that human ingenuity and technology will overcome the moderate problems that do exist (Simon and Kahn, 1984). This is a broad and complex debate, which we will revisit in Chapter 6.

Responding to Uncertainty

In view of the many uncertainties that surround environmental questions, Smil (1993: 36) writes:

> The task is not to find a middle ground: the dispute has become too ideological, and the extreme positions are too unforgiving to offer a meaningful compromise. The practical challenge is twofold. First, to identify and to separate the fundamental long-term risks to the integrity of the biosphere from less important, readily manageable concerns. The second task is to separate effective solutions to such problems from unrealistic paeans to the power of human inventiveness.

The first challenge is continually being addressed by geographers and other environmental scientists. Given our understanding of global environmental problems, there are three possible general responses to the second challenge:

1. We can attempt to develop new technologies to counter our deleterious impacts. For example, some scientists have suggested that dust might be deliberately spread in the upper atmosphere to reflect sunlight, replacing the depleting

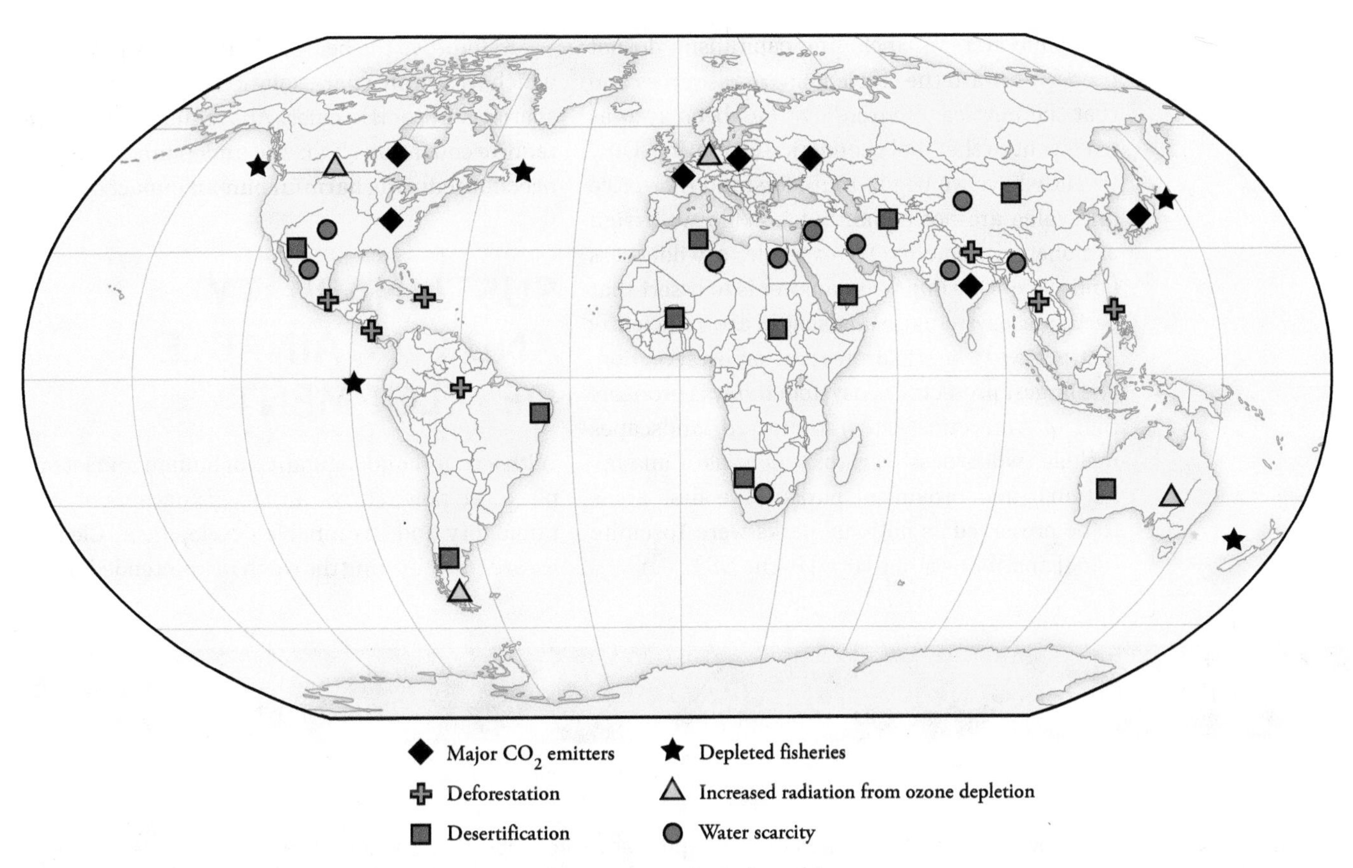

FIGURE 4.8 | Global distribution of some major environmental problems
This map shows the general regional occurrence of six current environmental stresses. Symbols are not intended to identify specific locations or countries but to indicate general regions affected by particular kinds of stress.

Source: Adapted from *Current History* 95 (604) (November 1996).

ozone layer. Overall, though, it seems inappropriate to rely too much on such technological solutions. Despite massive evidence that we are changing the earth's climate on a variety of scales—local to global—we still know too little to change it deliberately in ways we desire. The results of attempts at rain-making, hurricane modification, and fog dispersal have been mixed. But there are perhaps some simple changes we might make that will have positive consequences. Eating less meat will reduce methane emissions from animals and was advocated by the UN climate chief in 2008. Transforming dark urban surfaces into white will increase the reflection of sunlight and reduce warming. California recognized the value of this strategy and in 2005 passed legislation requiring commercial premises with flat roofs to repaint them white.

2. We can acknowledge that environmental impacts are inevitable and emphasize our need to adapt to such changes. In the case of possible climatic change, **adaptation** might involve new population movement, water-supply systems, and coastal defences.
3. The third response, probably the most popular and most logical, involves the conservation of resources and the prevention of harmful impacts. Further, even if we are uncertain about the details of change, the most logical way to approach the environmental crisis is to recognize the likelihood of significant negative consequences and to take immediate action designed to mitigate the anticipated consequences. **Conservation** refers generally to any form of environmental protection. Prevention involves limiting the increase of greenhouse gases, reducing the use of certain materials, and reusing and recycling. **Recycling** often makes sense economically as well as environmentally.

Prevention is central to many current moves to protect environments. It can be argued that

adaptation
The process by which humans adjust to a particular set of circumstances; changes in behaviour that reduce conflict with the environment.

conservation
A general term referring to any form of environmental protection, including preservation.

recycling
The reuse of material and energy resources.

economic systems, including capitalism, do not properly reward the efficient use of resources and that the physical environment has been seen as irrelevant in the final economic accounting. Only recently have we begun to understand that price and value are not equal. But how do we assign a monetary value to, for example, a wilderness landscape? One imprecise answer is to assert that certain ecosystems or landscapes are sufficiently distinct as to merit protection or preservation. The largest protected ecosystem today is probably that of Antarctica. Other protected landscapes include wilderness regions such as Canada's national and provincial parks. (The first areas to be preserved as national parks were Yosemite [1864] and Yellowstone [1872] in the US.)

To improve the health of the earthly patient of which we are all a part, solutions are needed at all spatial and social scales. Above all, perhaps, we require education about and understanding of the need for reducing harmful human impacts.

SUSTAINABILITY AND SUSTAINABLE DEVELOPMENT

With a sound understanding of human impacts in place, we now return to the key concepts of sustainability and sustainable development. Clearly, we are transforming the earth in unintended ways.

Focus on Geographers

Jobs in the Environment and Sustainability Sector | DAN SHRUBSOLE

Established by the federal government, ECO Canada (www.eco.ca) develops programs to help people find environmental jobs. It has provided the following information about current and future employment in this sector:

- There are over 682,000 environmental employees in Canada (4 per cent of total workforce) who spend 50 per cent or more of their time on environmental activities. Over 1.8 million workers (10 per cent of total workforce) spend some of their time on environmental activities.
- Over 318,000 organizations in Canada employ at least one environmental employee (approximately 17 per cent of Canadian organizations have one or more environmental employees).
- Retirements of environmental workers will create vacancies over the next decade as over 100,000 environmental employees (14 per cent of the environmental workforce) reach retirement age.

Thus, the opportunities for employment in the environmental sector in Canada, as well as in other parts of the world, are sizable and growing, and geographers will obtain many of these available positions for a variety of good reasons. First, the topics mentioned in this chapter—energy, deforestation, desertification, maintaining biodiversity, land management, water management, and climate change—are relevant and significant and are studied by many Canadian geographers around the world. Second, understanding and solving these problems are important because they reflect some of the transformational forces that are reshaping how the global community lives and interacts—with one another and with our shared environment. Since one of the geographer's strengths is integrating ideas, we are well placed to understand how these transformational forces are linked to environmental issues. Third, geography, as practised in many university departments, favours an integrated and interdisciplinary approach that combines social and physical sciences. We are thus well placed to research and help solve these types of issues and to team up with others from other academic disciplines. Fourth, as noted in the chapter, ecosystems range in size from a part of a community to the entire globe and are interconnected and interdependent. Since geographers work from a spatial framework, their approach is different from and uniquely useful to other environmental specialists and to understanding the complexity of problems and proposing solutions. Fifth, resource and environmental management is an area in which geography has had a long and strong tradition. Geographers who specialize in this type of training not only have the essential skills of a geographer (i.e. theories, concepts, and methods such as GIS, remote sensing, statistics, qualitative methods), but they also understand the planning techniques, such as environmental impact assessment and benefit–cost analysis as well as relevant international, federal, and provincial legislation. A complex web of acts and regulations guide environmental management in Canada. Some of these can be found on Environment Canada's website (www.ec.gc.ca/default.asp?lang=En&n=E826924C-1)

and include the Arctic Environmental Protection Act, Canada Water Act, and the Canadian Environmental Protection Act. Environmental planners and managers influence decisions, such as whether a development will occur in a specific location and, if so, what restrictions might be placed on it.

My own research has focused on water management in Canada and has dealt with topics such as flooding, wetland management, water quality as it relates to diffuse sources (e.g. from farms), water pricing, and watershed planning and management. My students have found employment with Ontario conservation authorities, environmental consulting firms, the federal and Ontario governments, and the non-governmental sector.

DAN SHRUBSOLE is a professor and the chair of the Department of Geography at Western University.

Equally clearly, we need to manage the earth along appropriate pathways. Management requires us to understand what kind of earth we want, to reach consensus, and to find an appropriate balance between the values of economic development and conservation. Although these aims will be difficult to achieve, if only because people in different areas live in very different circumstances and have very different value systems, it is imperative that a balance be established: the relationship between humanity and the land is such that environment and economics must both be central concerns.

The term *sustainability* was introduced in the late 1970s to refer to the idea that our way of life, based on ever-increasing consumption of resources, could not continue indefinitely and that we would have to find a more sustainable way of life. It was difficult to argue against the desirability of sustainability, but there was considerable disagreement as to what changes were needed to achieve the desired state. The debate on this theme advanced significantly with the introduction of the concept of **sustainable development**, loosely defined as development that accounts for social, economic, and environmental concerns. This term was introduced by *Our Common Future*, the World Commission on Environment and Development's influential 1987 report.

The report defined sustainable development as "development that meets the needs of the present without compromising the ability of future generations to meet their own needs." A more elaborate definition was put forward in 1989: "Sustainability is the nascent doctrine that economic growth and development must take place, and be maintained over time, within the limits set by ecology in the broadest sense—by the interrelations of human beings and their works, the biosphere and the physical and chemical laws that govern it" (Ruckelshaus, 1989: 167). The concept of sustainability can also be explained by reference to the systems concept we looked at earlier. The earth can be regarded as a closed system in that, although energy enters and leaves the system, matter only circulates within it. This type of system can reach a state of dynamic equilibrium—one that involves optimal energy flow and matter cycling in such a way that the

sustainable development
A term popularized by the 1987 report of the World Commission on Environment and Development; refers to economic development that sustains the natural environment for future generations.

Around the Globe

BOX 4.7 | Clayoquot Sound: The Case for Sustainability

Clayoquot (pronounced Klak-wat) Sound, on the west coast of Vancouver Island, is one of the largest (about 3,500 km² or 1,351 square miles) areas of coastal temperate rain forest in the world. It is a distinctive ecosystem, with the highest biomass (weight of organic matter to land area) of any forest type. It also contains at least 4,500 known plant and animal species and possibly several thousand other insects and micro-organisms. Several of the species, such as the sea otter, are on the Canadian endangered species list. Aboriginal groups, including the Nuu-chah-nulth (meaning "all along the mountains"), have lived in the area for more than 8,000 years without causing environmental degradation, but the area has become attractive to lumber companies in recent decades, and clear-cut logging is posing real threats to the ecosystem.

There are, then, several conflicting interests at stake. Conservationists, including groups such as the Sierra Club,

Continued

argue for preservation of the ecosystem; Aboriginal people ask that their values and lifestyles, which are intimately related to the forest, be recognized and respected; and lumber companies seek to make profits in a province that has 20 per cent of its economy in the forestry sector. In 1984 the first open confrontation occurred when local Aboriginal and other residents set up a blockade to prevent logging. The provincial government established a Wilderness Committee in 1986, then three mediating forums in 1986, 1989, and 1992. Meanwhile, in 1988, a second confrontation began in another location.

Discussions at the various forums failed to arrive at an agreement satisfactory to all interest groups. In 1993 the provincial government announced the Clayoquot Sound Land Use Decision, permitting logging in two-thirds of the area and protecting one-third. A massive civil disobedience movement followed, with about 900 people arrested by the end of the year. In 1994 the forum set up two years earlier (the Commission on Resources and the Environment) proposed arrangements for the area that would result in the loss of about 900 logging jobs. This prompted a mass demonstration by loggers.

In 1995 the provincial government endorsed a report of the Clayoquot Sound Scientific Panel, another advisory forum it had established following the 1993 protests; the report made a series of recommendations that included significant reductions in the rate of cutting but did not call for permanent protection. The recommendations represented a substantial change from previous reports, as they explicitly acknowledged the importance of maintaining the rain forest ecosystem. This change of attitude appears to be continuing. At their 1996 annual meeting, members of the World Conservation Union, including the BC government, endorsed a proposal supporting the designation of Clayoquot Sound as a UN Biosphere Reserve. Further, in 1999 the provincial government reached agreements with MacMillan Bloedel concerning compensation for the company's loss of cutting rights following the creation of new parks. In a move welcomed by environmental groups, the company announced new forest management policies involving the phasing out of clear-cutting, increased conservation of old growth, and independent validation of their forest practices. That same year, MacMillan Bloedel was acquired by Weyerhaeuser to create the largest North American forest products company. Currently, the annual total yield from the limited logging in Clayoquot Sound ranges between 0 and 100,000 cubic metres—a massive reduction from the late 1980s high of almost 1 million cubic metres per year—and Aboriginal people share in the logging and in the profit, a small but perhaps not insignificant instance of "sustainable livelihoods."

In the summer of 1993, more than 850 people were arrested for blockading a logging road into Clayoquot Sound. This non-violent civil disobedience campaign was designed to attract media coverage and pressure the provincial and federal governments into putting a stop to clear-cutting in the area.

This example highlights many important aspects of human impacts on land. Above all, it suggests that governments find it hard to appreciate the arguments for a sustainable economy—not surprising, given that governments need political support from both industry and trade unions and also benefit financially from economic activities (in this case from the sale of logging licences). But it also indicates that public education campaigns and intense public pressure can have an impact on government attitudes and policies. In 2006 the BC government announced that about one-third of the Great Bear Rainforest, which stretches along the Pacific coast between Vancouver Island and Alaska, is to be preserved. The remaining two-thirds will see some logging using sustainable practices. Environmental groups, including Greenpeace, continue to lobby for less logging.

system does not collapse. In a sense, sustainable development would represent a similar state of dynamic equilibrium.

The idea of sustainable development became a key focus at the UN Conference on Environment and Development held in Rio de Janeiro in 1992. The essence of sustainable development is the attempt to blend sustainability and development, which are often seen as opposites, into one process. For some critics the phrase is an oxymoron—a contradiction in terms. After all, to limit damage to environments, it might be necessary to limit growth. This contradiction is most easily resolved, in principle if not in practice, by suggesting that rich countries should strive for sustainability by limiting their own growth, while poor countries, which have not reached an adequate level of economic development and should have a chance to do so, should strive for sustainable development.

How do we move towards a sustainable world where environmental changes are in accord with sound ecological principles? Any such move clearly requires a significant—and deliberate—shift towards a new attitude. Four principles are essential to that new attitude:

1. We need to recognize that humans are a part of nature. To destroy nature is to destroy ourselves.
2. We need to account for environmental costs in all our economic activities.
3. We need to understand that all humans deserve to achieve acceptable living standards. A world with poor people cannot be a peaceful world.
4. We need to be aware that even apparently small local impacts can have global consequences. This is one of the basic themes of ecology—"think globally, act locally"—and it becomes ever more obvious as globalization processes unfold.

There are, of course, vast differences between acknowledging the need for a new attitude, seeing it adopted globally, and, finally, putting it into practice. The dilemma facing humans today was expressed forcefully some years ago: "To continue with the trial and error procedures of the past means to risk irreparable damage to [our] habitat. To accept responsibility for full environmental control means to anticipate change and decide in which direction to go" (Wilkinson, 1963: 32). Since these words were written, many strides have been made in the right direction. Clean air acts, environmental impact assessments, and environment ministries are now standard in many countries. Although it would be an exaggeration to assert that solutions are in sight, or even that they are being sought in all cases (Box 4.7), environmental issues are increasingly recognized as relevant at all levels, from individuals to governments. But this is not equally the case for all countries. Thus, it is the clear responsibility of the more developed world to demonstrate environmental concern by example:

> In creating the consciousness of advanced sustainability, we shall have to redefine our concepts of political and economic feasibility. These concepts are, after all, simply human constructs; they were different in the past, and they will surely change in the future. But the earth is real, and we are obliged by the fact of our utter dependence on it to listen more closely than we have to its messages. (Ruckelshaus, 1989: 174)

CONCLUSION

At the beginning of this chapter, we noted that everything is related to everything else, that one cannot change one aspect of nature without directly or indirectly affecting other aspects. This fact points to the difficulty of predicting the future with any degree of certainty: "The principal reason why even the cleverest and the most elaborate explorative scenarios are ultimately so disappointing is that they may get some components of future realities approximately right, but they will inevitably miss other ingredients whose dynamic interaction will create profoundly altered outcomes" (Smil, 2005: 202).

Here are four suggestive concluding comments.

- Through a combination of our large and growing population, our advancing technologies, and our improving living standards, we are changing the global ecosystem in many and varied ways, often with negative consequences. The powerful idea of the Anthropocene epoch suggests that we need to think and act differently from the past.
- Some of these ways are fairly easy to predict, at least in general terms, while others are likely unanticipated.
- Our global environmental future is uncertain, as evidenced by the multitude of contradictory reports that include results and forecasts that sometimes flow from preconceived outcomes.
- Because the authors of some research assume never-ending progress while other authors assume inevitable downfall, the best message is to read and think about environmental issues both thoughtfully and critically.

Summary

Use or Abuse?

There is considerable evidence that we are damaging our home—the earth—and increasing recognition of how fragile that home is.

Scale

The value of a global perspective in any discussion of human use of the earth is clear: everything is related to everything else.

The Increasing Human Impact

Two principal variables—sheer numbers of people and increasing use of technology and energy—explain why human activities have such great impact on the earth.

Ecosystems

Combining systems logic and ecological principles produces the concept of the ecosystem: any self-sustaining collection of living organisms and their environment is an ecosystem. A whole series of ecosystems together make up the global ecosystem—the ecosphere or biosphere—which is the home of all life on earth. The ecosphere includes air, water, land, and all life.

When humans change an ecosystem, the result is usually a simplification. We cause changes largely because of our numbers, our technology, and our energy use. Today, we number seven billion, and our technological ability—that is, our ability to convert energy into forms useful to us—is constantly increasing. As human culture (particularly technology) changes, so does the value attributed to resources. Different interest groups evaluate resources according to different criteria and hence have differing views. Stock resources are finite in quantity, whereas renewable resources are relatively unlimited.

Our simplification of ecosystems is now sufficiently evident that a new environmental ethic is needed—one involving co-operation with nature, not domination over it.

Human Impacts

The presence of humans on the earth changes the planet. Some of the changes brought about by humans are inevitable because they are necessary to human survival. Other changes, however, are the result of inappropriate actions motivated by human greed and misunderstanding. Over time, humans' ability to affect ecosystems at all levels has increased. Although it may be inappropriate to focus on individual parts of an ecosystem—which is by definition an inseparable whole—it is convenient to consider the impact of human activities on specific parts of ecosystems such as vegetation, animals, land and soil, water, and climate. In each case, we have caused many changes.

Two important issues are tropical rain forest depletion and desertification; in both cases, human activity is at least partially explained by the problems experienced by poor people in less developed countries. Impacts on animal life include those associated with domestication and species extinction. Extensive changes to land surfaces result from a host of human activities such as resource extraction and the creation of urban industrial complexes. Soil is particularly susceptible to abuses involving salinization, increased laterite content caused by removal of vegetation, and erosion. Water, an essential ingredient for life, is typically used in ways that result in shortages and contamination, and the oceans continue to be polluted.

The ecosystem concept of interrelatedness is especially useful with respect to climate. We are having a significant effect on the atmosphere by increasing the quantity of greenhouse gases and damaging the ozone layer. Climate change is undeniable and humans are a cause, but we also need to understand more about climatic change that is not caused by humans.

The Health of the Planet

Although debates continue concerning the consequences of human activity for the health of the planet, most agree that human-induced ills are many and serious. Probably the best solution is prevention—an argument advanced by many environmentalist groups.

Sustainable Development

Sustainable development serves our present needs but does not compromise the ability of later generations to meet their needs.

Links to Other Chapters

- **Globalization:** Chapter 2 (concepts); Chapter 3 (transnationals); Chapter 5 (population growth, fertility decline); Chapter 6 (refugees, disease, more and less developed worlds); Chapter 7 (popular culture); Chapter 8 (popular culture); Chapter 9 (political futures); Chapter 10 (agriculture and the world economy); Chapter 11 (global cities); Chapter 13 (industrial restructuring)
- **The interrelatedness of all things:** All chapters
- **Relations between humans and land:** Chapter 1 (Humboldt, Ritter, Vidal, and Sauer); Chapter 2 (environmental determinism; possibilism)
- **Energy use and technology:** Chapter 13 (especially the Industrial Revolution and fossil fuel sources)
- **More developed and less developed regions:** Chapter 6
- **Agricultural revolution:** Chapter 3 (beginnings of civilization); Chapter 10 (agriculture)
- **Natural resources and human values:** Chapter 2 (possibilism)
- **Human impacts:** Chapters 3, 10, 11, 12, 13 (economic geography); Chapter 10 (agriculture); Chapter 13 (industrial activity)
- **The catastrophist–cornucopian debate:** Chapters 5 and 6

Questions for Critical Thought

1. In what ways is the earth "fragile"? Do you agree with this description of the earth? Why or why not?
2. How and why have human impacts on the earth changed during the Holocene?
3. The discussion of the three contentious issues regarding the current debate about the environment states that "market forces are unlikely to solve environmental problems." Do you agree? Why or why not? What other factors (apart from market forces) might play a key role in our attempt to deal with environmental issues?
4. Are humans part of nature or are they separate from nature? Give reasons to support your answer.
5. Of the various global environmental issues discussed in this chapter (tropical rain forest destruction, desertification, biodiversity loss, pollution, and climate change), which one do you think is most serious? Why? What do you think should be done about this particular issue?
6. Are you a catastrophist or a cornucopian? How would you explain your position?

Suggested Readings

Visit the companion website for a list of suggested readings.

5 POPULATION GEOGRAPHY

This chapter focuses on the fundamentals of human population growth and spread through time, at both global and regional scales. The two factors on which all changes in population size depend are fertility and mortality. In this chapter, we introduce the various measures used to track fertility and mortality rates, some of the factors that influence those rates, and the temporal and spatial variations that occur in both cases. We look at recent evidence indicating that fertility is declining not only in the more developed parts of the world but in the less developed parts as well. We also pay attention to the way that many governments actively seek to influence fertility, with existing policies often being a major factor affecting fertility at the national scale. Our consideration of mortality includes factors affecting it and variations in mortality measures.

Some implications of aging populations are also debated. We look at the history of population growth and at five theories that have been proposed to explain the history of population growth and to predict possible scenarios for the future. The chapter concludes with discussions of global population distribution and density and of migration.

Here are three points to consider as you read this chapter.

- Is it appropriate for national governments to implement policies designed to either decrease or increase fertility or should people be left to determine for themselves how many children they have?
- In the early nineteenth century, Malthus argued that our human future would be characterized by famine, vice, and misery. Was he correct? Why or why not?
- Where do we live today, why there, and why does this matter?

The Rabindra Setu (also known as the Howrah Bridge) links the two cities of Howrah and Kolkata, India. Carrying over 100,000 vehicles and at least 500,000 pedestrians daily, this cantilever bridge is the busiest in the world. India is projected to become the world's most populous country by the end of the next decade.

vodafone
AIRCEL
per second
and STD @ 1paisa/sec.
HOWRAH DISTRICT POLICE
SGS
HOWRAH Stn.
CSTC

In 2015, the total world population was estimated to be 7.3 billion. By the time you read these words, that figure will be higher; the UN estimates 8.4 billion by the year 2025, 9.7 billion by 2050, and possibly a stable population of just over 10 billion by 2200. More significant than this numerical increase, however, is the fact that the increases are occurring in less developed parts of the world that are currently least capable of supporting them, whereas population in the more developed world is expected to remain essentially unchanged through to 2050. Hence, an ever-present theme in this chapter (and the one following) is the uncertainty that such increases in the less developed world imply for human well-being. A number of human geographers and other scholars have suggested links between population numbers and such problems as famine, disease, and (as discussed in the previous chapter), environmental deterioration. Others, however, are more optimistic about our ability to cope with increasing numbers, arguing that past increases have been accompanied by improvements in human well-being through technological change.

Much of this chapter draws on measures and procedures developed in **demography**, the science that studies the size and makeup of populations (according to such variables as age and sex), the processes that influence the composition of populations (notably fertility and mortality), and the links between populations and the larger human environments of which they are a part.

demography
The study of human populations.

fertility
Generally, all aspects of human reproduction that lead to live births; also used specifically to refer to the actual number of live births produced by a woman.

fecundity
A biological term; the ability of a woman or man to produce a live child; refers to potential rather than actual number of live births.

The UN celebrates the world's population reaching seven billion. In 1989 the organization designated 11 July as World Population Day, a day to focus on population issues.

UN Photo/Eskinder Debebe

FERTILITY

As mentioned in this chapter's introduction, all global changes in population size can be understood by reference to fertility and mortality. Thus,

$$P_1 = P_0 + B - D$$

where

P_1 = population at time 1
P_0 = population at time 0 (before time 1)
B = number of births between times 0 and 1
D = number of deaths between times 0 and 1

Fertility and mortality rates vary significantly according to time and location, and both rates are affected by many different variables.

At the subglobal level, a third factor becomes relevant: migration into and out of that particular subdivision of the world. Thus,

$$P_1 = P_0 + B - D + I - E$$

where

I = number of immigrants to area between times 0 and 1
E = number of emigrants from area between times 0 and 1

Fertility Measures

The simplest and most common measure of **fertility** is the *crude birth rate* (CBR), the total number of live births in a given period (usually one year) for every 1,000 people already living. Thus,

$$\text{CBR} = \frac{\text{number of live births in one year}}{\text{mid-year total population}} \times 1{,}000$$

The 2014 CBR for the world was 20. Historically, measures of CBR have typically ranged from a minimum of about 8 (the theoretical minimum is 0) to a maximum of 55, which is a rough estimate of the biological maximum. Clearly, the CBR is a very useful statistic, but it may be somewhat misleading because births are related to the total population and not to the subset that is able to conceive (it is called crude for this reason).

To more accurately reflect underlying fertility patterns, we measure female **fecundity**—the ability of a woman to conceive. The *total fertility rate* (TFR) is the average number of children a woman will have, assuming she has children at the prevailing age-specific rates, as she passes through the fecund years. This age-specific measure of fertility is useful because child-bearing during the fecund years varies considerably with age. It is calculated as follows:

$$\text{TFR} = 5 \sum_{A=1}^{7} \frac{\text{number of births to woman in age group A in a given period}}{\text{mid-year number of females in age group A}}$$

where A refers to the seven 5-year age groups of 15–19, 20–4, 25–9, 30–4, 35–9, 40–4, and 45–9. The 5 preceding the summation sign is necessary because each age group covers five years. The world TFR for 2014 was 2.5, meaning that the average woman has 2.5 children during her fecund years.

Although there are various other measures of fertility, with different measures used in different circumstances and with different aims in mind, we will concentrate on the CBR and the TFR. The former is a factual measure reflecting what has actually happened in a given time period—x number of children per 1,000 members of the population were born. The latter, on the other hand, predicts that a woman will have the same number of children as do other women her age.

The impact of a particular CBR or TFR on population totals is related to mortality. Generally, a TFR of between 2.1 and 2.5 is considered **replacement-level fertility**; that is, it maintains a stable population. Of course, if there were no early deaths, the replacement level would be 2.0—two parents replaced by their two children. But the level has to allow for the fact that females may die before giving birth. The lower figure applies to areas with relatively low levels of mortality (few females dying before having children) and the higher figure to areas with relatively high levels of mortality (more females dying before having children).

replacement-level fertility
The level of fertility at which a couple has only enough children to replace themselves.

Factors Affecting Fertility

The reproductive behaviour of a population is affected by biological, economic, and cultural factors. It is not difficult to identify many of these aspects; it is challenging, however, to assess their specific effects. But perhaps the most surprising fact to emerge as the discussion of population unfolds throughout this chapter is that human beings might be the only species that reproduces less when they are well fed.

Biological Factors

Age is the key biological factor in fecundity. For females, fecundity begins at about age 15, reaches a maximum in the late twenties, and terminates in the late forties. The pattern for males is less clear, with fecundity commencing at about age 15, peaking at about age 20, and then declining but without a clear termination age. Some females and males are sterile (i.e. incapable of reproduction).

Reproductive behaviour is also affected by nutritional well-being; populations in ill health are very likely to have impaired fertility. Thus, periods of famine reduce population growth by lessening fertility (as well as by increasing mortality). Another diet-related biological factor is that women with low levels of body fat tend to be less fecund than others. Members of nomadic pastoral societies, who are particularly likely to have low levels of body fat because they eat a low-starch diet, are characterized by low fertility.

Fertility in Tropical Africa

Economic Factors

Until recently, fertility changes in more developed countries were essentially part of the economic development process. With increasing industrialization and urbanization, fertility declines. This economic argument suggests that traditional societies, in which the family is a total production and consumption unit, are strongly *pro-natalist*; that is, they favour large families. By contrast, modern societies emphasize small families and individual independence. In economic terms, children once were valued for their contributions to the household but today represent an expense. The economic argument is that the decision to have children is essentially a cost–benefit decision. In the traditional (often, extended-family) setting, children are valuable both as productive agents and as sources of security for their parents in old age—hence large families. Neither of these factors is important in modern societies.

This economic claim implies that any reductions in fertility are essentially caused by economic changes. The idea is central to the demographic transition theory—one of the five explanations of population growth to be discussed later in this chapter.

Cultural Factors

A host of complex and interrelated cultural factors also affect fertility; it is frequently suggested that the reasons behind current reductions in fertility are primarily cultural rather than economic. This suggestion is central to the fertility transition theory (Box 5.1).

Most cultural groups recognize *marriage* as the most appropriate setting for reproduction. There are several measures of nuptiality, the simplest of which is the nuptiality rate:

nuptiality
The extent to which a population marries.

$$\text{nuptiality rate} = \frac{\text{number of marriages in one year}}{\text{mid-year total population}} \times 1{,}000$$

The age at which females marry is important because it may reduce the number of effective fecund years. A female marrying at age 25, for example, has "lost" 10 fecund years. Until recently, perhaps the most obvious instance of such loss was Ireland. In the 1940s, the average age at marriage for Irish women was 28 and for Irish men 33; moreover, high proportions of the population—as many as 32 per cent of women and 34 per cent of men—lost all their fecund years as a result of not

Around the Globe

BOX 5.1 | Declining Fertility in the Less Developed World

The less developed world is experiencing a decline in fertility that began in the 1970s. The reasons behind this decline were not apparent until recently because data on such matters were not readily available. However, information derived largely from 44 surveys of more than 300,000 women, carried out during the 1990s, is substantial and well documented. It shows that birth rates are falling rapidly without any prior improvement in economic or living conditions. Therefore, this change appears to have little to do with economic factors and everything to do with cultural factors (Robey, Rutstein, and Morris, 1993).

The magnitude of this decline is evident in Thailand, which had a 1975 TFR of 4.6 and a 2014 TFR of 1.8; similar reductions have occurred in many other countries in Asia, Latin America, and North Africa. The fact that several sub-Saharan African countries—including Kenya, Botswana, Zimbabwe, and Nigeria—also are experiencing reduced fertility is especially significant, as this area has always appeared to be resistant to any such trend. These fertility declines are in addition to the case of China, where (often compulsory) government policies have contributed to a below-replacement fertility level of 1.4 only about 30 years after the introduction of the regulations.

The importance of fertility declines in the less developed world is difficult to exaggerate, both practically and in terms of our understanding of fertility change. The conventional wisdom in the 1970s was well expressed by Demeny (1974: 105): countries in the less developed world "will continue their rapid growth for the rest of the century. Control will eventually come through development or catastrophe." It now appears that rapid growth is slowing down but not for either of these two reasons. As a result, many earlier accounts of anticipated population growth are being revised.

The most powerful influence on fertility in the less developed world has been improvement in the education and lives of women and the related extent to which modern contraceptive methods are employed (Robey et al., 1993). This logic is labelled the fertility transition model. Education leads to family planning; hence, fertility drops. In the less developed world (not including China), about 46 per cent of married females practise modern methods of family planning. Information about family planning is widespread because of the influence of the mass media.

According to the fertility transition model, large families have fallen out of favour because of the obvious problems associated with rapid and substantial population increases, such as pressure on agricultural land and poor quality of urban life. The evidence is clear: females—who are generally experiencing a rise in social status—favour later marriage, smaller families, and more time between births.

Thus, the principal reason fertility is declining so rapidly in the less developed world is that women are better educated. Also, effective contraception is available to meet the demand as soon as the necessary cultural impetus is in place; this was not the case when the more developed world was first culturally predisposed to smaller families, a time when abstinence, withdrawal, and abortion were the only techniques available. The current fertility decline may be best described as a "reproductive revolution" because it is rapid and substantial.

Economic development does help to create a climate conducive to reductions in fertility, but the key point of the fertility transition argument is that these reductions are caused by a new cultural attitude related to educational advances—a desire for smaller families and the willingness to employ modern contraceptive methods—and by the ready availability of these methods.

marrying. Late marriage and non-marriage are usually explained by reference to social organization and economic aspirations. Interestingly, since about 1971 the Irish people have tended to marry earlier and more universally. Other cultures actively encourage early marriage. Some Latin American countries, such as Panama, have legal marriage ages as low as 12 for females and 14 for males. Overall, delayed marriage and celibacy are uncommon throughout Asia, Africa, and Latin America. Some of the recent success in reducing fertility in China can be explained by the government requirement that marriage be delayed until age 25 for females and 28 for males.

As discussed in Box 5.1, fertility is also affected by *contraceptive use*. Attempts to reduce fertility within marriage have a long history. The civilizations of Egypt, Greece, and Rome all used contraceptive techniques. Today, contraceptives are used around the world but especially in more developed countries. The less developed countries typically have lower rates of contraceptive use, although most evidence shows that the rates are increasing (Table 5.1). The practice of contraception is closely related to government attitudes and to religion. Most of the world's people live in countries that actively encourage limits to fertility, yet as recently as 1960 only India and Pakistan had active programs to reduce fertility. The complex issue of government policies designed to impact fertility is discussed later in this chapter.

TABLE 5.1 | Contraceptive use by region, 2014

	% Married Women 15–49 Using Contraception	
	All Methods	Modern Methods[1]
World	63	56
More developed world	70	60
Less developed world	61	55
Less developed world (excluding China)	54	46

1. "Modern" methods include the pill, IUD, condom, and sterilization.
Source: Population Reference Bureau. 2014. *2014 World Population Data Sheet*. Washington, DC: Population Reference Bureau. www.prb.org/pdf14/2014-world-population-data-sheet_eng.pdf.

Another important cultural factor affecting fertility is *abortion*, the deliberate termination of an unwanted pregnancy. It is estimated that, for every three births in the world, at least one pregnancy is deliberately terminated. Romania has the highest rate of abortions, with about 75 per cent of pregnancies terminated, while in Russia about 66 per cent are terminated. Only about half of the estimated 45 million abortions performed annually in the world are legal.

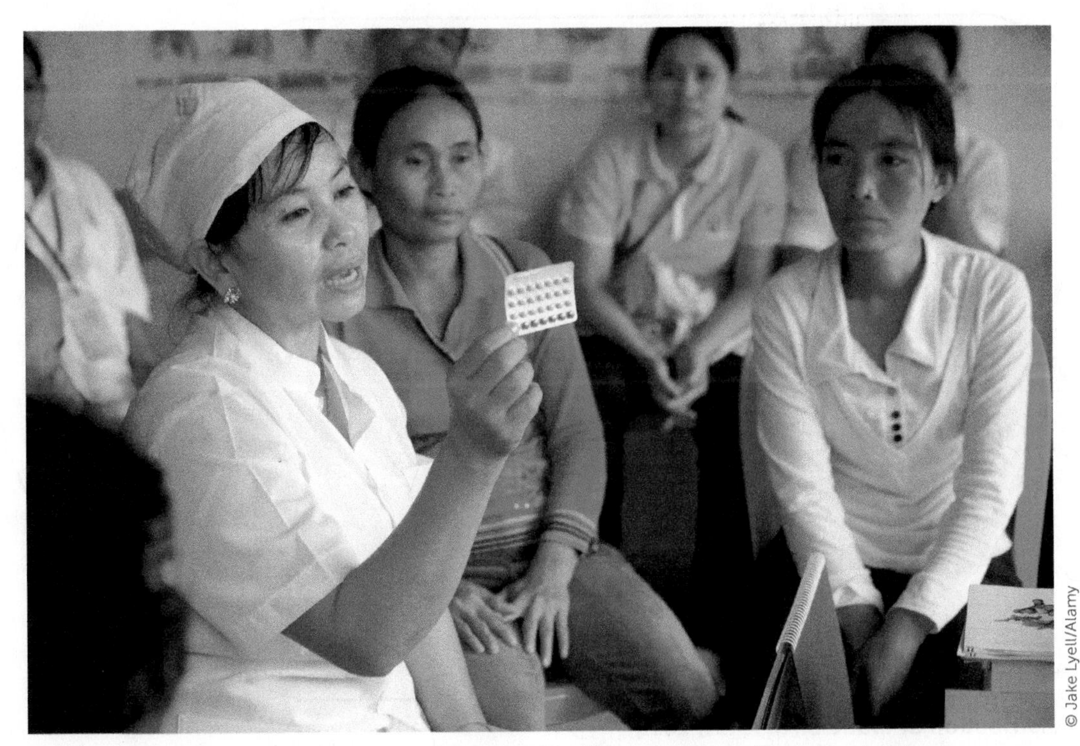

Women talk about birth control at a reproductive health clinic in Kampong Cham, Cambodia.

In many countries, governments actively try either to promote abortion or to discourage it for their own pragmatic reasons. China's liberal abortion law reflects the need for population control; indeed, abortion is obligatory for women whose pregnancies violate government population policies. Sweden has a liberal abortion law on the grounds that abortion is a human right. In other countries, particularly those where the dominant faith is either Catholicism or Islam, access to abortion is either denied or limited to instances in which the woman's life is threatened—about 40 per cent of women live in countries with very restrictive abortion laws.

Abortion is an even more complex moral question than contraception. Although it is a long-standing practice, it is still subject to condemnation, usually on religious and moral grounds. In countries such as Canada, Australia, the United Kingdom, and (especially) the United States, abortion is one of the most hotly debated biological, ethical, social, and political issues. Notably, UN documents do not refer to abortion as a right because of the opposition of some member countries.

Variations in Fertility

Spatial variations in fertility correspond closely to those in levels of economic development. But this does not necessarily mean that economic factors are the key cause. Modernization and economic development have been related to lower levels of fertility, mainly since the onset of the Industrial Revolution. At the broad regional level, the more developed world had a CBR of 11 and a TFR of 1.6 in 2014, while the less developed world had a CBR of 22 and a TFR of 2.6. More specifically, in the more developed world, Italy was among the group of countries with low fertility, with a CBR of 9 and a TFR of 1.4, while in the less developed world, Liberia had a CBR of 35 and a TFR of 4.7. Figure 5.1 maps CBR by country. In general, variations of this type reflect economic conditions.

As noted in Box 5.1, fertility rates in the less developed world are decreasing, and the reasons are more cultural than economic. Rather more surprisingly, there is evidence of declining fertility in several countries in the more developed world that may be prompted by cultural factors (Box 5.2).

Of course, fertility also varies significantly within any given country. There is, for example, usually a clear distinction between urban areas (relatively low fertility) and rural areas (relatively high fertility). This distinction can be found in all countries, regardless of development levels.

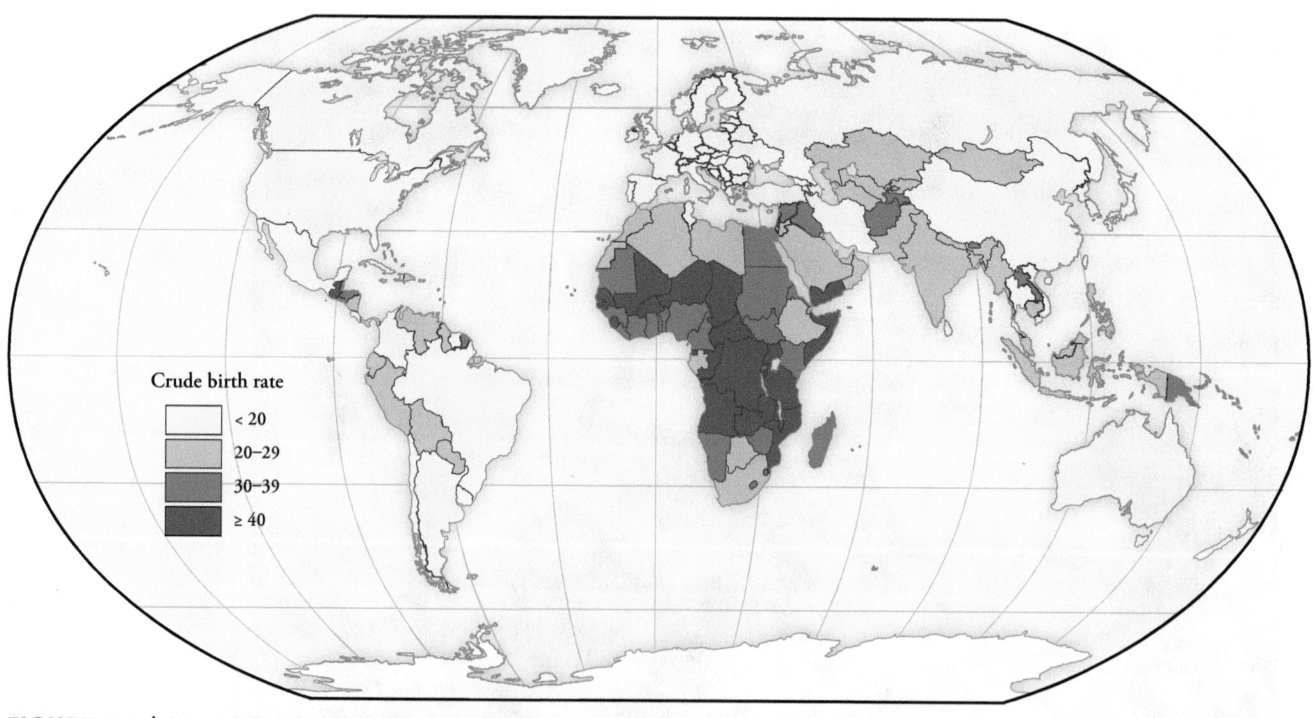

FIGURE 5.1 | World distribution of crude birth rates, 2014

Source: Population Reference Bureau. 2014. *2014 World Population Data Sheet*. Washington, DC: Population Reference Bureau.

Around the Globe

BOX 5.2 | Declining Fertility in the More Developed World

"The last European will die on 6 August 2960" (Nelson, 2006: 24). This "prediction" assumes that existing demographic trends continue and is the outcome of a European world where sex is separated from child-bearing and people are too busy to have large families. Indeed, the lowest fertility rates in the world are found in several European countries: the TFR is as low as 1.2 in Moldova and Poland and 1.3 in Hungary, Romania, Bosnia-Herzegovina, Greece, and Spain. The extraordinarily low figures for much of the continent reflect both a desire to postpone starting families and a desire for smaller families. For many Europeans, the ideal family now includes only one child. Such preferences may be prompted by uncertain economic circumstances and the growth of employment opportunities for women (Hall, 1993; King, 1993).

The TFR for Germany, 1.4, reflects an unprecedented trend in the former East Germany, which appears to have come as close to a temporary suspension of child-bearing as any large population in the human experience. Eastern Germans have virtually stopped having children. The explanation is not an increase in abortions, which have also fallen abruptly. One possible explanation is the trauma associated with the transition from communism to capitalism and, specifically, concern about employment opportunities for future generations in a region with high levels of unemployment. It may be appropriate, then, to interpret the low fertility rate in Germany—actually a form of demographic disorder—as a temporary outcome of the transition from the "old" to the "new" political order. This suggestion appears to have some merit, as fertility also declined between 1989 and 1993 in Poland by 20 per cent, 25 per cent in Bulgaria, 30 per cent in Romania and Estonia, and 35 per cent in Russia (*The Economist*, 1993: 54). These low fertility rates have continued to the present and are resulting in net losses in population (excluding the effects of immigration), since the replacement level in such countries is a TFR of about 2.1.

Both Germany and Poland have placed fertility high on the political agenda. Sweden and Norway in particular have a range of policies intended to encourage couples to have children, notably generous maternity and paternity leave arrangements, heavily subsidized daycare, and flexible work schedules; France has the most extensive state-funded child-care system in Europe. Several Asian countries are also concerned: both Japan and Hong Kong are actively promoting fertility.

Although Canada does not belong to the group of lowest-fertility countries, both CBR and TFR statistics have been declining or stable in recent years. Most European countries have slightly lower birth and fertility rates than Canada, but the United States have slightly higher levels (CBR is 13 and TFR is 1.9). Table 5.2 summarizes the demographic situation in Canada.

TABLE 5.2 | Population data, Canada, 2014

Total Population	34.5 million
CBR (crude birth rate)	11
CDR (crude death rate)	7
RNI (rate of natural increase; in per cent)	0.4
IMR (infant mortality rate)	4.8
TFR (total fertility rate)	1.6

Source: Population Reference Bureau. 2014. *2014 World Population Data Sheet*. Washington, DC: Population Reference Bureau.

Similarly, fertility within a country is higher for those with low incomes and for those with limited education.

MORTALITY

Mortality Measures

Like fertility, **mortality** may be measured in a variety of ways. The simplest, which is equivalent to the CBR, is the *crude death rate* (CDR), the total number of deaths in a given period (usually one year) for every 1,000 people living. Thus,

$$\text{CDR} = \frac{\text{number of deaths in one year}}{\text{mid-year total population}} \times 1{,}000$$

The CDR for the world in 2014 was 8. Measures of CDR have typically ranged from a minimum of 5 to a maximum of 50.

mortality
Deaths as a component of population change.

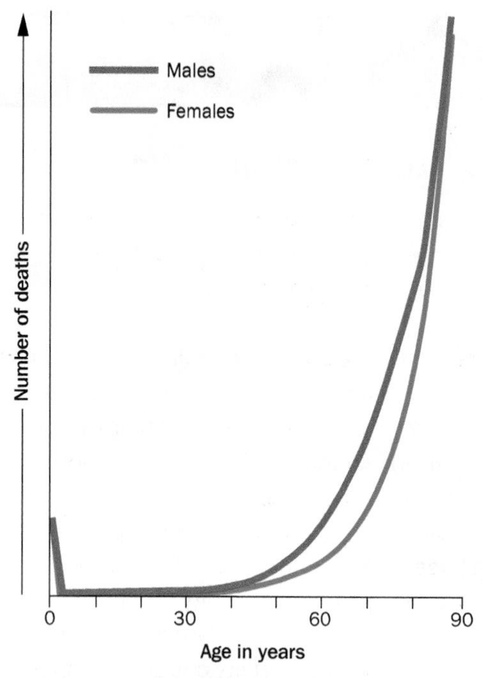

FIGURE 5.2 | Death rates and age

The CDR does not take into account the fact that the probability of dying is closely related to age (it is called crude for this reason). Usually, death rates are highest for the very young and the very old, producing a characteristic J-shaped curve (Figure 5.2). Other mortality measures consider the age structure of the population.

The most useful of these for our purposes is the *infant mortality rate* (IMR), the number of deaths of infants under 1 year old per 1,000 live births in a given year. Thus,

$$\text{IMR} = \frac{\text{number of infant deaths under one year old}}{\text{number of births in that year}} \times 1{,}000$$

The world IMR for 2014 was 38, although figures for individual countries ranged from as low as 1.8 for Finland and Iceland to as high as 116 for the Central African Republic. The IMR is sensitive to cultural and economic conditions, declining with improved medical and health services and better nutrition; reductions in the IMR usually precede overall mortality decline. Infant mortality, and child mortality more generally, are declining, largely because of measles vaccinations, mosquito nets, and increased rates of breast-feeding. This is detailed in Chapter 6 in the discussion of the Millennium Development Goals.

Although not a mortality measure, another useful statistic that reflects mortality is *life expectancy* (LE), the average number of years to be lived from birth. World life expectancy in 2014 was 71, although for Japan and some European countries the LE was in the low eighties and in the forties in Botswana, Lesotho, Sierra Leone, and Swaziland.

Factors Affecting Mortality

Humans are mortal. Whereas it is theoretically possible for the human population to attain a CBR of zero for an extended period, the same cannot be said for the CDR. Unlike fertility, which is affected by many biological, cultural, and economic variables, mortality is relatively easy to explain—despite the fact that the World Health Organization has recognized some 850 specific causes of death! As the CDR and LE figures cited above suggest, mortality generally reflects socio-economic status: high LE statistics are associated with high-quality living and working conditions, good nutrition, good sanitation, and widely available medical services—and vice versa. When disease is prevalent, mortality increases: the most dramatic example in recent years is AIDS. (Diseases are discussed in Chapter 6.) Mortality is also higher when countries experience conflict. In recent years, these include Afghanistan, DR of Congo, Iraq, Sudan, and Syria.

Variations in Mortality

The study of death statistics tells a great deal about how people lived their lives because many die not as a result of aging but rather as a result of environmental conditions and/or lifestyle. The world pattern for CDR shows much less variation than does that for CBR (Figure 5.3). This is a reflection of the general availability of at least minimal health-care facilities throughout the world. Figure 5.4 maps LE by country. In this case, major variations are evident: as already noted, LE figures are more sensitive to availability of food and health-care facilities than are death rates. The map of LE is a fairly good approximation of the health status of populations. Low LE statistics are found in tropical countries in Africa and in South and Southeast Asia. High LE statistics are particularly common in Europe and North America.

Mortality measures also vary markedly within countries. In countries such as the United Kingdom, Canada, the United States, and Australia, certain groups have higher CDRs and IMRs and lower LEs than the population as a whole. These differences reflect what we might call the

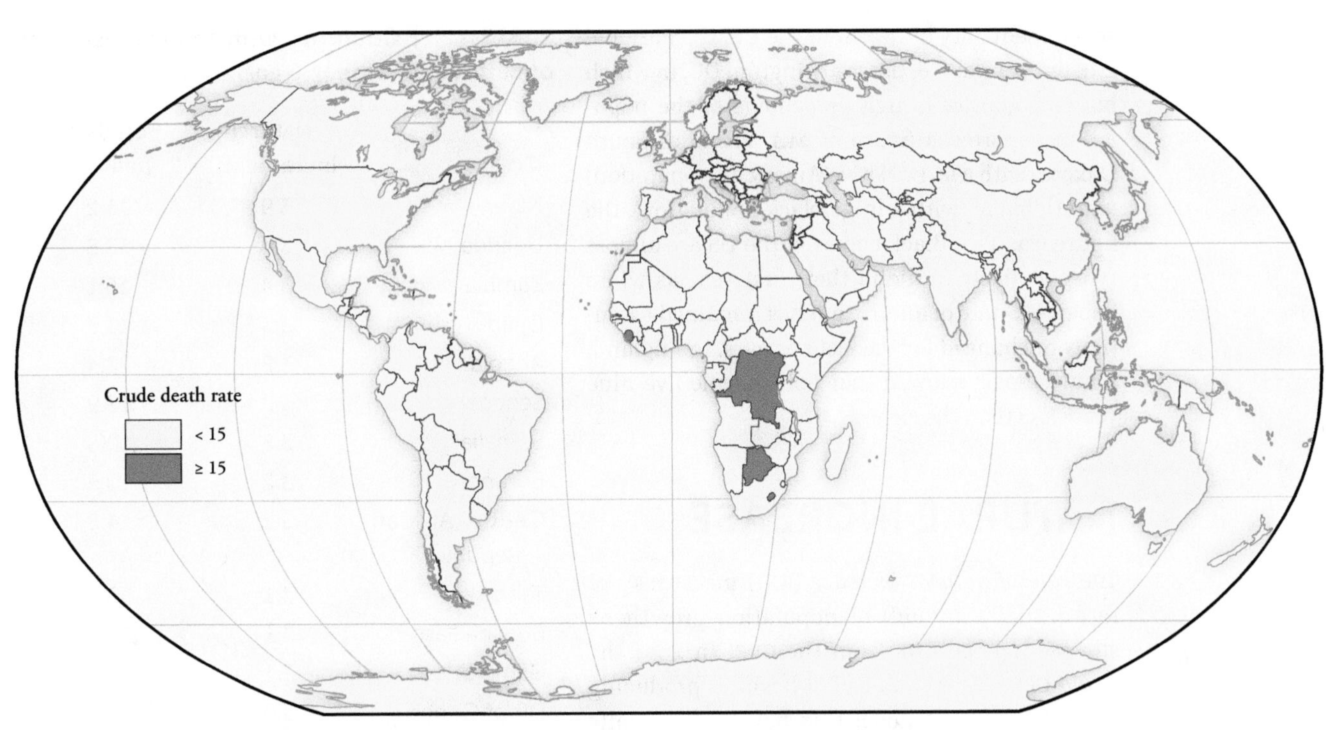

FIGURE 5.3 | World distribution of crude death rates, 2014

As this map shows, the world pattern of death rates varies relatively little. Most countries have rates of 15 or less, with the striking exception of several sub-Saharan African countries. Generally speaking, rates lower than 15 are acceptable. It is notable that, in some countries, rates are increasing because of the increasing percentage of elderly people in a national population; the greater the proportion of elderly people, the higher the death rate. For example, both Finland and the United Kingdom have a rate of 9, whereas Malaysia and the Philippines have rates of 5 and 6, respectively. These different statistics do not reflect differences in health and the quality of life but rather different age structures—all four countries have "low" death rates.

Source: Population Reference Bureau. 2014. *2014 World Population Data Sheet*. Washington, DC: Population Reference Bureau.

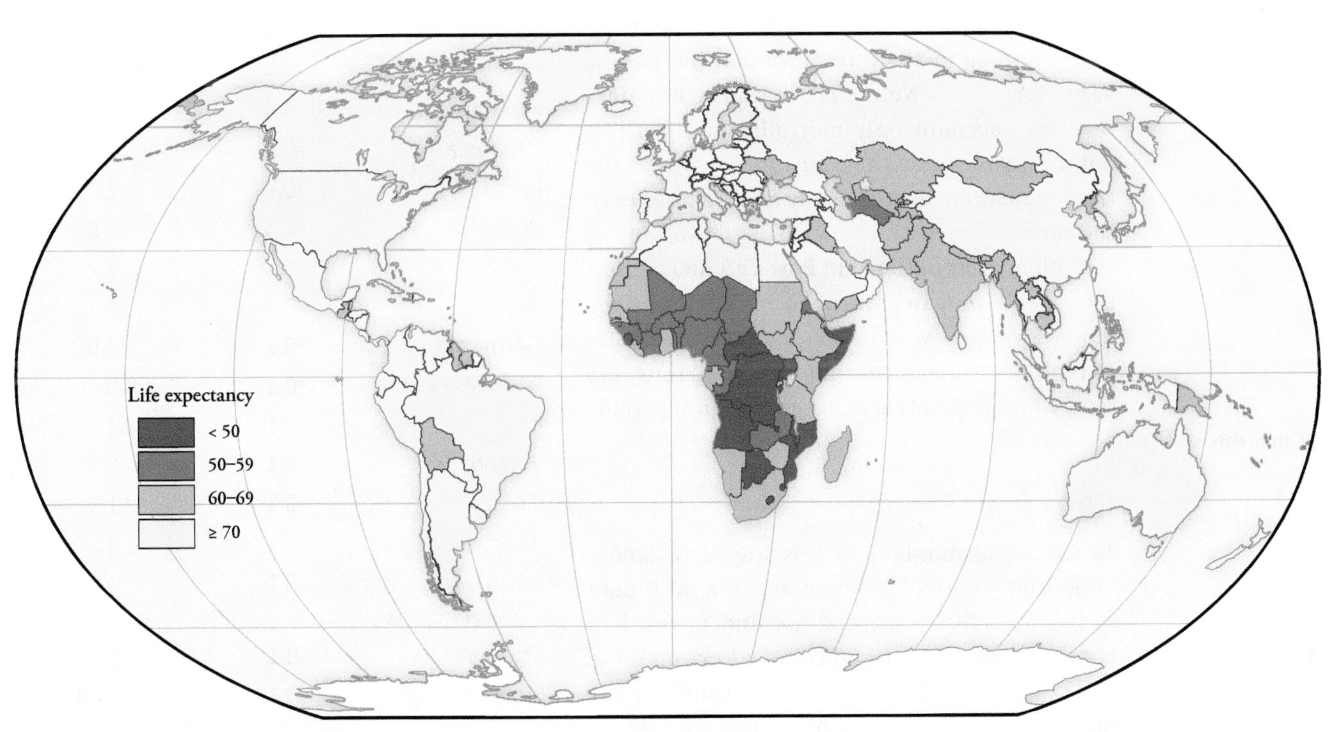

FIGURE 5.4 | World distribution of life expectancy, 2014

Source: Population Reference Bureau. 2014. *2014 World Population Data Sheet*. Washington, DC: Population Reference Bureau.

social inequality of death. In the US, the national IMR in 1993 was 8.6, but Washington, DC, in which blacks comprise a high percentage of the population, reported a figure of 21.1. For both South Dakota (with a large Native American population) and Alabama (with a large black population), the figure was 13.3. Black babies in the US are almost twice as likely to die in their first year as white babies—a state of affairs that has remained essentially unchanged for the last 50 years. In England, a 2014 report showed that poor people live nine years less than the richest.

NATURAL INCREASE

The *rate of natural increase* (RNI) measures the rate (usually annual) of population growth by subtracting the CDR from the CBR. In 2014 the world CBR was 20 and the CDR was 8, producing an RNI of 12 per 1,000; this figure is typically expressed as a percentage of total population (hence 1.2 per cent).

In 2013 the world's population increased by about 86 million people, with 143 million births and 57 million deaths. The size of the annual increase has been relatively constant in recent years, meaning that *world population is still increasing but at a slowly decreasing rate.* The reduction in the RNI began about 1990 because, as discussed earlier, world fertility is declining. This decrease in the rate of annual increase is very significant. Note that, because RNI data take into account only mortality and fertility, not migration, they generally do not reflect the true population change of any area smaller than the earth.

Although, as outlined in Boxes 5.1 and 5.2, fertility is declining in parts of both the less and the more developed worlds, the number of females of reproductive age continues to rise. Thus, the total world population continues to grow rapidly because of **population momentum**.

population momentum
The tendency for population growth to continue beyond the time that replacement-level fertility has been reached because of the relatively high number of people in the child-bearing years.

Regional Variations

To use world numbers in this way is to ignore important regional differences. The RNI data included in Tables 5.3 and 5.4 and mapped in Figure 5.5 provide a useful commentary on these variations. Low rates prevail in Japan and in many European countries, while high rates are concentrated in tropical African countries.

TABLE 5.3 | Countries with the highest rates of natural increase (3.0 and above), 2014

Country	Natural Increase (%)	Population (millions)
Niger	3.9	18.2
Uganda	3.4	38.8
Zambia	3.4	15.1
Chad	3.3	13.3
Angola	3.2	22.4
Senegal	3.2	13.9
Somalia	3.2	10.8
Burundi	3.2	10.5
Central African Republic	3.2	4.8
Tanzania	3.1	50.8
Burkina Faso	3.1	17.9
Gambia	3.1	1.9
DR of Congo	3.0	71.2

Source: Population Reference Bureau. 2014. *2014 World Population Data Sheet.* Washington, DC: Population Reference Bureau.

TABLE 5.4 | Countries with the lowest rates of natural increase (less than 0.0), 2014

Country	Natural Increase (%)	Population (millions)
Bulgaria	-0.5	7.2
Serbia	-0.5	7.1
Ukraine	-0.4	42.9
Hungary	-0.4	9.9
Lithuania	-0.4	2.9
Latvia	-0.4	2.0
Romania	-0.3	20.0
Croatia	-0.3	4.2
Japan	-0.2	127.1
Germany	-0.2	80.9
Portugal	-0.2	10.4
Italy	-0.1	61.3
Poland	-0.1	38.5
Greece	-0.1	11.0
Belarus	-0.1	9.5
Bosnia-Herzegovina	-0.1	3.8
Estonia	-0.1	1.3
Monaco	-0.1	0.04

Source: Population Reference Bureau. 2014. *2014 World Population Data Sheet.* Washington, DC: Population Reference Bureau.

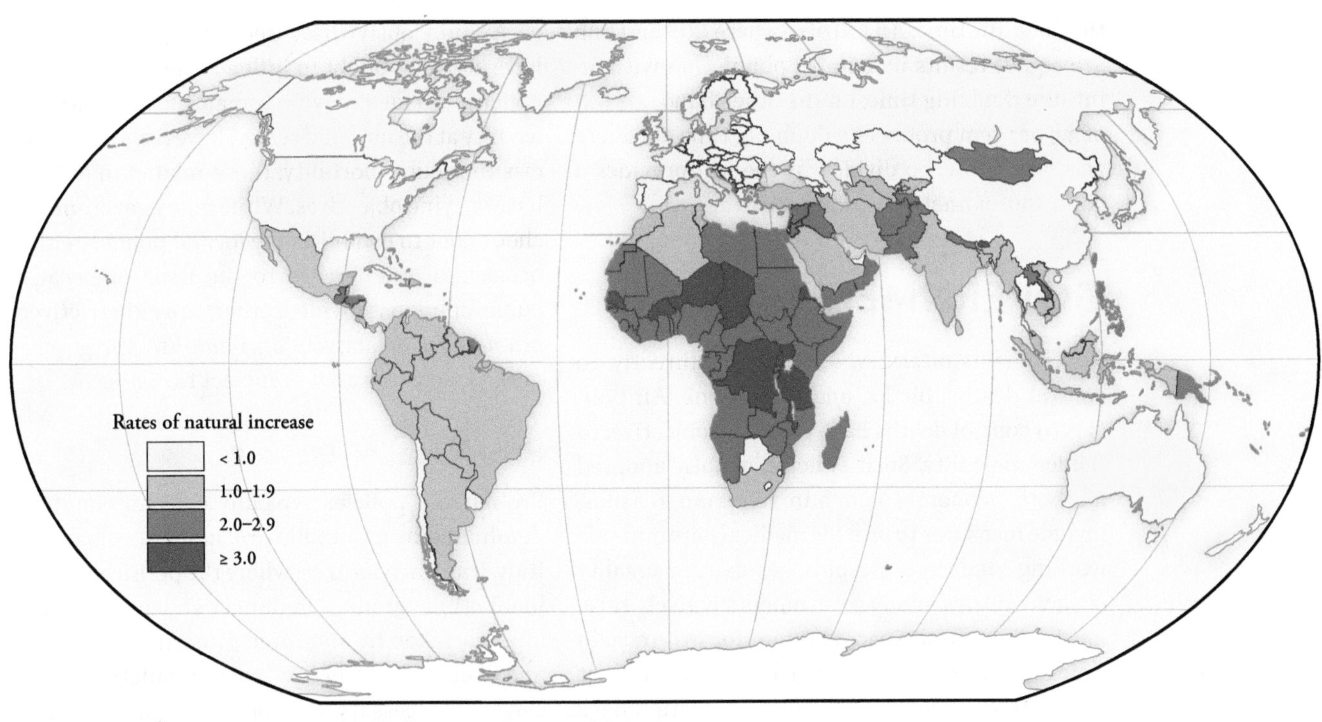

FIGURE 5.5 | World distribution of rates of natural increase, 2014

Source: Population Reference Bureau. 2014. *2014 World Population Data Sheet*. Washington, DC: Population Reference Bureau.

In addition to considering RNI data, it is important to recognize that large populations grow faster than small ones—even when they have lower fertility rates—simply because the base population is so much larger. At present, China and India are home to about 37 per cent of the world's population; hence, even a small increase in fertility in one of these countries will mean a significant increase in the total world population.

Table 5.5, showing projected data for selected world regions/countries, highlights the fact that much of the growth predicted between 2014 and 2050 will occur in sub-Saharan Africa and India. Indeed, of the anticipated world growth of 2.4 billion, sub-Saharan Africa will contribute 1.2 billion and India 361 million, together accounting for 62 per cent of the total growth. The rest will mainly be in other high-population Asian countries. These regional projections also have implications for the larger arena of political power and economic growth. For example, the population of Africa is projected to more than double by 2050, whereas that of Europe will decline. Precisely how such changes will play out politically and economically remains to be seen.

Doubling Time

Related to the RNI is the useful concept of **doubling time** (the number of years needed to double the size of a population, assuming a constant RNI). For example, in the unlikely event that the current 1.2 per cent growth rate continues, rather than declining as expected, the world population would double in approximately 58 years. Relatively minor variations in RNI can significantly affect

doubling time
The number of years required for the population of an area to double its present size, given the current rate of population growth.

TABLE 5.5 | Projected population growth, 2014–2050

Region or Country	Population, 2014 (millions)	Projected Population, 2050 (millions)	2050 Population as a Multiple of 2014
World	7,238	9,587	1.3
Sub-Saharan Africa	920	2,428	2.3
Europe	741	726	1.0
US and Canada	353	444	1.3
China	1,364	1,312	1.0
India	1,296	1,657	1.3

Source: Calculated from Population Reference Bureau. 2014. *2014 World Population Data Sheet*. Washington, DC: Population Reference Bureau.

the doubling time. An RNI of 0 (where CDR and CBR are equal) results in a stable population with an infinite doubling time; on the other hand, an RNI of 3.5 per cent produces a doubling time of a mere 20 years. (Note: 70 divided by the RNI provides a good approximation of doubling time.)

GOVERNMENT POLICIES

Governments often try, directly or indirectly, to control deaths, births, and migrations. All policies to control deaths have the same objective: to reduce mortality. Such policies, usually adopted for both economic and humanitarian reasons, include measures to provide medical care and safe working conditions. Despite the near universality of such policies, many governments actively raise mortality levels at specific times, for instance, in times of war. Further, many governments do not ensure that all members of their population have equal access to the same quality of health care. This is true throughout the world, regardless of levels of economic development.

As previously discussed, many governments have actively sought to influence fertility, and the policies in place may be a major factor affecting fertility at the national scale. However, unlike policies to reduce mortality, those related to fertility have varying objectives. While many governments choose not to establish any formal policies, either because of indifference to the issue or because public opinion is divided, others are either actively ***pro-natalist*** or actively ***anti-natalist***. Yet government policies are always subject to change.

Pro-Natalist Policies

Pro-natalist policies typically exist in countries dominated by a Catholic or Islamic theology (e.g. Italy, Iran), in countries where the politically dominant ethnic group is in danger of being numerically overtaken by an ethnic minority (e.g. Israel), and in countries where a larger population is perceived as necessary for economic or strategic reasons (Box 5.3).

Singapore and Malaysia, both of which belong in the Asian group of newly industrialized countries

Around the Globe

BOX 5.3 | Fertility in Romania, 1966–1989

In the October 1986 issue of the German magazine *Der Spiegel*, the president of Romania was quoted as saying that "the fetus is the socialist property of the whole society. Giving birth is a patriotic duty, determining the fate of our country. Those who refuse to have children are deserters, escaping the laws of national continuity."

Alarmed by a crude birth rate of 16 in 1966, the communist government in Romania ended all legal access to abortion and set a goal to increase the national population to between 24 and 25 million by 1980 (a 30 per cent increase). Contraceptives were banned, and all employed women under 45 had to take a monthly gynecological examination. Unmarried people over 15 and married couples without children or a valid medical reason for not having children were assessed an additional 30 per cent income tax. The TFR increased from 1.9 in 1966 to 3.6 in 1968, but since that time the rate has steadily dropped, and in 2014 it stood at 1.3. There is reason to believe that many women had illegal abortions despite the establishment of a special unit within the state security police whose job it was to combat abortion.

The government's principal reasons for attempting to increase fertility were probably to strengthen national security (more people equals more strength) and improve productivity (by alleviating labour shortages). Interestingly, all the strategies devised to increase the fertility rate were negative and coercive. The government punished people for not having children rather than rewarding them for having children with such positive inducements as maternity leave or family income supplements.

In general, it seems reasonable to say that pre-1989 Romania had the most stringent fertility policy in the world. Among other previously communist countries, the former USSR encouraged higher fertility but, at the same time, probably had the highest national legal abortion rate in the world (about one abortion for every birth).

Directly or indirectly, most states have an influence on fertility. What types of policies are evident in Canada? Does the Canadian government encourage or discourage births? What role, if any, does religion play?

(see Chapter 13), began to actively encourage fertility increases in the 1980s (Dwyer, 1987). The two countries had succeeded in lowering their TFRs by following an anti-natalist line but, in 1984, decided that new policies were required to reverse the trend and increase fertility again. In Singapore, this reversal was related to the economic difficulties perceived to be the result of reducing the TFR to 1.6; there was a perceived need to provide a larger market for domestic industrial production and a larger workforce. The Malaysian case was similar in that the motivation was economic, although the TFR remained at 3.9, well above replacement level. The extent to which such pro-natalist policies are successful is debatable; the 2014 TFR for Singapore was only 1.2 and for Malaysia 2.1.

Other countries are also greatly concerned about low fertility. In 2006 President Vladimir Putin stated that the situation in Russia was critical—in recent years, medals have been handed out to very large families, including in 2010 to a family with 17 children. In 2008 the president of Turkmenistan announced financial incentives for women who have more than eight children. Japan also has introduced financial incentives with the aim of encouraging a baby boom. The concern in this country is explicitly economic, as it is feared that an aging population supported by fewer working people would keep the country in a state of permanent recession. In addition, daycare facilities in Japan are being expanded and employers are making their companies more family friendly. Yet, although these policies have been in place for about 30 years, they do not appear to have prompted increased fertility. Canada has similar policies and practices are place, with subsidized daycare and with many employers offering generous maternity and paternity leaves.

One effective tactic, at least in the short term, was employed in Georgia by the head of the Georgian Orthodox Church. Government concern about low fertility prompted Patriarch Ilia II to issue a promise in 2007 that he would personally baptize any baby born to parents who already had two children. In a country where about 80 per cent of the population are adherents of this faith and the patriarch is held in great esteem, the immediate result was a 20 per cent increase in births in 2008.

It seems likely that more and more countries will adopt pro-natalist policies if their fertility declines to the point where not only the economic but also the political and cultural consequences come to be seen as unacceptable. For instance, such policies could be implemented if it is feared that immigrants entering a country to take up the slack in the labour force will lead to a smaller native-born population and a loss of national identity.

Commemorating Russia's twentieth-century baby boom, dancers create a "maternity ward" during the opening ceremony of the Sochi Olympic Games in Russia. The country's TFR is currently below the replacement rate.

Anti-Natalist Policies

Despite the pro-natalist policies of some countries, the most common policies relating to births are currently anti-natalist. Since about 1960, many of the less developed countries have initiated policies designed to reduce fertility. The argument behind such policies is that overpopulation is a real danger and that **carrying capacity** has been (or will soon be) exceeded. A fundamental problem with this argument is that it ignores the fact that carrying capacity is not static. Consider, for example, that without the early twentieth-century invention of a chemical process to produce nitrogen fertilizers, the carrying capacity of the earth would be about half the present population (see Chapter 10 for a discussion of nitrogen fertilizers).

carrying capacity
The maximum population that can be supported by a given set of resources and a given level of technology.

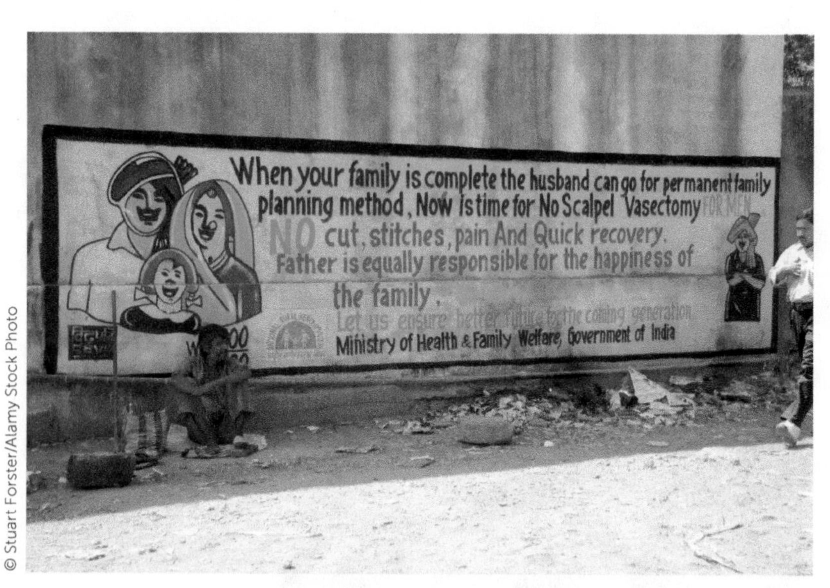

A government family planning poster in the city of Panaji, Goa, India

India was the first country to intervene actively to reduce fertility. Beginning in 1952, the government introduced a series of programs designed to encourage contraceptive use and sterilization. The first program offered financial incentives; others have been coercive or educational. Fertility rates have declined, but with a 2014 RNI of 1.5 per cent (meaning a doubling time of 47 years), it is clear that the various programs have not achieved the desired result.

China, on the other hand, has made great strides towards reducing fertility. The 2014 RNI was 0.5 per cent and the doubling time 140 years. However, the causes of this fertility decline are debated. Much of this success might be attributed to the programs introduced, mostly in the late 1970s, by the communist government. Families were restricted to having one child, and (as previously mentioned) marriage was prohibited until the age of 28 for men and 25 for women. Contraception, abortion, and sterilization were free. There were financial incentives for families with only one child and penalties for those with more than one. However, some Chinese academics contend that fertility was declining prior to the introduction of the one-child policy, making it a largely irrelevant factor. They say that fertility was declining because of the education of women, better health care, and social and economic development in general. As noted in Box 5.4, the policy was formally abandoned in 2015 with the government announcing that all couples could have two children.

The Best Policy?

Is it possible that the best government population policy is neither pro-natal nor anti-natal but no policy at all? The logic behind having no policy is simply that people ought to be allowed to determine how many children they have. Anti-natal policies in particular have destroyed lives (for example, through forced sterilization) and are derived from imperialist ideas of supremacy and the presumed ignorance of others.

Perhaps the most compelling reason not to have an anti-natal policy is that such policies may not be needed. The last few decades have made it clear that, with education and improved lifestyles, women choose to have fewer children and do not need to be told to or, worse, be coerced into doing so. Even in the case of China, as noted in Box 5.4, it can be argued that fertility was already declining prior to the introduction of the one-child policy and that improved educational opportunities for women, combined with the ongoing process of urbanization and industrialization, would have been sufficient to lead to the desired low fertility.

Viewed retrospectively, many anti-natal policies were put in place by government officials, or by presumed experts from more developed countries, on the arrogant grounds that people do not know what is best for themselves—it is perhaps a fine line between arrogance, assuming what is best for others, and inhumanity, requiring that people behave in ways deemed appropriate by others.

THE COMPOSITION OF A POPULATION

Age and Sex Structure

Fertility and mortality vary significantly with age. Inevitably, then, the growth of a population is affected by the population's age composition. Age composition is dynamic. As we have seen in Box 5.4, it is usual to represent age and sex compositions with a **population pyramid**. Three general patterns can be suggested (Figure 5.7). First, if a population is rapidly expanding, a high proportion of the total population will be in the younger age groups. Because fertility is high, each successive age group is larger than the one before it. Second, if a population is relatively stable, each age group, barring the older groups that are losing

population pyramid
A diagrammatic representation of the age and sex composition of a population; by convention, the younger ages are at the bottom, males are on the left, and females on the right.

Around the Globe

BOX 5.4 | Population in China

As of 2014, China continued to enjoy a declining CBR, one that is low by Asian standards (12), a CDR that is among the lowest in the world (7), an RNI of 0.5, a TFR of 1.6, and an LE that is approaching the average for the more developed world (75).

A useful way to depict the late twentieth-century history of Chinese population growth is with a population pyramid (Figure 5.6). The indentations in the 1990 pyramid show several changes in fertility and mortality during the 1960–90 period. The 1960s were a period of generally high fertility and low mortality, reflected by the bulge for the 10-year cohort aged 18–27 in 1990; these were people born in the high-fertility period, 1963–72. In 1964 more than 40 per cent of the population was under age 15. The high fertility of 1972 was followed the next year by dramatic decreases that resulted in a narrowing of the pyramid, and the 1990 percentage under 15 was reduced to 28.

The broadening at the base of the pyramid reflects the fact that fertility is highest for women in their late twenties; if the number of births to married women of this age remains constant, the CBR will remain high. As Jowett (1993: 405) noted, "China is currently faced with the 'echo' effect from demographic developments that occurred 20–25 years ago. The large-small-large-small cohorts of population aged eight, six, three and zero are a condensed form of the fluctuations from the previous generation aged 32, 29, 27, and 23."

Recent data for China, including results of the 2010 census, highlight three developments. First, it seems that, if existing fertility trends continue, China's population will begin to decline about the year 2042. For many commentators inside China, this prediction means the fertility rate is too low. This astounding prospect prompted authorities to informally relax the one-child policy, and it became common for rural couples to have two children, especially if the first was a girl. But the policy continued to be applied in some parts of the country by local officials. Several news items reported severe violations of human rights in the form of forced abortions and sterilizations designed to reduce births. On 29 October 2015 the government announced that all couples could have two children. Even though the one-child policy has been abandoned, most observers think it likely that the TFR will increase only slightly.

The second significant development concerns an increasingly skewed sex structure because of a marked disparity in the sex ratio of children born in recent years. Recent data show a ratio of 118 males to 100 females. This suggests that increasing numbers of female fetuses have been aborted. Female infanticide is practised but is likely much less evident than selective abortion. There is also evidence of a higher infant mortality rate for girls than for boys because of neglect or maltreatment. All of this happens because parents prefer sons over daughters—a daughter is responsible for her birth family only until she marries, whereas a son remains responsible after his marriage. The other country with a similar pattern of female abortion is India, where many parents still have to offer dowries in order to find husbands for their daughters.

The third significant development is an increasingly skewed age structure, specifically, the rapid aging of the population. It is estimated that the Chinese population's rate of aging will be the fastest in the world. Assuming current trends continue, the ratio of people of working age to retirees will fall from about 6:1 in the early twenty-first century to about 2:1 by 2040. Considering that China is on the way to becoming an economic powerhouse, one way to express what this aging means is to say that China will get old before it gets rich. Certainly, the huge elderly non-working population is expected to place great stress on the national economy. China is approaching what can be broadly described as a "4–2–1" scenario, that is, four grandparents, two parents, and one child.

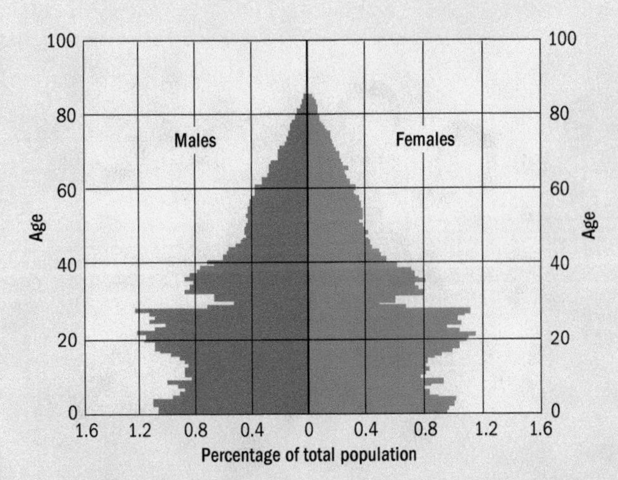

FIGURE 5.6 | Age and sex structure in China in the late twentieth century

Source: Jowett, J. 1993. "China's Population: 1,133,709,738 and Still Counting." *Geography* 78: 405. Reprinted by permission of the Geographical Association, www.geography.org.uk.

sex ratio
The number of males per 100 females in a population.

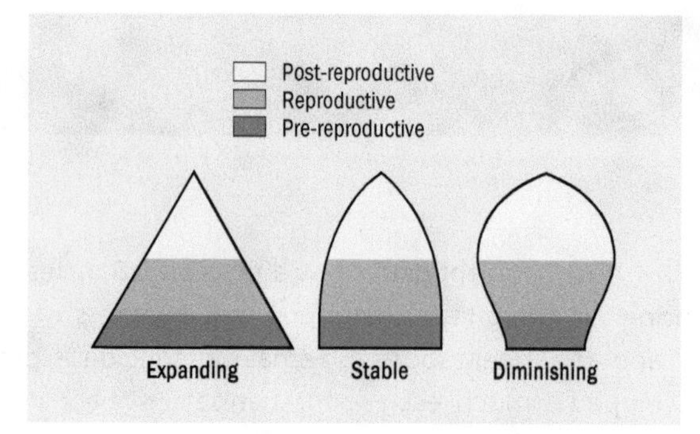

FIGURE 5.7 | Age structure of populations

Expanding populations have a high percentage in the pre-reproductive age group. Stable populations have relatively equal pre-reproductive and reproductive age groups. Diminishing populations have a low percentage in the pre-reproductive age group.

numbers, is similarly sized. Third, if a population is declining, the younger groups will be smaller than the older groups.

The three generalizations in Figure 5.7 are useful, but specific age and sex composition pyramids are more revealing. Usually, population pyramids distinguish males and females and divide them into five-year categories. Each bar in the pyramid indicates the percentage of the total population that a particular group (such as 30–34-year-old females) makes up. Figure 5.8 presents the population of Brazil for the period 1975 to 2025 in a series of three pyramids.

population aging
A process in which the proportion of elderly people in a population increases and the proportion of younger people decreases, resulting in increased median age of the population.

© Homer W Sykes/Alamy

Liufu village, Anhui Province, China. In China there is an abnormally high ratio of young males as a result of the one-child policy.

Discussions of the number of males and females in a population refer to the **sex ratio** (technically, a masculinity ratio as it relates the number of males to the number of females). Sex ratio data for individual countries frequently are estimated, but it is generally accepted that there are about 101 males for each 100 females. Under normal circumstances and in most parts of the world, about 104–08 boys are born for each 100 females. The major exceptions are those countries—especially India, Taiwan, South Korea, and China—where there is a preference for sons and the number of girls is reduced through abortion and female infanticide, as discussed in Box 5.4.

The surplus of males at birth is reduced through time as male mortality rates are generally higher and male life expectancy generally lower than that of females. In most countries the number of males in the population will be overtaken by the number of females by middle age, and the elderly population usually is predominantly female. In short, sex ratios vary with age.

Global Population Aging

The year 2000 was a watershed, as it was the first year that people under 14 were outnumbered by people over 60. The process by which older individuals come to make up a proportionally larger share of the total population, known as **population aging**, is occurring in most countries around the world, albeit at different rates.

The data are compelling. In 1900, about 1 per cent of the world's population was over age 60; by 2014 this figure had increased to about 12 per cent, and an estimate for 2050 is 21 per cent (Figure 5.9). Furthermore, global aging is occurring at an increasing rate: the older population is growing faster than the total population in most parts of the world, and the difference between the two growth rates is increasing. Table 5.6 provides data on the global aging trend, while Figure 5.10 shows the expected impact of aging on the Canadian population. As a result of these changes, the median age of the world's population is expected to increase from 29 years in 2014 to about 38 years in 2050.

Although population aging is indeed a global phenomenon, there are significant regional differences, most obviously related to differences between the more and less developed worlds in their patterns of fertility decline and increasing life expectancy. In the case of fertility, the

FIGURE 5.8 | Age and sex structure in Brazil: 1975, 2000, 2025

These three age–sex pyramids show the changes and predicted changes in Brazil's population over a 50-year period. The wide base and gently sloping sides in the 1975 pyramid indicate high rates both of births and deaths. Much of the population was under 20, ensuring that births would continue to be high in the future. The 2000 pyramid shows a slight reduction in the number of births and a decrease in deaths, essentially the second stage in the demographic transition. Throughout this period, the population increases rapidly because birth rates are high and death rates are dropping. This pyramid indicates that Brazil's population growth has not yet reached its maximum acceleration—a fact that does not bode well for a nation already experiencing difficulties supporting its current population. The predicted 2025 pyramid shows a significantly changed situation, with the narrowing base reflecting a reduction in births and evident signs of population aging.

Source: US Census Bureau. www.census.gov/ipc/www/idbpyr.html.

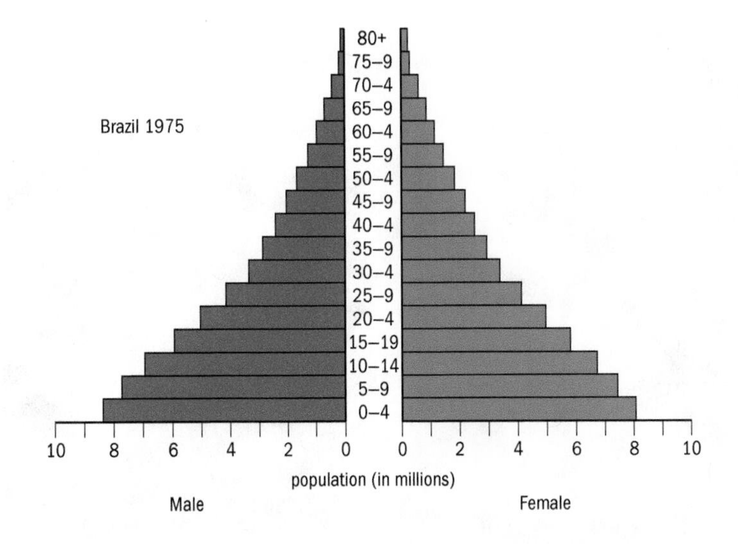

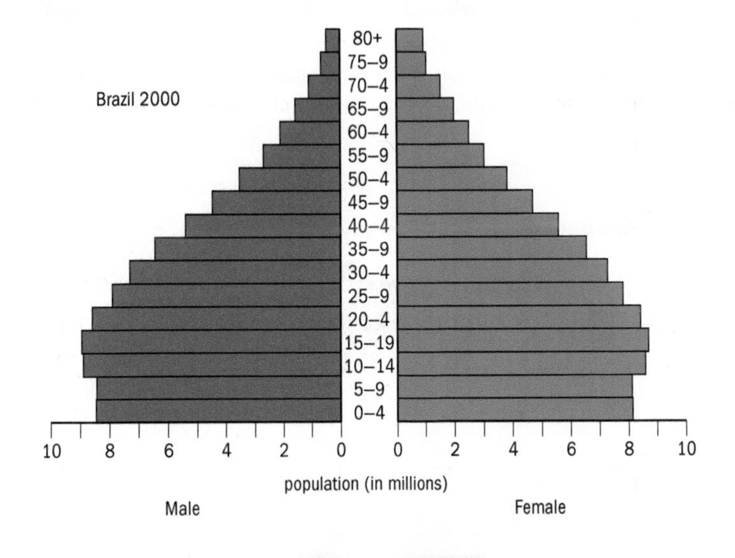

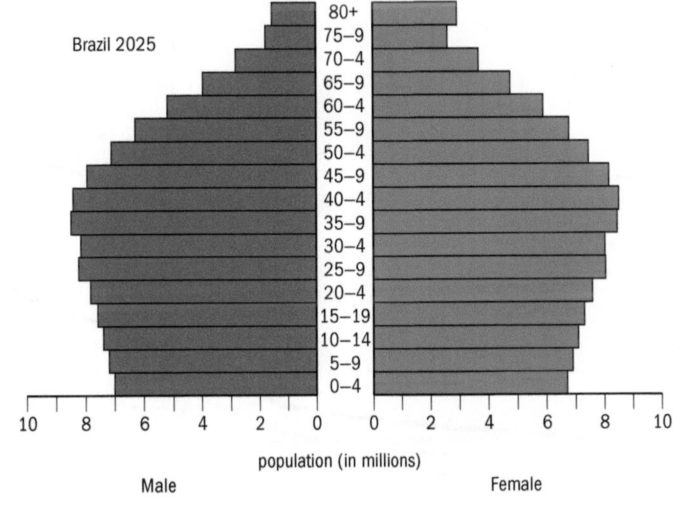

decline started later in the less developed world but is proceeding more rapidly. In the case of life expectancy, the increases are lower throughout the less developed world and have fallen in some places because of AIDS (see Chapter 6). Among specific countries, Italy and Japan have the highest median age (40 years), but it is estimated that, by 2050, 10 countries will have a median age above 50, with Spain heading the list at 55 years and Italy and Austria at 54 years. The two principal causes and some likely consequences of population aging are discussed in Box 5.5.

HISTORY OF POPULATION GROWTH

Like all animal populations, early human populations increased in some periods and declined in others. The principal constraints on growth were climate and the (related) availability of food. Unlike other animals, humans gradually increased their freedom from such constraints as a consequence of the development of culture. Cultural adaptation has enabled humans to increase in numbers and—so far—to avert extinction. For most of our time on earth, however, our numbers have increased very slowly.

Reasons for Growth

Among the cultural changes that permitted human numbers to increase before about 12,000 years ago were the evolution of speech, which facilitated co-operation in the search for food; the introduction of monogamy, which increased the chances that children would survive; and the use of fire and of clothing, which together

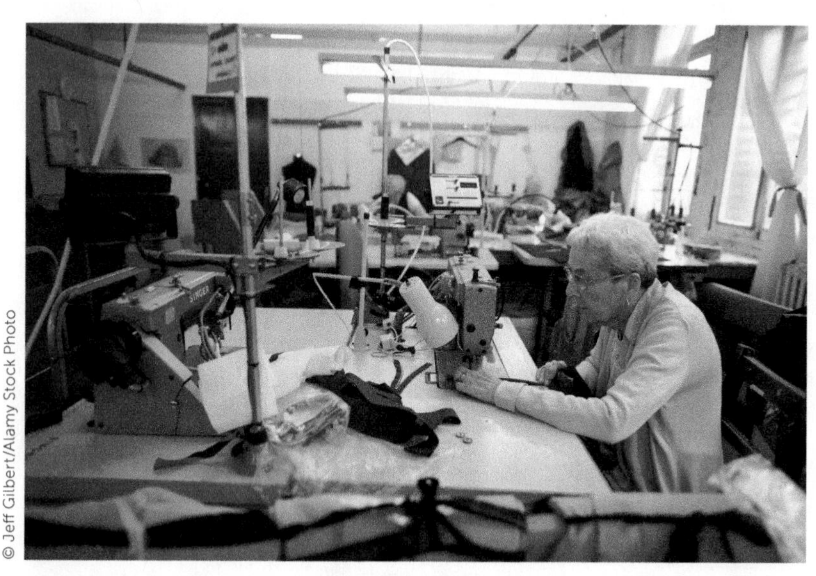

Jean Seddon, who is in her seventies, is factory supervisor at Cooper & Stollbrand in Manchester, England. Like Seddon, many retirees continue to work because they enjoy it; however, many others do so because the recent recession has made them unable to retire.

TABLE 5.6 | Global aging, 1950–2050

Region	Median Age (years)			% Aged 60 or Older		
	1950	1999	2050	1950	1999	2050
World	23.5	26.4	37.8	8.1	9.9	22.1
More developed world	28.6	37.2	45.6	11.7	19.3	32.5
Less developed world	21.3	24.2	36.7	6.4	7.6	20.6

Source: Updated from Population Division of the Department of Economic and Social Affairs of the United Nations Secretariat. 1999. *The World at Six Billion*. www.un.org/esa/population/publications/sixbillion/sixbillion.htm.

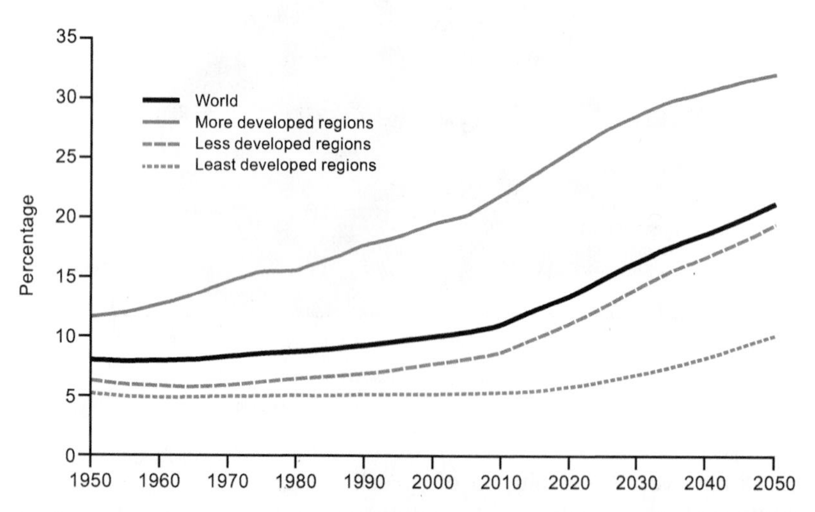

FIGURE 5.9 | Proportion of the population aged 60 years or over: World and development regions 1950–2050

Source: United Nations, Department of Economic and Social Affairs Population Division. 2013. *World Population Ageing 2013*. New York: United Nations. www.un.org/en/development/desa/population/publications/pdf/ageing/WorldPopulationAgeing2013.pdf.

made it possible to move into cooler areas. Nevertheless, the cumulative effect of these advances was not great; at this time, the human population totalled perhaps four million. The rapid growth since has not been regular. Rather, there have been several relatively brief growth periods and longer intervals of slow growth. Each growth period can be explained by a major cultural advance.

The Rise of Agriculture

The first such advance, beginning about 12,000 years ago, was the agricultural revolution (the gradual process, over thousands of years, by which humans domesticated animals and plants). As agriculture spread, the first human economic activities of hunting and gathering became marginal. This took place in various centres, beginning in the Tigris and Euphrates Valleys of present-day Iraq. Some 9,000 years ago, the first region of high population density appeared, stretching from Greece to Iran and including Egypt. By 4000 BCE agriculture and related population centres had developed on the Mediterranean coast and in several European locations, as well as in Mexico, Peru, China, and India. Around the beginning of the Common Era, new centres arose throughout Europe and Japan, and the total world population reached an estimated 250 million—a dramatic increase in the 10,000 years since the beginnings of agriculture.

Before the cultural innovation of agriculture, population numbers had changed little, increasing or decreasing depending on cultural advances and physical constraints. Birth and death rates were high, about 35 to 55 per 1,000, and life expectancy was short, about 35 years. Following the introduction of agriculture, birth rates remained high but death rates fluctuated; as the epidemics listed in Table 5.7 suggest, humans still had relatively little control over the environment. This state of affairs continued until the seventeenth century. The 250 million at the beginning of the Common Era had reached 500 million by 1650, but the pace of growth remained slow in the absence of any major cultural advance.

The Rise of Industry

From about 1650 onward, however, the world population increased rapidly in response to improved agricultural production and the beginning of a demographic shift to cities, especially after the onset of the second major cultural advance, the

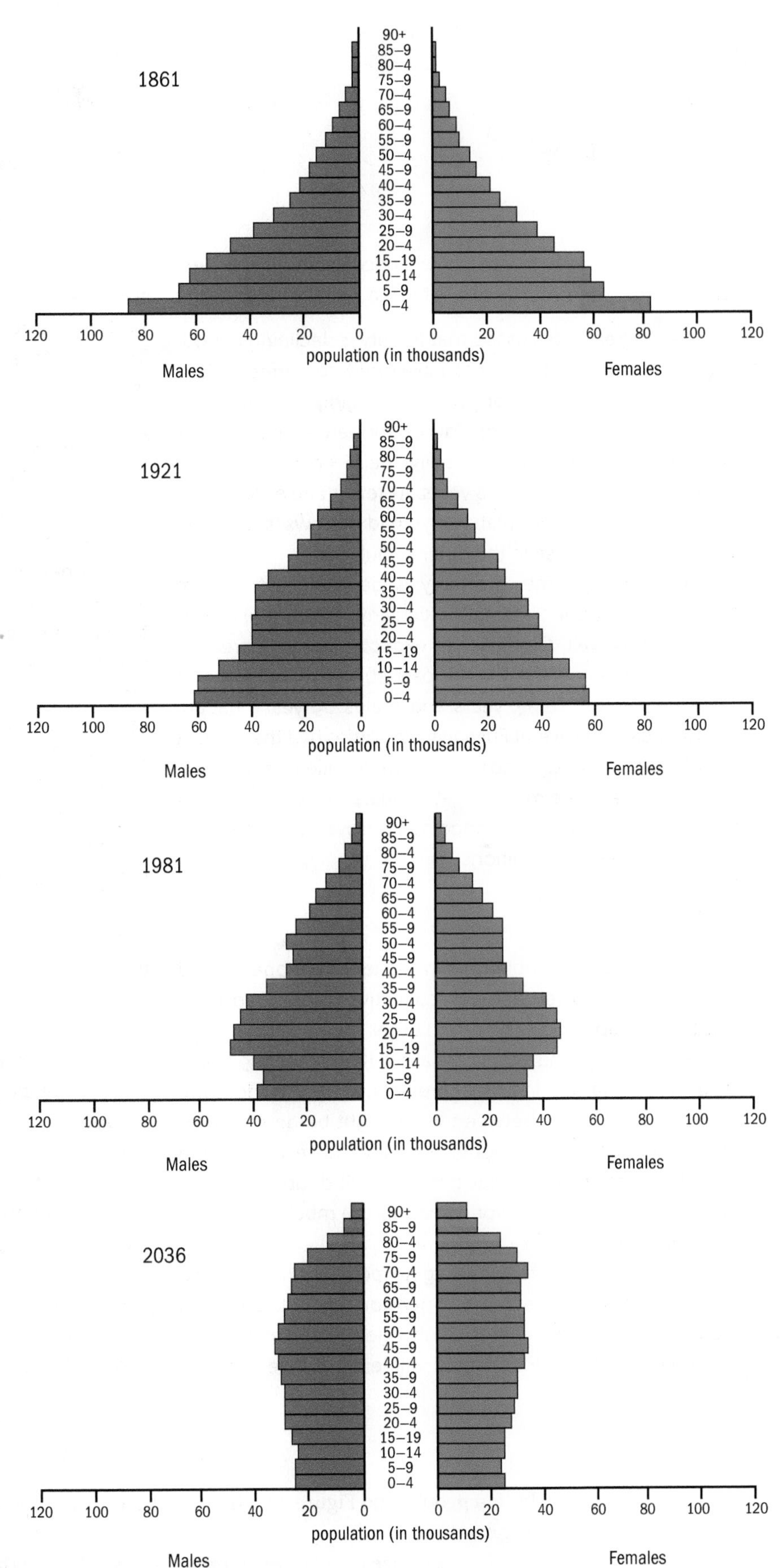

FIGURE 5.10 | Age and sex structure in Canada: 1861, 1921, 1981, 2036

These four population pyramids provide considerable information about the composition of Canada's population. The numbers along the base of each pyramid indicate the number of people in each age and sex group per 1,000 of the total population. In 1861 immigration and high fertility combined to produce a pyramid that shows cohorts steadily decreasing in size with age. The 1921 pyramid shows the effects of reduced fertility in a narrowing of the base, but the effects of immigration still are apparent. In 1981 the Canadian population was characterized by extremely low birth and death rates; the age–sex structure exhibits a beehive shape. The pyramid bulge between the ages of 15 and 35 represents the postwar baby boom, which accelerated population growth. The birth rate has long since returned to its normal low; without the baby boom, the pyramid would have had almost vertical sides. A continuing low birth rate and an aging population are reflected in the projection for 2036. This age–sex structure may well represent the future for most countries in the more developed world.

Note: As Figures 5.6, 5.8, and 5.10 demonstrate, population pyramids may be presented in various styles.

Source: Dumas, J., with Y. Lavoie. 1992. *Report on the Demographic Situation in Canada, 1992*. Catalogue no. 91-209E. Ottawa: Minister of Industry, 132–3.

Industrial Revolution. This development initiated a rapid growth and diffusion of technologies, and industry replaced agriculture as the dominant productive sector. The agricultural revolution had involved more effective use of solar energy in plant growth. This second revolution involved the large-scale exploitation of new sources of energy—coal, oil, and electricity. The result was a rapid reduction in death rates, with a delayed but equally significant drop in birth rates. World population increased to 680 million in 1700, 954 million in 1800, and 1.6 billion in 1900.

The Current Situation

The growth initially spurred by industrialization has ceased in the more developed world, but rapid growth continues in the less developed world. As a result, the total world population continues to increase substantially. Table 5.8 places the recent and expected growth in context by noting the number of years taken to add each additional

In the News

BOX 5.5 | Causes and Consequences of Population Aging

Causes

Population aging is the result of ongoing changes in two areas: fertility and mortality (United Nations, 2002).

1. We've already discussed that fertility is declining in both more developed and less developed countries. This is the primary reason for population aging: the fewer the young people in a population, the larger the proportion of elderly people. Globally, the TFR has declined by about half in the past 50 years. As fewer babies are born, the base of the population pyramid narrows, producing a relative increase in the older population.
2. At the same time, mortality is declining; hence, life expectancy is increasing (particularly for women) everywhere as well. Globally, life expectancy has increased by about 20 per cent in the past 50 years, with females expected to live 73 years and males 69 years. Japan, Canada, and several European countries had the highest life expectancies in 2014, with the female rate at about 84 years and the male rate at 80 years. Life expectancy is increasing because of ongoing improvements in health care and living conditions.

Consequences

The greying of the population in the less developed world may do more to reshape our collective future than even the proliferation of chemical and other weapons, terrorism, global warming, and ethnic tribalism (Peterson, 1999: 42). The reasoning behind this prediction has to do with changes in the ratio between what might be described as the dependent elderly population and those of working age. Estimates suggest that this ratio will double in much of the more developed world and triple in much of the less developed world.

Perhaps most notably, an aging population will place increasing stress on retirement, pension, and related social benefits, necessitating radical changes in social security programs. Further, global aging will lead to quite different patterns of disease and disability; for example, degenerative diseases associated with aging, such as cancer, heart problems, and arthritis, will become increasingly common. The need for adjustments to national health-care programs is apparent.

At the same time, national economies will face enormous strain as the numbers of workers available to support the growing population of non-working elderly gradually decline. It is possible that a decrease in the number of dependent young will free up some resources for the dependent elderly; however, the amount of public spending required to support an elderly person is two to three times that required to support a young person. Countries with a strong tradition of children supporting elderly parents, such as China, will face particular stress as increasing life expectancy makes this more and more difficult (Box 5.4).

Inevitably, the problems of aging will be exacerbated in those parts of the world that already lack financial and other resources. Many babies born in less developed countries can expect to live into their seventies. In the two very large population countries of China and India, the number of people aged over 80 is expected to increase about five times by 2050. The fact that the more developed world became rich before aging whereas the less developed world is aging before it has a chance to get rich is a very significant difference between the two.

It might be argued that aging is threatening national security in more developed countries because of a declining workforce and related loss of entrepreneurial initiative. Aged societies may favour consumption over investment and the elderly over the youthful. One very interesting argument is that population aging in more developed countries facilitates the economic globalization discussed in Chapter 3. Because people in these countries are living longer and having fewer children, they necessarily rely on the more affordable labour of people in other countries.

billion people, and Figure 5.11 displays these data graphically.

To summarize the current situation, the world population growth rate—the rate of natural increase—has fallen from a late 1960s high of 2.04 per cent per year to 1.2 per cent per year. Because of this decline, the world population will grow less rapidly in this century than it did in the twentieth, although it will continue to increase substantially because of population momentum.

This brief account explains the basis of population growth. Numbers have increased in response to cultural, specifically technological, advances. The dramatic increases since about 1650 reflect the fact that the death rate began to fall before the birth rate did. In what is now the more developed world, populations thus grew rapidly after about 1650; elsewhere, numbers increased rapidly in the last century, particularly since the 1940s. An interesting question about world population is raised in Box 5.6.

TABLE 5.7 | Major epidemics, 1500–1700

Date	Event
1517	Smallpox in Mexico caused the death of one-third of population, followed by other diseases in Mexico and other American countries
1522–34	Series of epidemics in Western Europe resulted in significant decrease of population
1563	Plague in Europe
1580–98	Epidemics in France
1582–3	Epidemics and famines in southern Sudan
1588	Epidemics and famines throughout China
1618–48	German population declined from 15 million to 10 million because of epidemics and war
1628–31	Disastrous plague epidemics in Germany, France, and northern Italy
1641–4	Epidemics, famines, and conflicts reduced Chinese population by 13 million
1648–57	Plague in Spain; one million died
1650–85	Plague and conflict in Britain; one million died
1651–3	Plague and conflict in France; one million died
1690–4	Food shortages and epidemics throughout Europe

Population Projections

Predicting population growth is hazardous. In the early 1920s, Pearl predicted a stable population of 2.6 billion by 2100. A second example is the 1945 prediction by an eminent American demographer, Frank Rotestein, of 3 billion by 2000. But despite some unimpressive precedents, there is good reason to suggest that we are in a position to make better forecasts. The principal reason is that both fertility and mortality rates now lie within narrower ranges than was previously the case.

Current UN projections suggest a world population of 9.7 billion for 2050 (Table 5.5). This projection assumes that the mortality transition will be complete at that time, such that the CDR is approximately equal throughout the world. As we have stated, a look further ahead might see the fertility transition in the less developed world completed by the year 2050, with a relatively stable world population of about 10 billion by 2200. This estimate has remained essentially unchanged for about 20 years, reflecting a broad consensus about our future growth. Nevertheless, as noted, it is still a projection and may not be correct.

There are two complicating issues regarding this prediction. First, the assumption of a continually declining TFR may be questioned. Some evidence suggests that this decline has stalled in parts of the less developed world because contraceptive needs are not being met and because large families are still preferred; as an example, research published in 2014 claimed that Nigeria's population will increase from 178 million today to 900 million by 2100.

Second, it is possible that population numbers will collapse because of **limits to growth** (this phrase became popular after it was used as the title of a report published by the Club of Rome [Meadows et al., 1972]). Many environmentalists and ecologists argue that there are definite limits to the growth both of populations and of economies. The earth is finite, and many resources are not renewable. The classic argument in this genre is that of Malthus (see below), but the Club of Rome report broadened the debate to include natural resources and environmental impacts. Judging by trends in

limits to growth
The argument that both world population and world economy may collapse because available world resources are inadequate.

TABLE 5.8 | Adding the billions: Actual and projected

Billions	Year Reached	Years Taken
1	1800	
2	1930	130
3	1960	30
4	1974	14
5	1987	13
6	1999	12
7	2011	12
8	2023	12
9	2042	19

Source: Updated from Population Division of the Department of Economic and Social Affairs of the United Nations Secretariat. 1999. *The World at Six Billion*. www.un.org/esa/population/publications/sixbillion/sixbillion.htm.

Examining the Issues

BOX 5.6 | How Many People Have Ever Lived on Earth?

This thought-provoking question was asked, and answered, in a short article first published by the Population Reference Bureau in 1995 and updated in 2011. The article acknowledges that the question was necessarily approached less scientifically than other demographic questions and that an answer was somewhat speculative. The reason is that there are no population data for 99 per cent of the time that humans have been on earth. We must estimate when humans first appeared and the likely size of the population at particular times, with the latter estimation requiring assumptions about birth rates and life expectancy.

The date 50,000 BCE is used as a reasonable approximation of the first appearance of humans. Even if this estimate is shorter or longer than has been the case, the impact on the answer will be slight as the early years of human life involved very few people. Estimates of the total population at particular times, including estimates of CBR, are shown in Table 5.9. The resulting guesstimate is that almost 108 billion have lived on earth as of 2011. The 2011 population represents 6.8 per cent of that total.

TABLE 5.9 | Estimating how many people have ever lived on earth

Year	Population	Births per 1,000	Births between Benchmarks
50,000 BCE	2	-	-
8,000 BCE	5,000,000	80	1,137,789,769
1 CE	300,000,000	80	46,025,332,354
1200	450,000,000	60	26,591,343,000
1650	500,000,000	60	12,782,002,453
1750	795,000,000	50	3,171,931,513
1850	1,265,000,000	40	4,046,240,009
1900	1,656,000,000	40	2,900,237,856
1950	2,516,000,000	31–8	3,390,198,215
1995	5,760,000,000	31	5,427,305,000
2011	6,987,000,000	23	2,130,327,622
Number who have ever been born			107,602,707,791
World population in mid-2011			6,987,000,000
Percent of those ever born who are living in 2011			6.5

Source: Haub, C. 2011. *How Many People Have Ever Lived on Earth?* Washington, DC: Population Reference Bureau. www.prb.org/Publications/Articles/2002/HowManyPeopleHaveEverLivedonEarth.aspx.

the early 1970s, the authors of the report argued that world population was likely to exceed world carrying capacity, leading to a population and economic collapse. A well-publicized earlier work, *The Population Bomb* (Ehrlich, 1968), similarly anticipated widespread famine, raging pandemics, and possibly nuclear war by about 2000 as a result of worldwide competition for scarce resources.

As we saw in Chapter 4, this thesis is articulated by catastrophists, but another, more positive view, is articulated by cornucopians. According to this second view, technological advances will make new resources available as old resources are depleted: "There are limits to growth only if science and technology cease to advance, but there is no reason why such advances should cease. So long as technological development continues, the earth is not really finite, for technology creates resources" (Ridker and Cecelski, 1979: 3–4).

Indeed, some believe people to be the ultimate resource. Japan, for example, has few resources apart from people, and that resource is slowly decreasing. According to some neo-liberal commentators, a slowdown in population growth will lead to a slowdown in economic growth and therefore is to be avoided. The ideological debate between the left-leaning catastrophists and the right-leaning cornucopians is far from over. Perhaps the best we can do at present is to recognize the wide range of views on the subject, seek to set aside opposing ideologies, and weigh the demographic and scientific evidence with extreme care.

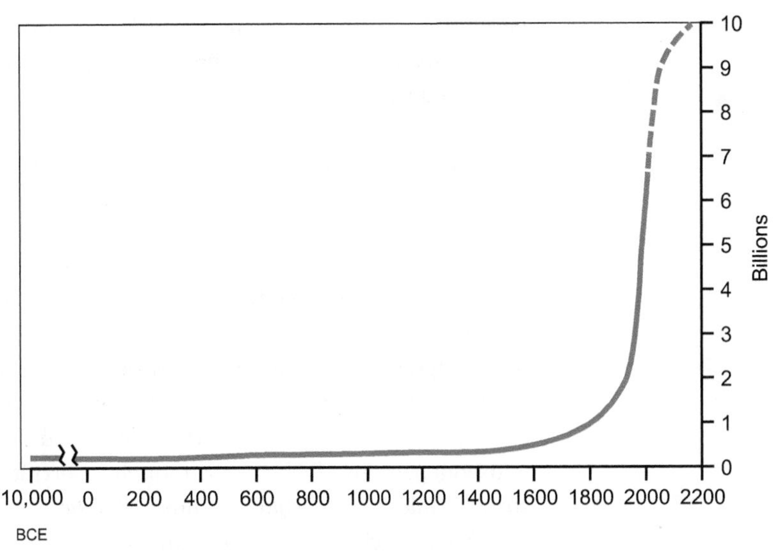

FIGURE 5.11 | World population growth

EXPLAINING POPULATION GROWTH

Making sound predictions about future population numbers is not easy, nor is it easy to explain past population growth. In this section, five different theories and models are described and evaluated.

The S-Shaped Curve Model

All species, including humans, have a great capacity to reproduce, which is regulated by constraints such as space, food supplies, disease, and social strife. Our first explanation proposes a simple and natural statement of population growth known as the S-shaped curve (see Figure 2.12).

The S-curve is produced under carefully controlled experimental conditions. The growth process begins slowly, then increases rapidly (exponentially), and finally levels out at some ceiling. It seems probable that such growth curves never actually occur in nature. It may not be unusual for growth to be first slow and then rapid, but it does appear to be unusual for any population (plant or animal) to remain steady at some ceiling level. A more characteristic final stage would involve a series of oscillations. Various scientists nevertheless predicted that world human population growth would correspond to the S-shaped curve. However, using the curve as a predictive tool is risky because it does not take into account the variety of cultural and economic factors that affect human populations. As seen in regard to fertility, stable populations result not from some natural law but from human decisions to control birth rates knowingly and willingly. On the other hand, it is still possible, as suggested in Figure 5.11, that the history of human population growth will reflect an S-shaped curve by 2200.

Malthusian Theory

Thomas Robert Malthus (1766–1834) was born in England at a time of great technological and other changes associated with industrialization and rapid population growth. He opposed the prevailing school of economic thought, **mercantilism**, which was explicitly pro-natalist; for mercantilism, more births meant more wealth since a large labour force was needed both in England to increase national productivity and overseas to increase English strength in the colonies.

mercantilism
A school of economic thought dominant in Europe in the seventeenth and early eighteenth centuries that argued for the involvement of the state in economic life so as to increase national wealth and power.

Malthus expressed his views in *An Essay on the Principle of Population*, first published in 1798. Basic Malthusian logic is straightforward. It can be usefully presented in terms of two axioms and a hypothesis:

Axiom 1: Food is necessary for human existence; further, food production increases at an arithmetic rate, i.e. 1, 2, 3, 4, 5 . . .

Axiom 2: Passion between the sexes is necessary and will continue; further, population increases at a geometric rate, i.e. 1, 2, 4, 8, 16 . . .

Given these two axioms, which Malthus regarded as reasonable assumptions, the following hypothesis is deduced:

> Population growth will always create stress on the means of subsistence.

According to Malthus, this conclusion—the inevitable result of the different growth rates of food supplies and population—applies to all living things, plant or animal. As rational beings (that is, with culture), humans theoretically have the capacity to anticipate and therefore avoid the consequences by deliberately reducing fertility through adoption of what Malthus called *preventive checks*, such as moral restraint and delayed marriage. However, he believed that, in practice, humans are incapable of voluntarily adopting such checks: they would do so only under the pressure of extreme circumstances such as war, pestilence, and famine—what he called *positive checks*. Hence, the human future would be one of famine, vice, and misery.

Malthus's central concern was imbalance between population and food. His arguments are intriguing, if not especially prophetic. In considering his theory, we must first note that there is no particular justification for the specific rates of increase proposed in the two axioms. Indeed, Malthus acknowledged in later editions of his work that the rates of population and food increase could not be definitely specified. Malthus failed to anticipate that food supplies could be increased not only by increasing the supply of land (which he correctly saw as finite) but also by improving fertilizers, crop strains, and so forth. Nor could he have predicted that contraception would become normal and accepted. He was aware of contraception such as existed at the time but saw it as an immoral preventive check (homosexuality was another), unlike the moral preventive checks of delayed marriage and self-restraint, of which he approved.

Events have proven Malthus incorrect in his predictions, and his theory lost favour in the mid-nineteenth century as birth-rate reductions and emigration eased population pressures in Europe. Today, what might be called neo-Malthusian theory is frequently purported to be relevant in the less developed world, particularly in several sub-Saharan African countries.

Marxist Theory

One of the earliest and most powerful critics of Malthusian theory was Marx, who objected to its rigorous axioms and hypothesis and believed that population growth must be considered in relation to the prevailing mode of production in a given society. For Marx, Malthus represented a bourgeois viewpoint in which the primary aim was to maintain existing social inequalities. Malthus saw population growth as the primary cause of poverty, whereas Marx saw the capitalist system as the primary cause. The difference is fundamental: for Marx, the problem was not a population problem at all but a resource distribution problem caused by capitalism.

A related distinction can be made between the concepts of overpopulation and surplus population. These terms are not synonyms. Malthus was concerned with overpopulation: he believed that a society could be said to be overpopulated when food and other necessities of human life were in such short supply that life-threatening circumstances arose. The key concept for Marx was surplus population: surplus—unemployed—workers represented a "reserve" labour force. According to Marx, capitalism depends on the existence of those workers to keep wages low and profits high; surplus population is, then, an inevitable consequence of capitalism. Both concepts are valuable to geographers.

Boserup Theory

A fourth theory is available in Ester Boserup's (1965) writings on historical changes in agriculture in subsistence societies. According to this argument, subsistence farmers select farming systems that permit them to maximize their leisure time and will change these systems only if population increases and it becomes necessary to increase the food supply accordingly.

Boserup argued that, in subsistence societies, population growth requires that farming be intensified, first by reducing the amount of land left fallow and then by employing multi-cropping, so that the supply of food is increased to feed the additional population. Thus, though population increase prompts an increase in gross food output, food output per capita decreases. Further, a series of negative changes ensues because most of the agricultural areas with high and increasing populations are already areas of poverty and often limited agricultural technology.

The contrast with Malthus is clear. Malthus saw population as dependent on food supply; Boserup reversed this relationship by proposing population as the independent variable. Evidence suggests that this theory is applicable in a subsistence context, especially throughout much of Africa, but not in the more developed world, where technology plays a much larger role and agricultural populations are declining. For many geographers, the principal attraction of these ideas is the questioning of Malthusian claims.

The Demographic Transition

So far, our discussion of theories about population growth has considered one natural law (the S-shaped curve) and three influential writers (Malthus, Marx, and Boserup). Each has something to offer; each is at least partially flawed. Clearly, population growth is not easily explained. Our fifth explanation is in a somewhat different category.

The **demographic transition** model is a descriptive generalization; it simply describes changing levels of fertility and mortality, and hence of natural increase, over time in the contemporary, more developed world. Because it is based on known facts rather than specific axioms or general assumptions, it has a major advantage: although simplified, it is clearly more realistic than the other explanations we have looked at. Inevitably, it also has a major disadvantage: because it is descriptive, it does not offer a bold, provocative perspective or hypothesis. Figure 5.12a presents the demographic transition in conventional graphic form; Figure 5.12b shows that the transition can be graphed as an S-shaped curve.

The first stage of the demographic transition is characterized by a high CBR and a high CDR—the two rates are approximately equal. The CDR fluctuates in response to war and disease, but the principal reason for the high death rates is the lack of clean drinking water and an effective sewage system. Throughout this stage, infant mortality is high and life expectancy low. The high birth rates are characteristic of a situation where there is no incentive to reduce birth. This stage involves a low-income agricultural economy. Population growth is limited, with doubling times as long as 5,000 years. This stage applied to all human populations around the world until about the mid-eighteenth century.

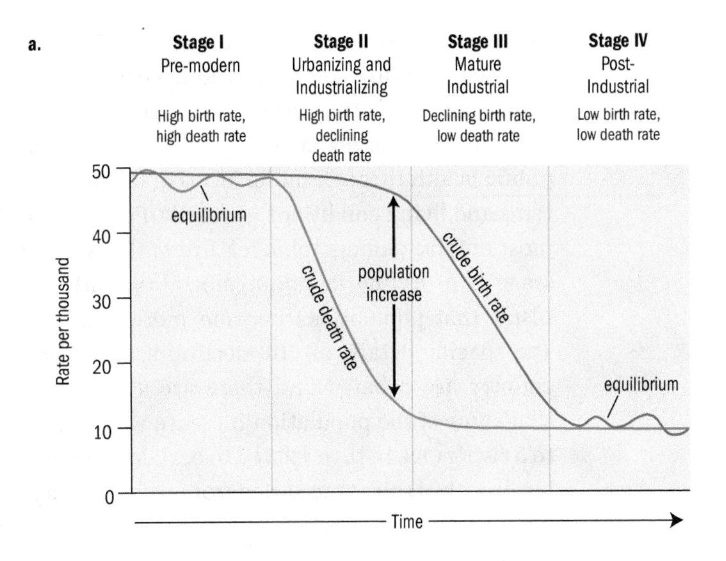

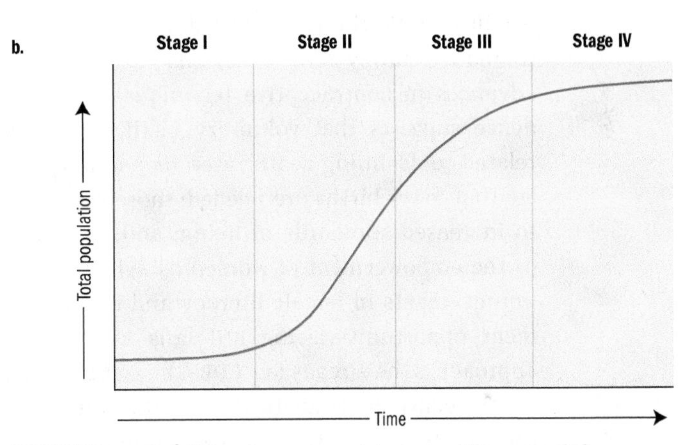

FIGURE 5.12 | The demographic transition model

a. The demographic transition model. This model describes population change over time. It is based on the observed changes, or transitions, in birth and death rates in industrialized societies, with its focus on about the past 250 years.

b. The corresponding growth of population, highlighting the fact that the transition in births and deaths can be graphed as an S-shaped curve.

In the second stage, CDR is reduced dramatically as a result of the onset of urbanization and industrialization and related sanitation, medical, and health advances. This reduction is not accompanied by a parallel reduction in CBR; consequently, the RNI is high and population growth is rapid. This second stage occurred first in Northern and Western Europe and soon spread east and south throughout most of the continent and then to other industrializing countries. Within the larger context of urbanization and industrialization, there are several specific reasons for the decline in deaths. Improvements in agricultural technologies, especially crop rotation and selective breeding, preceded the Industrial Revolution,

demographic transition
The historical shift of birth and death rates from high to low levels in a population; mortality declines before fertility, resulting in substantial population increase during the transition phase.

as did the introduction of new crops from the Americas, mainly the potato and corn. As the Industrial Revolution progressed, city planners became increasingly concerned with improving public health through more effective sewage systems and living conditions generally. Perhaps the most notable demographic feature of this second stage is a decline in infant mortality, with the result that populations become more youthful. The specific details of CDR decline in any given country are debated and there are suggestions that some of the population increase was related to a rising CBR, in turn related to earlier marriage. The length of this stage varied from place to place, but it usually lasted several decades.

The principal feature of the third stage is a declining CBR, the result of voluntary decisions to reduce family size—decisions facilitated by advances in contraceptive techniques. The evidence suggests that voluntary birth control is related to declining death rates, as parents realize that fewer births are needed; more generally, to increased standards of living; and, critically, to the empowerment of women as evidenced by improvements in female literacy and in employment opportunities. The RNI falls as the CBR approaches the already low CDR. The fact that this stage occurs contradicts the Malthusian claim that death rates are the key to explaining changes in population. The first clear signs of the onset of this third stage were in the late nineteenth century, again in Northern and Western Europe.

In the fourth (and final?) stage, the CBR and CDR are once again approximately equal but are low, as is the RNI. The process of population momentum ensures that population continues to increase for some years following the onset of this stage. Key demographic features include fertility rates below replacement level and an aging population.

Growth rates are similar in the first and fourth stages, but the ways in which the rates are generated are very different. The transition involves two central stages during which birth and death rates are unequal and a high rate of increase prevails. The three population pyramids in Figure 5.8 exhibit the characteristics of a country moving through the transition. The four population pyramids in Figure 5.10 show a country that has passed through the transition, with the added complications of the baby boom and an aging population.

The demographic transition model accurately describes in a simplified form the experience of the more developed world. Most of that world reached the fourth stage during the first half of the twentieth century, but what about the less developed world? Throughout much of Asia, Africa, and Latin America, the situation resembles the third stage of the model. Can we assume that it is only a matter of time before economic development guarantees passage through the third stage and entry into the stable fourth stage? To put the question another way: Is the demographic transition a valuable predictive tool? The answer appears to be "yes" but for rather different reasons. As we have seen earlier in this chapter, compelling evidence suggests that the fertility decline in the third and fourth stages of the demographic transition is indeed occurring in the less developed world.

FOTOGRAFIA INC./iStockphoto

An indigenous family from Tena, Ecuador, poses for a photo. As the demographic transition continues in Latin America and other areas, such families will be smaller in future generations.

DISTRIBUTION AND DENSITY

The preceding account of population growth through time has paid little specific attention to either the present-day spatial pattern of population or to the long history of human migration that has contributed to this pattern. We consider these matters in the final two sections of this chapter.

Determining population distributions and densities—where people are located and in what numbers—is one of the human geographer's central concerns. Establishing these basic facts, normally through a census, is not as easy as you might think. There are three related problems. The required data are not available for all countries—let alone all regions within countries—and even the data that are available may not always be reliable. The available data have been collected for various purposes and hence may reflect different divisions from the ones we are interested in. Since distribution and density statistics are closely related to spatial scale (recall Figure 2.10), they can often suggest an inaccurate picture of reality. World population density is discussed in Box 5.7.

census
The periodic collection and compilation of demographic and other data relating to all individuals in a given country at a particular time.

Evaluating the Available Explanations

Examining the Issues

BOX 5.7 | World Population Density

Conventionally, population **density** is arithmetic density, the total number of people in a unit area. Maps of population regularly combine two characteristics: distribution and density. Figure 5.13, showing population density and distribution for the world, is a good example.

density
A measure of the number of geographic facts (for example, people) per unit area.

Many data sources use a single statistic to identify the population density of a country, but that practice can be seriously misleading. Canada is a prime example: because its population is not evenly distributed—in some areas the density is very high, in others it is very low—a single density statistic for the entire country would be meaningless. The 2014 population density was 3.5 people per square kilometre, but this single statistic is merely an average produced by combining large areas that are virtually unpopulated and relatively small areas that are densely populated. In most cases, the local scale is the only one where density measures are appropriate; the smaller the area, the less likely it is to include significant spatial variations.

More refined density measures relate population to some other measure, such as cultivated or cultivable land. For example, **physiological density** is the relation between population and that portion of the land area deemed suitable for agriculture.

physiological density
Population per unit of cultivable land.

Although the world map of population distribution and density (Figure 5.13) is valuable, it presents a static picture of a dynamic situation and provides no indication of how the distribution developed or of how it might change in the future. It shows three areas of population concentration: East Asia, the Indian subcontinent, and Europe. Both Asian areas are long-established population centres, locations of early civilizations and early participation in the agricultural revolution. Both include areas of high rural population density (especially in the coastal, lowland, and river valley locations) and large urban centres. Densities are high in Europe but lower than in the high-density areas of Asia, and these are related primarily to urbanization. In all three areas, population densities are clearly related to land productivity.

Other scattered areas of high density are northeastern North America, around large cities in Latin America, the Nile Valley, and parts of West Africa. Perhaps the most compelling impression conveyed by Figure 5.13, however, is that a large proportion of the earth's surface is only sparsely populated.

There is a simple—but potentially misleading—correlation between Figure 5.13 and basic physical geography. Certainly humans have favoured some physical environments and not others. Even a cursory comparison of Figure A1.8 (our online map of global environments) and Figure 5.13 suggests that three particular environments—monsoon, Mediterranean, and temperate forest areas—are associated with high population densities, while three others—desert, tundra, and polar areas—are associated with very low densities. Once again, though, it is important to remember that physical geography does not cause human geography; the correlation between the two simply reflects the fact that humans have recognized the relative attractiveness of certain areas, specifically with respect to productivity or carrying capacity.

At the global scale, four physical variables are relevant: temperature, availability of water, relief (the physical contours of the land), and soil quality. High-density areas typically have temperatures that permit an agricultural growing season of at least five to six months a year. Water is essential, whether it comes as precipitation or in the form of irrigation water. Optimum temperature levels and precipitation amounts are not easy to specify, since technologies such as irrigation, dams and reservoirs, and heating and cooling systems can help to

Continued

ameliorate natural conditions, but it is clear that extremely high temperature and precipitation together, as in tropical rain forests, are not associated with dense population.

A second set of factors related to global population densities has to do with cultural organization. In many of the high-density areas shown in Figure 5.13, a form of state organization was established relatively early—China, India, Southern Europe, Egypt, Mexico, and Peru all were centres of early civilizations. A strong state organization facilitates the concentration of population; conversely, the collapse of such organization (usually with the onset of war or climate change) works against continued concentration. However, not all current areas of high density experienced early state development: in Western Europe and northeastern North America, high densities developed much later, in conjunction with the Industrial Revolution and the urbanization associated with it.

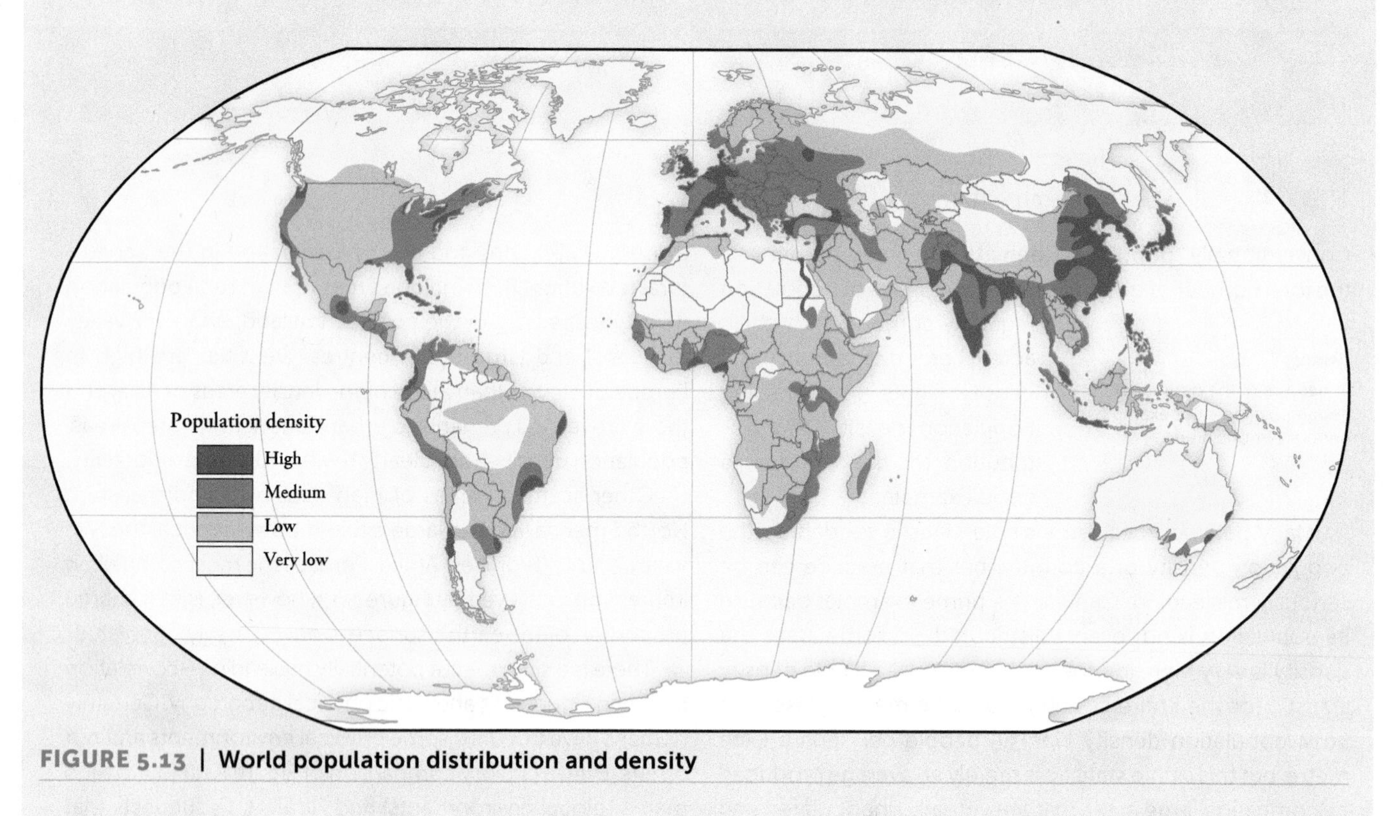

FIGURE 5.13 | World population distribution and density

Distribution of World Population

Table 5.10 provides information about population distribution by major world region, highlighting the dominance of Asia with 60.1 per cent today and 58 per cent predicted for 2050. Perhaps the most significant data in the table are those for Africa, which has 15.7 per cent today but a predicted 20.7 per cent by 2050. Table 5.11 identifies the 10 countries with the largest total populations in 2014 and includes projections for 2050. Although China and India continue to dominate, India is projected to overtake China in 2028 as the world's most populous country. African countries will also be more prominent by 2050, with Nigeria moving up from seventh to third place and the DR of Congo and Ethiopia, occupying ninth and tenth places, respectively. The two countries slipping off the list by 2050 are Russia and Japan, both of which are expected to lose population or perhaps grow only very slowly between 2014 and 2050 because of low fertility.

Table 5.12 shows the number of people per square kilometre in each of the top 10 countries; note that the countries with the largest populations are not those with the highest densities. Of the 10 most populous countries, Bangladesh is by far the most densely populated, followed by India and Japan.

MIGRATION

Together, the distribution and density of world population represents the dynamic outcome of the long historical process of population increase already discussed. But it is also a result of population movement. We now consider population movement, both as it has contributed to the current pattern and as it occurs today. As we will see, some areas of the world have grown largely as a consequence of European overseas exploration and resultant large scale migration.

Humans have always moved from one location to another—early pre-agricultural humans moved out of Africa to populate all major land masses of the world except Antarctica. Such movements expanded the resource base available, facilitated overall population increases, and stimulated cultural change by requiring ongoing adaptations to new environmental circumstances. Migration has continued to be important since these early movements.

For our purposes, migration may be defined as a particular kind of mobility that involves a spatial movement of residence. We do not regard the journey to work, the trip to the store, or the seasonal movements of some pastoralists and agriculturalists as migration. Focusing on those movements that involve a change of residence, we are particularly concerned with the distance moved, the time spent in the new location, the political boundaries crossed, the geographic character of the two areas involved, the causes of the migration, the numbers of persons involved, the cultural and economic characteristics of those moving, and some consequences of movement.

Why People Migrate

To ask why people migrate is not to suggest that a single reason is behind the multitude of migrations that have taken place in human history, although it is appropriate to note that migration often responds to the fact of spatial differences as people strive to improve their quality of life. We identify five approaches to this topic.

Push–Pull Logic

The first approach might best be described as a useful generalization—people move from one location to another because they consider the new location to be more favourable, in some crucial respect, than the old location. This is the key idea of inequality from place to place. How much more favourable the new location needs to be is a matter of individual judgment.

In mid-nineteenth-century Ireland, for example, there was virtually unanimous agreement that some overseas location such as the United States, Canada, or Australia was preferable to famine-stricken

TABLE 5.10 | World population distribution by major area (percentage): Current and projected

Region	2014	2050
Africa	15.7	20.7
Asia	60.1	58.0
Europe	10.2	7.3
Latin America and Caribbean (including Mexico)	8.5	8.3
North America (Canada and US only)	4.9	5.1
Oceania	0.5	0.5

Source: Calculated from Population Reference Bureau. 2014. *2014 World Population Data Sheet*. Washington, DC: Population Reference Bureau.

TABLE 5.11 | The 10 most populous countries: Current and projected

2014		2050	
Country	Population (millions)	Country	Population (millions)
China	1,364	India	1,6576
India	1,296	China	1,312
United States	318	Nigeria	396
Indonesia	251	United States	395
Brazil	203	Indonesia	365
Pakistan	194	Pakistan	348
Nigeria	177	Brazil	226
Bangladesh	158	Bangladesh	202
Russia	144	DR of Congo	194
Japan	127	Ethiopia	165

Source: Population Reference Bureau. 2014. *2014 World Population Data Sheet*. Washington, DC: Population Reference Bureau.

TABLE 5.12 | Population densities of the 10 most populous countries, 2014

Country	Population Density per km^2 (millions)
Bangladesh	1,101
India	394
Japan	336
Pakistan	244
Nigeria	192
China	143
Indonesia	132
United States	33
Brazil	24
Russia	8

Source: Population Reference Bureau. 2011. *2011 World Population Data Sheet*. Washington, DC: Population Reference Bureau.

Eastern European immigrants travel to North America, c. 1910–14.

Ireland. In this case, the advantages of a new location appeared to be so overwhelming that the decision to stay behind may have been just as difficult as the decision to leave, if not more so. In other cases, though, only a small percentage of the population will decide to migrate, suggesting that the perceived difference between old and new locations is not so substantial. Thus, in simple terms, migration decisions can be conceptualized as involving a push and a pull. Being located in an unattractive area is a push; being aware of an attractive alternative area is a pull.

Table 5.13 summarizes some common push and pull factors. Typical elements can be sorted into four categories: economic, political, cultural, and environmental. A simple *economic* explanation is that migration is a consequence of differences in income, with people moving from low- to high-income areas. Relatively low income is a push; relatively high income is a pull. Another economic explanation involves the relative availability of agricultural land. From perhaps the sixteenth to the twentieth centuries, land was less readily obtainable in Europe than in temperate areas overseas. A third economic explanation, and perhaps the most fundamental, applies in situations where life is threatened because of inadequate food supplies; in this case, any viable agricultural alternative area will be more attractive.

Another respect in which areas can be seen as unequal has to do with *politics*. In some extreme cases, people have felt obliged to seek refuge in a country other than their own. In recent years, the political environments in Afghanistan, Iraq, Ethiopia, the former Yugoslavia, Rwanda, Somalia, Syria, and Ukraine have prompted refugee movements. This topic is so important to human geographers that it will be addressed separately in the discussion of less developed world circumstances in the following chapter.

Culturally, migrants are most likely to move to areas that they perceive as similar in terms of language and religion. Much of the movement from former colonies to the former colonial power relates to these considerations and also to the relative ease of such movement in terms of the legal requirements of entry. Migration to a new location is also more likely when friends or relatives have migrated to that location previously.

Relative attractiveness can also be measured *environmentally*. For example, migration may be induced by flooding and desertification, and the historical experience indicates that most migrations have been towards the temperate climatic areas.

Push factors at one place depend on pull factors at others. Understanding migration requires both sets of factors to be considered, as well as the relationship between these factors. The balance between push and pull factors is complex. For example, high unemployment might be a push factor and low unemployment in another place a pull factor. But the difference in employment rates needs to be considered in the context of the economic cost of migration and the cultural upheaval associated with leaving family and friends.

As obvious as push–pull reasoning may seem, it does have a serious limitation: it assumes that people invariably behave according to the logic of a theory. Specifically, it is unable to explain why some people decide to stay in an unfavourable area when a more favourable alternative is available.

Laws of Migration

A classic attempt to formulate specific "laws" of migration was made in the late nineteenth century by E.G. Ravenstein (1876, 1885, 1889). Based on analyses of population movements in Britain, Ravenstein's laws are not laws in a formal, positivistic sense but generalizations with varying degrees of applicability. The 11 generalizations that Ravenstein developed (Box 5.8) still are among the most valuable concepts we have for understanding migration. However, like push–pull logic generally, they fail to take into account individual differences.

The Mobility Transition

A third effort to explain why people migrate was made by Zelinsky (1971). This theory is labelled the mobility transition. The term is derived from the concept of demographic transition discussed earlier in this chapter. Zelinsky proposed five phases of temporal changes in migration:

1. the pre-modern traditional society
2. the early transitional society
3. the late transitional society
4. the advanced society
5. a future, super-advanced society

TABLE 5.13 | Some push–pull factors

Push Factors	Pull Factors
Localized recession because of declining regional income	Superior career prospects and increased regional income
Cultural or political oppression or discrimination	Improved personal growth opportunities
Limited personal, family, career prospects	Preferable environment (climate, housing, medical care, schools)
Disasters, such as floods, earthquakes, wars	Other family members or friends

Source: After Bogue, D. J. 1969. *Principles of Demography*. New York: Wiley, 753–4.

Each of these five societies has particular migration characteristics. Phase 1 features minimal residential migration and only limited human mobility. Zelinsky sees this phase as temporally parallel to the first stage of the demographic transition (high birth rates, high and fluctuating death rates). In phase 2, numerically significant migration begins in the form of rural-to-urban and overseas movements. This phase is temporally parallel to the second stage of the demographic transition (continuing high birth rates, rapidly falling death rates) and includes the mass movements associated with industrialization and European overseas expansion. Rural-to-urban mobility declines somewhat in phase 3 but remains numerically significant, while overseas migration is drastically reduced. This phase

Legal migrant workers from Mexico harvest lettuce in California's Imperial Valley. Thousands of people who cannot find employment at home cross the US–Mexico border every day to work in the valley's fields. The US farmers are able to fill these low-paying positions that are undesirable to most.

Examining the Issues

BOX 5.8 | The Ravenstein Laws

More than a century ago, E.G. Ravenstein wrote three articles on migration that have been highly influential in much subsequent research on the subject. On the basis of information contained in the British censuses of 1871 and 1881, Ravenstein identified 11 laws—or, more correctly, generalizations. In the following list, each generalization is followed by a brief comment.

1. The majority of migrants travel only a short distance. The idea of distance friction, noted in Chapters 2 and 3, is a fundamental geographic concept.
2. Migration proceeds step by step. Thus, a migrant from Europe to Canada might go first to a port city such as Montreal and then to rural Quebec.
3. Migrants moving long distances generally head for one of the great centres of commerce or industry. This reflects the fact that large centres are usually better known than small ones to people from far away.
4. Each current of migration produces a compensating countercurrent. Any such countercurrent is usually relatively small.
5. The natives of towns are less migratory than those of rural areas. This law reflects the frequency of rural-to-urban migration.
6. Females are more migratory than males within their country of birth, but males more frequently venture beyond. Females often move within a country in order to marry. International migrants usually are young males.
7. Most migrants are adults. Families rarely migrate out of their country of birth.
8. Large towns grow more by migration than by natural increase. Remember that Ravenstein was writing at a time of dramatic industrial and urban growth.
9. Migration increases in volume as industries and commerce develop and transport improves. Such developments make urban centres more attractive and reduce distance friction.
10. The usual direction of migration is from agricultural areas to centres of industry and commerce. This is still the most common direction in the early twenty-first century, as evidenced by the ongoing rural depopulation in the Canadian prairies.
11. The major causes of migration are economic. This law, too, reflects Ravenstein's time: the present text includes several examples of migrations undertaken for social or political reasons.

Although these statements have been modified by subsequent research, they have not been disproven. Clearly, they are somewhat time specific, but all of them have some validity in different places and at different times (Grigg, 1977). Probably Ravenstein's principal limitation is his neglect of various forms of forced migration.

What is your impression of these laws? Do they make sense to you, both intuitively and with reference to the ideas introduced in this chapter? Or are they too general?

developmentalism
Analysis of cultural and economic change that treats each country or region of the world separately in an evolutionary manner; assumes that all areas are autonomous and proceed through the same series of stages.

is temporally parallel to the third stage of the demographic transition (declining birth rates, low death rates). In phase 4, residential mobility continues apace; rural-to-urban movement lessens but continues; urban-to-urban movement is significant; and international migration increases, with both unskilled and skilled workers moving from the less developed to the more developed world. This phase is temporally parallel to the fourth stage of the demographic transition (low birth rates, low death rates). Finally, in phase 5—Zelinsky's attempt to predict future trends—most migration is between urban centres. There is no equivalent stage in the demographic transition model.

Zelinsky's mobility transition theory is a useful summary of temporal changes in mobility, but it is not without its critics, especially because it can be seen as an example of **developmentalism**—a "geography of ladders" in which "the world is viewed as a series of hearth areas out of which modernization diffuses, so that the Third World's future can be explicitly read from the First World's past and present in idealized maps and graphs of developmental social change" (Taylor, 1989: 310). An alternative way of looking at the mobility history of a particular country would be to see it as reflecting global processes as well as those taking place within the country. The same issue arose, somewhat less formally, in our discussion of the demographic transition.

A Behavioural Explanation

Another way of approaching the question of why people migrate is from a more humanistic perspective. None of the three approaches outlined so far has focused primarily on the people themselves. Push–pull logic, the 11 laws of Ravenstein, and the mobility transition model all have positivistic overtones: they can be interpreted as implying that each individual human responds in an identical fashion to various external factors. Of course, any human geographer knows that this is not so, but many of us have been willing to sacrifice some reality for the sake of simplicity. Our fourth attempt to explain why people migrate shifts attention to the people themselves and away from the forces presumed to be affecting their decisions.

For convenience, we will label this approach the behavioural view of migration, since it centres on the behaviour of individuals rather than on aggregate, usually large-group, behaviour. Behavioural approaches usually employ a version of push–pull logic but typically do so at the level of the individual.

Place utility is a measure of the extent to which an individual is satisfied with particular locations. Typically, the place utility that people attribute to their current place of residence is much better grounded in fact than the place utility they attribute to other locations. Place utility is thus an individually focused version of push–pull logic. This concept was introduced by Wolpert (1965), who argued that it was necessary to research an individual's **spatial preferences**.

Clearly, such preferences are based on perceptions, not objective facts. All people have what we call mental maps: mental images of various places that can contribute to migration decisions. This approach has typically been applied at a relatively local scale and is discussed further in Chapter 8.

Moorings

The studies in migration that we have looked at are not conceptually sophisticated by the standards of much contemporary human geography; indeed, "migration research is in danger of being left behind by recent developments in social theory" (Halfacree and Boyle, 1993: 337).

One way of advancing studies in this area is to acknowledge that migration reflects a personal decision made within a larger political and economic framework, for example, by drawing on theories of human motivation as developed in social psychology and putting greater emphasis on the cultural influences on migration. An example is our fifth and final attempt to explain why people migrate. This is the idea of "moorings"—issues through which individuals give meaning to their lives—suggested by Moon (1995). Table 5.14 identifies some typical moorings.

The moorings approach centres on the idea that individuals' perception of their current location, and hence the likelihood of their either remaining there or migrating to another location, depends on the value they place on their various moorings. It seems probable that future efforts to explain why people migrate at various spatial scales, including the global, will increasingly focus on conceptual formulations that incorporate

place utility
A measure of the satisfaction an individual derives from a location relative to his or her goals.

spatial preferences
Individual (sometimes group) evaluation of the relative attractiveness of different locations.

TABLE 5.14 | Some typical moorings

Some typical moorings
Life-Course Issues
household/family structure
career opportunities
household income
educational opportunities
caregiving responsibilities
Cultural Issues
household wealth
employment structure
social networks
cultural affiliations
ethnicity
class structure
socio-economic ideologies
Spatial Issues
climate features
access to social contacts
access to cultural icons
proximity to places of recreation interest

Source: Moon, B. 1995. "Paradigms in Migration Research: Exploring 'Moorings' as a Schema." *Progress in Human Geography* 19: 515.

life cycle
The process of change experienced by individuals over their lifespans; often divided into stages (such as childhood, adolescence, adulthood, old age), each of which is associated with particular forms of behaviour.

both the personal and the cultural aspects of the migration decision.

The Selectivity of Migration

Migration is a selective process. Objectively speaking, areas in themselves are neither unattractive nor attractive. Relative attractiveness is a matter of subjective, individual perception: what is attractive or unattractive to one person may not be so to another, and migration decisions are, in the final analysis, made by individuals. We are not all equally affected by the general factors prompting migration. Who moves and who stays?

Among the factors that appear to influence individual decisions about migration are age (most migrants are older adolescents or young adults); marital status (today, as in the past, most people migrating from the less developed world are single adults); gender (males typically are more migratory, but there are many exceptions to this generalization); occupation (higher-skilled workers today are most likely to move); and education (migrants have higher levels of education than non-migrants). In regard to migration across national borders, all these factors are determined to some degree by the immigration criteria of the receiving country. In the most general sense, however, there is a useful relationship between life cycle and the likelihood of individual migration.

In addition, there is often a substantial difference between what people would like to do and what they are able to do. People who want to migrate may be unable to leave their home for political reasons, and people who want to move to a specific new area may not be able to do so because of the immigration policies of that area. Some striking examples of this circumstance are included in the account of racism in Chapter 8. Similarly, potential migrants need to consider the economic and personal costs; some may be unable to pay for the move; some may not be able to move because of health, age, or family circumstances (Lee, 1966).

Types of Migration

One of the most useful attempts to classify migration was made by the sociologist Petersen (1958), who identified four classes of migration: primitive, forced, mass, and free. Along with these four, it is useful to add *illegal* migration, which can take two forms: illegal exit occurs when a country prohibits out-movement; illegal entry occurs when people enter a country without official approval. Figure 5.14 provides an overview of the principal forced and free/mass migrations that have taken place at the global scale since about 1500.

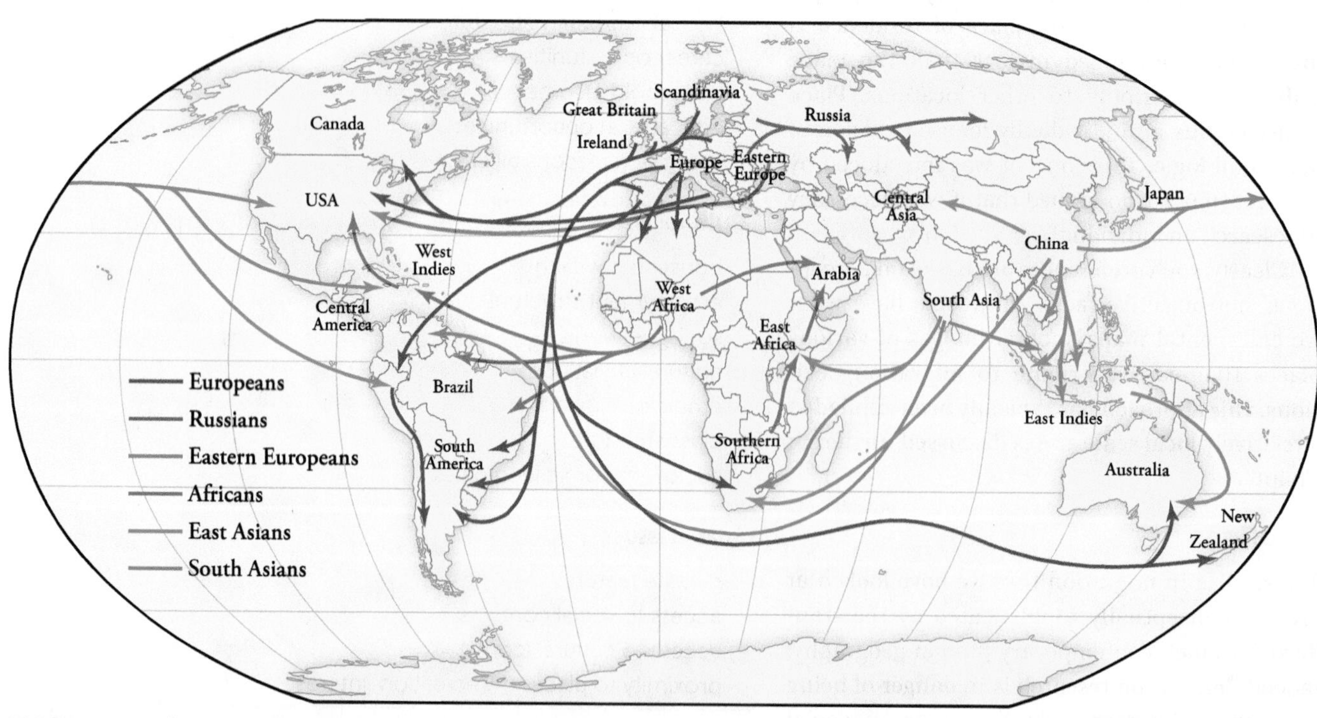

FIGURE 5.14 | Major world migrations, 1500–1900

Primitive Migration

Primitive migration is really a specific instance of adaptation to environment, in which people respond to an unfavourable environment by leaving it in search of a more favourable one. Pre-industrial societies tend to make such adaptation decisions on a group rather than on an individual basis. Thus, hunting and gathering groups might migrate regularly as the resources of an area are depleted or as game animals move on; some agricultural groups might move as soil loses fertility. Another instance of primitive migration occurs when populations increase in size so that additional land is needed. In pre-industrial societies, primitive migration was a normal part of the human search for appropriate environments—the search that was responsible for the human occupation of most of the earth.

Forced Migration

Forced migration has a long history. Slavery was a significant institution in early civilizations such as those of Greece and Rome. It appears that slaves forced to migrate from areas occupied by the Romans made up the largest portion of Rome's population at its peak. The Europeans who colonized the Caribbean and the warm coasts of North and South America—areas not conducive to large-scale European migration—also relied on slave labour. Perhaps as many as 11 million Africans were moved as slaves between 1451 and 1870.

A second example of forced migration can be seen in the late nineteenth century, when workers from China, Java, and India were shipped to the new European-controlled plantations of Malaysia, Sumatra (now Indonesia), Burma, Ceylon (now Sri Lanka), and Fiji. These workers were supposed to be engaged voluntarily, on the basis of contracts, but in fact force often was used. A third and quite differently motivated example was the post-1938 movement of Jewish populations in areas controlled by Nazi Germany. In each of these instances, movement was literally forced on people.

A variant of forced migration is the situation in which the migrant has some voice, however small, in the decision-making process. Examples of such *impelled* migration would include the many cases in which people have chosen to flee oppressive political regimes, war zones, and areas of famine. The dividing line between forced and impelled is not clear, nor is the line between impelled and free. Most of the contemporary refugee movements discussed in Chapter 6 qualify as impelled. Most forced and impelled migrations are related to the actions of others who directly influence the migration decision.

Free Migration

In free migration the person has the choice either to stay or to move. Much free migration takes place within rather than between countries. In the United States, the South and West are attractive destinations for a combination of reasons having to do with climate and job opportunities. Movement to destinations perceived as attractive is evident in England as well, where the southeast is the most attractive area, and in Canada, where Alberta is especially attractive. People also continue to move from rural to urban areas and from central to suburban zones within urban areas. Because these movements tend to be selective, reflecting the migrants' stage in the life cycle, gender, and ethnic background, they frequently lead to significant changes in the structure and composition of local populations. Most migration within countries, regional or more local, continues to be primarily economic in motivation, being related to employment, income potential, and the housing market.

Free migration between countries continues, but all of the more developed countries are popular destinations, including the European states that so many earlier migrants left behind. This situation confirms our observation that individuals' migration decisions often reflect their assessments of relative place utility. It is estimated that about 40 per cent of people in less developed countries would like to emigrate, which is one reason that immigration controls are in place in more developed countries. There are concerns about a flood of migrants taking jobs or collecting welfare benefits.

Mass Migration

Mass migrations are free migrations prompted by push–pull factors that are widely experienced and involve large numbers of people. Historically, mass migrations included movement from densely settled countries to less densely settled ones, as in the period between 1800 and 1914, when some 70 million people migrated from Europe to temperate areas such as the United States, Canada, Australia,

New Zealand, South Africa, and Argentina. As we saw in our discussion of the mobility transition, this migration was closely related to the demographic and technological changes that began in Europe after about 1750.

Thus, emigration relieved Europe of at least some of the population pressures that came with the second stage of the demographic transition. In 1800, people of European origin totalled 210 million; by 1900, the total was 560 million—a 166 per cent increase—and one in every three people in the world was of European origin. Most European countries participated in the nineteenth-century wave of migration. Irish, English, Scottish, Germans, Italians, Scandinavians, Austro-Hungarians, Poles, and Russians all moved overseas in large numbers; at the same time, many other Russians moved east to the Caucasus and Siberia. This historically brief period of movement had massive and wide-ranging effects in areas both of origin and of destination, redistributing a large number of people and bringing into contact many previously separate groups. Indeed, free and mass migration lies at the root of some of the most difficult political issues in the world, particularly those involving the territorial claims of various minority groups.

Illegal Migration

The term *illegal migration* covers a wide variety of situations. Although it is obviously not possible to determine the exact number of people involved, it is certainly significant. The most important category of illegal immigrants consists of those who consciously violate immigration laws, but other immigrants become "illegal" through no fault of their own, as a result of policy changes or the complexities of maintaining legal residency. One example of a policy change that created illegal immigrants involved several millions of West Africans who had legally immigrated to Nigeria during the 1970s and early 1980s; their status was changed after a new government changed the immigration regulations.

Other migrants, however, deliberately violate immigration laws. The simple explanation sees such illegal movement as the result of desperate push factors (overpopulation, political turmoil, economic crises) combined with irresistible pull factors (high wages, plentiful job opportunities). This explanation is accurate in many cases, but it is not adequate. Indeed, much illegal movement takes place between more developed countries; for example, a 1993 analysis concluded that Italians made up the largest group of illegal immigrants in New York City. Most illegal immigrants are young, clustered in urban areas, and involved in industries such as construction and hospitality. The illegal alien population in the US is estimated at more than 11 million.

One of the best-known cases of illegal immigration is that of Mexicans moving into the US; every night, the border sees many attempts at illegal immigration. But there are numerous other examples. Approximately 10,000 Chinese illegal immigrants enter the US each year, and a significant Chinese illegal movement finds its way into Canada. Australia is experiencing overstays of legal temporary admissions; the total number of overstays recorded by 1992 was 81,500—a significant number when compared with the legal immigration in that year of 35,100. The countries of the EU may have as many as three million illegal immigrants from Africa, Eastern Europe, and Asia. Major movements of illegal immigrants occur within Asia, where the favoured destinations are Japan and Singapore. Some of this movement occurs between more developed countries (for example, from South Korea to Japan). Before 1997, when it was returned to China, Hong Kong received many illegal immigrants from the mainland, while Singapore, despite very tight policies, has received many illegal immigrants from Malaysia.

These examples are only the best known. Although the motives for illegal immigration in many cases appear obvious, efforts at explaining other cases are complicated by the absence of precise data. What we can say with some certainty is that international migration, legal and illegal, reflects both global inequalities and globalization processes, specifically a growing interdependence among the world's countries and a web of international relations that is becoming ever more complicated.

CONCLUSION

The factual information included in this chapter is fundamental to any understanding of our world and how we live in the world. For many human geographers, an awareness of population is the key building block for any *human* geographic study. Without this information it would be very difficult to discuss such important topics as global inequality, political conflict, changing agriculture, urban growth, and industrial activity. But clearly we have not yet asked a fundamental question: What might be some of the consequences of the anticipated population increase in the coming decades? Our world of 7.2 billion people has many serious problems, as we saw in Chapter 4 with respect to impacts on environment, so what sort of world might the expected 9.7 billion people of 2050 live in? We think about some aspects of our global future in the following chapter.

In this chapter we have noted several key facts and identified some of the implications of these facts:

- World population continues to increase markedly but at a decreasing rate such that population may stabilize at about 10 billion people in 2200.
- Women in less developed countries are having fewer children, and the explanation appears to be that women are better educated and more empowered than in the past.
- Declining fertility and longer life expectancy are causing populations to age.
- It appears that government attempts to either decrease or increase fertility have had little impact (with the possible exception of the one-child policy in China). Rather, fertility varies because of the decisions that women are able to make.
- Population is distributed unevenly on the earth, with areas of high density related especially to agricultural potential.
- The current distribution of people is also an outcome of a long history of migration.

Summary

How Many?

The earth had a population of 7.2 billion in 2014, increasing to about 9.7 billion by 2050 and possibly stabilizing at just over 10 billion by 2200. Almost all future population increases will be in the countries of the less developed world.

Fertility

There are various measures of fertility, including the crude birth rate and the total fertility rate. In 2014 the world birth rate was 20 (20 live births per 1,000 members of the population), while the total fertility rate was 2.5 (the average woman has 2.5 children). A total fertility rate of about 2.1 to 2.5 is sufficient to maintain a stable population. Fertility is affected by many variables: age and related fecundity, nutrition, level of industrialization, age at marriage, celibacy, governmental policies, contraceptive use, abortion, and empowerment of women. Spatial variations in fertility are closely related to level of development; the total fertility rate in the more developed world is 1.6 and 2.6 in the less developed world. At present there is evidence of declining fertility in much of the less developed world (related to advances in women's education and widespread acceptance of family planning) and in much of the more developed world (possibly related to uncertain economic prospects).

Mortality

Three mortality measures are the crude death rate, the infant mortality rate, and life expectancy. In 2014 the global death rate was 8 per 1,000 people and life expectancy was 71 years. Major causes of death are old age, disease, famine, and war. Around the world, death rates show much less variation than birth rates; sub-Saharan Africa is the last region of high mortality. Data on infant mortality and life expectancy are good indicators of health. Many countries exhibit significant internal variations in mortality patterns.

Natural Increase

In 2014 the rate of natural increase was 1.2 per cent. If maintained, which seems highly unlikely, this rate would result in a doubling of world population in about 58 years. Natural increase is affected by the age composition of a population.

Population Aging

The age structure of the world's population is changing significantly as the proportions of elderly people increase relative to other age groups. This change is the result of declines in fertility and increases in life expectancy. The year 2000 was the first year in which people under 14 were outnumbered by people over 60. The social and economic implications of this shift are considerable.

Government Intervention

Most governments intervene directly and/or indirectly to influence growth. Policies aimed at limiting unnecessary early death are normal and designed to reduce mortality. Birth control policies can be laissez-faire, pro-natalist, or anti-natalist. A pro-natalist approach may be related to a dominant religion or economic and strategic motives. Anti-natalist policies, of varying degrees of success, are common in many of the less developed countries. The two most populous countries, China and India, have employed different approaches; China has been highly successful, India less so. Many anti-natalist policies are highly intrusive.

Predicting Growth

Past forecasts have often erred seriously. As of 2014, population is estimated to continue to increase rapidly for about another 40 years and then increase more slowly. There is much uncertainty concerning the relative merits of two arguments. The limits-to-growth, or catastrophist, thesis sees definite limits to population and economic growth because the earth is finite. The cornucopian thesis sees technology as continually making new resources available and hence enabling the earth to accommodate increasing numbers of people.

World Population Growth

The world's population increases in response to cultural change. An agricultural revolution began about 10,000 BCE and prompted increases from about 4 million to 250 million by the beginning of the Common Era. A second major change, the Industrial Revolution, made it possible for the world's population to increase from 500 million in 1650 to 1.6 billion in 1900.

Explaining Growth

The S-shaped curve is a useful biological analogy and may be applicable in the long term. Malthus, Marx, and Boserup each contributed usefully to our understanding of population growth—Malthus saw population as limited by food supplies; Marx saw population as a response to a particular social and economic structure; Boserup saw population increases as prompting food supply increases—but none is of general applicability. Malthus is the most-discussed theorist, and a contemporary version of his theory is known as neo-Malthusianism. The demographic transition model is a summary of birth and death rates over the long period of human history in what is now the developed world. There is a clear need for more and better theory.

Distribution and Density

Distribution refers to the spatial arrangement of a phenomenon; density refers to the frequency of occurrence of a phenomenon within a specified area. Population maps frequently combine the two characteristics. Asia has 60 per cent of the world's population; China is the most populous country, followed by India. A third area of population concentration is Europe. The Asiatic regions include some areas with very high rural population densities. Distribution and density of population are related to land productivity and cultural organization.

Causes of Human Mobility

Human populations have always been mobile. Migrations are often explained in terms of the relative attractiveness of different locations. This idea can be expressed by reference to simple push–pull logic, "laws" such as Ravenstein's, or the concept of place utility. Another explanation for migration, related to the demographic transition, is called the mobility transition, and other explanations consider individual preferences and social contexts.

Types of Migration

Primitive migration is the process by which humans originally moved over the surface of the earth, gradually adapting to new environments. Forced migration occurs when people have little or no choice but to move; slaves and refugees are examples of forced migrants. Free migration is the result of a decision made following evaluation of available locations. Mass migration is a specific form of free migration that involves a great many people making a specific migration decision at about the same time.

Links to Other Chapters

- **Globalization:** Chapter 2 (concepts); Chapter 3 (all of chapter); Chapter 4 (global environmental issues); Chapter 6 (refugees, disease, more and less developed worlds); Chapter 8 (popular culture); Chapter 9 (political futures); Chapter 10 (agricultural restructuring); Chapter 11 (global cities); Chapter 13 (industrial restructuring)
- **Demographic measures (CBR, LE, etc.):** Chapter 6 (inequalities)
- **Human impacts on earth:** Chapter 4
- **Inequalities between the more and less developed worlds:** Chapter 6
- **Population distribution and density:** Chapter 2 (environmental determinism and possibilism); Chapter 4 (population growth)
- **Explanations of migration:** Chapter 2 (positivism and humanism)
- **Forced migration:** Chapter 8 (types of society; landscapes and power relations; ethnicity; racism)
- **Free migration within countries:** Chapter 11 (rural settlements in transition; rural–urban fringe); Chapter 12 (urban sprawl)
- **The catastrophist–cornucopian debate:** Chapter 4 (environmental futures); Chapter 6 (refugees, the less developed world)
- **Explaining population growth:** Chapter 2 (philosophical approaches)

Questions for Critical Thought

1. How do culture and development affect fertility and mortality in the more developed and less developed parts of the world?
2. By virtually all population indicators, Africa faces more population-related challenges than any other continent. What can be done about the continent's high rates of natural increase, high mortality rates, low life expectancy, and other related issues?
3. To what extent can/should a government take measures to control the population of a country through placing limits on how many children an individual can have and encouraging (or limiting) immigration?
4. What are the social, economic, and political impacts of an aging population in countries of the more developed world?
5. Most experts agree that the rate of world population increase is slowing down and that the predictions made in the 1970s and 1980s for the twenty-first century were overestimates. Does this mean that world population is less important as a global concern now than it was three or four decades ago? How significant will population issues be in the future?
6. How are population trends (births and deaths) tied to levels of economic development as described in the demographic transition model? What are the alternative explanations to this model?

Suggested Readings

Visit the companion website for a list of suggested readings.

6 GLOBAL INEQUALITIES

Issues of spatial and social inequality were introduced in Chapter 3's account of globalization, while the terms *more developed world* and *less developed world*, defined in the introduction, were employed regularly in Chapter 5's discussions of population. In this chapter, we ask why the world is divided into more and less developed countries. To answer this complex question, we consider various factors, including physical geography, agricultural domestication, and, of profound significance, historical and current political and economic relationships between countries. We then discuss problems of defining and measuring development, paying particular attention to the distinction between economic measures and measures that focus on quality of life. An overview of major problems in less developed countries, such as food supply, refugees, disasters, diseases, and national debt, includes several detailed examples. The final sections of the chapter consider some of the strategies designed to further economic growth and improve quality of life in less developed areas, with emphasis on the question of political governance.

Here are three points to consider as you read this chapter.

- All too clearly, and tragically, the world comprises a collection of unequal places. Why is this the case?
- Is the distinction between more and less developed worlds meaningful and helpful? Or are the differences within each category sufficiently great to render it meaningless? Is it useful to identify a least developed world, mostly comprising countries in sub-Saharan Africa?
- How important is good government, especially democracy, to the well-being of a country and its citizens?

A gypsy girl searches for water in a slum area of Lahore, Pakistan. Access to clean water, proper sanitation, and sufficient food and opportunities are some of the discrepancies between the more developed and less developed worlds.

EXPLAINING GLOBAL INEQUALITIES

We begin by noting and rejecting an often-influential purported answer to the question of different levels of development and related inequalities: differences between groups of people in terms of their ingenuity and ability (i.e. their human potential) explain global differences in development. This answer, which has routinely been employed since it came to the fore in the nineteenth century, assumes the existence of groups of people called races and their inherent inequality. As Chapter 8's discussion of the myth of race conclusively demonstrates, this assumption is scientifically incorrect.

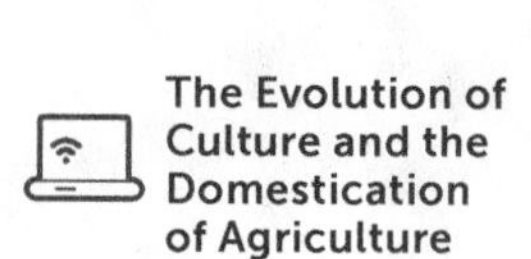
The Evolution of Culture and the Domestication of Agriculture

civilization
A contested term because it can be understood to mean that some groups are civilized while others are not; traditionally understood to refer to a culture with agriculture and cities, food and labour surpluses, labour specialization, social stratification, and state organization.

Nevertheless, there is no one easy answer as to why the world is characterized by inequalities from place to place such that it is routinely divided into two different and unequal parts. As we might expect, answers sometimes tend to be ideologically driven, reflecting a specific world view and related discourse. They are not necessarily wrong, but we must appreciate the larger intellectual context of the response. In this section, we consider two categories of answer that, importantly, can be thought of as sequential and complementary, and we place these answers in the needed larger contexts.

Our first answer—the shape of continents—requires that we think about why certain areas of the world experienced cultural development and the technology of agricultural domestication and, critically, why the practice of agriculture diffused away from some core areas of domestication. This explanation requires an understanding of physical geography and, for some critics, has overtones of environmental determinism. The first response can be seen as an essential prelude to the second, world systems and dependency theories. The second response, which considers relations between different parts of the world from the onset of European overseas expansion to the present, is rooted in Marxist logic of dominance and subordination.

The Shape of Continents

Civilizations developed in some areas and not others and were a consequence of the slow transition from hunting and gathering societies to agricultural societies (Figure 6.1 maps the most significant of these). But why did some civilizations, notably those in Eurasia, involve agricultural diffusion across great distances while others remained relatively limited in spatial extent, notably those in the Americas and Africa?

Here is one thought-provoking answer. Diamond (1997) contends that basic world geog-

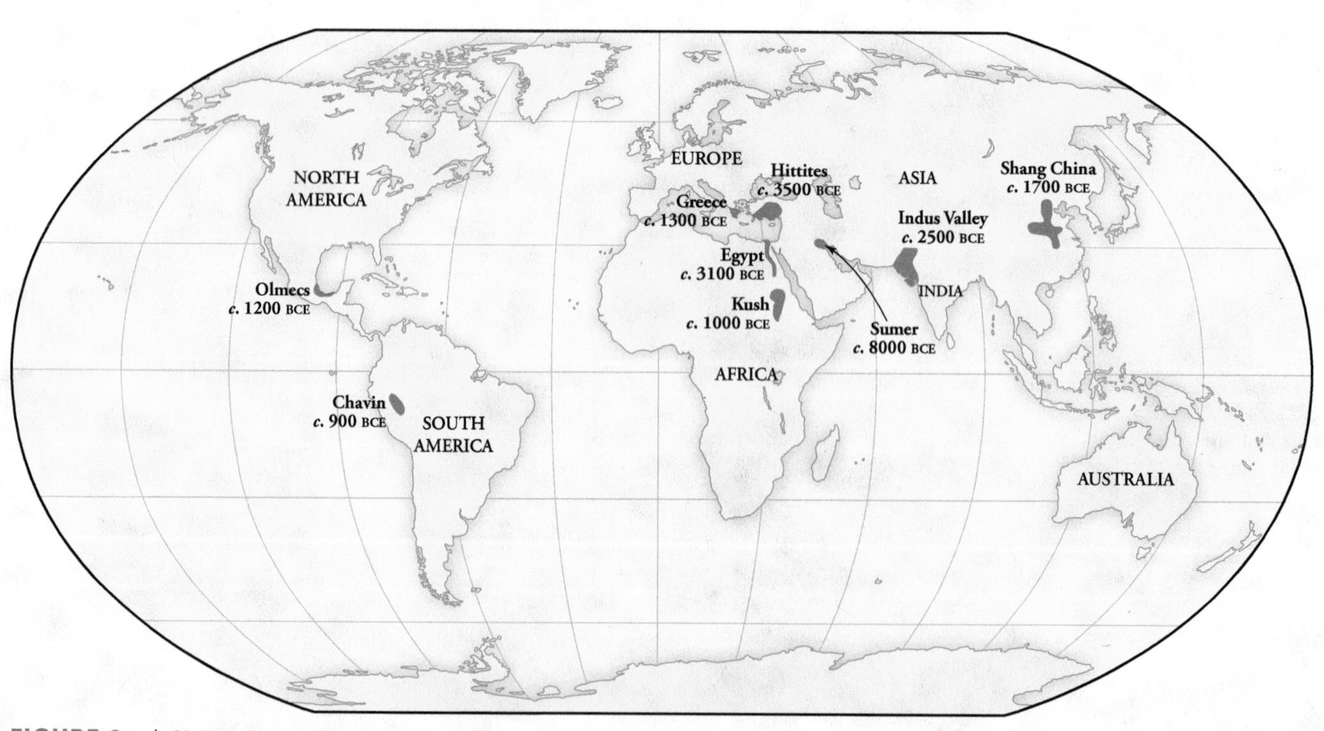

FIGURE 6.1 | **Civilizations of the ancient world**

Source: Stearns, P. N., M. Adas, S. B. Schwartz, and M. J. Gilbert. 2007. *World Civilizations: The Global Experience*. New York: Pearson, 3. © 2008. Printed and electronically reproduced by permission of Pearson Education, Inc., New York, NY.

raphy is critical to understanding not only the rise but also the subsequent spread of agriculture and, thus, the rise and spread of civilization. Figure 6.2 highlights the simple fact that the major axis of Eurasia is east–west while the major axes of the Americas and of Africa are north–south. Diamond proposes that these different continental shapes have been crucial because latitudinal (east–west) extent allows agricultural technologies to spread great distances over areas of similar climate, but lack of significant latitudinal extent means that spread is necessarily more limited.

It can be argued that, because agricultural technologies are the precursor to the rise of civilizations, this basic geographic circumstance is the ultimate factor in beginning a chain of causation that eventually leads to some societies' ability to spread globally and dominate other societies (Figure 6.3). It is possible to conceive of Diamond's ideas as aiding our understanding of why Europe (or possibly, as discussed in Box 1.2, China) was able to move overseas. The latitudinal extent of Eurasia is the first link in the chain because it permitted agricultural technologies to spread and, combined with suitable plant and animal species, led to a sophisticated agricultural region that became what we call civilization. The eventual outcome was a series of proximate factors—including the guns, germs, and steel that Diamond identifies in the title of his work—that enabled some people (Europeans) to move around the world and dominate most of those whom they contacted.

The shape of continents argument is not proven, but it is a highly suggestive argument. After all, if correct, it helps explain why Europe was able to move overseas.

World Systems and Dependency Theories

Our second answer builds on the shape of continents idea but has a quite different conceptual logic, relying as it does on a Marxist understanding of the colonial period and of relations between countries since about 1500. For many human geographers, the single most important factor in explaining the plight of countries in the less developed world is their history of **colonialism** and subsequent ongoing relationship with more developed countries. Most of the less developed countries have a colonial history; even those that do not (such as China, Thailand, Liberia, Saudi Arabia, Iran, and Afghanistan) have been affected by Europe's world dominance between about 1500 and 1945.

Why is this relationship so important? On the world scale, it can be argued that colonialism has led to **dependence**. In the past, many former colonies became economically dependent on the more developed countries; more recently, aid intended to promote development has served to encourage increased dependence. The indigenous cultures and social structures of former colonies have been largely relegated to secondary status, their place taken by European structures. Thus, in the broadest sense, the less developed countries lack power, including controlling and directing their own affairs.

An exciting contribution to human geography that addresses these issues of colonialism and the related subordination of less developed countries is the **world systems theory**, proposed by Wallerstein (1979), and related ideas of **dependency theory**. Describing the dynamic capitalist world economy from 1500 onward, world systems logic examines the roles that specific states play in the larger set of state interrelationships. To summarize briefly, capitalism emerged gradually from feudalism in the sixteenth century, consolidated up to 1750, and expanded to cover the world in the form of industrial capitalism by 1900. In 1917, however, the capitalist system entered a long period of crisis that some suggest may eventually bring the world closer to a socialist system. Although the changes that have taken place since 1989 appear to make further movement towards

colonialism
The policy of a state or people seeking to establish and maintain authority over another state or people.

dependence
In political contexts, a relationship in which one state or people is dependent on, and therefore dominated by, another state or people.

world systems theory
A body of ideas that suggests a division of the world into a core, semi-periphery, and periphery, stressing that the periphery is dependent on the core; has numerous implications for an understanding of the less developed world.

dependency theory
A theory that centres on the relationship between dependence and underdevelopment.

Climate and Proximity to Coastline

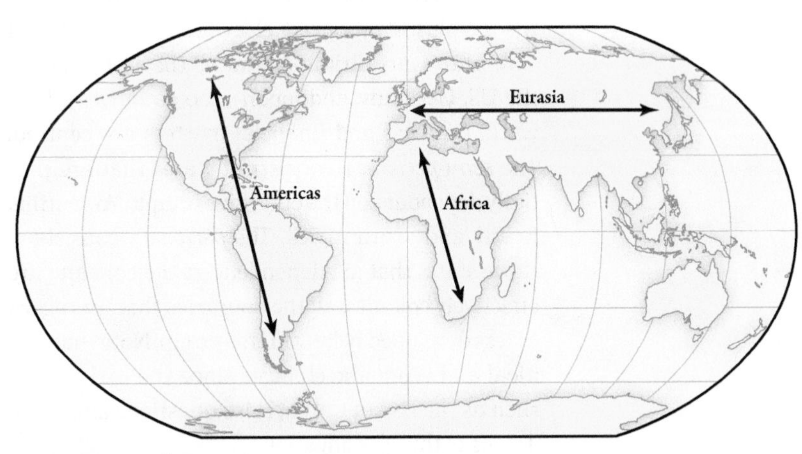

FIGURE 6.2 | The shape of continents

Source: Diamond, J. 1997. *Guns, Germs and Steel: The Fates of Human Societies*. London: Jonathan Cape. Figure 10.1, "Major Axes of the Continents." Copyright © 1997 by Jared Diamond. Used by permission of W.W. Norton & Company, Inc.

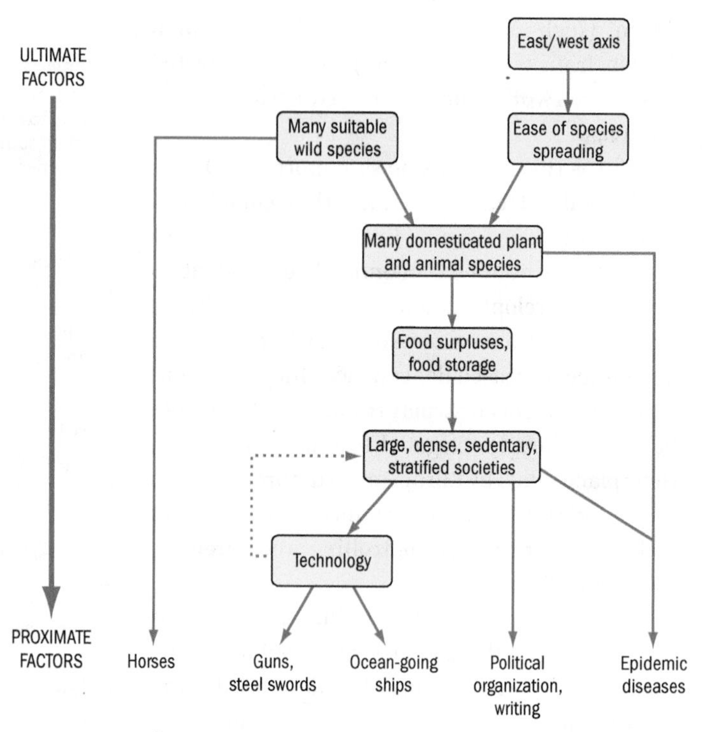

FIGURE 6.3 | Factors underlying the broadest patterns of history

Diamond suggests a chain of causation that begins with the advantage of latitudinal extent and leads up to factors such as guns, diseases, and steel that eventually enabled Europeans to conquer other people.

Source: Diamond, J. 1997. *Guns, Germs and Steel: The Fates of Human Societies*. London: Jonathan Cape. Figure 4.1, "Factors Underlying the Broadest Pattern of History." Copyright © 1997 by Jared Diamond. Used by permission of W.W. Norton & Company, Inc.

socialism less likely, especially in Europe, this prospect in no way detracts from Wallerstein's general observation.

The contemporary result of this historical process is a world divided into three principal zones: core, semi-periphery, and periphery (Figure 6.4). The *core* states benefit from this situation, as they receive the surpluses produced elsewhere. The principal core states are Britain, France, the Netherlands, the US, Germany, and Japan—the countries where world business and financial matters are centred. The *semi-periphery* consists of states that are partially dependent on the core, for example, Argentina, Brazil, and South Africa. The *periphery* consists of those states that are dependent on the core and are effectively colonies; all the countries that we regard as less developed belong in this group. Note that political and economic changes since the early 1990s, such as the rise of China, raise questions about the details of this classification.

Although this world system is dynamic, it is extremely difficult for a state to move out of peripheral status because the other states have vested interests in maintaining its dependency. The closely related logic of dependency theory, as developed by Frank (see, for example, Chew and Denemark, 1996), stresses that some areas have to remain underdeveloped for others to become developed. This is because economic value is transferred in one direction only: from periphery to core. For example, in simple terms, most European countries benefited from extracting resources from their colonies but provided no substantial benefits in return. Hence, the European colonizing countries grew economically while the colonies lost potential.

IDENTIFYING GLOBAL INEQUALITIES

Many of the topics introduced in earlier chapters were discussed in the context of an unequal world. As we saw in Chapter 1, the growth and institutionalization of geography in the nineteenth century was largely spurred by the perceived importance of the discipline as a source of information about the parts of the world "discovered" by Europeans during the period of overseas exploration. As we will see in Chapter 8, the myth of "race" played an important part in justifying Europeans' colonization and exploitation of those parts of the world inhabited by "inferior races." Chapter 3 acknowledged that globalization processes are driven by the more developed countries, and the prevailing view is that globalization benefits those areas and further disadvantages less developed countries, thereby exacerbating an already serious problem. Chapter 4 pointed out that a primary distinction between more and less developed countries concerns energy, both the quantities used and the sources exploited. Chapter 5 painted a vivid picture of global differences in such basic demographic measures as birth rates, death rates, rates of natural increase, life expectancy, and infant mortality. These and related ideas inform the present section.

Where Is the Less Developed World?

First, some comments on terminology. The term *Third World* was first used in the early 1950s, in the context of suggestions that former colonial territories might follow a different economic route from

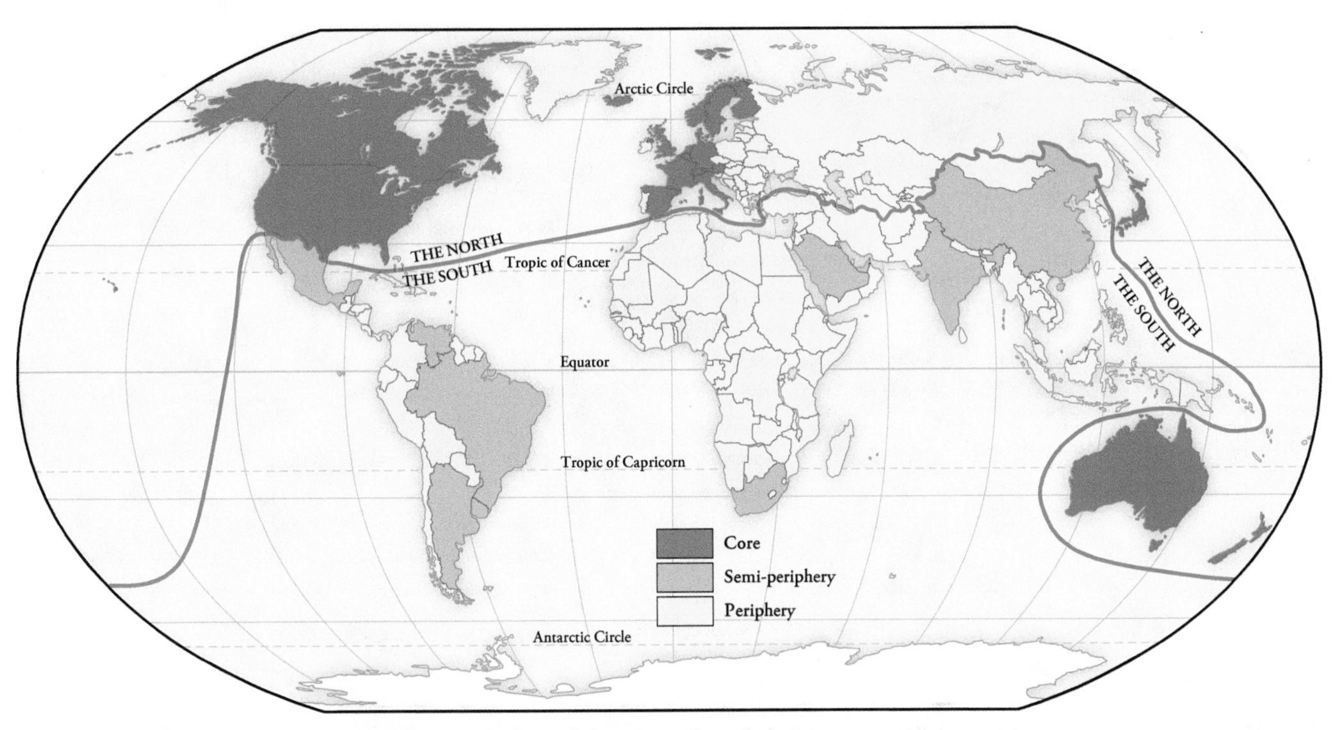

FIGURE 6.4 | The world system: Core, semi-periphery, and periphery
The line marks the 1980 Brandt Report division of North and South.

Source: Adapted from Know, P., and J. Agnew. 1994. *The Geography of the World Economy*, 2nd edn. London: Arnold, 2.

either the capitalist "First" or the socialist "Second" Worlds. By 1960, the term was being used to designate a group of African, Asian, and Latin American countries described by Prime Minister Lee Kuan Yew of Singapore as "poor, strife-ridden, [and] chaotic." Like all groupings based on broad generalizations, the Third World contained many more variations than it did similarities. Nevertheless, the classification had some value, and other terms have been used in much the same way. For example, the Brandt Report, *North–South: A Programme for Survival* (Brandt, 1980), explicitly distinguished between North and South and rich and poor. Other commonly used terms were *developed* and *underdeveloped*, or *developed* and *developing*.

All these terms are largely out of favour today. Because the geopolitical bipolarity of the Cold War era and the First and Second Worlds effectively ceased to exist with the demise of the Soviet bloc, the term *Third World* is anachronistic. Further, *North* and *South* are geographically misleading; *developed* and *underdeveloped* seem to imply that nothing can change; and *developing* may be factually incorrect.

The preferred terminology is that of *more developed* and *less developed*. These terms have the advantage of being relative rather than absolute and are used by organizations such as the United Nations, the World Bank, and the Population Reference Bureau. The more developed world comprises all of Europe and North America plus Australia, New Zealand, and Japan. All other countries in the world are classed as less developed. This simple twofold division is unsatisfactory in many respects, particularly because it disguises significant differences among the countries classified as less developed. Indeed, the UN and other organizations sometimes find it useful to exclude China from the less developed category and to identify a group of countries within the less developed world as comprising the least developed world: as of 2014, 48 countries were assessed as being in this group. Least developed countries have especially low incomes, high economic vulnerability, and poor human development indicators. Figure 6.5 maps the more, less, and least developed worlds. This book routinely employs the established distinction between more and

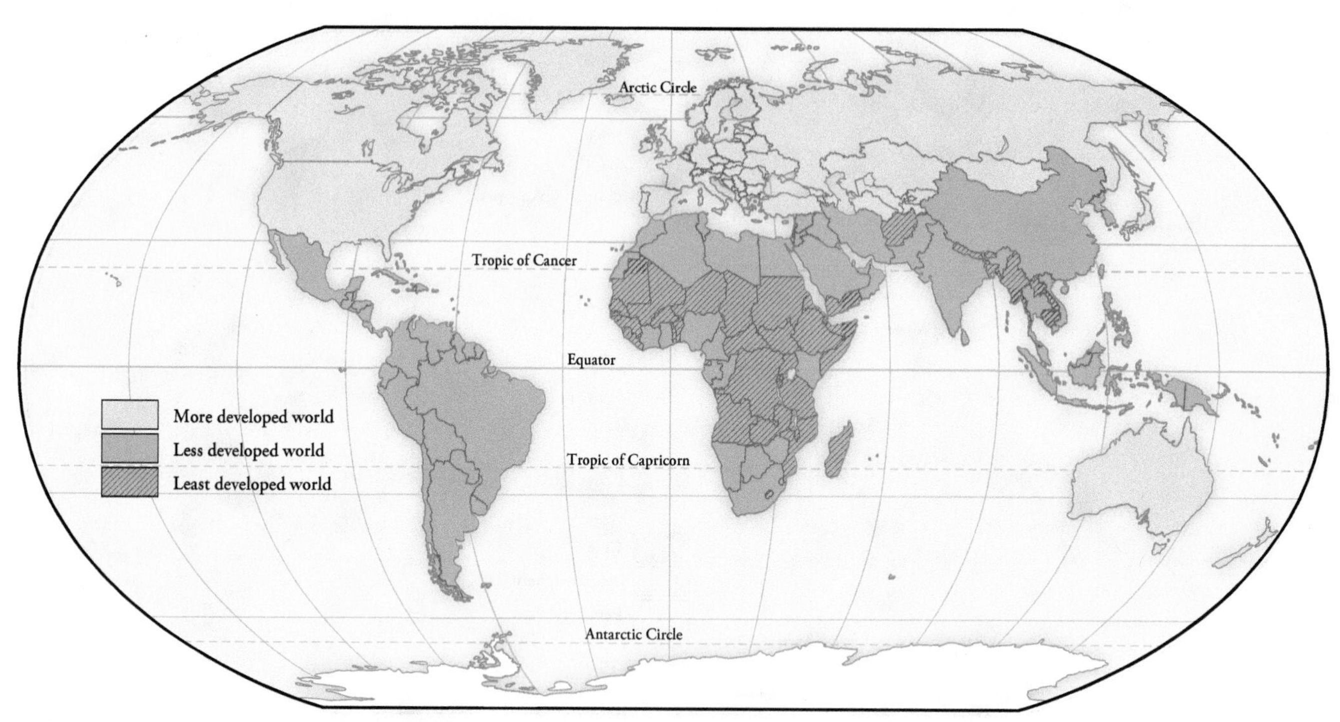

FIGURE 6.5 | More, less, and least developed countries

As noted, most sources of information use the terminology of *more* and *less developed countries* as an explicit recognition of two unequal worlds. The term *least developed* has been used since 1971 to identify the least developed of the less developed countries. The list is reviewed every three years: Botswana (1994), Cape Verde (2007), Maldives (2011), and Samoa (2014) have "graduated" (UN terminology) from this group, and Equatorial Guinea, Tuvalu, and Vanuatu are currently candidates for graduation. As of 2014, the 47 least developed countries are as follows:

Africa (34 countries): Angola, Benin, Burkina-Faso, Burundi, Central African Republic, Chad, Comoros, DR of Congo, Djibouti, Equatorial Guinea, Eritrea, Ethiopia, Gambia, Guinea, Guinea-Bissau, Lesotho, Liberia, Madagascar, Malawi, Mali, Mauritania, Mozambique, Niger, Rwanda, São Tomé and Principe, Senegal, Sierra Leone, Somalia, South Sudan, Sudan, Togo, Uganda, Tanzania, and Zambia.

Asia and Pacific (12 countries): Afghanistan, Bangladesh, Bhutan, Burma (Myanmar), Cambodia, Kiribati, Lao People's Democratic Republic, Nepal, Solomon Islands, Tuvalu, Vanuatu, Yemen.

Caribbean (1 country): Haiti.

Source: Adapted from Clarke, C. G., J. P. Dickenson, W. T. S. Gould, S. Mather, R. M. Prothero, D. J. Siddle, C. T. Smith, and E. Thomas-Hope. 1996. *A Geography Of The Third World*. London: Routledge. Copyright © 1996 Routledge.

less developed countries but also, when appropriate, makes reference to the least developed.

In general, countries in the less developed world (and, more specifically, in the least developed parts) have relatively high levels of mortality and fertility and relatively low levels of literacy and industrialization. In addition, they are often beset by political problems stemming from poor government and from ethnic or other rivalries. The basic demographic data are frequently unreliable; not only do the poorest countries have limited capital to conduct censuses, but low literacy levels may also affect the quality of the data collected. Box 6.1 provides capsule commentaries on three case studies of less developed countries, one from each of Africa, Asia, and Latin America. The African and Latin American cases belong to the least developed subgroup.

Development: Problems of Definition and Measurement

Defining Development

Traditionally, economic and social development have been measured by reference to gross domestic product (GDP) per capita or gross national product (GNP) per capita—usually called gross national income (GNI) per capita (see Chapter 3)—on the grounds that such macroeconomic indicators not only provide reliable data for comparing various countries' economic performance but also serve as reliable surrogate measures of social development in the areas of health, education, and overall quality of life. Others, however, believe that such measures are inappropriate because they do not take into account either the spatial distribution of economic benefits or the real-life

Around the Globe

BOX 6.1 | Less Developed World Case Studies: Ethiopia, Sri Lanka, and Haiti

© AP Photo/Elias Asmare

A recent photograph of the Merkato marketplace in Addis Ababa, Ethiopia's capital. Located at the approximate geographical centre of the country, Addis Ababa is home to over 2.7 million of the nation's inhabitants.

Ethiopia is one of several countries in the Horn of Africa suffering from a tragic combination of human and environmental factors. It is categorized as a least developed country.

Ethiopia was settled by Hamitic peoples of North African origin. Following an in-movement of Semitic peoples from southern Arabia in the first millennium BCE, a Semitic empire was founded at Axsum (it became Christian in the fourth century CE). The rise of Islam displaced that empire southwards and established Islam as the dominant religion of the larger area; the empire was overthrown in the twelfth century CE. Ethiopia escaped European colonial rule, but its borders were determined by Europeans occupying the surrounding areas, and the Eritrea region was colonized by Italy from the late nineteenth century until 1945. Attempts by the central government in Addis Ababa to gain control over both the Eritreans and the Somali group in the southeast generated much conflict. In 1974–5 a revolution displaced the long-serving ruler, Haile Selassie, and established a socialist state.

Ethiopia is slightly larger than Ontario—1.2 million km² (463,400 square miles). Much of the country is tropical highlands, typically densely populated because such areas have good soils and are free of many diseases. In 2014 the population was estimated at 95.9 million; the CBR was 28, the CDR 8, and the RNI 2.3. Although it is a leading coffee producer, Ethiopia is one of the poorest states in Africa, with many people reliant on overseas food aid. Population growth is highly uneven, and urban growth has actually declined since 1975 because of socialist land reform policies and the low quality of life in urban areas.

Most Ethiopians have little or no formal education, especially among the (predominant) rural population. The key social and economic unit is the family, in which women are subordinate, first to their fathers, then to their husbands and, if widowed, to their adult sons. Health care varies substantially between urban and rural areas.

The country also has a particular regional problem. People in the Lower Omo River Valley live in difficult circumstances. The Omo tribes are agricultural peoples who depend on the annual cycle of river flooding. Each year, communities on the riverbanks move to higher ground prior to the flood, then return to plant their crops (mostly sorghum) in the replenished soil when the waters recede. But this resource and their cattle are often insufficient. Furthermore, disputes between groups about water rights are common, with there being evidence that many of the males carry an automatic weapon.

The situation may deteriorate further. The Ethiopian government is constructing a huge dam (Gibe 111) upstream that, the Omo people claim, will mean the end of the annual flooding that forms the basis of their livelihood. The government argues otherwise, maintaining that it will be possible to regulate flooding as needed and that the Omo people will not be disadvantaged. Many scientists consider the Omo position legitimate, and European and British politicians have expressed concern. It is notable that the Omo were not involved in any consultations about construction.

Sri Lanka

Sri Lanka (Ceylon until 1972) is an island located in the Indian Ocean, south of India. Mostly low-lying, it has a tropical climate and a limited resource base. Minerals are in short supply, as are sources of power. The dominant economic activity is agriculture; about 36 per cent of the country is under cultivation.

Continued

© Lukasz Kulicki/iStockphoto

Sri Lankan people on the street of Colombo, the country's largest city and capital.

© Niko Guido/iStockphoto

Haitian children play together near an open bazaar street in Port-au-Prince. Behind them are colourful aluminum sheets, which were used to make temporary shelters after the January 2010 earthquake destroyed a major amount of the Haitian infrastructure and the lives of millions of people.

In 2014 Sri Lanka had a population of 20.7 million, a CBR of 18, a CDR of 6, and an RNI of 1.2. Although these figures are good by the standards of the less developed world, the country's 2010 per capita GNI was US$9,470—well below that of countries such as Singapore (US$76,850) and Malaysia (US$22,460). Interestingly, Sri Lanka and Malaysia had comparable GNIs in the 1960s. Three factors account for Sri Lanka's continuing problems. First, although it has effectively passed through the demographic transition (along with only a few other low-income countries such as China), its population increased rapidly in the past. Second, Sri Lanka has a long history of colonization—by the Portuguese from 1505 to 1655, the Dutch until 1796, and the British until independence was achieved in 1948. Problems of colonial dependency developed largely in the nineteenth century, when the ruling British established a system of plantation agriculture benefiting themselves rather than the local population. By 1945, tea plantations covered about 17 per cent of the cultivated area. Third, Sri Lanka includes two principal ethnic groups: the Sinhalese (74 per cent of the population), who moved down from north India and conquered the island in the sixth century BCE, and the Tamils (18 per cent of the population), who arrived from south India in the eleventh century. A long-running conflict between government soldiers and Tamil Tiger separatists, whose primary goal was the creation of an independent state (Tamil-Eelam), ended in 2009. This conflict is discussed in a broader South Asian context in Chapter 9.

Contemporary Sri Lanka is a far cry from the tropical Indian Ocean island called Serendip by the first Arab visitors and Paradise by many later Europeans.

Haiti

The "nightmare republic" of author Graham Greene, Haiti is a third example of a politically troubled country in the less developed world (Barberis, 1994). Like Ethiopia, it is categorized as least developed—the only such country in the Latin American region.

The basic demographic data for 2014 show a population of 10.8 million, a very high CBR (29), a moderate CDR (9), and a high RNI (1.9). The use of modern contraceptive techniques is low, although there are good reasons to believe that demand for them is high. There is a high population density of 387 per square kilometre.

Mostly mountainous with a tropical climate, Haiti has a long history of political turmoil since gaining independence (the first Caribbean state to do so) from France in 1804, following a 12-year rebellion. More recently, the first free election was held in 1990, after 30 years under the brutal rule of the Duvalier family. The victor was overthrown in 1992, and the country was again ruled by a military despot until democracy was re-established in 1994. However, the country has continued to be politically unstable.

Economic disparities are extreme, not only between the poor Creole-speaking black majority (95 per cent of the population) and the rich French-speaking minority (5 per cent) but also between the capital city of Port-au-Prince and the rural areas. It is estimated that the wealthiest 1 per cent of the population hold 44 per cent of the wealth. The rural population practise subsistence agriculture on soils that are generally poor, leading to erosion and declining soil fertility. Access to clean drinking water is difficult; hunger is widespread; and rates of infant mortality, tuberculosis, and HIV are all high. In 2009 Amnesty International initiated an online petition protesting government indifference to the high incidence of rape by gangs of armed men—rape was not a crime in Haiti until 2005. A collapsed infrastructure and drug trafficking and related corruption prevail. Finally, bad government and lax building standards mean that the country is tragically unable to cope with natural disasters, such as tropical storms or the January 2010 earthquake that killed many thousands and virtually destroyed Port-au-Prince.

conditions that less developed countries face, such as population displacement, inadequate food supplies, and vulnerability to environmental extremes. It can be argued that, for the less developed countries, GDP or GNI may indicate how the minority wealthy population is progressing but tell us nothing about the poor majority, just as they tell us little about poor populations living in wealthier countries. The different opinions on measures of development reflect a lack of agreement on what development means.

One problem is that definitions of development frequently are ethnocentric. Thus the standard of wealth and modernization. But some observers, for example, the Mexican activist Gustavo Esteva, argue that the entire concept of development is nothing more than Eurocentric nonsense invented by the Western world and imposed on other places and cultures. For most people, however, the fundamental error in this alternative way of thinking is that it appears to deny the need to eliminate poverty and disease.

Measuring Development

The *World Development Report*, an annual publication of the World Bank, measures development on the basis of selected economic criteria, grouping countries into four categories—low income, lower-middle income, upper-middle income, and high income—according to GNI per capita. Figure 6.6 illustrates this classification of national economies. One problem with this way of ranking countries is that it reflects a developmentalist (defined in Chapter 5) bias, suggesting that, as countries become more technologically advanced, they can—and should—increase their GNI. Moreover, as the World Bank (1998: 306–07) has admitted, this measure "does not, by itself, constitute or meas-

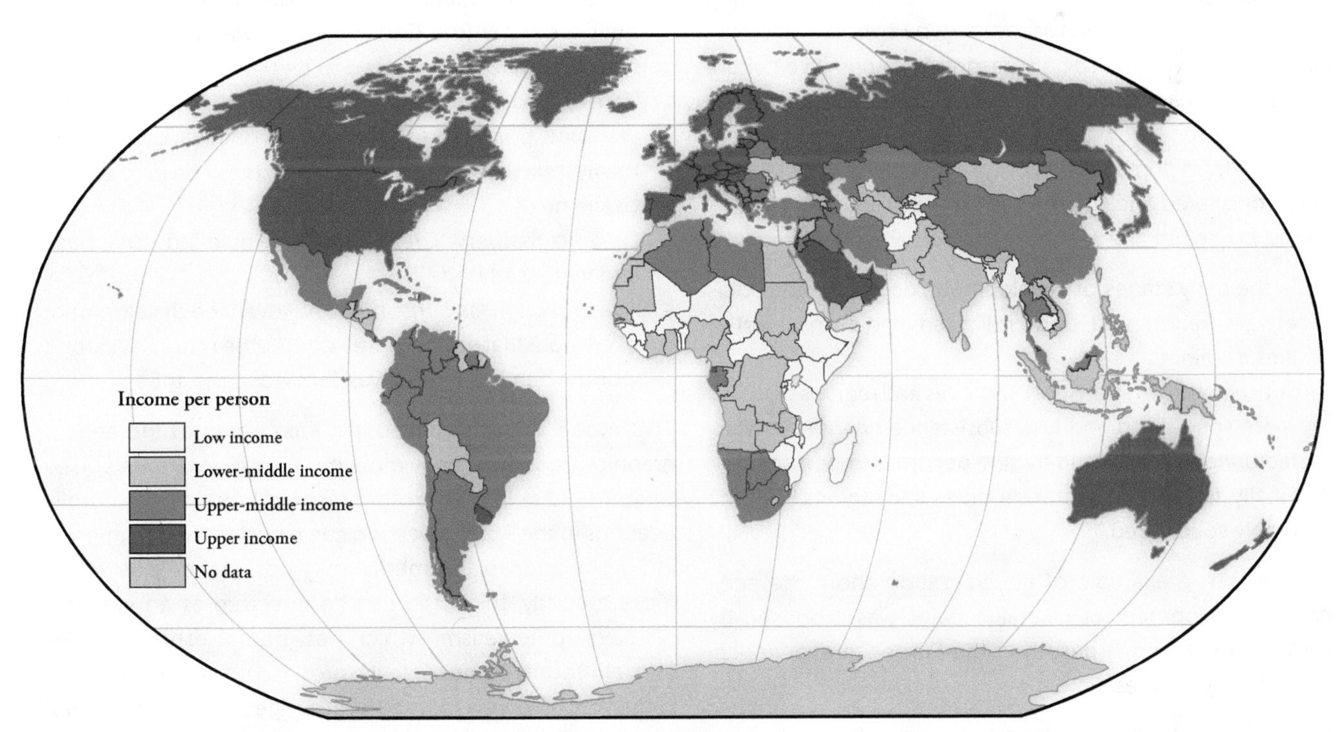

FIGURE 6.6 | National economies: Income per person, 2013
Low-income countries have a GNI per capita of US$1,045 or less; lower-middle-income countries have from US$1,046 to US$4,125; upper-middle-income countries from US$4,126 to US$12,745; and upper-income countries US$12,746 and above.

Source: databank.worldbank.org/data/download/GNIPC.pdf.

ure welfare or success in development. It does not distinguish between the aims and ultimate uses of a given product, nor does it say whether it merely offsets some natural or other obstacle, or harms or contributes to welfare." Notwithstanding these limitations, the mapped data provide a useful description of global differences. Box 6.2 comments on the problematic idea of growth following a series of developmental stages.

Measuring Human Development

The UN's *Human Development Report*, published annually since 1990, is intended to complement GNI measures of development. This publication has three distinctive characteristics. First, its underlying concept of development focuses on the satisfaction of basic needs and environmental issues. Second, it uses a wide variety of data to construct a Human Development Index (HDI) that is based on three primary development goals: life expectancy, education, and income. Beginning in 2010, the HDI has been modified to incorporate inequalities within countries by considering national disparities in gender, income, health, and education. The HDI does not measure absolute levels of human development but ranks countries in relation to one another. The value of thinking in terms of the HDI rather than solely in economic terms was reinforced by some UNDP research, citing data from Angola and parts of

Examining the Issues

BOX 6.2 | The Idea of Growth through Developmental Stages

Societies have frequently been seen as travelling along a sequential path of development. Several anthropologists have proposed sequences of cultural change, sometimes referring to stages using terms such as *savagery, barbarism*, and *civilization*. Marx envisaged society passing from primitive culture to feudalism, capitalism, and communism. Others have seen society as experiencing shifts in the dominant occupational category, from primary to secondary to tertiary. Geographers have considered and contributed to these various generalizations. In general, the sequence is as follows:

1. In the early stages of development, economy and society are fragmented, trade is limited, and primary activities dominate.
2. Subsequently, as transport improves and regions become more specialized and less subsistence oriented, manufacturing develops and trading becomes important.
3. Finally, the service sector develops and regions become highly specialized.

The classic example of generalization about the economic development of capitalist states was proposed by Rostow (1960) and is based on the European experience. There are five stages:

1. *Traditional society:* Subsistence agriculture, domestic industry, and a hierarchical social system; a stable population/resource balance.
2. *Preconditions for take-off:* Localized resource development because of colonialism or activities of a multinational corporation; export-based economy; often a dual economy (the Canadian Shield and the Canadian North might be considered to be in this stage).
3. *Take-off to sustained growth:* Exploitation of major resource; possibly radical and rapid political change (the newly industrialized countries [NICs] may be in this stage or the next).
4. *Drive to maturity:* Creation of a diverse industrial base and increased trade.
5. *Age of high mass consumption:* Advanced development of an industrial economy; evident in the more developed world when the model was formed in the 1960s.

This model is useful because it links easily to the demographic transition and mobility transition models (see Chapter 5). As with these other models, however, the simplifications in the Rostow version can be risky: what happened in Europe and North America may not happen elsewhere. More crucially, this logic can be criticized as an example of developmentalism, which, as noted earlier, implies that all countries necessarily proceed through the same sequence—the developmentalist or "geography of ladders" idea mentioned in Chapter 5. It is also worth reiterating that economic growth and development are not necessarily the same thing, as our discussion of the HDI emphasizes.

India and showing that economic growth does not necessarily translate into improvements in child mortality. Third, the UN report is explicitly concerned with how development affects the majority poor populations of the less developed world and recognizes a need to enlarge the range of individual choice.

Figure 6.7 maps the HDI in four general categories, indicating the countries in each category. Table 6.1 presents data on 20 countries (out of the 187 countries for which data are available): the 10 with the highest HDI values and the 10 with the lowest, using data from the 2014 *Human Development Report*. Again, as with GNI per capita data, African countries are the least developed. The HDI has a maximum value of 1.000 and a minimum value of 0.000; hence, Norway, with a score of 0.944, has a shortfall in human development of 5.6 per cent, whereas Niger, with a score of 0.337, has a shortfall of 66.3 per cent. It is also important to note that, although a general relationship between economic prosperity and human development is evident, there is no direct link—some countries are more successful than others in translating economic success into better lives for their people.

The HDI data in Table 6.1 paint a distressing picture of global inequalities. A positive, however, is that most countries have managed to reduce their shortfall. The world's average HDI has increased over time, with most countries showing an increase.

FEEDING THE WORLD

We all know that parts of the world, particularly in sub-Saharan Africa, seem to be especially vulnerable to hunger and famine. But it was not always like this. Until the nineteenth century, famine, hunger, and malnutrition were common elsewhere. For example, in 1229 a famine in Japan, probably caused by volcanic activity and damp weather, killed about one-fifth of the population in some areas; between 1315 and 1322 a famine in Northern Europe, most likely caused by flooding and cold weather, killed between one-fifth and one-third of the population; and in 1454 drought caused a famine in Mexico (Ó Gráda, 2009). A common

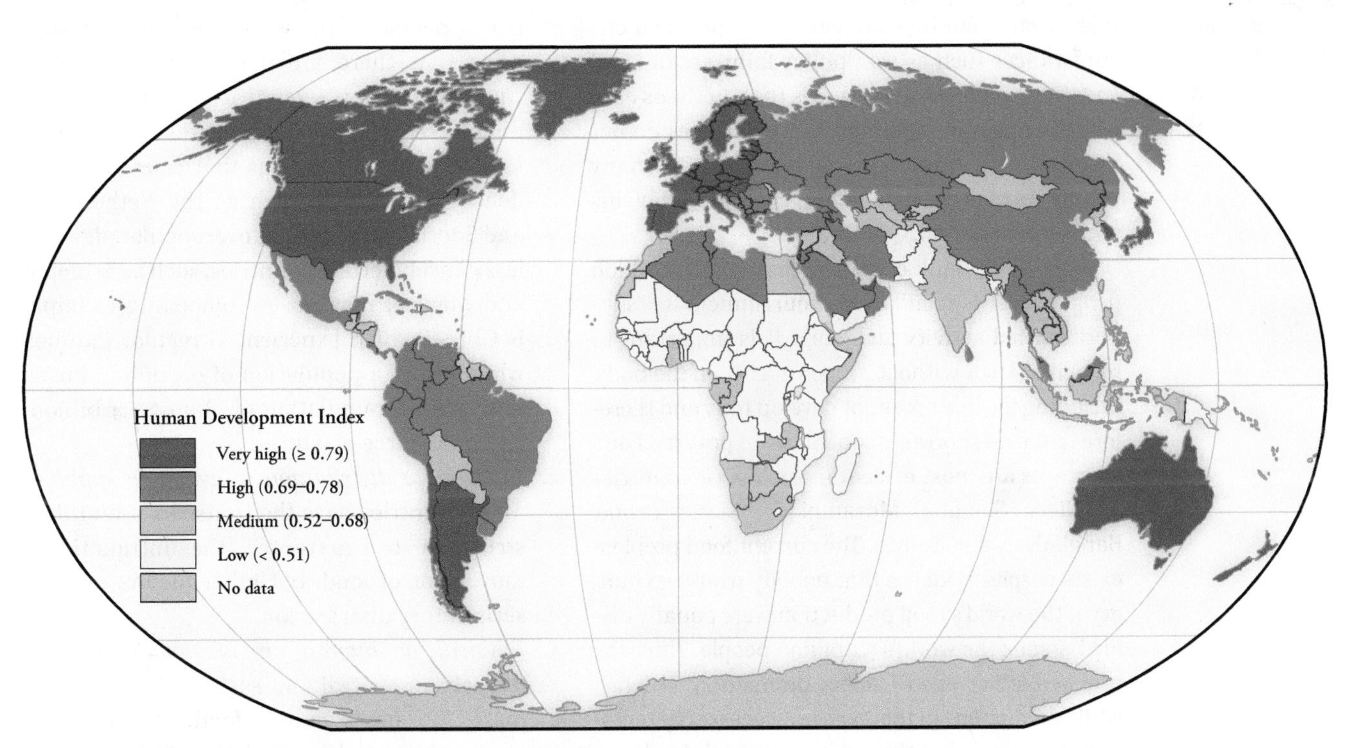

FIGURE 6.7 | Global distribution of human development, 2014

The UNDP divides the data set as follows: there are 49 countries with a very high level of human development, 53 with a high level, 42 with a medium level, and 43 with a low level. Data are not available for 8 countries, most notably North Korea, Somalia, and South Sudan.

Source: UNDP, *Human Development Report, 2014. Sustaining Human Progress: Reducing Vulnerabilities and Building Resilience.* Table 1. http://hdr.undp.org/en/content/table-1-human-development-index-and-its-components.

TABLE 6.1 | Extremes of human development, 2014

Top 10			Bottom 10		
Country	Rank	HDI Value	Country	Rank	HDI Value
Norway	1	0.944	Mozambique	178	0.393
Australia	2	0.933	Guinea	179	0.392
Switzerland	3	0.917	Burundi	180	0.389
Netherlands	4	0.915	Burkina Faso	181	0.388
United States	5	0.914	Eritrea	182	0.381
Germany	6	0.911	Sierra Leone	183	0.374
New Zealand	7	0.910	Chad	184	0.372
Canada	8	0.902	Central African Republic	185	0.341
Singapore	9	0.901	DR of Congo	186	0.338
Denmark	10	0.900	Niger	187	0.337

Source: UNDP. 2014. *Human Development Report, 2014. Sustaining Human Progress: Reducing Vulnerabilities and Building Resilience*, Table 1. http://hdr.undp.org/en/content/table-1-human-development-index-and-its-components.

undernutrition
Diet inadequate to sustain normal activity.

malnutrition
A condition caused by a diet lacking some food necessary for health.

denominator in these and other early famines is that they probably resulted directly from physical geographic circumstances.

As we will see, the situation has changed, with physical geography playing a secondary role as most famines are caused by political circumstances such as war, policy failures, and bad government. The bottom line is that, perhaps contrary to apparent common sense, famines do not result directly from food shortages and too many people (as Malthus contended) but from problems with access to food.

The World Bank estimates that about a billion people receive insufficient nourishment to support normal activity and work. It is important to recognize that, without proper nutrition, the body, including the brain, cannot develop fully and therefore that food shortages contribute to poverty. Food shortages are most evident in such poor countries as Niger, Somalia, Mozambique, Sierra Leone, Bangladesh, and Bolivia. The current food problem exists despite evidence that nobody would go hungry if the world's food production were equally divided among the world's 7.2 billion people. Thus, the problem is not one of global production. Further, while predictions of this type are necessarily tentative, it appears that there will be enough food produced by 2050 to feed all expected 9.7 billion people (*The Economist*, 2011). But producing sufficient food, whether today or in 2050, does not mean that everybody has the needed access to food.

A diet may be deficient in quantity, quality, or both. Requirements vary according to age, sex, weight, average daily activity, and climate, but an insufficient quantity of food (or calories) results in **undernutrition**. The best-known cases of acute undernutrition are the famines that attract media attention.

An adequate diet includes protein to facilitate growth and replace body tissue, as well as various vitamins. A diet deficient in quality results in **malnutrition**, usually a chronic condition. Among the health problems caused by undernutrition or malnutrition are kwashiorkor (too few calories), poor sight (inadequate vitamin A), poor bone formation (inadequate vitamin D), and beriberi (inadequate vitamin B1). The most extreme consequence of undernutrition and malnutrition is death.

Explaining Food Shortages

As we've previously stated, most explanations of the world food problem have, until recently, focused on three factors:

1. *Overpopulation:* The root cause of the food problem is often said to be the sheer number of people in the world. Yet evidence suggests that there are not too many people. *Overpopulation* is a relative term, and densely populated areas are not necessarily overpopulated. Table 6.2 confirms that countries with dense populations, such as the Netherlands and South Korea, are not overpopulated, while less densely settled countries, such as Ethiopia and Mexico, may be. A compelling example is China, which experienced regular famines when it had a population of 0.5 billion but is now, with a population of almost 1.4 billion, essentially free of famine.
2. *Inadequate distribution of available supplies:* Most countries have the transportation infrastructure to guarantee the intranational movement of food, but other factors prevent satisfactory distribution.
3. *Physical or human circumstances:* A 1984 drought in central and eastern Kenya is estimated to have caused food shortages for 80 per cent of the population. Other causes in specific cases include flooding and war. Circumstances such as these obviously aggravate existing problems, but they cannot be considered root causes.

TABLE 6.2 | Population densities, selected countries, 2014

Country	Population Density per km²
South Korea	507
Netherlands	406
Ethiopia	87
Mexico	61

Source: Population Reference Bureau. 2014. *2014 World Population Data Sheet*. Washington, DC: Population Reference Bureau.

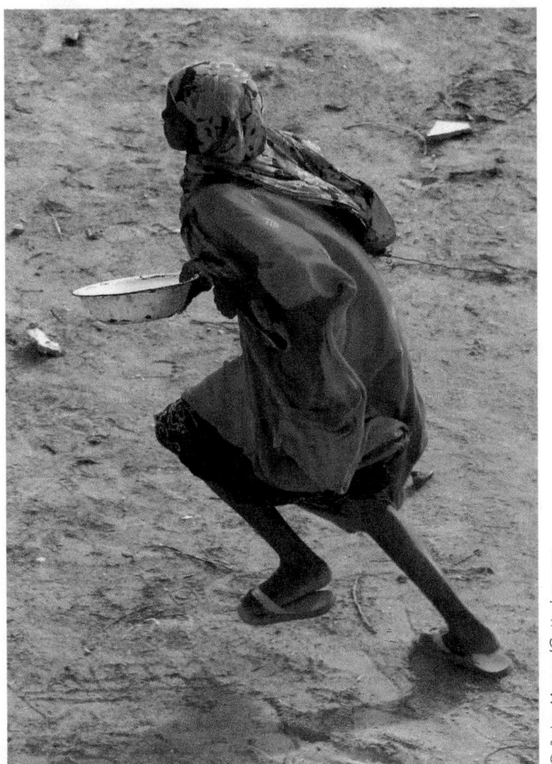

© John Moore/Getty Images

A girl runs to get in line for a cooked meal at a camp for Somalis displaced by drought and famine in Mogadishu, Somalia. The UN estimates that more than 100,000 people fled the countryside to Mogadishu in July and August 2011 due to the crisis. At the time, some 1.5 million Somalis were estimated to be displaced nationwide due to drought, famine, and war, creating extreme pressure on already scarce food resources.

Political and Economic Explanations

Most recent explanations of the food problem have tended to focus on the political and economic aspects. For example, the world systems perspective suggests that global political arrangements are making it increasingly difficult for many in the less developed world to grow their own food. Not only is the percentage of the population involved in agriculture in the less developed world declining (from more than 80 per cent in 1950 to less than 60 per cent today), but the vast majority of those remaining in agriculture are incapable of competing with the handful of commercial agriculturalists who can benefit from technological advances. Indeed, the vast majority has lost control over their own production because of larger global causes. For example, farmers in Kenya are actively encouraged to grow export crops such as tea and coffee instead of staple crops such as maize; peasant farmers in many countries are losing their freedom of choice because credit is increasingly controlled by large corporations; and many governments make it their policy to provide cheap food for urban populations at the expense of peasant farmers. The essential argument here is that the capitalist mode of production is affecting peasant production in the less developed world in such a way as to limit the production of staple foods, thus causing a food problem.

In the same way, global economic considerations make it increasingly difficult for people in the less developed world to purchase food. Throughout the world, food is a commodity; hence, production is related to profits. In other words, food is produced only for those who can afford to buy it. In 1972, a year when famine was widespread in the Sahel region of Africa, farmers in the US were actually paid to take land out of production to increase world grain prices.

These arguments suggest that the cause of the world food problem is the peripheral areas' dependence on the core area. Yet any attempt to correct that problem is likely doomed to failure because the more developed world is not about to initiate changes that would lessen its power and profits. If this argument is followed to its logical conclusion, the world food crisis can only get worse, regardless of technological change, because the cause of the problem lies in global political and economic patterns. Further support for the connection between famine and politics can be found in the historical record. Between 1875 and 1914, some 30–50 million people died in a series of famines in India, China, Brazil, and Ethiopia. Describing these famines, Davis (2001) recognizes the role played by meteorological conditions but also points to the complex politics of colonialism and capitalism as essential causes.

The fact is that feast and famine live side by side in our global village. There are rich and poor countries and, in the poor countries especially, rich and poor people. Increasingly, agriculture is a business, and it is in the nature of business that not all participants compete equally well. Food problems are not typically caused by overpopulation, and population density is a poor indicator of pressure on resources. We will not solve food problems merely by decreasing human fertility. A concerted and co-operative international effort is needed to improve the quality of peasant farming and to reorient it towards the production of staple foods. Meanwhile, undernutrition and malnutrition are not disappearing, and famines will continue to occur unless there is meaningful political change.

What makes the world systems perspective particularly valuable is that it emphasizes the need to focus on structural causes—not just the fairly obvious immediate causes, such as drought and crop failure, that tend to receive media attention. But the basic world systems logic can be usefully applied at various spatial scales below that of the world as a whole. Indeed, the role of power, in the sense of the politics that govern access to food, is relevant at various scales.

The Idea of Entitlements

A case in point is the explanation of world food problems suggested by Young (1996: 99) and based on the ideas of Sen: "Patterns of food distribution may be examined with reference to people's entitlements reflected in their ability to *command* food." Entitlements are the factors and mechanisms that explain people's ability to acquire food in terms of their power. In addition to entitlements at the international scale (addressed by world systems theory), this argument identifies entitlements at three other scales: national, regional, and household.

At the international scale, as world systems logic suggests, among the historical legacies of colonialism is an emphasis on export production at the expense of local production and therefore a vulnerability to global market changes. In other words, small changes in the price of a commodity can significantly affect entitlements. During the 1980s, high levels of debt led the International Monetary Fund (IMF) and the World Bank to require that countries in the less developed world adopt structural adjustment policies—usually involving even greater emphasis on export production—before additional loans would be issued or existing loans restructured. Contemporary globalizing trends aggravate this situation because the demand for food continues to increase in parts of the world that already are poor and lack power.

At the national scale, governments may not be committed to ensuring that economic growth is accompanied by the elimination of food shortages. Because of the distribution of power, national governments may support urban activities and commercial agriculture at the expense of the rural peasant sector. At the regional scale, governments may neglect areas inhabited by relatively powerless minority ethnic groups. Further, problems of various kinds are often regionally distinct: some areas may be more subject to conflict or environmental problems than other areas. At the household scale, the most vulnerable family groupings are those that are poorest, that include many dependants, that are isolated, and that are powerless. Even within households, there are differences in the ability to command food; females and the elderly are frequently the most vulnerable.

This multi-scale approach to understanding food shortages stresses the inequalities that exist in people's ability to acquire food. Food shortages are placed in context using broad geographic and historical frameworks with emphasis on entitlements and the related ability of people to command food. We will introduce a conceptual basis for this political economy approach in Chapter 10. Those global areas, countries, regions within countries, households, and even individuals within households whose entitlements are most limited are the ones least likely to command adequate amounts of food. The basic equation is this: lack of power equals lack of food.

The Role of Bad Government

In a similar vein, bad government is a cause of food problems and lack of economic and social development more generally. Sen (1981) argues that widespread hunger has nothing to do with food production and everything to do with poverty—which in turn is closely related to political governance. Sen points out that famine does not occur in democratic countries because, even in the poorest democracy, famine would threaten the survival of the ruling government. A cursory review of current global famine areas confirms that famine seems to have six principal causes, with bad government front and centre. Of the other causes, four are closely related to bad government: a prolonged period of underinvestment in rural areas, political instability related to conflict that causes refugee problems, HIV/AIDS and other diseases depriving families of productive members and damaging family structures, and continued population growth because of high birth rates. The fifth cause is bad weather.

Links between quality of governance and food production are emphasized in the 2008 *World Development Report*. Arguing that agriculture can play a key role in development, the report noted the need for sound agricultural policies and investment. Unfortunately, many of the less developed countries perform poorly on a range of governance measures.

For example, a 2005 World Bank report, *Doing Business in 2005*, noted many of the regulatory and bureaucratic obstacles to conducting business and achieving prosperity in some less developed countries. Consider, for example, that it takes two days to incorporate a business in Canada but 153 days in Mozambique or that it costs 1,268 per cent of average annual income to register a company in Sierra Leone and nothing in Denmark. In Lagos, the commercial capital of Nigeria, recording a property sale involves 21 procedures and takes 274 days. Obtaining a licence to conduct business in Kenya is an opportunity available only to those with government connections. One consequence of these circumstances is that, instead of creating wealth through private enterprise, many entrepreneurs resort to begging or criminal activity. These examples of inefficiency can be multiplied many times over and, together, such inefficiencies hinder economic and social development. Clearly, many of the solutions to poverty and related problems need to be addressed by the poor countries themselves. Doing so is difficult in many cases because, tragically, poor countries are prone to civil war, especially if there are many young, uneducated men and several minority ethnic groups (Collier, 2007).

Box 6.3 extends the consideration of the role of bad government in an account of the food crisis that began in 2007. The larger question of the global spread of democracy is considered in Chapter 9. Figure 6.8, employing a fourfold division of countries and six measures of quality of governance, shows a clear relationship between a poor governance score and emphasis on agriculture.

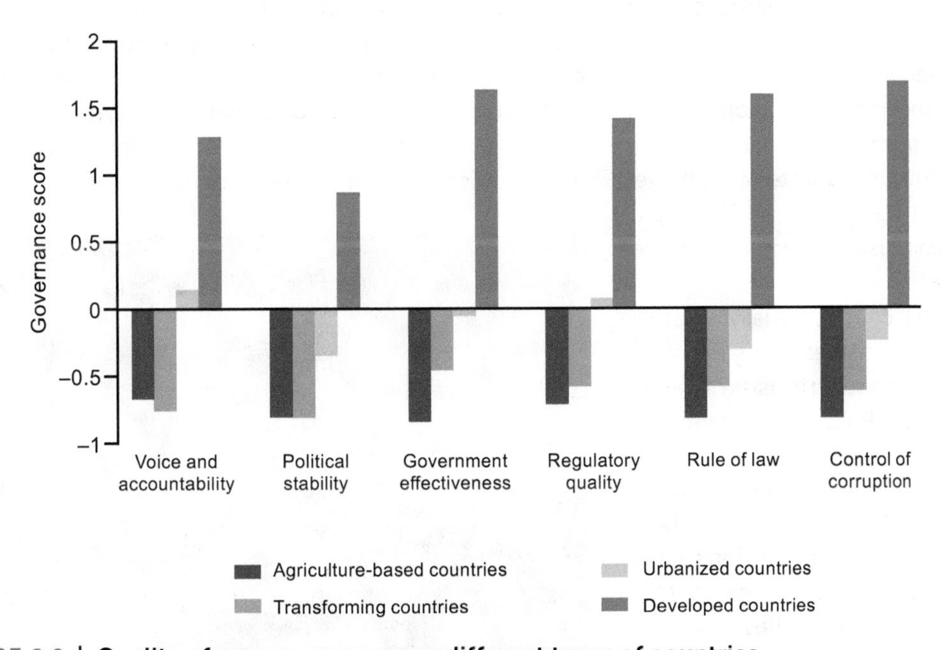

FIGURE 6.8 | Quality of governance among different types of countries

The four country groupings are not mutually exclusive: agriculture-based countries are mostly in sub-Saharan Africa; transforming countries are mostly in South and East Asia, the Middle East, and North Africa; urbanized countries include most of Latin America and much of Europe and Central Asia; and the developed countries (a grouping that includes many of the urbanized countries) are those in the more developed world. The six dimensions of governance—voice and accountability, political stability and absence of violence, government effectiveness, regulatory quality, rule of law, and control of corruption—used to calculate the governance score are taken from a substantial World Bank research project (see Kaufmann, Kraay, and Mastruzzi, 2006). For the purposes of this figure, scores below zero indicate poor governance and above zero indicate better governance.

Sources: World Bank, *World Development Report: Agriculture for Development* (Washington, DC: World Bank, 2008), 23, Figure 12.

Examining the Issues

BOX 6.3 | A Food Crisis in the Early Twenty-First Century

In 2007–08 many basic foodstuffs trading on international commodity markets experienced dramatic price increases after about 30 years of stability; prices for wheat, corn, soya, rice, coffee, and meat all increased, in some cases more than doubling. While these increases affected consumers everywhere, the less developed world suffered most.

Attempts to explain this situation referred to the classical economic logic of an increased demand combined with a reduced supply. But several other factors played roles:

- Some explanations referred to the growing world population, specifically to increased demand from the growing and wealthier populations of India and China.
- Other explanations centred on environmental pressure, for example, accelerating desertification that resulted in loss of agricultural land.
- Most observers agreed that increasing oil prices, which in turn raise the costs of agricultural production, were a key cause.
- There was general agreement that the shift to biofuels—related to concerns about global warming—shifted grains away from food to fuel. In 2007, for example, one-third of the United States corn harvest was directed to fuel. It is estimated that filling up an SUV fuel tank with ethanol uses enough corn to feed one person for a year.
- A variety of more local causes, usually weather-related, included poor rice harvests in some Asian countries and, in 2010–11, flooding in parts of Canada and Pakistan and drought in Russia and Argentina.

It can also be argued that bad government contributed to the crisis. In response to initial price rises and reduced surpluses, some governments ceased exports, thus increasing the likelihood of further price increases. More generally, especially in some African countries, poor governance may create barriers to distribution through inadequate transport infrastructure; impose internal levies and tax technologies such as refrigeration and packaging and basic inputs such as fertilizer; limit improvements in agriculture through landownership legislation, weak property rights, and a lack of the rule of law; permit political elites to promote their interests at the expense of agricultural progress for the majority; and operate marketing boards that oblige farmers to sell at below-market prices.

Knowing how best to respond to increases in the price of food is far from simple. For many, the solution is to repeat the successes of the "green revolution" (discussed in Chapter 10), which saw massive production increases in much of the less developed world through new technologies of production. Although African countries benefited little from this movement, partly because of the varied topography and small farms, they might benefit from technologies tailored for local circumstances (Conway, 2008). Alternatively, the head of the UN Environment Programme has argued for organic farming in Africa as the best way to increase yields, improve soils, and raise incomes. There is also a need to correct many of the government-related problems noted in this chapter, in particular through the provision of reliable and accessible input and output markets. Other observers argue that governments globally need to liberalize markets—eliminating export quotas and trade restrictions—not intervene in them.

© Scott Olson/Getty Images

Signs mark the location of various corn hybrids grown for use in ethanol production on a plot of farmland near Freeport, Illinois.

Providing Food Aid

Food aid provided by the more developed world helps the less developed world enormously but has not proven to be a solution to famine. This is not surprising, given our observation that the ultimate cause is political (which no amount of aid is likely to change) and that factors such as drought are proximate causes. Aid is often directed to urban areas, even though the greatest need is usually in rural areas; indeed, much donated food goes to governments, which then sell it for profit. It tends to depress food prices in the receiving country, reducing the incentive for people to grow crops and increasing their dependence. In many cases it is not effectively distributed, whether because of inadequate transport or because undemocratic governments in the receiving countries control the food supply and feed their armies before anyone else.

There are other concerns. In 2006, when suffering from a severe drought and food shortages, Eritrea expelled three charities working in the country in an apparent attempt to attract international attention to the long-running border disagreement with Ethiopia (see Box 6.1). Even more disturbing, a 2006 report from Save the Children stated that humanitarian workers and peacekeeping soldiers in Liberia were providing food aid in return for sex with young girls. Again, during the 2011 famine in Somalia, Islamist militias that controlled parts of the country prevented food aid reaching some of those in need.

These comments about food aid may also apply more generally to attempts to help countries emerge from poverty. It has been said, only half-jokingly, that foreign aid merely moves money from poor people in rich countries to rich people in poor countries. There is, for example, much debate about the merits of the Make Poverty History campaign, initiated in 2003 and strongly supported by the Irish rock singer Bob Geldof, other popular entertainers, and many politicians. A major focus of this campaign was collecting money from individuals and pressuring governments to provide more aid. Although such attempts to help poor countries are laudable and well motivated, their effectiveness is another matter. A country such as Malawi has received huge amounts of aid in recent decades but without any indication of meaningful improvement in quality of life for the vast majority of the population. Raising this critical point is not to suggest that many countries do not require humanitarian aid, disaster relief, or medicine. The criticism is that merely pouring in money does not achieve the desired end to poverty. A 2005 report by two major humanitarian agencies, Oxfam and ActionAid, accused Western governments of using aid to reward strategic allies and to support favoured

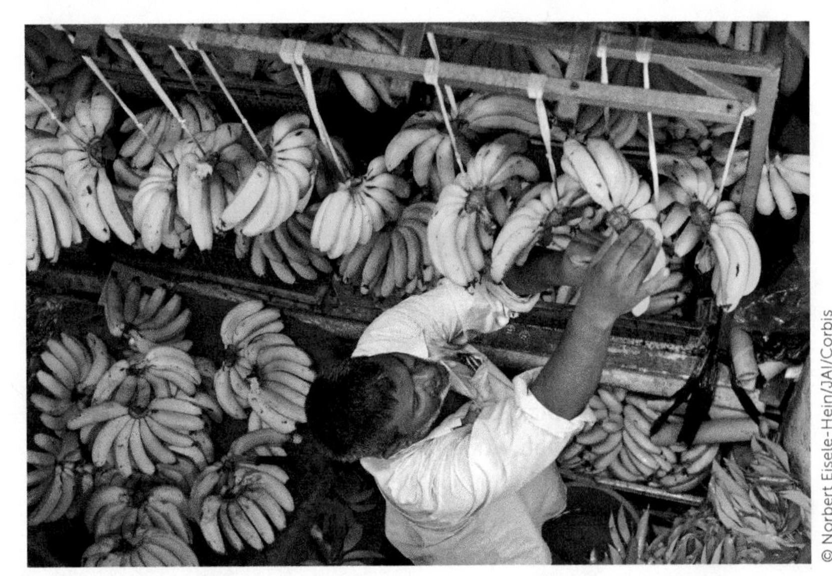

A market worker sets up his stall at the Big Vegetable Market in Port Louis, Mauritius. Situated about 800 kilometres east of Madagascar, the island is Africa's only full democracy. Severe poverty is rare in the country; however, it is vulnerable to rising global food prices because it relies on imports for 70 per cent of its food need. For example, the main fruit crops (banana, pineapple, litchi, and mango) yield only half the requirements.

economic development projects. The document also suggested that only about one-fifth of aid money went to the countries most in need, with only half of that one-fifth being invested in health and education.

Perhaps more important, aid funds are commonly sent to countries known to be governed by corrupt politicians. Botswana is an example of a country that has been governed well and has steadily improved in terms of quality of life; Malawi, Zimbabwe, and Zambia are contrary examples, mired in poverty and not always well governed. In other countries, wealth achieved from oil or other natural resource revenues is not used to benefit the population but a select few. Indeed, Collier (2007) argues that natural resources lead to corrupt government. A recent notorious example involved the son of the president of Equatorial Guinea spending more than CA$2 million on luxury cars and owning expensive homes in several countries around the world, all while the vast majority of the country's population is poverty-stricken. More generally, about 40 per cent of Africa's private wealth is held overseas, much in Swiss bank accounts. Further, foreign companies are typically quite content to exploit local ignorance in order to maximize their profits.

Indeed, a 2014 report authored by a group of 13 United Kingdom and Africa-based NGOs asserted that providing aid money acts as a smokescreen to hide the money being "looted" out of Africa. Therefore, while Africa receives about US$30 billion in aid each year, more than six times that amount is effectively removed from the continent by transnational companies based in the same countries that provide much of the aid. Weak African governments facilitate this situation.

Food shortage and related problems are complex. There are many undernourished and malnourished people in the world, yet enough food is being produced to feed everyone. As we have noted, there seem to be good reasons to believe that current food shortages are not caused by inadequate supplies and that supplies will be adequate in the future. On the basis of a scientifically detailed analysis of the complete food cycle, Smil (2000: 315) arrived at a conclusion he described as "encouragingly Malthusian." In other words, the future may not be as bright as we might hope, but neither is it totally bleak. Thus, with reference to the competing catastrophist and cornucopian positions, we might suggest that neither extreme is a realistic portrayal of our future. Rather, like several other problems (for instance, the issue of refugees), the food problem appears to be associated with specific areas and a multitude of complex interrelated causes—physical, economic, and cultural but primarily political.

Examining the Issues

BOX 6.4 | Emerging Africa: The Hopeful Continent

The two phrases used in the title of this box are taken from a special report published in *The Economist* on 2 March 2013. Although the report regularly stresses uncertainties, the overall tone is most positive. The argument is that, while Africa faces numerous problems related to poverty, nutrition, health, and civil strife, the recent past shows substantiated improvement. Indeed, it might be said that the perception of the continent from outside is both partial and warped, as the media tend to cover only wars, disasters, and famines and continually stress the need for aid and intervention. At least implicitly, these accounts suggest that Africa needs to be saved from Africans (Dowden, 2008).

The good news is that several developments over the past decade or so are beginning to have a significant impact. Large amounts of targeted aid, including US$17 billion from the Bill and Melinda Gates Foundation (since 2006) are showing results. There is much new infrastructure, notably schools that are improving education standards and roads that are making it easier to move goods and people. The World Bank predicts that most African countries will be classed as middle income by 2025, whereas only about half are today.

Also, the number of major conflicts has declined, with the end of wars in Ethiopia, Mozambique, Angola, and Chad. Reasons include the end of the Cold War, which meant the United States and the former Soviet Union began to reduce their involvement in propping up regimes or supporting anti-government groups; intervention by Western countries to actively facilitate a peace process; and the fact that many civil wars simply end when the warring groups become exhausted. Significant conflict is evident only in Sudan, DR of Congo, and Somalia, although Islamic extremism has increased in several Saharan countries. Political violence undermines state stability, limits the ability of governments to function effectively, weakens state institutions, forces school and hospital closures, and deters foreign investment. When conflicts end, the potential for growth and for a better quality of life is greatly enhanced. Significantly, democratic ideas and institutions become more evident.

In recent years, many African countries—including Ghana, Niger, Gambia, Malawi, Guinea, and Kenya—have shown enthusiasm for democratic institutions and increasing personal freedom. A 1999–2001 survey of more than 21,000 people in 12 African countries showed that more than 70 per cent of respondents favoured democracy over other forms of government. In 2008 Festus Mogae, a former president of Botswana, won the US$5 million Ibrahim Prize for Achievement in African Leadership, which recognizes good governance in Africa. Botswana has been one of the most politically stable countries in Africa since independence in 1966, never having had a coup and holding regular multi-party elections. As of 2015, the repressive regime of Robert Mugabe in Zimbabwe, which has contributed to a food crisis affecting millions, is more the exception than the norm in sub-Saharan Africa.

Less dramatic, but no less important, are some changes in the approach to combatting poverty. Since about 2000, some African countries, including Uganda and Kenya, have adopted a strategy that sees poor households given a small stipend to spend as they choose. Initiated in Latin America, this program clearly helps pull people out of poverty. The money is not typically wasted by recipients but used to, for example, build a new roof or buy seed.

There are good reasons to be increasingly confident about the future of Africa, with the ongoing development, decreased conflict, enhanced democratic institutions, and grassroots initiatives all having an impact on people's everyday lives.

REFUGEES

Since the end of World War II, distressingly large numbers of people have been forced to move from their homes. Specific reasons vary but include political changes following a period of conflict, political repression, shortage of food, drought, or often a combination of these circumstances. Some people move within their country, while others cross international borders. Some of the major early movements were in the more developed world, but less developed countries are now the primary source.

Globally, the first major movements of refugees took place after World War II in response to changing political circumstances. Immediately after 1945, about 15 million Germans relocated. The partition of India in 1947, which created the Muslim state of Pakistan, caused the movement of about 16 million people—8 million Muslims fled India for the new state, while 8 million Hindus and Sikhs fled Pakistan for India. A rather different migration took place prior to the construction of the Berlin Wall in 1961, when some 3.5 million moved from what was then communist East Germany to democratic West Germany. Even after the wall was in place, about 300,000 succeeded in fleeing west before it was taken down in late 1989 (Germany was reunified in October 1990).

From about 1960 to the mid-1970s, the annual total number of refugees in the world was relatively low at between two and three million (Figure 6.9). However, in 1973 the end of the Vietnam War gave rise to one of the first large transcontinental movements of refugees; some two million people fled Vietnam, necessitating a major international relief effort. Because of its involvement in the war, the US received about half of these refugees. The flow of Vietnamese refugees continued until the early 1990s. More generally, the 1980s saw a huge increase in numbers of refugees for various political, economic, and environmental reasons. Total refugee numbers began to fall in the first decade of the twenty-first century, standing at 10.5 million in 2010, but have increased since then largely because of the civil war in Syria and the subsequent emergence of the Islamic State; the major origin countries are listed in Table 6.3.

The Problem Today

It is generally agreed that refugees are people forced to migrate, usually for political or food shortage reasons. There is, however, no agreement on how to determine the legitimacy of refugee claims; there are considerable logistical difficulties in counting refugees, and governments may have a vested interest in providing incorrect information. Accordingly, estimates of total numbers of refugees vary.

The most reliable source of information on refugee numbers is the United Nations High Commissioner for Refugees (UNHCR). A few definitions are important at the outset. According to the 1951 Convention Relating to the Status of Refugees, a refugee is a person who, "owing to a well-founded fear of being persecuted for reasons of race, religion, nationality, membership in a particular social group, or political opinion, is outside the country of his nationality, and is unable to, or unwilling to avail himself of the protection of that country." Although most countries offer protection to their citizens, others choose not to or are unable to do so, and these prompt refugee movement. Most current refugee movements are prompted by civil wars and forms of ethnic conflict, as was the case, beginning in 2012, of Syrians fleeing the civil war in their country. By late 2014, more than three million refugees had left Syria for camps in Turkey, Lebanon, Jordan, and Iraq, while others had crossed borders but were not in camps and still others remained internally displaced within Syria. A country that receives refugees is known as a country of asylum. In addition to refugees, the

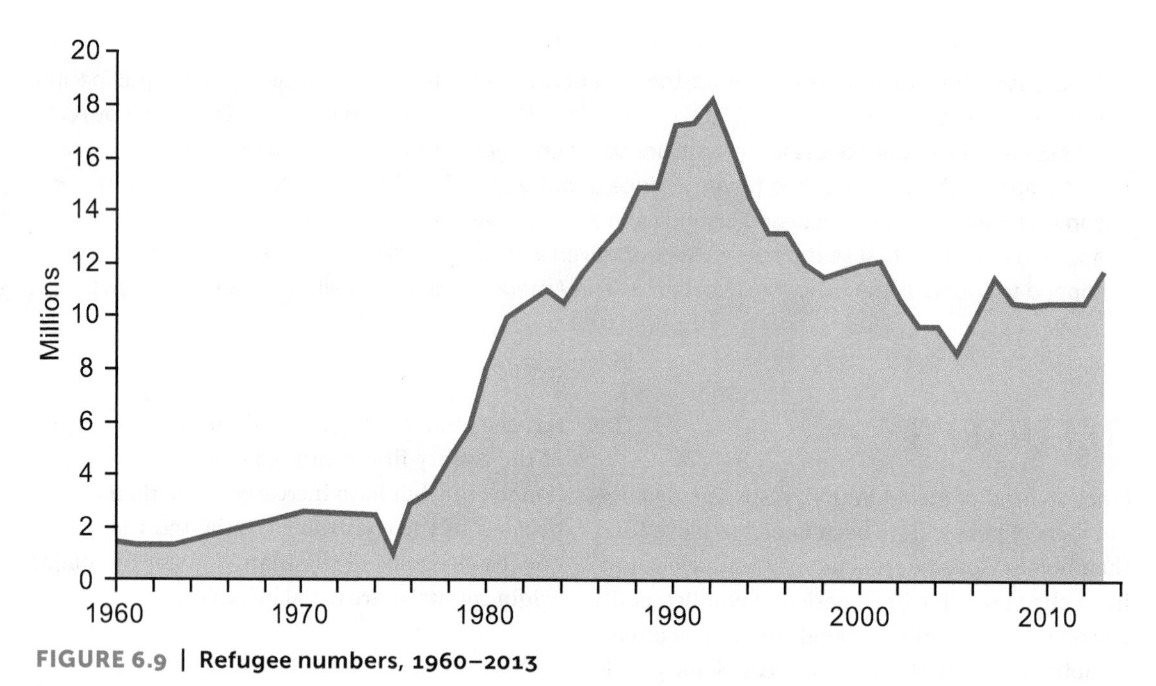

FIGURE 6.9 | Refugee numbers, 1960–2013

Source: Updated from UNHCR. 2014. *Global Trends 2013: War's Human Cost.* www.unhcr.org/cgi-bin/texis/vtx/home/opendocPDFViewer.html?docid=5399a14f9&query=refugees%20and%20IDPs%201993-2013%20(end%20year).

UNHCR includes three other classes of "persons of concern":

1. Asylum seekers are people who have left their home country and applied for refugee status in some other country, usually in the more developed world.
2. Returnees are refugees who are in the process of returning home (the desired goal for most refugees); in recent years the UNHCR has been involved in major repatriations in many places, including Afghanistan, Iraq, Burma, Cambodia, and several African countries.
3. Internally displaced persons (IDPs) are people who flee their homes but remain within their home country; unlike refugees, they do not cross an international boundary. A recent example is Sri Lanka's Tamil population, many of whom fled their homeland in the north of that country as civil war escalated from 2006 to its bloody conclusion in 2009.

SOEREN BIDSTRUP/AFP/Getty

Refugees arrive on the shores of Lesbos, Greece. In 2015 over half a million people fled Syria, Iraq, and Afghanistan for Europe. Eastern European countries such as Slovenia, Croatia, Turkey, and Hungary have struggled to handle the influx of people, with some closing their borders. The EU and the German government have sought more co-operation from central and southeast European states in processing and accepting refugees.

TABLE 6.3 | Main origins of refugees, 2013

Country of Origin	Total
Afghanistan	2,556,600
Syria	2,468,400
Somalia	1,121,700
Sudan	649,300
DR of Congo	499,500
Burma	479,600
Iraq	401,400
Colombia	396,600
Vietnam	314,100
Eritrea	308,000

Source: Data from UNHCR. 2014. *Global Trends 2013: War's Human Cost.* www.unhcr.org/cgi-bin/texis/vtx/home/opendocPDFViewer.html?docid=5399a14f9&query=refugees%20and%20IDPs%201993-2013%20(end%20year).

Today, the total number of persons of concern is 42 million. But even this figure seriously underestimates the problem. According to UNHCR estimates, there are probably as many as 60 million people worldwide who have been forced to leave their homes—the difference between that and the official figure reflects many distressing situations for which reasonably detailed data are simply not available, as well as a variety of situations that are more difficult to categorize. For example, about 258,000 Urdu-speaking Muslims, the Biharis, have been stranded in Bangladesh (formerly East Pakistan) since the 1947 partition. The Biharis are in a difficult situation because of their past support for the former West Pakistan (now Pakistan). Similarly, about 800,000 Iranians in Turkey are not recognized as refugees. Figure 6.10, which maps major source countries of refugees in 2013, highlights the key problem areas (East and Central Africa, West Asia, and Colombia in South America).

It is sometimes suggested that the vast majority of refugees and other persons of concern are women and children, but this is not the case. Globally, about 48 per cent are female and about 12 per cent are under the age of five. These percentages do not differ radically from standard gender and age distributions. It seems that, when a population is displaced en masse, its demographic structure remains relatively balanced.

Most refugees move to an adjacent country, which can result in some very complicated regional patterns when the same nation is both a country of asylum for some refugees and the home from which others have fled. Such situations often arise in the context of ethnic conflict and civil war, and those affected are usually in serious need of aid and support. Current examples include the East African countries of Sudan, Somalia, Ethiopia, and Djibouti; the Central African countries of Burundi, Rwanda, Uganda, and DR of Congo; and the West Asian countries of Afghanistan, Iran, Iraq, and Syria. Some other refugees are able to move to countries in the more developed world. Table 6.4 lists the principal countries of asylum. This table confirms that, although most refugees are in countries adjoining their homelands, many others have obtained legal status in the more developed world, where most countries receive locally significant numbers of refugees.

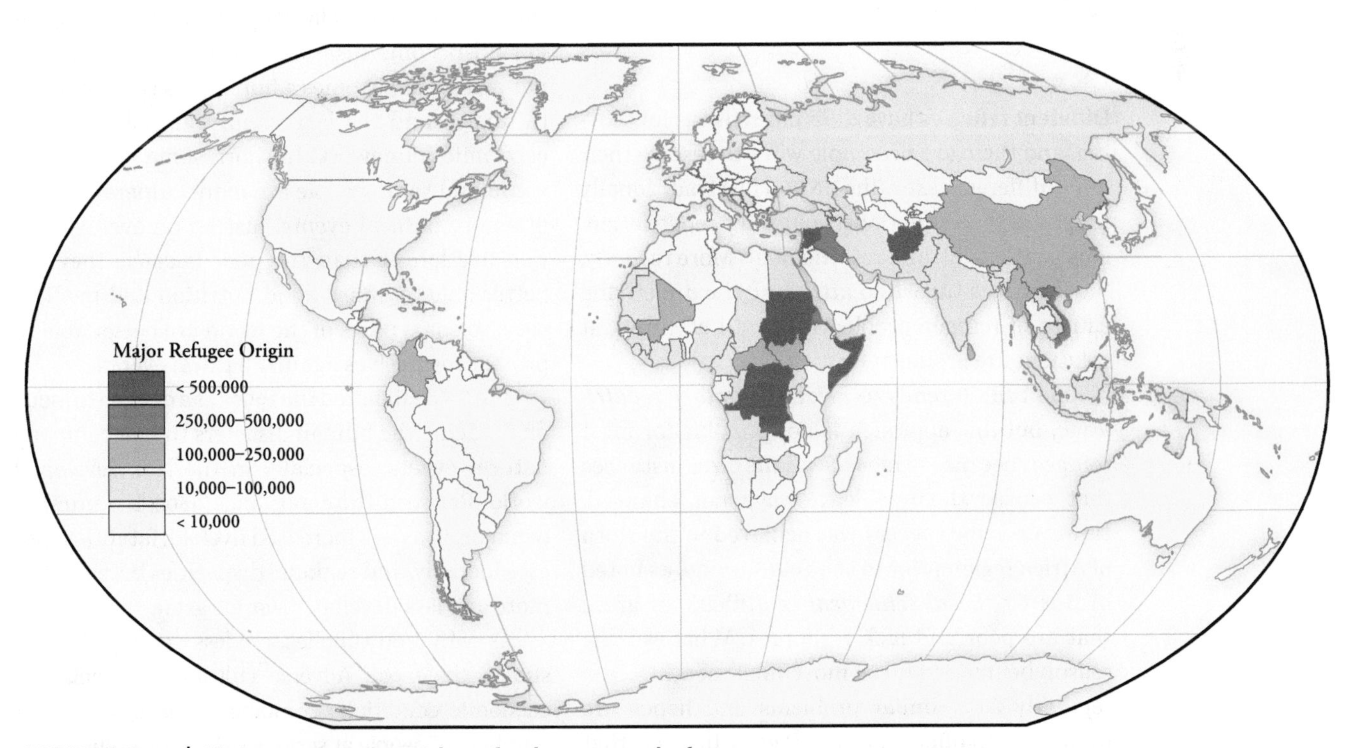

FIGURE 6.10 | Major source countries of refugees, end of 2013

Source: UNHCR. 2014. UNHCR *Statistical Yearbook 2013: Statistical Annexes*, Table 1. www.unhcr.org/cgi-bin/texis/vtx/home/opendocPDFViewer.html?docid=54cf9bc29&query=Major%20source%20countries%20of%20refugees%20|%20end-2013.

TABLE 6.4 | Main countries of asylum, 2013

Country of Asylum	Total
Pakistan	1,616,500
Iran	857,400
Lebanon	856,500
Jordan	641,500
Turkey	609,000
Kenya	534,900
Chad	434,500
Ethiopia	433,900
China	301,000
United States	263,600

Source: UNHCR. 2014. *Global Trends 2013: War's Human Cost*. www.unhcr.org/cgi-bin/texis/vtx/home/opendocPDFViewer.html?docid=5399a14f9&query=refugees%20and%20IDPs%201993-2013%20(end%20year).

Refugee and related problems are greatest, however, in the less developed world, where few asylum countries have the infrastructure to cope with the additional pressures that refugees bring. In the more developed world, the collapse of communism in several European countries in the late 1980s, the breakup of the former USSR in 1991, and the conflict in the former Yugoslavia after 1991 all gave rise to mass refugee movements.

Solutions

Different refugees have different reasons for moving, and there are no simple ways to resolve their many different cases. The UNHCR has traditionally proposed three solutions: voluntary repatriation, local settlement, and resettlement. More recently, the focus has turned to attacking the underlying causes of refugee problems. But all attempts at solutions are problematic.

The UNHCR tends to favour *voluntary repatriation*, but this approach is not possible for most refugees because in most cases the circumstances that caused them to leave have not changed. However, some success was achieved in the Horn of Africa region prior to the 2011 famine, as noted in Box 6.5. *Local settlement* is difficult in areas that are poor and lack resources. Whatever the reason behind a refugee movement, nearby areas regularly face similar problems and hence are rarely able to offer solutions. People in areas that are already poor and subject to environmental problems, such as drought in parts of eastern and southern Africa, will be hard-pressed to provide refugees with the food, water, and shelter they need. *Resettlement* in some other country is an option for only a few. No country is legally obliged to accept refugees for resettlement, and only about 20 countries do so regularly.

Attacking the root causes of refugee problems is a mammoth challenge. Not only are those causes often a complex mixture of political, economic, and environmental issues for which no simple solutions are available, but refugee problems are also greatest in parts of the less developed world that already face huge challenges. A key first step in solving the many refugee and related problems is working towards national political stability.

DISASTERS AND DISEASES

Disasters

Human geographers have long used the term *natural disasters* to refer to physical phenomena such as floods, earthquakes, and volcanic eruptions and their human consequences. In recent years, however, we have recognized that this label is inappropriate. Such events are natural in the sense that they are part of larger physical processes, but not all become disasters. To understand why a natural event becomes a human disaster, we need to understand the larger cultural, political, and economic framework. In short, some parts of the world and some people are more vulnerable than others to natural events. Just as, on average, the rich live longer than the poor because they are better able to afford good nutrition and medical care, so some parts of the world are better able to protect themselves against natural events.

The UN designated the 1990s as a decade to focus on reducing the human disasters that accompany natural events, especially in the less developed world. Between the 1960s and 1980s, the number of major disasters increased fivefold, fatalities rose considerably, and regional disparities between the more and less developed worlds grew.

As with food supplies, the less developed world suffers the most. Adverse cultural, political, and economic conditions combine to place increasing numbers of people at serious risk in the event of any environmental extreme. "The problems of staggering population/urban growth and crippling overseas

Examining the Issues

BOX 6.5 | War, Famine, and Refugees in the Horn of Africa

Tragically, the four countries that comprise the Horn of Africa—Somalia, Ethiopia, Eritrea, and Djibouti (Figure 6.11)—are rarely out of the news. Along with neighbouring Sudan, South Sudan, and Kenya, the Horn is a land of refugees, with areas that are war-torn and lacking food. This region is home to several protracted refugee problems. The reasons behind this situation are a tragic mix of human and physical geographic factors. Eritrea was a province of Ethiopia from 1962 to 1993, but there were ongoing tensions between the two. For example, when Eritrea and the neighbouring Ethiopian province of Tigray experienced severe droughts for several years in the 1980s, the Ethiopian government refused to allow international aid efforts access to these areas. Following a nearly unanimous vote to secede in 1993, Eritrea became a separate state—and Ethiopia became a landlocked nation. Border disputes simmered after 1993, and about 80,000 people died during a 1998–2000 war that was the largest conventional conflict in Africa since World War II. UN peacekeeping forces withdrew from the border zone in 2008, and the dispute continues.

Politics is also a factor in conflict in Somalia. Even though it is one of Africa's poorest countries, it has placed economic and social development second to the political ideal of integrating all Somali people into one nation. Since independence in 1960, this has involved conflict with Ethiopia (in the Ogaden) and, particularly, with Kenya. A full-scale war between Somalia and Ethiopia in 1977–8 coincided with drought. Since 1991, however, Somalia has not had an effective central government, being plagued by clan warfare and lawlessness. It is a country in crisis. A new government was established in 2004 but continued clan conflict, and the rise of Islamist militias in the south of the country are preventing reconciliation. The ongoing absence of an effective national government has also resulted in Somali pirates becoming a major threat to international shipping off the nation's coast. Further, the continuing political instability has prevented any effective response to several severe food shortages, which have exacerbated an already tragic scenario.

The situation in Somalia became much worse in 2011. The UN declared a famine in some areas of the country, the first such declaration in Africa since the 1980s. *Famine* is a highly emotive word that is used only when three circumstances are met:

1. At least 20 per cent of the population has access to fewer than 2,100 kilocalories of food a day.
2. Acute malnutrition exists in more than 30 per cent of children.
3. Two deaths per 10,000 people, or four deaths per 10,000 children, occur every day.

In recent years, the Horn, and East Africa more generally, has experienced drought conditions, the worst in about the past 60 years, but Somalia was least able to cope. The reason parts of Somalia suffered more than neighbouring countries is that, as noted, it has no effective government and is in a state of civil conflict—in short, it is a failed state (see Chapter 9). Drought added onto more than 20 years of conflict and ongoing lack of effective central authority resulted in famine. There is a simple but useful way to understand this situation: if drought conditions occur when a conflict situation is already prevailing, famine results for the most vulnerable members of the population. It is important to stress that only parts of Somalia were in a state of famine in 2011. Those who suffered were members of weak or minority clans. In 2012 the UN declared that the famine was over

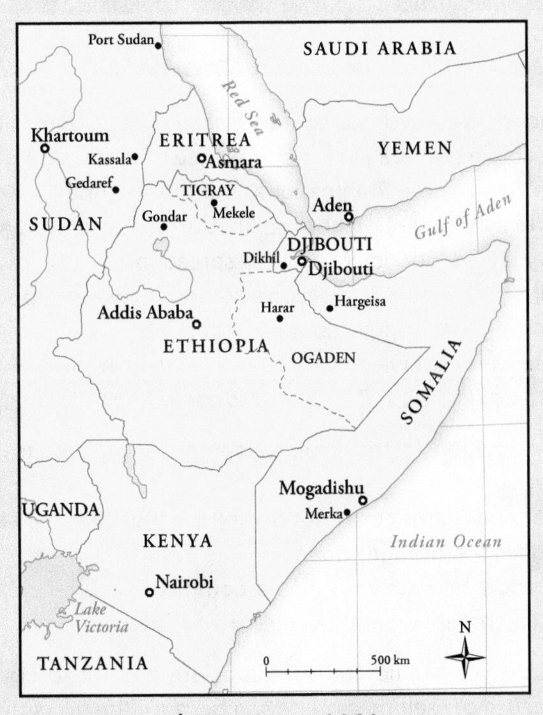

FIGURE 6.11 | The Horn of Africa

Continued

but cautioned that over two million people in the country continued to need emergency aid to survive.

It is also significant that the political conflicts in the Horn are not merely regional but are closely related to European colonial policies. France, Britain, and (especially) Italy had a presence in the region during the colonial era; subsequently, the United States and the USSR (more recently Russia) have been involved in the Horn of Africa. Despite the region's continuing political instability, the UNHCR achieved much success in repatriating refugees prior to the famine. As of 2014, however, there are almost a million Somali refugees in neighbouring countries, with almost half of these in Kenya.

debt that face many less developed countries have often become manifested in poor construction standards, poor planning and infrastructure, inadequate medical facilities and poor education—all of which exert a direct influence on vulnerability to natural hazards" (Degg, 1992: 201).

Box 6.6 provides details on Bangladesh, which is regularly subject to devastating floods.

Earthquakes, Volcanic Eruptions, and Tsunamis

If we were able to peel away the thin shell of air, water, and solid ground that is the outer layer of the earth, we would find that our planet is a furnace. Many of our major cities are a mere 35 km (22 miles) above this furnace, and in the deep ocean trenches the solid crust above it is as little as 5 km (3 miles) thick. Not surprisingly, earthquakes and volcanic eruptions are regular occurrences. Between 200 and 300 earthquakes occur each year in Canada alone, while several major volcanic eruptions and earthquakes have made headlines recently (Mount Pinatubo in the Philippines and Mount Unzen in Japan, both in 1991; the earthquakes in Pakistan, China, and Haiti in 2005, 2008, and 2010, respectively).

The impact of an earth movement on the Indian Ocean seabed resulted in a devastating tsunami

Around the Globe

BOX 6.6 | Flooding in Bangladesh

Bangladesh is a low-lying country (mostly only 5–6 m/16–20 feet above sea level) at the confluence of three large rivers: the Ganges, the Brahmaputra, and the Meghna. Floods are normal in this region; indeed, they are an essential part of everyday economic life, as they spread fertile soils over large areas. During the monsoon season, however, floods often have catastrophic consequences, especially if they coincide with tidal waves caused by cyclones in the Bay of Bengal (see Figure 4.7). Three factors contribute to flooding:

1. Deforestation in the inner catchment areas results in more runoff.
2. Dike and dam construction in the upstream areas reduces the storage capacity of the basin.
3. Coincidental high rainfall is a common occurrence in the catchment areas of all three rivers.

Regardless of specific causes, floods are difficult to control because of their magnitude and because the rivers frequently change channels.

The most extreme consequence of flooding, of course, is death; in a 1988 flood, over 2,000 died. Impelled migration is another serious consequence: in 1988, over 45 million people were uprooted. Disasters occur regularly. In a normal year, over 18 per cent of Bangladesh is flooded, and even normal floods result in shifting of river courses and erosion of banks that cause population displacement. In an already poor country, such displacement aggravates landlessness and food availability. Most displaced persons move as short a distance as possible for cultural and family reasons. During these impelled migrations, the poorest suffer the most, and women-headed households are particularly vulnerable. Some migrants move to towns, hoping to become more economically prosperous. Typically, however, such rural-to-urban migrants become further disadvantaged, clustering together in squatter settlements generally regarded as unwelcome additions to the established urban setting.

that hit parts of Asia and Africa in December 2004. The tsunami moved across the ocean with little loss of energy, resulting in much destruction when it reached land. Figure 6.12 shows the origin and movement of the tsunami and the areas affected. There were about 283,000 deaths and a further 1.1 million people displaced. An even greater tragedy was avoided as aid agencies moved in quickly to provide clean water in order to combat disease.

Scientists are continually attempting to improve our ability to predict earthquakes and volcanic eruptions. We know that most earthquakes occur at the margins of tectonic plates, when the plates move against each other; we also know some of the signs that can indicate a quake is likely in a particular area. To date, however, more exact predictions—predictions that might save lives—remain elusive.

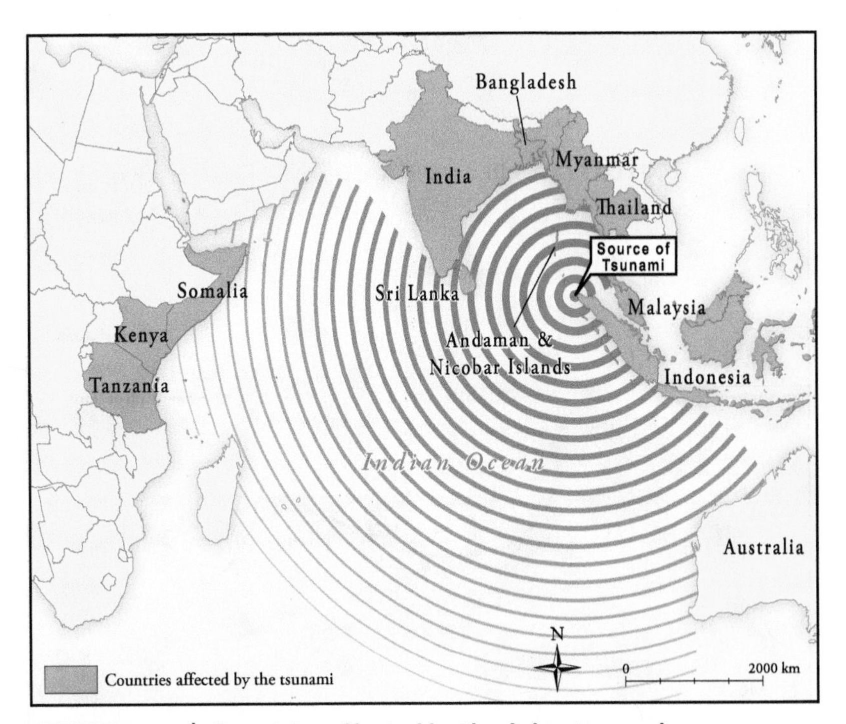

FIGURE 6.12 | Countries affected by the Asian tsunami, December 2004

Tropical Cyclones

Tropical cyclones regularly kill people, but how many die may be far from natural. This fact was starkly highlighted when Burma was hit by a typhoon in May 2008 (*typhoon* is a regional term for a tropical cyclone; the term used in the North Atlantic region, including the Gulf of Mexico, is *hurricane*). At the time, Burma was ruled by a secretive, isolated, non-democratic, military junta that chose not to respond effectively to the disaster (as of 2015 there is some evidence of political change). Some people died immediately, but the vast majority of deaths occurred in the following days as corpses spread disease and as victims were unable to obtain food and fresh water. Even six days after the cyclone struck, the number of people affected was unclear. Other countries tried to send aid workers and aid, but visas were refused and the military authorities insisted on handling, or in some cases rejecting, all aid shipments. At a time when food was scarce for victims, Burma continued to export rice to other countries. One month after the typhoon struck, the UN estimated that 2.4 million people were in need of food, shelter, or medical care. Even one year after the cyclone, the UN reported that several hundred thousand people were still in need of assistance, many continued to live in flimsy shelters, water supplies were contaminated, and reconstruction had barely begun. The inhumane government response might be explained by the fact that the government was more concerned with minimizing outside influences than it was with the welfare of its citizens. Put simply, the government chose not to save lives.

But disasters also can have devastating impacts in the more developed world, as evidenced by the hurricane that hit New Orleans in August 2005 (Figure 6.13). With a sustained wind speed of about 200 km/h, Hurricane Katrina

Residents of Chautara, Nepal, sift through the rubble of their homes, which were demolished by earthquakes in spring 2015. The UN reports that more than 8,000 people in the country died in the earthquakes and over 16,000 were injured. The devastation affected 5.6 million people, with almost half a million homes destroyed.

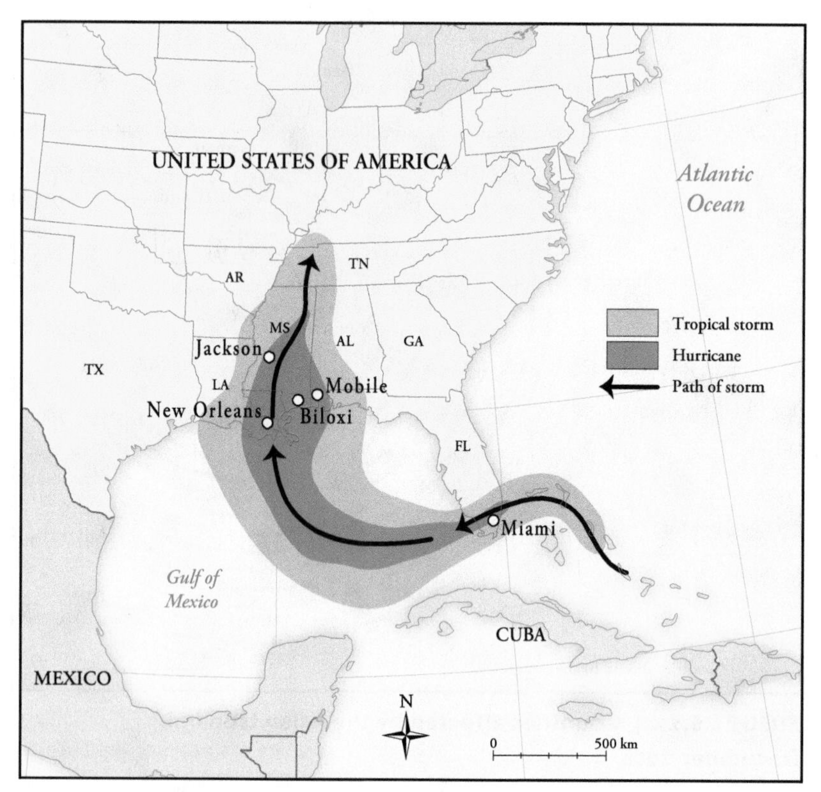

FIGURE 6.13 | The path followed by Hurricane Katrina, August 2005

passed directly through New Orleans, flooding much of the city and creating conditions that most people would have thought impossible in a major American city. As the city was evacuated, it was clear that there was a close relationship between vulnerability and poverty, and it is possible to interpret the effects of this hurricane as a selective disaster affecting the poorest most severely.

Diseases

Easily preventable diseases continue to kill large numbers of people. Malaria, which is carried from person to person by mosquitoes, is a prime example. Figure 6.14 maps malaria in Africa, which has about 90 per cent of all cases, and large areas of South and Southeast Asia and of Latin America also are vulnerable to the disease. Note the extensive area of Africa where the disease is endemic (i.e. constantly present in the population). In Africa, malaria accounts for an estimated US$12 billion in lost productivity each year. On average, there are about 500 million cases of clinical infection and between 1.5 million and 3 million deaths from malaria each year. Young children are most vulnerable; malaria kills some children and keeps others from developing.

These facts are difficult to understand because malaria can be prevented and can be treated. Of particular significance is that it is not difficult to predict a malaria outbreak, often several months in advance, because the disease is so closely related to local weather circumstances. Recent medical advances in the understanding of the disease, evidence from Kenya that the free distribution of mosquito nets dramatically reduces child deaths, and increased funding have prompted the UN-backed Global Malaria Action Plan to identify the goal of reducing malaria deaths to almost zero in the near future.

Other diseases that kill many poor people, typically several hundred thousand each year, include the following, most of which are preventable.

- Pneumonia and other lung diseases cause many deaths of young children.
- Polio, once believed to have been defeated, has returned to Africa, and a major health campaign is underway to eradicate this disease permanently.
- Tuberculosis, a respiratory disease, might be the greatest killer in history, with perhaps a billion deaths since about 1800. Poverty, poor nutrition, overcrowding, poor sanitation and hygiene, and HIV/AIDS all make tuberculosis infection more likely.
- A cholera epidemic that began in the summer of 2008 and peaked in March 2009 affected about 90,000 people in Zimbabwe. Cases can be reduced if latrines and taps for washing hands are available; it is estimated that about a billion people lack such facilities and have to defecate outdoors.
- Whooping cough is still a major killer, with most cases in sub-Saharan Africa.
- Tetanus kills many other people, with most deaths in sub-Saharan Africa, South Asia, and Southeast Asia.
- Meningitis is an ongoing threat, with a "meningitis belt" stretching across Africa from Senegal to Ethiopia.
- Syphilis is especially prevalent in Southeast Asia, sub-Saharan Africa, and Latin America.
- Lung cancer, heart disease, and other illnesses are closely linked to smoking, and it is estimated that tobacco products cause about 5.4 million deaths annually. About 50 per cent of the world's 1.3 billion smokers live in China, India, and Indonesia.

Several of these diseases are being tackled. One recent success is measles, with the number of deaths globally falling by half between 1999 and 2004. Vaccination is cited as the principal reason for the drop.

The Tragedy of AIDS

The origin of the AIDS (acquired immune deficiency syndrome) pandemic has been traced to 1920s Kinshasa in the DR of Congo, where population growth, an active sex trade, use of unsterilized needles in health clinics, and rapid railway expansion combined to spread the disease. AIDS was first identified in 1981, and HIV (the human immunodeficiency virus) was recognized as its cause in 1984. The immune system of an individual infected by HIV weakens over time, leaving the body less and less able to combat infection. The majority of HIV-infected people develop AIDS within a few years. Currently, in the more developed areas of the world, drugs make AIDS relatively manageable, but the worst-affected regions are in less developed areas—notably sub-Saharan Africa—where drugs are less likely to be available. In this part of the world, it has not been unusual for a person developing AIDS to die just six months after infection.

AIDS is able to spread through a population with astonishing rapidity. Consider that the affected population in South Africa was 1 per cent in 1990 but about 22 per cent by 2006. By 2009, it was estimated that some 33 million people globally were infected with HIV and that more than 20 million had died of AIDS since the early 1980s. The news has improved, as the incidence of new infections peaked in the late 1990s and the global percentage of people living with AIDS has stabilized—obviously at an unacceptably high level (Box 6.7).

Given these numbers, and the fact that AIDS has spread to all parts of the world, the disease is appropriately described as a pandemic. Although it is especially prevalent in sub-Saharan Africa and among the poorest members of individual populations, AIDS honours no social or geographical boundaries. It is therefore very difficult to predict its spatial spread. Because AIDS is most often transmitted sexually, the most vulnerable population is the 15–49 age group—the very group that ought to be most highly productive and is most likely to have dependants, both young and old.

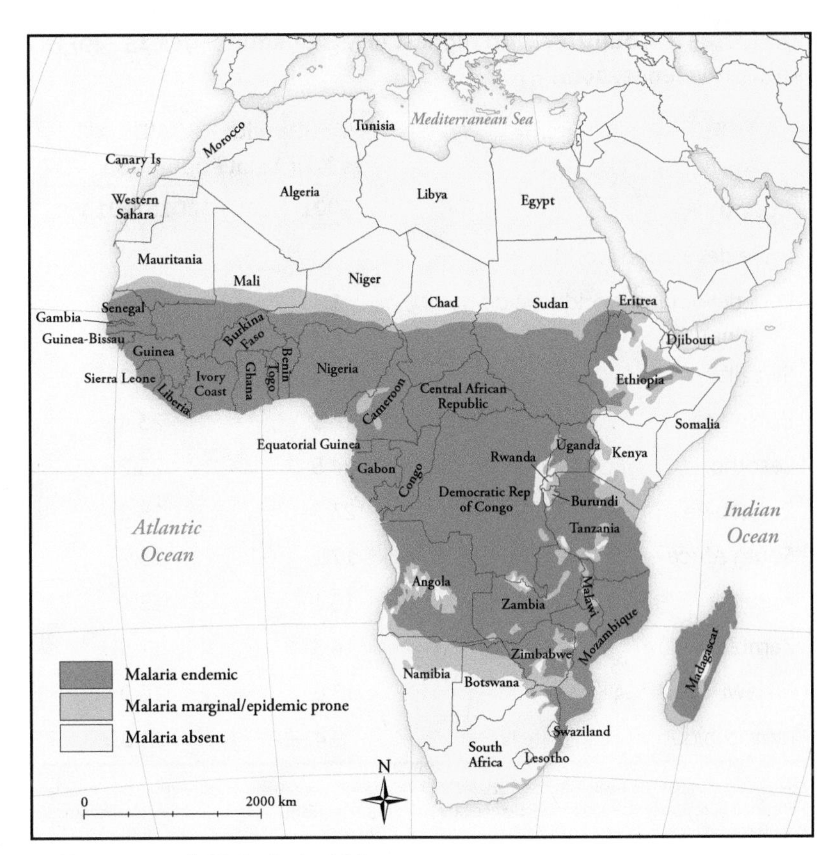

FIGURE 6.14 | Malaria in Africa

As mentioned, sub-Saharan Africa is the area of the world most devastated by AIDS. Some national death rates doubled; rates of infant mortality increased sharply; and life expectancy was reduced by as much as 23 years. Table 6.5 provides data on the 10 countries with the highest percentages of adults living with HIV/AIDS. Although some of these rates are declining, they remain devastatingly high. In sub-Saharan Africa, HIV/AIDS is most common in two age groups: infants and adults aged between 20 and 40 years. Women are at greater risk of contracting AIDS than men, by a ratio of about 1.5:1 (Daniel, 2000: 47), and frequently infect their infants, either before birth or after, through breast-feeding.

No amount of data can do justice to the human suffering caused by AIDS. Writing about one clinic in Malawi, Toolis (2000: 29) stated:

> If you have no money, then there is no choice; you must queue in the morning heat of the Boma's outpatients clinic and hope to be admitted. Demand outstrips supply. There are 130 beds, and every morning another 250 would-be patients appear at the gates. All but the sickest are turned away, as well as the medically

pandemic
A term used to designate diseases with very wide distribution (a whole country, or even the world); epidemic diseases have more limited distribution.

TABLE 6.5 | Countries with highest levels of adult (ages 15–49) HIV/AIDS prevalence, 2014

Country	Adults with HIV/AIDS as % of Total Population	
	2001	2011/2013
More developed world	0.3	0.4
Less developed world (excluding China)	1.4	1.1
Swaziland	23.6	26.0
Botswana	26.3	23.4
Lesotho	24.5	23.2
Zimbabwe	23.7	15.2
South Africa	17.1	15.9
Namibia	16.1	13.4
Zambia	14.3	12.5
Malawi	13.8	10.0
Mozambique	9.4	11.3

Sources: Population Reference Bureau, *2011 World Population Data Sheet* (Washington, DC: Population Reference Bureau, 2011); Population Reference Bureau, *2013 World Population Data Sheet* (Washington, DC: Population Reference Bureau, 2013).

hopeless, who are sent back home to die. There is not enough staff, not enough drugs, not enough needles, not enough anything.

It seems clear that poverty is the biggest single factor contributing to the spread of AIDS. Poverty means that infections remain untreated, greatly increasing the risk of further transmission. Poverty also keeps children out of school, increasing the likelihood that the sale of sex will be their primary means of supporting themselves. But AIDS only serves to exacerbate poverty. Yet, in many countries, the disease is still seen as so shameful that it cannot even be named. In Mozambique, for example, death certificates for AIDS patients often bear the words "cause unknown." As a result of these attitudes, the frank discussion needed to raise public awareness is still lacking in many cases.

Although most accounts of AIDS today focus on sub-Saharan Africa, the disease is evident elsewhere. According to a 2008 UN report, most of those outside this area who are affected are injecting drug users, men who have sex with men, and sex workers. As is so often the case with

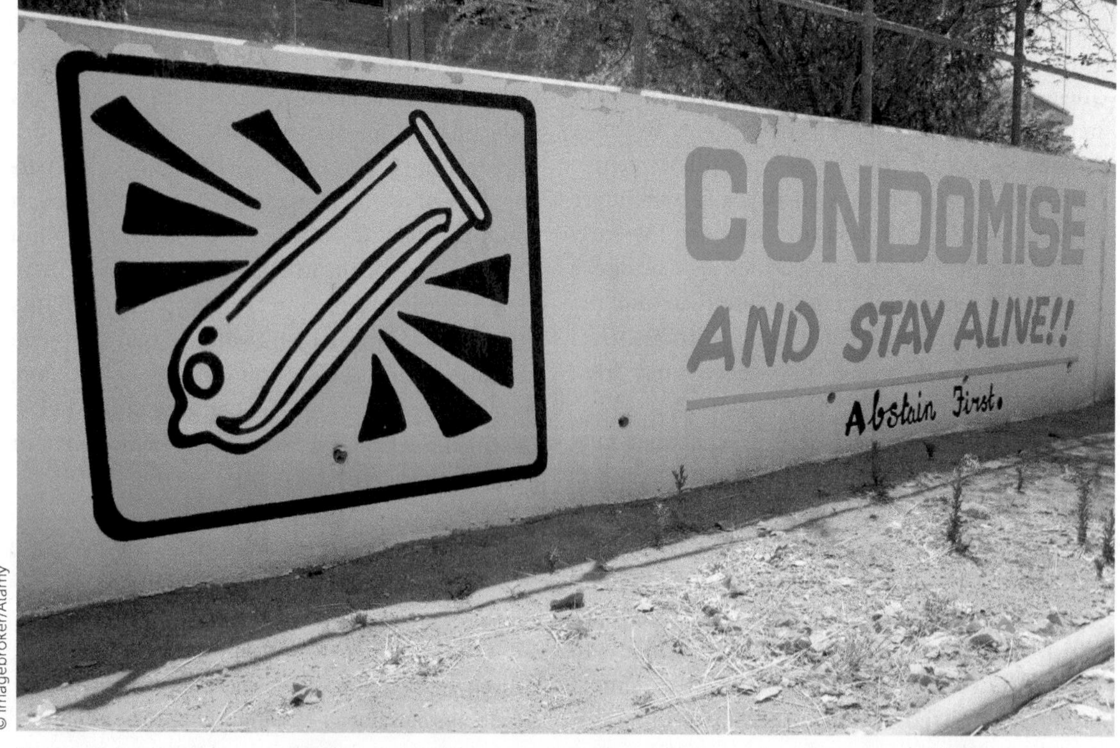

Much of the effort to combat HIV/AIDS focuses on education, such as this HIV/AIDS sex education poster in Francistown, Botswana, Africa.

In the News

BOX 6.7 | Defeating AIDS?

Talking about the "end of AIDS" seems reasonable. Although many factors can influence the course of the battle against the disease, positive signs are emerging, and the UN AIDS agency is now talking about the epidemic being under control by 2030. Crucially, the number of new infections has fallen from 3.4 million in 2001 to 2.1 million in 2013; also, the number of deaths from AIDS and related illnesses has fallen from 2.4 million in 2005 to 1.5 million in 2013.

The reason for these positive changes appears to be a combination of education, changing sexual behaviour, and increased availability of antiretroviral drugs. Countries such as Thailand, the Philippines, and Brazil have been fighting the disease with much success since the 1990s through prevention programs designed to educate people about such matters as condom use. In sub-Saharan Africa, the principal reasons for these positive changes appear to be that condom use is more widespread, people are engaging in sexual behaviour at a later age, and the more developed world has intensified efforts to combat the disease, including helping to make drugs more readily available.

In Uganda the adult infection rate has dropped from 30 per cent in 1992 to less than 6 per cent. Perhaps surprisingly, this reduction is attributed not only to increased use of condoms but also to a government campaign stressing abstinence and fidelity. In many other countries, including South Africa, people are increasingly well educated concerning the causes of AIDS, and there are clear signs that attitudes towards sexual behaviour are changing. In Botswana, some mining companies that employ large numbers of adult males provide hospital facilities for those already infected and are working to prevent AIDS not only through education but also by making free condoms readily available.

Of course, condom use does meet with opposition. A controversy erupted in 2009 following remarks by Pope Benedict XVI prior to a visit to Africa. Stating that condoms did not help solve the problem but rather exacerbated the effect of HIV/AIDS, he advocated the traditional church teachings of abstinence and fidelity. In a response that was unprecedented in its forcefulness and condemnation, the prestigious medical journal *Lancet* wrote that the pope was publicly distorting well-established scientific facts in order to promote Catholic doctrine.

Another important factor in the worldwide fight against AIDS is the response of the more developed world. Here, too, the signs are positive. A 2001 UN summit agreed to set up a global fund to fight AIDS, along with tuberculosis and malaria. Despite initial funding problems, there is reason to believe that the errors of the 1990s, when the more developed world quite literally looked away, are being corrected. In 2003 the United States committed up to $15 billion over five years to combat AIDS in Africa and the Caribbean, although much of this funding is targeted to programs advocating abstinence.

Changing attitudes in both less and more developed countries and increased use of antiretroviral drugs appear to be both limiting the spread of AIDS and reducing the number of deaths.

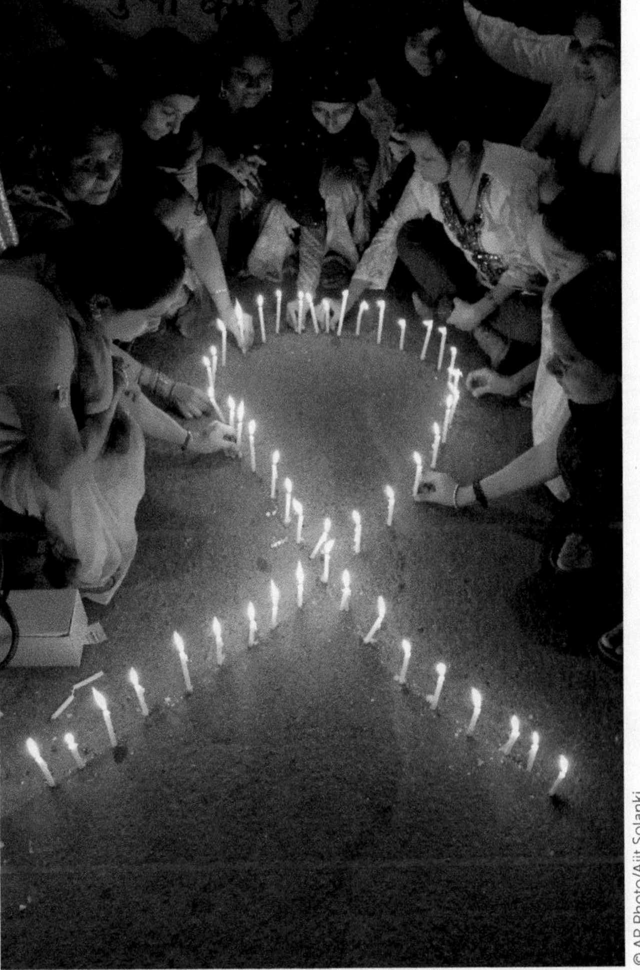

Women make a ribbon formation with candles to mark World AIDS Day in Amadabad. The World Health Organization established the day in 1988 to further global awareness and put a focus on prevention.

HIV/AIDS, the data are unreliable. Nevertheless, it appears that the infection rate in Russia may be three times that in the United States, in part because of the social and economic changes that have been underway since the fall of communism. Together, increasing poverty, increasing social freedom, and (especially) increasing illegal drug use are making Russians particularly vulnerable to infection. In the case of China, there are perhaps 10 million cases. One secondary cause of AIDS transmission in the country is the sale of blood by impoverished peasants to illegal brokers (*The Economist*, 2002: 75). India may have about three million people infected with HIV. Most of the epidemic there is concentrated in major cities and is associated with heterosexual behaviour. Because there is little open discussion of AIDS, the risks of infection are not widely recognized.

The Diffusion of Disease

Ebola

Since it was first discovered in 1976, near the Ebola River in the DR of Congo, there have been sporadic outbreaks of Ebola, all in Africa. There are normally about 500 cases reported each year, but none were reported between 1979 and 1994. Although rare, it is a deadly disease and is caused by infection with one of four virus strains. Details are unclear, but it is likely that the virus is animal-borne, with bats as the probable reservoir.

An outbreak of epidemic proportions was reported in March 2014, though it probably appeared in late 2013. By early 2015, this incident was responsible for more deaths (about 9,000) than all previous outbreaks combined. This was also the first outbreak in West Africa. The first reported death was in Guinea, followed by cases in Liberia, Sierra Leone, Nigeria, and Senegal. In June, Médecins Sans Frontières/Doctors Without Borders reported that the disease was out of control; that same month, the UN Health Agency declared an international public health emergency; and, in September, the director of the Centers for Disease Control and Prevention compared the outbreak to the emergence of HIV/AIDS. By early 2015, it was clear that the outbreak was under control.

The disease was able to spread so rapidly because the illness was initially seen as a spiritual rather than a medical problem; family members cared for those already sick. Also, it took time to be diagnosed as Ebola, partly because it was thought to be Lassa fever, which has similar symptoms but is much less deadly. More generally, there was a mistrust of government and ignoring of advice from officials. Other contributory factors were the movements of people from place to place and crowded cities with limited health-care facilities and personnel. Hopefully, lessons have been learnt by local populations, national governments, and the international community such that another outbreak will not become another epidemic.

During this most recent Ebola outbreak, about 3.6 million died from AIDS, malaria, and tuberculosis combined. One way of addressing disease problems in the less developed world is to move away from targeting specific issues, such as Ebola or AIDS, and instead to emphasize the creation of strong health systems that are capable of tackling all diseases, although this approach is likely to cost much more than is currently being spent.

PROSPECTS FOR ECONOMIC GROWTH

As this chapter demonstrates, there are poor countries and wealthy countries. But there are so many variations, especially within the former group, that generalizations are often misleading. Much of the discussion so far has focused on the principal problem areas: the least developed of the less developed countries. The following accounts of debt and of prospects for industrial growth serve to broaden the discussion and suggest there is much hope for the future. In particular, there are indications that industrial growth is proceeding apace in some poorer countries.

Solving the Debt Crisis

The significance of a country's debt to a country depends on its economy. Some countries—such as South Korea, which borrowed heavily to finance industrialization—have been able to repay their loans because of their industrial and export success. In many less developed countries, however, the cost of servicing the foreign debt accounts for much of the income from exports. For those countries, foreign debt is such a crushing burden that some, such as Colombia, have actually exported food to help repay it, even though their own populations are malnourished. Not surprisingly, there is a concentration of severely indebted low-income countries in Africa.

The external debt of the less developed world—mostly money owed to international lending agencies and commercial banks in the more developed world—increased from US$59.2 billion in 1970 to $445.3 billion in 1980, $1.2 trillion in 1990, more than $2 trillion by the early twenty-first century, and about $4 trillion today. The 1970s saw high lending to less developed countries by the commercial banking sector, development agencies, and governments in the more developed world. These loans, intended for the establishment and support of economic and social programs, had long payback terms because it was generally expected that, in the long run, the less developed economies would boom and eventually provide good returns on the investments.

Unfortunately, the 1980s brought a recession, with rising interest rates, declining world trade, and steadily increasing debt. As a result, many countries are so poor and owe so much that they need to borrow more money just to keep up the interest payments on their existing debt—a situation that first arose in 1982 in Mexico (Sowden, 1993). But the news is not all bad. In 1996 the World Bank and the IMF launched the Heavily Indebted Poor Countries (HIPC) Initiative to assist countries encumbered by debt. Also, agreement was reached in 2005 to reduce some debts, and the Multilateral Debt Relief Initiative (MDRI) was launched the next year to provide additional resources to HIPCs to meet Millennium Development Goal number 8 (discussed later in this chapter). By 2010, $76.4 billion in HIPC debt relief had been committed to 36 countries, of which 30 have received an additional $45.8 billion under the MDRI. In 2006 Nigeria became the first African country to pay off its **Paris Club** debt, of about US$12.4 billion, leaving a remaining debt to the World Bank and other lenders of about US$5 billion. This repayment means that Nigeria is able to invest significantly in health care and education and might also attract increased foreign investment.

Loans intended to help impoverished countries can have quite the opposite effect. Box 6.8 outlines one alternative approach, an innovative program in which loans are provided directly to poor individuals rather than to their country and from within countries rather than from outside. It is also possible to rethink the whole question of world debt and argue that the debt owed by the less developed world is only one side of the coin. For example, a 1999 report published by Christian Aid, a leading British charity, asserted that the more developed world owes a huge debt to the less developed because of the disproportionate environmental damage that the former causes. Similarly, discussions at the 2001 World Conference against Racism, Racial Discrimination, Xenophobia, and Related Forms of Intolerance, sponsored by the UN and held in Durban, South Africa, included arguments that rich countries should pay poor countries compensation for the abuses (such as slavery) inflicted on them in the past. The likelihood that rich countries would have heeded such a proposal was minimal; a few days later, the 9/11 terrorist attacks on the US shifted global priorities far away from poor country debt.

Industrial Growth

As noted, the distinction between more and less developed is never straightforward, and there is much variation within the less developed world. Because of their recent industrial success, a group of newly industrialized countries (NICS; see Chapter 13) are frequently referred to as economic miracles, yet they are still generally regarded as belonging to the less developed world. More typical less developed countries have been much less successful industrially, and no miracles are expected of them in the near future.

An obvious distinction between the more and the less developed worlds is the degree of industrialization: many countries in the latter group have not yet experienced anything like an industrial revolution. But it is widely hoped that industrialization might lead to employment for

Paris Club
An ideologically neo-liberal grouping of financial officials from 19 of the biggest developed economies—15 European countries plus the US, Canada, Japan, and Australia—loosely formed in 1956 and more formally structured in the 1970s, which provides financial services and organizes debt restructuring, debt relief, and debt cancellation for indebted countries and their creditors.

Around the Globe

BOX 6.8 | The Grameen Bank, Bangladesh

Bangladesh is one of the poorest countries in the world. As of 2013 it has a large population of 158.5 million, an exceptional population density of 1,101 per km², a CBR of 20, a CDR of 6, and an RNI of 1.5. The per capita GNI is among the lowest in Asia, at US$2,810.

But a remarkable transformation has occurred in parts of the country, a transformation brought about by a credit program that focuses exclusively on improving the status of landless and destitute people. The Grameen ("village") Bank program began in 1976, under the leadership of Muhammad Yunus, as a research project at the University of Chittagong to explore the possibility of providing credit for the most disadvantaged of the Bangladeshi people and of encouraging small-scale entrepreneurial activity. In 1983 the project became an independent bank, which had 571 branches covering 17 per cent of the villages in Bangladesh by 1988.

Essentially, the Grameen program seeks to enhance the social and economic status of those most in need, especially women, by issuing them small loans (the first was $27 made to 43 women; at present, the institution's loans average $300). The bank requires that loan applicants first form groups of five prospective borrowers and meet regularly with bank officials. Two of the five then receive loans and the others become eligible once the first loans are repaid. The focus is on the collective responsibility of the group. As the recovery rate—a remarkable 98 per cent—shows, the Grameen Bank's innovative approach works.

This successful technique is being replicated elsewhere in the world (see Mahmud, 1989; Todd, 1996), and there are now about 10,000 microfinance institutions in more than 100 countries serving about 100 million people. Yunus and the Grameen Bank were awarded the 2006 Nobel Peace Prize.

the unemployed or underemployed rural poor. In addition, many in the less developed world believe that industrialization would demonstrate economic independence, encourage urbanization, help to build a better economic infrastructure, and reduce dependence on overseas markets for primary products. At the same time, governments and development agencies in most poor countries need to address the issue of significant regional differences within these countries.

Potentially, industrialization has many advantages. So why is it so difficult for many countries to achieve it? According to dependency theory logic, a major obstacle is the legacy of former colonial rule. Historically, European colonial countries saw their colonies both as producers of the raw materials needed for their domestic manufacturing industries and as markets for those industries. This attitude can be a difficult legacy to move beyond. Most countries in the less developed world have neither the necessary infrastructure nor the capital required to develop it. Furthermore, fundamental social problems resulting from limited educational facilities do not encourage the rise of either a skilled labour force or a domestic entrepreneurial class. Finally, the domestic markets in many of these countries lack the spending capacity to make industrial production economically feasible.

Perhaps the lack of development throughout most of sub-Saharan Africa is explained in terms of low population density relative to other major world areas, long distances between the major settlement locations, and the political partitioning of Africa by the colonial powers, all of which has resulted in post-colonial states composed of deeply divided ethnic nations of different cultures and religions. Consider, too, that Africa has more countries (54) relative to the total land mass than any other major world region and that each of these countries has an average of four neighbours (in Latin America the average is 2.3) (World Bank, 2009: 283).

Despite the many difficulties faced, some industrialization has taken place in sub-Saharan African countries since World War II. Successes have been achieved in import substitution—manufacturing goods that were previously imported, often with the help of protective tariffs. Examples include light industries that are not technologically advanced, such as food processing and textiles. Heavy industry has been more difficult to develop, even though many of the raw materials used in the more developed world come from the less developed world. There is clear evidence of a growing middle class (defined in terms of daily spending) in many of the rapidly expanding urban areas; the African Development Bank estimated that 34 per cent of the urban population were in this category in 2010, up from 27 per cent in 2000, and is expected to reach 42 per cent by 2060.

More generally, since about 1980, many less developed countries have been under sustained pressure from both the IMF and the World Bank to adopt top-down free-market economic policies that will result in structural adjustment. Specifically, governments are encouraged to open their economies to increase international trade, privatize previously state-owned enterprises, and reduce government spending. The logic is that such policies will generate wealth and development that will trickle down through all sectors of the economy to benefit all the people. This approach has been criticized on the grounds that, in fact, most of the wealth stays with the elite and any trickle-down benefits are minimal.

We will see in Chapter 10 that some failings of the green revolution have been blamed on the preponderance of top-down strategies. An alternative approach favoured by many national development agencies and by most NGOs working in poor countries, such as Oxfam, is to initiate development from the bottom, building on local community strengths and being sensitive to particular local circumstances. Many poor countries employ both methods to regional economic and social development, and it may be that both are necessary to achieve meaningful growth.

STRIVING FOR EQUALITY, FAIRNESS, AND SOCIAL JUSTICE

Discussion concerning the many challenges faced by less developed world countries continues unabated. Earlier in the chapter, we asked why there is a less developed world, but so far we have had little to say about possible solutions. What pragmatic approaches are taken by the principal organizations that have responsibility for these matters, such as the various branches of the UN, the World Bank, and the IMF? In this section, we review some of the ideas about ongoing uneven development, describe strategies taken by major organizations, and assess the progress currently being made towards a more equal world.

Explaining spatial variations in economic development and quality of life is not easy. Disparities are common at all spatial scales, from the basic global division into two worlds all the way to local differences between, for example, residential areas in a city. One influential recent contribution is called the new economic geography.

The Ideas of New Economic Geography

Since about 1990, economists, not geographers, have initiated this influential research interest. Its impact is evident in the fact that one of the leading advocates, Paul Krugman, was awarded the 2008 Nobel Prize in economics for his work on trade patterns and location of economic activity. Also, the World Bank employed this approach in the 2009 edition of its flagship annual publication, the *World Development Report*. Subtitled *Reshaping Economic Geography*, the report argues that the best way to promote long-term growth is to acknowledge the inevitability of uneven development and to develop policies that enable geographic concentration of economic activity. Challenging the assumption that economic activities are best spread spatially evenly, the report asserts that attempting to prevent concentration only results in limiting prosperity and, at the same time, stresses that development can still be spatially inclusive through government promotion of economic integration. The report explains uneven development in terms of differences in population density, distances between places, and political and other divisions.

Overall, human geographers have not been attracted to this new economic geography, at least partly because it appears to ignore issues of spatial injustice with its failure to recognize that many places are peripheral and undeveloped because of a history of exclusion. As we have seen, it can be argued that global inequalities are a consequence of the historic and present patterns of relations between peripheral and core countries, while some instances of regional inequality within countries also result principally from social and power relations, as evidenced, for example, in the spatial pattern of racism and related landscapes (Rigg et al., 2009). Notwithstanding these criticisms, the new economic geography is able to provide numerous insights into why some areas develop more than others and is attractive to policy-makers.

It is possible that free migration within countries, specifically rural-to-urban migration, will contribute to improving the quality of life in less developed countries. An interesting way to think about reducing inequality is to observe that people are living in the wrong place and that movement into cities is helping solve that problem.

Millennium Development Goals

In 2000 the UN's Millennium Summit of world leaders identified eight Millennium Development Goals (MDGs) aimed at improving the lives of people in poor countries by 2015. The Millennium Declaration, approved by all UN member states and by 23 international organizations, stated that "every individual has the right to dignity, freedom, equality, a basic standard of living that includes freedom from hunger and violence, and encourages tolerance and solidarity." The eight goals, listed below, comprise a blueprint for development. As the brief comments on each show, must progress has been made in achieving them.

1. *Eradicate extreme poverty and hunger.* Extreme poverty has been reduced by more than half, and is now about 20 per cent of the population of less developed countries. This progress was achieved through investments in agriculture, job creation, expanded social safety nets, expanded nutrition programs targeting very young children, improved education, improved gender equality, and the protection of vulnerable countries during crises. A notable improvement is in Ethiopia, where the number of people with access to safe water increased from 60 per cent in 1990 to 94 per cent in 2008. Much success has also been achieved in Rwanda, Nepal, and Bangladesh.
2. *Achieve universal primary education.* Ninety per cent of children are attending primary school. Education is essential both to the reduction of poverty and inequality and to the building of successful economies and democratic institutions. In particular, educating girls contributes significantly to the elimination of poverty as it results in later marriages, fewer children, reduced risk of AIDS infection, improved employment prospects, better medical care, increased political participation, and better access to credit. Most less developed countries reached gender parity in primary schools as of 2015.
3. *Promote gender equality and empower women.* Here again, education is the key. Also important are increasing women's labour force participation and strengthening labour policies affecting women, promoting women's political rights and participation, and expanding reproductive health programs and family support policies. Notable achievements include a 15–20 year increase in average life expectancy for women since 1970. However, the fact that only 1 per cent of global agriculture credit goes to African women suggests that much remains to be accomplished.
4. *Reduce child mortality.* Child mortality has been halved. Deaths of infants and children have fallen dramatically in recent years, largely because of improved health care. Yet the mortality rate for children under five remains far too high, with the majority of deaths resulting from preventable diseases (e.g. dysentery, malaria) and/or malnutrition. Child mortality can be further reduced by strengthening national health systems, expanding immunization programs, ensuring the survival and improved health of mothers, supporting better nutrition for child and mother, and investing in improved reproductive health. One example of a specific improvement is Rwanda, where

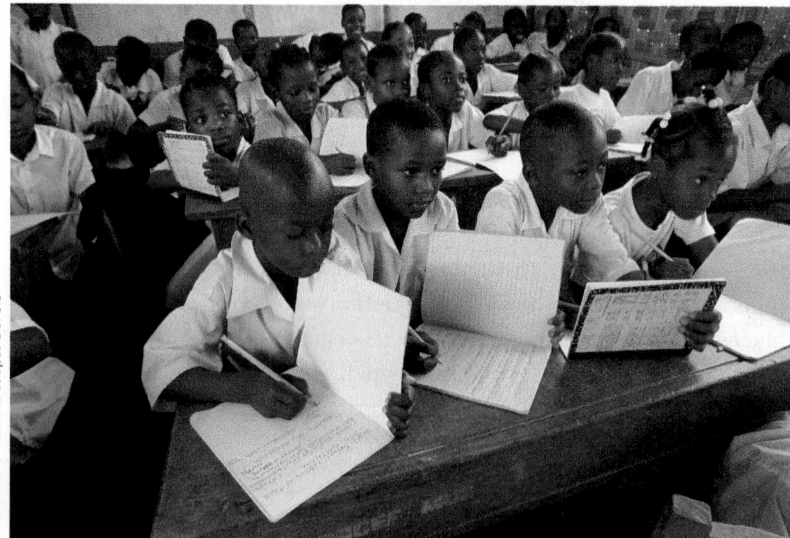

Students take notes during class at Harper School in Maryland County, Liberia. Reports claim that the Liberian government and activists are campaigning for more parents to enroll their daughters in school.

health visits for children ages 2–5 increased 133 per cent between 2006 and 2008.

5. *Improve maternal health.* Limited progress has been made towards achieving this goal. The maternal mortality ratio fell by 45 per cent (1990 to 2013) but remains far too high. Complications related to pregnancy and childbirth are a leading cause of death among women of reproductive age in the less developed world, where a rate of 1 maternal death per 100 births is not unusual. The corresponding figure in the more developed world is about 1 in 100,000. Again, these deaths are preventable. Strategies to achieve this goal include reducing teenage pregnancies, empowering women to space births and choose family size, improving maternal nutrition, ensuring trained attendance at deliveries, and providing better access to needed care.
6. *Combat disease.* The spread of AIDS appears to be under control, but many areas suffer from malaria and tuberculosis, and the threat of new diseases, such as Ebola, is always present. There is an ongoing need to provide effective prevention, care, and treatment services for a number of diseases.
7. *Ensure environmental sustainability.* Wherever resources are depleted, the poorest suffer first and most severely. Sustainable practices are especially urgent with regard to fresh water, forests, and soils. There is a need to invest in clean energy, make infrastructure improvements, increase access to sanitation, offer technical assistance, and promote ecosystem-based solutions. The target of halving the number of people without access to clean drinking water was met in 2012, but there are still 2.5 billion people lacking basic sanitation.
8. *Develop a "global partnership for development."* An open and non-discriminatory trading and financial system must take into account the special needs of the less developed world. It is particularly important to address the debt problem faced by many countries. Strategies include promoting debt relief (discussed earlier), developing information technology infrastructure, and expanding trade agreements.

Achieving these goals was never going to be easy, and the global economic crisis that began in 2008 had negative impacts. However, it is now widely accepted that achieving goals such as these is a global responsibility and that international aid is most effective when it goes to countries with good economic policies and sound governance. It is clear that the more developed world has become increasingly committed to reducing poverty and improving lives in the less developed world.

Moving beyond the Millennium Development Goals

At the time of writing, the UN is working to develop an ambitious post-2015 agenda, including new goals. Many of the same problems need to be addressed again, but there are other considerations to take into account. There will certainly be increased focus on the possible impacts of global warming. Our discussion of food shortages noted that the key causes are political and mainly concern bad government. Hence, better government (meaning democratic institutions) is required. Democracies are much less likely to wage war and to instigate ethnic rivalries. But establishing a democracy is no simple task. Dictators are not easily toppled, and authoritarian regimes are difficult to replace. A major obstacle to political change is that, as discussed further in Chapter 9, many boundaries in African countries were determined by colonial Europeans for reasons related to European, not African, interests. The World Bank acknowledged this need for better government in 2011, when it recommended that aid money should be targeted at building stable government, working towards a just society, and establishing an efficient police force. Most civil wars occur in countries that have had a similar conflict during the previous 30 years. Thus, there is a need to prevent conflict rather than simply spending money to alleviate the consequences.

This point leads to a second requirement: ensuring security for the poor. A former Kenyan politician, John Githongo, set up the Inuka Kenya Trust in 2008, which works with local partners to initiate security-building schemes throughout the country. Githongo argues that increasing inequality, mainly in resource-rich countries, only makes matters worse as large sections of the population are quite literally excluded from any development that is occurring. Closely associated with this need for security is the need to eliminate corruption; Githongo estimates that one-third of Kenyan government revenue disappears into private bank accounts.

So what helps a less developed country to develop? We have noted that money, while frequently essential (especially in emergency situations), is not a panacea. According to a UK think tank, the Overseas Development Institute, there are four prerequisites: smart leadership, smart policies, smart institutions, and smart friends.

Debating the Future

Much of this chapter's discussion of the less developed world paints a grim picture. Yet there are also promising signs. What will our global future be? The answer is far from clear—and, as our discussion of world systems analysis demonstrates, it depends at least partly on ideological perspective. As we have seen, there is an ongoing debate between pessimists and optimists or, as they are commonly known, catastrophists and cornucopians. Similarly, antithetical perspectives on the future of the world have existed for centuries. In the mid-1700s, just a few years before Malthus predicted a future of famine, vice, and misery, Berlin vicar and statistician Johann Peter Süssmilch claimed that the earth could feed 10 billion people.

The differences of opinion in the contemporary debate often seem almost equally extreme. Pessimists and optimists base their forecasts on very different kinds of assumptions. Pessimists such as the Club of Rome, with its 1972 analysis *The Limits to Growth*, and Paul Ehrlich, whose 1968 book *The Population Bomb* stirred both interest and controversy regarding overpopulation, usually assume an existing set of conditions, "other things being equal." Optimists, by contrast, argue that "other things" are not equal; they maintain that circumstances change and that human ingenuity is always increasing, finding new ways to cope with problems.

One writer who has focused attention on specific problem regions in several African and Asian countries is Robert Kaplan (1994, 1996). Reporting in a style that one reviewer described as "travel writing from hell" (Ignatieff, 1996: 7), he details a nightmare scenario of population growth, ecological disaster, disease, refugee movements, ethnic conflict, civil war, weakening of government authority, empowerment of private armies, rural depopulation, collapsing social infrastructure, increasing criminal activity, and urban decay. He also identifies West Africa as the region in the most severe trouble. Kaplan writes convincingly about the interrelatedness of problems such as population growth, environmental degradation, and ethnic conflict—all characteristic in many parts of the less developed world. The evidence that he finds, particularly concerning refugees and food supply, leads him to predict an immediate future of instability and chaos.

Others disagree with both the details of Kaplan's findings and his conclusions. Pointing to steady improvements in health, education, life expectancy, and nutrition, both globally and specifically in the less developed world, Gee (1994: D1) asserts that "By almost every measure, life on earth is getting better." It is definitely true that, by some measures, life is improving. As we have seen, fertility is declining and life expectancy and adult literacy are increasing throughout much of the less developed world. One especially interesting development concerns Botswana, a country that has seen rapid growth in income per person over the past 35 years. Attempts to explain this surprising fact are varied, but it appears that Botswana has succeeded in aligning the interests of the elite with those of the masses and has used aid money to improve the legal and court system and to fight disease. Elsewhere in Africa, Uganda is educating nearly all its children of elementary school age. There is also evidence that income poverty is declining in percentage terms, but this needs to be interpreted with care. Reducing income poverty is a result of both economic growth and the extent to which the very poor benefit from economic growth. The most notable declines in income poverty are in the East Asia and Pacific and South Asia regions and result from economic growth in China and India, respectively. Indeed, a global decline in income poverty from 40.4 per cent in 1981 to 20.7 per cent in 2001 was largely the result of poverty reduction in China.

Critics like Gee have compared Kaplan to such earlier doomsayers as Malthus and the Club of Rome, whose most dire predictions have not come to pass. But there are differences, and they may be significant. Unlike most earlier predictors of nightmare futures, Kaplan was looking at the relatively short term—perhaps two decades—and basing his judgments on specific circumstances in the mid-1990s. As of 2015, the overall picture looks brighter than a pessimist might have predicted. For example, the proportion of the world's people living in extreme poverty is falling. On the other hand, regional circumstances may well confirm the worst fears.

Some evidence shows improved standards of living and levels of education globally, while a recent book argues that Africa will be able to feed itself within a generation through improvements in infrastructure, mechanization, and genetically modified crops (Juma, 2010). But there are also indications that the very poorest countries are falling further behind. An ambitious goal to halve the number within 10 years was set at the 2011 UN conference on least developed countries. Recall that there are currently 48 such countries and only 4 have moved out of the category since 1971. Clearly, the debate between cornucopianism and catastrophism remains unresolved.

CONCLUSION

The fact that our world comprises very poor and very rich countries is one of the most important of human geographic circumstances. This fact was introduced in earlier chapters and resurfaces in later ones simply because it is a necessary consideration in any account of global geographies. This chapter attempts to explain why there are many differences between countries, paying particular attention to ideas about the early spread of agriculture and the facts and consequences of colonial rule. The account of circumstances in the less developed world focused on problems of food supply, refugees, disasters and diseases, and some of the prospects for economic growth. It is noted that aid, while often essential in emergency situations, does not necessarily assist countries to improve their circumstances. Consideration of the eight MDGs permitted some conclusions to be made about the prospects for social and economic change.

There is a continuing debate between those who argue that most major problems can and will be resolved and those who argue that things are bad and likely to get worse. This debate takes place against a background of continuing population increases and uncertain political developments. But one fundamental conclusion cannot be overemphasized: the key to improving the lives of people in less developed countries is political change. There is compelling evidence that the common denominator in such major problems as food shortages, civil wars, and refugee problems is bad government, meaning a lack of democracy.

Summary

Explaining and Identifying the Less Developed World

It can be argued that early agricultural areas in Eurasia that had room to expand latitudinally (east and/or west) resulted in civilizations that were able to move overseas in due course. Wallerstein's world systems approach posits a world organized in such a way that core countries benefit at the expense of peripheral countries. Less developed countries may be identified according to various criteria: some, such as GDP or GNI, are solely economic; others reflect a number of measures of human development. All these countries have similar relationships with the more developed world. Many have been colonies and most are in a dependent situation; virtually all have massive debt loads; most are vulnerable to environmental extremes such as droughts, floods, and earthquakes; and most are prone to disease. Sub-Saharan Africa is the location of many less, and even least, developed countries.

The World Food Problem

Undernutrition is caused by insufficient food; malnutrition is caused by low-quality food lacking the necessary protein and vitamins. One explanation of the global food problem is overpopulation (too many people for the carrying capacity of the land); another is inadequate distribution of food. Some scholars have argued that the world food problem is a consequence of the imposition of a capitalist mode of production, which has made it impossible for increasing numbers of people in the less developed world either to grow or to purchase food. Much evidence also suggests that food shortages and related problems can be traced to a lack of democratic institutions and to bad government more generally.

Refugees

Refugees are forced migrants. Major movements first occurred after World War II, but a marked increase in numbers began in the 1980s in less developed countries for various political, economic, and environmental reasons. Afghanistan and Syria are the source countries of large numbers of refugees. Most refugees move to adjacent countries.

Disasters and Diseases

Disasters such as earthquakes are much more likely to have adverse impacts on people in the less developed world than on people in the more developed because of a general inability to cope. The impact of diseases is also spatially unequal, with Africans especially suffering from several preventable diseases.

Developing Societies and Economies

For several decades many less developed countries have faced unbearable debt loads, but these are lessening somewhat. A major obstacle to industrial growth is the legacy of former colonial rule. Historically, European countries saw their colonies both as producers of the raw materials needed for their domestic manufacturing industries and as markets for those industries. Progress made towards the eight Millennium Development Goals is significant but uneven. A positive sign is that the more developed world has become increasingly committed to reducing poverty and improving lives in the less developed world. A particular need is the promotion of good government, as it is widely agreed that this is a prerequisite to development. The debate between cornucopianism and catastrophism continues unabated.

Our Global Future

There is much debate concerning how best to eliminate poverty in the less developed world, with some favouring huge inputs of aid money and others favouring helping poor countries to help themselves. There is also debate concerning what, if any, improvements in lives and livelihoods are taking place.

Links to Other Chapters

- **Globalization:** Chapter 2 (concepts); Chapter 3 (all of chapter); Chapter 4 (global environmental issues); Chapter 5 (population growth, fertility decline); Chapter 8 (popular culture); Chapter 9 (political futures); Chapter 10 (agricultural restructuring); Chapter 11 (global cities); Chapter 13 (industrial restructuring)
- **Refugees:** Chapter 9 (political geography)
- **The less developed world:** Chapter 4 (energy and technology, earth's vital signs); Chapter 5 (fertility and mortality); Chapter 9 (colonialism); Chapter 10 (agriculture); Chapter 12 (settlement); Chapter 13 (industry)

Questions for Critical Thought

1. The history of humanity can be viewed as an ongoing series of migrations spanning six million years (see Table A3.2 online). Why did our ancestors move throughout this history? Why do we continue to move? Does it make sense to think in terms of "laws" of migration?
2. What role does government policy play in migration today?
3. How should governments respond to refugee situations? For example, who should be responsible for providing assistance for political and environmental refugees?
4. Why is development uneven (i.e. why is there a more developed world and a less developed world)?

5. What, if anything, can/should be done by the more developed world to address issues such as malnutrition and debt in the less developed world? Justify your position.
6. From a moral point of view, is there an obligation on the part of the more developed world to send aid to the less developed world? Justify your position.
7. In your view, are the Millennium Development Goals realistic? Attainable? Worth striving towards? Justify your position.

Suggested Readings

Visit the companion website for a list of suggested readings.

7 GEOGRAPHIES OF CULTURE AND LANDSCAPE

This and the following chapter focus on three closely related subdisciplines of human geography—historical, cultural, and social—that share concerns with cultural identity, the behaviour of humans as individuals and group members, and the landscapes and places that humans create. We begin this chapter with some provocative ideas about cultural identity in the contemporary world. We then look at cultural geographic regions at various scales, the idea of landscapes as places, and the making of cultural landscapes. We continue with substantial discussions of the two cultural variables that are perhaps most important with respect to both human identity and the human creation of landscapes as places with meaning: language and religion.

Here are three points to consider as you read this chapter.

- Do you agree that our greatest human achievement—our culture—has been responsible for erecting barriers and for sparking misunderstanding and even conflict between peoples?
- Are cultural regions, especially at the world scale, best seen as useful classifications rather than as insightful applications of a geographic method?
- Several different religions sometimes value, perhaps even lay claim to, the same places on the earth's surface. Why is this the case, and why does it matter?

A UNESCO World Heritage Site since 1985, Quebec City is the only city in North America that has preserved its colonial fortifications. Quebec is one of the world's many cultural landscapes, with its own language and traditions.

Boutique
Eclipse
DE CE CÔTE
Parmesan
Parmesan
VOITURIER
gratuit

A WORLD DIVIDED BY CULTURE?

> Humans have "brought into being mountains of hate, rivers of inflexible tradition, oceans of ignorance." (James, 1964: 2)

In 1964 the eminent American geographer Preston James published an introductory geography textbook with an inspired title: *One World Divided*. We have already recognized divisions related to aspects of population geography such as fertility and mortality and have identified, described, and attempted to explain the fundamental division of the world into more developed and less developed regions. Our concern now is the human divisions that derive from differences in culture. In fact, the barriers created by human variables are often far more difficult to cross than any physical barrier. The divisions we noted in our accounts of population and inequality are only the tip of the iceberg.

culture
A complex term that typically refers to the way of life of a society's members; also usefully understood as referring to our ability first to analyze and then to change the physical environments that we encounter.

Humans differ from all other forms of life in that they have developed not only biologically but culturally. Other forms, largely limited to biological adaptation, have become so highly specialized that most are restricted to particular physical environments; indeed, environmental change has resulted in species extinction in many cases. Humans have avoided such a fate primarily because of our culture, our ability first to analyze and then to change the physical environments that we encounter. Unlike other animals, humans can form ideas out of experiences and act on the basis of these ideas; we are capable of not only changing physical environments but also changing them in directions suggested by experience. This is just one meaning of the term *culture*.

Humans in Groups

The world we live in features tremendous cultural diversity, but it also, as discussed in Chapter 3, has increasing interaction between cultures. As a result, in many areas traditional groups are struggling to protect their established ways of life against "foreign" influences; in others, the frequently uncontrolled passions that surround language, religion, and ethnicity are erupting in cultural and sometimes political conflicts. Ironically, as James argues, our greatest human achievement—our culture—has been responsible for both erecting barriers and sparking conflict between peoples. We consider this idea in our discussions of language and religion in this chapter and of ethnicity in Chapter 8.

cultural regions
Areas in which there is a degree of homogeneity in cultural characteristics; areas with similar landscapes.

Why has culture tended to have the unfortunate consequences noted by James? A large part of the reason is that each cultural change, each new idea prompted by experience, brings with it new knowledge and responsibilities that, at least initially, do not fit easily into the cultural framework of existing attitudes and behaviours. The value that humans generally continue to place on tradition and the past means that our cultural frameworks rarely are conducive to the efficient use of change in culture. Once in place, differences in attitude and behaviour tend to be self-perpetuating and can be resolved only with the development of new, overarching values. Developing and establishing such new values is a challenging task, as it involves human engineering—literally changing ourselves in accord with these new values.

We can summarize this important introductory argument as follows:

1. Our world is divided, especially because of spatial variations in culture.
2. By culture, we mean the human ability to develop ideas from experiences and subsequently act on the basis of those ideas.
3. Once cultural attitudes and behaviours are in place, an inevitable tendency is for them to become the frame of reference—however inappropriate—within which all new developments in culture are placed and evaluated.
4. One task humans may choose to tackle is to create new sets of values—in effect, to re-engineer ourselves.

FORMAL CULTURAL REGIONS

This and the following section examine how the concept of cultural regions is used in contemporary human geography. We first focus on formally defined regions; as discussed in Chapter 2, such regions are identified because of the presence of some particular cultural characteristic(s).

Human geographers, we have seen, are not concerned primarily with humans or human cultures but with human impacts on landscape. Yet these impacts can vary considerably, depending on the specific characteristics of the human group in question—in other words, depending on culture. Because different cultures have emerged in different areas and because the earth is a diverse physical environment, there is a wide variety of human landscapes. In addition to describing and explaining these landscapes, human geographers have sorted them into cultural regions.

Delimiting formal cultural regions requires decisions on at least four basic points:

1. criteria for inclusion, that is, the defining characteristic(s)
2. date or time period (since these regions change over time)
3. spatial scale
4. boundary lines

These four issues are interrelated and difficult to resolve. As you read the following examples of geographers identifying and mapping regions, appreciate that the goal is to enhance understanding, not to arrive at some ideal scheme.

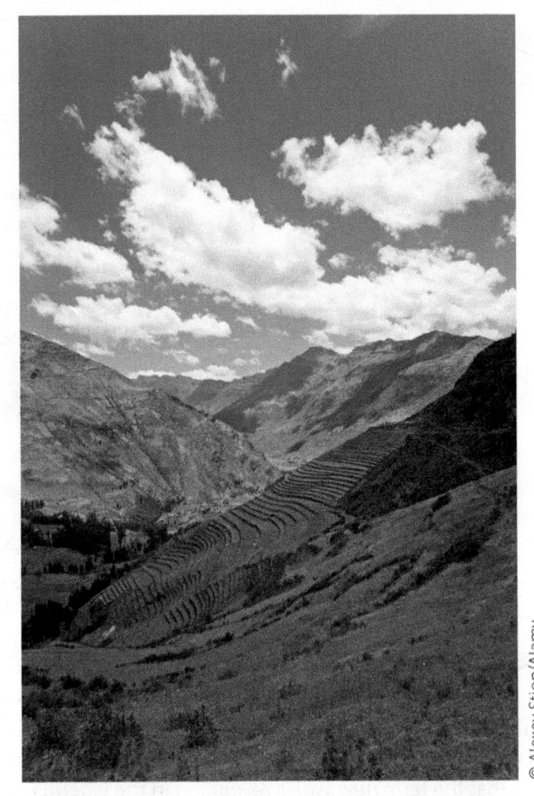

Irrigated terraces at the ancient Incan site of Pisac, Peru.

World Regions

On the world scale, many notable attempts have been made to delimit a meaningful set of cultural regions. The historian A.J. Toynbee (1935–61), for example, identified a total of 26 civilizations as responses to environment, past and present: 16 "abortive" and 10 surviving. He described three of the latter—the Polynesian, Nomad, and Inuit civilizations—as "arrested" because their overspecialized response to a difficult environment left them unable either to expand into different regions or to cope with environmental change. He identified the remaining seven living civilizations as Western Christendom, Orthodox Christendom, the Russian offshoot of Orthodox Christendom, Islamic culture, Hindu culture, Chinese culture, and the Japanese offshoot of Chinese culture. Toynbee considered most of Africa to be "primitive."

Although its limitations are clear, Toynbee's schema is instructive. His principal criterion obviously is religion, since he named most of his civilizations for their religions. But, in a broader sense, the most important feature of each is the manner in which it has responded to the environment—hence Toynbee's recognition of arrested and abortive civilizations.

The first geographers to attempt to delimit world regions were Russell and Kniffen (1951), who began by recognizing various cultural groups and then related them to areas so as to delimit seven cultural regions, each a product of a long evolution of human–land relations (Figure 7.1). With the addition of one "transitional" area, these regions effectively cover the world. Russell and Kniffen's regions are well justified, and their classification is a valuable contribution to our general understanding of the world. Nevertheless, it is important to emphasize that world regionalizations of this type do not represent the best use of the culture concept in geography. The central problem is spatial scale: the larger the area to be divided, the more likely it is that the regions identified will be either too numerous or too superficial to be helpful. Like regionalizations based on physical variables (see online Appendix 1), cultural regionalizations are more appropriately seen as useful classifications than as insightful applications of geographic methods. The difficulties involved in delimiting Europe as one world cultural region are outlined in Box 7.1.

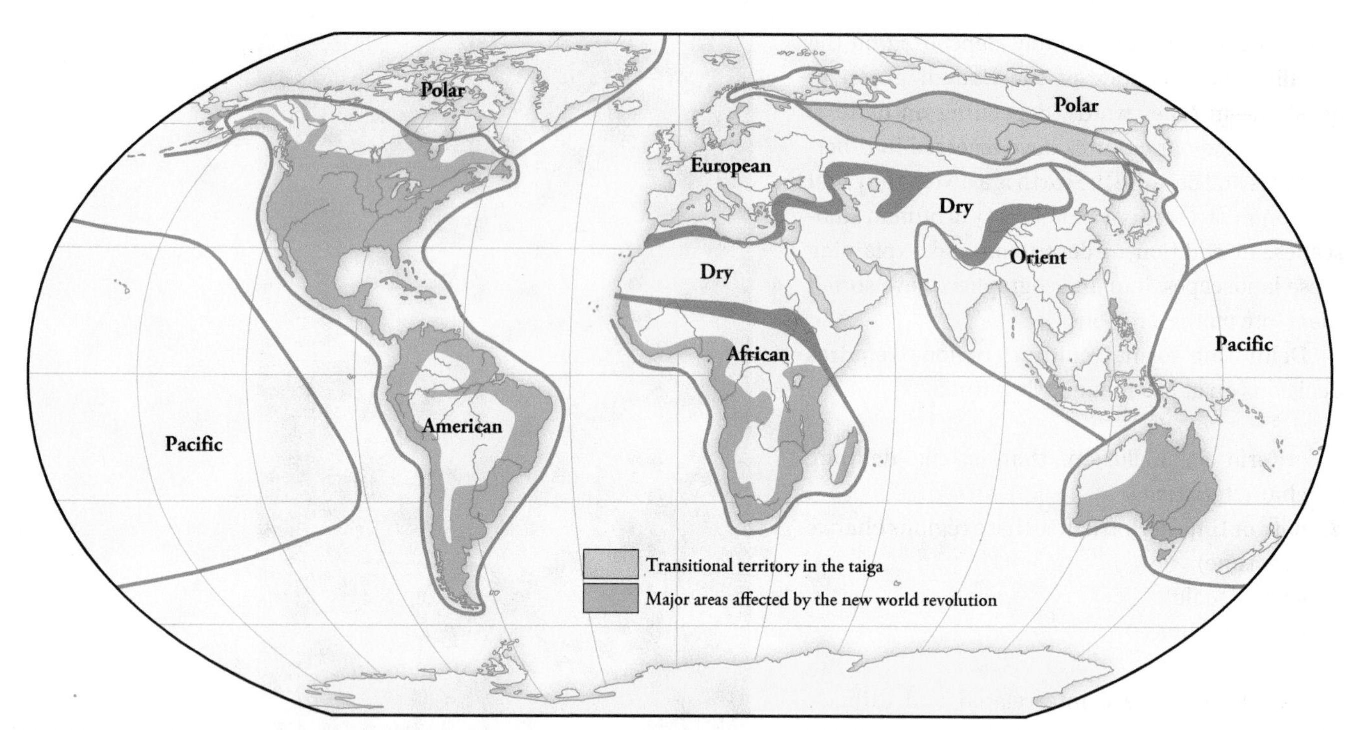

FIGURE 7.1 | Cultural regions of the world

The varying width of the lines between regions is one means of acknowledging that boundaries may be zones of transition rather than clear dividing lines. Note that the term *new world revolution* in the legend refers to the diffusion of European ideas and technologies with European expansion.

Source: Adapted from Russell, R. J., and F.B. Kniffen. 1951. *Culture Worlds*. New York: Macmillan.

first effective settlement
A concept based on the likely importance of the initial occupancy of an area in determining later landscapes.

Cultural regions and the distinctive landscapes associated with them are more effectively considered on a more detailed scale. If we take North America as one example of a world region, we find that further subdivision greatly enhances our appreciation of the relationship between culture and landscape.

North American Regions

North America is relatively easy to subdivide because the development of its contemporary cultural regions is recent enough that we are able to trace their origins. Figure 7.3 delimits 16 regions in North America on the basis of several criteria, such as physical and economic differences. Each region is likely to be generally recognizable, and the authors of the map attempt to convey the "feeling" of each region in their discussions. This regionalization is interesting because it focuses on broad regional themes rather than on specific criteria and recognizes that the border between the United States and Canada is not a fundamentally geographic division.

A Regionalization of the United States

Figure 7.4 is a regionalization of the United States explicitly based on the interesting concept of **first effective settlement**: "Whenever an empty territory undergoes settlement, or an earlier population is dislodged by invaders, the specific characteristics of the first group able to effect a viable, self-perpetuating society are of crucial significance for the later social and cultural geography of the area, no matter how tiny the initial band of settlers may have been" (Zelinsky, 1973: 13).

Accordingly, Zelinsky demarcates five regions—West, Middle West, South, Midland, and New England. Only New England has a single major source of culture (England); the others are further divided into various subregions, all of which have multiple sources of culture. The Midland is divided into two subregions; the South and the Middle West are divided into three; and the West is divided into nine. A number of sub-subregions are noted as well, along with three regions of uncertain status (Texas, peninsular Florida, and Oklahoma). As a cultural regionalization, this example is especially useful

Around the Globe

BOX 7.1 | Europe as a Cultural Region

Although clearly an area of considerable diversity, Europe generally is seen as a single world region—"a *culture* that occupies a *culture area*" (Jordan, 1988: 6). Before the sixteenth century, it was widely believed that the continent was physically separated from Asia. Even after this error had been corrected, the image of a separate continent was so powerful that a new divide was needed: the idea that the Caucasus, the Urals, and the Black Sea represented a meaningful boundary became generally accepted. The concept of a separate Europe, an area that was culturally distinctive, was allowed to remain intact.

Using the political map of the mid-1980s, T.G. Jordan defined Europe on the basis of 12 traits measured for each country (Figure 7.2):

1. The majority of the population speak an Indo-European language.
2. The majority of the population have a Christian heritage.
3. The majority of the population are Caucasian.
4. More than 95 per cent of the population are literate.
5. The infant mortality rate (IMR) is less than 15.
6. The rate of natural increase (RNI) is 0.5 per cent or less.
7. Per capita income is US$7,500 or more.
8. Seventy per cent or more of the population are urban.
9. Fifteen per cent or less are employed in agriculture and forestry.
10. There are 100 km (62 miles) of railway plus highway for 100 km² (39 square miles).
11. Two hundred or more kg (441 lbs) of fertilizer are applied annually per ha of cropland.
12. Free elections are permitted.

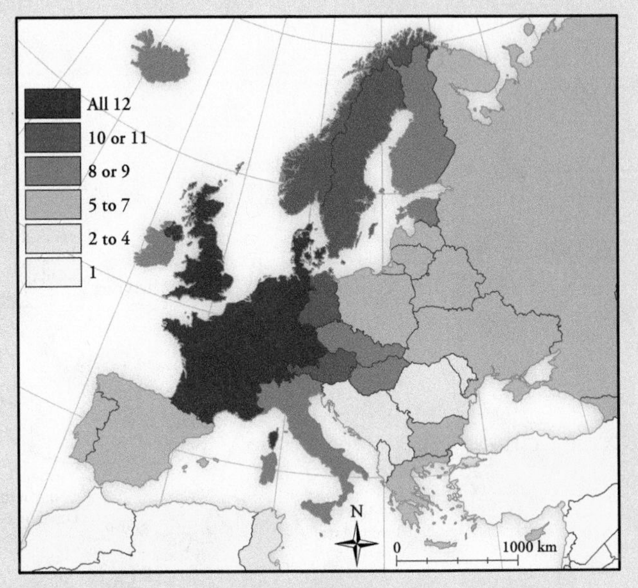

FIGURE 7.2 | Europe defined

Source: Based on Jordan, T. G. 1988. *The European Culture Area: A Systematic Geography*, 2nd edn. New York: Harper & Row, 14.

Figure 7.2 implies a regionalization; to proceed further would require a more explicit focus on variables such as language or religion. Jordan (1988: 402) devised a particularly innovative delimitation of regions within the larger European region, one that combines three distinctions: north/south, east/west, and core/periphery. Recognizing such divisions is really to admit that Europe cannot be defined in terms of a single culture, an idea that is supported by the fact that Europeans have a long history of fighting each other on the basis of religion or language.

because of the single clear variable employed. Yet the simplicity of the approach does not disguise the complexity of the cultural landscapes revealed by the many subregions.

A Regionalization of Canada

Canada is divided into six regions (Figure 7.5). This regionalization employs several variables, but a basic cultural division is evident. The North is delimited according to political boundaries and can be subdivided into a Northwest that is relatively forested and has a mixed Aboriginal and European population and an Arctic North that is treeless, is populated mostly by Inuit, and has limited resource potential by southern Canadian standards. The British Columbia region is characterized by major differences in physical and human geography and has a resource-based economy; the same is true of the Atlantic and Gulf region, which also has the problems of low incomes and high unemployment rates. The Interior Plains is an agricultural region (cattle, wheat, and mixed farming), and its dispersed farmsteads and often regularly spaced towns reflect government surveying and

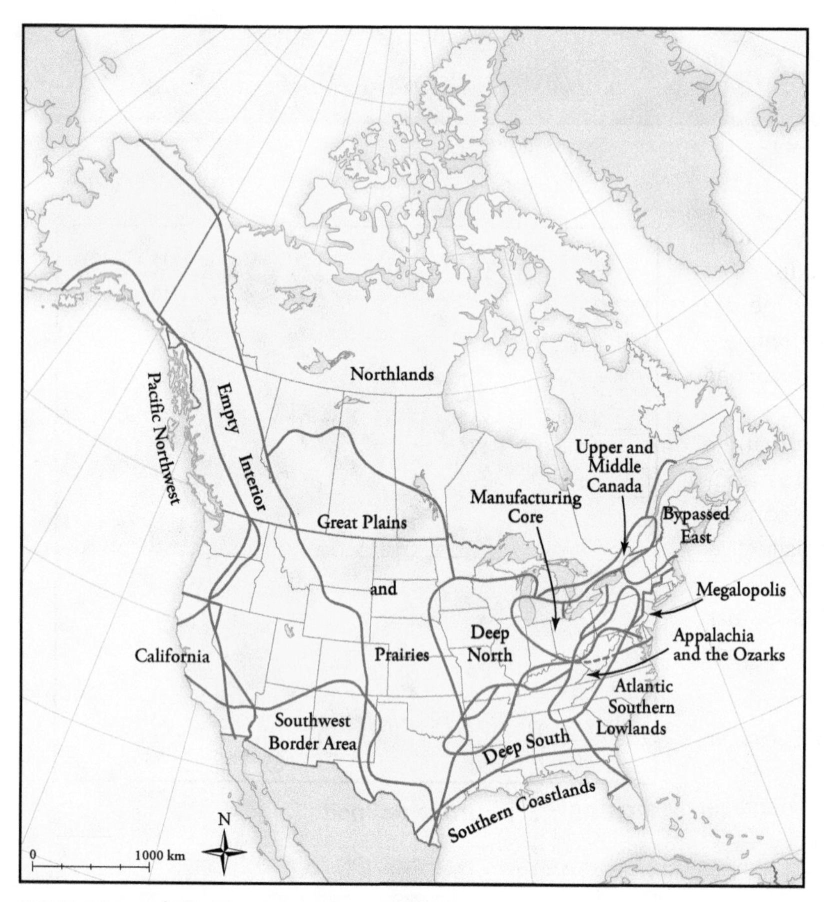

FIGURE 7.3 | Regions of North America

Overlapping boundaries are included to emphasize the uncertain status of areas that might be seen as belonging to more than one region. This is another way of acknowledging that many of the boundaries between regions are zones of transition rather than firm divisions (see also Figure 7.1).

Source: Adapted from Birdsall, S. S., and J. W. Florin. 1981. *Regional Landscapes of the United States and Canada*, 2nd edn. New York: Wiley, 18.

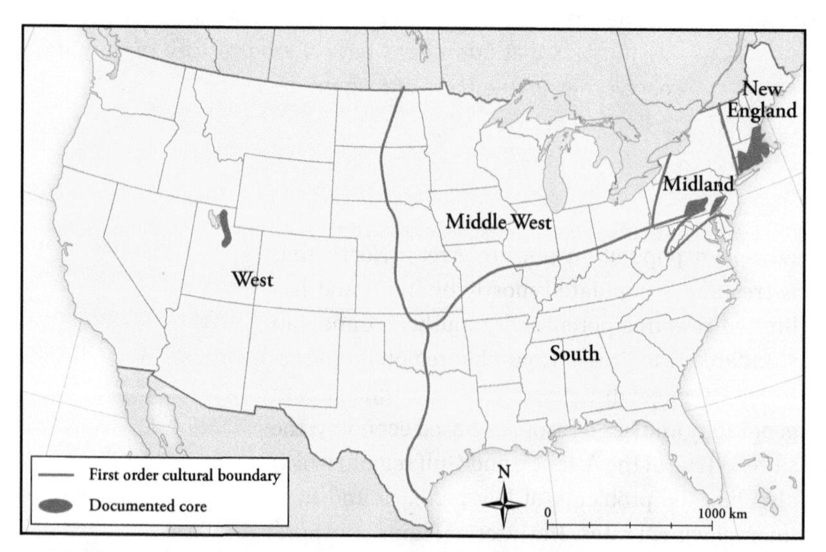

FIGURE 7.4 | Cultural regions of the United States

Source: Adapted from Zelinsky, W. 1973. *Cultural Geography of the United States*. Englewood Cliffs, NJ: Prentice-Hall, 118.

planning before settlement. The core region of Canada—the cultural and economic heartland—is the Great Lakes–St Lawrence Lowlands, with most of the country's people, the largest cities, major industry, and intensive agriculture. The largest region is the Canadian Shield, a sparsely inhabited area in which most of the settlements are resource towns.

These accounts of regions in the United States and Canada are enhanced by consideration of the term **homeland**. According to Nostrand and Estaville (1993: 1), a homeland has four basic ingredients: people, place, sense of place, and control of place. "Sense of place" refers to the emotional attachment that people have to a place, while "control of place" refers to the requirement of a sufficient population to allow a group to claim an area as their homeland. Most homelands are associated with groups defined on the basis of common language, religion, or ethnicity.

VERNACULAR CULTURAL REGIONS

Regions, and the landscapes that characterize them, are not simply locations; they are also places in the sense that they convey meaning. The meaning that a place has is a human meaning, dependent on social matters. There is an important circularity here. Humans, as members of groups, create places; in turn, each place created develops a character that affects human behaviour.

Our experience of place and the meaning we attach to it are not simply individual matters: they are intersubjective (shared). The most compelling example of the intersubjective character of place is the idea of home. Whatever the word specifically refers to—a dwelling, other people, the earth—for most people, a home is a shared place.

Further, as noted in Chapter 2, our experience of any place may be characterized either by topophilia, a positive attitude towards place, or by topophobia, a negative attitude towards place. These terms refer to the emotional attachment that exists between person and place, so that the feelings are related to certain characteristics of both the individual and the environment in question. Individual characteristics include personal well-being and familiarity with place. Environmental characteristics include the attractiveness of places as judged by individuals;

for example, a pastoral setting may be attractive, while an urban slum may be unattractive.

The cultural regions discussed so far are formal regions; areas with one or more cultural traits in common, they are the geographic expressions of culture. Such regions are essentially the product of the geographer who analyzes people and place and proceeds to delimit regions. Vernacular regions are rather different in that they may or may not be formally defined: what matters is that the region is perceived to exist by those living there and/or by people elsewhere. Such regions are most clearly characterized by a sense of place. Of course, a formal region may possess a distinct sense of place and thus may also be a vernacular region.

Vernacular regions are often viewed more positively by those living within the region than by those outside. For many living in the American Bible belt, the name is a source of great pride; yet for some outsiders it is a term of derision. In other cases, if the name of the vernacular region has been imposed for purposes of tourism or commercial promotion, with no real roots in the region, it may mean little to either residents or outsiders.

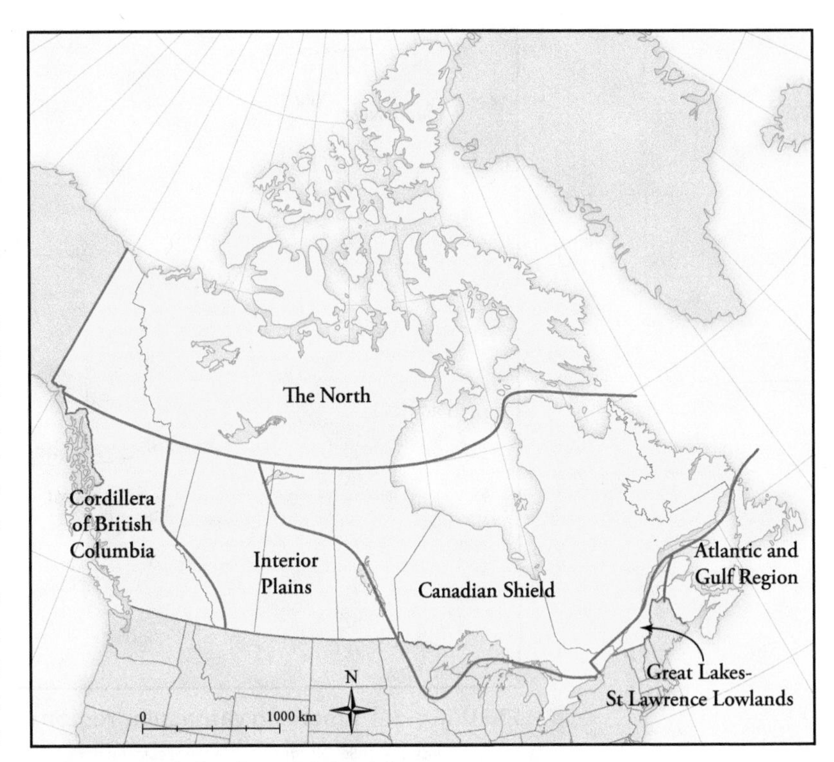

FIGURE 7.5 | Regions of Canada

Source: Adapted from Robinson, J. L. 1989. *Concepts and Themes in the Regional Geography of Canada*, rev. edn. Vancouver: Talon Books, 17.

Vernacular Regions in North America

North American geographers have delimited many vernacular regions, typically by collecting information on individual perceptions. Probably the most elaborate survey was conducted by Hale (1984); it gathered 6,800 responses from such people as local newspaper editors. In order to identify true vernacular regions, any survey has to ensure that the responses received in some way represent those of average people.

Zelinsky (1980) avoided the survey problem by studying the frequency of a specific regional term and a more general national term in metropolitan business usage; for example, if the region in question is the American South, the incidence of the term *Dixie* might be compared with that of *American*. Figure 7.6 shows Zelinsky's findings: 15 named vernacular regions, an area that lacks regional affiliation, and two areas that are included in either one or both of two regions (these are the areas between the West and Southwest and between the Northwest and West). It is instructive to compare this map to Figure 7.3 to see the extent to which the cultural region and the vernacular region are spatially related.

Newspaper journalist Joel Garreau made an insightful contribution to the geographic understanding of North America, first with a short article that proved very popular and then with his book, *The Nine Nations of North America*. Describing the inspiration for his work as "a kind of private craziness," he collected information from fellow journalists who spent much time on the road concerning their impressions of places (Garreau, 1981: ix). Gradually, it became clear that most political divisions lacked meaning and that North America was better understood as a collection of nine nations, places that are full of meaning but that do not appear on any map. Garreau produced a regionalization that is a useful way of thinking about North America. The nine nations are shown in Figure 7.7 and justified as follows (Garreau, 1981: 1):

> Forget the pious wisdom you've been handed about North America.
> Forget about the borders dividing the United States and Canada, and Mexico, those pale barriers so thoroughly porous to money, immigrants, and ideas.

homeland
A cultural region especially closely associated with a particular cultural group; the term usually suggests a strong emotional attachment to place.

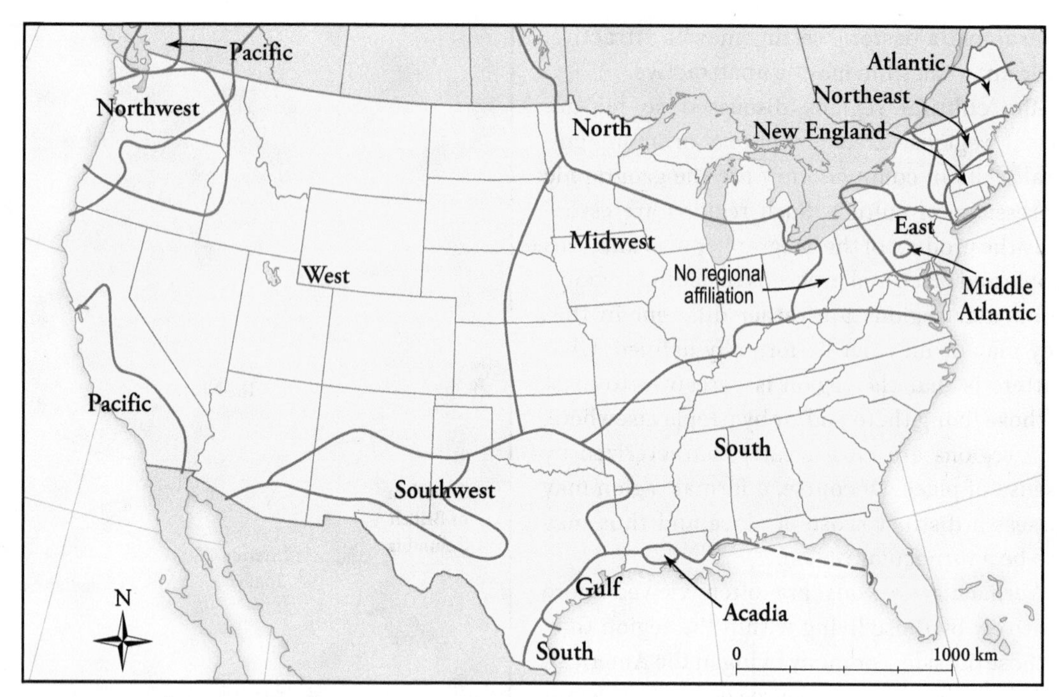

FIGURE 7.6 | American vernacular regions

Source: Zelinsky, W. 1980. "North America's Vernacular Regions." *Annals of the Association of American Geographers* 70: 14, Fig 9 (MW/RAAG/P1889).

> Forget the bilge you were taught in sixth-grade geography about East and West, North and South, faint echoes of glorious pasts that never really existed save in sanitized textbooks.

Having forgotten these and other ideas, Garreau invites the reader to consider how North America really works and notes that each nation is characterized by a particular way of looking at the world. Consider just two examples. First, it is not helpful to talk about Colorado because it is three different places, being part Bread-Basket, part Empty Quarter, and part MexAmerica. Second, San Francisco and Los Angeles are not best seen as two cities in California; they are the capitals of two of the nations, San Francisco of Ecotopia and Los Angeles of MexAmerica.

French Louisiana

The area of French Louisiana, the only surviving remnant of the vast French Mississippi Valley empire, is perceived as a distinctive vernacular region in which the population is associated with one group identity, namely Cajun. According to Trépanier (1991), however, this perception is not an accurate reflection of cultural reality. Four important French subcultures inhabit the area: white Creoles, black Creoles, French-speaking Indians, and the descendants of the Acadians who were expelled from Nova Scotia in 1755, now known as Cajuns. Furthermore, until recently, outsiders' perceptions of the Cajun group identity were largely negative, and only since the late 1960s has a process of image modification, which Trépanier (1991: 161) called "beautification," been underway. The Creole identity, on the other hand, carried a positive image for both black and white French speakers. How can we explain this anomaly?

The decision to identify the area with the Cajun group was made by the state government. In 1968 the Louisiana legislature created the Council for the Development of French in Louisiana (CODOFIL) to gain political benefits by cultivating a French image, but this did not garner any real popular support. Once created, CODOFIL was determined to unify French Louisiana using the then-negative Cajun label rather than the more positive Creole label. Trépanier (1991: 164) interpreted this effort as a way to guarantee a white identity for the French-speaking area. Once this decision was made, it became necessary to improve the popular image of the Cajun identity, which was achieved through publicity campaigns, including the organization of Cajun festivals. It helped enormously that a new governor who identified with the Cajun group was elected.

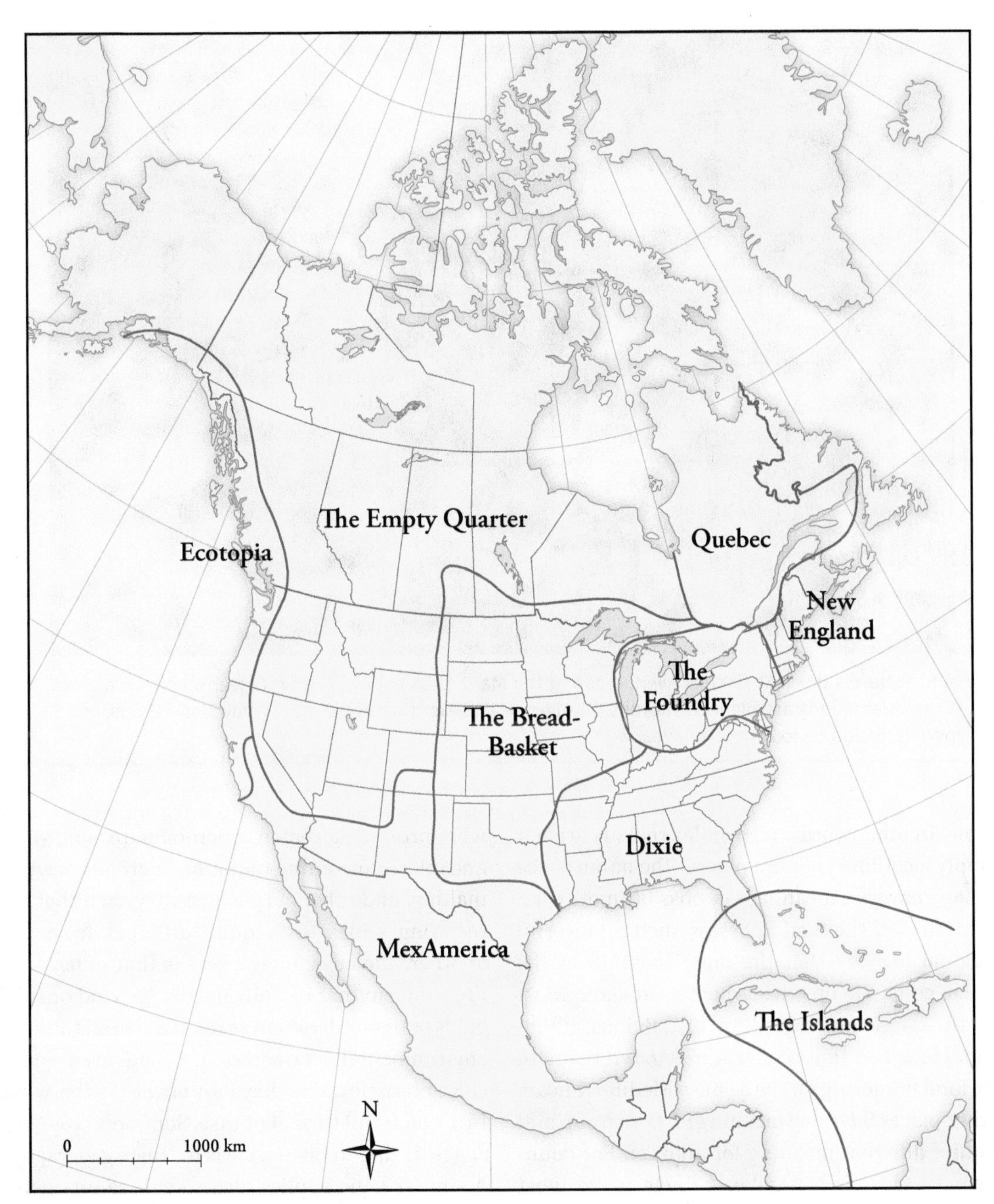

FIGURE 7.7 | The nine nations of North America: A journalist's perception
Geographers often refer to this thoughtful map when discussing vernacular or perceived regions. It is instructive to compare this map not only to Figure 7.6 but also to the divisions of North America suggested in Figure 7.3.

Source: Adapted from Garreau, J. 1981. *The Nine Nations of North America*. Boston: Houghton Mifflin, following 204.

For many, French Louisiana exists as a vernacular region—Cajun Country—with a Cajun identity that disguises the variety of French-speaking people and the fact that both the black Creoles and the French-speaking Indians reject the Cajun identity. As this example vividly illustrates, vernacular regions are much more than areas perceived to possess regional characteristics. They are also regions to which precise meanings and values are attached—they have social and symbolic identity.

Regional Identity

If an area has an identity specific enough to be named, as with Cajun Country, this suggests that the identity is meaningful to those using the

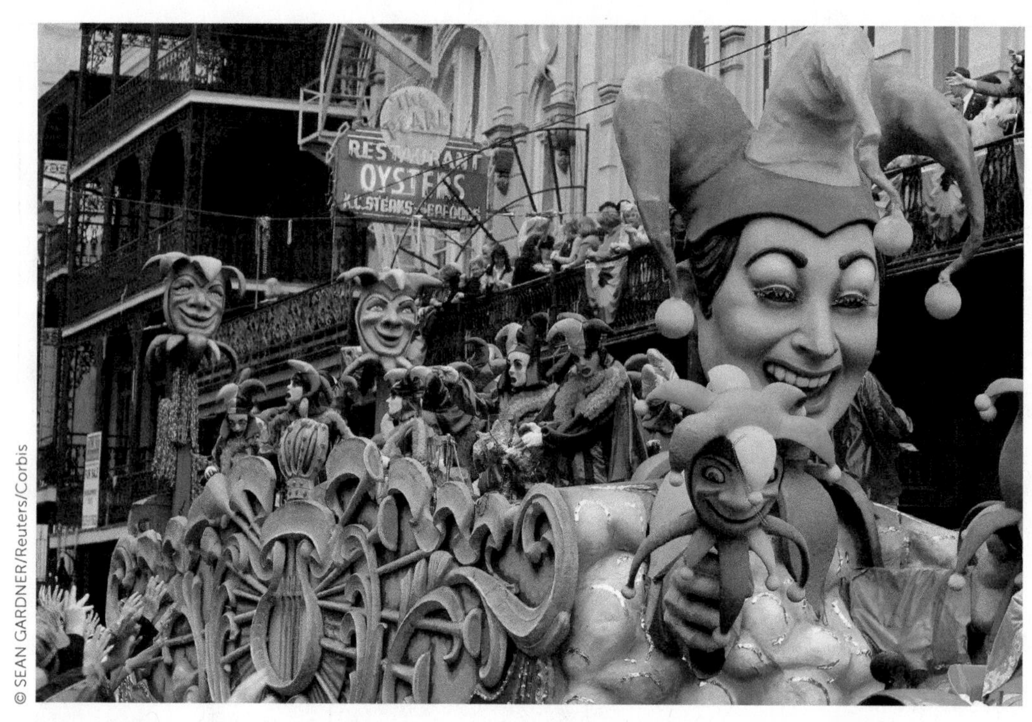

© SEAN GARDNER/Reuters/Corbis

A float travels down St Charles Avenue during the Mardi Gras parade in New Orleans, Louisiana. Mardi Gras celebrations are planned months in advance and last for several days; they draw thousands of people, including tourists, every year.

name. In other words, vernacular regions are not simply locations: they are places. The name of the region conveys a meaning—or possibly more than one. Parts of the world possess such a powerful regional identity that the mere mention of the name conjures up vivid mental images. Most North Americans, regardless of religious affiliation, recognize that the term *Holy Land* refers to the land bordering on the eastern Mediterranean. Other places have one meaning for one group and a quite different meaning for another. For country-music lovers, Nashville is likely to be most meaningful as the home of the Grand Ole Opry; for others it may be simply the capital of Tennessee. Similarly with literature, where vernacular regions associated with particular novelists or poets are created and then marketed to interested tourists, as in the well-known British examples of Wordsworth country (the Lake District), Brontë country (the Yorkshire moors), and D.H. Lawrence country (rural Nottinghamshire). In such cases, place blends with identity.

An extremely vivid example of the links between regional identity and sense of place is wine-producing regions. More so than most crops, wine grapes and the wines produced from them vary from place to place. Even within a wine-producing region, microclimate, soil, slope and related incoming radiation, methods of wine-making, and other details can vary such that the resulting wines taste quite different from one another. The key concept here is that of *terroir*, a French term that literally means "ground or soil," but is better understood as referring to all the local environmental characteristics, and even social characteristics, that have an effect on the wine. In a wonderful turn of phrase, Sommers (2008: 19) suggests that to taste wine is to "taste geography." A similar logic applies also to cheese-producing regions, especially those in Europe but also in parts of southern Ontario and Quebec.

THE MAKING OF CULTURAL LANDSCAPES

So far, our discussion of cultural regions has been descriptive rather than explanatory. We will now consider in more detail how regions of distinctive cultural landscape arise, beginning with a look at cultural variations over space and the ability of culture to affect landscape.

Although all cultures share certain basic similarities—all need to obtain food and shelter and to reproduce—they differ in the methods used to achieve those goals. As humans settled the earth, culture initially evolved in close association with the physical environment. Over time, however, cultural adaptations brought increasing freedom from environmental constraints. Gradually, as humans' ties to the physical environment loosened, their ties to their culture increased. One of the most important changes in human history was the shift (associated with the Industrial Revolution) to a capitalist mode of production, which largely destroyed our ties to the physical environment but created a new set to culture. Some human geographers see this change as part of an ongoing process of diffusion that is effectively homogenizing world culture by minimizing regional variations; this view is derived most clearly from Marx (see Box 2.3).

Focus on Geographers

Family Geographies | BONNIE C. HALLMAN

What attracts us to certain places/landscapes for family leisure activities? How do we "read" such places and what do we (desire to) experience in them? Spaces of family leisure (e.g. zoos, museums, theme parks) are landscapes physically and socially constructed as destinations for quality family time spent engaged in culturally valued behaviours thought to build family cohesiveness and connection (Hallman and Benbow, 2007).

Family photography is a tool for documenting the high points of family life and developing a narrative of remembrance. Photographic analysis informs understanding of the experience of family leisure landscapes and how they are used to facilitate a place-centred narrative of family life. Photos—tangible memories—help us recall times, places, people, and activities. The desire to remember, to create a family story, is the primary motivation for taking and sharing family photographs, keeping scrapbooks, and creating photo displays. Planning and engaging in family leisure activities where family photography is a primary activity are integral to the narrative of remembrance. Looking at and sharing our family photographs both creates and enacts our place-grounded family narratives, reinforcing specific understandings of family relationships. A sense of well-being and fulfillment may be gained via the stories of what we did, which are triggered by family photos. A family "collectivity" evolves from the shared events or experiences, contextualized via the people and places depicted in the shared photos, and through which we feel interconnected spatially (van Dijck, 2007).

An illustrative example of these ideas can be found in Disney's "Let the Memories Begin" media campaign and park theming, begun in 2011. Incorporating online photo and story sharing, park décor, and "The Memories, The Magic and You" evening shows, the campaign reflects an understanding that families want to deeply and personally connect to their holiday destination. This is not only through what they take away (e.g. personal family photos, videos, and souvenirs) but also by leaving a trace of their family story at the website, a communal scrapbook that mixes social media with photo and video sharing. Through adding text and images, Disney Park guests add their memories, contributing to a shared dialogue centred on the idea that "behind every memory is an unforgettable story" (memories.disneyparks.disney.go.com/).

Connecting with others who share their interests, family members cement their emotional connection to place through participation in cultural exchange with not quite strangers who also value and wish to share their Disney experiences. Through the site, individual family expressions of their experiences of place are articulated as part of a larger shared cultural commonality. The individual family story is now part of a larger Disney narrative celebrating a particular understanding of family togetherness and well-being—a powerful, culturally embedded narrative on what constitutes family happiness and fulfillment. However, private memories are on public display—the visitors who post to the website gain a sense of being connected to and sharing with others, while Disney's commodification of family togetherness attracts more of their target audience.

Another geographic expression of private family narratives embedded in the family leisure landscape is the daily "The Memories, The Magic and You" show at both Disneyland and Walt Disney World. In a mosaic of lighting, animation, music, and daily guest photos (collected via Disney Photo Pass), images of families and children are interwoven with classic Disney stories and music. Family photographs—the narratives of remembrance—are literally embedded into the parks themselves, however temporarily, projected onto iconic Disney landmarks—the "It's a Small World" facade in Disneyland and Cinderella's Castle in Walt Disney World.

The campaign evokes the desire of families to be immersed in the values associated with these sites of family leisure. Such connection reinforces desired and enacted self-identities as parents, children, and family members. Families become quite literally embedded in the Disney story of happy and cohesive family life. Disney Parks are places evoking, for guests, desired sets of values and practices associated with meaningful and positive family interaction. Family members can literally "see" themselves as "good families" in a "family place," with place-based narratives reinforcing their desired understanding of themselves and their most intimate relationships. Family photography is an integral practice for telling the story of good families spending quality time together "making memories" that can then be packaged, shared, told, and retold, further reinforcing a certain type of family life narrative.

BONNIE C. HALLMAN is an associate professor in the Department of Environment and Geography at the University of Manitoba.

REFERENCES

Hallman, B. C., ed. 2010. *Family Geographies: The Spatiality of Families and Family Life*. Toronto: Oxford University Press.

——, and S. Mary P. Benbow. 2007. "Family Leisure, Family Photography and Zoos: Exploring the Emotional Geographies of Families." *Social & Cultural Geography* 8 (6): 871–88.

van Dijck, J. 2007. *Mediated Memories in the Digital Age*. Stanford, CA: Stanford University Press.

Cultural Adaptation

As humans settled the earth, different cultures evolved in different locations. Despite their many underlying similarities, each developed its own variations on the basic elements of culture: language, religion, political system, kinship ties, and economic organization. As we seek to understand how different cultures create different landscapes, we must also ask how sound relationships develop between humans and their physical environment. This question is central to human geography (witness environmental determinism, possibilism, and related viewpoints). One answer centres on the concept of **cultural adaptation**. Humans are adapting continuously to the environment, genetically, physiologically, and culturally. Cultural adaptation may take place at an individual or group level—human geographers are interested in both—and consists of changes in technology, organization, or ideology of a group of people in response to problems, whether physical or human. Such changes can help us to deal with problems in any number of ways, such as by permitting the development of new solutions or by improving the effectiveness of old ones and by increasing general awareness of a problem or by enhancing overall adaptability. In other words, cultural adaptation is an essential process by which sound relationships evolve between humans and land.

cultural adaptation
Changes in technology, organization, and ideology that permit sound relationships to develop between humans and their physical environment.

We noted earlier the idea that culture represents the human ability to develop ideas from experiences and subsequently to act on the basis of those ideas. Cultural adaptation thus includes changes in attitudes as well as behaviour. One current example of attitudinal change is our increasing awareness of the importance of environmentally sensitive land-use practices; an example of behavioural change is the implementation of such practices. Needless to say, behavioural changes do not always accompany attitudinal changes—or even follow them.

Human geographers have proposed a number of concepts to explain more precisely what cultural adaptation is and how it works. Among them are the following three ideas.

First, in their analysis of the evolution of agricultural regions, Spencer and Horvath (1963: 81) suggested that such regions could be seen as the "landscape expression of . . . the totality of the beliefs of the farmers over a region regarding the most suitable use of land." According to this view, cultures are particular beliefs, psychological mindsets, that result in a culturally habituated predisposition towards a certain activity and hence a certain cultural landscape; for example, the corn landscape of the early European-settled American Midwest reflected the commercial aims of incoming settlers.

Second, according to another view, a cultural group moving into a new area may be pre-adapted for that area: conditions in the source area are such that any necessary adjustments have already been made prior to the move or are relatively easy to make thereafter. Following this logic, Jordan and Kaups (1989) proposed that the American backwoods culture had significant Northern European roots, specifically in Sweden and Finland.

Third, and perhaps most useful, is the core, domain, and sphere model (Meinig, 1965). According to this view, a cultural region or landscape can be divided into three areas—core (the hearth area of the culture), domain (the area where the culture is dominant), and sphere (the outer fringe)—and cultural identity decreases with increasing distance from the core. Figure 7.8 shows what most observers would agree is one of the easiest regions to delimit: the Mormon region of the United States (Box 7.2). But Meinig has successfully extended these ideas, in modified form, to other regions such as Texas and the American Southwest. Few attempts have been made to apply these ideas outside North America.

Each of the three views is, however indirectly, very much in the Sauer landscape school tradition (see the online content for this chapter). There is a consistent concern with the landscapes created by cultural groups and particularly with the material manifestations of culture: the visible landscape.

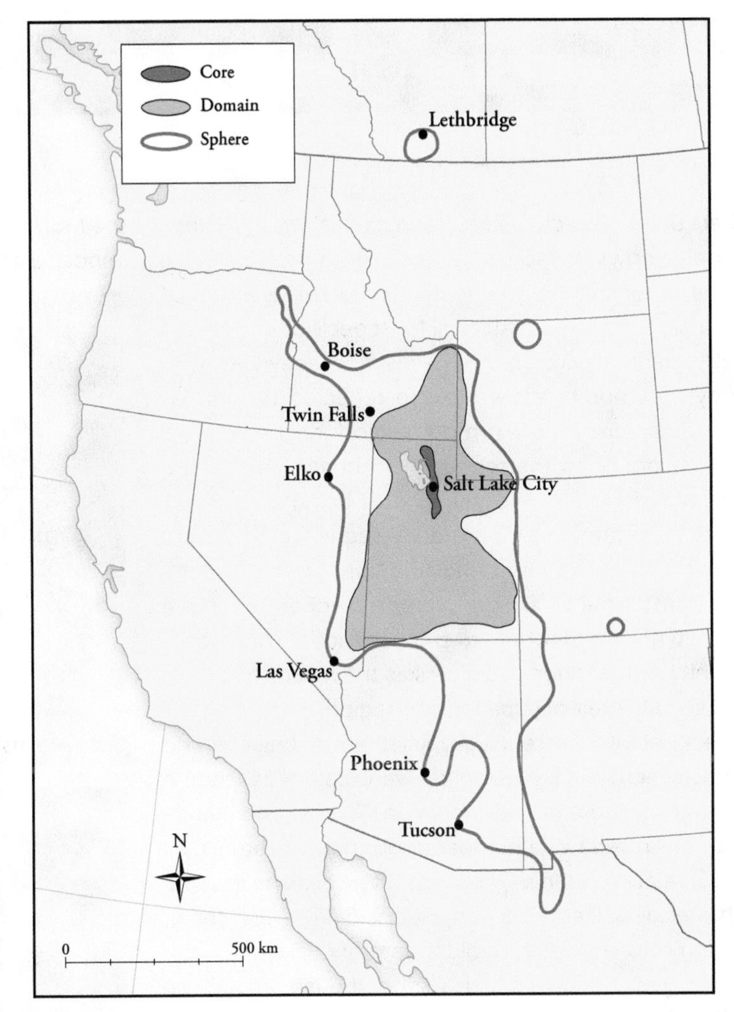

FIGURE 7.8 | Core, domain, and sphere

Source: Adapted from Meinig, D. W. 1965. "The Mormon Culture Region: Strategies and Patterns in the Geography of the American West, 1847–1964." *Annals, Association of American Geographers* 55: 214, Fig 7 (MW/RAAG/P1887).

Cultural Diffusion

As discussed, regions comprising areas of relatively uniform landscapes may evolve through a process of adaptation. But it is also evident that many landscape changes are related to the diffusion of new ideas and practices from elsewhere. Diffusion is best interpreted as the process of spread in geographic space and of growth through time. Migration can be regarded as a form of diffusion, although the term is more often used in the context of some particular innovation, such as a new agricultural technique. Diffusion research has a rich heritage in geography and has been associated with three approaches: cultural geography, spatial analysis, and political economy.

A Traditional Cultural Geographic Emphasis

Until the 1960s, most diffusion research took place under the general rubric of historical and cultural geography in efforts to understand cultural origins, cultural regions, and cultural landscapes and was specifically associated with the landscape school initiated by Sauer. Typical studies focused on the diffusion of certain material landscape features, such as housing types, agricultural fairs, covered bridges, place names, or grid-pattern towns. The usual approach was to identify an origin and then describe and map diffusion outwards from it; the work of Kniffen (1951) is a good example.

Various issues emerged from this research. In many cases, debate centred on the question of single or multiple invention, the numbers and locations of hearth areas, or the problem of determining source areas for agricultural origin and subsequent diffusion. The importance of ethnic or social characteristics in relation to a

Around the Globe

BOX 7.2 | The Mormon Landscape

Members of the Church of Jesus Christ of Latter-day Saints, commonly known as Mormons, first arrived near Salt Lake in 1847 and subsequently settled in an extensive area that included part of southern Alberta. Most decisions concerning settlement expansion and landscape activities were made by the church leadership and reflected the fundamental character of the Mormon religion: decisions were made in response to the requests of church leaders and as one part of the larger experience of being a Mormon. Thus, settlement followed an orderly sequence, in accord with larger church concerns and ambitions, and landscape change demonstrated a high degree of uniformity. More than most groups, Mormons emphasized co-operation, community, and economic success as the means to establish and solidify their occupation of a region.

Settlement sites were usually selected by leaders, and the settlers either volunteered or were called by church leaders to move to each new location. The Mormon landscape that evolved has a number of distinctive features, all intimately related to the religious identity. Speaking in 1874, church president George Smith observed: "The first thing, in locating a town, was to build a dam and make a water ditch; the next thing to build a schoolhouse, and these schoolhouses generally answered the purpose of meeting houses. You may pass through all the settlements from north to south, and you will find the history of them to be just about the same" (*Church News*, 1979).

Francaviglia (1978) describes the Mormon landscape in detail, based on extensive travels within the region. Mormon towns were laid out in approximate accord with the City of Zion plan as detailed in 1833 by the church's founder, Joseph Smith: a regular grid pattern, with square blocks, wide streets, half-acre lots that included backyard gardens, houses of brick or stone construction, and central areas for church and educational buildings. Towns laid out along these lines continue both to reflect and to enhance the Mormons' community focus and represent a distinctive town landscape within the larger region of the mountain West.

The rural landscape of the Mormons is also distinctive, emphasizing arable agriculture rather than pastoral activities (which necessarily involve greater movement and lower population densities). Among the notable features of the farm landscape are networks of irrigation ditches, Lombardy poplars, unpainted fences and barns, and hay derricks. Together, these features serve both to distinguish the Mormon landscape from the surrounding areas and to impose a sense of unity on it.

William Norton

The wide main street in Manassa, Colorado.

William Norton

The central temple in Cardston, Alberta.

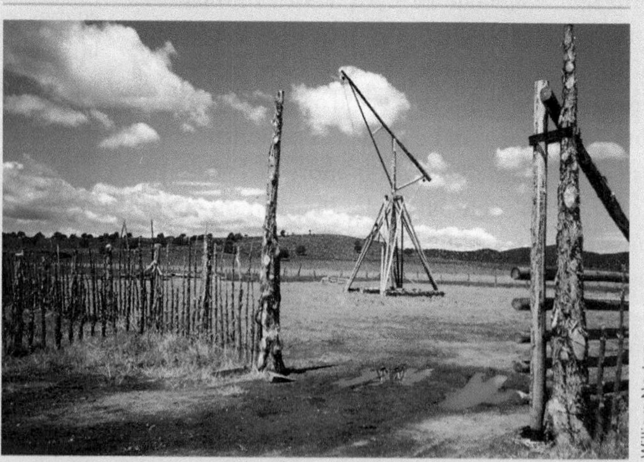
William Norton

A hay derrick at Cove Fort, Utah.

group's receptivity to innovations was frequently acknowledged. For example, a detailed study of cigar tobacco production in the United States that focused on the significance of ethnicity—"In each of the tobacco producing districts, tobacco culture came to be identified with hard work, clever farming techniques, economic independence, and with an ethnic group" (Raitz, 1973: 305)—found, in Wisconsin, a very close relationship between people of Norwegian descent and tobacco production.

A Spatial Analytic Emphasis

The character of diffusion research changed substantially following the work of the innovative cultural geographer Hägerstrand (1951, 1967), who pioneered the use of models and statistical procedures. This shift was associated in North America with the rise of spatial analysis as a major approach in human geography after about 1955 (as discussed in Chapters 1 and 2). Although the exact interests of these researchers were quite different from those of cultural geographers working in the Sauerian landscape school tradition, the central concern was unchanged: to study diffusion as a process effecting change in human landscapes. The importance of chance factors was explicitly acknowledged by Hägerstrand in his use of a procedure known as Monte Carlo **simulation**, which allowed for the likelihood of any given acceptance of an innovation to be interpreted as a probability. Spatial analysis–oriented diffusion research also introduced a number of themes that we might call empirical regularities because they are consistently observable in geographic analyses. Four of these empirical regularities are outlined below.

An innovation such as a new farming technology will likely first be adopted close to its source and later at greater and greater distances. This is the neighbourhood effect—a term that describes the situation where diffusion is distance biased. In general, the probability of new adoptions is higher for those who live near the existing adopters than it is for those who live farther away. The simplest description of this situation is a diagram showing a series of concentric circles of decreasing intensity with increasing distance—similar to the ripple effect produced by throwing a pebble into a lake.

The neighbourhood effect occurs in those circumstances where an individual's behaviour is strongly conditioned by the local social environment. It is most common in small rural communities with limited mass media communication, where the most important influences on behaviour are personal relationships—in traditional or ***Gemeinschaft*** societies (see Chapter 11). The neighbourhood effect is likely to be least evident in urban settings and in circumstances of improving transport and communication technologies that reduce distance friction.

The hierarchical effect is evident when larger centres adopt first and subsequent diffusion spreads spatially and vertically down the urban hierarchy. Thus, the innovation jumps from town to town in a selective fashion rather than spreading in the wave-like manner associated with the neighbourhood effect. The hierarchical effect is likely to occur in circumstances where the receptive population is urban rather than rural, and it is increasingly evident as transport technology improves. Box 7.3 describes a compelling example of the circumstances associated with both effects.

Reception of an innovation does not guarantee acceptance. Resistance to innovation is a sociological phenomenon. The greater the resistance, the longer the time before adoption occurs. Resistance also shows a spatial pattern: urban dwellers typically are less conservative than their rural counterparts.

Probably the best-supported empirical regularity is the S-shaped curve. When the cumulative percentage of adopters of an innovation is plotted against time, this curve is the typical result (see Figure 2.12). It describes a process that begins gradually and then picks up pace, only to slow down again in the final stages.

A Political Economy Emphasis

For many researchers, the emphasis outlined in the preceding section represents an example of the dehumanizing approach that is characteristic of the spatial analysis school. After about 1970, a third approach to diffusion research was developed, focusing not on the mechanics of the diffusion process but on the human and landscape consequences of diffusion. Although there are various aspects to this approach, it is most frequently associated with Marxist perspectives.

Any diffusion process adds to existing technological capacity, but it also affects the use of resources in some way. For example, some innovations result in significant time savings and hence change individuals' daily time budgets; others actually demand more time, as in the case of the

Gemeinschaft
A term introduced by Tönnies; a form of human association based on loyalty, informality, and personal contact; assumed to be characteristic of traditional village communities.

simulation
Representation of a real-world process in an abstract form for purposes of experimentation.

Around the Globe

BOX 7.3 | Cholera Diffusion

In the nineteenth century, North America experienced three major cholera pandemics: in 1832, 1848, and 1866. Cholera is generally spread by water contaminated with human feces, but it can also be carried by flies and food. Climatically, its diffusion is promoted by warm, dry weather.

Each pandemic diffused differently as the urban and communication systems changed. The 1832 outbreak spread in a neighbourhood fashion, the 1866 outbreak in a hierarchical fashion. The 1848 outbreak demonstrated aspects of both types (Pyle, 1969). Figure 7.9 maps the spread and growth of the disease for the two earlier pandemics.

The first pandemic occurred at a time when both urbanization and the communication system were limited. Urban centres were few and small, and movement was largely by water—coastal, river, and canal. Data on cholera spread show a neighbourhood process. After initial outbreaks near Montreal and New York City, the disease moved along distance-biased,

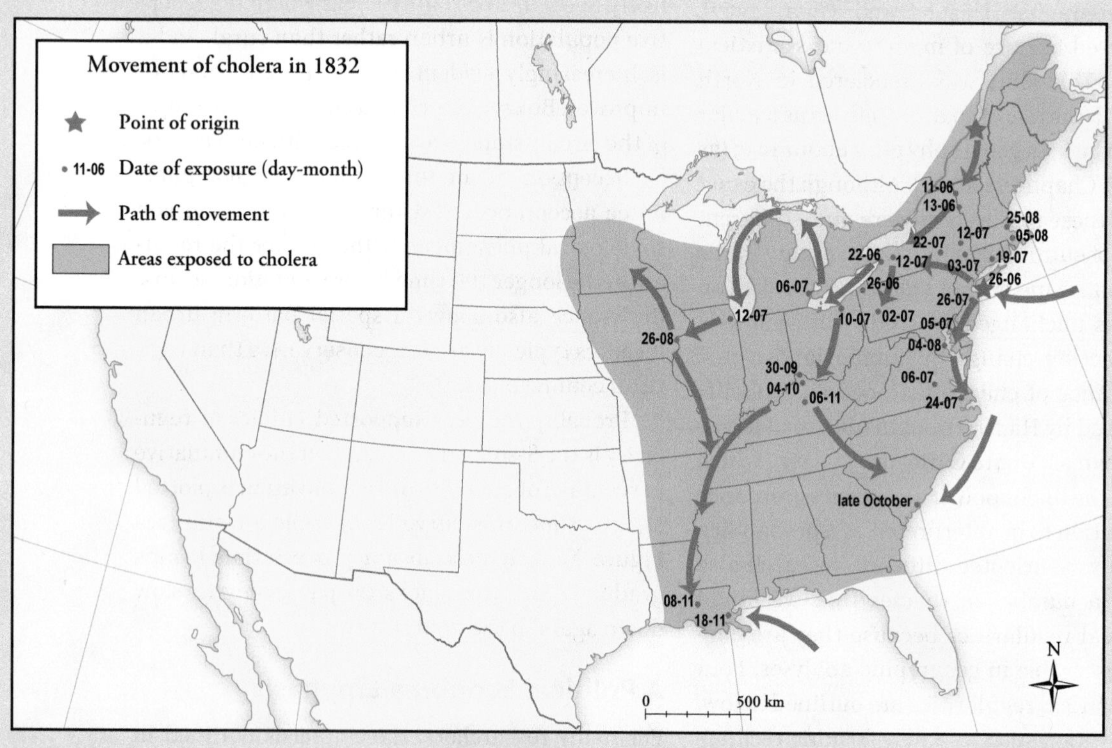

FIGURE 7.9 | Diffusion of cholera, North America, 1832 and 1866

Source: Pyle, G. 1969. "The Diffusion of Cholera in the United States in the Nineteenth Century." *Geographical Analysis* 1: 63, 72. Published by Blackwell Publishing Ltd.

introduction of formal schooling. Analyzing such innovations means analyzing cultural change.

Much current research into innovation diffusion also considers the extent to which there may be a spatial pattern of innovativeness related to a wide range of social variables such as relative wealth, level of education, gender, age, employment status, and physical ability. These variables go far beyond the broad spatial variables we have noted, such as rural–urban or ethnic differences. Not surprisingly, such research also tends to consider overarching social, economic, and political conditions over which most individuals exercise little or no control.

In a study of the diffusion of agricultural innovations in Kenya since World War II, Freeman (1985) discovered that the diffusion process could be greatly affected by the pre-emption of valuable innovations by early adopters. The difference between the two curves shown in Figure 7.10

not city-size-biased, paths. The second pandemic occurred in the context of a very different urban and communication system: eastern North America was by then well served by rail, and an integrated urban system was developing. These changing circumstances allowed cholera to diffuse hierarchically. From New York City, it travelled along major rail routes to distant second-order centres, as well as along a number of other minor paths. In many cases, cholera took much longer to reach areas close to the urban areas where it started than it did to reach more distant large cities.

This example is a useful reminder of the importance of economic and other infrastructures to any diffusion process. In general, a neighbourhood effect is more likely in circumstances of limited technology and on a local scale, while a hierarchical effect is more likely in a developed technological context and in a larger area.

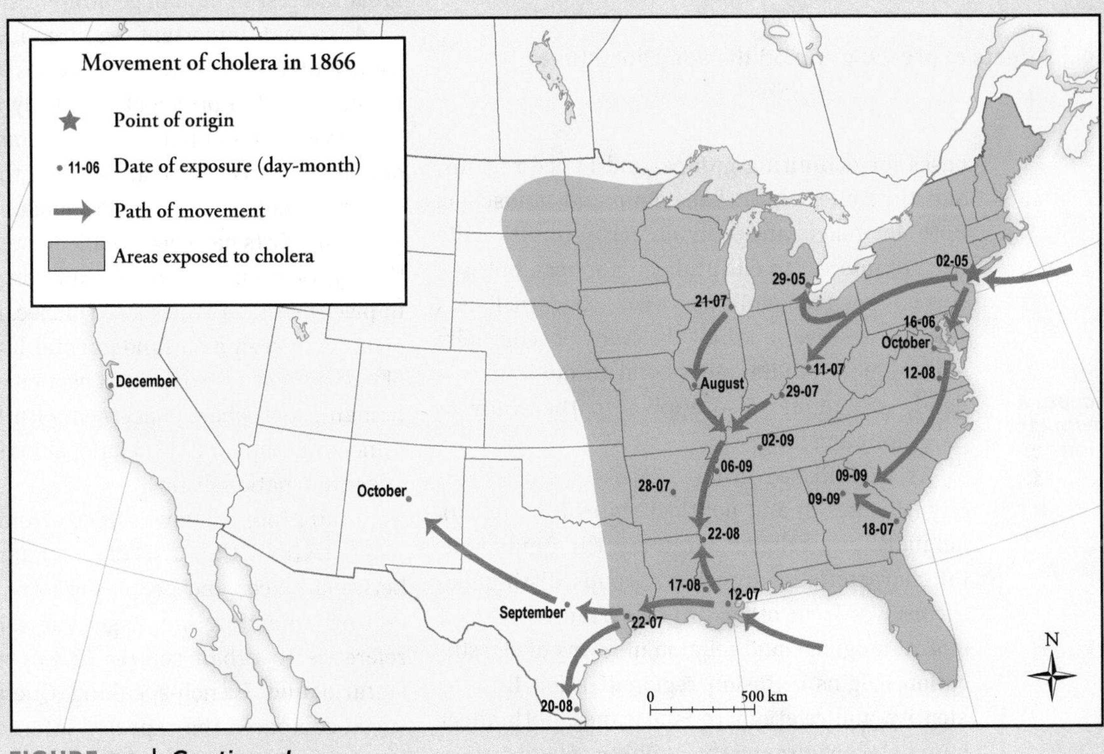

FIGURE 7.9 | *Continued*

reflects the influence of an entrenched elite who, as early adopters, saw advantages in limiting adoption by others. From their positions of social authority, members of the elite were able to take political action, lobbying for legislation to prevent further spread of an innovation or limiting access to essential agricultural processing facilities. This pattern was observed in Kenya for three agricultural innovations (coffee, pyrethrum, and processed dairy products). This example makes it clear that the process by which an innovation spreads can be just as important as the innovation itself.

LANGUAGE

Two aspects of culture—language and religion—are particularly important in understanding people and their landscapes. Not only are they important in themselves, but they are also good

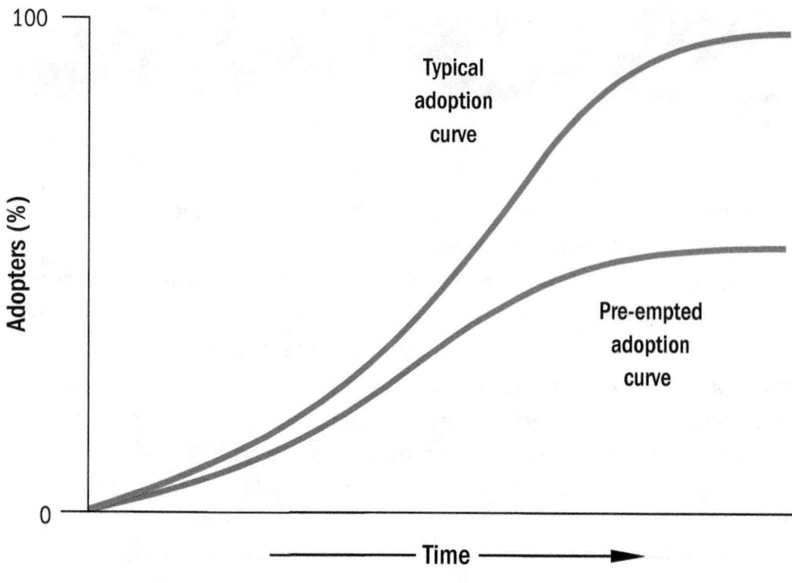

FIGURE 7.10 | Effects of pre-emption on the adoption curve

bases for delimiting groups, and hence regions, and they affect both behaviour and landscape. Both language and religion have traditionally been of interest to cultural geographers, but neither has been recognized as a core concern. Park's (1994: 1) observation that "the study of geography and religion remains peripheral to modern academic geography" also applies to the study of geography and language.

A Conceptual Framework for Diffusion

As we will see in our discussions of ethnicity (Chapter 8) and political states (Chapter 9), language and religion often are the fundamental characteristics by which groups distinguish themselves from others. In this chapter, we will look at language and religion in terms of classification, origins, diffusion, regionalization, links to identity, and relations to landscapes. Both affect human behaviour and the building of landscapes, and both are inherently communal, helping members of language and religious groups share a sense of togetherness and belonging, as well as being inherently categorical, emphasizing differences and thus helping keep groups apart.

A Cultural Variable

Language, probably the single most important human achievement, is of interest to human geographers for several reasons. First, language is a cultural variable, a learned behaviour that initially evolved so that humans could communicate in groups, probably for the purpose of organizing hunting activities. It is possible that early humans, concentrated in the area of origin, all spoke the same early language. According to McWhorter (2002), this brief phase of human unity occurred about 150,000 years ago in sub-Saharan Africa. Recent research on phonemes—the smallest sounds that differentiate meaning, such as *th*, *f*, or *s*, which result in words with different meanings when placed in front of, for example, *in*—in 504 languages has confirmed this location. However, as people moved across the earth, different languages developed in areas offering new physical environmental experiences. From the beginnings of cultural evolution, then, language has been a potential source of group unity (and therefore of total-population disunity), a topic of great interest to human geographers.

A second important feature of language is its usefulness in delimiting groups and regions. Language is the principal means by which a culture ensures continuity through time—the death of a language is frequently seen as the death of a culture. It is for this reason that such determined political efforts have been made to ensure the continuity, or even revival, of traditional languages in places such as Wales and Quebec. Because it is commonly seen as a fundamental building block of nationhood, language is of particular interest to human geographers concerned with the plight of minority groups and the relationships between language and nationalism.

A third interest flows directly from the second. As suggested at the outset of this chapter, divisions between places and peoples are one outcome of culture, including language. For example, with reference to urban centres in English-speaking Ontario and French-speaking Quebec, Mackay (1958) compared the expected interaction (calculated using accepted spatial analytic ideas that assume interaction relates to settlement size and the distance between settlements) and the actual interaction. Results showed much less interaction between the two regions than within each region, and it is clear that language plays a role in lessening the expected interaction.

Our fourth interest in language concerns its interactions with environment, both physical and human. Spatial variations in language are caused in part by variations in physical and human environments, while language is an effective moulder of the human environment. The importance of language to the symbolic landscape links it to group identity, returning us to our first interest.

How Many Languages?

Like many cultural variables—indeed, like culture itself—language began as a single entity (or at most a few different ones) and diversified into many. Over the long period of human life on earth, many languages have arisen and many others have died out. During the past several hundred years, however, a new language is a rarity and the disappearance of a language commonplace. Of the roughly 7,000 distinct languages (not counting minor dialects) that existed 400 years ago, approximately 1,000 have disappeared. Many more may vanish over the next few hundred years. Current estimates by UNESCO suggest that about 3,000 languages are endangered and that one language dies about every two weeks.

Most languages are spoken by relatively few people. Indeed, about 96 per cent of the world's population speak only 4 per cent of the world's languages. Table 7.1 identifies the nine languages that have more than 100 million native speakers. Different sources focus on all speakers rather than native speakers; in that case, the number of English speakers increases markedly from that noted in the table.

Disappearing Languages

Languages die for two related reasons. First, a language with few speakers tends to be associated with low social status and economic disadvantage, so those who do speak it may not teach it to their children. In some cases this results in people choosing to speak a different language associated with economic success and social progress. The specific reason for the disappearance of a language with few speakers might be a natural disaster, such as drought or the spread of disease. Second, because globalization depends on communication between previously separate groups, it is becoming essential for more and more people to speak a major language such as English or Chinese. At present, a very few dominant languages effectively control global economics, politics, and culture. About 2,500 are in danger of extinction, with about 1,000 of these seriously endangered.

There are varying reactions to language loss. For many observers, it represents culture loss and is just as serious a threat as loss of biodiversity. This is because each language might be understood as a window on the world, a distinct

© GYI NSEA/iStockphoto

The global diffusion of products and business such as McDonald's is facilitated by the more general diffusion of American cultural and economic characteristics. In this photo, Ronald McDonald and Chinese lion dancers perform during the opening ceremony for a new McDonald's drive-thru facility in Beijing. McDonald's opened its first restaurant in mainland China in 1990 and now operates 760 restaurants countrywide, employing over 50,000 people.

TABLE 7.1 | Languages with more than 100 million native speakers, 2014

Language	Number of Speakers (millions)
Mandarin	848
Spanish	399
English	335
Hindi	260
Arabic	242
Portuguese	203
Bengali	189
Russian	166
Japanese	128

Source: *Ethnologue: Languages of the World.* www.ethnologue.com/statistics/size.

and different way of seeing and thinking, a point highlighted by the fact that most languages include words that are effectively untranslatable. A survey of 1,000 linguists identified the most untranslatable word as *ilunga*, from the Tshiluba language spoken in the southeast of the DR of Congo. In English this word means "a person who is ready to forgive any abuse for the first time, to tolerate it for a second time, but never a third time." Thus, losing a language might be described as a process of cultural forgetting. Loss of languages is also an important practical issue because most languages include detailed knowledge—about local environments, for example—not available elsewhere. In Canada, as in many countries, the usual reaction to possible language loss is to identify it as a problem that requires addressing.

Others see the loss of linguistic diversity as a sign of increasing human unity, countering the divisive tendencies identified at the beginning of this chapter. In communist China the promotion of Mandarin has long been official government policy and language loss is not typically seen as a problem. From a pragmatic perspective, it is notable that an endangered language is likely to be saved only if speakers are prepared to be bilingual or multilingual. Three examples of the many disappearing languages are Eyak (Alaska), the last speaker having died in 2008; Haida (Haida Gwaii, British Columbia), with about 40 speakers; and Tofa (Siberia), with fewer than 30 speakers. Most Haida and Tofa speakers are elderly.

The concern with endangered languages means that there is a tendency to overlook questions about why just a few languages have been so spectacularly successful. A first explanation is migration; as noted below, several of the European languages spread around the world as a component of colonial expansion. Second, and closely related to migration, is that some languages become successful when people find that speaking a particular language is to their economic advantage. A third explanation, as hinted above, is prestige, with the number of speakers increasing if the language is associated with a culture that is in some way impressive, perhaps in military, artistic, economic, or religious terms. However, prestige is necessarily temporary, suggesting that the current popularity of English (as discussed later in this chapter) may be a temporary state of affairs.

Classification and Regions

By studying language families—groups of closely related languages that suggest a common origin—scholars have found evidence of language evolution dating from perhaps 2.5 million years ago and evidence of language distribution patterns dating back some 50,000 years. Nevertheless, it appears that the distribution just prior to the changes initiated by European overseas expansion (Figure 7.11) is a product of migrations during the last 5,000 years or so. Note that this map would be significantly different today because of the expansion of several Indo-European languages since the sixteenth century.

As human groups moved, their language altered, with groups that became separated from others experiencing the greatest language change. Something similar has happened over the past several hundred years as Europeans and others have moved overseas. Although obvious differences in dialect have developed (compare the varieties of English spoken in southern England, the southern United States, and South Africa, for example), few new languages have come into being as a result of these more recent movements because not enough time has elapsed for this to occur; and the groups that moved have remained in close contact with their original groups. This is not to say that languages cannot change significantly over a relatively short period of time, however: just compare Shakespearean and modern English.

Table 7.2 provides a summary of the number of speakers in each of the 19 language families. The largest family is the Indo-European, followed by the Sino-Tibetan. Interestingly, the Indo-European family is spatially dispersed, while the Sino-Tibetan is essentially limited to one area. The difference reflects the fact that many Indo-European–speaking countries colonized other parts of the world, whereas Sino-Tibetan–speaking countries did not. Thus, the numerical importance of Mandarin and Hindi is attributable simply to the large population of China and India, respectively, whereas the numerical importance of English, Spanish, and Portuguese is attributable to, among other things, colonial activity.

As shown in Figure 7.12, an interesting feature of the spatial distribution of languages is that temperate areas of the world have relatively few languages with many speakers, whereas tropical

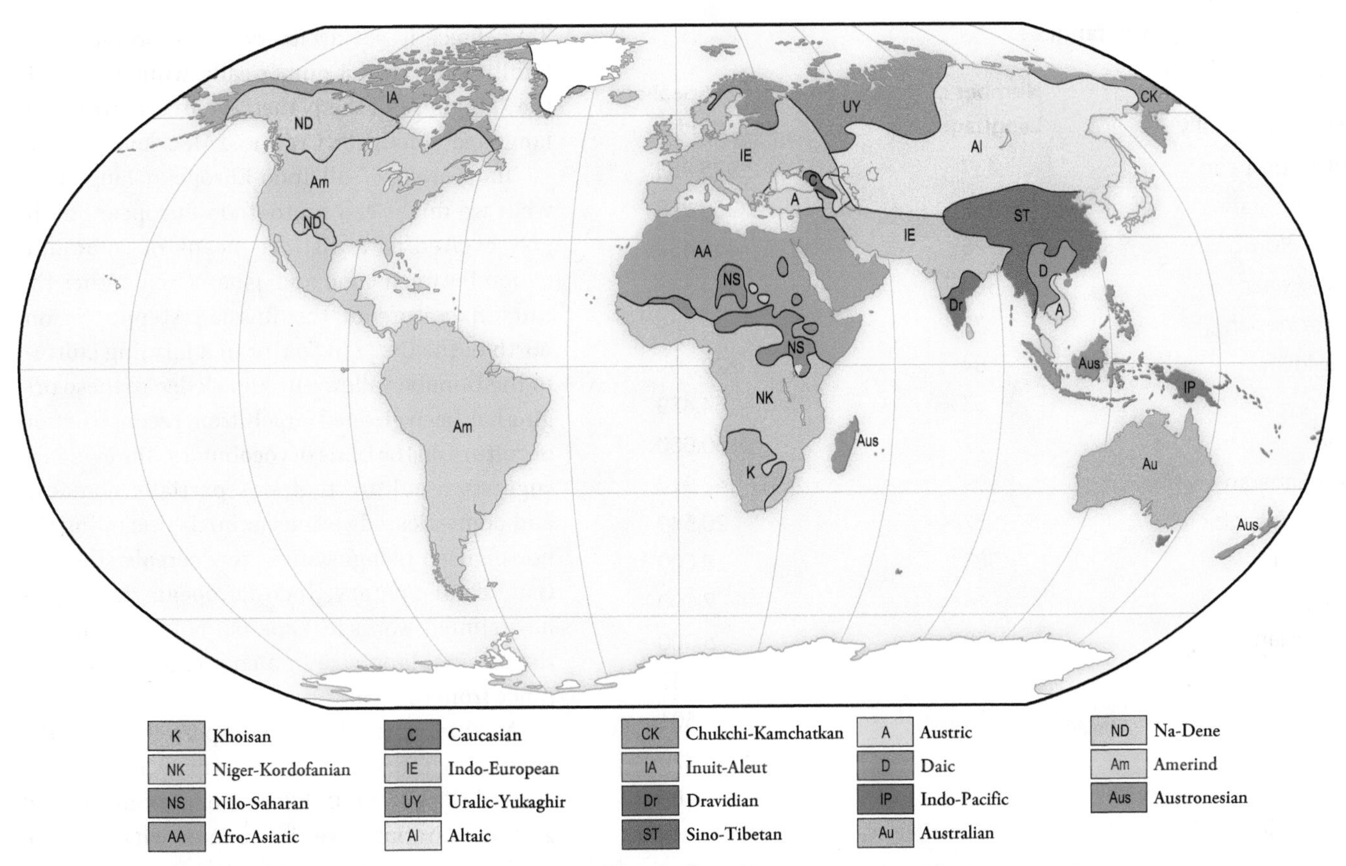

FIGURE 7.11 | World distribution of language families before European expansion

Source: Adapted from Ruhlen, M. C. 1987. *A Guide to the World's Languages: Volume 1, Classification.* Stanford, CA: Stanford University Press. Copyright © 1987 by the Board of Trustees of the Leland Stanford Jr University.

areas have many languages with relatively few speakers. One reason for this distribution is that agriculture spread throughout temperate areas, leading to the resultant emergence of large-scale societies that gradually replaced smaller groups. More specifically, two features of landscape may predict linguistic diversity, namely, river density and landscape roughness (variations in topography). This is because rivers and mountains have historically formed natural barriers.

Counting languages is fraught with problems, but we can estimate that Europe has about 200, the Americas about 1,000, Africa about 2,400, and the Asian-Pacific region about 3,200. Perhaps the most remarkable example of linguistic diversity is Papua New Guinea, which has about 800 languages.

The Indo-European Language Family

Rather than review each language family in detail, we will investigate one family, Indo-European, and one component language, English.

Tijana Martin Photography

Khelsilem (also known as Dustin Rivers), centre, is one of fewer than 10 fluent Squamish language speakers in the world. He has been working to preserve the Squamish language by teaching classes on the Capilano reserve in North Vancouver and developing an English–Squamish dictionary.

TABLE 7.2 | Language families

Language Family	Number of Languages	Number of Speakers (thousands)
Indo-European	437	2,925,253
Sino-Tibetan	455	1,268,218
Niger-Kordofian	1,526	435,432
Afro-Asiatic	367	374,573
Austronesian	1,222	346,489
Dravidian	84	229,294
Altaic	63	164,439
Daic	57	50,000
Nilo-Saharan	199	42,917
Uralic-Yukaghir	37	20,549
Amerind	583	18,000
Austric	4	7,000
Caucasian	38	5,000
Indo-Pacific	731	2,735
Khoisan	24	501
Na-Dene	41	180
Inuit-Aleut	9	85
Australian	201	40
Chukchi-Kamchatkan	5	7

Sources: Adapted and updated from Ruhlen, M. C. 1987. *A Guide to the World's Languages: Volume 1, Classification*. Stanford, CA: Stanford University Press; *Ethnologue: Languages of the World*. www.ethnologue.com/statistics/family.

This choice is not arbitrary. The Indo-European family is the largest numerically, while English is the "language of the planet, the first truly global language" (McCrum, Cran, and MacNeil, 1986: 19).

The parent of all Indo-European languages, what we might call proto–Indo-European, probably evolved as a distinct means of communication between 6000 and 4500 BCE in either the Kurgan culture of the Russian steppe region, north of the Caspian Sea, or in a farming culture in the Danube valley. Our knowledge of these origins has been gleaned largely from reconstruction of culture on the basis of vocabulary. The evidence suggests a culture that was partially nomadic, had domesticated various animals (including the horse), used ploughs, and grew cereals. The fact that contemporary Indo-European languages have similar words for *snow* but not for *sea* places the original language in a cold region some distance from the sea.

As this cultural group dispersed, using the horse and the wheel, the language moved and evolved (Figure 7.13). Different environments and group separation gave rise to many changes. The earliest Indo-Europeans to settle in what is now England were Gaelic-speaking people who arrived after 2500 BCE. They were followed by a series of Indo-European invaders: the Romans in 55 BCE

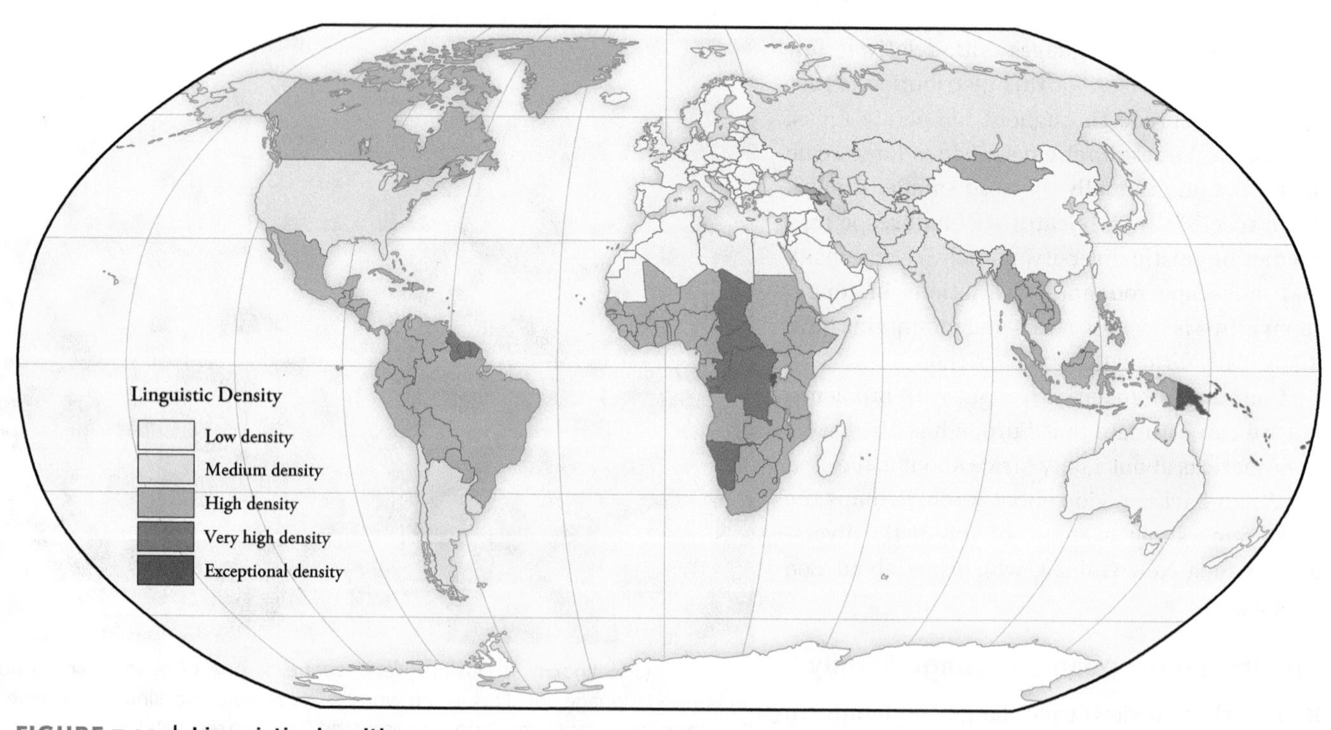

FIGURE 7.12 | Linguistic densities

Source: *The UNESCO Courier*, April 2000. www.unesco.org/courier.

and the Angles, Saxons, and Jutes in the fifth century CE (Figure 7.14).The latter three groups introduced what was to become the English language, forcing Gaelic speakers to relocate in what is currently known as the Celtic fringe. Surprisingly, little mixing took place between Gaelic and the incoming Germanic language. Old English, as it is called, was far from uniform and was subsequently enriched with the introduction of Christianity (Latin) in 597, the invasions of Vikings between 750 and 1050, and the Norman (French) invasion of 1066. Each invasion led to a collision of languages, and English responded by diversifying. Today, it has an estimated 500,000 words (excluding an equal number of technical terms); by contrast, German has 185,000 words and French 100,000.

Beginning in the fifteenth century, the English language and other Indo-European languages spread around the globe. This spread encouraged the diversification of English, including the development of many American varieties. Not only is English the first language of many countries, but it is also the second language of many others. In India, for example, perhaps 70 million people speak English as a second language, a vital unifying force in a country of many languages. In China, English is increasingly a requirement for admission to university. English is often considered a global language, entirely replacing some minority languages and becoming increasingly necessary as a second or additional language in many parts of the world. For example, it has become the de facto working language of the EU. It is an accident of history that, at the time when the need for a common language became apparent in the twentieth century, an English-speaking country, the United States, was dominant. South Sudan, which was recognized as a new country in 2011, chose English as the official language.

Language and Identity

For many groups, language is the primary basis of identity—hence the close links between language and nationalism, the desire to preserve minority languages, and even the various efforts that have been made to create a universal language. A common language facilitates communication; different languages create barriers and frictions between groups, further dividing our divided world.

Language and Nationalism

At first glance, the link between language and **nationalism** seems clear, but it is fraught with problems. The boundaries of language regions are rarely clearly defined, resulting in many difficulties for those countries aspiring to declare their national territory on the basis of a common language. But more significantly, for many people, the language they speak is a result of power struggles for cultural and economic dominance. In Europe

nationalism
The political expression of nationhood or aspiring nationhood; reflects a consciousness of belonging to a nation.

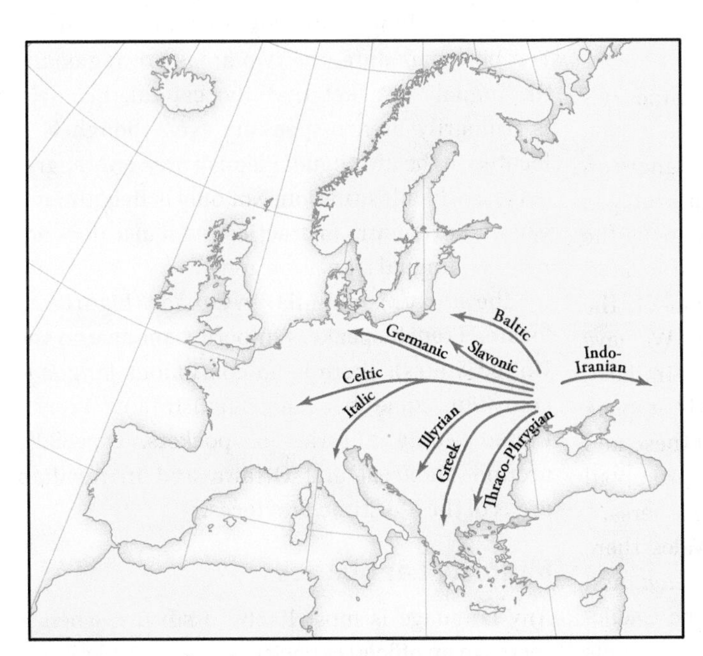

FIGURE 7.13 | Initial diffusion of Indo-European languages

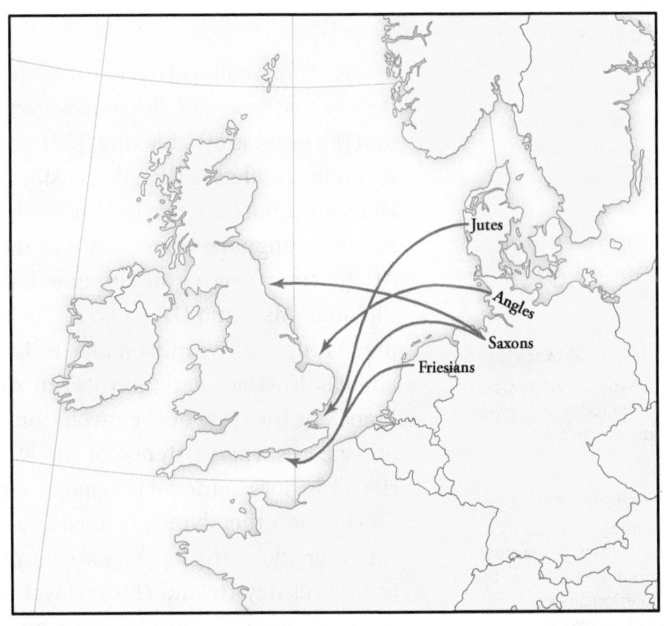

FIGURE 7.14 | Diffusion of Indo-European languages into England

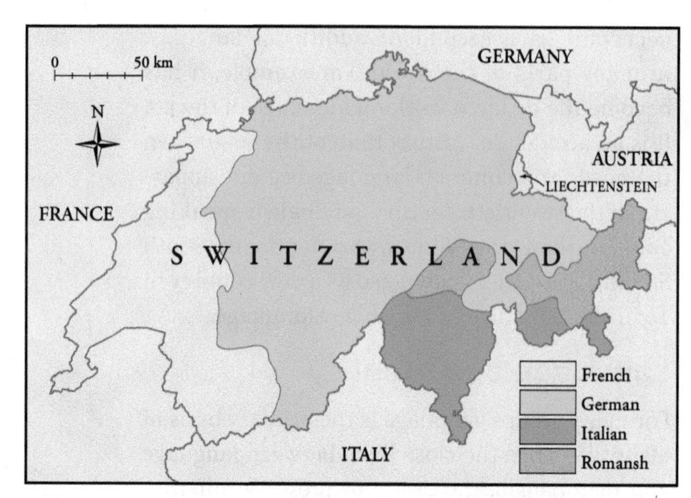

FIGURE 7.15 | Four official languages in Switzerland

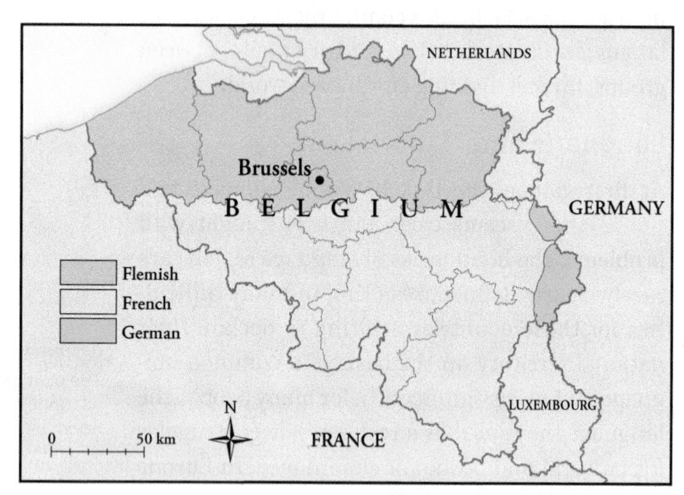

FIGURE 7.16 | Flemish, French, and German in Belgium

before the nineteenth century, the boundaries between states and languages rarely coincided and the process of achieving this concordance can be understood as artificial. Consider, for example, that only about 2.5 per cent of Italians spoke the national language when Italy was unified in 1861. At the first meeting of the new parliament, the diplomat Massimo D'Azeglio remarked, "We have made Italy, now we must make Italians." Similarly, about half of the population of France did not speak French at the time of the revolution. In these and other cases, the existence of the state prompted the desire for a national language, not vice versa.

multilingual state
A state in which the population includes at least one linguistic minority.

minority language
A language spoken by a minority group in a state in which the majority of the population speaks another language; may or may not be an official language.

On the other hand, in medieval Wales there was a single word for *language* and *nation*, and in present-day Ireland the survival of the Gaelic-speaking area is considered to be "synonymous with retention of the distinctive Irish national character" (Kearns, 1974: 85). Language is often seen as the basis for delimiting a nation for two reasons. First, a common language facilitates communication. Second, language is such a powerful symbol of "groupness" that it serves to proclaim a national identity even where, as in Ireland, it does not serve a significant communication function.

Multilingual States

Other areas in Europe, however, did not follow the trend to build a state with a single dominant language. Switzerland is a prime example of a viable political unit with several official languages: 70 per cent of the people speak German, 19 per cent French, 10 per cent Italian, and 1 per cent Romansh (Figure 7.15). The political stability for which Switzerland is known reflects several factors, including the pre-1500 evolution of the Swiss state, its long-standing practice of delegating much governmental activity to local regions, and its neutrality in major European conflicts, the latter factors aided by the extreme geography.

More typical examples of **multilingual states** are Belgium and Canada (Box 7.4). Despite conscious efforts to follow the Swiss example, Belgium is politically unstable. Created as late as 1830 as an artificial state—a move that pleased other European powers but not necessarily the people who became Belgians—Belgium has failed to remain detached from European conflicts and is divided between a Flemish-speaking north and French-speaking south (Figure 7.16). Although it is a bilingual state, the two areas are regionally unilingual. The fact that the capital, Brussels, is primarily French-speaking even though it is located in the unilingual Flemish area only aggravates a difficult situation. Not only is Belgium not a bilingual country in practice, but it also does not have a bilingual transition zone.

The situation is similar in Canada (Figure 7.17 locates French speakers in both Canada and the United States). There is no continuous language transition zone between English and French Canada—only a series of pockets, especially around Montreal and Ottawa and in Acadian areas of the Maritime provinces.

Minority Languages

Any language is most likely to survive when it serves in an official capacity. **Minority languages** without official status typically experience a slow

Around the Globe

BOX 7.4 | Linguistic Territorialization in Belgium and Canada

Although the specific causes are quite different, the Belgian and Canadian language situations are basically similar. Belgium is clearly divided into two linguistic territories, while Canada appears to be approaching the same situation.

Since about the fifth century CE, the area that is now Belgium has been divided into a northern area of Flemish speakers (Flemish is the ancestor of modern Dutch) and a southern area where Walloon (French) is spoken. Thus, Belgium lies on both sides of the linguistic border separating the Germanic and Romance subfamilies of the Indo-European family. To expect a viable state to emerge in this context as late as 1830 was quite ambitious, particularly when the French speakers explicitly wished to join France. In fact, Belgium is an artificial state purposely created by the British and Germans, neither of whom favoured territorial expansion for France. It comprises two distinct language areas with a minimal transition zone; the country has attempted to follow Switzerland's example but without success.

Canada, settled by both French and English during its formative years, is also a bilingual state. With Confederation in 1867, the French formed a majority in Quebec and the English a majority in Ontario, Nova Scotia, and New Brunswick. Most of the remaining land west of Ontario was settled either by English speakers or by immigrants from other countries who settled in English-speaking areas; immigration of French speakers has been very limited. Over centuries, interaction between French and English gave rise to distinct transition zones in western New Brunswick, southern Quebec, and eastern Ontario. However, analysis of local migration and interaction between the two groups (Cartwright, 1988) suggests that the zone is disintegrating into a series of pockets and that the linguistic territorialization evident in Belgium since 1830 is becoming a reality in Canada. This change is a consequence of spatially delimited language differences that have occurred despite the 1969 passing of the Official Languages Act, which institutionalized French as one of two official languages in Canada.

but inexorable demise. Consequently, minority-language speakers often strive for much more than the mere survival of their language: they may press for the creation of their own separate state, using the language issue as justification.

Examples of minority languages include Welsh and Gaelic in Britain, Spanish in the United States, French—despite its official status—in Canada, Basque in Spain, Hausa and other languages in Nigeria, and Cantonese in China. This is a highly condensed list; most countries in the contemporary world have at least one minority language. The consequences are varied. Welsh and Irish Gaelic continue to strive for survival and indeed independence (Box 7.5). The Basque language—one of very few languages in Europe that do not belong to the Indo-European family—was for many years at the centre of a powerful and violent Basque independence movement, although the principal terrorist group declared a cease-fire in 2006. In Nigeria the official language is English—a result of colonial activity and the fact that, although the country has several indigenous languages, none is predominant.

Communications between Different Language Groups

Historically, some languages have played important roles even when they are not the first language of populations. Following independence in India, English joined Hindi as an official language. In parts of East Africa, Swahili is an official language; combining the local Bantu with imported Arabic, it is an example of a **lingua franca**, a language developed to facilitate trade between different groups (in this case, Africans and Arab traders). In some other colonial areas, **pidgin** languages have developed as simplified ways of communicating between different language groups and are especially common in areas that developed as trading centres, with imported slave populations, and in areas of plantation economies. One such area is Southeast Asia, where pidgins are usually based on English, with some Malay and some Chinese. A pidgin that becomes the first language ("mother tongue") of a generation of native speakers is known as a **creole**. Creoles have larger vocabularies and a more sophisticated grammatical structure

lingua franca
An existing language used as a common means of communication between different language groups.

pidgin
A new language designed to serve the purposes of commerce between different language groups; typically has a limited vocabulary.

creole
A pidgin language that assumes the status of a mother tongue for a group.

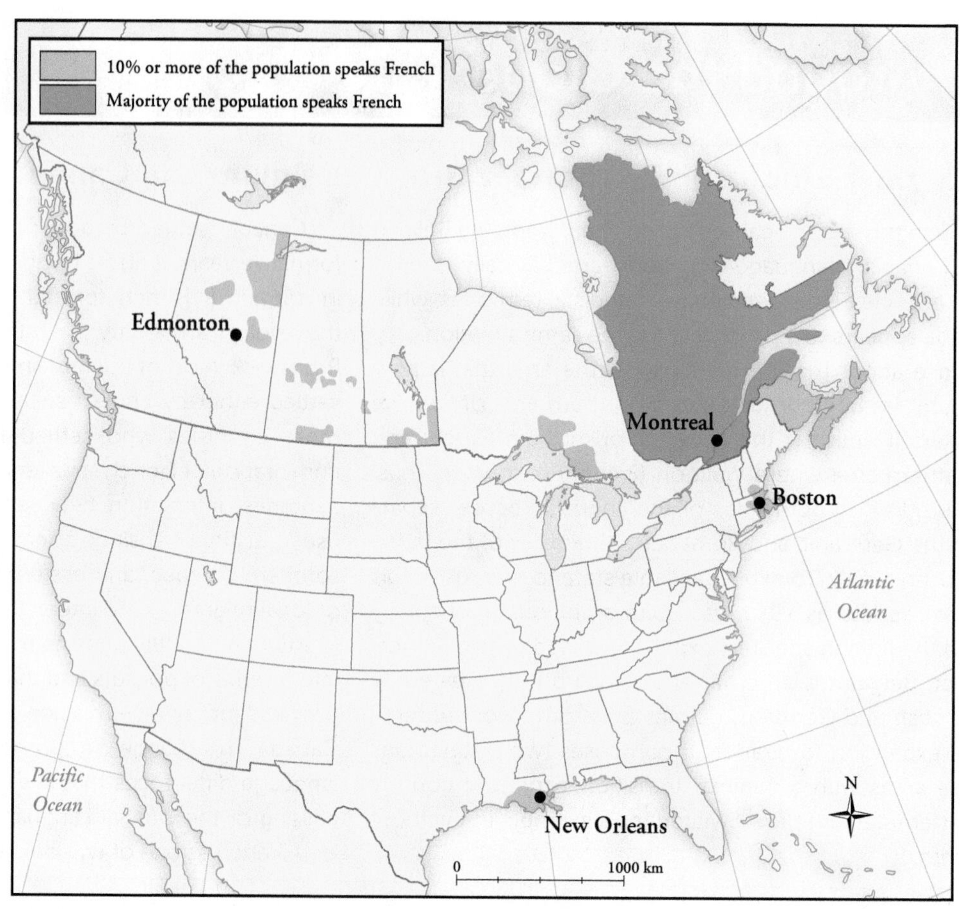

FIGURE 7.17 | French and English in North America

than pidgins. They are relatively common in the Caribbean region and vary according to whether the principal European component is English, Spanish, or French; often the other components are the native Carib and imported Bantu.

Some countries are uncertain which of many languages spoken merit official status. In the South American country of Suriname, with about half a million people, languages spoken include Dutch (it was a Dutch colony), Portuguese, English, and variants of Chinese, Hindi, Javanese, and six creole languages. Dutch is the official language and is taught in schools, but the main language of everyday communication is a creole, Sranan Tongo (meaning Suriname tongue). This creole is based largely on English. In general, people speak Dutch in formal settings and Sranan Tongo in informal settings.

toponym
Place name; evidence provided by place names can be crucial in a historical study of movement and settlement if other sources of information are unavailable.

A number of attempts have been made to promote the use of a single universal language; artificial languages have been invented for this purpose. In 1887 the most popular such language, Esperanto, was introduced but failed to make a significant impact largely because, although it was based on roots common to several European languages, it had no ties to any specific tradition, culture, or environment. For the same reason, most human geographers reject the idea of a universal language—even though they recognize that, in principle, a universal language could promote communication and understanding between groups, thereby minimizing division and friction. Regardless, it seems unlikely that any artificial language will succeed in playing such a role. The best hope for a universal language rests with English.

Language and Landscape

Naming Places

Language plays a key role in landscape, in the form of place names (**toponyms**). We name places for at least two reasons. The first is in order to understand and give meaning to landscape. A landscape without names would be like a group of people without names; it would be difficult to

Around the Globe

BOX 7.5 | The Celtic Languages

The Celts were one of the most important early groups to diffuse from the Indo-European core area. Beginning about 500 BCE, they spread across much of Europe. But after some 500 years of expansion, during which they came into contact with other, more organized groups, the Celts were pushed to the western limits of Europe by the Anglo-Saxon in-movement and gradually retreated into some of the more inhospitable and isolated areas. Over time, various Celtic languages, such as Cornish (southwest England) and Manx (Isle of Man, in the Irish Sea), disappeared. The remains of the once-large Celtic group live in four small pockets: western Wales, western Scotland, western Ireland, and northwest France. Each area still has some Celtic speakers: in Wales about 500,000 people speak Welsh; in Scotland about 80,000 speak Gaelic; in Ireland about 70,000 speak Erse (Irish Gaelic); and in northwest France about 675,000 speak Breton.

These languages are in precarious positions because they are not associated with political units. The Irish-language area, called the Gaeltacht, covered about 33 per cent of Ireland in 1850 and had perhaps 1.5 million speakers; today it covers about 6 per cent and has lost 95 per cent of its speakers. Since 1956, when the first concerted efforts to save the Gaeltacht were made, the Irish government has actively encouraged retention of the language and cautious social and economic development for the rural, remote regions. The basic assumption is that language is a key requisite for any group that aspires to retain a traditional culture. So far, the various efforts to foster development have been quite successful and, unlike some had feared, the Irish language has continued to thrive in the area.

Welsh, a minority language intimately tied to traditional Welsh culture and a rural way of life in an increasingly anglicized environment, is in a similar position. Welsh-speaking Wales is restricted to the extreme northwest and southwest, and the Welsh language is threatened despite vigorous local efforts to preserve it. As a part of the United Kingdom, Wales is not in as strong a position as Ireland when it comes to implementing local development and language retention policies.

distinguish one from another. Second, naming places probably serves an important psychological need—to name is to know and control, to remove uncertainty about the landscape. For these reasons, humans impose names on all landscapes that they occupy and on many that they do not (the moon is a prime example).

Place names, then, are a significant feature of our human-made landscapes, often visible in the form of road signs and an integral component of maps—our models of the landscape. Many place names combine two parts, generic and specific. Newfoundland, for example, has *Newfound* as the specific component and *land*—the type of location being identified—as the generic component.

Analysis of place names can provide information about both the spatial and the social origins of settlers. In addition, names such as Toivola (Hopeville) in Minnesota or Paradise in California tell us something about their aspirations. Place names typically date from the first effective settlement in an area. Thus, groups such as the French and Spanish were able to place their languages on the landscape at the regional scale; witness the profusion of French names in Quebec and

A Mayan family gathers in a doorway in Guatemala to share a meal. There are 69 Mayan languages, which are spoken by about six million people in Central America. Guatemala has one of the largest populations of Maya speakers.

Louisiana and of Spanish names throughout the American Southwest. Other groups who arrived later, such as Finns in Minnesota, were restricted to naming places at a local scale.

Place names can be an extremely valuable route to understanding the cultural history of an area. This is especially clear in areas where the first effective settlement was relatively recent but is also the case in older settled areas. In many parts of Europe, for example, former cultural boundaries can be identified through place-name analysis. Jordan (1988: 98) provides an example from before 800, when the Germanic–Slavic linguistic border roughly followed the line of the Elbe and Saale rivers in present-day Germany. Although German speakers began moving east after 800, evidence of the earlier boundary remains: place names are German west of the line and Slavic east of it. Similarly, throughout much of Britain it is possible to identify areas settled by different language groups by studying place names; in north-east England, for example, place-name endings such as *by* are evidence of Viking settlement.

exonym
A name given to people (or a place) by a group other than the people to which the name refers (or who are not native to the territory within which the place is situated).

Renaming Places

Place names also are about power. The former republic of Yugoslavia now called Macedonia came close to war with neighbouring Greece over the choice of name because Macedonia is the name of a Greek province. In response to this political issue, the UN elects to call Macedonia by the much more convoluted name, the Former Yugoslav Republic of Macedonia (or FYROM). One of the greatest confusions in today's world concerns international uncertainty over the names of countries and of many landscape features. Especially throughout much of Africa, Central Asia, and India, place names were changed as one part of a larger rejection of the colonial era. For example, Mumbai has replaced Bombay in India and Iqaluit in Canada was named Frobisher Bay until 1987. The case of Burma is particularly complex. In 1989 the ruling military junta changed the English version of the country name to Myanmar, a change that was endorsed by the UN but has not been widely accepted elsewhere, including Canada. The key reason for many countries not accepting the new name is that the democracy movement in Burma, which is supported by most Western countries, does not accept the legitimacy of the junta, arguing that it has no authority to change the name of the country.

In some multilingual countries the situation is extraordinarily complex. In South Africa, which has 11 official languages, the same city is called Cape Town (English), Kaapstad (Afrikaans), and eKapa (Xhosa). Elsewhere in South Africa, white residents have opposed some name changes, such as changing the name of the town Lydenburg, which was named by early Dutch settlers, to Mashishing, an African name. In Israel the city of Jerusalem (English) is called Yerushalayim in Hebrew and either Urshalim or Quds in Arabic. Sometimes the consequences of place-name ambiguity can be very damaging, as evidenced during the first Gulf War, when bombs were dropped in an unintended location because of misunderstanding about place names. The UN even has a committee of geographers to discourage the use of **exonyms**, names given to a group or place by a group other than the people/place to which the name refers.

Renaming also occurs because previous names may now cause offence. A cape on Lake Ontario known as Niggerhead Point, because it was on the route of the underground railway aiding slaves to escape to Canada, was renamed Negrohead Point and is currently marked on New York state maps as Graves Point. For a somewhat different reason, Gayside in Newfoundland was renamed Baytona.

The "Great American Desert"

Just as language is everywhere in landscape, it can also make landscape. A striking example is the nineteenth-century use of the term *desert*—as in "Great American Desert"—to describe much of what is now the Great Plains region. Most European North Americans knew nothing of the Plains until several expeditions in the early nineteenth century, including those of Zebulon Pike (in 1806) and Stephen Long (in 1823), returned with reports of a "desert" that would surely restrict settlement. This misconception arose because individual explorers recorded what impressed them most, the small areas of sand desert were indeed impressive, and the absence of trees did make the Plains a desert in comparison with the heavily forested East. Among the consequences was delayed settlement as migrants bypassed the Plains in favour of the west coast. Thus, the language used to describe a specific environment affected human geographic changes in the landscape.

Landscape in Language

Not only is language in landscape and capable of making landscape, but landscape is also in language. Because physical barriers tend to limit movement, there is a general relationship between language distributions and physical regions (this is evident on a world scale in Figure 7.11). The vocabulary of any language necessarily reflects the physical environment in which its speakers live; hence, there are many Spanish words for features of desert landscapes and few comparable English words. Similarly, many words are available to discriminate between different types of snow in the Inuit language and between the colourings of cattle in the Masai language. As languages move to new environments, new words are added to help describe these new environments; *outback*, for example, entered English only with the settlement of Australia. Finally, the human landscape is in language to the extent that language reflects class and gender; within a given language, word choices and pronunciations say a great deal about social origins. This area of research is labelled sociolinguistics.

RELIGION

A second fundamental cultural element is religion. Although the precise origins of religious beliefs are no less difficult to trace than those of language, the universality of religion suggests that it serves a basic human need or reflects a basic human awareness. Recent research suggests that the traditional understanding of the rise of civilization, namely, that plant and animal domestication set the scene for organized religion and political systems, may not have applied in all cases. A suggested alternative sequence of events, based on investigations of several Middle Eastern sites, sees the rise of organized religion preceding domestication.

Essentially, a religion consists of a set of beliefs and associated activities that are in some way designed to facilitate appreciation of our human place in the world. In many instances, religious beliefs generate sets of moral and ethical rules that can have a significant influence on many aspects of behaviour. Typically, a religion has a core set of beliefs that find expression in many forms, including texts, rituals, everyday behaviour, symbols, and, of course, landscapes.

There are often major distinctions between women and men with regard to religious behaviour. Women usually make up the majority of a religion's followers and play a critical role in teaching religion to their children. Further, most major religions—with the exception of Islam—have at least some female figures of worship (although the most important figures are generally male). Yet in many religions women are excluded from serving in a formal, structured role; in the Christian tradition, women are admitted to the ministry in many Protestant churches but remain largely excluded from the Catholic and Orthodox traditions. Men play the dominant official roles in Hinduism, Buddhism, Christianity, and Islam, to name only the largest organized religions.

Figure 7.18 identifies the origin areas of the four largest religions, as measured by numbers of adherents, while Figure 7.19 maps the contemporary distribution of these and other religions. These maps make two things clear. First, there are two major religious "hearth" areas: Indo-Gangetic (a lowland area of north India that drains into the Indus and Ganges Rivers) and Semitic (the Near East). The Indo-Gangetic hearth has given rise to Hinduism, Buddhism, Jainism, and Sikhism, while the Semitic hearth has given rise to Judaism, Christianity, and Islam. Second, two of the four

A Hindu devotee making the painful pilgrimage to Batu Caves, Kuala Lumpur, Malaysia.

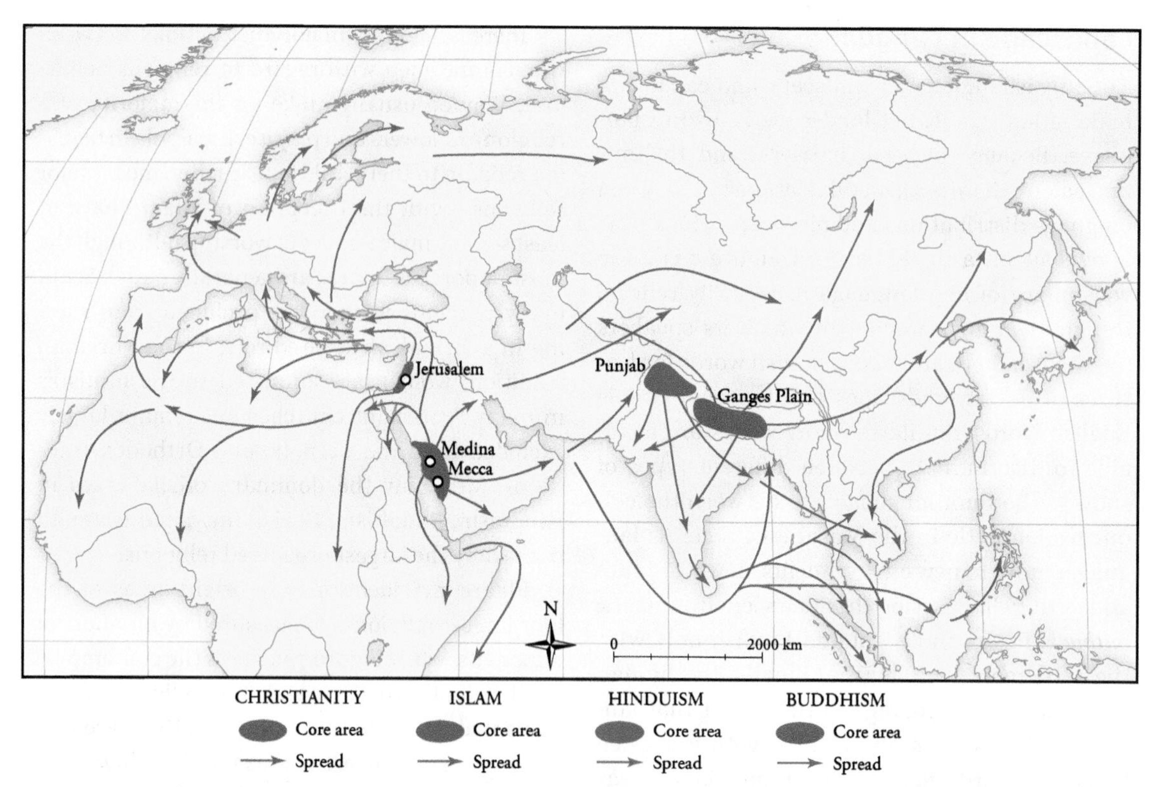

FIGURE 7.18 | Hearth areas and diffusion of four major religions

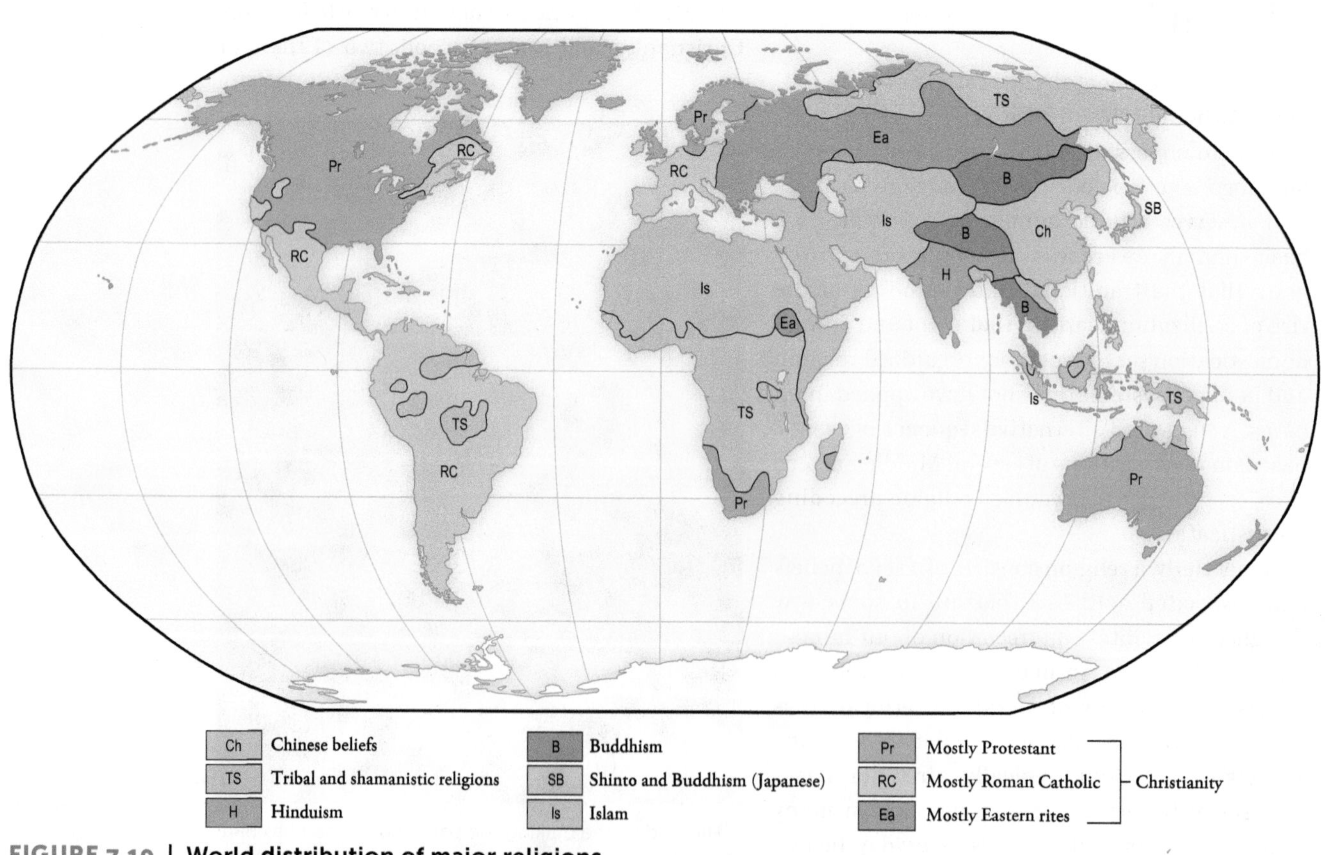

FIGURE 7.19 | World distribution of major religions

major religions, Christianity and Islam, have diffused over large areas, whereas Hinduism has not experienced significant spread and Buddhism has effectively relocated to China and other parts of Asia. These conclusions are confirmed by Table 7.3, which provides detailed data on the distribution of religious groups by selected major world regions. This table also shows that Christianity is the leading religion numerically, followed by Islam, Hinduism, and Buddhism. A high 13 per cent are classed as non-religious (agnostic or atheist), and this percentage appears to be increasing. The Czech Republic reports 60 per cent of its population as non-religious, while a recent mathematical analysis suggests that religion, as measured by numbers of people who declare a religious affiliation in national censuses, is set for extinction in nine countries, including Canada.

Using a single name to describe a religious identity can be misleading as all four of the major religions are divided into often very different competing branches with differing beliefs and practices. These schisms arose as religions moved through landscapes, coming into contact with different cultural contexts so that new understandings emerged; also, oral traditions and ancient texts are easily subject to differing interpretations.

A useful classification of religions, favoured by geographers because it relates closely to spatial distributions, distinguishes between ethnic religions, which are closely identified with a specific cultural group, and universalizing religions, which actively seek converts. Although very useful, this system necessarily excludes hundreds of numerically small religions. For example, many cultures embraced a form of **animism**—in which a soul or spirit is attributed to various phenomena, including inanimate objects—especially before the diffusion of universalizing religions, notably Christianity and Islam.

animism
A general name for beliefs that attribute a spirit or soul to natural phenomena and inanimate objects.

Origins and Distribution of Ethnic Religions

The principal ethnic religions emerged earlier than their universalizing counterparts. They are associated with a particular group of people and do not actively seek to convert others.

Hinduism

Of the several hundred religions in this category, the largest is Hinduism, which evolved in the Indo-Gangetic hearth about 2000 BCE. Initially more correctly described as Vedism, this faith was the first major religion to evolve in the area, from which it spread east down the Ganges and then south through India, eventually to dominate the entire region. As it diffused during the last few centuries BCE and much of the first millennium CE through an already diverse cultural landscape, core features absorbed and blended with other local religious beliefs. There were certain distinctions between

TABLE 7.3 | Major world religions: Number of adherents (thousands) by continental region, 2010

Religion	Africa	Asia	Europe	Latin America	North America	Oceania	World	%
Christianity	488,880	350,822	584,809	544,592	283,308	28,205	2,280,616	33.0
Islam	421,938	1,083,354	40,174	1,599	5,598	524	1,553,188	22.5
Hinduism	2,945	935,753	991	789	1,867	526	942,871	13.6
Agnosticism	5,995	504,352	84,652	16,941	43,211	4,629	659,781	9.6
Buddhism	258	455,412	1,777	760	3,845	573	462,625	6.7
Chinese folk religions	133	452,762	438	189	781	101	454,404	6.6
Ethnic religions	109,592	153,565	1,150	3,802	1,246	368	269,723	3.9
Atheism	594	116,204	15,390	2,901	2,013	462	137,564	2.0
New religions	117	59,611	364	1,744	1,747	101	63,864	0.9
Sikhism	74	22,496	400	7	613	49	23,738	0.4
Judaism	134	5,980	1,914	963	5,720	113	14,824	0.2
Spiritism	3	2	143	13,330	247	8	13,732	0.2

Source: *Time Almanac, 2011*. 2011. Chicago: Encyclopedia Britannica, 508–09. Reprinted with permission from Encyclopædia Britannica Almanac 2011, © 2010 by Encyclopædia Britannica, Inc.

Indo-Aryan cultures to the north, Dravidian cultures to the south, and numerous local hill cultures in central and eastern India. The diffusion and ongoing evolution of Hinduism means it might be better described as a variety of related religious beliefs rather than as a homogeneous religion. Indeed, variations are evident even at the scale of the local village community.

Hinduism was carried overseas, but it has not retained significant numbers of adherents outside India. It remains an Indian religion associated with a country and a broadly defined cultural group. Hinduism has no dogma and only a loosely defined philosophy by religious standards. It is polytheistic (worshipping more than one god) and has close ties to the rigid social stratification of the caste system. Like many other religions, Hinduism has spawned numerous offshoots. Jainism, developed from the teachings of a sixth-century BCE holy teacher, rejects Hindu rituals but shares many basic tenets of Hinduism, including the belief in reincarnation and *ahimsa* (the ethical doctrine that humans ought to avoid hurting any living creature). A more recent offshoot is Sikhism, a hybrid of Hinduism and Islam that arose about 500 years ago in India's Punjab region.

caste
A social rank, based solely on birth, to which an individual belongs for life and that limits interaction with members of other castes.

Judaism

The first monotheistic religion (worshipping a single god), Judaism originated about 2000 BCE in the Semitic hearth area, initially in the form of proto-Judaism. Following the Romans' destruction of Jerusalem in 70 CE, the Jews were driven out of their homeland and eventually dispersed throughout Europe; the entire body of Jews living outside Israel, in Europe and elsewhere, is known as the Diaspora.

Judaism contains significant internal divisions reflecting theological and ideological differences. These differences developed as Jews moved, with Sephardic Judaism concentrated in Spain and Ashkenazic Judaism elsewhere in Europe. Sephardic Jews easily adapted to different areas, but Ashkenazic Jews tended not to integrate with larger non-Jewish, Christian, society. This lack of integration was undoubtedly a twofold process, with Jews wishing to maintain their traditions and with Christians being antagonistic towards Jews. Ashkenazic Judaism experienced a number of divisions, especially with the eighteenth-century rise of Hasidic Judaism in western Ukraine and the emergence of ultra Orthodox Judaism, initially in Hungary.

Much Jewish movement, including a general eastward migration in Europe from about the fourteenth century onward and the late nineteenth- and early twentieth-century movements from Europe to North America, can be understood as a response to persecution. Anti-Jewish sentiment was most evident when it became a part of larger state policies in Germany in the 1930s. Only in 1948 was the long-term goal of a Jewish homeland achieved with the creation of the state of Israel (see Chapter 9).

Other ethnic religions include Shinto, the indigenous religion of Japan, and Taoism and Confucianism, both of which are primarily associated with China.

Origins and Distribution of Universalizing Religions

Both Hinduism and Judaism have given rise to major universalizing world religions.

Buddhism

Buddhism was the first universalizing religion, an offshoot of Hinduism founded in the Indo-Gangetic hearth by Prince Gautama, who was born in 644 BCE. During his lifetime he preached in northern India, but after his death Buddhism was spread by missionary monks into other parts of India and then throughout much of Asia. There is debate concerning whether this diffusion of Buddhism occurred mostly before or after two versions of the religion evolved in the first few centuries CE—Mahayana Buddhism emerged in the Upper Indus Valley, while Theravada Buddhism first displayed a clear identity in Sri Lanka. Some scholars contend that, by the third century BCE, Buddhist scriptures were codified, that the religion was effectively the state religion of a large India-wide empire, and that it was then carried to China after about 100 BCE and throughout most of Asia. But these dates are uncertain. What is clear is that the Mahayana version, which is more inclusive (syncretistic), diffused north from India into Central and East Asia, including China, Korea, Japan, Tibet, and Mongolia. The conservative Theravada version, which does not seek to reconcile with other belief systems, moved into Southeast Asia, including Burma, Thailand, Cambodia, Laos, and Vietnam.

Perhaps surprisingly, this spread and growth did not result in Buddhism's acceptance as the religion of China, as both Taoism and Confucianism wielded important influences. Also, Buddhism in India merged with Hinduism, which also competed with it in some parts of Southeast Asia. Only in Burma and Thailand did Buddhism achieve the status of a state religion. The patterns of diffusion remain evident in the current distributions, with Mahayana Buddhism prevailing in Central and East Asia and Theravada Buddhism in Southeast Asia. Like Hinduism, Buddhism is characterized by numerous local variations, as it blended easily with pre-existing local beliefs.

Christianity

Christianity developed about 600 years after Buddhism as an offshoot of Judaism. It began when disciples of Jesus of Nazareth accepted that he was the expected Messiah. The religion spread slowly during his lifetime and more rapidly after his death as missionaries carried it initially to areas around Jerusalem and then through the Mediterranean to Cyprus, Turkey, Greece, and Rome. Although local variations soon appeared, these never assumed the same significance as was evident in Hinduism and Buddhism. This was because Christianity included the idea of a unified Church, because written details of the founding of the religion were soon available as the four gospels, and because there were deliberate efforts to formalize universal doctrines at a series of councils, the first of which was held in Nicea in 325 CE.

Christianity spread rapidly through the Roman Empire and was formally adopted as an official religion. However, the collapse of the empire and resultant political fragmentation created competing Germanic and Roman influences, with many Germanic people practising a version of Christianity known as Arianism. Also, a division was evident between a Latin-speaking western region and a Greek-speaking east. This distinction became more significant with debate about the status of the bishop of Rome, later known as the pope. Christians in the east rejected popes' claims that they had authority over the entire Church. As a result of these differences, a major division between the Roman Catholic (western) and Eastern Orthodox forms of Christianity occurred in 1054, and the line dividing these two remains the most basic religious boundary in contemporary Europe (see Figure 9.8). The two versions of Christianity competed for converts throughout the Slavic areas of Europe.

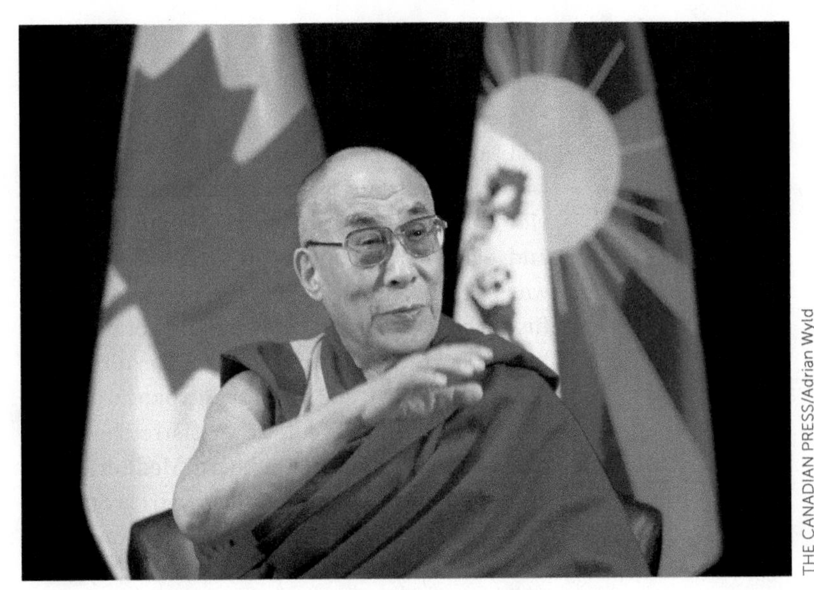

THE CANADIAN PRESS/Adrian Wyld

The Dalai Lama, the spiritual head of Tibetan Buddhism, speaks at the Sixth World Parliamentarians' Convention on Tibet in Ottawa in 2012. The Dalai Lama was also the spiritual and temporal ruler of Tibet until a revolt against Chinese communist rule in 1959, when he fled the country.

The Protestant Reformation of the 1500s produced a third major version of Christianity. The Reformation was an effort to reform Roman Catholic dogma, teachings, and practices and also an attempt by some Northern European leaders to challenge the wealth of the Catholic Church. Protestants established a number of independent churches, notably Lutheran, Zwinglian, Calvinist, and Anglican. Several of these churches, most notably Anglican, had strong national connections. Northern Europe became largely (though by no means exclusively) Protestant with all the religious diversity that this implied, but the Reformation did not affect Southern Europe, which remained predominantly Catholic.

Both Catholic and Protestant versions of Christianity spread to other parts of the world—the Americas, Africa, much of Asia, and Australasia—in the course of European movement overseas. The importance of Christianity in the contemporary world reflects not only the large number of its adherents and their wide spatial spread but also the fact that Christian thinking has been a cornerstone of Western culture, affecting attitudes and behaviours at both group and individual levels. Christianity has proven to be the most influential religion in shaping the world because it was carried overseas by European colonial powers and because it has been very adaptable and therefore effective at encouraging converts.

Islam

Islam also arose in the Semitic hearth area; it is related to both Judaism and Christianity but has additional Arabic characteristics. Rather like Christianity, Islam is an all-encompassing world view, shaping both group and individual attitudes and behaviours. Founded by Muhammad, who was born in Mecca in 570, the religion had diffused throughout Arabia by the time of his death in 632. Further diffusion was rapid as a result of Islamic political and military expansion. Arab Muslims created an empire that stretched west to include parts of the northern Mediterranean as far as Spain and much of North Africa and east to include the areas of modern-day Iraq, Iran, Afghanistan, and Pakistan. Islam has also spread into much of Southeast Asia. In India, Muslims added to the already diverse religious identities and landscapes of Hinduism. Essentially, this historical expansion was limited to a vast west–east belt that barely penetrated into temperate climates. The limited expansion of Islam into Europe resulted in conflicts with Christians; in the sixteenth and seventeenth centuries, for example, there was a massive expulsion of Muslims from Spain, with many relocating in North Africa.

For several reasons, Islam has not experienced the same degree of divisiveness as the other two major universalizing religions. The basic structure and content of Islam was widely accepted from the outset, as were the ideas of the Islamic *umma*, a group sharing basic beliefs, and of the obligatory *Hajj*, or pilgrimage. The most notable division in Islam is between Sunni and Shia versions, which came into being with the death of Muhammad. Sunnis believed that the *umma* was responsible for electing Muhammad's successor, while Shiites believed that Muhammad had designated a specific person (his cousin and son-in-law). The break between these two groups, formalized in 680, thus revolved around the question of human leadership. Other schisms in Islam have involved the reassertion of fundamental beliefs and practices, discussed in the following section.

Sunnis, who represent about 90 per cent of the current Muslim population, dominate in Arabic-speaking areas, in Pakistan, and in Bangladesh; Shiites are a majority in Iran and Iraq. For most of the history of Islam, the Shiite–Sunni divide has been peaceful, at least partly because the two groups usually occupied different places. But in recent years, both Iraq, which was under the political and military control of the minority Sunnis, and Iran especially have experienced conflict.

Although Islam is frequently considered a religion of the Middle East, today's largest Muslim populations are in Indonesia, India, Pakistan, and Bangladesh. Islam was carried overland and overseas as part of a larger political expansion, but it has not been as adaptable as Christianity because Muslims see their religion as a way of life that encompasses social and political affairs, meaning that it is well-suited to some societies but not to others. Islam is a more prescriptive religion than Christianity as it comprises a universal code of behaviour that adherents accept and pursue. The body of Islamic law, the *shariah*, serves as a basis for the religion and the political state. A significant tension in many parts of the Islamic world concerns disagreements about state observance of the *shariah*, with revolutionary movements in some countries asserting that any departure from the Islamic law indicates that the government lacks legitimacy. This is a very sensitive issue in a globalizing world where ideas about universal human rights, as evident in UN documents, derive primarily from a Christian tradition.

Religion, Identity, and Conflict

The preceding account of the origins, spread, and growth of major religious groups has highlighted an important general fact about religious identity, religious territory, and conflict between religious groups: a person's sense of identity and community, and all that this implies, can often be closely tied to religion. Religions have competed, directly or indirectly, with each other and with different versions of the same religion as they have spread from source areas and become established in particular places. Indeed, such competition has sometimes resulted in conflict and the expulsion of the members of one group. In the case of competing universalizing religions and their many versions, this competition has included attempts at conversion. These circumstances can be sources of tension.

Most notably, our human tendency to identify with a specific religion has given rise to many military conflicts, from the medieval Crusades and the European religious wars of the sixteenth and seventeenth centuries to recent conflicts in Pakistan (Muslim–Hindu), Lebanon (Christian–Muslim), and Northern Ireland (Catholic–Protestant Christian). In some of these conflicts, religious differences might have been used as excuses for aggression that actually had other motives; many commentators see this as the case in Northern Ireland, arguing that the real cause of conflict was British involvement in Ireland. But there is no doubt that, in many instances, religion promotes mistrust of non-believers, an attitude that, combined with a general lack of understanding, may lead to hostility and conflict. Christian attitudes to, and sometimes forced expulsion of, Jews and Muslims were referred to in the preceding section.

The example of the area of Palestine is informative. Locations here are sacred to Jews, Christians, and Muslims, and the region has long been contested space. Beginning in 1095 and continuing for about the next 200 years, the Crusades were attempts by Christian Europe to wrest control of Palestine from Muslim empires. More recently, especially since the creation of the state of Israel in 1948, conflict has intensified between Jews and Muslims concerning the legitimacy of the state as well as access to and ownership of many locations within it.

As discussed in Box 7.6 and in Chapter 9, some contemporary scholars point to hostility between Islam and Christianity as a major cause of conflict in the contemporary world, although other scholars consider this a gross oversimplification. Not surprisingly, this debate intensified following 9/11 and has been exacerbated by subsequent incidents such as the 2015 murder of French journalists in Paris. What is clear is that just a few religious fanatics can take actions that have drastic consequences on very large populations. Only 19 people were physically involved in the terror attacks of 11 September 2001, but their actions contributed to American-initiated invasions of Afghanistan and Iraq. In turn, these invasions prompted many Muslims elsewhere to worry about the possibility of American invasion and to numerous expressions of anti-American sentiment. More generally, it appears that mutual misunderstanding contributes to disagreements and even outrage.

Religious Landscapes

Religion and Identity Today

Religion and landscape are often inextricably interwoven, for three principal reasons:

In the News

BOX 7.6 | Islamic and Christian Identities

Some of the differences between the two closely related religions of Islam and Christianity are assuming major importance in the opening decades of the twenty-first century. Indeed, especially since the 2001 attacks on the United States, discussions of fundamentalist Islam, frequently called jihadism, are rarely out of the news. But it is important to place these tensions in perspective (something that is not necessarily a part of many news stories) by stressing that only a small fraction of the about 1.5 billion Muslims in the world are involved in violent activities against the Western world. The few Muslims who do subscribe to extremist views reflect an Islamic reformist ideology, a jihadist movement propagated in Arabia in the 1740s by Muhammad ibn Abd al-Wahhab and known as Wahhabism, which is based on the claim that an adherent has a religious duty to kill someone who does not convert to his way of thinking. Wahhabism did not prosper in its extreme form, although it played a role in the 1932 creation of the Saudi state; opponents of Wahhabism, along with many Shia Muslims, were massacred in the decades prior to state creation. Wahhabism has also influenced jihadist movements in Mali, Niger, northern Nigeria, Cameroon, and Sudan. In many of these countries, Islamic and Christian traditions are becoming increasingly polarized. Wahhabism has survived to this day, being evident in the Taliban in Afghanistan, in the Al-Qaeda movement led by Osama bin Laden (who was killed in 2011), and in the Islamic State movement.

To understand these religious outlooks, it is crucial to appreciate that most Muslim countries are profoundly Muslim, whereas most Christian countries are no longer profoundly Christian. In the Muslim world, values that might compete with religion have not been as evident as in the Christian world. The influential and often controversial scholar of Islam, Bernard Lewis (2003), argues that the West sees the world as a system of nations, whereas Islam sees the world as a system of religions. A key idea in Islam

Continued

is that religious unity is more important than tribal loyalties. Consequently, Muslim countries have found it difficult to create successful democracies, although some Muslim countries with large populations are democratic, notably Indonesia, Malaysia, Bangladesh, and Turkey.

Part of the difference in viewpoint is that democracy implies that humans make laws, whereas more fundamentalist interpretations of Islam contend that the laws dictated by God to Muhammad cannot be changed. This belief creates close ties between religion and state in some Islamic countries, while these countries are seen by the Western world as overly rigid and inflexible. Saudi Arabia, for example, bans the practice of all religions except Islam and, as a Sunni Islam state, even looks unfavourably on other versions of Islam. At the same time, however, most Saudis are unsympathetic to fundamentalism and strongly oppose violence conducted in the name of Islam. For many Western observers, these differing points of view within Islam appear to be contradictory. An unfortunate tendency is to treat the apparent illogic in dehumanizing terms.

On the other hand, Muslims sometimes view the West as practising double standards. The 2006 publication (initially in a Danish newspaper and subsequently in other European papers) of cartoons deemed by some Muslims as offensive to Muhammad was justified by many European commentators as freedom of speech. But many Muslim observers were quick to point out that European countries typically have limits to freedom of speech on the grounds of national security and prevention of disorder. Most notably, 11 countries—Austria, Belgium, the Czech Republic, France, Germany, Israel, Lithuania, Poland, Romania, Slovakia, and Switzerland—have laws against denying the Holocaust. Coincidentally, British historian David Irving was jailed in Austria for denying the Holocaust at about the same time as the cartoon controversy erupted.

The cartoon issue was just one specific expression of an underlying tension that also received expression in, for example, the French ban on head scarves and other displays of religious symbols in state schools. More recently, the production and Internet distribution of an anti-Muslim film, apparently by a Coptic Christian of Egyptian origin residing in the US, led to the deaths of four American diplomats in Libya and anti-American demonstrations and riots throughout the Muslim world in September 2012.

Improved understanding of and respect for each other within the Islamic and Christian worlds are necessary for a more stable and secure cultural and political world. A better understanding of Islam might develop in the Western world if there was fuller appreciation of some of the contributions made by Muslims. Here are just four examples of many: the first pin-hole camera was invented by Ibn al-Haitham after he realized that light entered the eye rather than leaving it; the crankshaft, which translates rotary into linear motion, was invented about 1200 by al-Jazari to raise water for irrigation; the windmill was invented in the seventh-century Islamic world; and soap was perfected in the Islamic world.

Religions, like languages, not only divide our world but also regularly encourage people to engage in what may appear, from a detached perspective, to be inappropriate behaviour; recall James's comment at the beginning of this chapter. In some cases, religion serves as an even more potent force for group unification than language.

1. Beliefs about nature and about how humans relate to nature are integral parts of many religions.
2. Many religions explicitly choose to display their identity in landscape.
3. Members of religious groups identify some places and load them with meaning—these are called sacred spaces.

Religious Perspectives on Nature

In many cases, an important function of religion is to serve an intermediary role between humans and nature—although the type of relationship favoured varies. Judaism and Christianity place God above humans and humans above nature. In other words, they have traditionally incorporated an attitude of human dominance over the physical environment, which is reflected in numerous ways. Christians in particular have seen themselves as fulfilling an obligation to tame and control the land. Other religions, however, take a very different view. Eastern religions in general, as well as many Aboriginal belief systems, see humans as a part of, not apart from, nature and both as having equivalent status under God. The result is a quite different relationship between humans and land.

Religious beliefs about human use of land, plants, and animals can have significant impacts on regional economies. For example, the pig is a common domesticated animal in Christian areas but absent in Islamic and Jewish areas. The traditional Catholic avoidance of meat one day per week prompted European fishermen to sail across the North Atlantic to fish the teeming Grand Banks off the coast of Newfoundland long before Europeans settled in the New World. Viticulture diffused with Christianity because of the use of wine as a sacrament. Hindus regard cows as sacred animals that are not to be consumed. These are among the most familiar examples of how beliefs influence human behaviour.

Different religions incorporate different beliefs and attitudes, and what is important to one group may not be important to another. When religion affects the way we use land, it can be a powerful cultural factor operating against economic logic.

Religious Displays of Identity in Landscape

Many religions, especially when they are a minority group, may choose to reinforce their identity through settling in close proximity. The Mormon example discussed in Box 7.2 is a prime case, and there are many others in North America among smaller Protestant denominations such as the Amish and Mennonites. In Europe the Jewish ghetto was commonplace in many cities, and this preference was brought to North America.

Landscape is a natural repository for religious creations, a vehicle for displaying religion. Sacred structures, in particular, are a part of the visible landscape. Hindu temples are intended to house gods, not large numbers of people, and are designed accordingly; Buddhist temples serve a similar function. By contrast, Islamic temples (mosques) are built to accommodate large congregations of people, as are Christian churches and cathedrals; yet these different faiths can affect landscape differently in large, modern cities such as Toronto. Islamic temples, because they do not use the conventional seating of Christian churches and can accommodate many more people in the same amount of space, require larger parking lots outside the building (Hoernig and Walton-Roberts, 2006: 414–15). The size and decoration of religious buildings often reflect the prosperity, as well as the piety or devotion, of the local area at the time of construction.

When a religious building is proposed for an area where there are no similar buildings, established residents might interpret it as a threat to traditional and generally accepted religious identity, causing friction. Europe has seen many recent disputes over the building of mosques needed by the large immigrant Muslim population. A particularly sensitive case erupted in London when a large and secretive Islamic sect, Tablighi Jamaat, proposed construction of what would be the largest mosque in Europe (housing 70,000 worshippers) close to the financial district and the main site of the 2012 Olympic Games. As of 2015, the future of a downsized version of this proposed mosque is uncertain.

Religious identity is frequently displayed by adherents at the personal scale of the body. Sikh men and Ultra Orthodox Jewish men, to cite just two examples, are readily visible to others through choice of dress and beards. Muslim women often favour the wearing of veils or burkas, some Christians choose to wear crosses, and some Hindus are marked on the forehead.

Religion and Sacred Spaces

Most religions recognize a holy land: there is the promised land of Israel for Jews, all of India for some Hindu fundamentalists, western Arabia (including Mecca and Medina) for Muslims, and the larger Palestine area for Christians. As

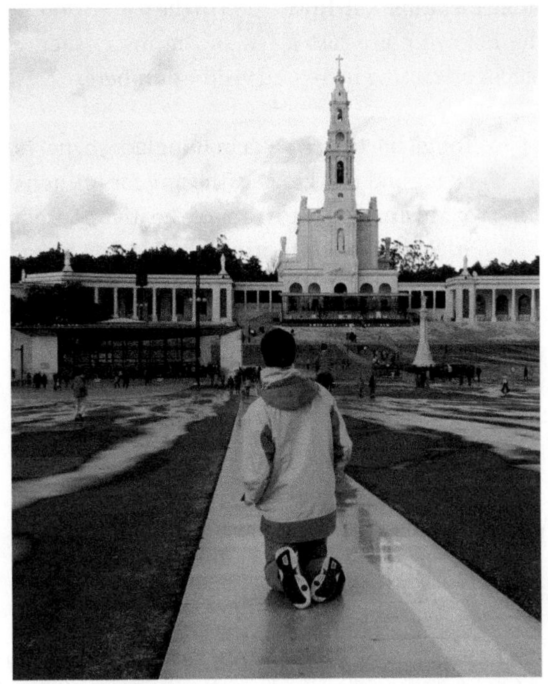

Penitents approach the shrine of Our Lady of Fatima, in Portugal. Roughly four million people visit the shrine each year.

discussed earlier, several different religions sometimes value the same places, many times with unfortunate results. Indeed, sacred places are not always treated with respect, especially when there are competing claims. The traditional site of Christ's crucifixion and resurrection, the Church of the Holy Sepulchre in the old city of Jerusalem, has been the site of several recent brawls between Greek Orthodox and Armenian monks, each group blaming the other for the incidents. Six Christian sects share control of the church, and disagreements are not unusual.

Many religions ascribe a special status to certain features of the physical environment. Rivers and mountains may be sacred places, including the Ganges for Hindus, the Jordan for Christians, and Mount Fuji in Shintoism. Human environments may also achieve sacred status, including Mecca for Islam, Varanasi for Hinduism, and the Vatican for Catholicism. Some sites attract diverse visitors: the Golden Temple at Amritsar, India, is sacred to Sikhs but also attracts many others, as do Lourdes in France and Westminster Abbey in London.

More generally, almost any religious addition to landscape—church, temple, mosque, cemetery, shrine—is sacred. With sacredness come tourists. What might be called faith tourism is no longer a small niche market. In addition to the places noted, the following are just a few of the many sacred places attracting increased visitor numbers.

- Tongi, north of Dhaka in Bangladesh, hosts an annual three-day gathering for millions of Muslims on the banks of the River Turag.
- Sri Pada, a mountain in southern Sri Lanka, is sacred to Hindus, Buddhists, Muslims, and Christians, with each group having a specific reason; Buddhists, for example, believe that Buddha ascended the mountain, leaving a footprint.
- Uman, in central Ukraine, is where Rabbi Nachman of Breslov, an influential Hassidic Jew, died and was buried; the site attracts thousands of pilgrims during Rosh Hashanah.
- Mount Kailash, in the remote western Tibetan Plateau of China, is sacred to Buddhists, Hindus, and other groups; it is believed that walking around this holy mountain erases a lifetime of sins.
- Medjugorje, a small village in Bosnia, is where local youth claimed to see the Virgin Mary in 1981; subsequent sightings have been claimed, with the result that this is now a major location of faith tourism.
- Djenne, in Mali, has long been a centre of Islamic learning and pilgrimage; the Grand Mosque is the tallest dried-earth building in the world.
- Mount Athos, in Greece, is the oldest surviving monastic community in the world and receives a strictly limited number of male visitors for brief stays.

In Calcutta (since 2001, officially spelled Kolkata), traditional Hindu temple architecture reflected the belief that temples are the homes of gods. Because mountains are also traditional dwelling places of gods, temple towers were built to resemble mountain peaks, while small rooms inside the temples resembled caverns. In the late eighteenth century, these temples were replaced by flat-topped two- or three-storey buildings built next to the homes of the very rich. Finally, in this century, a series of new temple styles appeared as a result of the pressures of urbanization and related institutional processes. These styles are described in detail by Biswas (1984) and are a clear example of the effects of changing power relations.

Places to house gods or to gather for worship are typical features of most landscapes. In many Christian communities, the church is a religious and social centre serving many extra-religious functions. Although some religious groups actively reject such external expressions of religion as churches and create landscapes devoid of religious expression (Box 7.7), such cases are unusual.

Other religious practices that create a distinctive landscape include the construction of roadside shrines and the use of land for burying the dead. Hindus and Buddhists cremate their dead, but Muslims and Christians traditionally opt for burial, a practice that requires considerable space.

These brief comments do little more than highlight some of the many ways in which religious beliefs may be expressed in landscape. Additional examples of religion and landscape symbolism will be noted in the following chapter.

Around the Globe

BOX 7.7 | Religious Landscapes: Hutterites and Doukhobors in the Canadian West

Most landscapes include evidence of religious occupation. Places of worship in particular are visible even in an increasingly secular developed world. Unlike the Mormons (Box 7.2), the Hutterites and Doukhobors, in parts of the Canadian West, have not settled across a broad region or aspired to express themselves in landscape. Therefore, the impact of their religion on landscape is not quite so visible.

A number of Protestant Anabaptist groups emerged during the Reformation in Europe. All shared a belief in adult baptism and a literal interpretation of the Bible, but they varied in other areas of religious belief and practice. Three such groups, all of which immigrated to North America because of persecution in Europe, are the Hutterites, Mennonites, and Amish. Each expressed some aspects of their religious belief in landscape. Hutterites immigrated beginning after World War I and continue to live in the Canadian prairies, creating a landscape that reflects their belief in community and their desire for isolation from the larger world (Simpson-Housley, 1978). At the centre of each communal settlement are a kitchen complex and long houses, and around them are buildings used for economic functions; the two types of building are painted different colours. Hutterite communal settlements do not include commercial stores or bars. The farming landscape typically involves a greater diversity of activities than do neighbouring farms.

Doukhobors, like Hutterites, are a Christian sect. They broke from the Orthodox Church in the eighteenth century and were banished to the Caucasus region, where they built a flourishing community. As a result of persecution, they immigrated to Canada beginning in 1898. Doukhobors reject the "externalities" of religion; their settlements are thus without houses of worship and there are no crosses or spires in the landscape. In short, the Doukhobor areas lack any religious symbolism. Doukhobors also believe in the equality of life; hence, their settlements are communal and all work and financial matters have a group rather than an individual focus. Communal living arrangements required the construction of distinctive double houses that accommodate up to 100 people (Gale and Koroscil, 1977).

Although most Christian landscapes show evidence of religious symbolism and beliefs, such as houses of worship, cemeteries, and roadside shrines, some areas settled by community-based religious groups reflect group organization even when they do not include obvious symbolic features. Both Hutterites and Doukhobors belong in the second category.

AP Photo/Aman Sharma

Five baptized Sikhs lead a procession at the Golden Temple in Amritsar, India, in 2003, marking the 337th anniversary of the birth of the tenth guru, Gobind Singh.

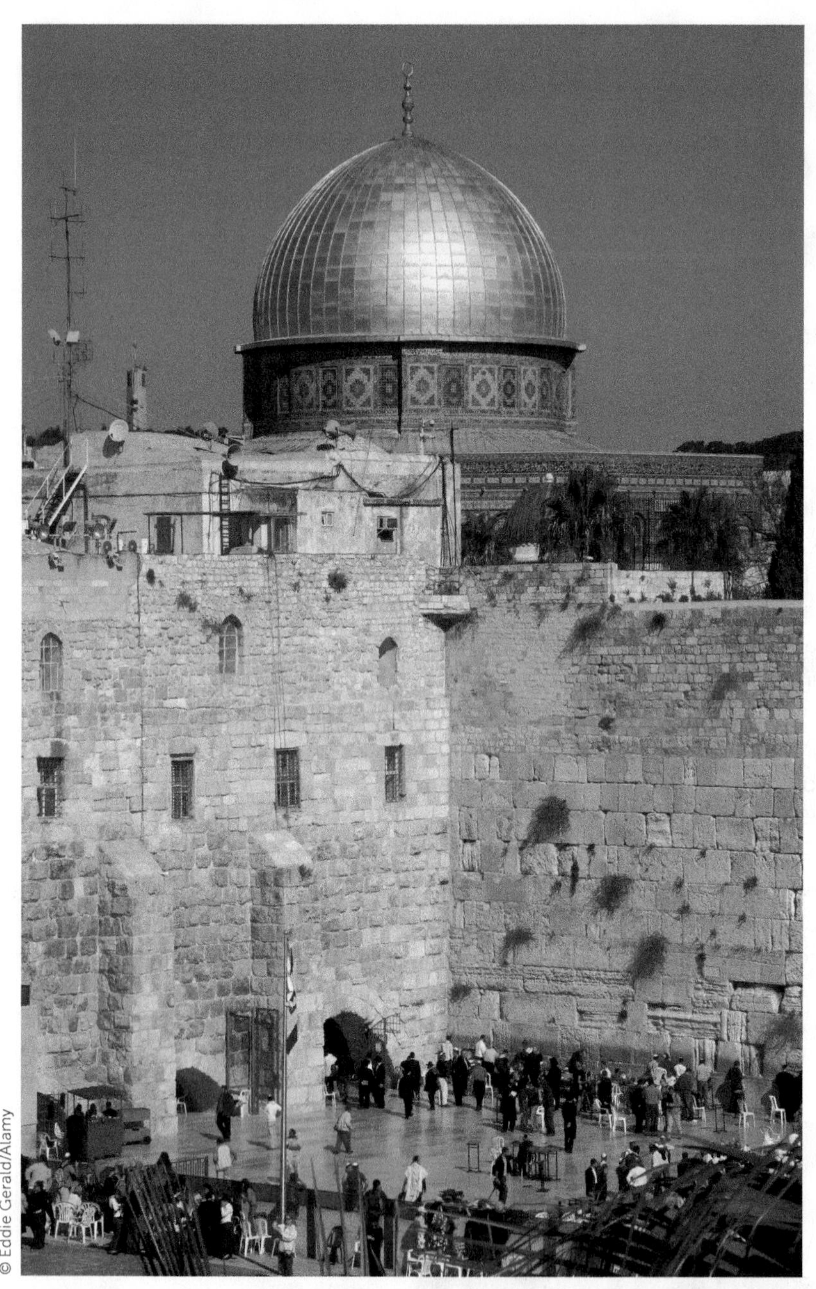

The Western or "Wailing" Wall and Dome of the Rock mosque in Jerusalem.

CONCLUSION

The contents of this chapter on cultural geography reflect its importance to human geography both of regions—one of our three recurring themes—and of the culturally based divisions described by James at the beginning of the chapter. Delimiting regions based on landscape and identity is an instructive exercise, but the resulting divisions must always be interpreted cautiously as they are essentially just one means of description rather than a necessary reflection of reality.

Although in some respects culturally based divisions between groups and places are being reduced as part of the cultural globalization discussed in Chapter 3, it is clear that some linguistic and religious differences are potent divisive forces in the contemporary world. Indeed, some religious differences are playing increasingly critical roles in global political life, especially because of the increasing popularity of fundamentalist interpretations of some major religions. It is also notable that some places are highly contested because they are understood to be sacred by two or more religious groups. Conflicts relating to access to and ownership of such places are commonplace today.

The continuing decline of the number of languages spoken in the world is routinely understood as unfortunate because it can be seen as a process of cultural forgetting, involving the loss of an identity and of a particular way of viewing our world. Clearly, however, it can be interpreted rather differently as involving the removal of culturally imposed barriers. It might also be claimed that some religious differences are disappearing, as evidenced by the increasing number of people who identify themselves as non-religious.

Perhaps the single most important conclusion is to acknowledge that cultural variables are human constructions. While these constructions are an integral part of our humanity, they do serve to promote divisions between people and landscape that have often led to conflict between groups. This fundamental point reappears in both Chapters 8 and 9.

Summary

Our Divided World

The human world is divided physically and, more important, culturally.

Cultural Regions

Regions can be delimited on various scales. The world scale can provide a useful overview but lacks precision. North America can be usefully regionalized, from a European settler point of view, using the concept of first effective settlement. Europe is usefully defined by reference to specific traits.

Cultural Landscapes

Cultural regions have distinct cultural landscapes because of the impact of culture on land and the variations in human–land relationships. "Cultural adaptation" is cultural change in response to environmental and cultural challenges. Geographers studying cultural adaptation have proposed a variety of explanations, including (along with first effective settlement) culturally habituated predisposition, cultural pre-adaptation, and the core, domain, and sphere model. Landscapes change also because of the diffusion of ideas and innovations, and geographers have employed three general approaches to diffusion research: a Sauerian culture trait emphasis, spatial analysis, and political economy.

Place

Landscapes are recognized as signifying systems that have meanings; we used the term *place* in this sense. In addition, they are interpreted as the continually changing product of ongoing struggles between groups and between different attitudes and behaviours. The distinction between place and placelessness is particularly significant, as is that between the related ideas of local and global environments.

Vernacular Regions

Vernacular regions are perceived to exist by people living inside and/or outside them; they may be created institutionally and typically possess a strong sense of place. In North America, the map of vernacular regions closely resembles the map of regions delimited using the concept of first effective European settlement. A related area of interest is known as psychogeography. Some vernacular regions may qualify as homelands if they are closely identified with a distinctive cultural group.

Languages

Probably the single most important human achievement, language, is an essential key to understanding human groups, their attitudes, beliefs, and behaviours. Indeed, it is not far-fetched to assert that language is the single most appropriate indicator of culture.

As the basis for group communication, language is the earliest source of group unity and the means by which cultures continue through time. For many groups, language is argued to be culture and presented as the principal basis for a national identity. There are currently as many as 6,000 languages in the world. Individual languages can be grouped into families—languages that share a common origin. The language family with the most speakers is Indo-European. Mandarin has more speakers than any other single language, but English is the most widespread and the nearest to a world language. Many languages are likely to disappear.

Languages create barriers between groups. A characteristic of multilingual states is that they are less stable than unilingual states.

Place names—toponyms—are the clearest expression of language on landscape. Place-name studies help us understand early settlement. But places also are renamed, often for political reasons. Language also helps to make landscape in the sense that a place may become what it is named—as in the case of the Great American Desert. Both physical and human landscapes are reflected in language.

Religions

For some people, religion is the basis of life. Thus, religion is a useful variable for regionalizing and may be even more powerful than language in reinforcing group identity. Because different religions affect attitudes and behaviour in different ways, they help to distinguish one group from another and to promote group cohesiveness. Religions are usefully classified as ethnic or universalizing. A feature of today's world is the increasing tension between Islam and Christianity.

Many religions function as intermediaries between humans and nature. Specific physical environments are regularly ascribed a special status as sacred spaces. Some religious beliefs can affect regional economies. Landscape also is a natural vehicle for religious expression.

Links to Other Chapters

- **Landscape:** Chapter 1 (Humboldt, Ritter, Vidal, and Sauer)
- **World cultural divisions:** Chapters 8 and 9
- ***Culture* and *society* as terms in human geography:** Chapter 1 (disciplinary evolution, especially Sauer and Vidal)
- **Social scale:** Chapter 2 (concept of scale)
- **World regions:** Chapter 2 (concept of region); Chapter 4 (environmental regions); Chapter 5 (demographic regions); Chapter 6 (more and less developed regions); Chapter 9 (political and cultural regions); Chapter 10 (agricultural regions); Chapter 11 (urban regions); Chapter 13 (industrial regions)
- **Cultural adaptation:** Chapter 2 (environmental determinism and possibilism); Chapter 4 (natural resources and human impacts in general)
- **Vernacular regions:** Chapter 2 (concepts)
- **Language and identity; religion and identity:** Chapter 8 (myth of race, ethnicity); Chapter 9 (nationalism)
- **Language and landscape; religion and landscape:** Chapter 2 (humanism); Chapter 8 (symbolic landscape)
- **Women and religion:** Chapters 5, 6, and 8 (gender)
- **Religion and land use:** Chapter 10 (agricultural land use)

Questions for Critical Thought

1. Why is culture our "greatest human achievement" on the one hand and the source of so much conflict on the other?
2. It is stated that humans might choose to "create new sets of values—in effect, to re-engineer ourselves." Can this be done? Should this be attempted? Justify your position.
3. To what extent do vernacular regions exist in the minds of people who live in them? Are such regions merely constructs of geographers that bear little or no meaning to those who live in them?
4. Discuss the rationale underlying the "regions of North America depicted in Figure 7.3. Do these regions exist in the minds of the people who live in them or are they figments of the geographer's imagination?
5. Why is language so important to culture? If a language disappears or becomes extinct, will that culture also disappear?
6. Why has religion been the basis of so much conflict throughout the course of human history?
7. Do you think cultural globalization (i.e. the homogenization of culture on a global scale) is a good thing? What would be the advantages and disadvantages of there being a single culture?

Suggested Readings

Visit the companion website for a list of suggested readings.

8 GEOGRAPHIES OF IDENTITY AND DIFFERENCE

This chapter continues the Chapter 7 exploration of identities and landscapes. However, the emphasis here is rather different. Specifically, although some of the previous chapter's material had conceptual, social, or symbolic overtones, the overall approach was empirical. This chapter is more conceptually informed and relies mostly on Marxist, feminist, and postmodern thought to inform discussions of the myth of race, ethnicity, gender, and sexuality, as these establish and maintain group identity and related landscapes. These accounts of geographies of difference are followed by an examination of contested landscapes, well-being, folk and popular culture, and tourism, each of which also highlights differences between peoples and places. The chapter concludes with a brief look at the controversial idea of social engineering.

Here are three points to consider as you read this chapter.

- Do you agree that there is no single, fixed entity called culture, but rather a plurality of cultures, understood as those values that members of human groups share in particular places at particular times?
- Traditionally, human geographers focused on describing landscapes and explaining how they came into being, but today they are also concerned with how those landscapes are lived in, experienced, contested, and resisted. What is your reaction to this broadening of interest?
- Do you agree that, although gender differences are not always as apparent in the landscape as, for example, ethnicity differences, landscapes are indeed shaped by gender?

The Paris Las Vegas is one of several expansive hotels on Nevada's famous Las Vegas Strip. With nearly 40 million visitors each year, Las Vegas is among the most popular tourist attractions in the world. Tourism not only plays a major role in the global economy but also allows groups to create and share their identities.

BALLYS
EIFFEL TOWER RESTAURANT

THE CULTURAL TURN

One way to think about the study of human geography is to suggest that a "cultural turn" has taken place. This phrase—which has been used in the context of recent changes in many of the social sciences and humanities—suggests both a different understanding of culture and an increased appreciation of the importance of culture in understanding humans and their political and economic activities.

Rethinking Culture

Chapter 7 mostly employed a single view of culture—albeit a traditional and important one with strong roots in anthropology. In this traditional interpretation, mainly for Sauer and other members of the landscape school, culture was a given. Accordingly, these geographers analyzed the impact of culture on landscape, especially as it was reflected in the material and visible landscape and the formation of regions. This view of culture as cause remained largely unquestioned in geography until about 1980; in recent years, however, several other perspectives on culture developed with new interests in both Marxism and humanism, as well as debate over the merits of the landscape school approach. The term *new cultural geography* (admittedly no longer particularly new) distinguished the revised concepts of culture from the traditional landscape school view.

critical geography
A collection of ideas and practices concerned with challenging inequalities, as these are evident in landscape.

As we saw in Chapter 2, the rise of humanist, Marxist, feminist, and postmodern approaches involved acceptance of the idea that human geography is an interpretive endeavour—in other words, questions of meaning and communication are relevant. Thus, a symbolic interpretation of culture became prominent in the discipline. This view broadens the concept of culture to more fully embrace non-material culture and to emphasize that individuals create groups through communication. This broader view of culture allows human geographers to consider topics beyond landscape, specifically topics that fall under the general heading of what might be called the spatial constitution of culture.

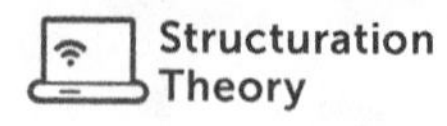
Structuration Theory

According to this logic, then, there is no single, fixed entity called culture but a plurality of cultures, understood as those values that members of human groups share in particular places at particular times. Considered from this perspective, cultures are not objects but mediums or processes, what Jackson (1989: 2) described as maps of meaning: the "codes with which meaning is constructed, conveyed, and understood." This interpretation has led geographers to study many topics besides landscape, including previously ignored groups, new cultural forms, Eurocentrism, and ideologies of domination and oppression.

Another theoretical approach, structuration theory (see, e.g. Giddens, 1984) developed from numerous earlier contributions to social theory, including Marxism, and focuses on the capacities that permit people to institute, maintain, and alter social life. But this approach has proven to be of less value to human geography.

Rethinking Identity

Human geographers now pay much attention to inequalities with respect to such variables as language, religion, ethnicity, class, gender, and sexuality, as these are reflected in landscape. Much of the work conducted in this tradition is identified as **critical geography**, a term referring to the need for geographic studies to move beyond description and explanation and become emancipatory.

Further, human geographers tend to base their questions about human identity on the logic of constructionism rather than the more traditional logic of essentialism (see Chapter 2). Whereas the essentialist view sees the characteristics that constitute identity as inherited and largely unchanging, the constructionist view stresses that those characteristics are socially made or acquired and that they are contested in the sense that there are no unequivocal meanings: different characteristics are important in different places and at different times. The distinction is important philosophically since the essentialist view is associated with empiricism, while the constructionist view is closely related to feminism and postmodernism. This shift in perspective was evident in Chapter 7, where we discussed the identity characteristics of language and religion from both viewpoints, and it is evident here as we consider the myth of race, ethnicity, gender, and sexuality and in Chapter 9, which looks at nationality from a primarily constructionist perspective.

Rethinking Landscape

Along with these new human geographic views of culture and identity has come a new understanding of landscape itself, including the natural landscape, as something that is socially constructed.

Studies in this area focus on two aspects: symbolic (specifically, the meaning contained in landscape) and represented (landscapes as represented in literature and art as well as the more usual visible and material landscapes). Such work often treats landscape as a text that is open to interpretation and, recognizing the importance of images, includes among its research methods **iconography** (the description and interpretation of images to uncover their symbolic meanings; see Box 8.1). The idea that nature is socially constructed, that it is effectively part of culture—has several important implications. We need to recognize that our understanding of nature is filtered through human representations of it and that these representations vary with time and place. We also need to be aware that the representation of nature is never neutral. Any such representation—and there may be several of them for any part of the natural world—is ideologically loaded, and it is part of the geographer's task to interpret them.

iconography
The description and interpretation of visual images, including landscape, in order to uncover their symbolic meanings; the identity of a region as expressed through symbols.

A Focus on Difference

Clearly, the cultural turn invites us to explore more than cultures as ways of life and the visible landscapes related to them (the geographic expressions of culture); it invites us to explore culture as a process in which people are actively involved and landscapes as places constructed by people (the spatial constitution of culture). Thus, a first reason for the interest in difference and inequality is the rethinking of culture, identity, and landscape. But it is also notable that difference and inequality did not emerge as topics of major interest before about 1970 because, like most other

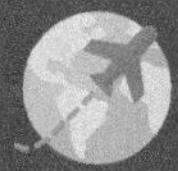

Around the Globe

BOX 8.1 | Iconographic Analyses of Canadian Landscapes

Iconography is the description and interpretation of visual images to uncover their symbolic meanings. To the extent that landscapes can be regarded as depositories of cultural meanings, it is possible to subject them to this process. A successful iconographic analysis will reveal how a landscape is shaped by and at the same time shapes the regional culture of which it is a part. Two Canadian examples are briefly summarized here.

Osborne (1988) focused on the development of a distinctive national Canadian iconography both in artistic images of lands and peoples (notably the paintings of Tom Thomson and the Group of Seven) and in the various responses these images provoked. The declared aim of these artists was to assist the development of a national identity, largely through their paintings of the Canadian North. During the first half of the twentieth century, they helped to create a distinctive image of Canada: "rock, rolling topography, expansive skies, water in all its forms, trees and forests, and the symbolic white snow and ice of the 'strong North'" (p. 172). Although this image is clearly limited and is just one among a vast range of national images, it remains a compelling example of iconography.

A different application of iconographic logic can be seen in Eyles and Peace's (1990) examination of images of Hamilton, Ontario, particularly their societal context and why they appear as they do. Derived primarily from newspaper accounts, the images were varied, but they were dominated by a single negative theme: Hamilton as Steeltown. This image most frequently arose in comparisons with neighbouring Toronto, in which Hamilton appeared as the blue-collar city of production and Toronto as the city of consumption. Because Hamilton is an industrial city, it was seen as a polluted landscape representing the past rather than the future, as an area of economic decline.

Hamilton at dusk.

academic disciplines, human geography was dominated by white, middle-class, heterosexual, able-bodied males. In the era when the dominance of patriarchal discourse was near-universal, questions of difference and inequality were rarely acknowledged.

Indeed, the cultural turn's impact has been expressly notable in studies of human identity and human difference and of the politics related to these matters. There is a persistent questioning of traditional concepts, classifications, and categories such as those used in our discussions of language and religion in Chapter 7. Acknowledging that identity is socially created and therefore subject to ongoing change rather than something predetermined and fixed has prompted human geographers to examine the power relations between dominant and other groups, as well as the politics of difference. Some key ideas about how geographic knowledge is constructed are noted in Box 8.2.

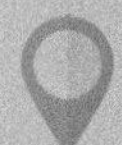

Examining the Issues

BOX 8.2 | Constructing Geographic Knowledge

Numerous theorists from various disciplines have contributed to the cultural turn in geography and have introduced important concepts for understanding this turn towards a constructionist view. Most of the material included in this box relates closely to the account of postmodernism in Chapter 2, and all of the concepts discussed are employed in this chapter.

Hegemony is a central idea. According to Antonio Gramsci (1891–1937), any group that desires to attain **power** in a larger society needs to achieve a degree of cultural and intellectual hegemony, that is, the ability to determine the ruling discourse, including what questions are or can be asked. This allows the group to express its world view and to structure social and other institutions in accord with its goals. Gramsci also introduced the term *subaltern* to refer to those socially subordinate, marginalized, exploited, and oppressed groups that lack both the unity and the organization of more dominant groups that exercise **authority** and control. The term *subaltern* has been used in the context of writing about peoples and places from "below," such as post-colonial literature produced by the colonized.

hegemony
A social condition in which members of a society interpret their interests in terms of the world view of a dominant group.

power
The capacity to affect outcomes; more specifically, to dominate others by means of violence, force, manipulation, or authority.

authority
The power or right, usually mutually recognized, to require and receive the obedience of others.

Closely linked to these ideas is the concept of discourse. Michel Foucault (1926–84) saw subjectivity as constructed within and through discourses, with discourse defined as a system, comparable to a language, that enables the world to be made intelligible. Discourses are important because they legitimize a particular view of the world that then becomes part of the taken-for-granted world. Power is practised through discourse, and dominant discourses affect how members of a society understand the world. It can be argued, for example, that the discipline of geography evolved within the late nineteenth-century discourses of imperialism, colonialism, and racism.

Building on these ideas, the cultural turn has highlighted the idea that research conducted by geographers may not be free from bias. For example, much previous work might be best described as colonial research essentially exploitive of those being studied. Post-colonial research explicitly premised on the rejection of colonial attitudes, ideas, and values is advocated instead; such research is inclusionary, seeking to empower those whose voices previously were unheard.

Another influential writer, Edward Said (b. 1935), introduced the related idea of **Orientalism**, a fundamental post-colonial concept contending that the Orient is an invention of those who study it from outside and that North America and Europe employ ideas from within their cultures (such as freedom, democracy, and individualism) to facilitate their conquest and domination of other regions. Thus, the Orient is a necessary European image of the **Other**, a construct of a dominant European discourse. The contemporary importance of these ideas is evident in Chapter 7's

Orientalism
Western views of the Orient, implying a view of the periphery from the centre; closely associated with post-colonial theory, especially the work of Edward Said.

Other
Subordinate group as seen by and contrasted to dominant groups; implies both difference and inferiority.

account of Islam and Christianity, where it was noted that parts of the Christian world sometimes see the Islamic world in dehumanized terms, a tendency exacerbated by terrorist activities and by the ongoing conflict between Palestinians and Israelis. More than one point of view may be legitimate; it is important to be attentive to multiple voices.

The concept of **contextualism** acknowledges that a specific discourse is employed in human geography. We need to be aware of, and sensitive to, precisely how knowledge is being constructed—in other words, readers need to know who is conducting the research and what their agenda is. For this reason, some current work in human geography, especially that inspired by feminist thought, explicitly acknowledges the **positionality** and **situatedness** of the authors, in short, who they are and what they believe. Expressed rather differently, the impossibility of producing universally valid and accurate representations of the world is increasingly accepted because any **representation** is actually an interpretation. Traditionally, it was assumed that a real world could be mirrored by geographers in their writings; this assumes a neutrality on the part of the author that is not feasible. Rather, geographers interpret the world through particular lenses, particularly in the context of prevailing power relations and discourses. Furthermore, representations do not simply interpret the world but also help to shape it.

contextualism
Broadly, the idea that it is necessary to take into account the specific context within which any research is conducted.

positionality
The ideological preference and the identity of the researcher as they relate to the subjects of the research.

situatedness
An idea that rejects notions of researcher authority and impartiality—knowledge is not neutral and cannot be acquired in some detached and disembodied manner but is partial and located somewhere.

representation
A depiction of the world, acknowledging the impossibility to be exact as all such depictions are affected by the researcher's identity.

Building on ideas about hegemony and power introduced in Box 8.2 and also on the postmodern focus on previously repressed experiences, Tables 8.1 and 8.2 identify some of the main issues relating to geographies of difference. Table 8.1 suggests that difference is evident at various geographic scales, from the body through to the entire world, and Table 8.2 provides examples of inclusion and exclusion at different geographic scales. Together, these two tables suggest the need for geographers to think critically about how landscapes are lived in, experienced, contested, and resisted. The need to focus on difference results from recognition that, put simply, "space and place are intimately connected to . . . gender, class, sexuality and other axes of power; all geographic knowledges are situated, and location matters" (McKittrick and Peake, 2005: 43).

The cultural turn has opened a number of new directions for human geographers. Much traditional work in cultural geography was informed by the Sauerian landscape school tradition and now seems partial. This tradition saw humans modifying physical geography into a cultural landscape and emphasized the regional mapping of material and visible features of landscape on the grounds that such mapping assumes the existence of a single distinctive and unchanging cultural group constituting a culture. Contemporary cultural geography, on the other hand, stresses the fact that some groups of people are able, because of their position of dominance, to exercise power over other groups and hence that different groups have markedly different abilities and opportunities to build landscapes that reflect their interests and needs. The following accounts of geographies of difference first focus on four important components of human identity: the myth of race, ethnicity, gender, and sexuality.

TABLE 8.1 | Examples of scales of difference

	Body	Home	Nation	Globe
Geographies of domination	Racialization, racism, heterosexism	Domestic violence, domestic labour	Colonization, genocide, apartheid	Imperialism, globalization, language, religion
Experiential geographies	Bodily geographies, such as transgendered bodies	Geographies of fear, geographies of fleeing or staying put	Geographies of diaspora and migration; critiques of nation	Geographies of fair trade, refugees; anti-globalization activism

Source: Adapted from McKittrick, K., and L. Peake. 2005. "What Difference Does Difference Make to Geography?" In *Questioning Geography*, edited by N. Castree, A. Rogers, and D. Sherman, 45. New York: Blackwell.

TABLE 8.2 | Examples of scales of inclusions and exclusions

	Home	Neighbourhood	Nation	Globe
Spaces of assimilation and/or exclusion	Master bedrooms, den	Ethnic neighbourhoods, gated communities	Public spaces such as parks, shopping malls	Trading blocs, language, religion
Spaces of containment/ internment/exile	Homeless shelters, homelessness	Ghettos	Concentration camps, refugee camps, prisons	Apartheid-like systems
Spaces of objectification	Women in home seen as housewives	Youth in malls seen as delinquent	Immigrants seen as drain on national welfare system	Women in less developed world seen as victims of development processes

Source: Adapted from McKittrick, K., and L. Peake. 2005. "What Difference Does Difference Make to Geography?" In *Questioning Geography*, edited by N. Castree, A. Rogers, and D. Sherman, 45. New York: Blackwell.

THE MYTH OF RACE

Scientific research is finally dispelling three long-standing myths about human evolution. A first myth concerns the idea of human evolution, or evolution generally, as some sort of ladder of progress. Evolution is not a history of steady progress to a finished product; it is not correctly seen as a ladder. Rather, to borrow a metaphor popularized by the distinguished evolutionary biologist Stephen Jay Gould, evolution is best seen as a bush with endlessly branching twigs, each representing a different **species**. In the terms of this metaphor, the human species is "a fragile little twig of recent origin" (Gould, 1987: 19).

species
A group of organisms able to produce fertile offspring among themselves but not with any other group.

race
A subspecies; a physically distinguishable population within a species.

phenotype
Any physical or chemical trait that can be observed or measured.

genocide
An organized, systematic effort to destroy a group defined in ethnic terms; usually the targeted group is seen as living in the "wrong place."

A second myth concerns the supposed existence of several different human **races** (Gould, 1985; Montague, 1964). As modern humans, *Homo sapiens sapiens*, moved from Africa to different physical environments and splintered into distinct spatial groupings, adaptations developed not only in culture but also in some body features, or **phenotypes**. Relatively distinct phenotypes emerged: different skin colours, head shapes, blood groups, and so forth. Certain characteristics became dominant in certain areas, either because they were environmentally appropriate or as a result of chance genetic developments. The resulting variations in human physical characteristics are quite minor and do not in any way represent different types of humans (Box 8.3).

A third myth is the idea that races not only exist but also can be classified according to their level of intelligence. We are fortunate in that we have clear scientific evidence of the unity of the human race. Such knowledge was not available until relatively recently. The characteristic response to physical variations in humans, particularly different skin colours, has been to regard these as evidence of different types of humans with different levels of intelligence. Box 8.4 summarizes the history of the erroneous idea that the human population can be divided into distinct and unequal groups, usually called races. The reality is that all humans are members of the same species; the only differences are of secondary characteristics, such as skin colour (Kennedy, 1976).

Regardless of the fact of human biological unity, the term *race* is commonly used to set apart outsiders whose physical appearance does not accord with some generally accepted norm. The key divide is the most visible one: skin colour. The fact that races do not exist does not prevent the label from being applied, frequently with tragic consequences.

Racism, Identities, and Genocide

Tragically, groups seen as different from the majority within a society may be labelled negatively and treated unequally by the majority group that has cultural and intellectual hegemony and is able to construct the prevailing discourse. One extreme consequence of this circumstance is the targeting of a group for **genocide**. The UN Convention on Genocide, written following the Holocaust, sees victim groups in national, ethnic,

Examining the Issues

BOX 8.3 | Species and Races

Homo sapiens sapiens is the only species of humans in the world. That all humans are members of the same species is biologically confirmed by the fact that any human male is able to breed with any human female.

All species, including humans, display variation among their members. Indeed, many plant and animal species exhibit sufficient variation that they can be conveniently divided into subspecies or races, geographically defined aggregates of local populations. Different races emerge within a species if groups are isolated from one another, in different environments, for the necessary length of time. As a consequence of selective breeding (the result of isolation), together with various adaptations to their respective environments, each group will eventually exhibit some genetic differences from every other group.

To what extent is it possible to delimit distinct races—geographically defined aggregates of local populations—in the human species? To answer this question, we need to identify three stages:

1. Current understanding suggests that *Homo sapiens sapiens* evolved by 150,000–195,000 years ago, probably only in Africa.
2. Over the course of many millennia, members of this species moved across the surface of the earth, replacing archaic *Homo sapiens* and gradually separating into various groups. Consequently, some selective breeding and adaptation to different environments occurred, giving rise to some minor genetic differences.
3. As a result of population increases and further migrations that began in the fifteenth century, groups that had been temporarily isolated began to mix with other groups.

During the second stage the groups of humans often known as Negroid, Mongoloid, Caucasoid, and Australoid appeared. But are these groups races? The answer is no, because none was ever isolated long enough or completely enough to allow separate independent genetic changes to occur. Use of the word *race* is therefore mistaken because, genetically speaking, humans cannot be divided into clear-cut stable subspecies. This fact was powerfully affirmed in 2000 when researchers from the Human Genome Project announced that all humans are 99.9 per cent identical at the genetic level and that there is hence no genetic basis for "race."

There are some small genetic differences between members of the human species, such as skin colour and shape of eyes, as a consequence of the earlier period of isolation of groups in different environments. Biologically, however, these differences are not sufficient to justify the identification of separate races. Thus, the differences among people throughout the world are very few and very minor. There are no subspecies—races—within our human species.

racial, or religious terms, although some contemporary scholars favour a broadening of the term to include groups defined politically, occupationally, or socially. Thus, genocide "is a form of one-sided mass killing in which a state or other authority intends to destroy a group, as that group and membership are defined by the perpetrator" (Chalk and Jonassohn, 1990: 23).

A necessary precondition for genocide is a symbolic, and sometimes spatial, distancing or separation of one group, the perpetrator "in-group," and another, the victim "out-group." The victim group is regularly given a derogatory label, further emphasizing that they do not belong. Identities frequently are linked to place, and in most cases perpetrators justify genocide in terms of their "right" as a group to occupy a particular place without others present. For genocide to occur, the perpetrator group must be able to exercise power and authority over the victim group. Totalitarian states have been especially prone to commit genocide because they function without real constraints on their ability to exercise power. The worst mass killings of the twentieth century were committed by dictatorial states in which most aspects of cultural, political, and economic life were controlled by the government. Genocide requires the participation of many people in the perpetrator group, and this level of participation is more likely

Examining the Issues

BOX 8.4 | A History of Racism

Box 8.3 explained the unity of the human species: there are no distinct subspecies or races within that species. Notwithstanding this biological fact, the concept of distinct racial groupings has long been popular in both lay and scientific circles. Ideas about the existence of races and the relative abilities of the supposed races have been central to many cultures.

Racism is the belief that human progress is inevitably linked to the existence of distinct races. Notions of racial variations in ability appear to have been common in most cultures. They flowered especially in Europe beginning about 1450 with the overseas movement of Europeans, which brought them into contact with different human groups, and new theories were later developed according to which of these groups came from different origins. This notion of multiple origins is incorrect.

By the eighteenth century, racial explanations for the increasingly evident variety of global cultures were standard. Philosophers such as Voltaire and David Hume perceived clear differences in ability between what were commonly seen as racial groups. Undoubtedly one of the major racist thinkers was Joseph Arthur de Gobineau, whose *Essay on the Inequality of Human Races* was published in 1853–5. Gobineau ranked races as follows: whites, Asians, Negroes. Within the whites, the Germanic peoples were seen as the most able. A second influential racist was Houston Stewart Chamberlain, but racist logic was apparent among many of the greatest scientists. Darwin wrote of a future when the gap between human and ape would increase because such intermediaries as the chimpanzee and Hottentot would be exterminated (see Gould, 1981: 36). As we have seen, the racial categories employed in all such discussions have no biological meaning.

One explanation for the popularity of racist thought is the fact that most cultures classify other cultures relative to themselves. The consequences of racist thinking are varied and considerable. Belief in the inferiority of specific groups has led to mass exterminations, slavery, restrictive immigration policies, and, most generally, unjust treatment. The American Anthropological Association Statement on Race (1998) summarizes the development and consequences of thinking based on race:

racism
A particular form of prejudice that attributes characteristics of superiority or inferiority to a group of people who share some physically inherited characteristics.

> "Race" . . . evolved as a worldview, a body of prejudgments that distorts our ideas about human differences and group behavior. Racial beliefs constitute myths about the diversity in the human species and about the abilities and behavior of people homogenized into "racial" categories. . . . Given what we know about the capacity of normal humans to achieve and function within any culture, we conclude that present-day inequalities between so-called racial groups are not consequences of their biological inheritance but products of historical and contemporary social, economic, educational, and political circumstances.

Both past and present discussions about racism generate heated debate. For example, as noted in Chapter 6, discussions at the 2001 United Nations World Conference against Racism, Racial Discrimination, Xenophobia, and Related Forms of Intolerance included arguments that rich countries should pay poor countries compensation for the abuses (such as slavery) inflicted on them in the past and disputes about whether or not Zionism is racism. On both sides of the colonial divide, of the "privileged" West and the "othered" Third World, politics is often central to discussions of racism. In 2009 a follow-up UN conference held in Geneva was boycotted by many Western countries, including Canada, over concerns about anti-Western and anti-Israel bias. Of particular concern was that the then president of Iran, who has denied the Holocaust and made homophobic statements, was to address the conference.

to occur when the actions taken are formally authorized by the state, such that participating in mass killing is legitimized. Further, much evidence suggests that most of those who participate in genocide are "relatively ordinary people engaged in extraordinary behaviors that they are somehow able to define as acceptable, necessary, and even praiseworthy" (Alvarez, 2001: 20). Most cases of genocide occur at times of severe national tension.

Twentieth-century examples of genocide include forced movements and murder of Armenians in Turkey during World War I; population purges and deportations in the USSR from the 1920s to the 1950s; the Holocaust perpetrated by Nazi Germany against the Jews, as well as against the Roma (gypsies), the disabled, and homosexuals; mass killings in Cambodia by the communist Khmer Rouge between 1975 and 1979; killing of Kurds in Iraq in the 1980s; killing of Muslims by Serbs in the former Yugoslavia in the 1990s; and Hutu killing of Tutsis and moderate Hutu in Rwanda in 1994. In all these cases, genocide is part of larger efforts by one group to dramatically reshape the geography of peoples and places. Racist and nationalist ideas form the basis for categorizing populations into groups. Typically, victim groups are blamed for any and all social and economic problems and are seen as impediments to progress and prosperity. The perpetrators, using powerful metaphors of "purifying" and "cleansing," rationalize genocide on the grounds that victim groups are something less than human.

Racism, Identities, and Apartheid

In areas of European overseas settlement during the colonial era, it was usual for the dominant European group to see others as different, to treat them differently, and to impose some form of spatial separation. Once in place, identity and spatial separation often continued. The most extreme case was in South Africa, where the first permanent European settlement was established in 1652 to grow food to supply Dutch ships. The colony became British in 1806; from then until 1948 it was predominantly British in political character.

The history of South Africa was characterized by a series of conflicts between several relatively distinct groups, namely British, Afrikaners (descendants of early Dutch settlers), and Africans. Segregation of different groups, as determined by the British, was characteristic of the landscape prior to 1948, but in that year an Afrikaner-dominated government passed legislation to institutionalize a much more rigid system known as **apartheid**, or "separate development." Hence, apartheid had roots in long traditions of social conflict and of relatively informal segregation. Understanding apartheid, however, clearly requires further consideration of the Afrikaners, the social group that formally institutionalized it.

As their name suggests, the Afrikaners separated themselves from their European heritage, although their language, Afrikaans, is derived from Dutch and they have traditionally belonged to the Dutch Reformed Church. By the twentieth century they identified themselves fully with Africa, often being called "the white tribe." In 1948 the Afrikaners gained political control of a state made up of whites, blacks (of various linguistic and tribal groups), coloureds (people of mixed descent), and Asians. On gaining power, the Afrikaners instituted apartheid on the grounds that it was the best way to allow a socially fragmented country to evolve; their argument was that integration of different groups leads to moral decay and racial pollution, whereas segregation leads to political and economic independence for each group. As shown in Figure 8.1, 10 African homelands were created, 4 of which were theoretically independent (their independence was never recognized by the UN or by any government outside South Africa).

According to the Afrikaners, then, the social and spatial separations imposed by apartheid provided blacks with independent states and rights comparable to those of whites. In practice, they simply enabled the whites to maintain their own cultural identity and political power while exploiting black labour. Interestingly, the dismantling of the apartheid system, which began in

apartheid
The South African policy by which four groups of people, as defined by the authorities, were spatially separated between 1948 and 1994.

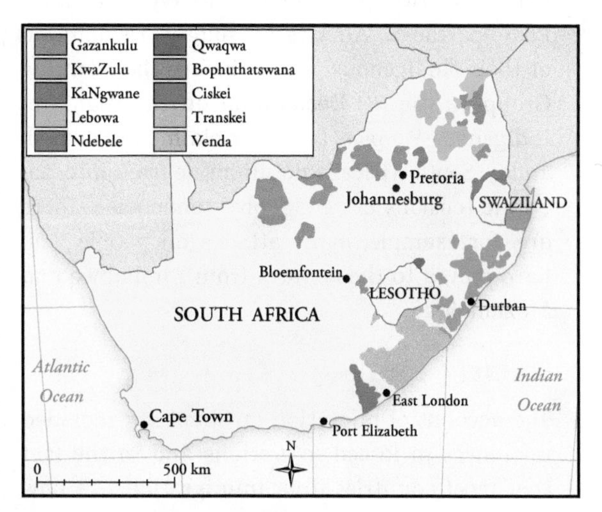

FIGURE 8.1 | South African "homelands"
Apartheid on the national scale: the "homelands" planned by the South African government in 1975.

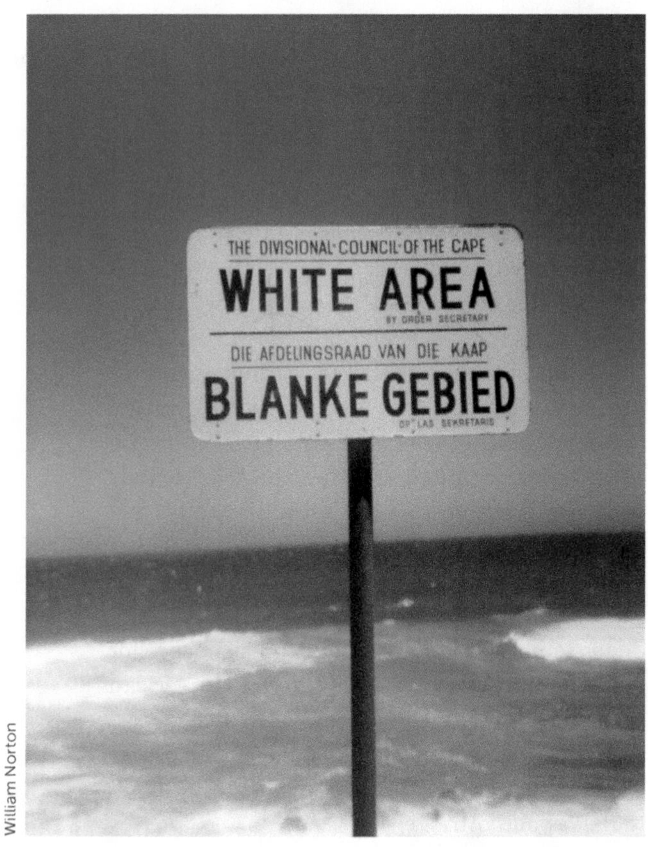

William Norton

During the apartheid era, desirable landscapes, such as this Indian Ocean beach north of Port Elizabeth, South Africa, were designated for "whites only." Similar signs were placed in many urban parks, while official buildings, such as post offices, had separate entrances for whites and other groups.

the late 1980s and concluded with the 1994 election of the first government representative of the black majority, proceeded relatively peacefully. Post-apartheid, Afrikaners sought recognition of their "indigenous" status before the Working Group on the UN Declaration on the Rights of Indigenous Peoples, but their claim was denied. Today, South Africa is not immune from internal ethnic tensions or xenophobic tendencies. There are, for example, many attacks on people who have moved to the country from Zimbabwe and Mozambique.

Racism and Migration

The account of migration in Chapter 5 included references to forced migrations and to the fact that most countries limit immigration. We now consider these matters in greater detail, with specific reference to the role played by racist attitudes and policies.

Indentured Labour

From a broad global and historical perspective, forced (or unfree) labour has been normal. Free labour, in which the labourer has the right to choose an employer, did not become usual in Europe until the late eighteenth century with the emergence of a capitalist mode of production. Indeed, slavery persisted into the nineteenth century in parts of Europe; serfdom (an essential social relationship in feudalism that involved the legal subjection of peasants to a lord) was not abolished in Poland until 1800 and in Russia until 1861. Forms of forced labour also were central to the evolution of the European colonial world. To produce tropical products and precious metals overseas, European powers relied first on slavery. Following the abolition of slavery by the British in 1834, the French in 1848, and the Dutch in 1863, these powers turned to indentured (that is, contracted) labour. The migration of indentured labourers became a key component in the global system by which Europeans combined cheap labour and abundant land to make big profits in their colonial areas. Both slavery and the use of indentured labour had explicit racist overtones.

The majority of indentured labourers—perhaps as many as 1.5 million—came from India between the years 1830 and 1916. The British were able to establish and maintain this system as a result of their penetration into the Indian economy and society, a penetration that brought commercialization of agriculture, payment of rents in cash rather than kind, a decline in traditional crafts, and discriminatory taxation. These changes had a severe impact on the lower agricultural classes and served to make the prospect of emigration attractive. In the course of the nineteenth century, more than 500,000 indentured labourers left India for the island of Mauritius (a British colony in the Indian Ocean), where most worked on sugar cane plantations.

In 1837 British legislation established rules for organizing the emigration of indentured labourers. In principle, the labourers contracted to work for a certain period (normally five years) in exchange for their passage. In reality, however, forced banishment, kidnapping, and deception were common parts of the emigration process. Once overseas, indentured workers were not slaves, but their freedom was severely constrained. Movement outside the plantations was restricted, and the opportunities available after contracts expired were limited by measures such as taxes and vagrancy laws. Overwork, low wages, poor food, illness, low-quality housing, and inadequate medical and educational facilities were usual. Typically, any attempt at resistance on the part of workers was met by stringent labour legislation. Although offering all workers transport back to India after 10 years was a condition of indenture, this responsibility was neglected in Mauritius after 1851. Very few indentured labourers ever returned home, and their descendants have become the majority population in Mauritius.

Tragically, forced labour, frequently involving forced migration, remains a real problem. The UN judges the traffic in human beings to be the third most lucrative in the world, after drugs and arms. The International Labour Organization suggests a minimum of 21 million people work as forced labourers, effectively slaves; about one-sixth of these people are victims of human trafficking. A Global Slavery Index published by the Walk Free Foundation suggests a higher number of 30 million, with 14 million in India, 2.9 million in China and 2.1 million in Pakistan. Young children and women are especially vulnerable, with many forced to work as soldiers or in the sex trade. In a few cases—notably Burma and Sudan but also China—the state is directly involved. It is noteworthy that most developed countries are indirectly involved as many companies rely on forced labour, although this is often not clear because of the complex webs of subcontracting and supply chains.

Restrictive Immigration Policies

In the mid-nineteenth century, Chinese people moved to North America and Australia as free migrants, prompted in both cases by the discovery of gold. Many others worked in transcontinental railroad construction, first in the United States in the mid-1860s and then in Canada in the 1880s. But their presence was quickly opposed by local populations motivated by some combination of racist and economic fears.

In British Empire countries, the British government had to agree to restrictive immigration policies before they could be implemented. Had this not been the case, policies would likely have been in place much earlier in some colonies. In Australia restrictions were imposed following the immigration of Chinese labourers and gold-field workers. The first legislative action, in the state of Victoria in 1855, was followed by similar legislation in other states. But the most explicit restriction—namely, that all new immigrants had to be proficient in a European language—was introduced in the South African province of Natal in 1897. Such a restriction is "colour-blind" in principle but not in practice. New Zealand introduced a restrictive immigration policy in 1881 but replaced it with a literacy requirement in 1899. In Australia the in-movement of all coloured peoples was similarly limited: the requirement that all immigrants be literate in a European language marked the beginning of the "White Australia Policy."

Similar tactics were used in British Columbia, where Chinese immigrants had initially been welcomed as cheap labour. Discrimination began in 1885 with the introduction of a federal head tax, and in 1923 Chinese immigration was virtually prohibited by an Act of the federal Parliament that was not repealed until 1947. Immigration from India began about 1900 but was largely restricted in 1908 by the requirement that immigrants arrive by a "continuous voyage"—this at a time when there were no direct voyages between India and Canada. Japanese immigration, however, was not severely curtailed during the period of the Anglo-Japanese alliance (1902–22).

In the United States the Chinese Exclusion Act was in place from 1882 until 1943 and restrictions were imposed on most other Asian groups after 1917. These policies were greatly expanded in 1924 to include some European groups, particularly from Eastern Europe, because of widespread concerns about political radicalism as well as the various post–World War I nationalisms and economic dislocations. The 1924 Act included an intentionally discriminatory system and set quotas on the numbers of immigrants from specified groups. One term used to identify the outlook that prompted such policies is ***nativism***, the protection of the interests of the native-born population against those of "foreign" minorities. It was not until after World War II that these restrictions were gradually relaxed.

nativism
Intense opposition to an internal minority on the grounds that the minority is foreign.

By the time of World War I, then, many of the countries of European overseas settlement had restrictive immigration policies regarding Asians, policies that continued for several more decades. Of the motivations behind such policies, racist attitudes are the most obvious; British immigrants were typically regarded as the most desirable. In addition, however, there was fear of the economic competition that Asian immigrants—as an abundant source of cheap labour—would represent.

ETHNICITY

Ethnicity is generally regarded as one of the most confusing terms in social science. The greatest confusion occurs when terms such as *race* or *minority* are used interchangeably with *ethnic*. Some groups are generally regarded as ethnic—either by themselves or by themselves and others—because they are in some way different from (and sufficiently visible to be excluded by) the majority. Most groups who self-identify as ethnic base their ethnicity on one or both of the two principal cultural variables discussed in the previous chapter: language and religion. Other common bases for delimiting an ethnic group are perceived racial identity, recent immigrant status, and way of life as evidenced by particular culture traits such as music, dance, food, and drink preferences.

ethnic group
A group whose members perceive themselves as different from others because of a common ancestry and shared culture.

ghetto
A residential district in an urban area with a concentration of a particular ethnic group.

chain migration
A process of movement from one location to another through time sustained by social links of kinship or friendship; often results in distinct areas of ethnic settlement in rural or urban areas.

assimilation
The process by which an ethnic group is absorbed into a larger society and loses its own identity.

acculturation
The process by which an ethnic group is absorbed into a larger society while retaining aspects of distinct identity.

The geographer K.B. Raitz (1979: 79) argued that an **ethnic group** is any group with a common cultural tradition that identifies itself as a group and constitutes a minority in the society where it lives: "Ethnics are custodians of distinct cultural traditions.... [T]he organization of social interaction is often based on ethnicity." Thus, the group may be delimited according to one or more cultural criteria, but it must not live in its national territory. Swedes in Sweden do not constitute an ethnic group, whereas Swedes in the United States do because they identify themselves as one. This understanding of ethnicity accorded with Sauerian understandings of culture and was regularly used in geographic studies of European settlement areas overseas, but it is less popular today. The favoured current understanding is of a collective identity through a shared history in a shared space.

Like language and religion, ethnicity is both inclusionary and exclusionary. Some people are defined as insiders because they share the common identity of the group, while others are seen as outsiders because they are different. Yet it is quite possible for an ethnic group to change its identity and behaviour over time. Human geographers often talk about ethnicity when discussing cultural identities and cultural regions. Generally speaking, an ethnic region or neighbourhood is an area occupied by people of common cultural heritage who live in close spatial proximity. Defined in this way, much of the Chapter 7 discussion of regions, language, and religion dealt with ethnicity, as does much of the Chapter 9 account of nations and nationalism. For these reasons, the discussion of ethnicity in this chapter is limited to an account of migration and related culture change.

Ethnic Migration

Ethnic groups frequently have shared attitudes and behaviours, including shared perceptions of place. To determine these shared perceptions, geographers use questionnaires asking individuals for their perceptions of different areas and then proceed to construct individual and group mental maps. The pioneering work in this area was accomplished by Gould and White (1986), who gathered the data on group perceptions (spatial preferences) mapped in Figure 8.2. In principle, we can use maps such as these to predict free migration behaviour.

Most immigrant ethnic groups, specifically those moving into urban areas, experience an initial period of social and spatial isolation that may lead to low levels of well-being, relative deprivation, and the development of an ethnic colony, enclave, or **ghetto**. We can interpret these common initial experiences of deprivation and residential segregation as social and spatial expressions of outsider or Other status. Often a local group identity is reinforced by **chain migration**, the process whereby migrants from a particular area follow the same paths as friends and relatives who migrated before them.

Assimilation, Acculturation, and Multiculturalism

Despite negative initial experiences, most new immigrant groups eventually experience either **assimilation** or **acculturation**. Some groups that move into cities steadily lose their ethnic traits and assimilate, eventually becoming part of the larger culture. Other groups, however, are more likely to acculturate, to function in the larger culture but retain a distinctive identity. In both Canada and the United States, acculturation has been more common than assimilation. One main factor determining whether or not assimilation

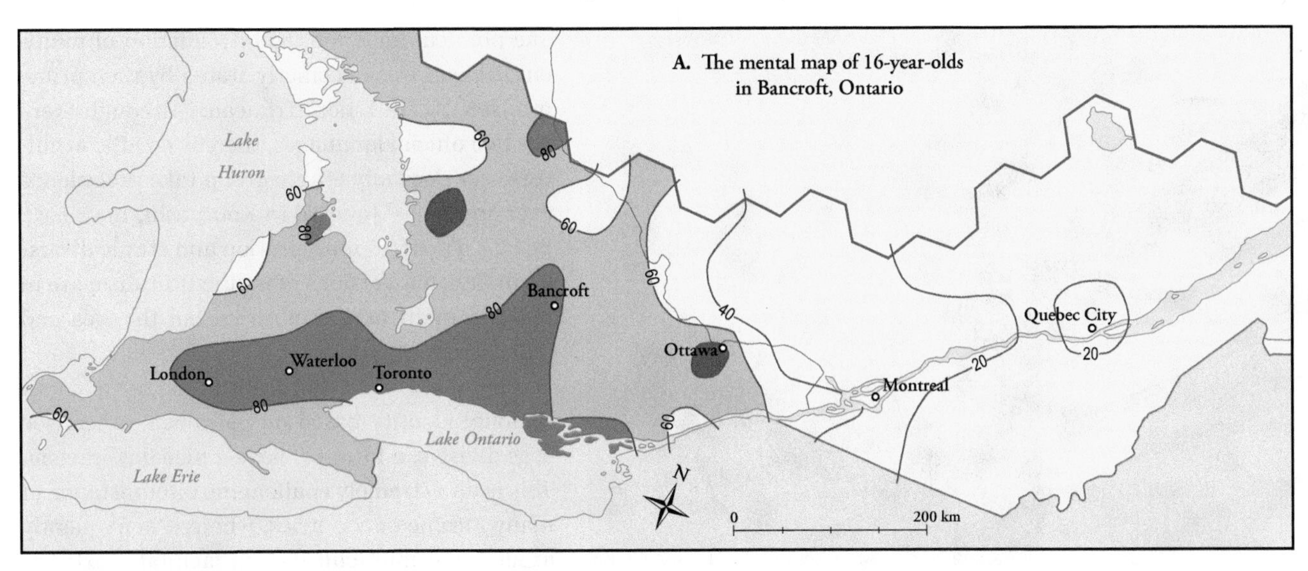

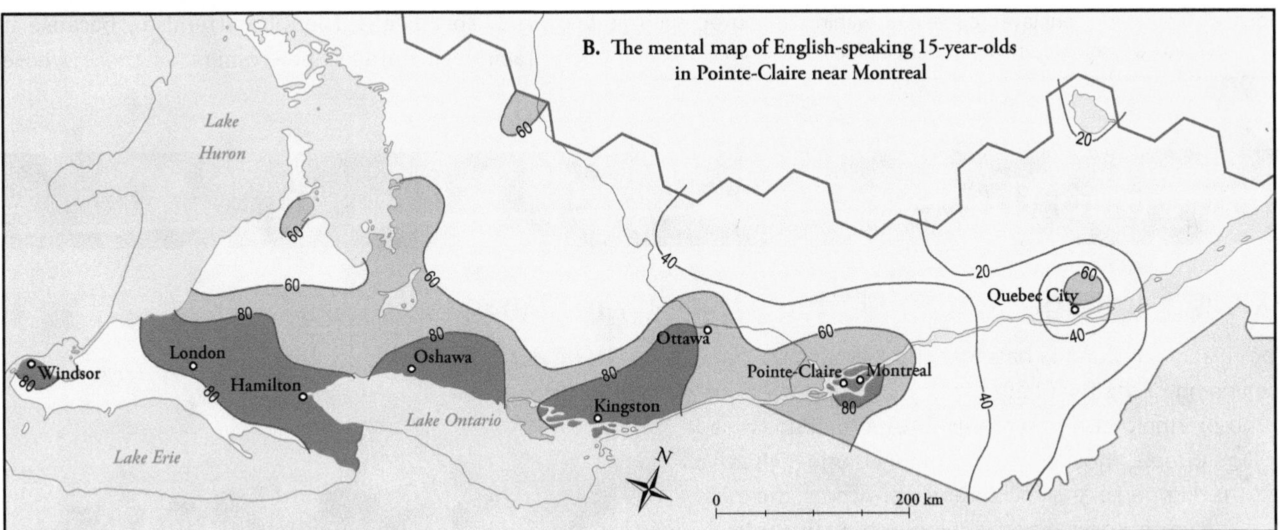

FIGURE 8.2 | Mental maps

Numbered "isolines" represent spatial preferences on a scale from 1 to 100. These particular examples demonstrate that, in some cases, ethnic identity is more important than distance. Map A represents the mental map of 16-year-olds in English-speaking Bancroft, Ontario. Note the preference for larger urban areas and the sharp drop in ratings across the border with Quebec. Map B represents the mental map for English-speaking 15-year-olds in Pointe-Claire, near Montreal, in the predominantly French-speaking province of Quebec. Instead of favouring their local area, these young anglophones favour various areas in English-speaking Ontario. Relevant concepts in discussions of mental maps include those of place utility and spatial preferences, introduced in Chapter 2.

Source: Adapted from Gould, P., and R. White. 1986. *Mental Maps*. 2nd edn. Boston: Allen and Unwin, 77–9.

occurs is the degree of residential propinquity: if group members live in close spatial proximity, social interaction with the larger culture is limited and assimilation unlikely.

Immigrants moving into rural areas, especially as a group or in a chain migration, tend to retain aspects of their ethnic identity longer than do those moving into urban areas. Many regional landscapes, particularly in areas of European expansion, are characterized by distinct ethnic imprints because of the relative recency of the occupation, the cohesion and homogeneity of the group, and the relative isolation of rural areas. Historical and cultural geographers often study the extent of cultural change that accompanies and follows settlement and the creation of landscapes (see McQuillan, 1993).

The extent to which ethnic groups assimilate is also related to state policies. Canada, for example, has had a policy of **multiculturalism** since 1971.

multiculturalism
A policy that endorses the right of ethnic groups to remain distinct rather than to be assimilated into a dominant society.

The various vibrant ethnic enclaves contained within Vancouver, such as the Punjabi Market area and Chinatown, are evidence of acculturation.

The political logic for the introduction of multiculturalism was succinctly stated by then prime minister Pierre Elliott Trudeau: "Although there are two official languages, there is no official culture, nor does any ethnic group take precedence over any other" (quoted in Kobayashi, 1993: 205). Box 8.5 discusses immigration and ethnic diversity in Canada. Versions of multiculturalism are in place in many other countries, but the idea and policy are contentious.

Some see multiculturalism as a vision of national identity based on pluralism; others see it as divisive, a form of "have a nice day" racism. This is an extremely challenging circumstance in many European countries, where it is frequently argued that multiculturalism facilitates extremism, specifically Islamic extremism, because it tacitly encourages newcomers or those whose

Around the Globe

BOX 8.5 | Immigration and Ethnic Diversity in Canada

The population of Canada has changed dramatically since the beginning of the twentieth century, both in total numbers and in ethnic composition. In 1901 the country had 5.3 million people, most of British or French origin; in 2014 Canadians numbered 35.5 million and were one of the most ethnically diverse populations in the world. Four relatively distinct waves of migration can be identified during the twentieth century. The years 1901 to 1914 brought dramatic growth—many immigrants came from the new source areas of Eastern and Southeastern Europe and headed for the prairies. For some years during this period, the annual population growth rate was almost 3 per cent, but these rates slowed with the onset of World War I and remained low, at about 1.4 per cent, until the end of World War II.

The next major increase in immigration, between 1951 and 1961, joined with high fertility to produce an average annual growth rate of 2.7 per cent. In this period, during which the Canadian government was anxious to fill labour shortages in both the agricultural and industrial sectors, immigration policy continued to show a strong preference for British and other European settlers. As noted earlier in this chapter, potential immigrants from Asia were subject to quotas.

Since 1961 both immigration and fertility in Canada have declined, resulting in annual growth rates of about 1.3 per

New Canadians swear an oath of citizenship during a citizenship ceremony in Edmonton, Alberta.

cent. In 1962 Canada's immigration rules dropped all references to supposed race and nationality, and in 1967 a points system was introduced that reflected economic needs and gave particular weight to education, employment qualifications, English- or French-language competence, and family reunification. Thus, policies in this most recent phase have been relatively liberal, permitting significant diversification in the ethnic mix of immigrants. Immigrants from Europe and the United States accounted for about 95 per cent of the total in 1960 but presently make up only about 20 per cent. Application of the points system has resulted in an increase in the number of qualified professionals, many from Asia and the Caribbean. More generally, there have been huge increases in immigrant numbers from less developed countries. A new Immigration Act in 1978 had three principal objectives: to facilitate family reunification, to encourage regional economic growth, and to fulfill moral obligations to refugees and persecuted people. Although subsequent legislation has modified policy details concerning refugee claimants and illegal immigration, the overall thrust remained the same until the Conservative election victory in 2006. Today, there is increased emphasis on admitting immigrants who already have offers of employment.

Canada's liberal immigration policies over the last four decades have benefited the country in many ways. Yet increasing ethnic diversity has made the establishment of a coherent cultural identity an elusive goal. Moreover, some Canadians' reactions against that diversity make it clear that we still have a long way to go if we are to free ourselves of racist attitudes.

ethnicity is "different" not to assimilate. Policymakers in several European countries currently are seeking to identify and implement some alternative policy that promotes pluralism without creating divisiveness. One strategy favoured by conservative thinkers involves the promotion of a stronger national identity.

GENDER

As discussed in the Chapter 2 account of feminism, our identities are gendered, a fact that introduces another category of difference. Gender is traditionally understood to involve power relations between dominant males and subordinate females. Indeed, so widespread is this pattern of relations that the few exceptions are well known. For example, the Indian state of Meghalaya has a matrilineal system with all wealth passing from mother to daughter and with routine discrimination against men. The system is so deeply embedded in local culture that many nouns assume a feminine form only if the thing referred to becomes useful; tree is masculine, but wood is feminine.

Human geographic recognition of gendered identities and related differences owes much to the work of mostly female geographers inspired by various bodies of feminist theory. Feminist research has also been inspired by Marxist ideas about inequality and need for social justice and by postmodern ideas about excluded groups.

Feminist Geography

Building on Pratt (2004: 128), the history of feminist geography can be summarized as follows. In the 1970s a critique of the sexism and patriarchy that prevailed, both institutionally and intellectually, in the discipline of geography was initiated. At that time, geography was exceptionally male-dominated, at least partly because of the long-standing links with fieldwork and because any form of outdoor activity typically was presumed to be within the male domain. The new interests in Marxism and humanism did little to correct the traditional gender blindness of the discipline. The feminist geographic critique employed the concept of contextualism to focus on the varied ways that geography ignored the everyday lives of women, including identification of research topics and analytical strategies.

A significant body of feminist geographic research soon emerged with the aim of making space for women, in the everyday world of the university and society more generally and also in the knowledge produced by geographers. This involved acknowledging that people were active agents, living their lives in households and other places.

As this research tradition evolved, it became evident that it was not necessarily appropriate to treat women as a universal category and by the 1990s feminist geographers were considering additional bases for difference besides gender, such as the myth of race, ethnicity, sexuality, class, and relationship to colonialism. Feminist geographers have been at the forefront of geographic research about the production of difference and the way that some groups are marginalized and excluded; there is a focus on gender, but many other categories of difference are incorporated.

Further, there is increasing recognition that it is not always appropriate to conceive of such simple binaries as the taken-for-granted norms of woman/man, female/male, and feminine/masculine. We must acknowledge the diversity of gender and sexuality to include such identities as lesbian, gay, transsexual, and queer/questioning. Recognizing that there are several different ways of being women and men presents another dimension to the practice of human geography.

Finally, feminist geographers pay much attention to the body and embodiment, reflecting the claim that the "body is the touchstone of feminist theory" (Nelson and Seager, 2005: 2). Indeed, bodies can be understood like places, as geographies that are created and recreated. In particular, the transition from production to consumption and the related rise of a service economy in more developed countries have resulted in a new emphasis on the idea of bodily performance, especially how the self is presented to others in terms of, for example, body shape, facial expression, and dress. There is also interest in the bodily performance of particular sexual identities. Research on the body is informed by a diverse literature on social theory, including Foucault's ideas about power and the disciplining of bodies and psychoanalytic work suggesting identity derives from bodily differences.

In addition to challenging human geographers to rethink their approach and then proceeding to regularly introduce new ideas and research topics, feminist geography has affected both teaching and research in human geography in two rather fundamental ways. First is the increased awareness that researchers cannot be authoritative and impartial and that all knowledge is partial (i.e. the concept of situatedness). Second, our discipline is now more personal. It has become routine for many researchers to explicitly acknowledge their positionality—stating whether or not they are separated from the subjects of their work in terms of gender, sexuality, ability, and other aspects of identity. If they do not share the same identity, an explanation may be offered concerning how they strive to overcome this difference. Sharing the same identity is often seen as an advantage, legitimizing both teaching and research. This recognition of positionality, outsider or insider status, is one form of self-reflection.

The practice of feminist geography as described here presents an important challenge to the larger discipline of human geography and to the idea of a feminist geography itself. A seminal text, *Feminism and Geography*, noted: "To think geography—to think within the parameters of the discipline in order to create geographical knowledge acceptable to the discipline—is to occupy a masculine position" (Rose, 1994: 3). The statement implies that it may not be possible to conduct feminist research within geography because such research is necessarily a different type of knowledge. This challenge has been the subject of much debate but remains unresolved. Intellectually, most feminist-inspired geography has been framed within the parameters of the larger discipline of human geography and, practically, most feminist geographers work in university departments of geography, where their teaching interests are part of the curriculum, and participate in academic meetings of geographers. Of course, specific feminist geography "spaces" have been created, most notably new journals devoted to gender, study groups, and teaching texts, but these are within rather than outside geography. In a related context, concerns that gender studies might become ghettoized within the larger discipline have not been realized.

Feminist-inspired studies continue to challenge the practice of human geography, being variously understood as both a part of and yet somehow against that practice. But despite these tensions, feminist geography has contributed significantly to the human geographic enterprise, particularly to our understandings of hegemony, difference, the power of exclusion, and the social construction of identity. Most notably, it is widely accepted within human geography that gender is a fundamental component of human identity and that it intersects with other components of identity. This understanding enhances appreciation of the causes and consequences of social and spatial exclusion.

Gender in the Landscape

Interestingly, traditional gender differences are not always as readily apparent in the landscape as are some other differences, such as ethnicity and social class. Yet landscapes are shaped by gender and provide the contexts for the reproduction of gender roles and relationships.

Numerous examples of landscapes, both visible and symbolic, reflect the power inequalities between women and men and demonstrate the embodiment of patriarchal cultural values and related sexism. In urban areas, statues and monuments reinforce the idea of male power by commemorating male military and political leaders: "Conveyed to us in the urban landscapes of Western societies is a heritage of masculine power, accomplishment, and heroism; women are largely invisible, present occasionally if they enter the male sphere of politics or militarism" (Monk, 1992: 126).

sexism
Attitudes or beliefs that serve to justify sexual inequalities by incorrectly attributing or denying certain capacities either to women or to men.

Focus on Geographers

Urbanization, Gender, and Everyday Life in Botswana | ALICE J. HOVORKA

As a feminist geographer and gender researcher, I am interested in the differences and inequalities that exist between men and women in society. I investigate gender dynamics through a two-tiered process based on gender-disaggregated data collection followed by gender analysis. First, I collect data on men's and women's circumstances, experiences, interests, needs, and access issues. This allows me to establish an accurate overview of the context at hand. Vital questions at this stage of research are who, what, when, where, and how particular dynamics function with regard to gender. Second, I analyze why exactly such gender dynamics occur. It is not enough to document differences between men and women, as this has the potential to simply reproduce inequalities on the ground. Instead, it is important that we probe deeper to examine key factors that create and reproduce different opportunities and constraints for men and women at local, regional, and global levels. Many times, what we see happening at the household or community level is a reflection of what goes on at broader scales. Economic systems, legal structures, policy decisions, or cultural traditions, for example, contain many assumptions and norms regarding what it means to be a man or woman. The resulting gender system creates substantial imbalances and inequities—its impacts are felt by both men and women. A multi-faceted approach helps expose the gender relations of power deeply embedded within social, cultural, economic, political, and ecological processes and structures.

Equipped with these gender tools, I have conducted numerous research projects focused on urbanization and everyday life in Botswana. One significant project focuses on whether men and women farm differently in the capital city of Gaborone. My gender-disaggregated data, based on 51 female and 48 male commercial urban farmers, reveal clear differences and inequalities within social variables: women have lower levels of education than men; women are employed in "female" jobs (e.g. within administrative, education, and retail sectors) that offer lower wages and in part-time work; this female employment carries less status than is the case for men employed in "male" jobs (e.g. within government and construction sectors); and women earn less monthly income than men do.

Gender differences and inequalities also emerge through spatial variables whereby women have poorer access to land than men for their agricultural activities. Women are located in peri-urban villages on small residential plots and on tribal land allowing for user rights only, while men access larger, agriculturally zoned plots through land rental or outright purchase throughout the greater Gaborone area. The type of agricultural operations also reflects gender differences and inequalities: men's operations are large-scale, mechanized operations with substantial investment in fixed assets and labour within a wide range of agricultural subsectors; women's operations are on a small scale with little investment in fixed assets or labour and are focused on broiler production. Finally, the output of men's and women's commercial urban agriculture activities reveal a discrepancy in terms of quantity of foodstuffs produced per month for the urban market (men produce much more) and type of foodstuffs produced (men produce a wider range).

Why do such stark differences and inequalities exist between men and women in the Gaborone commercial urban agriculture sector? To answer this question, I embarked on an analysis exploring gender power relations across numerous scales and realms. In Botswana the education system perpetuates culturally driven gender inequalities whereby boys are privileged over girls in sheer numbers (particularly at higher levels of schooling) and in terms of male-based subjects (e.g. sciences or business) compared to female-based subjects (e.g. social sciences or arts). Powerful conventions restrict women's domain to the household and women's autonomy under male guardianship. This means

Continued

that women's right and access to productive resources, including land, are significantly curtailed relative to men's. The labour market entrenches these gendered ideas such that women are seen as secondary earners compared to men as breadwinners, and female entrepreneurs are the exception rather than the norm. These ideological and institutional norms mean that female commercial urban farmers are socially marginalized (lower education, lower income), spatially marginalized (less land access), and production marginalized (small-scale operations). Women face challenges expanding their agricultural activities given that they are "fixed" in space (through the land tenure system) and in socio-economic class (through their inability to generate enough profit from their small plots of land).

Ultimately, gender analysis helps us not only unravel how gender "works" but also how gender matters in a particular context. In Gaborone, gender dynamics matter because they affect men's and women's everyday lives within the commercial urban agriculture sector, with individual women unable to realize their full potential as producers and entrepreneurs. Further significance of this is that, at a broader scale, gender dynamics influence the quantity and type of foodstuffs produced for the urban market, thus inhibiting food availability and economic growth potential for everyone in the city.

ALICE J. HOVORKA is a professor in the Department of Geography and Planning at Queen's University. Her research focuses on human–environment relations with specific emphasis on gender and environment, critical animal studies, urban geography, and Southern Africa.

The design of domestic space is strongly influenced by ideas about gender roles. In many cases, including traditional Chinese and Islamic societies, areas for women and for men are separated, and those for women are commonly isolated from the larger world. In Western societies, the favoured domestic design has centred on the home as the domain of women and as a retreat from the larger world. It has been usual in the Western world to locate the kitchen, designed as a separate area for the unpaid work of women, at the rear of the home, and some feminist scholars have interpreted this as devaluing women's work. Following this logic, a home can be understood as a site of unpaid domestic work, oppression, and sometimes violence.

According to some feminist arguments, this same patriarchal logic influenced the expansion and morphology of city suburbs: men were seen as commuters and women as homemakers/consumers requiring ready access to shops and schools. It is argued that this arrangement further disadvantages women, who may become isolated because of distance from city centres or struggle to cope with work at home and limited opportunities for paid employment in the suburbs. The morphology of the city therefore works to reinforce traditional gender roles and identities and serves as an obstacle to change. This is evident in the shopping mall. Because of increasing personal mobility, additional discretionary income, and conformity with gender expectations, women are the principal consumers at shopping malls. As a result, the typical shopping mall landscape reflects the gendered identity of consumers in the way the mall is laid out, in the way many individual stores display merchandise, and in advertisements directed at women.

Gender and Work

The concept of work as a social construction developed within a masculinist discourse that explicitly distinguished between work and home, public and private. As such, it is necessarily partial. In the capitalist mode of production, work is conventionally understood as waged employment in the public sphere. Activities in the private sphere of the home are thus not work. Although attitudes have changed in many parts of the world, domestic, reproductive, and caring (usually for children, aged relatives, or a spouse) activities performed by women remain undervalued compared to waged employment.

This understanding of work is challenged in at least four ways. Paid domestic and caregiving work is usually performed by low-income women for middle- and upper-class families. Many women are involved in unpaid volunteer and activist work, often with the intent of enhancing the local community. Rather differently, there is an increasing tendency for both women and men to conduct businesses in their home, thereby confusing the classic work/home dichotomy. A new spatial division of labour is seeing more and more women in paid employment, especially in urban areas. This new spatial division of labour has involved some firms relocating to suburban areas in order to attract women workers who need to work locally, while other firms relocate to city centre areas that are hubs of public transportation services.

The idea that work in the home is not work has had implications within the sphere of waged employment, with women seen as suited to domestic and caregiving work outside the home. Because of the low status of such work, it is typically relatively undervalued and lower paid than other work. When identity characteristics such as ethnicity are factored in, the wage gap widens further. Research has also shown that it is usual for women to encounter sexist practices and gender stereotyping in paid workplaces, particularly as they move into positions widely understood to be men's work. This topic is considered further in Chapter 13.

Gender Relations in Agriculture

In the less developed world, there is increasing recognition of the roles played by women in agriculture, past and present. In the West African country of Gambia, for instance, women play an important role in rice cultivation, although this work does not generate income. They have also been enthusiastically involved in recent government initiatives in dry-season horticultural production of commercial crops such as chilies, eggplants, and okra and subsistence crops such as potatoes and tomatoes. If such initiatives increase women's incomes, they are clearly of great value; in fact, several of them yield well and generate needed income. Typically, however, such efforts face two problems.

First, the new schemes are based not on the local knowledge of horticulture but on imported (or at least non-local) knowledge. Second, they may not be environmentally or economically sustainable as this production frequently involves tree removal that in turn leads to reductions in the soil's fertility and moisture-holding capacity. Once again, there is evidence of government efforts to improve agricultural circumstances, but such efforts are probably misguided. In Bhutan and Gambia long-established agricultural practices are ignored and new imported practices are favoured. It is not difficult to understand why governments are tempted by the new, but it is also important that the new learn from the old. The agricultural practices that have survived, perhaps little changed, for generations are usually grounded in economic and cultural realities; they may need to be improved, but this does not mean that they need to be replaced.

Gender and Health

Despite women being the principal users of health-care systems and predominating in the caregiving professions, there has been relatively little attention given to gender in geographic studies of health and health-care systems. Traditional health geography research tended to be gender blind. However, since about the mid-1990s, the feminist geography research agenda has prompted increased interest in the complex links between health, place, and culture. In common with other topics investigated within the general rubric of feminist geography, researchers are concerned with how the life experiences of women and men differ, how the key ideas are socially constructed from a male perspective, and how women are typically disadvantaged or even excluded.

Much of this work emphasizes the need to listen to the voices of women. Examples of research topics include health and beauty as women move from school to employment or university, the body spaces of breast-feeding, the role of the mall for women struggling with agoraphobia, health and social isolation, and health and environmental risk.

Gender and Human Development

Typically, expressions of gender in landscape, in work, and in health display disadvantage. Not surprisingly, then, there is also clear evidence of gender inequality in the larger context of global human development. While it is correct to note that women's education and health levels are improving markedly in many less developed countries, in many places far too many women still die in childbirth, are excluded from playing important roles in public life, lack a strong voice in families and communities, and have limited employment opportunities. It should go without saying that all such inequalities are wrong. Speaking on the UN's International Day for the Elimination of Violence against Women in 2008, the UN Secretary-General stated that one in every three women in the world will be abused, beaten, or forced into sex in her lifetime. The principle of equity is enshrined in the UN Charter, and the "promotion of gender equality and the empowerment of women" is the third Millennium Development Goal. The fundamental point regarding human development is clear:

> Gender equity is an intrinsic dimension of human development. If girls and women are systematically denied freedoms and opportunities, this is not consistent with human development. Gender equality also has instrumental value for human development—there is much country level evidence showing how investments in women and girls can be a vehicle to promote long-term prospects for growth prospects and human development. (Gaye et al., 2010: 1)

Eliminating gender inequality not only ought to be a fundamental goal of any civilized society, but is also essential to winning the war on global poverty. This second point was central to discussions in Chapter 5 concerning population growth, where the empowerment of women, especially improved education, was identified as critical for reducing fertility.

sexuality
In some feminist and psychoanalytic theory, interpreted as a cultural construct rather than as a biological given; aligned with power and control.

TABLE 8.3 | Gender Inequality Index

Top 10			Bottom 10		
Country	GII Rank	GII Value	Country	GII Rank	GII Value
Slovenia	1	0.021	Ivory Coast	143	0.645
Switzerland	2	0.030	Central African Republic	144	0.654
Germany	3	0.046	Liberia	145	0.655
Sweden	4	0.054	Mozambique	146	0.657
Denmark	5	0.056	DR of Congo	147	0.669
Austria	5	0.056	Mali	148	0.673
Netherlands	7	0.057	Afghanistan	149	0.705
Italy	8	0.067	Chad	150	0.707
Norway	9	0.068	Niger	151	0.709
Belgium	9	0.068	Yemen	152	0.733

Source: UNDP. 2014. *Human Development Report, 2014. Sustaining Human Progress: Reducing Vulnerabilities and Building Resilience*, Table 4. http://hdr.undp.org/en/data. Countries not included in the source data set are North Korea, Somalia, South Sudan, and a number that are very small.

Measuring Gender Inequality

Recall from Table 6.1 that the Human Development Index (HDI) measures average development in a country but says little about gender inequality as it is evident in development. Because of this failing, the Gender-related Development Index (GDI) was first calculated by the UN in 1995. This statistic used the same criteria as the HDI except that it also captured inequalities between women and men for each criterion. A revised measure, the Gender Inequality Index (GII), was introduced in 2010. This index focuses on reproductive health, on empowerment as indicated by women's participation in political decision-making, and on participation in the labour market. Specifically, the GII measures the negative human development impact of social and economic disparities between men and women by calculating national HDI losses from gender inequities. The minimum GII value is 0.0, meaning there are no losses; the greater the GII value, the greater the inequality.

Table 8.3 lists the 10 countries that rank highest and the 10 that rank lowest on the GII. It is notable that all the highest-ranked countries are located in Europe, mostly in the west and north. Canada ranks twenty-third. Also notable is that the largest losses due to gender inequality are in Afghanistan, Yemen, and eight countries in Africa. There is a close correlation between gender inequality and low education levels and poverty. The global pattern of gender inequality is mapped in Figure 8.3.

Human rights groups regularly highlight Saudi Arabia as a country that severely limits women's rights in order to maintain male control. In 2008 the New York–based Human Rights Watch reported that women have limited access to justice and may have to obtain permission from male relatives to work, travel, study, marry, or receive health care. Here again, some change is taking place with women being given the right to vote, at least theoretically, in 2011. China, the country with the largest population, outlawed sexual harassment and established gender equality as a national goal as recently as 2005.

SEXUALITY

Sexuality is increasingly studied as an expression of identity and as one way that the dominant heterosexual landscape can be challenged. Work on diverse sexual identities and related

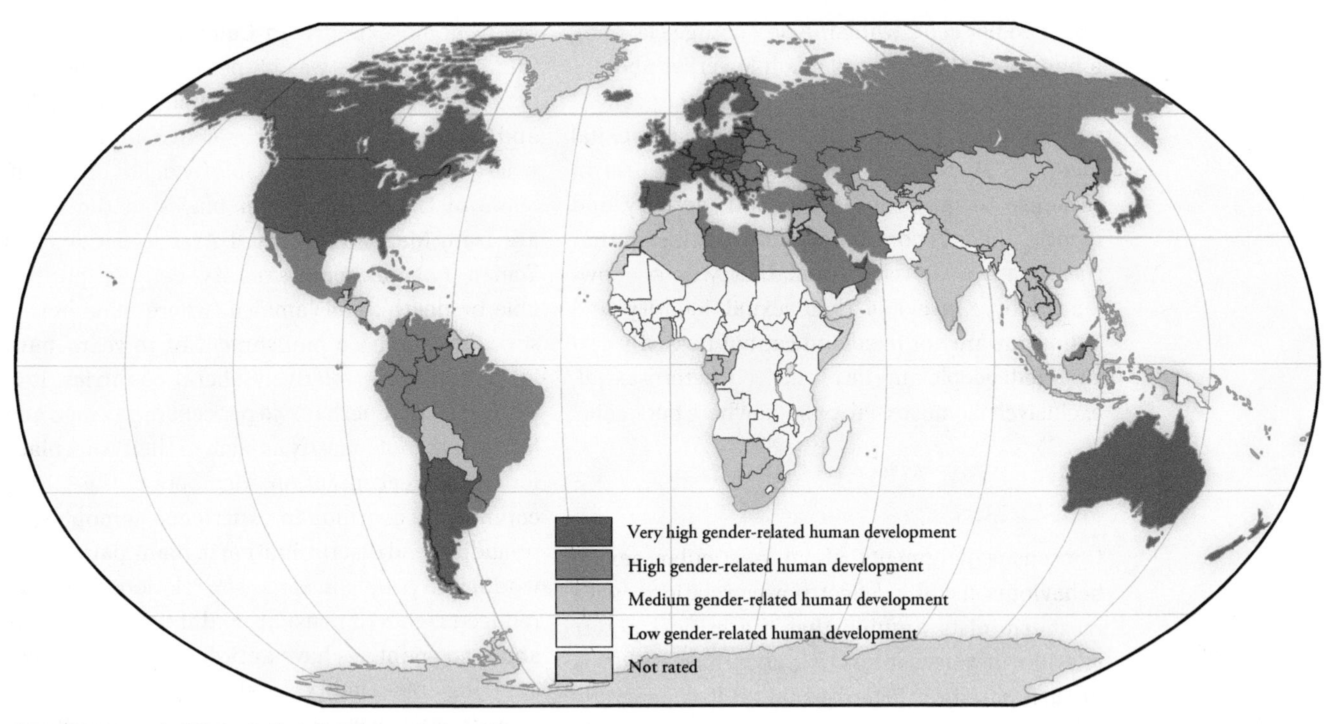

FIGURE 8.3 | Global distribution of the GII

Source: Based on UNDP. 2011. *Human Development Report, 2011. Sustainability and Equity: A Better Future for All*, Table 4. New York: Macmillan.

landscapes has been conducted mostly within the larger framework of feminist geography, but researchers also employ **queer theory**, a rather controversial term that refers to a concern with all people who are seen as and/or who have been made to feel different because of their sexual identity and who share a common identity on the fringes of hetero-normative society. Queer theory also emphasizes the fluidity and even hybridity of identities and is concerned with empowering those who lack power. Terminology is important in discussions of sexuality, as certain terms tend to be loaded with emotive meanings. The terms *gay* and *lesbian*, for example, are commonly preferred by male and female homosexuals, respectively.

But what percentage of the population are homosexual? Debate on this question has been intense since the 1940s surveys of American sexual behaviour conducted by Alfred Kinsey, which suggested that 13 per cent of men and 7 per cent of women had predominantly homosexual activity. Hence the often stated estimate of 10 per cent. This claim was highly controversial and was challenged in 1991, when a major American survey concluded that the percentage for men was, at 2.3, much lower. Recently, surveys have become more nuanced, recognizing that there are three different issues at stake: identity, what sexual orientation is claimed by an individual; attraction, who an individual considers attractive; and behaviour, whether or not an individual has engaged in same-sex activity. Not surprisingly, the percentages for these are quite different. A general conclusion reached by recent surveys is that the percentage of people with same-sex experience is

queer theory
Ideas developed in gay and lesbian studies and concerned with oppressed sexualities in terms of both social rights and cultural politics.

Women and men wearing traditional dress in Riyadh, Saudi Arabia.

about 10 per cent, while the percentage claiming a homosexual identity is much lower, at about 1 for women and 1.5 for men (about 1.5 per cent of women and 1 per cent of men claim a bisexual identity). There are also notable differences in response to questions about both activity and identity if age is considered. One conclusion that these data make clear (and that was acknowledged by Kinsey) is that sexual identity and behaviour are not fixed and cannot be easily categorized: people may be exclusively heterosexual, exclusively homosexual, or somewhere between.

Changing, and Unchanging, Attitudes

Consensus judgments about particular sexual behaviours usually reflect power relations, with most feminists arguing that the critical power relation in this context is patriarchy, which is in turn based on heterosexuality. It is generally accepted that sexual behaviour is not simply the product of some instinctual drive designed to ensure the continuity of the species; sexual activity is a multi-functional behaviour that can be fully understood only in a larger cultural context. This is why a particular sexual behaviour may be unacceptable in one cultural context and acceptable in another and why cultural attitudes towards sexual orientation change through time. By the early twentieth century, the dominant view of homosexuality in the Western world was negative, a view reinforced by both the legal and medical establishments.

Attitudes are changing in the Western world. For example, relationships involving same-sex partners are being legally recognized as legitimate and socially acceptable lifestyle choices, although many contrary views are evident, especially where religious fundamentalists are an influential presence. Countries in Northern Europe were the first to recognize same-sex unions. Increasingly, same-sex marriage is becoming institutionalized and, as of 2015, it is legal in 17 countries and some jurisdictions within countries. An additional indication of changing attitudes to sexual identity is the introduction by a few countries, including Australia, Germany, and the UK, of a third gender option—"indeterminate"—to accommodate transgendered people and those of ambiguous identity.

But such acceptance is not universal. A major 2010 survey of African attitudes to religion and morality reported a huge majority disapproving of homosexuality; in Cameroon, Kenya, and Zambia the figure was 98 per cent. Homosexual acts remain criminalized in about 80 countries, and about 2.8 billion people live in countries where being gay is punishable by death or prison. Some of the least tolerant places in the world are Iran, Mauritania, Saudi Arabia, Sudan, and Yemen (where homosexuality can be punishable by death), and Jamaica (where male homosexuality carries a punishment of 10 years' hard labour). Even in relatively liberal countries, it is estimated that perhaps 50 per cent of gay men and lesbians do not identify as such in their workplace for fear of recrimination. Members of the LGBTQ community continue to experience homophobia, prejudice, and discrimination in many parts of the world, and people in same-sex relationships have reduced rights to pensions and inheritance. Only about 50 countries have anti-discrimination laws applying to sexuality. In short, people are punished in many ways for their homosexual identity.

Two relationships regarding the acceptance of homosexuality stand out. First, if religion is central to people's lives, homosexuality is less likely to be accepted. Second, wealthy countries are more accepting than poor countries.

Sexuality in the Landscape

One way that sexuality is expressed in landscape, especially in large Western cities, is a part of the city being identified as a residential and commercial area where lesbians and gays dominate. Early examples include Greenwich Village in New York, the Castro district in San Francisco, West Hollywood in Los Angeles, and Soho in London. Another expression is a gay pride parade, march, or demonstration. These are now routine in many cities and are often major celebratory occasions attracting large tourist audiences. But they can be unsettling for some, and controversy frequently surrounds these displays of identity.

For example, the 2005 parade in Jerusalem was opposed by some Jewish, Christian, and Muslim religious leaders. In Auckland, New Zealand, the first two gay pride parades took place outside the area understood to be gay and therefore were seen as challenges to the larger heterosexual community as well as expressions of resistance in landscape. The third parade, in 1996, took place within an area widely perceived to be gay and was less controversial; this change suggests that the identical behaviour may be interpreted by the majority differently, depending on whether it represents a landscape challenge to the heterosexual majority (see Johnston, 1997). In both Boston and New York, annual St Patrick's Day Parades have been an influential strategy promoting and celebrating Irish identity but have been plagued by controversy in recent years because of organizers' decisions to exclude openly gay and lesbian participants. (However, shortly before the 2015 parade, organizers in Boston announced that both gay and lesbian groups were welcome to participate.) These few examples of the many cases of controversies surrounding parades are explicit illustrations of identity politics at play in the urban landscape.

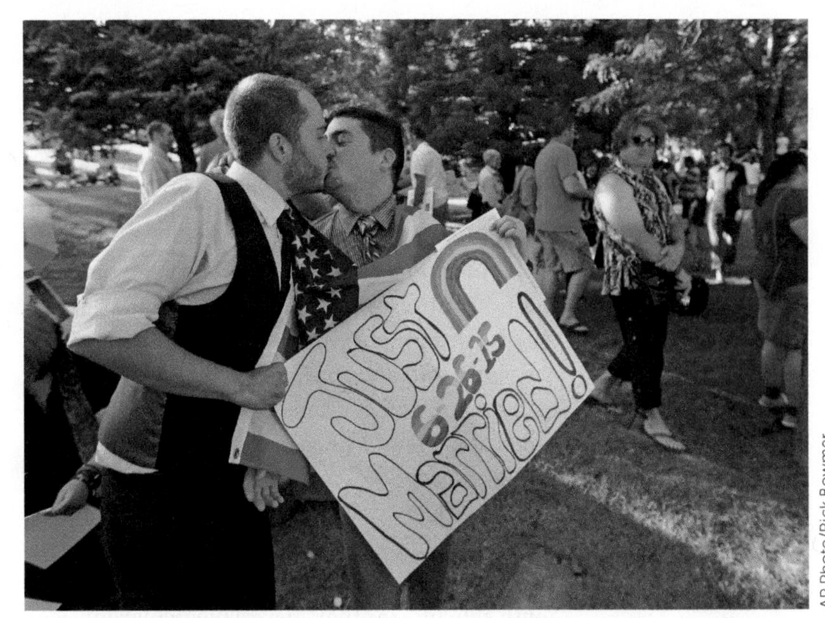

AP Photo/Rick Bowmer

Ryan Jones (left) and Ruben Vellejo kiss after their wedding at a same-sex marriage rally in Salt Lake City, Utah. In 1995 the state was the first to enact a Defense of Marriage Act (DOMA), which allowed states to refuse to recognize gay marriage. On 26 June 2015 the US Supreme Court ruled such refusals non-constitutional and declared that same-sex couples can marry anywhere in the country.

IDENTITIES AND LANDSCAPES

Four important general ideas are evident from our discussions of difference: imposing identities on others, landscape as a reflection of identity, microculture, and contesting identity and landscape.

Imposing Identities on Others

Groups of people often understand themselves in relation to other groups. In the case of apartheid, the dominant Afrikaners imposed their views of themselves and of the indigenous black population on the latter. Like all imperialists, they found their justification in a logic that, in the context of Asia and the Middle East, has come to be known as Orientalism (see Box 8.2). According to this logic, which implies a relational concept of culture, the indigenous people of areas that Europeans wanted to colonize were Others—not only different from but also less than the Europeans, who as "superior beings" had a natural right to use them and their lands as they chose. The same logic has been extended beyond the colonial context, as conventionally understood, to operate in more locally defined situations. In many cases, those colonized have responded by creating landscapes of resistance.

Landscape as a Reflection of Identity

Dominant groups have constructed landscapes as places that have meaning for them and that take certain characteristics—those of the dominant group—for granted. Thus, urban areas in the Western world have been constructed assuming such characteristics as heterosexual nuclear families, women dependent on men, and able-bodiedness. Because of these assumptions, those individuals and groups who do not conform to these societal expectations are perceived as different and are frequently excluded or disadvantaged as members of society. When such people intrude into landscapes that were not constructed for them, the result is often controversial, both culturally and spatially. Indeed, as evident in the account of sexuality, a distinctive feature of the contemporary world is the unsettling effect that the expression of other identities and the crossing of spatial boundaries has on dominant groups.

The Concept of a Microculture

This point refers back to an important question raised in Chapter 2 in regard to social scale: Do we as human geographers study individuals or groups of people and, if groups, what size? There is no one correct answer. Human geographers employ the social scale of analysis that seems appropriate, given the research topic. On the one hand, much of the discussion of population issues considered the entire human population or very large subsets such as those people living in less or more developed countries. On the other hand, much of our discussion in Chapter 7 and this chapter has employed a cultural group scale of analysis, such as groups defined by a shared ethnicity, that has been much favoured by geographers. Human geographers display increased interest in smaller groups, usually defined in terms of a particular characteristic rather than some collection. This interest is one component of the cultural turn; it reflects the facts that group identities are not fixed and unchanging but fluid and malleable and that people regularly identify meaningfully with a small subgroup rather than some larger, broadly based cultural identity.

The concept of a microcultural or subcultural identity, understood as relatively small groups of people within a larger culture who differ in some way from the majority, also reflects this interest. Youth subcultures that share a particular musical interest, clothing preference, or gang membership are prime examples. Such groups typically distinguish themselves from a larger culture by means of language, employing words in new ways or coining new words. This is one strategy for highlighting and applauding difference and for exercising power through emphasizing the exclusion of those who do not belong. The vocabulary associated with rap is one example. Much of the discussion in the following section is of microcultural identities.

Contesting Identity and Landscape

As previously suggested, it is often the case that dominant groups have endowed their landscapes with meaning; that is, they have created places. But because the understanding of place held by one group may be different from that held by another, the identity and ownership of those places may be contested. It is not surprising, then, that places frequently become sites of conflict between different groups of people, as in the apartheid landscape, gay pride marches, or the Chapter 7 examples of places claimed by more than one religious group.

Further, religion, the myth of race, ethnicity, gender, and sexuality are not the only grounds for exclusion from the landscape created by the dominant group. The homeless, the unemployed, the disabled, and the elderly all may find themselves in some way incapable of fitting into the dominant landscape. Some disadvantaged groups are more visible than others, and some are more controversial, but all are in some way oppressed because the landscapes created by dominant groups presume a set of identity characteristics that they do not possess. For instance, dominant groups "presume able-bodiedness, and by so doing, construct persons with disabilities as marginalized, oppressed, and largely invisible 'others'" (Chouinard, 1997: 380).

Other examples of these landscapes of resistance are associated with microcultures in the form of new social movements—in support of the environment, social justice, ethnic separatism, and so on—and they are best interpreted as expressions of opposition to power. The Occupy protests that began in New York in 2011 and subsequently spread to many other cities, including several in Canada, were explicitly designed to challenge established ideas about how space was occupied, that is, what it was used for and what it meant to people. Hence the need to occupy major symbolic spaces and use them for purposes other than the usual. The London protest centred on the symbolic open space adjacent to St Paul's Cathedral and in the heart of the financial district.

Some subcultures, especially those associated with youthful populations, express opposition to mainstream lifestyles and places. As noted earlier, such groups may choose to display their difference through the adoption of alternative lifestyles, including music, clothing, and linguistic preferences, and through their attachment to places such as inner-city neighbourhoods.

A little-known example of a means of communication developed in response to oppression from a majority is the form of language known as

The Canadian Press Images—Mario Beauregard

As part of worldwide protests, Occupy protesters in Montreal took over Victoria Square, a public area directly between the Montreal World Trade Centre and the Montreal Exchange, which trades derivatives.

Polari. Associated with gays and, to a lesser degree, lesbians during about the first two-thirds of the twentieth century, Polari was an underground vernacular used in large British cities. This form of language had many inspirations, borrowing words from such early sources as travelling entertainers, costermongers, and beggars and from more recent slang sources such as the British Navy, American Air Force, and drug users. It also used rhyming and backwards slang words. Effectively a secret, or anti, language used only by insiders, Polari arose because of the need for people to communicate in an environment of both legal and routine informal discrimination. This is one example of a group creating a form of language that allowed it to discuss its reality without others understanding. Some Polari words mocked those considered unfriendly; for example, police were disparagingly referred to as "Betty bracelets" or "Hilda handcuffs." The need for Polari disappeared with the decriminalization of homosexuality and the rise of a more mainstream gay identity.

GEOGRAPHIES OF WELL-BEING

The term *well-being* is used to refer to the overall circumstances of a group of people and to specific components of security and comfort—economic, social, psychological, and physical. Landscapes at various spatial and social scales differ all too widely in terms of the well-being of their occupants. On the global scale, as discussed in Chapter 6, there are areas of feast and areas of famine; some children are born into poverty and others into affluence. On the regional scale, many landscapes in areas of European overseas expansion have continued to evolve in response to the needs of the former imperial powers and the capitalist imperative to seek profit over equity. Needless to say, the results have been to the detriment of most indigenous inhabitants. On a more local scale, the landscapes of disadvantaged groups also reflect the fact that there are dominant and subordinate groups in any culture. Human geographers are paying increased attention to these landscapes, at a variety of scales, and to the social and power relations that lie behind them.

Much of the work done in these areas is labelled **welfare geography**; this is a general approach to a wide range of issues but with an emphasis on questions of social justice and equality. Concern with the geography of well-being and welfare geography has encouraged research into such important issues as the geography of education (focusing on the often inequitable spatial distribution of services and facilities) and the geography of justice (focusing on the variations in the spatial availability of social benefits).

Measuring Well-Being

In the 1960s concern about social problems in the United States prompted the development of spatial social indicators as measures of well-being. In a pioneering study, Smith (1973) identified seven sets of indicators representing seven different components of well-being:

1. income, wealth, and employment
2. the living environment, including housing
3. physical and mental health
4. education
5. social order
6. social belonging
7. recreation and leisure

Examining the extent to which quality of life differs for different groups in different places, Smith found extreme inequalities at all spatial levels in the country.

well-being
The degree to which the needs and wants of a society are satisfied.

But do all people have the same needs and the same ideas of what is beneficial and what is harmful? One attempt to conceptualize along these lines argues that all of us share a basic goal, to avoid harm (Doyal and Gough, 1991). Achieving this basic goal requires satisfaction of two needs: physical health and the ability to make informed decisions concerning personal behaviour. These two in turn require that the following conditions be satisfied:

1. adequate supply of food and water
2. availability of protective housing
3. safe workplace
4. safe physical environment
5. necessary health care
6. security while young
7. relationships with others
8. physical security
9. economic security
10. safe birth control and child-bearing
11. required education

welfare geography
An approach to human geography that maps and explains social and spatial variations.

It is one thing to list such needs and another to measure them effectively in such a way that the integrity of different groups of people is not lost. As we have seen, the UN has led the way in the measurement of well-being, variously called quality of life or human development. An important issue in this work is the existence of spatial inequalities, something that most of us take for granted. An atlas of British identity shows that where an individual lives can be a strong predictor of identity, including class, health, life expectancy, and family structure (Thomas and Dorling, 2007). In North American cities, there are often clearly defined areas occupied by ethnic groups that are relatively disadvantaged. In short, place provides advantages or disadvantages, life opportunities or barriers.

Indeed, our capitalist society and economy are predicated on inequalities. Capitalism allows a few to be very wealthy, many to be comfortable, and large numbers to live in poverty. This should not be the case. It is possible to strive for greater spatial equality so that at least basic human needs are met for all people in all places. Two important areas where spatial differences are clearly evident are crime and health and health care.

Crime

Human geographers have addressed the study of criminal behaviour and activities with the central concern of relating crime to relevant spatial and social contexts. Mapping the incidence of crime has helped identify places where criminal activity is most prevalent, such as inner cities and other areas of poverty, low-quality housing, high population mobility, and social heterogeneity. These essentially spatial, and frequently empiricist or positivistic, studies are accompanied by analyses of the social environments of criminals, the victims of crime, and the geography of fear, all of which are more clearly Marxist or humanist in focus.

Two topics of particular interest at present are the relationship between areas of criminal activity and urban decline and how our use of space is affected by our concerns about the likelihood of a criminal act (Box 8.6).

Health and Health Care

Susceptibility to disease, morbidity (illness), and mortality and the need for health-care provision are closely related to questions of well-being and the quality of life, and all have been analyzed by human geographers (see Chapters 5 and 6). Regardless of spatial scale, there has been a tendency to focus research on the less privileged members of society to demonstrate how health issues relate to a wide range of social and economic conditions. Clearly, health problems are related to the physical environment, to social factors such as housing and living conditions, and to access to health services. Several studies have emphasized inequalities in availability of services.

Canada's Aboriginal peoples offer a compelling example, as their health status and their access to care are much poorer than those of other Canadians. This situation is one aspect of a way of life characterized by poverty and low self-esteem, which result at least partly from a long period of oppression and, more specifically, from a residential school system that removed children from their parents and imposed an alien way of life. Today, many Aboriginal communities have high rates of physical and sexual abuse and of alcohol and substance abuse. As we see in Box 8.7, Aboriginal Canadians have higher rates of infant mortality and a lower life expectancy than do other Canadians. Major causes of death include suicide and tuberculosis (a disease related to poverty), while a common chronic condition is diabetes mellitus (indicative of poor diet).

The Geography of Happiness

An interesting recent focus is on the idea of happiness. Notoriously difficult to measure, happiness is clearly related to, but different from, the idea of well-being. Essentially, it is how we perceive ourselves and thus cannot be measured objectively. It is not unusual in some countries for surveys to ask, to put it simply, whether we are very happy, just happy, or not so happy. There is even a regularly updated World Database of Happiness housed at Erasmus University, Rotterdam. Those who conduct these surveys and those who then interpret and publish their thoughts on happiness are at the forefront of what *The Economist* (2006: 13) humorously described as the "upstart science of happiness," blending economics, psychology, and geography. Two general points that emerge from this work are, not surprisingly, that the rich report greater happiness than do the poor and, perhaps surprisingly, that people in affluent countries have not become happier as they have become richer. The latter finding might be because: (1) for many people, happiness is having things that others

Examining the Issues

BOX 8.6 | The Geography of Fear

Geographers have made considerable advances in the analysis of social and spatial variations in the distribution of fear. Compelling evidence suggests that the fear of crime is a major problem affecting our perceptions of certain areas and our behaviours in them (Pain, 1992; Smith, 1987). The problem is particularly acute for women, the elderly, and other social groups who are relatively powerless and vulnerable.

Fear of domestic violence, sexual harassment, and rape can limit women's access to and control of space within and outside the home; it also imposes limitations on social and economic activities. Pawson and Banks's (1993) study of the geography of rape and fear in Christchurch, New Zealand, demonstrates the extent of the problem. Christchurch has a population of 292,000, of whom 91 per cent are of European origin. Pawson and Banks had two goals: to determine the spatial and social incidence of rapes that were publicly reported in the press and to map the spatial and social distribution of the fear of crime. Their results showed a spatial distribution that emphasized the inner-city area and a general correlation with patterns of other criminal activity (such as burglary), areas of youthful population, and rental homes. They demonstrated that younger women under 25 years are especially at risk from rapists; that about 50 per cent of the rapes occurred in the victim's home, not a public place; that most occurred at night during or immediately after the hours of social contact and at a time of minimal community surveillance; and that there was a marked pattern of seasonality, with fewer rapes in the winter months.

To map the spatial and social distribution of the fear of crime, a survey of about 400 persons was conducted. The results showed that, although fear was widespread, there was a distinct spatial pattern, with the low-income, least stable, and high-rental areas reporting the greatest levels of fear. Socially, women and the elderly displayed the greatest concern, and fear was much greater at night than during the day. The evidence clearly showed that women and the elderly are often reluctant to use either their neighbourhoods or the city centre at night.

The conclusions of this research are generally in accord with those of several other studies. As this book has already emphasized in other contexts, we live in an unequal world. For women, the city at night may well be a landscape dominated by males.

do not have, so as others become wealthier the already wealthy become less happy; and (2) when people achieve a better standard of living, they are unable to appreciate its pleasures.

Of particular interest to human geographers is research that focuses on the spatial distribution of happiness. The first world map of happiness at the country scale, published in 2006, showed Denmark first, followed by Switzerland, Austria, Iceland, Bahamas, Finland, Sweden, Bhutan, Brunei, and Canada. Least happy of the 178 countries included were DR of Congo, Zimbabwe, and Burundi (Figure 8.4). This ranking illustrates a close relationship between happiness and measures of health, prosperity, and education, which are all high in these 10 nations. These results are as expected, and much of the interest in this work is on the detailed differences between countries, on how happiness might be more scientifically measured, and on whether or not individual responses reflect reality. With this last point in mind, a geography of happiness project has been set up in Britain, involving geography teachers travelling through Europe to determine how happy people are and whether these informal surveys accord with the published data.

Elitist Landscapes

Much of the discussion in this chapter, and elsewhere in this text, focuses on the different types of landscapes occupied by different types of people. As we have seen, there are landscapes for the privileged and the less privileged, the advantaged and the disadvantaged, which are generally labelled elitist landscapes and landscapes of stigma, respectively. There are obvious links between the fallacy of race, ethnicity, gender, and sexuality and these contrasting landscape types.

The existence of elitist landscapes is obvious, although they have seldom been analyzed. Geographers studying cities have traditionally recognized that all include a range of areas associ-

In the News

BOX 8.7 | Less Developed Canada?

Although Canada is a part of the developed world, as measured by various cultural and economic criteria, the benefits are not distributed equally among all Canadians. Throughout the country, many Aboriginal people live different lives from most other Canadians. Aboriginal communities are characterized by more youthful populations, shorter life expectancies, and higher birth rates that result in a high rate of natural increase (about double the national average and comparable to rates in many less developed countries). More generally, Aboriginal people in Canada are disadvantaged; they are more likely than other Canadians to be unemployed, to rely on social assistance, to live in poor-quality housing, to have limited access to medical services, and to have little formal education.

In 2011 the small Cree community of Attawapiskat on James Bay in northern Ontario made headline news when the federal government appointed a third-party manager to oversee the band's finances. The related publicity highlighted the squalid living conditions that many of the inhabitants experienced, including crowded ramshackle homes and lack of bathrooms and running water. One opposition politician described the conditions as Third World.

Another notorious setting was Davis Inlet, a government-created Innu community on the Labrador coast of about 680 people that lacked even running water and had deep-seated social problems. In late 2002 and early 2003 the community moved—at a cost to the federal government of some $152 million—15 kilometres across the Labrador Sea to the new village of Natuashish, in the hope that a new place would allow for a fresh start. Suicide and crime rates are much reduced, but serious social problems are still evident despite a 2008 ban on the sale, purchase, and possession of alcohol.

The situation on reserves in the north of many provinces is also deplorable. A survey of housing on northern Manitoba reserves estimated that 1,800 (29 per cent) of the 7,200 units were in need of major repair and that an additional 1,400 units were required. More than 50 per cent of reserve residents in Saskatchewan are dependent on welfare, and the rate is close to 90 per cent on some isolated reserves. Further, many communities lack piped water and sewage systems—a situation that has clear health implications. For residents of these reserves, the rate of tuberculosis is eight times the provincial average. Frustration, hopelessness, depression, and apathy are common, with high rates of violent death, family breakdown, and substance abuse.

Another tragic circumstance came to light in late 2005 when many residents were evacuated from the Kashechewan reserve on the shore of James Bay because a polluted water supply caused serious health problems. More generally, a 2007 newspaper article noted that the new territory of Nunavut is "racked by violence, with rates of homicide, assault, robbery, rape and suicide stunningly above the national average" (Harding, 2007). That such conditions should persist in relatively close proximity to some of the most affluent communities in the world is a scandal; yet most Canadians are still able to ignore it because Aboriginal people are typically marginalized, both spatially and socially.

Explaining this state of affairs is far from easy. Writing with general reference to the Canadian North, Bone (1992: 197) noted that "the notion of 'ethnically blocked mobility' arises; that is, Native Canadians may value educational and occupational achievements less than other Canadians because of cultural differences such as the sharing ethic and because of perceived or experienced discrimination." Other important reasons involve the history of relations between indigenous people and European Canadians and the nature of capitalism as a social and economic system.

Canada is not alone in this respect. In Australia the death rate for Aboriginal children is three times higher than the rate for other Australians, and life expectancy is about 20 years shorter. As in Canada, Aboriginal people often live in overcrowded housing and have unsafe drinking water, poor sanitary conditions, and higher rates of diabetes, heart disease, cancer, and violence.

ated with different social classes; it is not unusual for the higher-class areas to be at a higher elevation than other areas; to be close to a lake, river, or ocean; or to be located so that they are not affected by the pollution generated by industry. In British cities, it is usual for higher-status areas to be spatially segregated and to include such features as golf courses and specialty retail areas. In Melbourne, Australia, an elite residential area has been a consistent feature of the urban landscape,

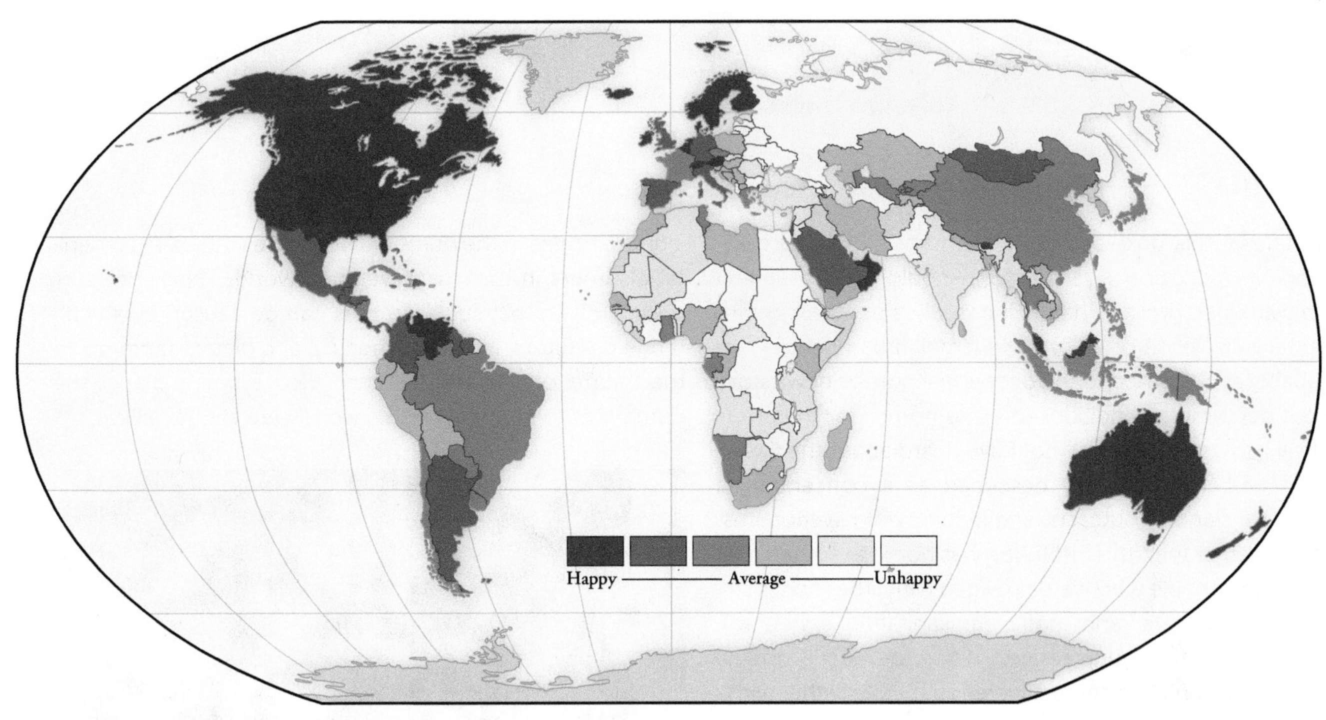

FIGURE 8.4 | World distribution of happiness

one that moved steadily south as the city grew. In an important sense, identity and landscape are closely interwoven in these elite areas. A 2008 survey listed the richest streets in the world as measured by property value. Ranked first was Avenue Princess Grace in Monaco, with apartments costing about US$50 million. Among the other rich streets are Severn Road in Hong Kong, Fifth Avenue in New York, Kensington Palace Gardens in London, and Avenue Montaigne in Paris.

Elite regions can also be identified. A long-standing distinction between south and north in Britain clearly reflects a distinction between privilege and lack of privilege. More obvious examples of privileged landscapes are tourist areas, especially in less developed countries (discussed later in this chapter).

Landscapes of Stigma

Human geographers have recently identified pariah landscapes, landscapes of despair, and landscapes of fear. Examples of pariah landscapes include many ghetto areas in cities and Aboriginal reserve lands in Canada and the United States. The experiences of discharged mental patients prompted the introduction of the term *landscapes of despair*, a vivid descriptor that applies equally well to pariah landscapes. One of the most publicized in recent years is the inner-city landscape of despair of homeless people. All such landscapes are expressions of exclusion, spatial reflections of social injustice (Box 8.8).

As already noted, violence is closely identified with specific locations. Many people must regularly cope with landscapes of fear; elderly people and women are most susceptible to violence, usually by males. Most people have mental maps identifying some areas as safe and others as dangerous—though certain "safe" areas may actually be so only for certain groups and/or during daylight hours. Large, open spaces, such as parks, are often perceived as unsafe, as are closed areas with limited exits, such as trains. The recent emergence of Neighbourhood Watch schemes is a clear reflection of increased fear and a corresponding social interaction in many city areas. Landscapes of fear result in restricted use of public space.

FOLK CULTURE AND POPULAR CULTURE

The distinction between "folk" and "popular" culture adds another dimension to our discussions of identity and difference. The relative homogeneity of

Examining the Issues

BOX 8.8 | Consigned to the Shadows

The title of this box—borrowed from an article in a geographical magazine (Evans, 1989)—could refer to many disadvantaged groups. In fact, it describes people who are mentally ill. Those of us living in larger urban centres are probably accustomed to the occasional horrific news story about deplorable conditions in some homes for the mentally ill. However, you may not have heard about the Greek island of Léros, variously described as a concentration camp, an island of outcasts, and a colony of psychopaths. Established in 1957, this institution was home to some 1,300 mentally ill people who were provided with little food, minimal medical care, and inefficient sanitation (when the odour became too unpleasant, they were hosed down). People with various disorders were grouped together without proper care, let alone hope of recovery. The EU began a project to close the institution in 1990, but it took several years before the last inmates were deinstitutionalized.

Conditions on Léros may appall us, but the island is not an extreme case historically. Mentally ill people have typically been separated from the majority with no actual attempt to cure them. Indeed, until very recently, there has been no real understanding of mental illness. In the United States homosexuality was regarded as a mental disorder until 1973; private mental hospitals in Japan house about 350,000 people (one of the highest per capita rates in the world), at least partly because it is relatively easy to have a person so detained.

The problem of properly housing the mentally ill is increasing. The World Health Organization (WHO) predicts a dramatic rise in the numbers of people with serious mental disorders in the less developed world. This is not surprising, given that such illnesses can be caused by a wide range of factors, including medical and dietary factors and the trauma of warfare. A recent WHO estimate places the number of mentally ill people worldwide at 100 million.

A homeless man collects bottles in Vancouver. Scenes such as this are increasingly common as many cities struggle to cope with growing numbers of people living on the street.

people in traditional folk cultures survives mainly in rural areas among groups linked by a distinctive ethnic background and/or religion, language, and occupation. Folk cultures tend to resist change and to remain attached to long-standing attitudes and behaviours. By contrast, popular culture is largely urban and tends to embrace change—even to the point of seeking change for its own sake, as in the case of fashion. Geographic studies of both folk and popular culture are plentiful in North America and Europe.

At present, folk studies still tend to be closely associated with the traditional landscape school of Sauer; however, the more theoretically informed methods outlined and employed in this chapter are becoming increasingly prominent. Studies of popular culture are being conducted both in the traditional landscape school and in the newer arena of cultural analyses enriched by social theory. Although the boundary between folk and popular is far from clear-cut, it is possible to identify some basic characteristics of each.

Folk Culture

In general, folk cultures are more traditional, less subject to change, and, in principle, more homogeneous than popular culture. (This third identifying characteristic is of limited value, however, since popular culture landscapes also can be homogeneous.) Folk cultures have pre-industrial origins and bear relatively little relation to class. Religion and ethnicity are likely to be major unifying variables, and traditional family and other social traditions are paramount. Folk cultures tend to be characterized by a rural setting and a strong sense of place.

This is not to say that elements of folk culture cannot be found in urban as well as rural areas. In large multicultural cities, attitudes and behaviours specific to many different ethnic groups not only survive but also flourish and in some cases have even become features of the broader popular culture. Prime examples include the Caribbean festivals held every year in Toronto and London and the dragon boat races mounted by Chinese communities in various Canadian cities.

The Red River dragon boat races at the Forks in Winnipeg.

In rural areas, domestic architecture, fence types, and barn styles vary from one folk culture to another. However, the landscapes of individual folk culture groups, such as the Amish or Mennonites, are uniform and largely unchanging—a reflection both of central control and of individuals' desire to conform to group norms. Diet, too, reflects a preference for long-established habits, as well as the rootedness in an agrarian or maritime economy and lifestyle. The same is true of musical preferences: traditional styles associated with a particular region (e.g. western swing in north Texas and Oklahoma) are strongly preferred over the products of North American/global popular culture. Indeed, folk traditions in musical landscapes and cultural regions have been studied extensively.

Popular Culture

Trends in popular culture, including new attitudes and behaviours, tend to diffuse rapidly, particularly in more developed areas where people have the time, income, and inclination to take part. The diffusion of popular culture is an important part of the cultural globalization discussed in Chapter 3. Box 8.9 discusses links between consumption and identity formation.

Jackson (1989) describes the rise of popular culture in nineteenth-century Britain, showing how the social activities of the working-class populations were constrained by more powerful classes. New places of entertainment (such as music halls) became major contexts of class struggle, as did activities such as prostitution. Many present-day issues similarly involve the control of space, and Jackson argues that the "domain of popular culture is a key area in which subordinate groups can contest their domination" (p. 101). One of the better-known instances of social inequality that has expression in landscape relates to matters of sexuality. As noted earlier, gay/lesbian neighbourhoods serve as powerful examples of the spatial constitution of culture.

Other geographic analyses of popular culture focus on specific landscapes (shopping malls, tourism, sports, gardens, urban commercial strips) and regions (musical regions, recreational regions). Shopping malls—artificial landscapes of consumption, located in most urban areas in the more developed world—provide a vivid illustration of the impact of popular culture on landscape. Usually enclosed, windowless, and climate-controlled, catering to the commerce-driven consumer tastes of the moment, malls are characterized above all by their sameness. In fact, they exemplify the placelessness that is one of the by-products of globalization. When completed in 1986, the West Edmonton Mall in Edmonton, Alberta, was the largest in the world, with 836 stores, 110 restaurants, 20 movie theatres, a hotel, and a host of recreational features (see Jackson and Johnson, 1991a, 1991b). But nothing compares to the Dubai Mall that was visited by about 80 million people in 2014, with that number expected to increase to 100 million by 2020. Indeed, the Dubai Mall is visited by more people than any other place in the world, more than the Eiffel Tower, Niagara

Geophagy

Examining the Issues

BOX 8.9 | Geography, Consumption, and Identity Formation

> While I am watching television, wearing trainers, eating at a restaurant, shopping at the local supermarket, going out clubbing, or jetting off on holiday, I am not often consciously thinking about geography. Indeed, until recently, geographers largely ignored such activities, deeming them frivolous and peripheral, to be left to the disciplines of sociology or cultural studies (or left outside academia altogether). This neglect is now being redressed, however, and cultural geographers in particular are increasingly arguing that the taken-for-granted activities which fill our everyday lives should be seen as very important and therefore worthy of serious academic inquiry. (Jayne, 2006: 34)

Some human geographers occasionally express surprise and even concern at the seemingly never-ending proliferation of subject matter, including advertisements, yard sales, clothing preferences, shopping malls, and much more. But in an important sense, such concerns reflect a misunderstanding. Geographers, especially cultural geographers, have long studied the everyday lives of people and place, with many studies in the Sauerian tradition focusing on such seemingly "esoteric" topics as covered bridges, agricultural fairs, log building construction, geophagy, and much more.

Indeed, it is not so much the choice of subject matter that is different, as evidenced by the fact that Sauerian and new cultural geographers have shared interests in many aspects of consumption—food, drink, sports, and music to name a few—and in many aspects of identity formation as these relate to consumption. What is new is the philosophical underpinning of research. Whereas much earlier work on everyday lives and landscapes, including consumption habits and places of consumption, was essentially empiricist, current work is informed by a disparate body of social theory, much of which is postmodern in character. Consider the construction and reconstruction of shopping mall cultures. The best-known example of this phenomenon was the emergence of the much ridiculed Valley girls, who identified themselves with the Glendale Galleria in California.

Most notably, present work typically acknowledges that the "consumption" of places (for example, tourist heritage sites or resort settings), the "consumption" of peoples (for example, indigenous or ethnic peoples), and the links between identity and consumption (as evidenced, for example, by musical, clothing, and food and drink preferences) are worthy of study because such consumption is constituted by economic, political, and cultural processes that vary from place to place. Most notably, many researchers are interested in disentangling local processes, regional processes, and global processes. In short, human geographers recognize that what, when, where, and how we consume are inherently geographic and therefore merit study.

Falls, and Disney World combined. The mall, currently being expanded, has 1,200 stores and 200 restaurants. As products of popular culture, shopping malls encourage the further acceptance of popular culture. Tourist landscapes, discussed more fully in this chapter's final sections, are another expression of popular culture.

Music

As with folk culture, there is much interest in studying musical landscapes and regions in popular culture contexts, as well as a focus on links between music and identity. As identities evolve, cities sprawl, and landscapes change, so music changes. Much recent and contemporary music is closely tied to certain places, as with the Detroit Motown sound of the 1960s and the Seattle grunge music of the 1990s.

According to a 2008 study of United Kingdom musical tastes that analyzed sales, regional charts, and live performances, there are striking differences in musical preference from place to place. Folk is particularly popular in Scotland, jazz in the southwest, heavy metal in the Nottingham area, world music in Bristol, handbag house in Northern Ireland, reggae in Birmingham, and Eurodisco on the south coast. Some of these preferences are linked to immigration patterns and ethnic identity, while others are more difficult to explain. Of course, some musical preferences are far more widespread, even global, reflecting the globalization processes that contribute to cultural homogeneity.

Sport

Sport is another important part of popular culture. In fact, major organized sports, such as baseball, hockey, and soccer, originated in the late nineteenth century in association with the social changes initiated by the Industrial Revolution. Regions can be delimited in terms of sporting preferences among spectators and participants alike. Rooney (1974) discovered clear regional variations in the United States: for example, prior to the 1970s American professional football players came primarily from a cluster of southern states (Texas, Louisiana, Mississippi, Alabama, and Georgia). Geographers frequently see links between such preferences and vernacular region (four of these five states can be identified as the American Deep South).

For some people, supporting a particular sports team is an important component of their everyday identity. College and high school sports teams in the United States often receive huge support from local people and from those who have graduated from these institutions. In Europe and Latin America affiliation with a soccer team can be a significant part of many people's lives, and fans have a passionate interest in how well their team performs. Identity at the national scale comes to the fore when there are major international competitions, most notably the FIFA World Cup in association football (soccer).

© nik wheeler/Alamy

People watch a dolphin show at the West Edmonton Mall. Built in three phases from 1981 to 1986, the mall is a prime tourist destination—a massive complex of department stores, shops, restaurants, recreation areas, amusements, and services, including a luxury hotel.

TOURISM

Our final discussion relating to questions of identity and difference is concerned with tourism, an activity generated by the more developed world but that operates in both the more and the less developed worlds. Tourism has experienced dramatic growth since about 1960; according to the World Travel and Tourism Council, it is now the largest economic activity in the world and is growing at a phenomenal rate—about 20 per cent faster than the overall world economy. This group claims that travel and tourism account for about 10 per cent of global GDP. Globally, the number of tourists is estimated at more than 700 million annually, and World Travel and Tourism (somewhat optimistically perhaps) expects that number to increase to some 1.5 billion by 2020. That said, tourist expenditures and tourist numbers are very difficult to determine precisely and these numbers need to be viewed with caution.

Regardless, the importance of tourism cannot be measured simply in numerical terms; tourism is explicitly bound up with matters of identity and difference. "Tourism is a significant means by which modern people assess their world, defining their own sense of identity in the process" (Jakle,

© GLYN KIRK/AFP/GettyImages

A Belgian fan cheers as he waits for the start of the international friendly football match between England and Belgium at London's Wembley Stadium.

1985: 11). Because it is "one of the most penetrating, pervasive and visible activities of consumptive capitalism, world tourism both reflects and accentuates economic disparities, and is marked by fundamental imbalances in power" (Robinson, 1999: 25).

In other words, tourism is a means by which both tourist and host communities create their respective identities and emphasize their difference from one another. Because tourism offers significant prospects for cultural contact, it creates opportunities for both cultural understanding and misunderstanding. In short, tourism is one means of creating and recreating human geographies.

The Rise of Tourism

In seventeenth-century Europe, the elite visited spas for medical purposes; later, the supposed health benefits of seaside resorts attracted a broader clientele. The idea that travelling might be for pleasure and also for broadening the traveller's horizons first emerged in Europe during the eighteenth century, especially as European nations developed their overseas empires and travellers and colonial officials published accounts of exotic **locales** and peoples.

locale
The setting or context for social interaction; a term that has become popular in human geography as an alternative to *place*.

Mass tourist travel, however, depended on the introduction of the annual vacation (as opposed to holidays for religious observance) negotiated between employer and workforce, a product of the Industrial Revolution. By the late nineteenth century, the vacation proper was established and seaside resorts in Europe, specifically Britain, were becoming playgrounds for the working classes. Beginning about 1960, changes in employment patterns that allow for more leisure time, additional discretionary income for many people, decreasing travel costs (mainly air travel), and increasing numbers of retirees have helped to diversify tourism and make it a year-round activity. By the 1990s, the Internet allowed people to arrange all aspects of the vacation experience without using a professional travel agent, thus reducing overall costs.

Tourist Attractions

As already noted, the most visited place in the world today is the Dubai Mall, a fact that highlights the importance of consumption to tourist activity. But there are six other, more traditional, attractions for contemporary tourists: good weather (usually meaning warm and dry); attractive scenery (coastal locations are particularly favoured); amenities for such activities as swimming, boating, and general amusement; historical and cultural features (old buildings, symbolic sites, or birthplaces of important people); accessibility (increasingly, tourists travel by air; in general, costs increase with the distance travelled); and suitable accommodation. Cruise lines, a major part of the tourism industry, both take their passengers to such destinations for brief stopovers and provide most of these attractions on board their massive ocean liners, in a sense becoming placeless destinations.

Locations with an appropriate mix of these six features tend to be major tourist areas. In the more developed world, such areas usually are urban or coastal. In Europe many of the most popular coasts for tourism are peripheral (for example, in Spain and Greece), although this is not the case in countries such as the United States and Australia. Also, some landscape features have assumed prominence as tourist attractions because of their identification as World Heritage Sites as defined by UNESCO. There are currently more than a thousand such sites, and this list is regularly reviewed. Most sites are identified as cultural, some are physical, and others are a mix of the two. As of 2015, Canada has 17 World Heritage Sites, with a further 7 under consideration. The most recent Canadian addition, in 2013, was the Red Bay Basque Whaling Station on the shore of the Strait of Belle Isle.

Mass Tourism

Mass tourism is a form of mass consumption: it involves the purchase of commodities produced under conditions of mass production; the industry is dominated by a few producers; new attractions are regularly developed; and many supposedly different sites are essentially similar. Together, these characteristics make the mass tourist experience much the same everywhere, regardless of site. Further, as with other cases of mass production and consumption, the producers of tourist sites effectively determine the consumers' options, having the power to direct what is sometimes called the tourist gaze.

While commercial producers create such attractions as coastal resorts, governments can direct the tourist gaze by creating parks and other favoured locations, although the tourist element in such cases is sometimes quite small.

For example, the 10 new national parks and park reserves that the Canadian government has created in the Canadian North since 1972 were intended primarily to preserve habitats and landscapes.

Alternative Tourism

The impact of tourism on local ways of life is contradictory. On the one hand, tourism encourages local crafts and ceremonies; on the other hand, it can destroy local cultures and dramatically change local environments. One commentator suggested that "we are more prone to vilify or characterize mass tourism as a beast; a monstrosity which has few redeeming qualities for the destination region, their people and the natural resource base" (Fennell, 1999: 7). Indeed, the Asian tsunami of December 2004 had such a devastating impact in places such as Phuket, Thailand, because protective mangrove forests and coral reefs had been cleared away to create playgrounds for visiting tourists and shrimp farms to feed European demand. Increasing awareness of the problems associated with conventional mass tourism has stimulated development of various "alternative" types of tourism centred on unspoiled environments and the needs of local people. Table 8.4 lists some of the advantages of such forms.

The Cefalù Cathedral is part of the Arab-Norman Palermo on Sicily's northern coast. Named a World Heritage Site in 2015, Arab-Norman Palermo consists of two more cathedrals, two palaces, three churches, and a bridge. These buildings demonstrate the co-existence of different cultures and religions; the amalgamation of Western, Islamic, and Byzantine cultures is evident in the use of space, structure, and decoration.

Focus on Geographers

The Paradox of Polar Bear Ecotourism in Churchill, Manitoba | CHARLES GREENBERG

A 2004 study by the Arctic Climate Impact Assessment (ACIA) team, made up of an international group of 300 scientists, concluded that the polar regions are warming almost twice as fast as the rest of the planet and that the retreat of sea ice has particularly devastating consequences for polar bears. Canada has 65 per cent of the world's estimated 20,000 polar bears. The western Hudson Bay population is about 600, and these bears obtain almost all their calories from eating seals, which they hunt on sea ice. Ian Stirling, a University of Alberta biologist, has studied the animal for nearly four decades, and is concerned that the rapidly dwindling population in this area will be too low to sustain a continued breeding community. This could lead to a 20–30 per cent annual population decline (Stirling and Parkinson, 2006). It is not hard to see how the population, at least in this region, may be wiped out in just a few decades.

Despite this problem, approximately 6,000–10,000 international tourists from as far away as Europe and Australia travel to Churchill, Manitoba—which calls itself the polar bear capital of the world—each year to view these magnificent mammals. This type of tourism is called ecotourism, which "involves travelling to relatively undisturbed or uncontaminated natural areas with the specific object of studying, admiring, and enjoying the scenery and its wild plants and animals" (Ceballos-Lascurain, 1988). Ecotourism, which is synonymous with the term *sustainable tourism*, is often associated with travel to undeveloped and remote locations involving modest accommodations and contrasts

Continued

sharply with "mass" and high volume tourism. Almost all ecotourism visits to Churchill occur in the late summer and early fall, when the bears congregate along the shores of Hudson Bay to wait for the sea-ice floes to form (Dawson et al., 2010). The bears are largely inactive, in a state of "walking hibernation," and are highly visible. Due to longer annual ice-free periods in Hudson Bay in recent years, the waiting period is extended, which lengthens the ecotourism season in Churchill.

Overall, tourism's contribution to climate change is significant. If tourism were a country, its emissions would be eclipsed by only China, the US, Russia, and the EU. The largest proportion of tourism's CO_2 emissions comes from transportation, and of that, aviation is the big menace. The paradox of polar bear ecotourism is clear. Long-haul air travel is necessary to reach remote ecotourism destinations such as Churchill, which means that polar bear viewing contributes a disproportionate share of emissions and therefore contributes to an accelerated extinction of the very species the tourists come to see. The paradox may be even deeper in that the ecotourists consider themselves to be "environmentally aware" and concerned about the long-term health of the species. As ecosystems around the world rapidly change in response to global warming, tourists are beginning to realize that there is a limited amount of time left before many endangered destinations vanish altogether. Habitats are shrinking and species are becoming extinct. This has prompted a travel movement that has been dubbed doomsday tourism—a surge in tourists visiting the places most threatened by global warming. In Churchill and other similar locations, this brand of ecotourism is also referred to as last chance tourism.

Meanwhile, Churchill is preparing for a surge in ecotourists. There is broad support for what Ernie Small (2011), writing in the journal *Biodiversity*, calls "marquee and poster species," which include polar bears, whales, elephants, and pandas. Polar bears, in particular, have been featured prominently in corporate ads that profess a "green" philosophy. Coca-Cola, which features animated versions of the animal in its ads, has announced that it will contribute $2 million to the World Wildlife Fund to protect polar bears and will match consumers' donations up to an additional $1 million. Polar bears have become recognized worldwide as living symbols of northern polar regions and have recently become prominent symbols in international campaigns to combat global warming. Who hasn't seen polar bears in TV commercials or the famous picture of two polar bears seemingly stranded on a tiny ice floe? The absurdity inherent in this discussion is the role ecotourism plays in undermining precisely what it champions to protect.

CHARLES GREENBERG teaches geography and tourism at Capilano University.

References

Arctic Climate Impact Assessment. 2004. *Impacts of a Warming Arctic*. Cambridge: Cambridge University Press.

Ceballos-Lascurain, H. 1988. "The Future of Ecotourism," *Mexico Journal*, 27 Jan., 13–14.

Dawson, J., et al. 2010. "The Carbon Cost of Polar Bear Viewing Tourism in Churchill, Manitoba," *Journal of Sustainable Tourism* 18, 3: 319–36.

Small, E. 2011. "The New Noah's Ark: Beautiful and Useful Species Only. Part 1. Biodiversity Conservation Issues and Priorities," *Biodiversity* 12, 4: 232–47.

Stirling, I., and C. L. Parkinson. 2006. "Possible Effects of Climate Warming on Selected Populations of Polar Bears (*Ursus maritimus*) in the Canadian Arctic," *Arctic* 59, 3: 261–75.

ecotourism
Tourism that is environmentally friendly and allows participants to experience a distinctive ecosystem.

Recent years have seen a striking upsurge of interest in "environmentally aware" or "sustainable" tourism, sometimes called **ecotourism** (see the Focus on Geographers box on page 291). Belize, in Central America, is one of the best-known ecotourist destinations and has hosted two major conferences on the topic. Tourism accounts for 26 per cent of Belize's gross national product and is becoming more important as prices for traditional cash crops, especially sugar cane, drop. Awareness of the fragility of coastal ecosystems led the government to encourage ecotourism; however, several observers have seen little difference between the developments designed for ecotourists in Belize and other, more traditional developments (Wheat, 1994: 18). In fact, specialized alternatives such as ecotourism are unlikely ever to attract the market that mass tourism does, although they may be able to make some contribution to local economies.

Creating Places and Peoples for Tourists

Culturally, the tourist industry can be seen as one more example of the dominance of the more developed world over the less developed. It can be argued that the activities of tourists both reflect and reinforce, perhaps even legitimize, existing patterns of inequality and the dominance of some groups over others. Increasingly, one reason

TABLE 8.4 | Characteristic tendencies: Conventional mass tourism vs alternative tourism

Variable	Conventional Mass Tourism	Alternative Tourism
Accommodations		
Spatial pattern	High density/concentrated	Low density/dispersed
Size	Large scale	Small (local) scale
Impact	Obtrusive (modifies local landscape)	Unobtrusive (blends into local landscape)
Ownership	Non-local/big business	Local/small business
Attractions		
Emphasis	Commercialized cultural/natural	Preserved cultural/natural
Character	Generic, contrived	Local, authentic
Orientation	Tourists only	Tourists and locals
Market		
Volume	High volume	Low volume
Frequency	Seasonal (cyclical)	Year-round (balanced)
Segment	Mass tourist	Niche tourist
Origin	A few dominant markets	Diverse origins
Economic impact		
Status	Dominant sector	Complementary sector
Linkages	Non-local	Local
Leakages	Profit expatriation, high import level	Low import level
Multiplier	Low multiplier	High multiplier
Regulation		
Control	Mainly non-local	Mainly local
Amount	Minimal	Extensive
Basis	Self-regulating free-market forces	Public-sector emphasis
Priority	Develop, then plan	Plan, then develop
Motives (holistic/integrated)	Economic (sectoral)	Economic, social, environmental
Time frame	Short term	Long term
Development and tourist ceilings	No ceilings	Imposed ceilings that recognize carrying capacities

Source: Weaver, D. B. 1993. "Ecotourism in the Small Island Caribbean." *Geojournal* 31: 458.

for people to visit "exotic" peoples and places is to experience cultural difference. Yet—not surprisingly—those peoples and places are rarely authentic; in many cases they have been constructed to satisfy the tourist gaze as described earlier. Travel companies frequently play on the desire for novelty by using "exotic" images and descriptions in their brochures. In fact, for many tourists, the places and peoples they visit are commodities, and—as with many other commodities—the advertising used to present them to prospective consumers relies more on fancy than on factual information. It can be argued that group tours to visit indigenous people are a form of cultural voyeurism, with visitors feeling uncomfortable and local people putting on a necessarily inauthentic display for money. Only tour operators benefit from these situations.

Human geographers analyze images and representations such as those used in tourist brochures in order to unpack and understand the meanings they convey. In many cases they find that such advertising stresses the power of the tourist in some dependent, often former colonial, area. "Tourist" places and peoples are clearly being socially constructed—in terms of ethnicity, gender, food, and drink—to stress the difference between "them" and "us" and enhance the elements of mystery and exoticism that attract tourists. Figure 8.5 shows one way of understanding the relationship between the places produced by the tourist industry and the places consumed by the tourist, suggesting that the social construction of tourist places comprises two subsystems: place production by the tourist industry and place consumption by the tourists. Where these constructions are in agreement, a zone of convergence exists.

As this discussion suggests, recreation and tourism are increasingly attracting the attention of human geographers. In particular, tourism is of interest as a process of consumption, a product of place construction, and a reflection of difference.

Falsifying Place and Time

Among the most distinctive tourist attractions are spectacles, such as sporting events or world fairs; theme parks, such as Disney World (Ley and Olds, 1988); and large shopping centres, such as the aforementioned Dubai Mall and the West Edmonton Mall (Shields, 1989; Hopkins, 1990; Jackson and Johnson, 1991a, 1991b). Many such manifestations of popular culture are created by developers to appeal to tourists as consumers, and they lend themselves to a wide range of human geographic interpretations.

spectacle
A term referring to places and events that are carefully constructed for the purposes of mass leisure and consumption.

Attractions created where there was previously nothing to draw visitors are often labelled artificial. It is probably correct to say that what matters in this kind of venture is innovation, creating something that people want—sometimes even creating a demand for something that no one had thought of wanting before. Two important attractions of artificial sites are that they offer a controlled and safe environment while purporting to be exotic, providing a glimpse into other cultures and places.

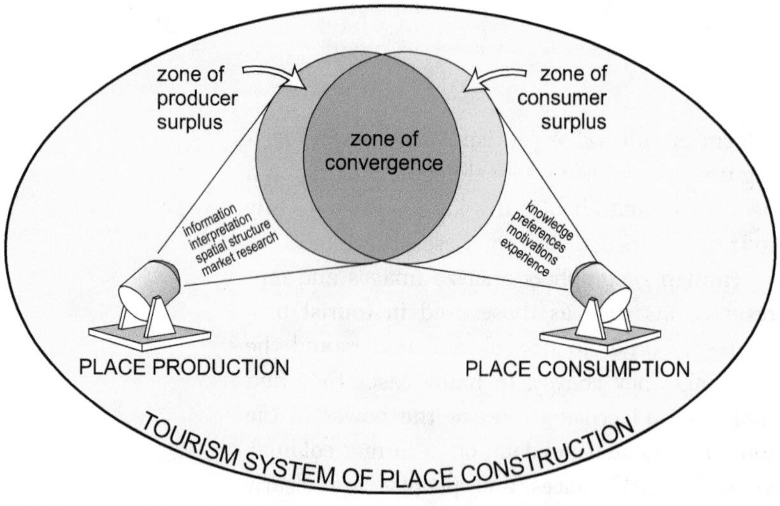

FIGURE 8.5 | The tourism system of place construction

Source: After Young, M. 1999. "The Social Construction of Tourist Places." *Australian Geographer* 30: 386, Figure 3 (MW/CAGE/P1892).

In many parts of North America, tourist attractions such as gambling facilities capitalize on a previously unsatisfied demand. In many respects, Las Vegas is the pioneer in this area, and it regularly reinvents itself to continue as a prime tourist destination. Most of the early hotels in Las Vegas offered little more than basic accommodation for gamblers and relatively tame nightclub performances, but in 1966 the first of the large theme hotels, Caesars Palace, opened, setting the pattern for subsequent development. Today the city attracts many families and business conferences in addition to gamblers.

One of the more interesting aspects of these essentially artificial tourist sites is that the experiences they offer are inauthentic—more about myths and fantasies than about reality. Most observers see this form of tourism as bound up with postmodernism, the emphasis on consumption rather than production, and the commodification of people, place, and time. Perhaps nowhere is this seen more clearly than in Branson, Missouri, on the edge of the Ozarks, which has been transformed from a quaint town with one significant tourist attraction—Silver Dollar City—into a country music (and other performances) destination for millions of visitors. Branson now offers dozens of single-purpose theatres owned, operated, or leased by the formerly famous or near famous, from Chubby Checker and Chinese acrobats to Shoji Tabuchi, Yakov Smirnoff, 3 Redneck Tenors, 12 Irish Tenors, and the ubiquitous Elvis Presley impersonators.

Even many supposedly "real" tourist sites may have more to do with myth than reality. Heritage and other historical sites, for example, may be so thoroughly reconstructed and packaged that it is difficult for tourist consumers to know how true to the past their experience is. Colonial Williamsburg in Virginia sometimes is criticized on these grounds, as are many battlefields and heritage sites, at least partly because the meanings attached to them are necessarily plural and contested. For some critics, our understanding of ourselves, our places, and our pasts increasingly is formed by the tourist industry.

© Paul Hakimata/Alamy

The Governor's Palace, Colonial Williamsburg.

Tourism in the Less Developed World

Many of the favoured tourist destinations today are in the less developed world—Mexico, various Caribbean and Pacific islands, and some Asian and African countries. Some of these places offer an appropriate mix of the six principal attractions noted earlier, but they may also be attractive for other reasons. For example, low labour costs usually mean that these areas can offer competitive rates, and some may represent—at least for Europeans or North Americans—an "exotic" cultural experience. Tourism is certainly growing in the less developed world, specifically in coastal areas, in response to increasing demand from the more developed world.

For many less developed countries, tourism generates the largest percentage of foreign exchange. Kenya has an especially well-developed tourist industry, and other African countries are seeking to follow a similar path. Hoping to capitalize on a long Atlantic coastline, lush forests, waterfalls, and wildlife, Angola has begun construction of several luxury hotels only a few years after the end of a civil war that left much of the country devastated.

This trend poses several problems for the less developed countries, however. Most important from the economic perspective is that dependence on tourism makes a country vulnerable to the changing strategies of the tour companies in more developed countries. Many destinations offer sun, sand, and surf, and those that have no distinctive attractions must keep their rates low if they are to draw clients away from their competitors. Even so, most less developed countries appreciate the benefits that tourism can provide: local employment, stimulation for local economies, foreign exchange, and improvements in services such as communications. In short, in less developed countries the tourist areas are generally much better off than other parts of the country. Today, tourism is a major growth industry in the tertiary sectors of several less developed countries, and current evidence suggests continued growth (Box 8.10).

Shangri-La

Not surprisingly, China's tourist industry is becoming increasingly similar to that of countries in the more developed world. Along with attracting tourists from other countries, domestic tourism is emerging. Consider the example of Yunnan

Province on the edge of the Tibetan Plateau in southwest China.

Designated a World Heritage Site in 2003, three of the great rivers of Asia—Yangtze, Mekong, and Salween—here run parallel through dramatic gorges in a region that may be, according to UNESCO, the most biologically diverse temperate region in the world. Particularly interesting is that one small nearby town—Zhongdian—has been completely transformed to become a major tourist centre, attracting visitors especially from the cities of eastern China. Capitalizing on the symbolism of the fictitious place in James Hilton's *Lost Horizon* (1933), Zhongdian was renamed Shangri-La in 2001. The once-derelict buildings were rediscovered as examples of traditional Tibetan architecture, and the town has become a commercialized centre of Tibetan culture. Urban

Around the Globe

BOX 8.10 | Tourism in Sri Lanka

Sri Lanka is representative of tourist destinations in the less developed world in that its tourist industry is problematic and uncertain. Long described as a paradise by Europeans (see Box 6.1), from a European perspective it is a logical area to be marketed as a tourist destination. There are scenic landscapes, including beaches, tropical lowlands, and mountains, as well as a rich and diverse cultural heritage. During the 1960s, the Sri Lankan government created the necessary infrastructure of hotels, roads, and airline facilities, and improvements and additions to this infrastructure have continued. The industry experienced spectacular growth, and by 1982 tourism was second only to tea as an earner of foreign exchange.

A period of decline between 1983 and 1989 reflected the ethnic conflict described in Box 6.1. Tourist numbers dropped by more than 50 per cent; northern and eastern Sri Lanka experienced the greatest losses in tourist numbers and hence the greatest economic damage. After 1989 the industry began to recover and attract visitors from elsewhere in Asia. By the mid-1990s, the industry seemed more secure because of strong government support and careful regulation (O'Hare and Barrett, 1993), but the hoped-for stability has not been achieved. Damage caused by the 2004 tsunami had an immediate negative impact with more than a hundred hotels destroyed or badly damaged. More significant is ethnic conflict, meaning that the island is not seen as a desirable destination (ethnic tensions are not conducive to a successful tourism business). By 2006 beach resorts catering to wealthy tourists from North America and Europe were functioning at only 20 per cent capacity largely because of the ongoing conflict, which came to a bloody end three years later.

Another problem for many tourist businesses is a government regulation passed in 2005 that created a buffer zone within 100 metres of the coast. Within this zone, businesses cannot build any new structures or repair damaged ones. However, this regulation was welcomed by environmentalists, and there is hope that an ecotourism industry focused on the shoreline environments might develop.

Euan White

The Sinhalese ruler Parakramabahu (1153–86) built the city of Polonnaruwa, the site of this 14-metre-long figure depicting the Buddha entering nirvana.

development proceeded rapidly, including hotel and airport construction and road improvements. Also nearby is Mount Kawagebo, one of the most sacred sites in Tibetan Buddhism. Many pilgrims walk around the mountain, a challenging trek that takes about two weeks. The combination of an outstanding nearby physical geography, a pilgrimage site, and the commodification of Tibetan cultural attractions encouraged three million tourists to visit Shangri-La in 2008.

CONCLUSION

The previous chapter began with a series of powerful metaphors: "mountains of hate, rivers of inflexible tradition, oceans of ignorance" (James, 1964: 2)—physical geographic terms used to describe aspects of our human condition. Language, religion, the myth of race, ethnicity, gender, and sexuality all have at some time and in some place been used as justifications for war, cruelty, hypocrisy, or dogma. The goal of a single universal language remains elusive, although English approaches that need; the prospects for uniting diverse religious beliefs are negligible; racism regularly rears its ugly head; ethnic groups continue to value traditions that in some cases may be inappropriate to modern societies; and the likelihood of altogether eliminating the cultural implications of gender and sexuality seems slight. All these variables are human constructions, and they are consequences of cultural evolution. Not only does each involve divisions within the human population, but divisive attitudes and behaviours are also actively encouraged by many cultural groups. For religious groups in particular, dogma may be fact.

Human diversity, then, has a dangerously divisive aspect. We are at a point in human cultural evolution when we might give serious thought to social engineering, designing ourselves and hence our future. It is therefore appropriate to ask whether such a project—to devise new, overarching values and moral standards that might bridge the divisions between us—is desirable. These are heady issues.

For many, a world without human diversity at both group and individual levels is unimaginable. Perhaps what we really need to achieve is a diversity that involves mutual respect between groups and individuals. Only then will our divided world be free of the circumstances that James noted.

Summary

New Cultural Geography

A welcome addition to Sauerian cultural geography, new cultural geography recognizes the existence of a plurality of changing cultures located in specific places at specific times and introduces new approaches and much new subject matter, especially relating to issues of dominance and subordination. The new cultural geography often implies constructionist rather than essentialist interpretations of such topics as human identity and landscape.

Geographies of Difference

Geographers understand difference through socially produced markers, such as ethnicity, gender, sexuality, able-bodiedness, and age (and also language and religion), and commonly focus on contested landscapes. Overall, these studies mark a dramatic break with the traditional focus initiated by Vidal and Sauer in particular. In brief, we are seeing increasing interest in Marxism, humanism, feminism, and postmodernism; decreasing concern with the landscape per se; and growing acknowledgement of human agency's role in both social and power relations. These newer concerns first emerged in the 1970s.

The Unity of the Human Race

Humans are members of one species—*Homo sapiens sapiens*. Early spatial separations of groups of humans facilitated the development of physical variations. These are of minimal relevance to our understanding of either people or places. The concept of race among humans has no basis in fact.

Racism

Race may be an illusion, but racism is a fact. Groups seen as different from the majority within a society often find themselves labelled negatively and treated unequally; genocidal circumstances and the policy of apartheid as previously practised in South Africa are extreme examples.

Ethnicity

Ethnic groups are frequently loosely defined; the key linking variable may be language, religion, or common ancestry. They are minorities and may be seen as set apart from the larger society, but over time many experience acculturation or assimilation. Canada is an officially multicultural society.

Gender

Beginning in the 1970s, feminist geographic research aimed to make space for women, both in the everyday world of the university and society more generally, and in the knowledge produced by geographers. Feminist-inspired studies challenge the practice of human geography, being variously understood as both a part of and yet somehow against that practice. A central idea is that landscapes are gendered to reflect the dominance of patriarchal cultures: dominant men and subordinate women. Gender differences in well-being and the quality of life in general are evident at the local scale, for example, in the development of suburbs in North American cities, and at the global scale using measures of human development.

Sexuality

Sexuality—particularly that other than heterosexuality—is studied as an expression of identity and as one way in which the dominant heterosexual landscape can be challenged. Acceptance of homosexuality is far from universal, with countries such as Nigeria, Saudi Arabia, and Iran being especially intolerant.

Microcultures

Some people often identify with a small subgroup rather than some larger, broadly based cultural identity. Youth subcultures are prime examples.

Well-Being

Human geographers are increasingly concerned with landscapes of resistance (created by those who are excluded from the landscapes constructed by and for dominant groups); topics such as crime and health as they reflect differences from place to place; and what might be called elitist landscapes and landscapes of stigma.

Folk Culture

Folk cultures tend to retain long-standing attitudes and behaviours and are conservative of traditions passed on from one generation to the next both orally and by example. A wide range of landscape features and folk activities are typically studied in the manner of the Sauer landscape school.

Popular Culture

The term *popular culture* is applied to groups whose attitudes and behaviours are constantly subject to change, in contrast to folk culture. Geographic studies of popular culture employ both traditional and more recent conceptual backgrounds and include analyses of landscapes and regions. Today, popular culture is usually discussed in the larger context of globalization processes.

Tourism

Tourism is explicitly bound up with matters of identity and difference. The largest industry in the world, it is generated by the more developed world, but many favoured destinations are in the less developed. Spectacles and theme parks are a distinctive form of tourist attraction. Many less developed countries take advantage of attractive climates and landscapes to cultivate a tourist industry, although their success can be affected by volatile political circumstances, as in the case of Sri Lanka. Tourist areas may benefit economically but may also experience negative cultural and environmental consequences. Some countries, such as Belize, are encouraging ecotourism. Human geographers are increasingly interested in how the places and peoples consumed by tourists are represented.

Future Cultural Identity

Geographers now ask questions about our future cultural identity and the possibility of human engineering—deliberately creating new cultures.

Links to Other Chapters

- **Globalization:** Chapter 2 (concepts); Chapter 3 (all of chapter); Chapter 4 (global environmental issues); Chapter 5 (population growth, fertility decline); Chapter 6 (refugees, disease, more and less developed worlds); Chapter 9 (political futures); Chapter 10 (agricultural restructuring); Chapter 11 (global cities); Chapter 13 (industrial restructuring)
- **Concept of place:** Chapter 2 (humanism)
- **Ethnicity:** Chapter 7 (language and religion)
- **Gender and sexuality:** Chapter 5 (fertility transition); Chapter 6 (global inequalities, indexes of human development); Chapter 9 (politics of protest)
- **Folk culture:** Chapter 7 (landscape)
- **Popular culture:** Chapter 12 (intra-urban issues)
- **Ecotourism:** Chapter 4 (sustainable development); Chapter 6 (less developed world)

Questions for Critical Thought

1. Outline the various theoretical perspectives that underlie the "new cultural geography." How can/should we make sense of these competing theories? Is one theory better than another? Justify your position.
2. How can/should we balance theory and empiricism?
3. Can there be too much attention paid to theory and, by implication, not enough attention paid to real-world social issues and concerns? Why or why not?
4. What do geographers mean when they talk of "landscape as place"? Why is this so important?
5. Why are topics such as gender, race, and ethnicity discussed under the heading of "geographies of difference"?
6. How can geographical perspectives on well-being be used to address issues of inequality at the local and global levels?
7. The last part of the chapter raises the issue of future cultural identities and the possibility of human engineering deliberately creating new cultures. Is this a good idea? Justify your position.

Suggested Readings

Visit the companion website for a list of suggested readings.

9 POLITICAL GEOGRAPHY

This chapter explores the subdiscipline of political geography—broadly understood as the study of the geographical manifestations of political phenomena, especially the effect states have on individual and group behaviour. One of the most basic divisions in the world is the division into political states. A focus on political states follows logically from our discussions of language and religion (Chapter 7) and ethnicity (Chapter 8).

To explain how today's world map came to look as it does, we begin by exploring nationalism, the nation-state, and colonialism. A consideration of geopolitics introduces some of the models that geographers have used to facilitate an understanding of world affairs. Some states are relatively stable and others subject to stress, and human geographers can make important contributions to understanding many of the problems that states encounter. Subsequent sections address the role of the state in everyday life, the rise of new political movements, and the geography of elections. We conclude with a discussion of the geography of peace and war and the possible waning of the nation-state.

Here are three points to consider as you read this chapter.

- Is the trend towards the creation of new states, reflecting ever more localized identities, something to be welcomed?
- Is the trend towards groups of states actively choosing to unite, sometimes even at the cost of sacrificing aspects of their sovereignty, something to be welcomed?
- Under what circumstances can electoral boundaries be considered fair? Some argue that districts ought to be meaningful human geographic regions; others argue that districts ought to be as diverse as possible.

A woman and her children wait at the Trnovec border crossing between Slovenia and Croatia. Conflicts in Syria, Afghanistan, Iraq, and other countries led to a migration surge in 2015; over 750,000 people crossed EU borders during the year. The influx of migrants caused several countries to close their borders and various governments and political organizations, individually and collectively, to develop methods of handling the situation.

If human geography is about the role played by space in the conduct of human affairs, politics is about struggles for power, specifically the power to exercise control over people and the spaces they occupy. Like many other animal species, humans have an inherent need both to delimit territory—an area inside which we feel secure against outsiders—and to apportion space among different groups. The creation of exact territories is the basis for political organization and political action. The political partitioning of space creates the most fundamental of human geographic divisions: the sovereign state. Most states are recognized as such by other states; their territorial rights are typically respected by others; they are governed by some recognized body; and they have an administration that runs the state. As of 2015, the UN has 193 member states.

Predecessors of the Nation-State

STATE CREATION

For most people, the most familiar map of the world is one that shows it divided into states. Our familiarity with the concept of international boundaries reflects our recognition that security, territorial integrity, and political power are enmeshed in this pattern. Each state, large or small, is a sovereign unit.

sovereignty
Supreme authority over the territory and population of a state, vested in its government; the most basic right of a state understood as a political community.

Defining the Nation-State

Nation is not an easy word to define. As we have seen in Chapters 7 and 8, humans are divided into numerous cultures based on such variables as language, religion, and ethnicity; cultural affiliation is important not least because it provides people with an identity.

nation
A group of people sharing a common culture and an attachment to some territory; a term difficult to define objectively.

Humans are also divided into formally demarcated political units known as states. The state is a set of institutions, the most important being the potential means of violence and coercion. A state also makes the rules that govern life within its territory, encouraging cultural homogenization.

state
An area with defined and internationally acknowledged boundaries; a political unit.

A nation-state, then, is a clearly defined group of people who self-identify as a group (a nation) and who occupy a spatially defined territory with the necessary infrastructure and social and political institutions (a state). Each nation-state is, in principle, a political territory including all members of one national group and excluding members of other groups. In practice, few nation-states fit that strict definition because the vast majority of them are not composed of just one national group.

nation-state
A political unit that contains one principal national group that gives it its identity and defines its territory.

Although it is often supposed that nations are in some way natural groupings of people, the reality is that most nations have been constructed. Most countries in the world engage in an ongoing process of nation construction and reconstruction. In this sense it is useful to conceive of nations as "imagined communities" (Anderson, 1983).

Nevertheless, the concept of the nation-state is important to us for two reasons. First, the phrase is part of a dominant discourse such that commentators frequently use the term *nation-state* to refer to the political unit of a state. Second, the concept of the nation-state has played, and continues to play, an important role in shaping the contemporary political world.

Nationalism

To understand the rise of nation-states, we need to refer to the concept of nationalism—the belief that a nation and a state should be congruent. Nationalism assumes that the nation-state is the natural political unit and that any other basis for state delimitation is inappropriate. Thus, an aspiring nation-state will usually make the following arguments:

1. All members of the national group have the right to live within the borders of the state.
2. It is not especially appropriate for members of other national groups to be resident in the state.
3. The government of the state must be in the hands of the dominant cultural group.

It is easy to understand how application of these arguments can lead to conflict, whether external (with other states) or internal (between cultural groups). Although this principle of nationalism is taken for granted in the modern world, it is a relatively recent idea. In retrospect, it seems that nationalism was more often a romantic yearning than a coherent political program and therefore must be a flawed—but vastly influential—concept. The English cleric W.R. Inge (1860–1954) summed up this contradiction rather acerbically: "A nation is a society united by a delusion about its ancestry and by common hatred of its neighbours."

Explaining Nationalism

The map of Europe as it was redrawn in 1815, following the Napoleonic Wars, depicted states demarcated on the basis of dynastic or religious criteria. No real attempt was made to correlate the national identities of populations with their states or their rulers. Indeed, such an exercise would have proved difficult because the cultural map of Europe was exceedingly complex and those cultural groups included many divisions. Such an exercise would also have been inappropriate because nationality was not of great importance at the time; people did not question boundaries based on dynastic, religious, or community differences. This raises an important question: Why, despite obvious problems of definition, did national identity emerge as the standard criterion for state delimitation during the nineteenth century? Here are five common suggestions.

1. Nation-states emerged in Europe in response to the rise of nationalist political philosophies during the eighteenth century.
2. Humans want to be close to people of similar cultural background.
3. The creation of nation-states was a necessary and logical component of the transition from feudalism to capitalism, as those who controlled production benefited from the existence of a stable state (a Marxist argument).
4. Nationalism is a logical accompaniment of economic growth based on expanding technologies.
5. The principle of one state/one culture arises from the collapse of local communities and the need for effective communication within a larger unit.

In Europe, Germany and Italy were created in the nineteenth century in response to centrally organized efforts to unify a national group politically. The transition came earlier for Britain and France. Britain evolved gradually throughout the eighteenth century, while monarchical absolutism was intact in France until the 1789 revolution; in these cases an early general correspondence existed between culture and state.

A vandalized Parks Canada sign in Quebec City. The political relationship between Quebec and the rest of Canada continues to be a controversial topic in the province.

Nationalism in the Contemporary World

Despite the transition to nation-states, our contemporary political map still includes many states that contain two or more nations. States that approach the nation-state ideal (such as Denmark) are found primarily in Europe, although immigration into the continent, often from former colonial territories, has transformed many of these countries. Among the multinational states are many African countries whose boundaries were drawn by Europeans without reference to African identities. Nationalistic ideology has been clearly articulated in the context of anti-colonial movements, but the only substantial change to the European-imposed boundaries is the 2011 creation of South Sudan.

Many multinational states are politically unstable, prone to changes of government and/or expressions of "minority nation" discontent. Canada and Belgium are examples of politically uncertain binational states: both include more than one language group and have experienced internal stresses related to the differing political aspirations of these groups (see Box 7.4). States that incorporate members of more than one national group are not necessarily unstable. The United States has largely succeeded in creating a single nation out of disparate groups, while Switzerland is a prime example of a stable and genuinely multinational state.

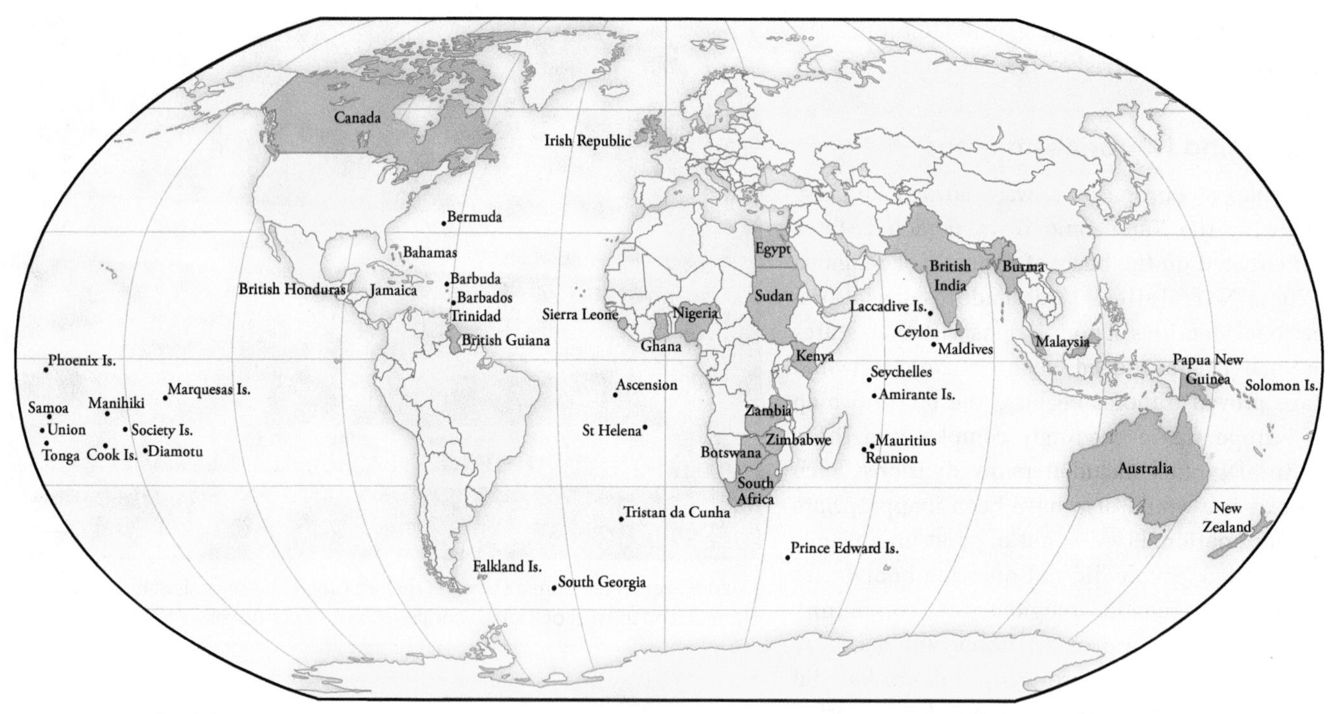

FIGURE 9.1 | The British Empire in the late nineteenth century

Exploration and Colonialism

Obviously, not all states in recent centuries have aspired to the nation-state model. European countries in particular attempted to expand their territories overseas to create empires similar to those that existed in earlier periods in Europe and elsewhere. Such empires are multinational by definition and were typically regarded as additions to rather than replacements for nation-states.

European countries began developing world empires about 1500, and most such empires achieved their maximum extent in the late nineteenth century. During the twentieth century, most were dissolved. Figure 9.1 shows the British Empire as it was in the late nineteenth century, when large areas of the world were under British control. The empire is currently limited to a few islands, most of which are in the Caribbean and South Atlantic (Box 9.1).

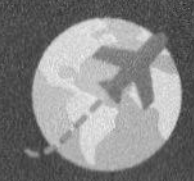

Around the Globe

BOX 9.1 | Remnants of Empire

In the late twentieth century, two former colonies changed their political status in accord with an agreed-upon timetable: Hong Kong was transferred from Britain to China in 1997, and Macao was transferred from Portugal to China in 1999. The UN identifies 16 other political units (representing some 1.2 million people) still best described as colonies. Although some colonies are likely to become independent as economic circumstances change, others are likely to remain politically dependent for some time to come.

One reason for continued dependency is that some areas are perceived as particularly valuable to the colonial power. Both France and the United States use islands in the Pacific for military purposes, and the US is known to be reluctant to relinquish control over Guam in the Pacific and Diego Garcia (a British colony mostly populated by US military) in the Indian Ocean. Both the UK and the US have refused to co-operate fully with the UN as it aims to end all colonial situations.

It also appears that some colonial areas would simply not be viable as independent states. A prime example is St Helena, an island of 122 km² (47 square miles) and a population of 5,700, located in the South Atlantic Ocean (see Figure 9.1). The island is perhaps best known as the location of Napoleon's exile and death.

First reached by the Portuguese in 1502, St Helena later became the object of competition between the Dutch and the British, and it remains part of the much-reduced British Empire. The competition reflected the island's location, which made it valuable as a stopover on the Europe–India route. Over time, movements of slaves from East Africa and indentured labour from China, as well as British and some other European settlement, created an ethnically diverse population that supported itself by providing food and other supplies for ships. By the late nineteenth century, however, technological advances such as steamships and refrigeration and the opening of a new route using the Suez Canal in 1869 had made St Helena unnecessary as a stopover, and the economic consequences soon became evident. With a very limited resource base, St Helena suffers from its small size and small population. For much of the twentieth century, the agricultural industry was dominated by flax; at times, half the working population was employed in the crop's production. Today, however, agriculture is limited to livestock and vegetables produced for only subsistence purposes.

The future of St Helena is uncertain. It is isolated, dependent on financial support from Britain, and unlikely to become a tourist attraction. From the British perspective, it is no longer a prestigious overseas possession but a financial liability (Royle, 1991).

Exploration

Most empires began as a result of exploratory activity (see Box 1.1 on the contentious term *exploration*). In general, geographers have paid little attention to exploration—defined as the expansion of knowledge that a given state has about the world—other than to list and describe events. One geographer, however, has derived a conceptual framework for the process of exploration, a framework that focuses on the links between events. As depicted in Figure 9.2, the process begins with a demand for exploration and may lead, in due course, to the "development" of the new area. Such development is likely to mean exploitation of the explored area as a source of raw materials needed by the exploring country. Britain, for example, viewed Canada as a source of fish, fur, lumber, and wheat and Australia as a source of wool, gold, and wheat.

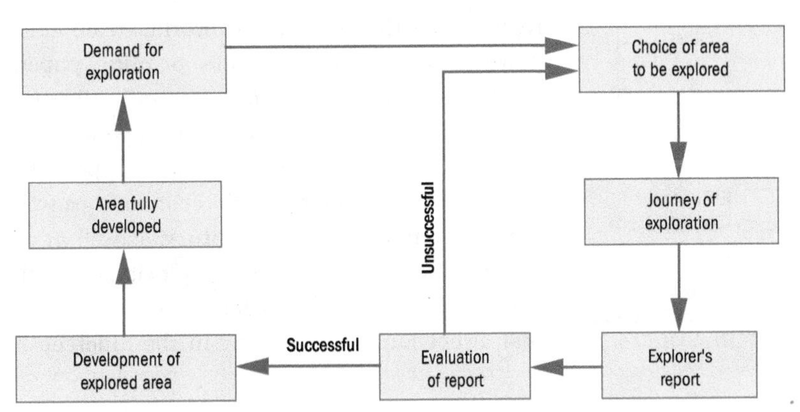

FIGURE 9.2 | Principal elements in the process of exploration

Source: Adapted from Overton, J. D. 1981. "A Theory of Exploration." *Journal of Historical Geography* 7: 57.

Colonialism

The exploring country's use of a new area was usually a function of perception and need. Economic, social, and political activity in explored areas that became colonies was determined by and for the exploring power. This was true of European empires and of the territories acquired by the United States from 1783 onward (Figure 9.3). In most cases, Aboriginal populations were eliminated or moved if their numbers represented a military or spatial threat. The economic activities that were encouraged benefited the central power. This process is known as colonialism (see Chapter 6).

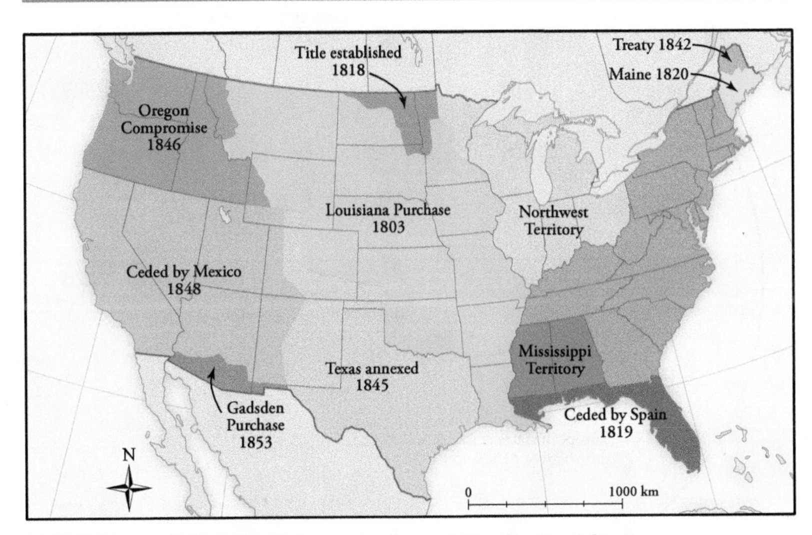

FIGURE 9.3 | Territorial expansion of the United States

The phrase "manifest destiny" reflected the belief that the new United States had an exclusive right to occupy North America. In 1803 President Jefferson purchased the Louisiana Territory from France. Spain ceded the Florida peninsula in 1819, while war with Mexico resulted in the acquisition of Texas and the Southwest. The Gadsden Purchase of 1853 completed the southern boundary. The northern boundary along the forty-ninth parallel was established in 1848. Subsequently, American interests expanded north to Alaska, purchased from Russia in 1867, and across the Pacific to include Hawaii.

In recent European history, colonialism took the form of territorial conquest throughout the Americas, much of Asia (including Siberia), and most of Africa. Competition for colonies was a major feature of the world political scene from the fifteenth to the early twentieth centuries. The earliest colonial powers were Spain and Portugal, followed by other Western European states, the United States, and Russia. The remarkable extent of European colonialism is shown in Figure 9.4.

The reasons behind colonial expansion are complex. Although often summarized as "God, glory, and greed," they may also have included various changes in Europe involving the prevailing feudal system and economic competition at the time. Recall that, shortly before the beginning of European colonialism, China and the Islamic region were the two leading world areas; gunpowder, the mariner's compass, printing, paper, the horse harness, and the water mill all were invented in China. The fact that Europe was the region that initiated global movements appears to be related to the demands of the economic growth that began in the fifteenth century, as well as to internal social and political complexity, turmoil, and competition. Reasons for the colonial fever that swept Europe, especially in the nineteenth century, included the ambitions of individual officials, special business interests, the value of territory for strategic reasons, and national prestige. But perhaps the most compelling motive was economic: colonial areas provided the raw materials needed for domestic industries and additional markets for industrial products.

Colonialism effectively ended following World War II. The defeat of Japan and Italy immediately removed them as colonial powers, and most of the remaining parts of the British Empire achieved independence over the next few years. Four European powers—France, the Netherlands, Belgium, and Portugal—tried to retain their colonies but were unsuccessful.

Decolonization

In recent years the number of widely or universally recognized states in the world has increased, from 70 in 1938 to 193 UN member states in 2015. Close to 20 other states (some fully functional and others less so), including Taiwan and Kosovo, do not have UN membership. Most of the new states achieved independence from a colonial power. Many reflect national groupings, but some occupy areas around which Europeans drew boundaries for their own reasons.

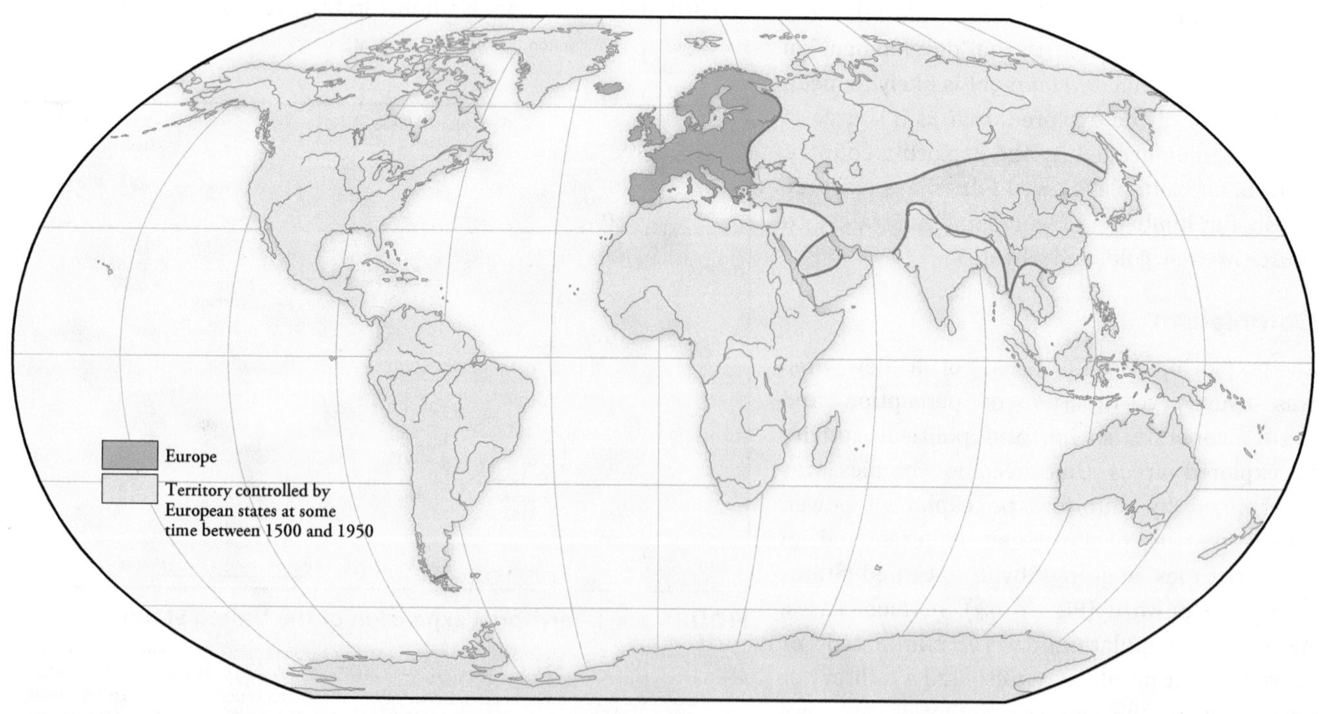

FIGURE 9.4 | European imperial coverage of the globe

Source: Murray, W. E. 2006. *Geographies of Globalization*. Copyright © 2006 Taylor & Francis UK. Reproduced by permission of Taylor & Francis Books UK.

As we have seen, the idea of a national identity originated in Europe, but it has been welcomed by colonial peoples who were discontent with colonial rule, aspired to independence, and suffered psychologically from foreign rule. The transition to independence has been violent in some cases and peaceful in others, depending in part on the way individual European powers responded as the independence movements in their colonies gathered momentum in the 1950s. Most of the former British colonies experienced a relatively smooth transition to independence, and almost all are now members of the Commonwealth, a voluntary organization of 54 countries. Some other former colonies, however, achieved independence only after long civil wars; examples include Algeria (France), Mozambique (Portugal), and the Congo region (Belgium).

INTRODUCING GEOPOLITICS (AND *GEOPOLITIK*)

One influential interest within political geography is **geopolitics**—the study of the relevance of space and distance to questions of international relations. Geopolitical discussions originated in the late nineteenth century, but manipulation of the concept by some German geographers and Nazi leaders in the 1930s and 1940s led others to reject it. Only since the 1970s has geopolitics revived and once again become a legitimate area of interest. Numerous discussions have centred on the "new geopolitics," particularly a critical geopolitics that places emphasis on the social construction of national and other identities.

The term *geopolitics* was coined early in the twentieth century by the Swedish political scientist Rudolf Kjellén, who expanded on earlier work by Ratzel to argue that territorial expansion was a legitimate state goal. This argument is only one aspect of the larger field, the relevance of space and distance to questions of international relations. It was, however, this narrow interest that dominated geopolitical discussions until they were largely abandoned in the 1940s.

Geopolitical Theories

Writing in 1904, leading British geographer H.J. Mackinder was the first to formulate a geopolitical theory. His **heartland theory**, which attempted to explain how geography and history had interacted over the past thousand years, has strong environmental determinist overtones that reflect British concerns about perceived Russian threats to British colonies in Asia, especially India. Mackinder (1919: 150) contended that the Europe–Asia land mass was the "world island" and that it comprised two regions: an interior "heartland" (pivot area) and a surrounding "inner or marginal crescent" (Figure 9.5):

> Who rules East Europe commands the Heartland;
>
> Who rules the Heartland commands the World Island;
>
> Who rules the World Island commands the World.

Mackinder argued that location and physical environment were key variables in any explanation of world power distribution. The theory is flawed because it overemphasizes one region (Eastern Europe) and because Mackinder could not anticipate the rise of air power, but his theory did influence other writers in general and may well have exercised a continuing influence on United States policy in particular.

These ideas influenced the rise of ***geopolitik***—a specific interpretation of more general geopolitical ideas—in the 1920s. It focuses on the state as an organism, on the subordinate role played by individual members of a state, and on the right of a state to expand to acquire sufficient *lebensraum* (living space). The individual most responsible for popularizing these ideas was Karl Haushofer, a German geographer in Nazi Germany who was bitterly disappointed by the territorial losses Germany experienced as a result of World War I. Haushofer's academic justification for German expansion coincided with Nazi ambitions. The extent of his influence on actual events is uncertain, but the German interest in *geopolitik* resulted in a general disillusionment in the Western world with all geopolitical issues.

Despite this disillusionment, several scholars continued to present their views on the global distribution of power. Nicholas Spykman, writing in the 1940s, built on Mackinder's "inner or marginal crescent" (rimland) to develop a **rimland theory**, arguing that the power controlling the rimland could control all of Europe and Asia and

geopolitics
The study of the importance of space in understanding international relations.

heartland theory
A geopolitical theory of world power based on the assumption that the land-based state controlling the Eurasian heartland held the key to world domination.

geopolitik
The study of states as organisms that choose to expand in territory in order to fulfill their "destinies" as nation-states.

rimland theory
A geopolitical theory of world power based on the assumption that the state controlling the area surrounding the Eurasian heartland held the key to world domination.

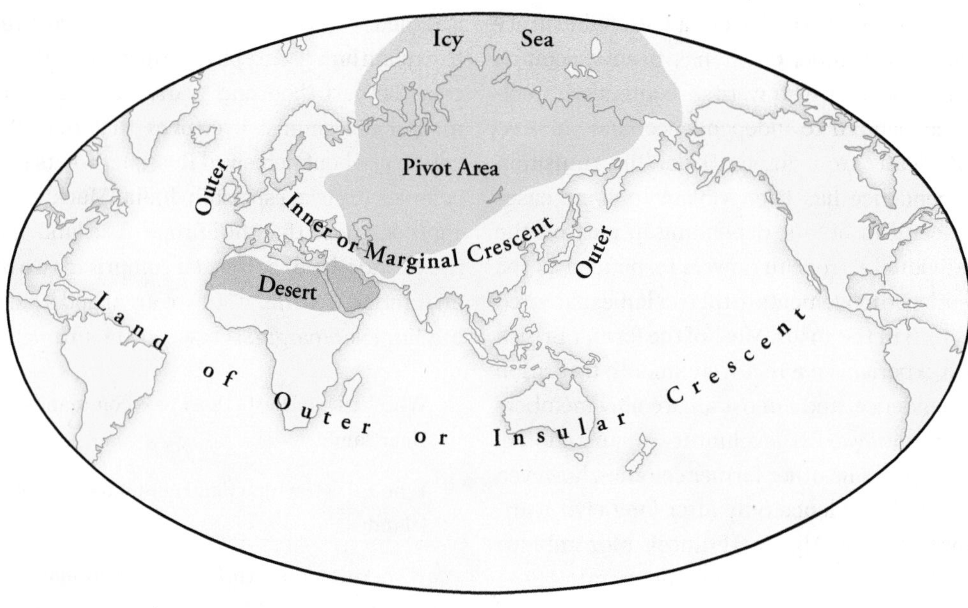

FIGURE 9.5 | Mackinder's heartland theory

therefore the world. As an American, Spykman saw considerable advantage in a fragmented rimland—a view that has influenced post–World War II American foreign policy in Asia (Korea and Vietnam) and in Eastern Europe. Also during this time, Alexander de Seversky stressed that the US needed to be dominant in the air to ensure state security.

Centrifugal and Centripetal Forces

In a consideration of the stability, or instability, of states, the geographer Hartshorne (1950) usefully distinguished between centrifugal and centripetal geopolitical forces. **Centrifugal forces** tear a state apart; **centripetal forces** bind a state together. When the former exceed the latter, a state is unstable; when the latter exceed the former, a state is stable.

centrifugal forces
In political geography, forces that make it difficult to bind an area together as an effective state; in urban geography, forces that favour the decentralization of urban land uses.

centripetal forces
In political geography, forces that pull an area together as one unit to create a relatively stable state; in urban geography, forces that favour the concentration of urban land uses in a central area.

federalism
A form of government in which power and authority are divided between central and regional governments.

The most common centrifugal forces are those involving internal divisions in language and religion that lead to a weak raison d'être or state identity. Other centrifugal forces include the lack of a long history in common (the case in many former colonies) and state boundaries that are subject to dispute. The fact that some countries, such as Australia, Canada, and the United States, are less centralized than others is not necessarily a centrifugal force as the key aim of **federalism** is to prevent one level of government from dictating to another.

The most common centripetal force is the presence of a powerful raison d'être: a distinct and widely accepted state identity. Other centripetal forces include a long state history and boundaries that are clearly delimited and accepted by others. The various centrifugal or centripetal forces are often closely related; for example, the presence of internal divisions generally indicates that groups do not share a long common history and that boundaries are not agreed upon. Stability and peace—and instability and conflict—are also closely linked within and between states.

Boundaries

Boundaries mark the limits of a state's sovereignty. They are "lines" drawn where states meet or where states' territorial waters end. A state's stability often reflects the nature of its boundaries.

The characteristics that give identity to a nation, such as language, are rarely as abruptly defined as are its boundaries. For this reason, many countries include at least one significant minority population. In principle, such situations are less common where the boundaries are antecedent (established before significant settlement began) because settlers moving into areas close to the boundary must acknowledge its presence. Boundaries of this kind frequently are geometric—as in the case of the US–Canada boundary west of Lake of the Woods, which follows the forty-ninth parallel.

Other boundaries are *subsequent*, or defined after an area has been settled and the basic form of the human landscape has been established. Such boundaries may attempt to reflect national identities (for example, the present boundary of France is an approximation of the *limites naturelles* of the French nation) or may totally ignore such distinctions (this is the case with most colonially determined boundaries). Many subsequent boundaries are continually subject to redefinition, as in Western Europe.

All boundaries are artificial in the sense that what is meaningful in one context may be meaningless in another. Rivers are popular boundaries, since they are easily demarcated and surveyed, but they are generally areas of contact rather than of separation and so tend to make poor boundaries. Rivers may also be poor boundaries if they are wide and contain islands (e.g. the Mekong River in Southeast Asia) or if they repeatedly change course (e.g. the Rio Grande along the Mexico–US [Texas] border).

In some cases, groups have erected barriers to serve as physical obstructions preventing others from entering. Well-known examples include the Great Wall of China, constructed in the third century BCE to prevent invasion from the north, and Hadrian's Wall, constructed across northern England in the second century CE to prevent invasion from Scotland. Along some parts of the US–Mexico border, fences are in place to restrict illegal immigration. In one exceptional instance—the Berlin Wall—the aim was to prevent people leaving the eastern sector for the west. In 2003 Israel began building an elaborate system of fences and military checkpoints along the border with the West Bank in the hope of reducing terrorist incursions (Box 9.2).

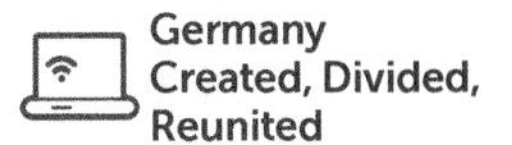

In the News

BOX 9.2 | The Jewish State

The closely related ongoing problems of Israeli security and the creation of a Palestinian state are routinely in the news. There is, though, no obvious resolution to the matter.

The Jews are a religious group with roots in the area that is now Israel. Jerusalem, in particular, is an important symbolic location for Jews, as it is for Christians and Muslims. During the time of Jesus, this area (known as Palestine), was under Roman rule. In 636 it was invaded by Muslims and from then until 1917 (with a brief Christian interlude during the Crusades), it was a Muslim country, specifically Turkish after 1517. Over the centuries, in a process known as the Diaspora, Jewish populations moved away to settle throughout much of Europe and overseas. At the end of World War I, Syria and Lebanon were taken over by France, and Palestine and Transjordan (now Jordan) were taken over by Britain, under League of Nations (the predecessor of the United Nations) mandates. The Palestine mandate—the Balfour Declaration of 1917, precisely—allowed for the creation of a Jewish national home but without damaging Palestinian interests. This was an impossible goal because a different national group, Palestinians, already inhabited the space being assigned a Jewish state.

In 1920 Palestine included about 60,000 Jews and 600,000 Arabs. Most of the Jews had arrived since the 1890s.

© AP Photo/Majdi Mohammed

Palestinian youths climb the Israeli separation barrier during a protest against the barrier in the West Bank village of Nilin near Ramallah, June 2012.

Jewish immigration increased under the mandate, especially as persecution of Jews in Hitler's Germany intensified. After World War II, survivors struggled to reach Palestine, which by 1947 included 600,000 Jews, 1.1 million Muslim Arabs, and about 150,000 Christians (mostly Arab). Arab–Jewish conflicts were by then commonplace, and Britain, unable to cope, announced that it intended to withdraw.

Continued

The UN then produced a plan to partition Palestine into a Jewish state and an Arab state, with Jerusalem as an international city. The Arabs did not accept the plan, and the Jews declared the independent state of Israel (1948). Conflict immediately flared between Israel and nearby Arab states and between Jews and Palestinian Arabs, which has continued to the present.

One long-standing issue of contention has been Israel's occupation of the West Bank (of the Jordan River) and the Gaza Strip, which it captured (from Jordan and Egypt, respectively) during the Six Day War of 1967. Although Israel has established relations with Egypt (in 1979), the Palestine Liberation Organization (in 1993), and Jordan (in 1994), serious obstacles to peace remain. Among them are the absence of diplomatic relations between Israel and Syria, which by 2015 was in the midst of a protracted civil war complicated by the rise of the Islamic State, the threat to Israel posed by Iran, the uncertain status of Jerusalem, and the continuing presence and expansion of Jewish settlements in the West Bank. Also, there are ongoing terrorist actions against Israel by Palestinian militants and Israeli military actions against the Palestinians, most recently in 2014. Finally, about 5 million Palestinians live in historic Palestine under Israeli control—2.5 million in the West Bank, 1.47 million in Gaza, and 1.13 million in Israel. There are also about 2.8 million Palestinians in Jordan and about 2 million more elsewhere. The result of this set of circumstances is that Israel's borders are vulnerable and the state is inherently unstable.

In recent years, a peaceful resolution to these conflicts seems no closer. In 2002 the United States acknowledged the need for an independent, democratic, and viable Palestinian state—an explicit acknowledgement of Palestinian identity—existing in peace and security with Israel. But progress in this direction has been unsure, particularly since the victory of Hamas, a militant Islamic group, in the 2006 Palestinian elections in the Gaza Strip. In 2011 the US formally acknowledged that a future Palestinian state should be based on pre-1967 boundaries, an idea strongly objected to by Israel.

Divided States

Some states have been divided into two or more separate parts; this situation increases the likelihood of boundary problems. A classic example is the partition of India as part of the decolonization of Britain's Indian empire. In some cases of partition, the boundaries between the newly formed states have been especially inappropriate. Examples include Germany (East and West from 1945 until 1990) and Korea (North and South since 1945). In both cases, one nation was divided into two states for reasons that had more to do with the political wishes of other states than with the wishes of the people. There are sound reasons to argue that artificial, externally imposed boundaries such as these are rarely long-lasting.

Building on these various geopolitical ideas, we next consider two seemingly contradictory trends in the contemporary world. First, there is much evidence to suggest that the number of states will continue to increase from the current 193 UN members because so many states have internal divisions, related primarily to claims of ethnic nationalism. Second, there is also evidence that some states are willing to sacrifice aspects of their independent national identity as they join with other states to form some degree of commonality, as has been the case with the EU. Both trends contribute to the creation of new maps of our political world.

irredentism

The view held by one country that a minority living in an adjacent country rightfully belongs to the first country.

UNSTABLE STATES

As the account of state creation explained, concordance between nation (a group of people) and state (a political territory) is rare. Internal ethnic (usually linguistic and/or religious) divisions frequently occur within a state. Although cases are rare, these can threaten a state's stability to varying degrees (Box 9.3).

In countries with significant internal divisions, any one of three general situations may threaten state stability. Secessionist movements arise when groups within multinational states want to create their own separate states; examples include the Québécois in Canada, the Flemish and Walloons in Belgium, and the Welsh and Scottish nationalist movements in the United Kingdom. In other cases, "nations within" may want to link with members of the same nation in other states to create a new state; the Basques in Spain and France and the Kurds in Iraq, Iran, Syria, Turkey, Armenia, and Azerbaijan are two examples (Box 9.4). **Irredentism** involves one

Examining the Issues

BOX 9.3 | Regional Identities and Political Aspirations: The Example of Canada

Many regional populations lack a sense of belonging as part of a political unit with which they are unable to identify, and the desire to be an independent political unit is at the head of the agenda for many groups. This desire is understandable, given the importance of states in the contemporary world: "They behave very much like Greek gods, so much so that, given the great powers which they wield, they need to be taken very seriously" (East and Prescott, 1975: 1).

Is this existence of local and regional nationalisms a good or bad thing? There is no correct answer, but the question is still worth thinking about. The current world political map, as we know all too well, is not sacrosanct. Changes occur, and often for very good reasons. It may seem that regionalism is in fashion. But is the trend towards the creation of new states, reflecting ever more localized identities, something that should be encouraged? The world as a whole is an uncertain and somewhat unpredictable environment. On one hand are the processes of globalization and movement towards the integration of some states; on the other are numerous groups arguing that they are different from others and therefore should be separate from them. Most states in which individual groups are seeking some degree of independence react negatively to the possible loss of territory, population, and prestige.

Is it better for a group to look outward and encourage cosmopolitanism or to look inward and encourage ethnicity? Or is it possible to accomplish both these goals simultaneously? Does support for a sub-state level of nationalism result in an ethnocentric world view and encourage lack of respect for others or does it allow for increased self-respect and a better understanding of others? Although there are no agreed-upon answers to such difficult questions, social psychology strongly suggests that simply classifying people into groups prompts a bias in favour of one's own group (Messick and Mackie, 1989). Such biases may arise even between groups that are interdependent and co-operative, as is the case with English- and French-speaking Canadians.

Canada is a country in which different groups hold different opinions of movements to achieve some degree of autonomy within the larger state. A huge country of 9.8 million km² (3.8 million square miles), Canada has 35.5 million people and large unsettled areas. Indeed, much of the land is inhospitable to humans; sixteenth-century French explorer Jacques Cartier described the southern coast of Labrador as the land that God gave to Cain. Further, Canada comprises 10 provinces and 3 territories, with marked regional variations. Ways of life range from traditional to very modern. It is a plural society made up of Aboriginal peoples, the two founding European cultures (English and French), and a host of other national groups. Together, these physical and human factors have produced strong regional feelings—a major cause of instability. In recent years, the unity of Canada has been threatened not only by Quebec (French-language) separatism but also by various western Canadian separatist movements and, more significantly, by the claims of national sovereignty of many Aboriginal nations.

state's seeking the return from another (usually neighbouring) state of people and/or territory formerly belonging to it. Irredentism is commonplace in parts of the Balkan region, the Middle East, and sub-Saharan Africa, all regions where state boundaries were largely determined by other powers.

To help clarify these general points, we consider some of the details of internal divisions through reference to a wide variety of examples from Africa, Europe, the former USSR, and South Asia, as well as the case of Canada.

Nation and State Discordance in Africa

Discordance between nation and state is especially evident in Africa (Figure 9.6). In only a few instances does a distinct national group correlate with a state; the southern African microstates of Lesotho and Swaziland provide examples. Perhaps ironically, it has been suggested that Lesotho might join South Africa (Lemon, 1996), and seeking this annexation is a popular idea in Lesotho because it is a dysfunctional country still

Around the Globe

BOX 9.4 | The Plight of the Kurds

According to some commentators, the biggest losers in the Middle East over the past hundred years have been the Kurds. Since the collapse of the Ottoman (Turkish) Empire at the end of World War I, other groups—Turks, Arabs, Jews, and Persians—have consolidated or even created their own states. Yet more than 20 million Kurds remain stateless, dispersed in Turkey (10.3 million), Iran (4.6 million), Iraq (3.6 million), Syria (1 million), and in smaller numbers in other neighbouring states. Their failure to achieve state identity is all the more surprising given that, in 1920, Britain, France, and the United States all acknowledged the need for a Kurdish state: "No objection shall be raised by the main allied powers should the Kurds . . . seek to become citizens of the newly independent Kurdish state" (Article 64, Treaty of Sèvres, signed 20 August 1920; quoted in Evans, 1991: 34).

Members of a Kurdish family prepare to leave their home in the Kurdish town of Chamchamal, near Kirkuk in northern Iraq, in March 2003. Local residents feared a chemical attack by Saddam Hussein's forces in the event of a US-led war.

Why have the Kurds not succeeded in creating a state? Part of the answer lies in an identity issue. Most Kurds are Sunni Muslims, but some belong to other groups; their society, located in a mountain environment, is typically poor and divided tribally; their dialects are varied; and there is no one political party to speak for all of them. One could argue that the Kurds are united more by persecution than by any sense of shared identity. A second part of the answer is that a Kurdish state would weaken the authority of all the states in which Kurds live. To discourage the development of Kurdish nationalism, Turkey banned the use of the Kurdish language in schools until 2002 (when it abandoned that policy as part of its unsuccessful effort to gain entry to the EU) and since the 1990s has resisted any suggestion of independence for Iraqi Kurds living next to its own Kurdish population.

The Kurds' preferred location for their own state would be in the mountainous border regions of Turkey, Iraq, and Iran. Unfortunately, this area is valued by existing states for its resources, which include oil deposits at Kirkuk. Iraq has used chemical warfare against the Kurds and, following the end of the 1991 Gulf War, mounted an extensive campaign to slaughter them, which caused massive population movements into Turkey. Ironically, Turkey is far from a safe haven: in 1979, the country's prime minister stated: "The government will defeat the disease (of Kurdish separatism) and heads will be crushed" (quoted ibid., 35). Political developments following the 2003 war in Iraq highlighted the precise divisions between Sunni, Shiite, and Kurdish groups and have led to a degree of Kurdish autonomy in that country, but the rise of the Islamic State makes the larger picture in the region unclear. It is possible that a weakening of states in the larger region might provide an opportunity for Kurdish independence.

suffering from HIV/AIDS. A principal argument in favour of the country sacrificing its identity in this way is that it is the only sovereign state in the world, apart from the microstates of San Marino and the Vatican, to be completely surrounded by a single neighbour.

Crucially, the states of contemporary Africa are not products of a long African history but creations of colonialism that subsequently achieved independence. Colonial activity erased many earlier states that were based on national identity. Among the many African nations that no longer have related states are the Hausa and Fulani nations in Nigeria, the Fon in Benin, and the Buganda in Uganda. States continued to grow and decline. Two early states in West Africa were Ghana and Mali (names later given to modern states covering different areas); both succumbed to Islamic incursions in the eleventh century. Figure 9.7 shows maps for the sixteenth, eighteenth, and nineteenth centuries—the colonial impact is particularly clear when we compare these maps with Figure 9.6. European neglect of African national identities is a prime cause of

African nationalism and contemporary instability in many African states.

State boundaries, shapes, and sizes, then, are colonial creations, reflecting past European rather than past or current African interests. In consequence, Africa is characterized by an inappropriate political fragmentation. Some states have a high degree of contiguity (when there are many states, most states have many neighbours); some are very small in size and/or population; some have awkward shapes and long, often environmentally difficult, boundaries; and some lack access to the sea. Each of these difficulties is related to the colonial past, which ignored national identity in the process of creating states.

The future of African political identities remains uncertain. The Organization of African Unity (OAU) was established in 1963 with the explicit goal of providing a common voice for the emerging independent African states. Sadly, it failed; most commentators found that it did more to impede than to promote democracy, human rights, or social and economic growth. In 2002 the OAU was replaced by the African Union (AU). There is hope that this organization will succeed in enhancing national identities, encouraging co-operation, and permitting Africa to speak with one voice in a global context. The AU is generally not sympathetic to secessionist movements on the grounds that independence for one group might lead to a drastic redrawing of boundaries. In a similar vein, some observers argue that the best way forward for Africa is through federalism, a linking of countries to facilitate co-operation.

And yet the one significant political change in recent years is the creation of the new state of South Sudan in 2011, representing a victory for a secessionist movement following a long conflict. South Sudan is thus the first explicit challenge to the colonially imposed boundaries. Sudan was obviously an instance of illogical colonial boundaries, with the north being Arabic-speaking and most of the population being Muslim, while the south includes a variety of African languages,

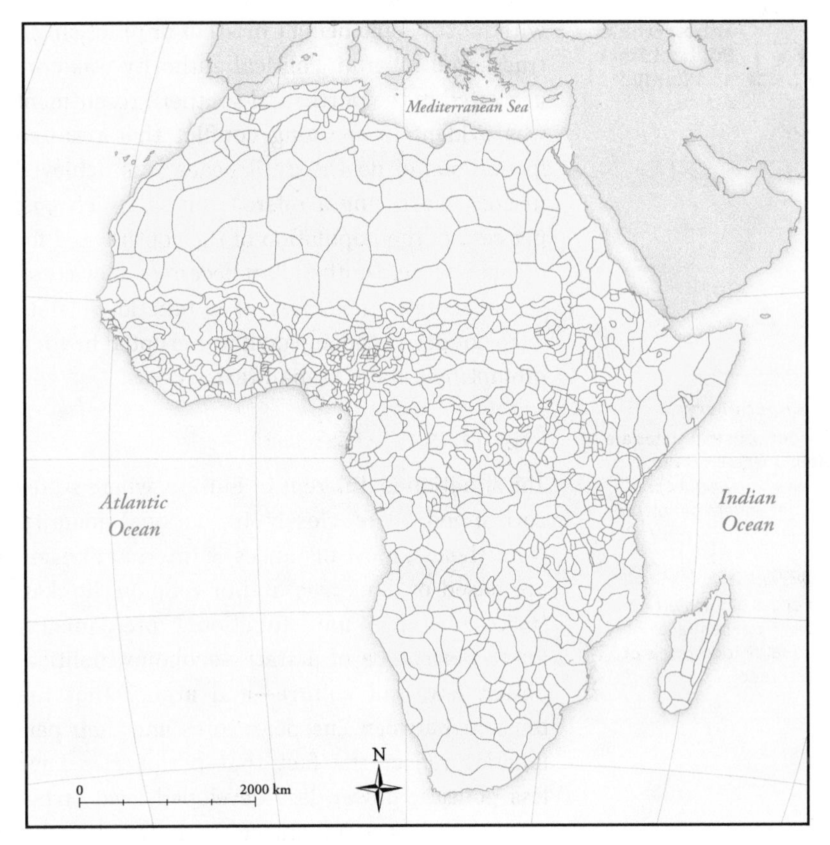

FIGURE 9.6 | African ethnic regions

Source: Stamp, L. D., and W. T. W. Morgan. 1972. *Africa: A Study in Tropical Development*, 3rd edn. New York: Wiley, 41. Copyright © 1972 John Wiley & Sons, Inc. Reprinted with permission.

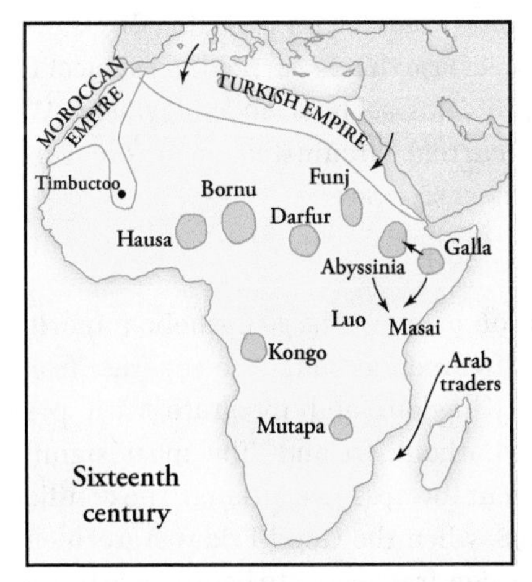

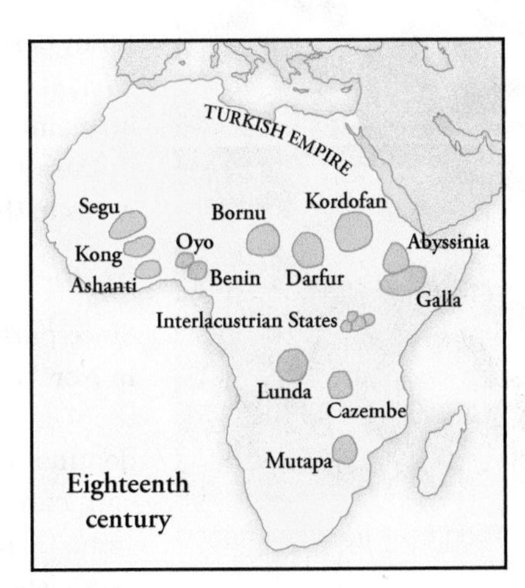

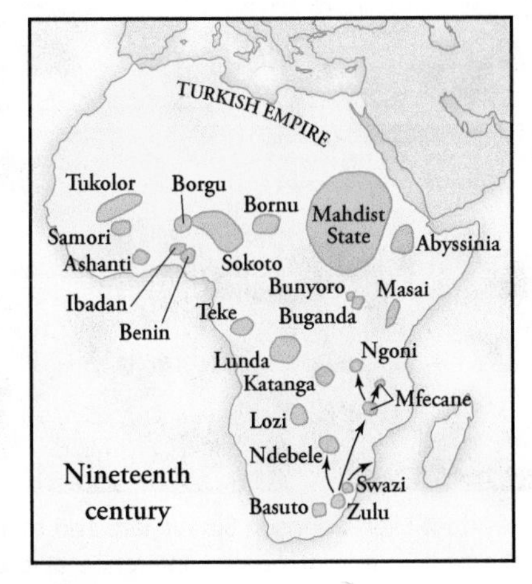

FIGURE 9.7 | African political areas in the sixteenth, eighteenth, and nineteenth centuries

Tribes, Ethnicity, Political States, and Conflict in Africa

with most people being Christian or practising a traditional religion. Political authority was concentrated in the north, and southern resentment was evident in the long conflict that resulted in millions of deaths until peace was achieved in 2005. Following a referendum in which 99.5 per cent of the population of the south voted for independence, South Sudan became a new state. Unfortunately, the bases of a functional state were not developed and civil war erupted in 2014, prompting a major humanitarian crisis.

core–periphery
The concept that states are often unequally divided between powerful cores and dependent peripheries.

devolution
A process of transferring power from central to regional or local levels of government.

Nations and States in Europe

The situation is different in Europe, where states correspond more closely to national identity. Nevertheless, most instances of internal division are based on language and/or religion. Rokkan (1980) identifies four functional prerequisites for the existence of a state—economy, political power, law, and culture—and argues that the tensions between European cores and their peripheries reflect the fact that peripheries have less political power, less developed legal structures, and less dominant cultures. He identifies two axes: a north–south cultural axis from the Baltic to Italy and an east–west economic axis. The north–south cultural axis was Protestant (a religion that favoured national aspirations) in the north and Catholic (a religion that cut across national boundaries) in the south. Along the east–west economic axis, the west had key commercial centres (an important factor in state formation), but in the east the urban centres were economically weak (and consequently unable to offer the resource base needed for the building of states). Arguing that the success of efforts at nation-building depended on cultural conditions and the success of efforts at state-building on economic conditions, Rokkan proposed that the north–south cultural axis determined the former, while the east–west axis determined the latter. Hence, the most favourable conditions for building nation-states existed in the northwest.

© CP/AP Photo/Pablo Martinez Monsivais

Liberian President Ellen Johnson Sirleaf, left, and then World Bank president Robert Zoellick take part in a joint news conference at the World Bank headquarters in Washington announcing that Liberia had significantly reduced its foreign debt.

Contemporary nationalist movements in Western Europe can therefore be seen as relics of a Rokkan-type **core–periphery** cleavage in areas not yet homogenized into an industrial society with its related class divisions. In our present context, this view is useful.

Separatist Movements in Europe

Regionalism and associated efforts to assert separate identities remain strong in Europe (Box 9.5). There are several areas with political parties whose policies are explicitly regional and nationalist. Some favour separation; others, self-government within a federal state structure, perhaps by means of **devolution** (the transfer of power from central to regional or local levels of government). In most of these cases, language is a key factor. However, it is worth noting that most of the areas asserting a distinct identity are peripheral and economically depressed. Indeed, one of the principal causes of social unrest in general is the core–periphery economic spatial structure—a common feature of modern states. Rich cores and poor peripheries often exacerbate separatist tendencies in the peripheral areas. Thus, most of these areas may feel disadvantaged not only in terms of national identity but also of economic well-being. Many peripheral areas have a specialized or short-lived economic base that is particularly subject to economic problems such as unemployment. We discuss the current circumstances of four such areas in the next sections.

Northern Ireland

Since partition in 1921, a large Catholic minority in Northern Ireland has sought to separate from the United Kingdom and integrate with predominantly Catholic Ireland. The most significant movement towards resolution of the conflict came in 1998, when the Good Friday Agreement established a new framework to help decide future constitutional change. The 2005 disarmament commitment from the Irish Republican Army (IRA) was a further positive step. Northern Ireland continues to be a spatially segregated society, with Catholics and Protestants living in different areas of the city of Belfast, a fact that is unfortunately highlighted when Protestant groups choose to march through or close to Catholic neighbourhoods, especially on the Glorious Twelfth (12 July, commemorating the 1690 Battle of the Boyne, where William III defeated the deposed James II's Catholic army).

Wales

Plaid Cymru, the Welsh nationalist party that favours self-government for Wales, was formed in 1925 to revive the Welsh language and culture. The devolution referendum of 1978 saw 12 per cent in favour and 47 per cent against a devolved assembly. However, in a 1997 referendum Welsh voters approved proposals—introduced by the British government—for a devolved National Assembly. Since 1999 the assembly has been responsible for the development and implementation of policy in a wide range of areas including agriculture, industry, the environment, and tourism.

Scotland

The Scottish National Party was formed in 1934 and achieved notable political success in 1974. By 1978, 33 per cent of voters favoured a devolved assembly, with 31 per cent opposed. As part of the same initiative that led to the creation of the Welsh National Assembly, proposals for a Scottish Parliament were introduced by the British government and a 1997 Scottish referendum approved its creation by a substantial majority. The Parliament's responsibilities are similar to those of the Welsh National Assembly. An independence referendum held in 2014 was rejected by Scottish voters.

Flanders and Wallonia

The two linguistically distinct regions of Belgium—Flanders is Flemish speaking and Wallonia is French speaking—were combined in 1830 but have never achieved effective national unity; the two areas legally became unilingual regions in 1930 (see Figure 7.16). Following a long-running constitutional crisis, Belgium seems even less unified than in the past. It is perhaps the best example of a state where the idea of national sovereignty has been weakened as a consequence of both economic and political globalization processes, especially by the increasing significance of European institutions.

The Former USSR

The Union of Soviet Socialist Republics (USSR) was never really a union; it was an empire tightly controlled by one of the republics, Russia (Figure 9.9). In this sense, it differed little from its predecessor, the pre-1917 Russian Empire. From 1917 until the collapse of the USSR in 1991, considerable numbers of ethnic Russians moved into other Union republics, which adopted policies favouring the Russian language and suppressing the practice of

Other Examples of Separatist Movements in Europe

Around the Globe

BOX 9.5 | Conflicts in the Former Yugoslavia

From the mid-fifteenth to the mid-nineteenth centuries, the area that was to become the federal state of Yugoslavia was part of the Ottoman Empire. By the beginning of the twentieth century, some parts—notably Serbia—were independent and others belonged to the Austro-Hungarian Empire. At the end of World War I, Yugoslavia was founded as a union of south (Yugo) Slavic peoples, although the name was not adopted until 1929. Following World War II, a socialist regime was established under the leadership of Tito, who dominated the political scene until his death in 1980, after which a form of collective presidency was established. The ethnically diverse area of Yugoslavia was organized as six socialist republics and two autonomous provinces.

The ethnic diversity evident in Table 9.1 is further complicated by the major European cultural division between eastern and western Christianity that runs through former Yugoslavia as well as through Romania, Ukraine, and Belarus (Figure 9.8). It is therefore not surprising that this federal state was dissolved in 1991, following Serbia's efforts three years earlier to exert a greater influence in the federation by assuming direct control of the two autonomous provinces (Kosovo and Vojvodina).

Continued

TABLE 9.1 | Ethnic groups in the former Yugoslavia

Republic	Ethnicity
Serbia	85% Serbs
Croatia	75% Croats, 12% Serbs
Slovenia	91% Slovenes
Bosnia-Herzegovina	40% Slavic Muslims, 32% Serbs, 18% Croats
Montenegro	69% Montenegrins, 13% Slavic Muslims, 6% Albanians
Macedonia Province	67% Macedonians, 20% Albanians, 5% Turks
Kosovo	77% Albanians, 13% Serbs
Vojvodina	54% Serbs, 19% Hungarians

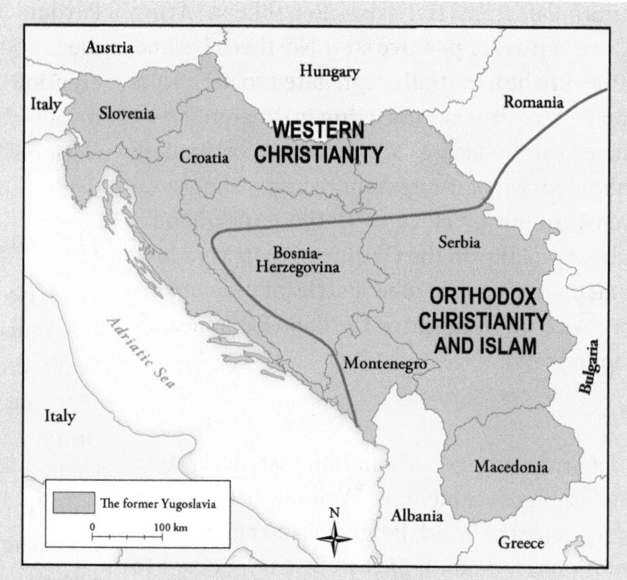

FIGURE 9.8 | The former Yugoslavia

The process of dissolution began in 1991 with the secession of Slovenia, which was quickly followed by that of Croatia. In Bosnia bitter conflict between Croatian Muslims and Serbs introduced that terrible euphemism "ethnic cleansing" (recall the discussion of genocide in Chapter 8). In 1999 violence flared in the southern province of Kosovo as Serbians and ethnic Albanians contested an area that has meaning to both groups; Western powers eventually intervened on behalf of the Albanian population.

Many commentators attributed that conflict to Kosovo's particular effect on the Serbian national psyche, for it was in Kosovo that the Serb Prince Lazar was killed by Turks in 1389—a defeat that would lead to Serbia's annexation by the Ottoman Empire in 1459. As a result of Prince Lazar's reported heroism in the face of adversity, many Serbs—or at least Serb leaders striving to reinforce Serbian identity—regard Kosovo as a sacred space. Moreover, the Serbs claim that the region's ethnic Albanian population had sided with the Turks. Not surprisingly, the Albanian people of Kosovo see things differently: according to their traditions, they were living in the region before the Serbs and fought with them against the Turks. When Serb forces entered Kosovo in 1999, purportedly to quell an armed independence struggle by the Kosovo Liberation Army, the conflict led to several instances of ethnic cleansing and massive refugee movements of ethnic Albanians.

Following this series of nationalist independence movements, several instances of ethnic slaughter, and a series of wars, Yugoslavia became five countries: Slovenia, Croatia, Bosnia-Herzegovina, Serbia-Montenegro, and Macedonia. Subsequently, in 2006 Montenegro voted to become independent from Serbia, a move that leaves the latter "landlocked and alone, a final nail in the malign nationalist dream of a 'Greater Serbia' that fuelled such bloodshed in the 1990s" (*The Times*, 2006).

Finally, Kosovo—with a 90 per cent ethnic Albanian population that has long sought independence—declared itself independent in 2008. The declaration was rejected by Serbia, essentially on the grounds that Kosovo was a province of Serbia, not a republic in the former Yugoslavia. But it was accepted by the United States and many other countries (as of 2015, 110 UN member states have given diplomatic recognition to Kosovo, and it was a member of the World Bank Group and the IMF), although described as a special case of separation not intended to set a precedent. The UN referred the matter to the International Court of Justice, which in 2010 concluded that Kosovo's unilateral declaration of independence did not violate international law. This ruling is expected to have significant implications for other territories demanding independence.

After the collapse of the Ottoman, Austro-Hungarian, and Soviet empires in the twentieth century, the demise of Yugoslavia might be seen as marking a symbolic end to efforts to bind multiple ethnic groups into artificial single entities. Identity—national, ethnic, linguistic, and religious—has been a central feature of recent conflicts in the former Yugoslavia. It was with this thought in mind that S.P. Huntington (1996) described the conflicts as one example of a "fault line" war, where West and East, Christianity and Islam, and Catholicism and Eastern Orthodox struggle to coexist and even to survive.

As noted, one exceptionally tragic component of the conflicts that erupted as Yugoslavia disintegrated was the practice of ethnic cleansing, or genocide. The most notorious example was carried out in 1995 at Srebenica—in a supposedly safe area under the control of UN peacekeeping forces—where 23,000 Bosnian Muslim women and children were expelled by Serbs and over 7,000 Muslim men and boys murdered, in some cases after being blindfolded and bound (Wood, 2001).

religion. The USSR's demise, part of the sweeping changes that occurred in Europe in the late 1980s and early 1990s, left a difficult and uncertain legacy of ethnic tensions and frequent ethnic conflict. Table 9.2 provides data on ethnic populations in the former republics and highlights the high percentages of Russians in most of these 15 independent states. The creation of the Commonwealth of Independent States in 1991 was an attempt by Russia to maintain links with the newly independent states, but it has proven ineffective.

As early as 1918, the three Baltic republics (Latvia, Lithuania, and Estonia), along with Ukraine, Byelorussia, Georgia, Azerbaijan, and Armenia, all fought for independence from the Russian empire. In 1990 conflict flared once again between Christian Armenians and Muslim Azerbaijanis in what was then the Soviet Republic of Azerbaijan; between 1988 and its collapse in 1991, the Soviet Union experienced considerable separatist pressure from the three Baltic republics and from southern republics with significant Islamic populations. Abkhazia and South Ossetia are, with Russian support, currently seeking to separate from Georgia. These two regions, situated on the Russia–Georgia border, were the sites of major conflict in the early 1990s and again in 2008. Abkhazia's independence is recognized only by Russia and three other countries, while the principal goal of South Ossetia's population is to unite with North Ossetians living in a neighbouring autonomous republic of the Russian Federation.

There are also problems in the Central Asian area, where the fall of the USSR resulted in the five republics of Kazakhstan, Uzbekistan, Tajikistan, Turkmenistan, and Kyrgyzstan emerging as independent countries. Each contains a significant minority population of Russians—in Kazakhstan it was as high as 38 per cent but is declining as a result of out-migration.

One group closely related to the majority populations in these new countries and with similar aspirations for independent state status seems unlikely to see those goals fulfilled in the immediate future. The Uighurs, who live in the northwestern Chinese province of Xinjiang, are part of the same larger group of people in the five new Central Asian countries. This group conquered much of Asia and Europe before converting to Islam in the fourteenth century; at that time the larger region was one of great wealth and included such important centres as Tashkent and Samarkand. But by the mid-nineteenth century,

FIGURE 9.9 | The former USSR

TABLE 9.2 | Ethnic groups in the former USSR

Country (former republic)	Population (millions)	Titular Nationality (%)	Principal Minorities
Russia	141.9	82	Tatars 4%
Ukraine	46.2	73	Russians 22%
Uzbekistan	24.4	71	Russians 8%
Kazakhstan	15.4	40	Russians 38%, Ukrainians 5%
Belarus	9.7	78	Russians 13%
Azerbaijan	7.7	83	Russians 6%, Armenians 6%
Tajikistan	6.2	62	Uzbeks 23%, Russians 7%
Georgia	5.4	70	Armenians 8%, Russians 6%, Azeris 6%
Moldova	4.1	65	Ukrainians 14%, Russians 13%
Turkmenistan	4.8	72	Russians 9%, Uzbeks 9%
Kyrgyzstan	4.7	52	Russians 22%, Uzbeks 12%
Armenia	3.8	93	Azeris 2%
Lithuania	3.4	80	Russians 9%, Poles 7%
Latvia	2.3	52	Russians 35%
Estonia	1.3	62	Russians 30%

the entire region had been conquered either by Russia or, in the case of the Uighur group, by China. Only the Uighurs have not achieved independence. They launched a new round of independence claims in 1997, following the death of long-time Chinese leader Deng Xiaoping. However, so many Han Chinese immigrants have moved into the region that the Uighurs make up only about half of the total Xinjiang population of some 16 million. There have been several recent instances of violence and in 2011 four men described by Chinese authorities as religious extremists and separatists who supported the creation of East Turkestan, a new independent state, were sentenced to death. The future of the Uighurs remains uncertain.

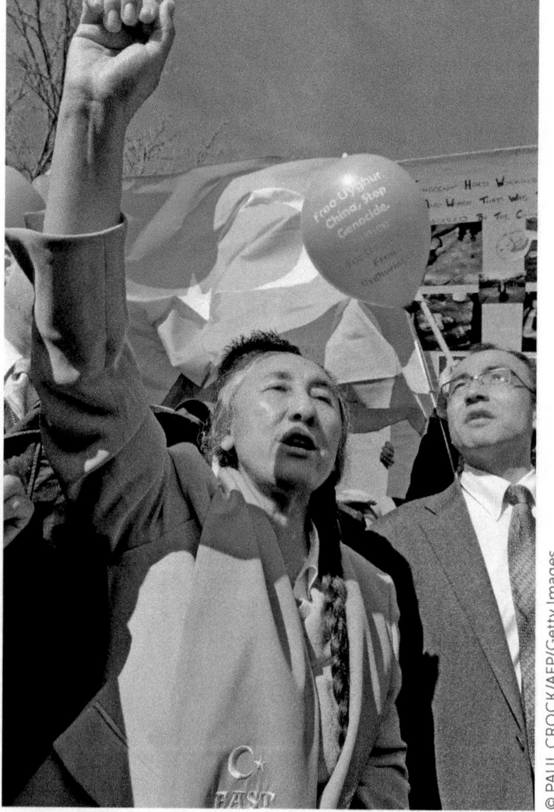

Exiled Uighur leader Rebiya Kadeer punches the air during a 2009 protest by Australian Uighurs outside the Chinese consulate in Melbourne. Kadeer is calling on the Australian government to push for an international inquiry into the alleged atrocities in Urumqi, the capital of Xinjiang province.

South Asian Conflicts

Ethnic and other tensions in South Asia are reflected in a highly unstable political landscape. Among the factors behind these tensions are a complex distribution of ethnic groups and associated religious rivalries (especially, but not exclusively, Hindu–Muslim), the caste system, and the 1947 departure of the British as colonial rulers (Zurick, 1999).

South Asia had been the "jewel in the crown" of the British Empire. With the withdrawal of the British in 1947, the two new states of India and Pakistan were born, and in 1948 Ceylon (renamed

Sri Lanka in 1972) became independent. However, the goal of creating national identities has proved elusive. The intention was for India to be predominantly Hindu and Pakistan predominantly Muslim, with the boundaries to be imposed (on the basis of 1941 census data) where none had existed before and dividing several well-established regions—primarily Kashmir and the Punjab in the northwest and Bengal in the northeast. One immediate result was the movement of about 7.4 million Hindus from Pakistan to India and 7.2 million Muslims from India to Pakistan. It is estimated that several million people died in the chaotic redistribution of population. The new state of Pakistan was divided into West and East, with roughly equal numbers of people separated by 1,600 km (1,000 miles) of Indian territory and with the capital located in West Pakistan, first at Karachi, then at Islamabad. Perhaps inevitably, this new state was unable to forge a cohesive national identity, and by 1971 a large part of the Pakistani army was in East Pakistan fighting Bengali guerrillas. The eastern section seceded that same year, creating the independent state of Bangladesh.

Conflict is presently endemic in much of South Asia (Figure 9.10). Five areas of tension are located in the Himalayan Mountains alone. Most notably, the northwestern region of Kashmir is the subject of ongoing struggles between India, which officially controls the entire state, and Pakistan, which in practice administers the upper portion of it and also lays claim to the predominantly Muslim Kashmir Valley to the south. It is widely feared that these tensions could erupt into war. At the same time, multiple Kashmiri separatist movements are demanding independence from both countries. Northwest of Kashmir in Pashtunistan, Pathan tribes are divided between Afghanistan and Pakistan but make strong claims for statehood. South of Kashmir, Pahari-speaking people favour creation of a separate state they call Uttarkhand, while in the eastern Himalayan region of Gorkhaland some 10 million ethnic Nepalese aspire to statehood and have been involved in several uprisings in recent years. Farther still to the east, in Bodoland—where India borders Burma and China—a number of groups live independent both of one another and of central state control.

Meanwhile, south of the Himalayas in the Punjab, Sikhs aspire to independence from India and the creation of their own state, Khalistan. The struggle for control led to serious violence in the 1980s, but recent years have seen Sikhs turning to more peaceful political activity. Farther east, numerous groups collectively known as the

The northeastern part of Sri Lanka, the disputed territory and the site of much of the fighting during the civil war.

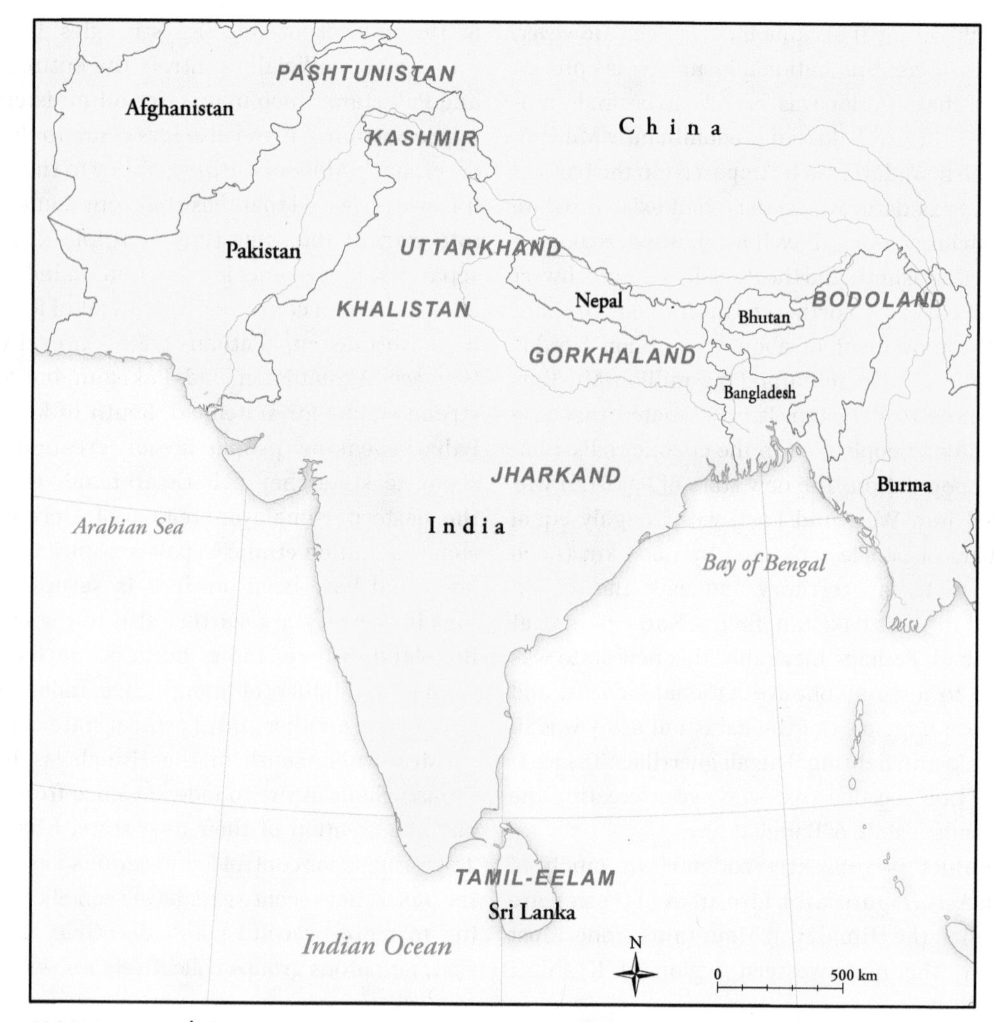

FIGURE 9.10 | Some areas of conflict in South Asia

Eight conflict areas are identified on this map: Kashmir, Pashtunistan, Uttarkhand, Gorkhaland, Bodoland, Khalistan, Jharkand, and Tamil-Eelam.

adivasi (indigenous peoples) assert their right to a state known as Jharkand. Finally, Sri Lanka was for many years in a state of civil war as Tamil Tiger separatists fought government soldiers (Sri Lanka has a Buddhist Sinhalese majority, who dominate the government, and a Hindu Tamil minority). The first Tamils arrived on the island around the eleventh century, but a second group, known as Indian Tamils, was taken there by the British in the late 1800s to work on tea plantations. Conflict first flared in the early 1900s, when Tamils moved onto Sinhalese land, prompting Sinhalese resentment and political and other discrimination. Sinhalese nationalism became more assertive over time, and the Tamil Tigers mounted many organized terror campaigns between the 1970s and 2009, when they finally surrendered to government forces. It is estimated that about 80,000 people died during the almost 30-year civil war. Although that war has ended, feelings of resentment and claims for a separate political identity are unlikely to go away (see Box 6.1).

This account covers only a smattering of the conflicts that plague South Asia. In some cases, ethnic groups war against the state; in others, ethnic groups war against one another; and in yet other cases the conflicts have international dimensions. There are, for example, continuing tensions between India and China (although a border agreement was signed in 2005) and between India and Pakistan (although recent developments suggest that these two nuclear-armed powers are increasingly able to co-operate, notwithstanding tensions relating to the Mumbai terror attacks in 2008). Further complicating the scenario are ongoing Hindu–Muslim differences, the discovery of a Hindu terror cell in 2008, and the complexities of the caste system.

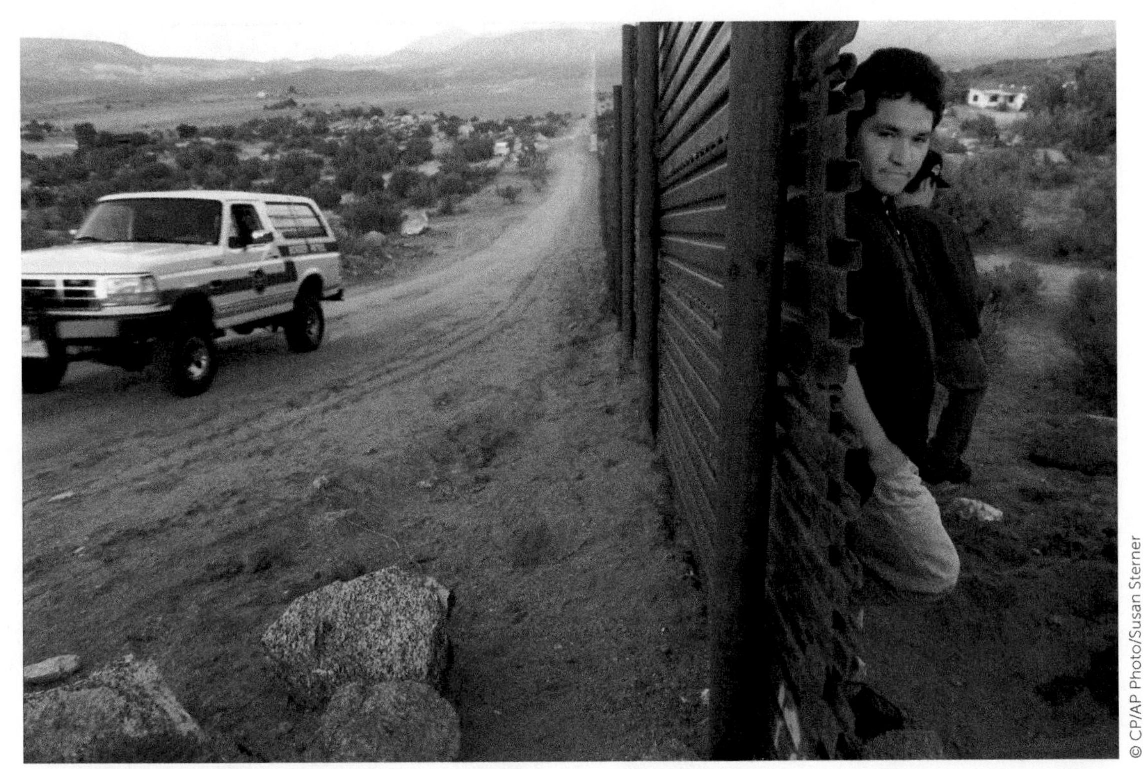

A US border patrol agent drives along the US–Mexico border in Jacumba, California, as prospective immigrants wait to attempt an illegal crossing.

GROUPINGS OF STATES

The contemporary world is characterized by two divergent trends. In addition to the many cases of regions and peoples actively seeking to create their own independent states, some groups of states are actively choosing to unite, sometimes even at the cost of sacrificing aspects of their sovereignty. Examples of this trend are detailed in Chapter 3, with the EU as the principal example of economic, political, and cultural integration.

Uniting Europe

Why have so many sovereign states been willing to sacrifice some components of their independence? In the case of the EU, probably the most important reason is the inherent appeal of a united Europe in a world dominated by the US and USSR and more recently by the US alone. The 28-member EU has a population of almost 510 million—larger than the total population of the three countries (Canada, the US, and Mexico) linked by the North American Free Trade Agreement (NAFTA). During the EU's formative years, the United States was seen as a real economic threat and the Soviet Union as a real military threat. Thus, it seemed that there was a place for Europe in the world but not for some patchwork of European states. As the proponents of union saw it, the two superpowers were large, occupied compact blocks of territory with a low ratio of frontier to total area, and had substantial east–west extent that increased the area of comparable environment. Both were also mid-latitude countries with large populations, densely settled core areas, and a wide variety of natural resources. Individual European countries did not share any of those crucial characteristics, but a united Europe would. Unlike the US, however, a united Europe would be multinational, and a voluntary multinational state might be more stable than an involuntary one such as the empire of the former USSR.

A radically different, indeed controversial, view of Europe's future was proposed by Leopold Kohr (1957). Arguing that aggression is a result of great size and power, he suggested that Europe would be a more peaceful place if, instead of uniting, it divided into many small ethnic states (Figure 9.11). Such an arrangement would minimize the problems associated with borders and national minorities and, in effect, return Europe to its medieval form. Although Europe appears to be moving in a different direction, evidence of regional authority within existing states is increasing.

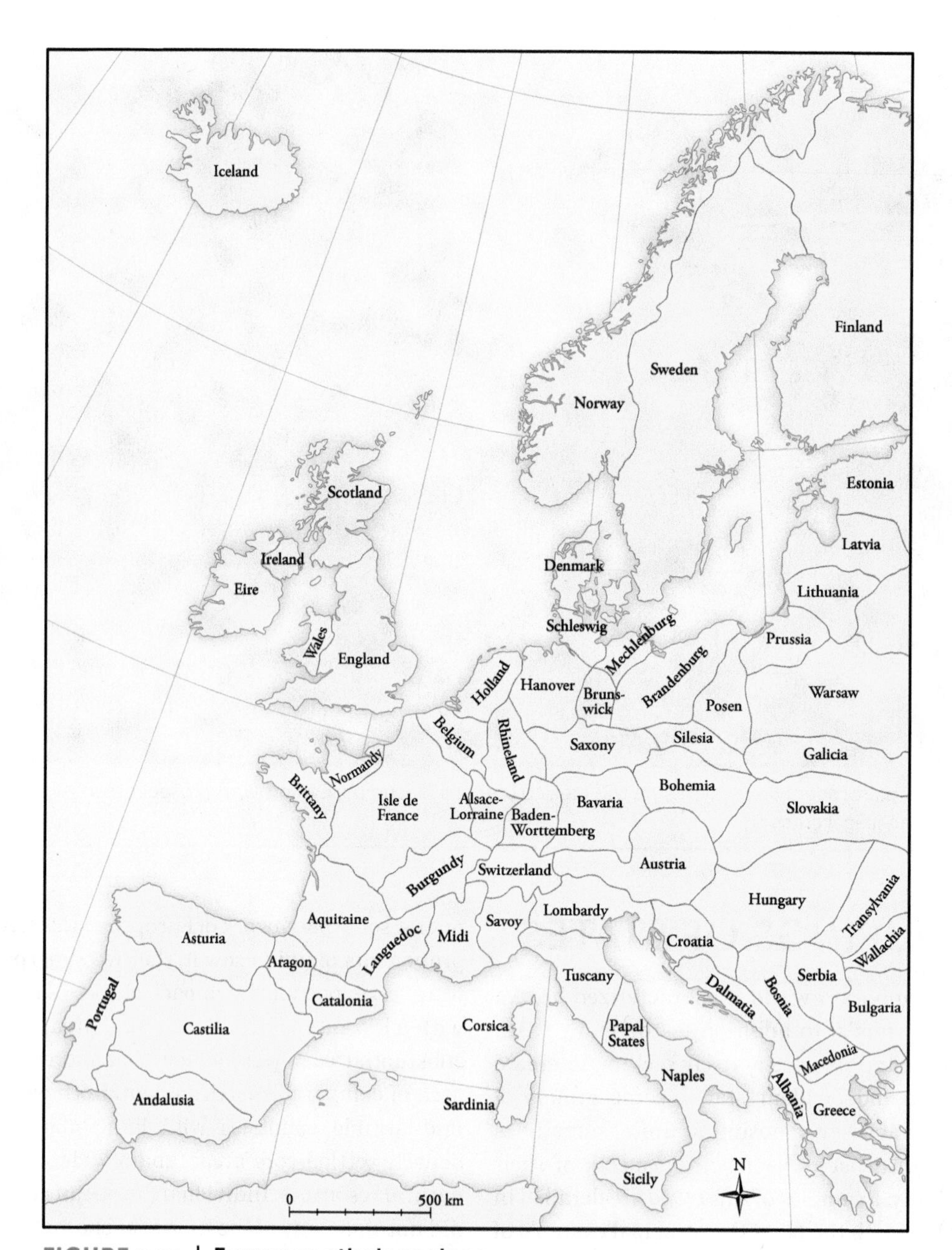

FIGURE 9.11 | European ethnic regions

This map needs to be read carefully and critically. As the text discussion notes, Kohr suggested that sometimes things do not work because they are too big and, a little over a decade after World War II, advocated a return to a pre-modern European world. He is sometimes described as a radical decentralist, unsympathetic to larger globalization tendencies. The map is not a definitive statement—indeed, it might be considered highly problematic in that whether these regions ever really existed in some meaningful way is debatable. It is included to encourage critical thought about the merits, or drawbacks, of a political world comprised of many small, ethnically based, political territories.

Source: Kohr, L. 1957. *The Breakdown of Nations*. Swansea: Christopher Davies.

Even as Europe is becoming increasingly integrated, the profusion of states and independence movements within it suggests that fragmentation might be more characteristic of its component parts. Since the end of World War II, political tensions in Europe have centred on intranational rather than international disputes. We are confronted with a complex scenario in which individual states are integrating with one another at a time when many of those same states are experiencing serious internal stresses.

Other Groupings of States

Apart from the EU, which has its own parliament, elected representatives from the member countries, and a rotating executive, the principal groupings of states—with the exception of the British Commonwealth, the African Union, and the Organization of American States—are essentially limited to establishing and maintaining trade blocs, such as NAFTA and ASEAN, noted in Chapter 3. Other such groupings include the Economic Community of West African States, the Latin American Integration Association, the Organization of Petroleum Exporting Countries (see Chapter 13 for a detailed discussion of this grouping of countries), Asia-Pacific Economic Cooperation (APEC), and the Maghreb Union. In Eastern Europe an economic group named Comecon (the Council for Mutual Economic Assistance) began meeting secretly in 1949 but was terminated in 1991 following the collapse of the communist bloc. Finally, there is the prospect of a single world government. Since 1945 the UN has striven to assist member states in limiting conflict.

THE ROLE OF THE STATE

Most of the world's people are citizens of a specific state and, as such, are subject to its laws. They have limited power—if any—to change the state significantly and are spatially tethered. Their everyday life is irrevocably involved with government, for all states "are active elements within society, providing services which are consumed by the public" (Johnston, 1982: 5). Political units correspond spatially to cultural, social, and economic units, and human geographers often find it useful to define regions on a political basis. The state is a factor affecting human life; it has evolved with society and is a key element in the mode of production. Because of the state's crucial role in the lives of people and places, it is important to recognize that there are many different ways in which a state can be governed.

Forms of Government

The current principal political philosophies are capitalism and socialism, but a number of related or alternative ideas are also important to us. Fundamental to the capitalist form of government is **democracy**, rule by the people. Democracy implies five features: regular free and fair elections to the principal political offices; universal suffrage; a government that is open and accountable to the public; freedom for state citizens to organize and communicate with each other; and a just society offering equal opportunity to all citizens. Democracy was important for a period in classical Greece, but only in the nineteenth century did it reappear as a major idea and only in the twentieth century was it generally approved and commonly practised. The possibility of increasing numbers of states accepting and implementing the key ideas of democracy is discussed later in this chapter.

Monarchy is rule by a single person. Constitutional monarchies survive in some countries but generally without any real power. Britain's monarchy legitimizes a hierarchical social order, while those of the Netherlands, Denmark, and Norway are more democratic. **Oligarchy** is rule by a few, usually those in possession of wealth. The term was introduced by Plato and Aristotle; the favoured term today is *elite*.

Government by **dictatorship** implies an oppressive and arbitrary form of rule established and maintained by force and intimidation. Military dictatorships are common in the contemporary world; the most notorious example is North Korea. Fascism is an extreme form of nationalism that, especially in Europe between 1918 and 1945, provided the intellectual basis for the rise of political movements opposing governments whether they were capitalist or socialist. In Italy and Germany, fascist parties achieved power through a combination of legal means and violence.

The political philosophy of **anarchism** may emphasize either individualism or socialism and was advocated by two prominent nineteenth-century geographers, Kropotkin and Réclus. It rejects the concept of the state and the associated division of society into rulers and ruled.

democracy
A form of government involving free and fair elections, openness and accountability, civil and political rights, and the rule of law.

monarchy
The institution of rule over a state by the hereditary head of a family; monarchists are those who favour this system.

oligarchy
Rule by an elite group of people, typically the wealthy.

dictatorship
An oppressive, anti-democratic form of government in which the leader is often backed by the military.

anarchism
A political philosophy that rejects the state and argues that social order is possible without a state.

Socialist Less Developed States

Socialism is an imprecise term, but we can identify two general characteristics of socialist regimes:

1. They aim to remove any and all features of capitalism, primarily private ownership of resources, resource allocation by the marketplace, and the class structure associated with them.
2. They have the power, in principle, to make substantial changes to society.

Although experiences in the pre-1990 socialist developed world of Eastern Europe made it clear that neither of these characteristics has met with popular approval, socialism has had a major impact in the less developed world. Several African countries have attempted to combine tradition with socialist ideals; perhaps the most successful has been Tanzania. In Latin America, Cuba's socialist system has survived since 1959 in defiance of the dominant capitalist model, although changes began in 2008 when ill-health prompted President Fidel Castro to step down in favour of his brother, Raul. Socialist movements have also played important roles in mobilizing the poor in various South American countries, notably in Uruguay, Bolivia, and Venezuela, with the election of leftist governments.

Maoism
The revolutionary thought and practice of Mao Zedong (1893–1976); based on protracted revolution to achieve power and socialist policies after power is achieved.

state apparatus
The institutions and organizations through which the state exercises its power.

public goods
Goods that are freely available to all or that are provided (equally or unequally) to citizens by the state.

Socialism has been able to exert its most significant and continuing influence in Asia (Hirsch, 1993). Four major Asian countries have socialist or communist governments: China, Laos, Vietnam, and North Korea. Of these, only North Korea still maintains a traditional hard-line and secretive system. China, in particular, seems to be moving towards more democratic institutions and a form of capitalism. Interestingly, North Korea is run by a communist "royal family"; before his death in 1994, Kim Il-sung groomed his son, Kim Jong-il, as heir, and when he died in 2011, his son, Kim Jong-un, was declared leader. North Korea remains a closed country, and its claim that it has nuclear weapons is being met with disapproval from the wider world.

Substate Governments

YAMIL LAGE/AFP/Getty Images

A woman watches a TV report announcing the re-establishment of full diplomatic ties between Cuba and the United States. The two countries had severed ties in 1962, following the Cuban Missile Crisis. The landmark agreement was marked by the re-opening of embassies in Havana and Washington in July 2015.

One version of socialism is based on the revolutionary thought and practice of Mao Zedong, the leader of the peasant revolution in China that led to the 1949 creation of the People's Republic of China. **Maoism** has two components: a strategy of protracted revolution to achieve power and the practice of socialist policies after the revolution succeeds. Both aspects continue to be influential in other parts of the world.

In many less developed countries, socialism has a strong anti-colonial, nationalist component. Unlike the former socialist states in the developed world, which are largely urban and industrial, most of these states are firmly rural in character. Although the details vary, it is fair to say that, even more than in capitalist states, individual behaviour in the less developed world is frequently determined by larger state considerations. Fertility policies, for example, are likely to be rigorous, as in China. Perhaps most important, central planning—that is, planning by the central government—means considerable state involvement in people's everyday lives.

Exercising State Power

In capitalist countries, state power is exercised through various institutions and organizations (Clark and Dear, 1984); this **state apparatus** includes the political and legal systems, the military or police forces to enforce the state's power, and mechanisms such as a central bank to regulate economic affairs. Significant spatial variations occur in the management of this state apparatus, as well as in public-sector income and spending and the provision of **public goods**, including services such as health care and education. Inequalities in the distribution of public goods often reflect government efforts to influence an electorate prior to an election (a popular American term for this behaviour is *pork barrelling*). The form of government and the political philosophy favoured directly affect the manner in which state power is exercised.

One of the most critical issues concerning the power exercised by individual states is the need for international co-operation in solving global environmental problems. As we saw in Box 4.3 on the tragedy of the commons, state power is needed to ensure that private-sector industries do not harm the environment. In the same way, some international authority is needed to ensure that individual states behave responsibly. No such international power exists yet—only the occasional meeting of states convened to address particular problems.

There are at least two dilemmas here (Johnston, 1993). First, governments in more developed countries may be unwilling to protect the global environment when such actions might result in short-term losses of jobs and wealth (and therefore votes) in their own states. Second, governments in less developed countries contend that they cannot afford to implement environmentally appropriate policies and practices, that their practices are not the principal causes of environmental problems, and that, as they develop, they should not be put at a disadvantage vis-à-vis the lack of stringent policies in place when the more developed countries first set out on the path of industrialization. As noted in Chapter 4, both of these dilemmas have been evident in recent meetings aimed at limiting global warming.

The Politics of Protest: Social Movements and Pressure Groups

The relationship between a state and its population is always subject to change. In extreme cases, a substantial proportion of people may reject the state's authority altogether—a situation that can lead to significant internal conflict. An example is Northern Ireland, where many citizens believe that the British state has no legitimate authority over them. In other cases, the people may accept the legitimacy of the state but oppose the governing body; in such situations, efforts to replace the people in charge with others deemed more appropriate may lead to civil war. Even in relatively stable and peaceful states, there are usually several groups striving to influence government policy in various areas.

In many parts of the more developed world, it is common for state governments to face protest from groups with specific agendas; such "beyond the state" social movements have been organized around a vast range of concerns, from racism and labour issues to women's issues, environmental conservation, LGBTQ rights, nuclear disarmament, housing conditions, and urban redevelopment. The 2011 Occupy protests were more generally aimed at established political and financial institutions for their failure to manage global affairs in what the protesters saw as an appropriate way.

Social movements are collective endeavours to instigate change, characterized by active participation on the part of their members and—unlike the state—a relatively informal structure (Carroll, 1992; Wilkinson, 1971). Details vary, but it is common for social protests to begin following the failure, from the perspective of the interested group, of earlier attempts at persuasion and collaboration. Nevertheless, social protest does not always lead to social change.

ELECTIONS: GEOGRAPHY MATTERS

Much attention has been paid to the question of voting since Edward Krebheil's (1916) analyses of geographic influences in British elections. The focus tends to be on spatial and temporal variations in voting patterns and on such causal variables as environment, economy, and society. Of particular importance is whether or not the boundaries of voting districts have been manipulated to create electoral bias (Box 9.6).

Legitimacy of Elections and Voter Turnout

Some elections are correctly described as free—meaning there is a multi-party system, universal adult suffrage, public campaigning, secret ballots, and no voter fraud—but others lack one or more of these characteristics and are better described as compromised. An analysis of recent elections in 154 countries identified 39 as compromised, mostly in African countries but also in Asia.

Voter turnout varies enormously. Recent compromised elections in Turkmenistan, Rwanda, and Russia reported voter turnouts in excess of 95 per cent. Free elections in Australia regularly have 95 per cent turnouts, at least partly because it is one of about 30 countries where voting is compulsory—but this requirement is not rigorously enforced in all cases and in others the penalty for not voting is insignificant (in Australia it is a $20 fine). Countries that often have a very low voter

Examining the Issues

BOX 9.6 | Creating Electoral Bias

In 1812 the governor of Massachusetts, Elbridge Gerry, rearranged voting districts (Figure 9.12) to favour his party. **Gerrymandering**—a word coined by Gerry's political opponents—refers to any spatial reorganization designed to favour a certain party. It aims to produce electoral bias either by concentrating supporters of the opposition party in one electoral district or by scattering them so that they cannot form a majority anywhere. Figure 9.13 shows how congressional district boundaries in Mississippi were manipulated in the 1960s: prior to 1966, blacks formed a majority in one of the five districts, but the redrawn boundaries ensured that they were a minority in all districts. Gerrymandering was especially prevalent in the early 1990s, when several US states fiddled with their electoral borders to create some districts dominated by black voters and others dominated by whites. In both cases the intention was to benefit the party in power (the Republicans) by reducing the impact of the black vote, which has traditionally gone mostly to the Democrats. The changes ensured that, in the first instance, many black votes would be wasted (since the winning candidate would be elected by a huge majority) and, in the second, black voters would be too few to elect the candidate of their choice.

gerrymandering
The realignment of electoral boundaries to benefit a particular political party.

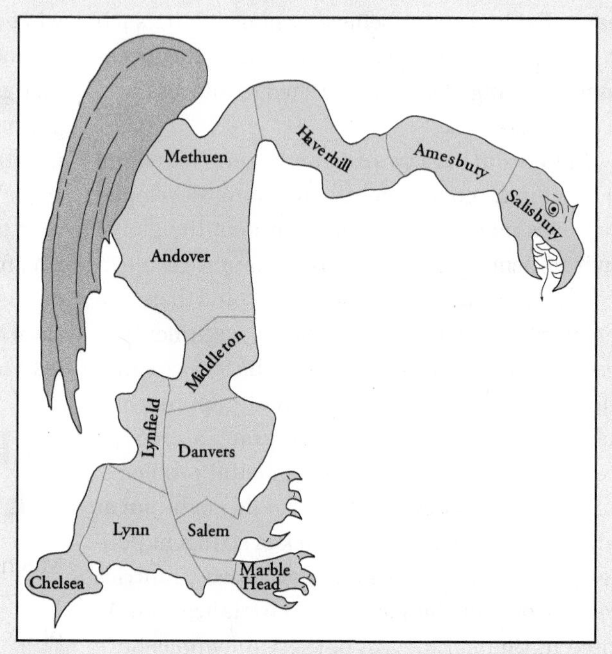

FIGURE 9.12 | The original "gerrymander"
This apportionment was intended to concentrate the vote for one party in a few districts. The term (introduced in the *Boston Gazette*, 26 March 1812) reflected the name of the governor who signed the law—Gerry—and the supposedly salamander-like shape of the new configuration.

Source: Silva, R. 1965. "Reapportionment and Redistricting." *Scientific American* 213 (5): 21.

turnout include several African and southeastern European countries. The lowest-ever voter turnout for a federal election in Canada was just under 60 per cent in 2008 (in 2011 it was 61.1 per cent).

Voting and Place

It is not uncommon to assert that class is a dominant influence on voting behaviour. Britain distinguishes between the Labour Party, traditionally representing the views of workers against those of the capitalist employers, and the Conservative Party, representing the employers. A similar distinction can be made in the US between the Democrats (worker-based) and Republicans (employer-based). In Canada the situation is more vague: although the New Democratic Party (NDP) and the Liberals are more oriented towards the working class than the Conservatives are, ethnic and regional divisions can play a more important role than class in Canadian politics.

Is class really a key explanation of voting behaviour? Do people from the same class but different regions vote similarly? We have already noted that nationalism can disturb this relationship. Place also matters. Geographic research has clearly demonstrated that where a voter lives is crucial, as the following two examples suggest.

First, consider the case of South Carolina, perhaps better described in voting terms as the many South Carolinas. A 2012 *Wall Street Journal* article mapped and described six regions of Republican primary voting preference that have been evident in general terms since before the Civil War (Figure 9.14).

Gerrymandering is a deliberate effort to produce an electoral bias, yet it has only recently been deemed a violation of the Constitution in the United States. A second method useful to parties that appeal to rural voters is **malapportionment**. Rural areas are typically much less densely populated than urban areas; boundaries are drawn so that rural electoral districts contain a relatively small number of voters who are thus able to exert a much greater influence than the same number of opposition voters in a large urban district. It is worth noting that attempts to create electoral bias continue. In 2014 a Florida judge ruled that Republicans illegally redrew the boundaries of two congressional districts to benefit their party and ordered that they be redrawn.

malapportionment
A form of gerrymandering, involving the creation of electoral districts of varying population sizes so that one party will benefit.

Electoral bias, then, is not difficult to produce. But it can also be produced quite unintentionally. Under what circumstances can electoral boundaries be considered fair? Some observers argue that a district ought to be a meaningful spatial unit in a human geographic context; others argue that districts ought to be as diverse as possible. Probably all that can be agreed on is that the number of voters in each district should be as close to equal as possible and that districts should be continuous.

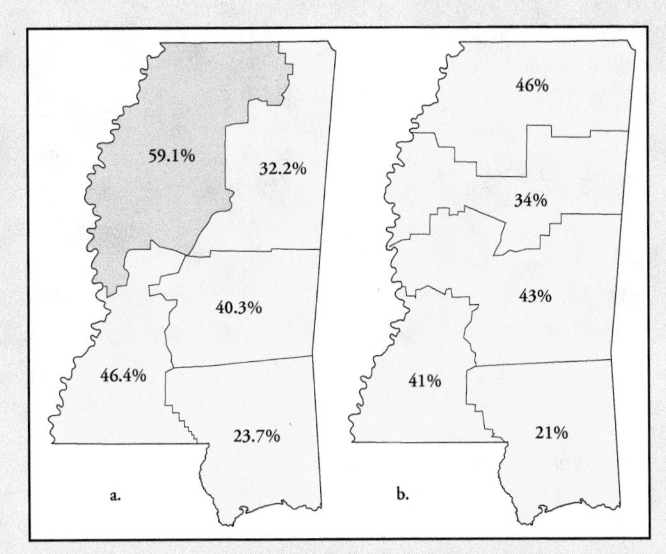

FIGURE 9.13 | Gerrymandering in Mississippi
Congressional districts in 1960s Mississippi with the percentage of black voters noted: (a) the pre-1966 districts with one of the districts having a majority of black voters and (b) the post-1966 districts that guaranteed a minority of blacks in all districts—an example of deliberate gerrymandering.

Source: Adapted from O'Loughlin, J. 1982. "The Identification and Evolution of Racial Gerrymandering." *Annals, Association of American Geographers* 72: 180, Figure 1 (MW/RAAG/P1885).

Second, the role of place is well documented by Johnston (1985), who used data from the 1983 British election to show that national trends cannot simply be transferred to the local scale. He identified four types of local influences on voting:

1. *Sectional effects:* "A long-standing geographical element of voting whereby differences in local and regional political culture produce spatial variations in the support given to the various political parties" (p. 279). Sectional effects have been clearly demonstrated in US presidential elections. Long-standing geographic cleavages require any successful presidential candidate to build an appropriate geographic coalition; these cleavages are so factored into the electoral college system that, to be elected president, votes alone are not enough—they must be in the right place.
2. *Environmental effects:* For example, in the 1983 British election it was found that, the higher the level of unemployment in an area, the more successful the Labour Party candidate. Candidate incumbency is another environmental factor: candidates who already hold their seats usually attract more votes than challengers do.
3. *Campaign effects:* "Vote-switching was more common in the safer seats and the longer-established parties (Conservative and Labour) did better in the marginal constituencies" (p. 287).
4. *Contextual effects:* Individuals may be influenced in their voting decisions (as they are in other behaviours) by their social contacts, such as friends, neighbours, relatives, and co-workers.

THE CANADIAN PRESS/Paul Chiasson

Liberal leader Justin Trudeau greets commuters at a Montreal subway station, one day after winning the 2015 federal election. The election featured one of the longest campaigns in the country's history and increased voter turnout, including a 71 per cent increase in advance poll voters.

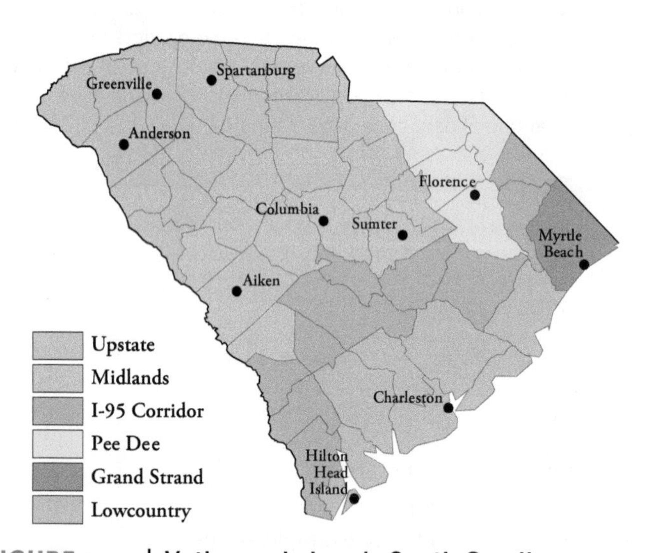

FIGURE 9.14 | Voting and place in South Carolina

Upstate: Evangelical Christians; whiter and wealthier than most of the state; region of manufacturing and foreign investment.

Midlands: Region of mostly moderate social conservatism but also some Tea Party support; Columbia is a major university town.

I-95 Corridor: Mostly Democrat; 43 per cent black population compared to 28 per cent statewide.

Pee Dee: Named after a river; roughly equal numbers of Republicans and Democrats.

Grand Strand: Economically thriving coastal region with many retirees; socially moderate.

Lowcountry: Home to social conservatives opposed to government intervention, for example, in health care; many retirees from the Northeast US.

Source: Based on Bauerlein, V. 2012. "Playing the Many South Carolinas." *Wall Street Journal*, 21 January.

Because voting is influenced by both class and place, any successful political party needs to develop a strong social and spatial base and any meaningful analysis of elections needs to consider both factors.

THE GEOGRAPHY OF PEACE AND WAR

> In geographical terms, this planet is not too small for peace but it is too small for war. (Bunge, 1988)

A 2005 report produced by the Human Security Centre at the University of British Columbia shows that wars are less frequent today than in the past and have had fewer casualties. Moreover, there has been a decline in all forms of political violence, except terrorism, since the early 1990s. Principal reasons for the reduction include UN intervention, the end of colonialism, and the end of the Cold War.

Conflicts

War and peace were not plainly distinguished in European thought until the eighteenth century, when capitalism and the concept of nationalism emerged. War became a state activity, with the creation of professional armies and the elimination of private ones and also became a temporary state of affairs, with long periods of peace between conflicts. Between the French Revolution and World War I, wars were waged between nation-states but were generally territorial; after 1918 conflicts became more ideological—variously between communism, fascism, and liberal democracy. Only since the end of the Cold War have non-Western cultures begun to play a major role in global politics. Since 1945 most conflicts have been centred in the less developed world. Before 1945 most casualties were caused by global wars; after 1945 most were caused by relatively local wars.

The contemporary world contains many traditional enmities. Most relate to ethnic rivalries and/or competition for territory. Conflicts between peoples, a feature of human life for eons, have become increasingly formalized and structured as the number of independent states has increased. Some authors believe that aggression is a natural human behaviour. Since 1945 the UN has offered member states the opportunity to work together to avoid conflict. In the Korean conflict (1950–3), several countries fought together under the UN flag to resist aggression.

Conflicts may be grouped into five categories: traditional conflicts between states, independence movements against foreign domination or occupation, secession conflicts, civil wars that aim to change regimes, and action taken against states that support terrorism. Category 1 conflicts since 1945 include three India–Pakistan wars, four wars involving Arab states and Israel, and the Vietnam War. Category 2 hostilities have arisen primarily as a consequence of decolonization; examples are the conflicts in Indonesia (1946–9), the Belgian Congo (1958–60), and Mozambique (1964–74). Category 3 types, over secession, include the struggles of Tibet (1955–9), Biafra (1967–70), and Philippine Muslims (1977–present). Category 4 includes the civil wars in China (1945–9), Cuba (1956–9), Bolivia (1967), and Iran (1978–9). Category 5 campaigns include the US-led invasion of Afghanistan in 2001 and the 2003 war in Iraq. However, in both of the latter cases terrorism was probably only one of several motivations for the US. In brief, the world may not be at war globally, but it is definitely not at peace.

Conflicts between states (Category 1) have the greatest potential to disrupt the human world. Geographers have focused on various theories of international relations to explain such conflicts. Among the principal variables in such theories are power, environment, and culture. Much of the related empirical work is strongly quantitative.

Are Humans Naturally Aggressive?

Civil Wars

Although the disruption caused by civil wars may be somewhat less than that caused by violent conflict between states, there is good reason to be especially concerned about them. Most wars now are civil wars—approximately 25 ongoing as of 2015—and many continue for years. Governments in the more developed world often assume that such wars are rooted in ancestral ethnic and religious hatreds and that little can be done to prevent them. Yet a 2003 World Bank study suggests that this assumption is not necessarily valid and that lack of development is the principal cause. As countries develop economically, they become progressively less likely to suffer violent conflict, and this in turn makes further development easier to achieve. By contrast, when efforts at development fail, a country usually is at high risk of civil war, which will further damage the economy and increase the risk of further war.

Focus on Geographers

Human Geography and Postwar Recovery of Land and Property Rights | JON D. UNRUH

Land rights have proven to be one of the most difficult issues in a peace process. The disintegration of land and property rights during armed conflict, combined with the importance of land, homelands, and territories to the cause and conduct of war, presents particular dilemmas for recovery. My work involves finding ways to re-establish legitimate, workable land and property rights systems after wars. I work with the UN and other components of the international community, together with countries recovering from war (governments, ethnic groups, ex-combatants, refugees, etc.), to move from war-related ways of pursuing land rights to more stable, peaceful, rule-of-law ways to engage in land rights (also called land tenure). Because of the spatial nature of both armed conflict and land tenure, geographers are uniquely suited to addressing the problem of land rights during postwar recovery.

An end to civil war in the developing world involves the return and reintegration of often several million dislocated persons into primarily agricultural pursuits. This is a vital process that re-engages populations in familiar land uses in order to establish food security and trade opportunities important for recovery and development. Unfortunately, land disputes then surge in frequency and volatility, especially between groups that fought each other, ethnic groups, victims and perpetrators of ethnic cleansing, or those engaged in exploitive or "warlord" interests in extractable

Continued

commodities such as diamonds, gold, and timber. Issues also arise with land used to conduct war, such as the placement of land mines. The number and severity of these land disputes inevitably overwhelm both state and traditional court systems and take a great deal of time to resolve, with impatience and disappointment pushing disputants to pursue alternatives, such as violence. With weapons still in wide circulation, there is frequently significant concern that the numerous land disputes could provide a flashpoint for a return to armed conflict.

One complicating issue is that the functioning of land and property rights (inheritance, buying and selling, access and use rights, renting, borrowing, etc.) cannot be stopped or suspended during times of war. In other words, because people conduct their livelihoods on the landscape in some form even during conflict, they derive forms of claim, access, and use rights even in the absence of state or customary laws and rules. After war, particularly long wars, "war-tenure" approaches become solidified. These can be hybrid, ad hoc, warlord, ethnic, tribal, or religiously based and regularly favour specific groups over others in abusive, exploitive ways.

My work in a country begins by meeting with in-country UN, government, civil society, and armed faction representatives to begin to understand the dynamics of land rights before, during, and after the war. We then start the process of land law reform, often involving both statutory and customary law. Because the misuse of land laws commonly contributes to the onset or escalation of armed conflict, such laws and the way they are implemented virtually always need to be reformed. This involves working with both the statutory side (lawyers, the relevant ministries, commercial interests, etc.) and the customary side (indigenous groups, squatters, ex-combatants, refugees) to see what the primary short-term problems are, as well as to chart the way forward for the longer-term transition towards a land tenure system that is legitimate and workable for all. Then I go to the rural areas of the country and find out what forms of land rights have been hybridized during the war (squatter, refugee, ex-combatant); what forms of indigenous, customary, and statutory land law might still be operational; and if any forms of warlord, religious, or ideological frameworks have emerged with regard to land rights. I do this to see how these ongoing systems of land rights "on the ground" can be stitched together with statutory law. We have learned that we must attend to these local realities after a war: trying to force a war-weary, semi-literate population to accept a statutory law that may have little to do with its reality will be ignored at best and face stiff resistance and confrontation at worst. Hence the need to include these forms of land rights in the process from the beginning and transition from them later. I also assess problems of title fraud, destroyed and missing documents, and transactions made under coercion, duress, or in bad faith during the war so as to derive remedies that have local legitimacy.

JON D. UNRUH is an associate professor in the Department of Geography at McGill University. He has worked on post-war land tenure problems in Latin America, Africa, the Middle East, and Asia.

The fact that many civil wars are linked to natural resources (as discussed later) suggests that they can be a curse rather than a blessing, with much national wealth that might be used to foster human development being directed to fund wars as groups within a country fight over access to oil, metals, minerals, and timber. Angola, where groups fought over oil and diamonds from 1975 until 2002, and Sudan, where fighting from 1983 to 2005 was related to oil, are prime examples. Civil wars may also result from attempts to overthrow an authoritarian leadership, as occurred in Libya in 2011; what might have become a prolonged conflict was undoubtedly shortened by the intervention of the North Atlantic Treaty Organization (NATO) air power on behalf of the rebel forces. The political unrest that had arisen in Syria as part of the Arab Spring of 2011 has boiled over into a brutal civil war.

Civil wars cause population displacement (as discussed in the Chapter 6 account of refugees), mortality, and poverty among local populations and have spillover effects on neighbouring countries. But they also have global impacts. For example, in a country engaged in civil war, some areas are impossible to control for illegal activities such as drug production or trafficking—about 95 per cent of the world's illicit hard drugs are produced in countries experiencing civil war. Tragically, civil wars may be prolonged because a few people benefit financially from such activities, which is one reason that a state might be appropriately described as failed.

Failed States

It has become commonplace, at least since the end of the Cold War, to suggest that some countries have failed because they are either critically weak or no longer functioning effectively—they are either ungoverned or misgoverned. One could argue that there are three types of state in the contemporary world: premodern states are those that have failed; modern states are governed effectively; and postmodern states are those where elements of national sovereignty are being voluntarily dissolved, as in the EU's member countries.

Failed states pose both national and global problems. Chapter 6 discussed several examples in the context of the discussion of refugees. Prime examples include Somalia (anarchy, civil war, piracy), Central African Republic (conflict and hunger), Chad (desertification, destitution, interfering neighbours), Zimbabwe (economic collapse, oppression, kleptocracy), Afghanistan (civil war, drugs, terrorism, limited infrastructure), and Iraq (ruined infrastructure, sectarian strife, terrorism). A particular concern in 2015 is the possibility of failed states emerging in West Africa because of the Ebola outbreak and the rise of Islamic extremism, along with the collapse of Libya into factional chaos, with at least three groups pledging alliance to Islamic State only a few years after the Arab Spring. The global security implications of state failure are apparent. Some failed states may serve as safe havens for terrorists and illicit drug production (Afghanistan) and/or fail to stop pirates from operating freely in busy shipping lanes (Somalia).

Several Latin American countries are struggling to cope with drug production and movement. Large quantities of cocaine are produced in and regularly shipped from the Andean region, forcing peasant farmers from the land, prompting gang wars, and compromising state institutions. In Mexico, over 12,000 people were killed in drug-related conflict in 2011. A few years earlier, when drug-war deaths were pegged at about 6,000, the US Joint Forces Command had suggested that Mexico was close to becoming a failed state.

A recent development in drug trafficking is movement from Colombia in South America to West Africa and then to Europe. This route is favoured because more intensive policing has made trafficking through the Caribbean more difficult. In West Africa, the small country of Guinea-Bissau might be labelled the world's first narco state after the arrival of Colombian drug cartels in 2005. This country, which has no prisons and few police, is already a failed state following a series of conflicts and therefore a logical location for storing cocaine prior to shipment to Europe. The demand in Europe and North America fuels most of the production, movement, and related conflict.

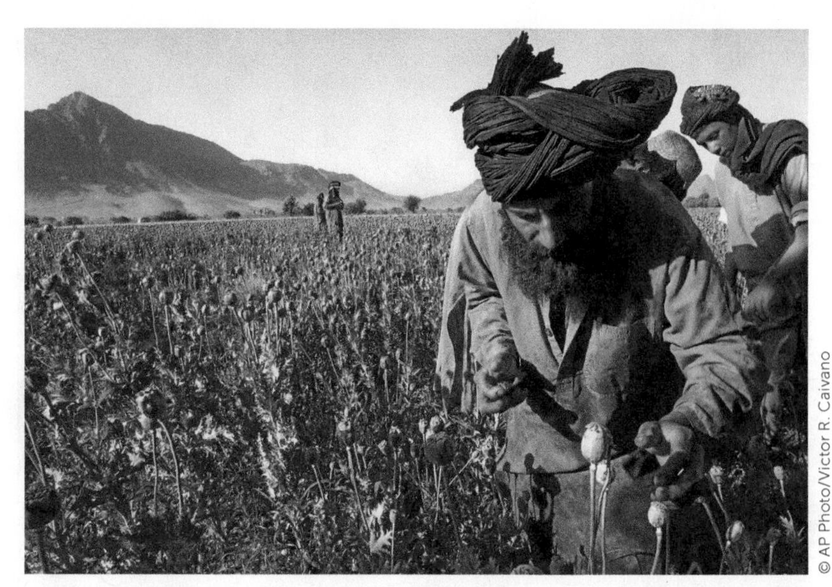

© AP Photo/Victor R. Caivano

Two farmers harvest opium from poppy plants near Kandahar, Afghanistan, in May 2002. According to the UN drug control agency, growers took advantage of a power vacuum created by the US-led war and the collapse of the Taliban to re-establish Afghanistan as the world's largest producer of opium.

Terrorism

Defining Terrorism

Terrorism is not quite so easy to define as might be expected. Attempts to reach international agreement on a definition date back to 1937, when the League of Nations proposed the following: "All criminal acts directed against a State and intended or calculated to create a state of terror in the minds of particular persons or a group of persons or the general public." This proposal was not formally accepted. In 1999 the UN's "Measures to Eliminate International Terrorism" proclaimed that the organization

1. strongly condemns all acts, methods, and practices of terrorism as criminal and unjustifiable, wherever and by whomsoever committed; and
2. reiterates that "criminal acts intended or calculated to provoke a state of terror in the general public, a group of persons or particular persons for political purposes are in any circumstance unjustifiable, whatever the considerations of a political, philosophical, ideological, racial,

Officers of the Royal Ulster Constabulary inspect the debris left by a bomb that killed 28 people and injured 220 in the market town of Omagh, Northern Ireland, in August 1998. A telephone "warning" had encouraged police to move people closer to the blast area.

> ethnic, religious, or other nature that may be invoked to justify them." (www.un.org/en/ga/search/view_doc.asp?symbol=A/RES/49/60)

Less formally, terrorism is the use or threat of violence by a group against a state or other group, with the general goal of intimidation designed to achieve some specific political outcome.

Terrorists or Freedom Fighters?

Difficulty in defining *terrorism* reflects a fundamental contradiction, namely, that many people labelled as terrorists by those whom they attack self-identify as freedom fighters and are considered as such by those whose interests they represent. Two notable examples are former prime minister of Israel Menachem Begin, who had been a terrorist/freedom fighter against the British in Palestine, and Nelson Mandela, the first post-apartheid president of South Africa, who had been a terrorist/freedom fighter against the apartheid regime and spent many years in prison as a consequence. Both received the Nobel Peace Prize.

In recent decades, most terrorist activities were a response to national policies. The IRA conducted a campaign against British rule in Northern Ireland, various Palestinian groups fought against Israeli occupation of certain territories, and ETA (Euskadi ta Askatasuna, meaning "Basque Fatherland and Liberty") fought for an independent Basque homeland in northern Spain and southwest France. Terrorism, then, is a well-established and popular strategy employed by many groups over the years, with numerous countries—including England, Ireland, France, Spain, Israel, Iraq, Pakistan, India, Indonesia, and Thailand—being especially vulnerable. Cohen (2003: 90–1) identifies roughly a hundred states that have been exposed to terrorist activities since the end of World War II. The contrast with peaceful civil disobedience, most famously employed by M.K. Gandhi in his opposition to British rule in India, is clear.

Global Terrorism

Terrorism is essentially a regional, not a global, phenomenon. But many commentators consider that its nature and goals changed with 9/11. These attacks and the organization behind them are of a different order to previous terrorist activities. Al-Qaeda was founded by Osama bin Laden in Afghanistan in 1988, moved to Sudan in 1991, and returned to Afghanistan in 1996. The organization has links with many other groups worldwide and is directly or indirectly responsible for the following acts:

- an explosion at the World Trade Center in New York (1993)
- attacks on US military in Somalia (1993) and Saudi Arabia (1996)
- explosions in the US embassies in Kenya and Tanzania (1998)
- an attack on a US warship in Yemen (2000)
- the 11 September attacks on the US (2001)
- an explosion at a synagogue in Tunisia (2002)
- attacks on nightclubs in Bali (2002)
- bombings in Saudi Arabia (2003)
- suicide attacks in Morocco (2003)
- separate attacks on synagogues and British interests in Turkey (2003)
- train bombings in Madrid (2004)
- killings in Saudi Arabia (2004)
- bombings in London (2005)

More generally, bin Laden issued a "fatwa" in 1998 calling for attacks on Americans (he was killed by US forces in 2011). Of these, the 11 September attacks were the most devastating, both in terms of the numbers killed and in terms of the response. The Islamic State appears to be pursuing similar goals to Al-Qaeda but with the significant additional goal of creating a caliphate, a state that is a political and religious entity.

Much of the immediate reaction to these developments was couched in the context of the long history of conflict between Islam and Christianity and employed the "clash of civilizations" terminology introduced by Huntington (discussed in the following pages). Others found the view superficial and argued that "Islamic" terrorism is not representative of Islam but the work of individuals and small extremist groups (see Chapter 7 for an account of the character of Christian–Islamic conflict). Organizations like Al-Qaeda and the Islamic State do not reflect larger religious identities, and associating Islam with terrorism is wrong. Indeed, 34 of the states identified by Cohen as being exposed to terrorist activities are Islamic states subject to attack from within by extremist groups—a fact that raises further questions about the logic of the clash of civilizations scenario.

Coping with terrorist organizations is especially difficult because, unlike states, they are not spatially tethered components of the political landscape. This does not mean, however, that they function in a geographic vacuum. Most terrorist groups are supported, at least informally, by states—hence the American-initiated invasions of Afghanistan and Iraq following the 9/11 attacks, although in the latter instance no linkages were ever proven, any more than weapons of mass destruction (another American government rationale for the invasion of Iraq) were found. The recent upsurge of extremist Islamic terrorism has dramatically changed global geopolitics, and it is not surprising that some commentators see this as a principal feature of the world and one likely to continue for the foreseeable future.

Competing for Resources

As we saw in Chapter 4, human activities can have serious implications for resource availability. Some commentators note that conflict over resources is central to several ongoing or recent civil wars and predict that resource shortages will play a key role in future conflicts (e.g. Klare, 2001). Perhaps most obviously, there may be increased competition for access to oil and gas, water, and commodities such as timber, copper, gold, and precious stones. In such a scenario, the areas containing those resources—many of which are located in contested and unstable areas of the less developed world—would be major sites of conflict.

In the case of oil and gas, areas vulnerable to conflict include the Caspian Sea basin, the South China Sea, Algeria, Angola, Chad, Colombia, Indonesia, Nigeria, Sudan, and Venezuela, not to mention the Persian Gulf region. The most vulnerable areas for water might be those where a major water body is shared between two or more states, as in the Nile (Egypt, Ethiopia, Sudan, and others), Jordan (Israel, Jordan, Lebanon, Syria), Tigris and Euphrates (Iran, Iraq, Syria, Turkey), Indus (Afghanistan, India, Pakistan), and Amu Darya (Tajikistan, Turkmenistan, Uzbekistan). In time the Great Lakes shared by Canada and the US could become a source of greater contention that it is at present. There may also be competition for diamonds in Angola, the DR of Congo, and Sierra Leone; for emeralds in Colombia; for gold and copper in the DR of Congo, Indonesia, and Papua New Guinea; and for timber in many tropical countries.

One cause of tension in many parts of the world is disagreement about precisely what territory belongs to a country and therefore who has the right to access its natural resources. The example of the Arctic region has come to the fore in recent years because of global warming.

Canada claimed sovereignty over the entire Arctic Archipelago of North America in 1895, but others see the matter differently. In particular, the United States and other shipping nations regard the Northwest Passage as an international waterway, while Russia, Norway, Denmark, and the US also have sectoral claims to parts of the High Arctic. These differing views were of relatively little importance until about 2005, when diamond mining, oil exploration, pipeline development, and navigation became more imaginable and feasible because of melting ice and resultant open water as a result of global warming. Some estimates suggest that the Northwest Passage will be a reasonable route for ocean-going ships within about a decade. Significantly, the journey between Asia and Europe is about 7,000 km shorter through the passage than through the Panama Canal.

Legally, Canada's claim to about one-third of the Arctic region appears well-founded, but this claim could be weakened if Canada does not maintain a significant presence there. Hence, several military exercises, flag-raising events, and scientific expeditions have been conducted in recent years. In 2008 Canada began a survey of parts of the Arctic seabed intended to demonstrate that its continental shelf extends through much of the Arctic region, and a diplomatic campaign was initiated the following year to forcefully and repetitively identify Canada as a major Arctic power. A positive development was the 2011 agreement involving all the Arctic claimants and concerning search-and-rescue responsibilities. There are hopes that it may serve as a blueprint for the more contentious issues of sovereignty and economic activity, although in late 2014 Denmark staked a claim to the North Pole. Reaching agreement, especially regarding oil drilling rights, is an urgent matter as there are extensive resources in the Arctic; the United States Geological Survey has concluded that about 25 per cent of the world total of undiscovered and technically recoverable hydrocarbons are in this region.

The Geography of Nuclear Weapons

As of 2015, there are nine nuclear powers. The United States, Russia, China, France, and Britain are also the five permanent members of the UN Security Council and are all committed to the Nuclear Non-Proliferation Treaty. The other powers are India, Israel, Pakistan, and (possibly) North Korea. There is much concern about North Korea's and Iran's nuclear intentions. The West suspects Algeria, Saudi Arabia, and Syria of having nuclear intentions, but no weapons programs have been identified. Iraq and Libya recently ended their nuclear programs, while Belarus, Kazakhstan, and Ukraine all scrapped the weapons inherited from their inclusion in the former USSR.

As discussed in Chapter 4, nuclear power is an increasingly important source of energy; unfortunately, however, it is also the first step along the road to the production of nuclear weapons. Thus, as more countries develop the technology for nuclear power, it seems likely that the world will confront more crises about their nuclear weapon intentions. We must ask: Do countries that currently have nuclear weapons have any moral authority to say that other countries cannot have those weapons?

Geographers have helped to clarify possible consequences of nuclear war. Openshaw, Steadman, and Greene (1983) published an extensive series of estimates of probable casualties following any nuclear attack. Bunge (1988) published a provocative yet penetrating *Nuclear War Atlas*, and other geographers have explored the possible climatological effects of a nuclear exchange (Elsom, 1985). Nuclear war would likely mean national suicide for any country involved, and a large-scale nuclear war would affect all environments. Large areas of the northern hemisphere would likely experience sub-zero temperatures for several months, regardless of the season. Low temperatures and reductions in sunlight would adversely affect agricultural productivity. A nuclear winter would result in many deaths from hypothermia and starvation. It is not necessary to belabour these points. Let us consider instead the role that geographers can play in influencing public awareness and alerting the makers of public policy to the folly of nuclear war.

OUR GEOPOLITICAL FUTURE?

One of the most interesting, and most important, questions we might ask concerns the character of the geopolitical world as it is unfolding. We have seen throughout this chapter, the world is a complex mix of relative stability and great uncertainty. Some of the political changes described here have followed quite predictable directions, while others have seemed unexpected. It would have been easy to predict the eventual decolonization of Africa, for example, as it took place over the decades after 1945. On the other hand, most observers were not prepared for the mass expressions of discontent throughout Eastern Europe that led to the end of the Cold War, for the peaceful termination of apartheid in South Africa, for the Arab Spring uprisings, or for the rapid rise of the Islamic State. With some trepidation, we ask: What might our geopolitical future have in store? Many answers to this question have been suggested; this section will discuss three intertwined possibilities.

Democracy and the Possibility of Perpetual Peace

Could we be approaching the beginning of what Kant called perpetual peace? One school of thought believes that we have seen the end of major wars between states, although local wars will continue. This "end of history" thesis is based on the idea that, with the termination of the Cold War, the principles of liberal democracy are increasingly accepted around the world and that liberal democracies do not wage war against one another (Fukuyama, 1992). This may be so, but the many local and regional conflicts that have erupted since the end of the Cold War make it difficult to see the phrase "perpetual peace" as applicable to the contemporary world. Nationalist, populist, and fundamentalist movements have proliferated in recent years, with resulting conflicts in many areas (e.g. the Persian Gulf region, the former Yugoslavia, Afghanistan, Somalia, and Central Africa). Evidence shows that the world has become less peaceful in recent years, especially with the rise of the Islamic State and competing claims in the South China Sea.

On the other hand, democracy continues to gain ground in recent decades, with many successful challenges to authoritarian rule. Yet the past few years have seen some setbacks (Table 9.3). Figure 9.15 maps democracy by country globally for 2014.

Around the world, citizens are increasingly unwilling to accept authoritarian rule: since 1980 alone, more than 30 military regimes have been replaced by civilian governments. Challenges to dictatorship may be internal, external, or both. Some of the most remarkable political events of recent years reflect the role played by "people power," with images of large crowds or single individuals peacefully protesting undemocratic governments becoming an icon of the contemporary world. The Philippines in 1986, much of Eastern Europe in 1989, Serbia in 2000, Ukraine in 2004, Nepal in 2006, and the several uprisings in Arab countries that began in 2011 all are examples of the power of mass protest. Rather differently, there are clear signs that the military junta in Burma is accepting the need for some political change.

TABLE 9.3 | Global trends in the spread of democracy, 1977–2014

	Free	Partly Free	Not Free
1977	43	48	64
1987	58	58	51
1997	81	57	53
2007	90	60	43
2014	88	59	48

Note: This table employs the standard terminology used by the Freedom House lobby group, identifying each country as free, partly free, or not free.
Source: Updated from Puddington, A. 2008. "Findings of *Freedom in the World 2008*—Freedom in Retreat: Is the Tide Turning?" www.freedomhouse.org/template.cfm?page=130&year=2008.

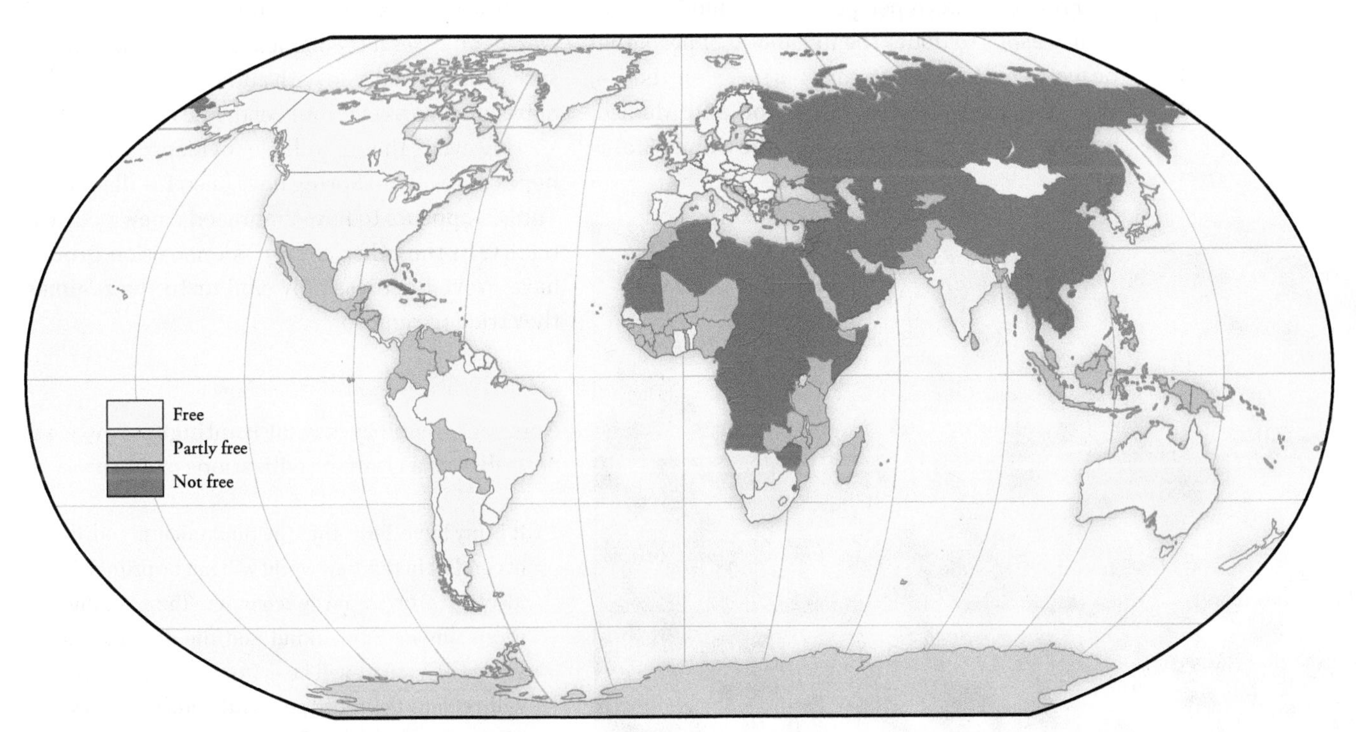

FIGURE 9.15 | Global distribution of freedom, 2014

Like Table 9.3, this map employs the standard terminology used by Freedom House, identifying each country as free, partly free, or not free. The details shown on this map need to be interpreted with caution. The concept of democracy is notoriously difficult to define, being understood differently in different parts of the world.

Source: www.freedomhouse.org/template.cfm?page=363&year=2008.

Since 1990 the number of countries that have ratified the six main human rights conventions and covenants has risen from about 90 to about 150.

But, according to some sources, this spread of democracy is showing signs of being arrested in some countries. The influential American lobby group Freedom House has identified some disturbing reversals, most notably weakening democracies in Russia, Hungary, Pakistan, Bangladesh, Sri Lanka, the Philippines, Venezuela, Kenya, and Nigeria.

The spread of democracy does not preclude conflict. Democratic states frequently experience challenges from within, particularly from marginalized ethnic or regional groups. Acceptance of democratic principles may encourage challenges to national identity. Identities—whether at the level of the nation-state or of a national group within the state—are not fixed but dynamic, continuously playing themselves out in a complex negotiation. For many states there is a fairly obvious basis in language and/or religion and also some sense of a shared history, but many individuals have several competing identities, such as belonging at once to a local community, a larger region, and a state. Democratic states also may experience the power of civil disobedience.

From a Western perspective, the most significant reasons to doubt the likelihood of perpetual peace were the 2001 terrorist attacks on the US, the subsequent attacks in London and Madrid, and the rise of the Islamic State. The United States responded first with a military campaign in Afghanistan designed to replace the Taliban government, which was sympathetic to Al-Qaeda, then with an invasion of Iraq that toppled the regime of Saddam Hussein, and most recently with air strikes against the Islamic State.

Egyptian anti-government protesters celebrate at Tahrir Square in Cairo on 11 February 2011. Cairo's streets exploded in joy when President Hosni Mubarak stepped down after three decades of autocratic rule and handed power to a junta of senior military commanders.

It is possible to interpret the aforementioned Arab Spring in the context of spreading democracy, similar to the political transformation of Eastern Europe in the late 1980s and early 1990s. But this idea may be too simplistic. The goal for Eastern European countries was apparent in all cases, namely to overthrow communist regimes and establish Western European–style democracies. In Arab countries the situation is much more complex because circumstances vary from country to country, especially regarding the role played by ethnic and tribal loyalties and the impact of Islamic understandings of the state's role (as discussed in Chapter 7).

Anti-government protests appeared first in Tunisia and then spread to Egypt, Libya, Syria, and Yemen but also to Algeria, Jordan, Kuwait, Lebanon, Morocco, and Oman. In 2011 authoritarian leaders were overthrown in Tunisia and Egypt through prolonged and often violent protests, while in Libya this was achieved by means of a brief civil war. In all cases it seems clear that there was widespread dissatisfaction with authoritarian rule, lack of basic human rights, and endemic corruption.

However, there is little evidence that the hopes of the Arab Spring have been fulfilled. Only Tunisia appears to have embraced a new political reality; in the other countries opposition groups have proved depressingly similar to the regimes they tried to remove.

Clash of Civilizations

Another possibility is what Huntington (1993: 22) describes as a clash of civilizations or cultures:

> It is my hypothesis that the fundamental source of conflict in this new world will not be primarily ideological or primarily economic. The great divisions among humankind and the dominating source of conflict will be cultural. Nation-states will remain the most powerful actors in world affairs, but the principal conflicts of global politics will occur between nations and groups of different civilizations. The clash of civilizations will dominate global politics. The fault lines between civilizations will be the battle lines of the future.

Huntington (1996) identifies nine major cultures: Western, Sinic, Buddhist, Japanese, Islamic, Hindu, Orthodox, Latin American, and African (Figure 9.16). It is instructive to compare these with the regions proposed by Toynbee and the cultural regions depicted in Figure 7.1 and to reflect again on the quotation from James (see Chapter 7). It can be argued that conflict will be based on cultural divisions for six basic reasons:

1. Cultural differences—of language, religion, and tradition—are more fundamental than differences between political ideologies. They are basic differences that imply different views of the world, different relationships with a god or gods, different social relations, and different understandings of individual rights and responsibilities.
2. As the world becomes smaller, contacts will increase and awareness of cultural differences may intensify.
3. The ongoing processes of modernization and social change separate people from long-standing local identities and weaken the state as a source of identity. Increasingly, various fundamentalist religions are filling the gap, providing a basis for identity.
4. The less developed world is developing its own elites, with their own—non-Western—ideas of how the world should be.
5. Cultural characteristics, especially religion, are difficult to change.
6. Economic regionalism is increasing and is most likely to be successful when rooted in a common culture.

The argument that cultural difference will be the principal basis for future global conflicts proved both controversial and influential. Many commentators interpret extremist Islamic movements in this way, while in this and the two preceding chapters we have discussed numerous regional and local conflicts rooted in cultural differences, specifically of language and religion. On the other hand, Huntington ignored the fact that many conflicts occur within rather than

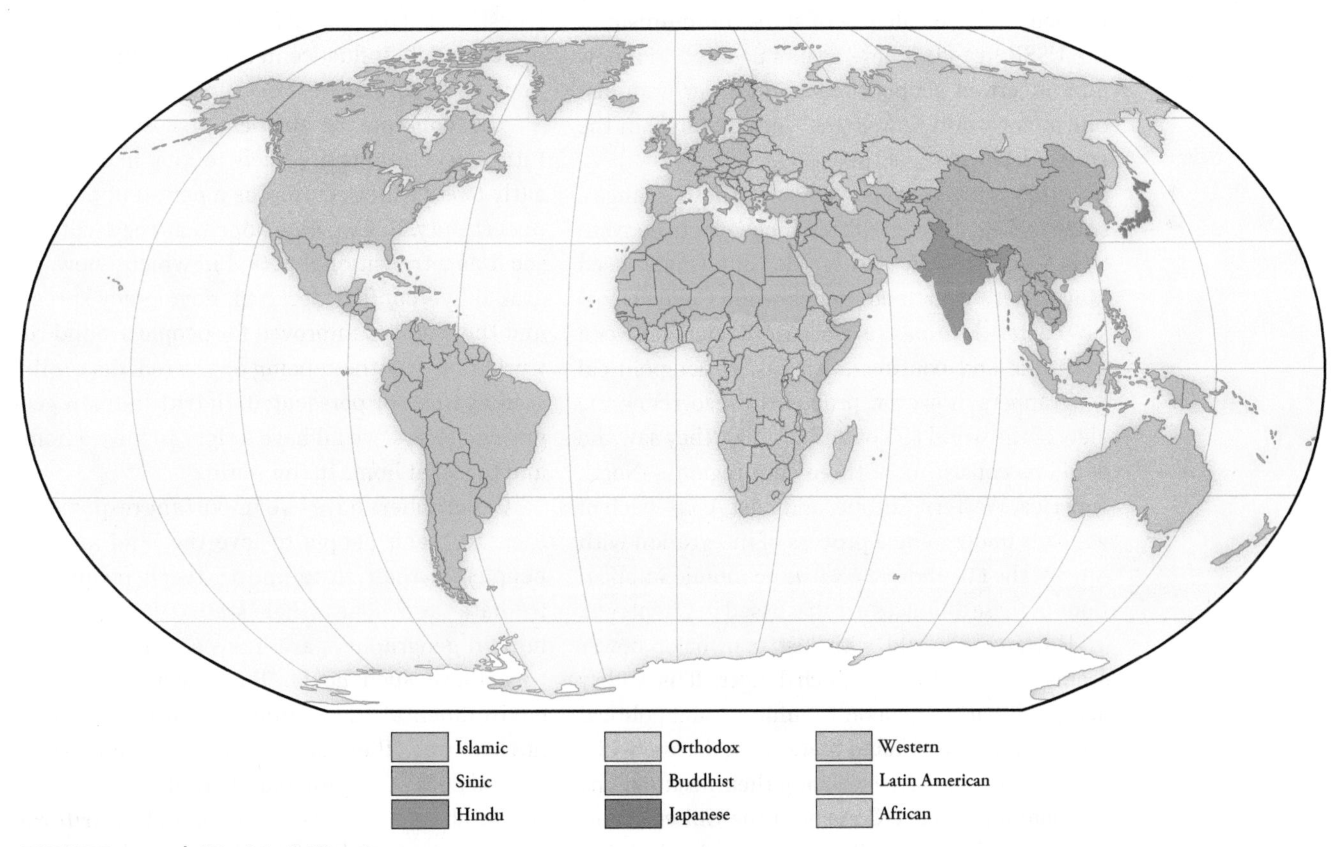

FIGURE 9.16 | World civilizations

Source: Huntington, S. P. 1996. *The Clash of Civilizations and the Remaking of the World Order.* New York: Simon & Schuster, 26–7. Copyright © 1996 by Samuel P. Huntington. © Base Map Hammond World Atlas Corp. All rights reserved. Reprinted by permission of Georges Borchardt, Inc., for the Estate of Samuel P. Huntington. Reprinted with the permission of Simon & Schuster, Inc., from *The Clash of Civilizations and the Remaking of the World Order* by Samuel P. Huntington. Copyright © 1996 by Samuel P. Huntington. Base map © Hammond World Atlas Corp.. All rights reserved.

between so-called civilizations; consider the many occasions when Europeans have fought each other and the instances of Muslims fighting other Muslims.

World Order—or Disorder?

During the Cold War period the world was ideologically divided between the democratic capitalist states belonging to the US-dominated NATO and the communist states belonging to the USSR-dominated Warsaw Pact. This bipolar division no longer exists: NATO has expanded from its original 12 member countries to a total of 28 in 2015, and all the new members are former communist states in Eastern Europe. Other new members may be admitted in the future.

The first stage in this geopolitical transition was Poland's installation of a non-communist government—approved by the USSR—in 1989. Events followed with remarkable speed: the collapse of communist governments elsewhere in Eastern Europe, the symbolic breaching of the Berlin Wall on 9 November 1989, the reunification of Germany in 1990, and the collapse of state communism in the USSR in 1991. These events ushered in a new and uncertain geopolitical world order with just one major world power (the United States) in the early twenty-first century.

Attempting to make sense of these changes, some political geographers focused on conventional geostrategic issues; de Blij (1992) used Mackinder's heartland model, while Cohen (1991) saw Eastern Europe as a "Gateway Region" between maritime and continental areas. Other political geographers, however, proposed a geo-economic view of the world (O'Loughlin, 1992). They saw the world as consisting of three core regions—North America, Western Europe, and East Asia—each of which is undergoing a process of integration with NAFTA, the EU, and ASEAN (the economic implications of these alliances are discussed in Chapter 3).

However, a world with just one major power seems unlikely to last much longer. It is widely accepted that the economic, military, and political dominance of the United States is in decline while other powers are strengthening their position. The National Intelligence Council in the United States sees China, India, and Russia as growing in influence. The power of the Islamic world is also increasingly evident. It seems unlikely that the UN will be able to assert more authority in this changing world.

CONCLUSION

This chapter has emphasized that the political map of the world and the power that parts of the world have exercised over others have been subject to much change through time. Most notably, it is possible that the long period of Western (European and American) dominance that has helped shape much of the world through colonial activity is coming to an end and that Asia may soon assume an equally important, or even a dominant, role. Since the mid-fifteenth century, the Western world steadily assumed ever-increasing importance until, by the onset of World War I, it controlled about 60 per cent of the global land mass and generated about 80 per cent of the wealth. But since then the West has declined relative to Japan and China especially. The reduced role played by the United States was obvious in the Arab Spring revolutions that proceeded, except in Libya, without any meaningful American support. It is also significant that neither the United States nor the EU seems able to resolve the Israeli–Palestinian conflict and that China is steadily increasing its influence in several countries, particularly in Africa.

The contents of this chapter suggest that future generations are likely to look back on the early twenty-first century as a period of political uncertainty. But we may hope that they will also see it as a transitional period in which—however gradually—conflict declined, democracy spread, and the chances improved for people around the world to live without being oppressed, discriminated against, or persecuted. In truly human geographic terms, we all have a right to have a home and to feel at home in the world.

Geographers have two important responsibilities: to teach people to love the land and the peoples of their state and to teach people not to hate and fear other states. Undoubtedly, human geographers are in an enviable position to achieve such goals. They have studied the environmental and human consequences of nuclear war. They have used cartography (long a highly effective propaganda tool) to teach and so are able to identify any deliberate territorial biases. Their study of landscapes, places, and their interrelationships can help to develop the global understanding so desperately lacking in our contemporary world.

Summary

Dividing Territory

Human beings partition space. The most fundamental of the divisions we create is the sovereign state. There are over 200 such states in the world, 193 of which are full members of the UN. Almost all people are subjects of a state. Loosely structured empires were typical of Europe before 1600. The link between sovereignty and territory emerged most clearly in Europe after 1600.

Nation-States

In principle, a nation-state is a political territory occupied by one national group. Despite the importance of nationalism and the territory–state–nation trilogy, there are many binational and multinational states. Nationalism is the belief that each nation has a right to a state; only in the twentieth century did nationalism become a prevalent view.

Empires: Rise and Fall

World history includes many examples of empire creation and collapse. European empires evolved after 1500 and typically disintegrated after 1900, greatly increasing the number of states in the world. Imperialism is the effort by one group to exert power via their state over another group located elsewhere. The colonies created in this way quickly became dependent on the imperial states. The most compelling motivation for colonialism was economic.

Geopolitics

Geopolitics is the study of the roles played by space and distance in international relations. Early geopolitical arguments by Ratzel and Kjellén included the idea that territorial expansion was a legitimate state goal. This idea, in the form of *geopolitik*, dominated geopolitics until the 1940s when it was discredited for being too closely associated with Nazi expansionist ideology. Other important geopolitical theories include Mackinder's heartland theory and Spykman's rimland theory. In the early 1990s the Cold War, which had started in the late 1940s, ended and a geopolitical transition began.

Boundaries of States

Stability may be closely related to boundaries. State boundaries are lines that can be depicted on maps, whereas national boundaries are usually zones of transition. Antecedent boundaries precede settlement; subsequent boundaries succeed settlement.

States: Internal Divisions

The stability of any state is closely related to the relative strength of centrifugal and centripetal forces within it. A key centripetal force is the presence of a powerful sense of shared identity; a key centrifugal force is the lack of a strong shared identity—perhaps because of the presence of minority groups that, in fact, may wish to secede. In Africa discordance between nation and state is common where state boundaries were imposed by former colonial powers without reference to national identities. European states typically have greater internal homogeneity, although many countries do experience conflict over minority issues, some of which might be explained by reference to core/periphery concepts.

The collapse of the former USSR, an ethnically diverse empire, is one example of instability; the disintegration of Yugoslavia, a country with several republics and a deep divide between Christians and Muslims, is another. Canada's internal divisions are related to a variety of physical and human factors. Many of the peoples asserting a distinct identity occupy areas peripherally located in their states and economically depressed.

Groupings of States

There are two opposing trends in our political world: minority groups increasingly aspire to create their own states, while states increasingly desire to group together, usually for economic purposes. The EU is the principal instance of the second trend. As of 2015 it includes 28 countries but is scheduled to expand.

The Power of the State

There are numerous types of states and related political philosophies. States perform many important functions, both internally with regard to the distribution of public goods and externally in their relationships with other states, especially regarding conflict and environmental concerns.

Elections

Geography is a central factor in elections. Some electoral boundaries are deliberately drawn to create an electoral bias using gerrymandering or malapportionment. The supporters of different political parties are often spatially segregated. Place has a significant influence on voting behaviour.

Peace and War in the Twenty-First Century

Peace and war, with their obvious implications for both people and place, are inherently geographic. Geographers can help us understand the causes and consequences of conflict and can contribute to increased understanding between states. Even countries not at war are heavily committed to military and related expenditures. Although the geopolitical future remains uncertain, most scholars agree that conflict will continue. The world may not be at war globally, but it is definitely not at peace with itself.

Links to Other Chapters

- **Globalization:** Chapter 2 (concepts); Chapter 3 (all of chapter); Chapter 4 (global environmental issues); Chapter 5 (population growth, fertility decline); Chapter 6 (refugees, disease, more and less developed worlds); Chapter 8 (popular culture); Chapter 10 (agricultural restructuring); Chapter 11 (global cities); Chapter 13 (industrial restructuring)
- **State creation, nation-states, nationalism, state stability:** Chapter 7 (language and religion); Chapter 8 (ethnicity)
- **Group formation and group identity:** Chapters 7 and 8
- **Colonialism, dependency theory, socialism in less developed states:** Chapter 1 (history of geography); Chapter 6 (the less developed world, world systems analysis)
- **European colonial activity:** Chapter 1 (European overseas movement)
- **Core and periphery:** Chapter 13 (uneven development)
- **State groupings (NAFTA, EU, ASEAN); globalization:** Chapter 3 (trade)
- **Nation–state discordance in Africa:** Chapter 6 (refugees)
- **Politics of protest:** Chapter 3 (Boxes 3.4 and 3.5); Chapter 4 (Box 4.7); Chapter 8 (contested landscapes)
- **Possible future conflicts:** Chapter 6 (future trends in population, food, and environment, the role of bad government); Chapter 7 (religious identities)

Questions for Critical Thought

1. What role did colonialism play in creating the current world political map?
2. Geopolitics is the study of the relevance of space and distance to questions of international relations. How has the "relevance of space and distance" changed over the past half-century? In other words, are space and distance more or less important today than they were 50 years ago?
3. It has been said that Canada has too much geography and not enough history, while the former Yugoslavia has too much history and not enough geography. What does this mean?
4. What similarities and differences are there among the various separatist movements in Europe? Why do you think there are so many such movements in Europe?

5. How important will "groupings of states" (e.g. the European Union) be for the geopolitical stability of the world in the future?
6. Are you optimistic or pessimistic about the future political stability of the world? Why?
7. To what extent is conflict due to the theory that humans are naturally aggressive?

Suggested Readings

Visit the companion website for a list of suggested readings.

10 AGRICULTURAL GEOGRAPHY

Agricultural geography as discussed in this chapter is generally considered part of the subdiscipline of economic geography, as are settlement and industrial geography (discussed in Chapters 11, 12, and 13). It may be helpful, then, to think of these four chapters as a unit. The study of economic geography obviously involves consideration of economic factors; however, cultural and political factors also play very important roles, so the content of these chapters is quite diverse.

This chapter begins with an examination of why agricultural activities are located where they are, with emphasis on a classic theoretical argument that many geographers have found useful. Next, we explain the origins and evolution of major agricultural activities, discuss major technological advances, and identify nine principal agricultural regions around the world. This is followed by a discussion of the global restructuring of agriculture. The chapter concludes with a look at changes in food consumption patterns and preferences. The final two sections reflect the current interest in agriculture as part of a wider production and consumption system that includes inputs, farming, product processing, wholesaling, retailing, and food preferences.

Here are three **points to consider** as you read this chapter.

- Is an agricultural location theory, which employs simplifications to produce an ideal world against which reality can be assessed, of value to agricultural geographers?
- Does understanding global patterns of agricultural production and consumption require an appreciation of the globalization processes detailed in Chapter 3?
- Will we be able to feed a global population that is expected to increase from 7.2 billion in 2014 to about 9.7 billion in 2050?

Farmers harvest brown mustard in Tryon, Prince Edward Island. Thousands of hectares of the crop have been planted to battle wireworm, a pest that destroys potatoes. Agricultural activities and where and why they occur are important elements in human geography, particularly regarding the three central themes examined in this book.

The human geographic study of agriculture, settlement, and industry—often collectively labelled economic geography—has broadened over the past century from the strictly economic (in the empiricist and spatial analysis traditions) to the economic and political (in the Marxist tradition) and the economic, political, and cultural (in the feminist and postmodern traditions). This and the following three chapters embrace this diversity of approaches and related subject matter.

When reading about the several positivist-informed location theories included in these chapters, it is helpful to remember that many human geographers see geographic theory and practice as social activities, inevitably influenced by the geographer doing the work. Thus, they demand that geographers situate themselves in relation to their work. Barnes (2001: 557) aptly describes this demand as reflecting the reality that all geographic writing is necessarily a "view from *somewhere*"—an observation that echoes the humanistic critique of positivism as dehumanizing and that is also evident in the accounts of postmodernism and feminism.

Inventing and Reinventing Economic Geography

THE AGRICULTURAL LOCATION PROBLEM

We begin our account of agriculture with a basic geographic question: Why are specific agricultural activities located where they are? Our answer will centre on economic issues, but it will

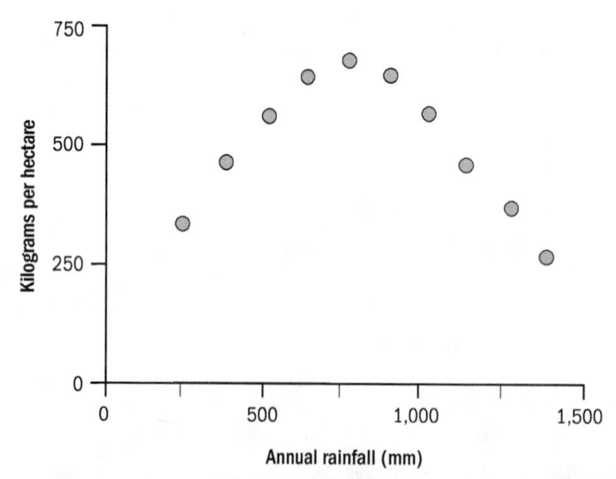

FIGURE 10.1 | Relationship between mean annual rainfall and wheat yield in the US, 1909

Source: Adapted from Baker, O. E. 1925. "The Potential Supply of Wheat" *Economic Geography* 1: 39.

first consider several other factors—physical, cultural, and political—that are all interrelated.

Physical Factors

Physical factors are important influences in agricultural decisions and hence in the creation of agricultural landscapes. Animals and plants are living things and require appropriate physical environments to function efficiently. Farmers have two options: ensure that there is a match between animal and plant requirements and the physical environment or create artificial physical environments by, for example, practising irrigation or building greenhouses. Because many environments are suitable for more than one crop, the agricultural decision can be based on other factors, which can also play a role when the demand for a product is lower than the quantity that can be produced.

Climatic factors are the main physical variables affecting agriculture. Plants have precise temperature and moisture requirements, although ongoing plant hybridization has succeeded in extending these needs. Optimum temperatures vary according to the plant cycle, but growing-season temperatures between 18°C (64°F) and 25°C (77°F) are often required. Low temperatures result in slow plant growth, while short growing seasons may prevent crops from reaching maturity. In many areas, frost may result in plant damage. Moisture is crucial to plant growth; too little or too much can damage plants. In some areas, especially semi-arid zones, rainfall variability is a problem. Many of the world's major wheat-growing areas (such as the Canadian prairies) are highly vulnerable in this respect—Figure 10.1 indicates the relationship between wheat yield and mean annual rainfall at a time before chemical fertilizer was widely used. Animals also have water requirements, particularly dairy cattle; sheep are much more adaptable to water limitations.

Two other important physical variables are soils and topographical relief. Soil depth, texture, acidity, and nutrient composition all need to be considered. Shallow soils typically inhibit root development; the ideal texture is one that is not dominated by either large particles (sand) or small particles (clay). Most crops require neutral or slightly acidic soils. The most crucial nutrients are nitrogen, phosphorus, and potassium. Soil fertility needs to be maintained if regular cropping is practised; methods include fallowing, manuring, rotating crops, and using chemical fertilizers.

Relief (the shape of the land) also affects agriculture, specifically through slope and altitude. A slope's angle, direction, and related insolation (exposure to sun) determines both the probability of soil erosion and the use of machinery. Generally, the flatter the land, the more suitable it is for agriculture. Altitude affects temperatures: in temperate areas, the mean annual temperature falls 6°C (11°F) for each 1,000 m (3,280 feet) above sea level.

© kevin miller/iStockphoto

View of vineyards in the Okanagan, British Columbia, Canada.

Technological, Cultural, and Political Factors

During about the past 300 years, advances in plant and animal breeding and, most recently, in biotechnology have improved agricultural productivity. A particularly effective approach uses pesticides and fungicides to reduce the competition between crops and various pests for both light and nutrients. Also important is increasing energy inputs through use of fertilizers. The dramatic consequences of the development of nitrogen fertilizers and current advances in biotechnology are considered in more detail later in this chapter.

Farmers are motivated not only by profit but also by cultural preferences. In the more developed world, they typically favour security and a relatively constant income over a life dedicated to the unlikely goal of profit maximization; in the less developed world, they typically maximize product output for subsistence rather than profit. Group religious beliefs may favour specific agricultural activities because of the value placed on either the activity or the product of the activity. Christianity, for example, values wine made from grapes, which is used in the sacrament of Holy Communion. The Church's demand for wine encouraged the spread and growth of viticulture, including its introduction to California by early Christian missionaries. Other agricultural activities are negatively affected by religious beliefs. Pigs are taboo in Islamic areas, while Hindus and Buddhists believe it is wrong to kill animals, especially cattle. Immigrants in a new land often continue to adhere to the agricultural practices of their homelands.

Political decisions affect much human behaviour, including that of farmers. Governments may have many reasons for deciding to influence farmers' behaviour: in the more developed world the most common motive is the desire to support an activity that, as measured by income, is in decline. In such cases, governments may intervene by fixing product prices and providing financial support for certain farm improvements such as land clearance (Box 10.1 and Figure 10.2). In the less developed world, governments may provide assistance that enables farmers to adopt new methods and products. Further, agricultural systems are affected not only by state policies but also by national and international trade legislation. Canada, the United States, Japan, and Europe all have marketing boards, quota requirements, government credit policies, and extension services to provide information to the farm community, all of which affect the agricultural landscape. Perhaps the best-known examples of interventionist policies come from the United States and the European Union, where in some cases farmers have been paid not to grow crops.

Supply and Demand

Spatial patterns of agricultural activities are affected by physical, cultural, and political variables. But human geographers agree that the most important variables generally—not necessarily in a specific location—are economic. Agricultural products are produced in response to market demand for them; thus, farming is subject to the basic laws of supply and demand. Figure 10.3 depicts characteristic supply and demand curves. Since supply increases and demand decreases when the price received rises, an equilibrium price (p) can be identified at the intersection of the two curves.

Farming Types in Illinois

Around the Globe

BOX 10.1 | Government and the Agricultural Landscape

The Canada–United States border region between the Great Lakes and the Rocky Mountains provides an excellent example of how agricultural landscapes evolve in response to government policies. Around 1960, despite similar climatic, soil, relief, and drainage conditions, the landscapes differed considerably because the US government's agricultural policy was much more interventionist than its Canadian counterpart. Figure 10.2 shows dramatic differences in crop and livestock combinations around 1960. The wheat allotment program in the United States meant that much wheat land was converted to raise barley, while the National Wool Act of 1954 encouraged sheep-raising through a guaranteed price.

The governments' approaches had changed by the 1970s. The United States had moved away from support programs towards relatively market-oriented policies. Canada had come to favour greater intervention to ensure adequate incomes for farmers. As a result, the differences between the two countries were reduced and the landscapes changed accordingly. The transboundary differences are significantly fewer today than in the 1950s and 1960s.

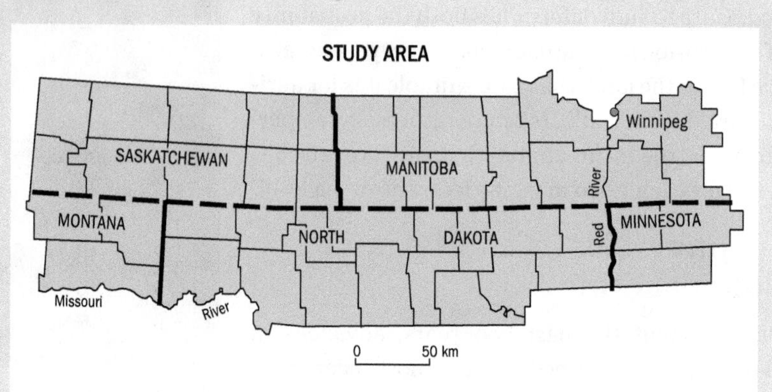

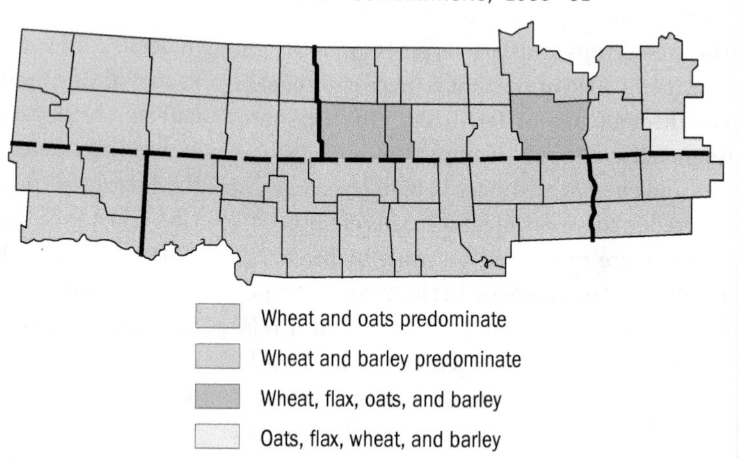

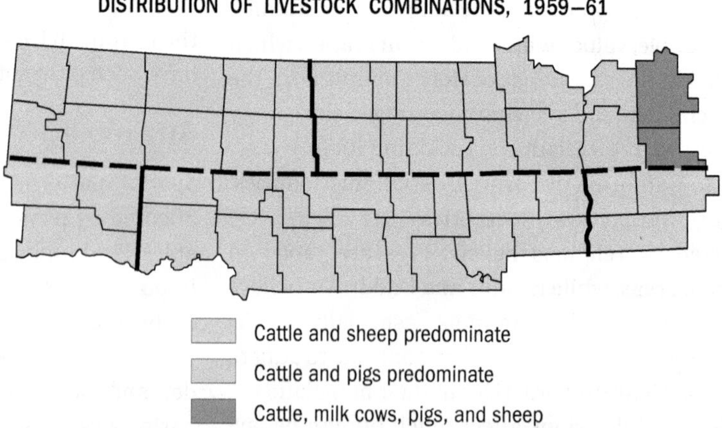

FIGURE 10.2 | Crop and livestock combinations along the US–Canada border

Source: Adapted from Reitsma, H. J. 1971. "Crop and Livestock Production in the Vicinity of the United States–Canada Border." *Professional Geographer* 23: 217, 219, 221, Figure 1 (MW/RTPG/P1891).

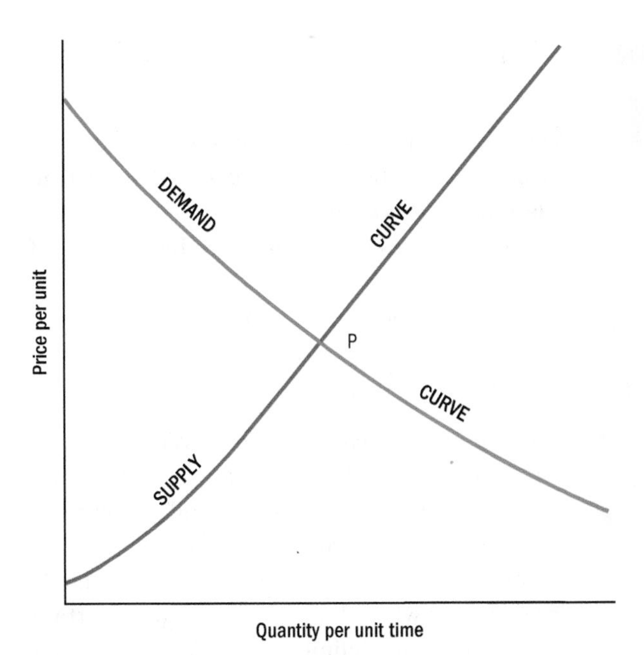

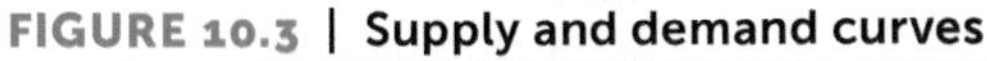
FIGURE 10.3 | Supply and demand curves

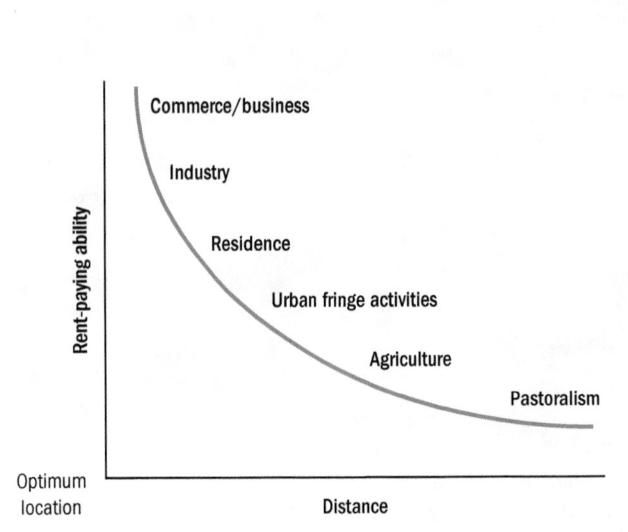

FIGURE 10.4 | Rent-paying abilities of selected land uses

The ideal economic world of commercial agriculture is occupied by farmers motivated solely by profit. This type of response is neither feasible nor desirable for this group. In the more developed world, farmers value stability and independence, and they may be willing to sacrifice profits accordingly or choose to respond to decreasing prices by increasing rather than decreasing supplies. Farmers in the less developed world tend not to be profit maximizers, but their behaviour may well be rational. The aim of subsistence agriculture is to produce the amount of product required to meet family needs; profit maximization has no meaning for such farmers. The number of subsistence farmers worldwide has decreased significantly since about 1800 and continues to fall. Therefore, human geographers studying agricultural location focus on commercial farmers.

Competition for Land

Although several theories have been developed to explain observed variations in the spatial patterns of farming activities, they all focus on the fact that different activities compete for use of any location. Such competition arises because it is not possible for all activities to be carried out at their economically optimal location, and different activities may have identical optimal locations. The obvious question is how to determine the most appropriate use of a particular piece of land. Conventionally, land is assigned to the use that generates the greatest profits; in a general sense, we can determine a hierarchy of land uses based on relative profits. But there is another way of viewing this situation. The greater the profits a certain use generates, the more that use can afford to pay for the land. The maximum amount that a given use can pay is called the ceiling rent. Figure 10.4 indicates the relative rent-paying abilities for a variety of land uses.

We can now identify the basic premise of much location theory, including agricultural: the competition among land uses, fought according to rent-paying abilities, results in a spatial patterning of those land uses. Agricultural location theorists use this idea in combination with the concept of economic rent, which explains why land is or is not used for production. Land is used for production if a given land use has an economic rent above zero (Box 10.2). Economic rent can be related to one or more variables. David Ricardo (1772–1823), the economist who formulated the concept, applied it to measures of fertility; however, the economist Johann Heinrich von Thünen (1783–1850) regarded distance as more important.

commercial agriculture
An agricultural system in which the production is primarily for sale.

subsistence agriculture
An agricultural system in which the production is not primarily for sale but is consumed by the farmer's household.

ceiling rent
The maximum rent that a potential land user can be charged for use of a given piece of land.

location theory
A body of theories explaining the distribution of economic activities.

economic rent
The surplus income that accrues to a unit of land above the minimum income needed to bring a unit of new land into production at the margins of production.

Commercial farmland in Saskatchewan. The province's agriculture and value-added sector of the economy includes approximately 44,000 farmers and ranchers and generates over $8 billion in economic activity.

Von Thünen's Agricultural Location Theory

Von Thünen was a German economist and landowner who published *The Isolated State* in 1826 (Hall, 1966). This seminal work reflected his interest in economics (he was inspired by the work of Adam Smith) and his 40 years as manager of an estate on the north German plain.

The Approach

Inspired by classical and neo-classical economics, von Thünen developed a location theory that aimed to explain agricultural patterns in any given place and at any time. This is our first example of a **normative** theory, meaning that it is concerned with what ought to be given certain assumptions. By definition, the theory does not purport to explain reality but what reality would

Examining the Issues

BOX 10.2 | Calculating Economic Rent

It is important to clarify the calculation method and spatial implications of von Thünen's use of the economic rent concept. Economic rent can be calculated as follows:

$$R = E(p - a) - Efk$$

where

- R = rent per unit of land (dependent variable)
- k = distance from market (independent variable)
- E = output per unit of land
- p = market price per unit of commodity
- a = production cost per unit of commodity
- f = transport rate per unit of distance per unit of commodity

(E, p, a, f: parameters)

This equation is a straightforward linear relationship between rent and distance. Rent is a function of distance with the details of the relationship determined by $E(p - a)$, which gives the intercept (on the rent axis), and Ef, which determines the slope of the line.

What are the spatial implications of this calculation? Because economic rent declines with increasing distance from market, it eventually becomes zero. When more than one agricultural activity is practised, a series of economic rent lines will appear, as in Figure 10.5. Where lines cross, one activity replaces another as the more profitable. The result is that agricultural activities are zoned around the central market and that a series of concentric rings emerges. In other words, the simple concept of economic rent as a function of distance results in a situation of spatial zonation.

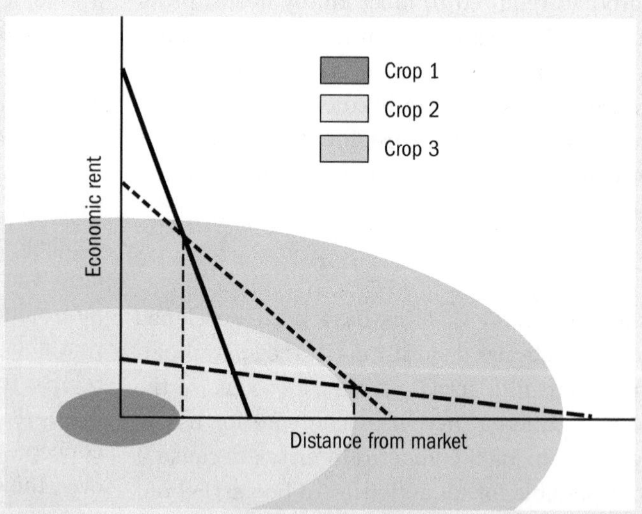

FIGURE 10.5 | Economic rent lines for three crops and related zones of land use

be like in some hypothetical ideal situation. For example, a key assumption is the unrealistic but useful concept of the economic operator, an individual who responds immediately to any price changes and aims to maximize profit (sometimes called economic man and one component of rational choice theory).

Rooted in classical economics, the economic operator concept is a normative theory according to which each individual (economic operator) minimizes costs and maximizes profits as a result of perfect knowledge and a perfect ability to use such knowledge in a rational fashion. There is, of course, no such person in the real world—none of us is blessed with the required "omniscient powers of perception" and "perfect predictive abilities" (Wolpert, 1964: 537). Nevertheless, this unrealistic concept is useful because it allows us to generate hypotheses that are not encumbered by the complexities of human behaviour. In fact, the economic operator represents one extreme on a continuum of possibilities that range from optimization to minimal adaptation. We know that most human behaviour falls between the two extremes and can be described as satisficing behaviour—aimed at satisfying the individual rather than optimizing the situation. The concept of satisficing behaviour is a basic alternative to rational choice theory.

The Problem

To tackle the complex question of which agricultural activities should be practised where, von Thünen used a highly original method of analysis (Johnson, 1962: 214). He deliberately excluded several factors known to be relevant and proceeded on the basis of a framework that he called the isolated state:

> Assume a very large city in the middle of a fertile plain which is not crossed by a navigable river or canal. The soil of the plain is uniformly fertile and everywhere cultivable. At a great distance from the city the plain shall end in an uncultivated wilderness by which the state is separate from the rest of the world.

The Assumptions

The isolated state model allowed for greatly simplified descriptions. Its assumptions may be elaborated as follows:

1. There is only one city, that is, one central market.
2. All farmers sell their products in this central market.
3. All farmers are profit maximizers—economic operators.
4. The agricultural land around the market is of uniform productive capacity.
5. There is only one mode of transportation by which farmers can transport products to market.

In effect, von Thünen was not so much excluding as holding constant such key variables as physical environment, humans, and transport. Only one—distance from market—was allowed to vary. This was a distinctive and highly original contribution and in accord with his definition of economic rent, which centres on distance from market.

The Answer

Given these assumptions, von Thünen asked, "How will agriculture develop under such conditions?" (Johnson, 1962: 214). The conditions are those of a controlled experiment isolating a single causal variable, distance from market. The answer is straightforward. Agricultural activities are located in a series of concentric rings around the central market, one zone for each product with a market demand. To determine the location and size of each product zone, von Thünen used data from his own estate on production costs, market prices, transport costs, and so on. The principle of concentric zones emerges from the concept of the isolated state, while the number, size, and content of zones is a function of particular places and times.

Combining the isolated state concept, his own data, and the economic rent concept, von Thünen came to two conclusions: zones of land devoted to specific uses develop around the market, and the intensity of each specific land use decreases with increasing distance from the market. The first conclusion, the crop theory, is summarized in Figure 10.6, which illustrates how product location is affected by perishability and weight as they affect transport cost. Remember that, while the activities depicted in this figure reflect north Germany in the early nineteenth century, the principle of zones applies generally.

In zone 1 are market gardening and milk production. Both fresh vegetables and dairy products are perishable, give high returns, and have high

Schools of Economics

normative
Focusing on what ought to be rather than what actually is; in normative theory, the aim is to seek what is rational or optimal according to some given criteria.

economic operator
A model of human behaviour in which each individual is assumed to be completely rational; economic operators maximize returns and minimize costs.

rational choice theory
The theory that social life can be explained by models of rational individual action; an extension of the economic operator concept to other areas of human life.

satisficing behaviour
A model of human behaviour that rejects the rationality assumptions of the economic operator model; assumes that the objective is to reach a level of acceptable satisfaction.

transport costs. Accordingly, they have a steep economic rent line reflecting a high Ef value (see Box 10.2) and a high intercept on the rent axis reflecting a high $E\ (p - a)$ value. In zone 2 are forestry products (used for fuel and building). Forestry could command a location close to market because of the high transport costs involved in moving such bulky products. Rye, the principal commercial crop, is located in zones 3, 4, and 5. The differences between the zones reflect differences in the intensity of rye cultivation: zone 3 uses a six-year rotation, zone 4 uses a seven-year rotation, and zone 5 uses a three-field system. Livestock ranching—producing butter, cheese, and live animals—is located in zone 6. Beyond zone 6 is wilderness that can be brought into production if spatial expansion is needed in the future to meet increased product demand.

Von Thünen's second conclusion is known as the intensity theory. Here von Thünen's own production data indicated that, for any given product, the intensity of production decreases with increasing distance from market. The easiest example to identify is the cultivation of rye in zones 3, 4, and 5, but the principle also applies inside specific zones.

Von Thünen was well aware that his work was a simplification. Accordingly, he modified several assumptions when elaborating on the basic hypotheses. The assumption of a single mode of transportation to market was relaxed with the introduction of a navigable waterway serving to elongate zones (Figure 10.7). Relaxing the assumption of a single market and allowing a transportation network including roads resulted in a more complex pattern of zones (Figure 10.8). Von Thünen also acknowledged that differential land quality would mean intensified use of the better areas.

DISTANCE, LAND VALUE, AND LAND USE

Numerous geographers tested the von Thünen hypotheses or the general relations with distance in the context of agricultural patterns. The seminal work in this area is Chisholm's (1962)

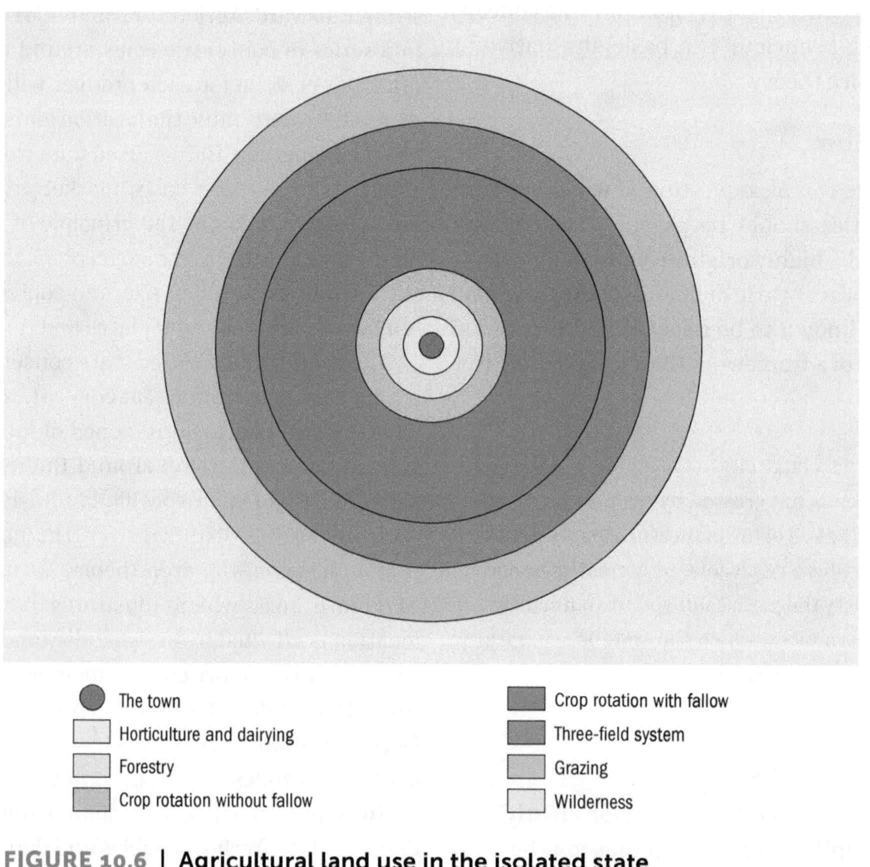

FIGURE 10.6 | Agricultural land use in the isolated state

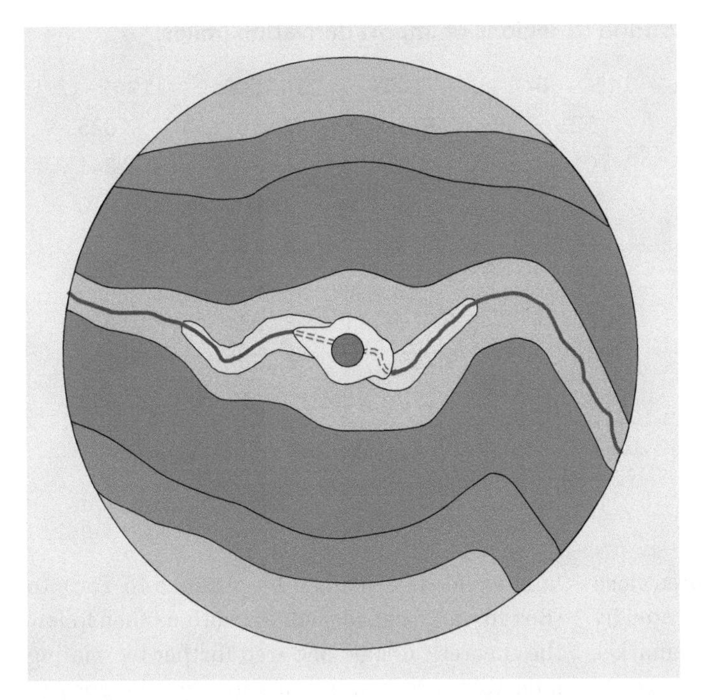

FIGURE 10.7 | Relaxing a von Thünen assumption
A navigable waterway elongates the agricultural zones.

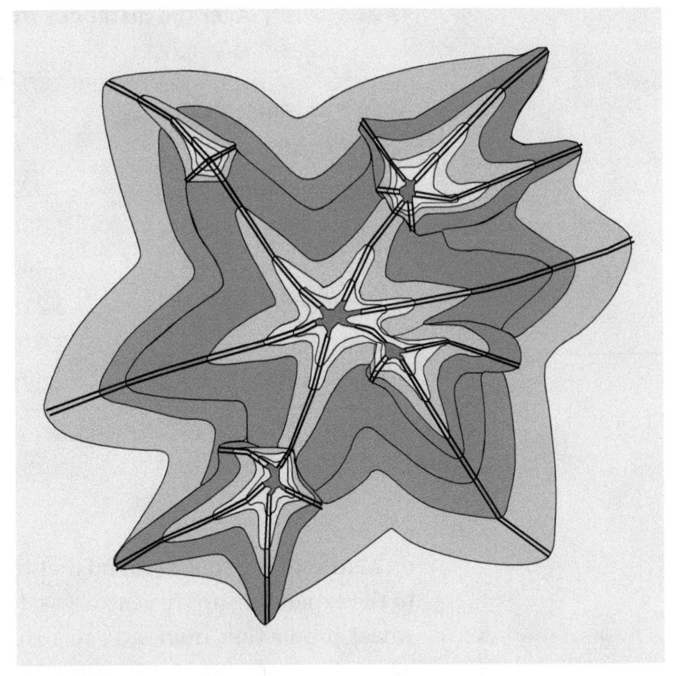

FIGURE 10.8 | Relaxing two von Thünen assumptions
Multiple markets and a transport network create a more complex model.

remarkably thorough pioneering study of the scale ramifications of the concepts, which is an excellent survey of world evidence pertaining to this topic.

Continental and World-Scale Studies

Von Thünen's logic is relevant at the continental and world scales. Several geographers have proposed concentric patterns of land use for North America, Europe, and the world. When Muller (1973) analyzed the nineteenth-century United States, he found an expanding system of concentric rings. Jonasson (1925) described European agriculture in terms of an inner ring of horticulture and dairying, followed by zones of less and less market-oriented activity. On the world scale, Schlebecker (1960) traced the evolution of a world city from Athens to Western Europe to Western Europe–eastern US.

The growth of such a world market and the related expansion of the supply area can be analyzed by reference to import data. Over time, the mean distances over which agricultural imports have been moved have increased (Table 10.1). This spatial expansion of the export system has not been haphazard: it is clearly zonal in character. The changing details of the import system reflect a modified von Thünen model. Peet (1969) has detailed this situation for wheat, showing that, in many instances, the arrival of wheat was preceded by extensive animal rearing (the Canadian prairies are an exception). In all cases, wheat cultivation was associated with major population growth, transport expansion, and the rise of an urban network. Spatially, wheat imports to Britain came first from Britain itself, then the Baltic region, followed by the United States, western Canada, and Australia.

Regional-Scale Studies

Von Thünen's own work is a regional historical study. Therefore, it is hardly surprising that it is applicable in a general historical context. In fact, Von Thünen was not the first to identify the existence of patterns in land-use zones. Two descriptions, one in 1576 and the other in 1811, identified four land-use zones close to London, namely, zones of clay pits, cattle pastures, market gardening, and hay.

Many regional analyses of agriculture in nineteenth-century North America highlight the von Thünen principles in a spatial and temporal framework. One study of an area centred

TABLE 10.1 | Average distances from London to regions of import derivation (miles)

	1831–5	1856–60	1871–5	1891–5	1909–13
Fruit and vegetables	0	521	861	1,850	3,025
Live animals	0	1,014	1,400	5,680	7,241
Butter, cheese, eggs	422	853	2,156	2,590	5,020
Feed grains	1,384	3,266	3,910	5,213	7,771
Flax and seed	2,446	5,229	4,457	6,565	6,275
Meat and tallow	3,218	4,666	6,018	8,093	10,056
Wheat and flour	3,910	3,492	6,758	8,286	9,574
Wool and hides	4,071	14,207	16,090	17,811	17,538

Note: The distances shown here are for imports only; British agricultural production continued to be a major source in the London markets.
Source: After Leonard, S. 1976. "Von Thünen in British Agriculture." *South Hampshire Geographer* 8: 28.

on Madison, Wisconsin, found that the area close to the expanding urban centre was dominated by wheat production from 1835 to 1870, by market gardening from 1870 to 1880, and by dairy farming after 1880, demonstrating that zonal change for any one period becomes temporal change as the urban centre increases in size (Conzen, 1971). A study of western New York state focused on the related idea that a prolonged period of falling transport costs is associated with increasing regional specialization; Leaman and Conkling (1975) showed that the expanding transport network of roads and canals allowed a frontier area (initially akin to von Thünen's wilderness) to rapidly become first an area of wheat cultivation and then an area of regional specialization related to market access.

The Southern Ontario Frontier

Both American studies found that areas experiencing increased settlement are transformed from non-agricultural areas into areas of subsistence agriculture and finally into a series of commercial agricultural types. Similar conclusions were reached in a study of nineteenth-century southern Ontario that focused explicitly on testing the von Thünen hypotheses (Norton and Conkling, 1974).

The area analyzed lay north of the then-emerging centre of Toronto and can be regarded as one segment of the market area of that centre. Using data for 1861, the relationship between distance and land value is measured at r being 20.6502, producing an r^2 of 42.64 and suggesting that 42 per cent of the spatial variation in agricultural land values is explained by distance to Toronto (Box 10.3 and Figure 10.9). The authors then follow the theoretical logic one step further by relating land values to a series of variables: distance to Toronto, distance to smaller regional market centres, distance to major lines of communication, and a measure of land capability for agriculture. The statistical result is $R = 0.7367$, producing an R^2 of 0.5427, which means that 54 per cent of the spatial variation in land values is explained by the set of independent variables.

Study of this pattern of land use showed an area divided into two basic zones: close to Toronto, a zone of fall wheat, peas, and oats, with production oriented to the Toronto market and an outer zone with much unoccupied land, used for spring wheat, potatoes, and turnips—an area anticipating commercial development. The incipient zonation in place by 1861 was to become fully realized in the later nineteenth century. This Ontario study also analyzed the intensity hypothesis and showed that the intensity of the key commercial crop (fall wheat) decreased with increasing distance from Toronto.

Uruguay

Griffin (1973) showed that Uruguay possesses many of the essential characteristics of the isolated state, and Montevideo has the main attributes of the theoretical central city. Actual land use there (Figure 10.10) closely resembles the von Thünen model of ideal land use. The first zone is used for horticulture and truck farming, followed by a zone of dairying, and finally a cereal zone, despite significant variations between model and

Examining the Issues

BOX 10.3 | Correlation and Regression Analysis

"A continuous theme in geographic research is that of analyzing the degree and direction of correspondence among two or more spatial patterns or locational arrangements" (King, 1969: 117). A traditional approach to the problem is to use map overlays—a procedure facilitated by geographic information systems. Another approach is to apply a set of quantitative procedures known as correlation and regression analysis.

By simple correlation and regression, we may determine the degree of association between two variables. The correlation coefficient, r, is a statistical measure of the relationship and may vary between −1 (a perfect inverse relationship) and +1 (a perfect direct relationship). Squaring the r value gives a coefficient of determination, r^2, which may be expressed as a percentage and interpreted as the percentage variation in one variable explained by the other variable. A related set of procedures, known as multiple correlation and regression analysis, allows one variable to be related to two or more variables. In this case, the correlation coefficient is identified as R.

The procedure by which these results are obtained is as follows:

$$y = a + bx$$

where
y = dependent variable
x = independent variable
a = y intercept
b = slope of the line

The economic rent equation is the simplest for predictive purposes. A method known as least squares is used to calculate the best-fit line on the graph of x plotted against y. Once the best-fit line is calculated, we can measure how well it fits. If all points fall on the line, it is a perfect fit and $r = -1$ or $+1$; more typically, points are not located on the line and r lies between −1 and +1. In cases where y does not vary with changes in x, r is close to 0 (Figure 10.9).

Procedures are also available to test the significance of r. All the results reported in the text are statistically significant.

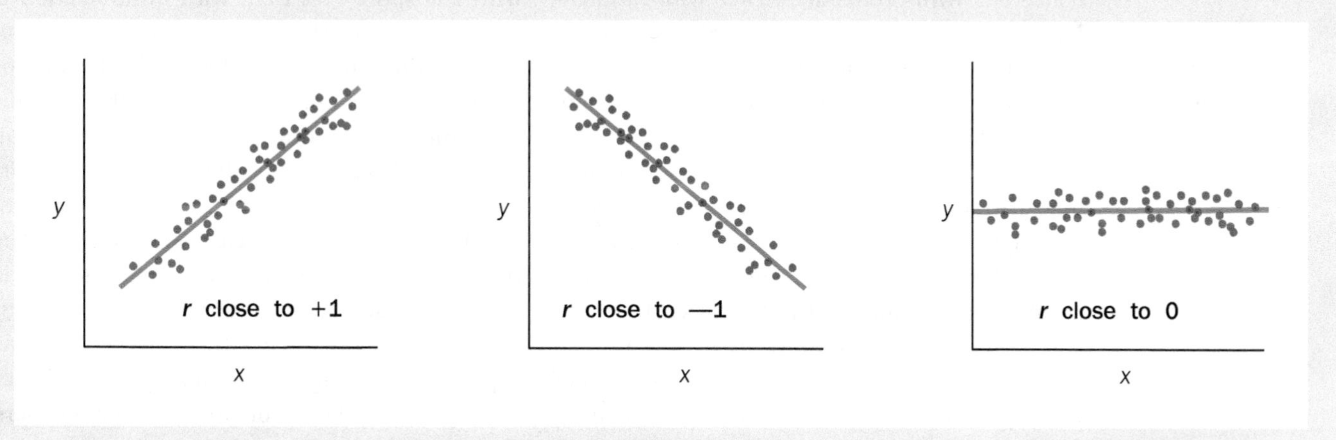

FIGURE 10.9 | Scatter graphs, best-fit lines, and r values

reality in variables such as soil fertility, transport efficiency, and ethnicity.

Ethiopia

More surprising than the previous examples is considerable evidence suggesting von Thünen-type spatial patterns in less developed countries, where much agricultural activity is subsistence in orientation. A regional pattern of land use is often a mix resulting from the different agricultural decisions made by relatively well-off commercial farmers and relatively less prosperous subsistence farmers—groups that necessarily have different economic outlooks. The commercial farmers are

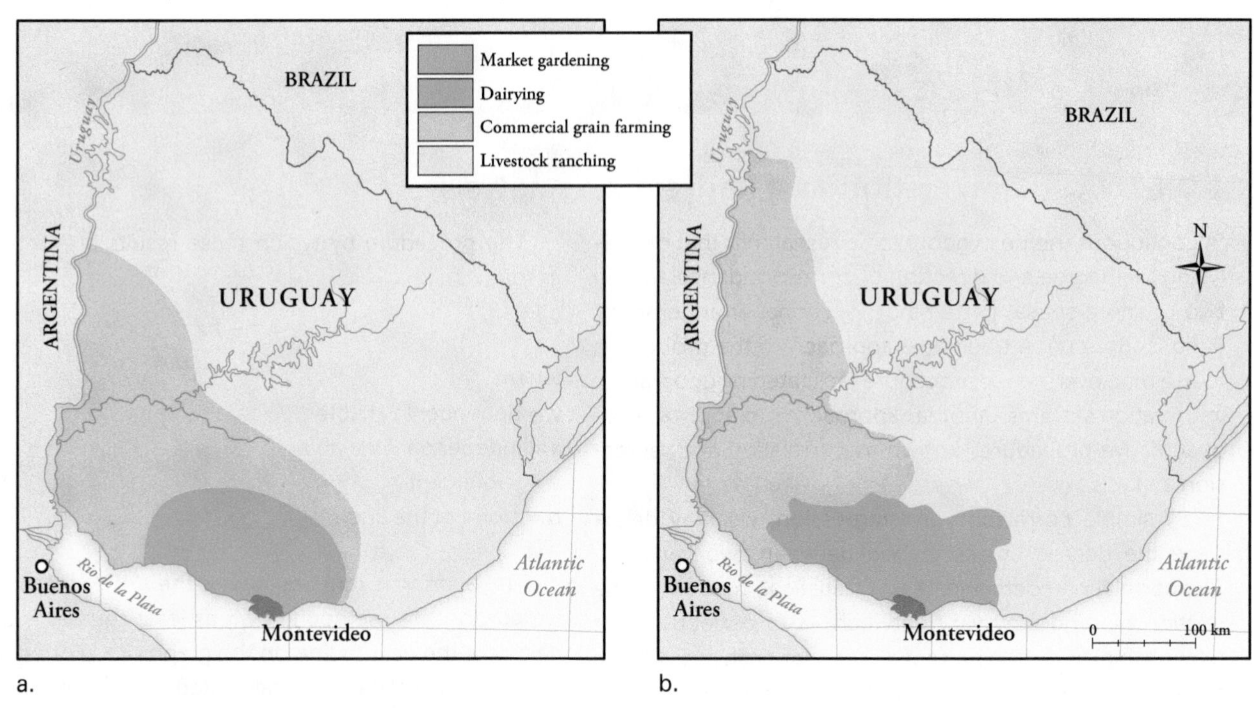

FIGURE 10.10 | Agricultural land use in Uruguay
(a) as predicted by von Thünen theory; (b) actual.

concerned with profits and hence with rationalizing land use according to economic constraints, whereas the subsistence farmers are limited in their activities by lack of capital and inadequate size of holdings.

Despite these complications, von Thünen-type patterns may emerge. Horvath (1969) analyzed land use *circa* 1964 in the vicinity of Addis Ababa, Ethiopia. Although this highland area is more physically and culturally varied than the von Thünen model assumes, basic von Thünen patterns were evident. These included an area of eucalyptus forest around the city that was used for fuel and building material (an orientation outward along roads was visible), vegetable cultivation within that forest rather than in a separate zone, and a zone of mixed farming that included commercial and subsistence activity beyond the first forest and vegetable zone. Horvath concluded that this was an example of incipient zonation, similar to the southern Ontario example.

Local-Scale Studies

Von Thünen's theory has also been tested on the scale of individual villages and farms. In such analyses, the concept of distance is equated less with transport cost than with minimization of movement in relation to time expended. Various investigations of villages have tested modified versions of the crop and intensity theories and confirmed their relevance. Von Thünen-type patterns are also evident in the immediate vicinity of large cities. Even though the considerations affecting land use close to cities are not those identified by von Thünen, the spatial consequences in terms of zonation are very similar (Sinclair, 1967). One such consideration is the prospect of urban expansion, specifically the anticipation of such expansion and hence of profits to landowners. The closer the land is to the city, the greater the anticipation of urban expansion and the lower the incentive for capital investment in the land. This argument helps us understand why so many areas on the outskirts of cities are either vacant or put to some clearly temporary use.

Overall, the message is clear. The method pioneered by von Thünen—especially the deliberate simplifications producing an ideal world against which reality can be assessed—is of value to agricultural geographers at a variety of spatial scales and, as we shall see in later chapters, also to urban and industrial geographers.

DOMESTICATING PLANTS AND ANIMALS

The contemporary pattern of world agriculture is the still-changing product of a long history: "The imprint of the past is still clearly to be seen in the world pattern of agriculture. To understand the present, it is essential to know something of the evolution of the modern types of agriculture" (Grigg, 1974: 1).

Early Domestication and Diffusion

It can be argued that animal and plant domestication was the most crucial event, or rather collection of events, in all human history. As noted in Chapter 4, agriculture was a technological advance that permitted population growth, new forms of social organization, the onset of urbanization, and the development of specialized occupations. By placing humans in close contact with animals (and with each other), it also led to many of the diseases that have helped shape our lives.

Agriculture originated in the domestication of plants and then animals. A domesticated plant is deliberately planted, raised, and harvested by humans; a domesticated animal depends on humans for food and, in many cases, shelter. As a consequence, domesticated plants and animals differ from their non-domesticated counterparts. From the human perspective, the domesticates are superior in that, for example, they bear more fruit or provide more milk: they have been deliberately engineered for these characteristics through selective breeding over long periods.

What was the first species to be domesticated? Where and when? How and why? In all likelihood agriculture began some 12,000 years ago—a 2006 discovery in the Jordan Valley of figs that were grown through human intervention is dated between 11,200 and 11,400 years ago. From several initial centres, agriculture diffused to other areas, gradually replacing the main pre-agricultural economic activities of hunting, gathering, and scavenging. Although it was traditionally thought that agriculture first emerged in a few Asian centres, most notably Southwest Asia (present-day Iraq), it has since been accepted that it evolved independently in several other places. Figure 10.11 maps the likely centres of domestication and suggests directions of early agricultural diffusion (compare this map with the ideas summarized in Figure 6.2, which shows the shape of continents). It seems likely that people began domesticating plants and animals in response to specific circumstances in each place.

Farmland and housing north of Toronto exist side-by-side along Major Mackenzie Drive, west of Highway 400. Scenes like this are increasingly common around the world as residential developments encroach on valuable agricultural land.

We are learning more and more about plant and animal domestication with advances in genetic research. Most notably, such research is providing support for the idea that many plant and animal species were domesticated more than once, in different places at different times. Cows, pigs, sheep, goats, yaks, and buffalo each were domesticated at least twice, dogs at least four times, and horses on even more occasions. Horses were first domesticated in what is now northern Kazakhstan, where they were ridden and used for meat and for milk, by about 3500 BCE. They represent a highly significant domestication as they greatly increase human mobility. The fact that many different groups of people came up with the idea of domesticating animals suggests that the reason some species are not domesticated is not that no one tried but that they proved unsuitable candidates.

It is important to appreciate that there is nothing inherently difficult or complicated about the process of domesticating plants and animals; no knowledge of genetics is needed. The process simply makes use of artificial selection as opposed to natural selection. Natural selection results in the survival and reproduction of those plants and animals that are best able to cope in a particular

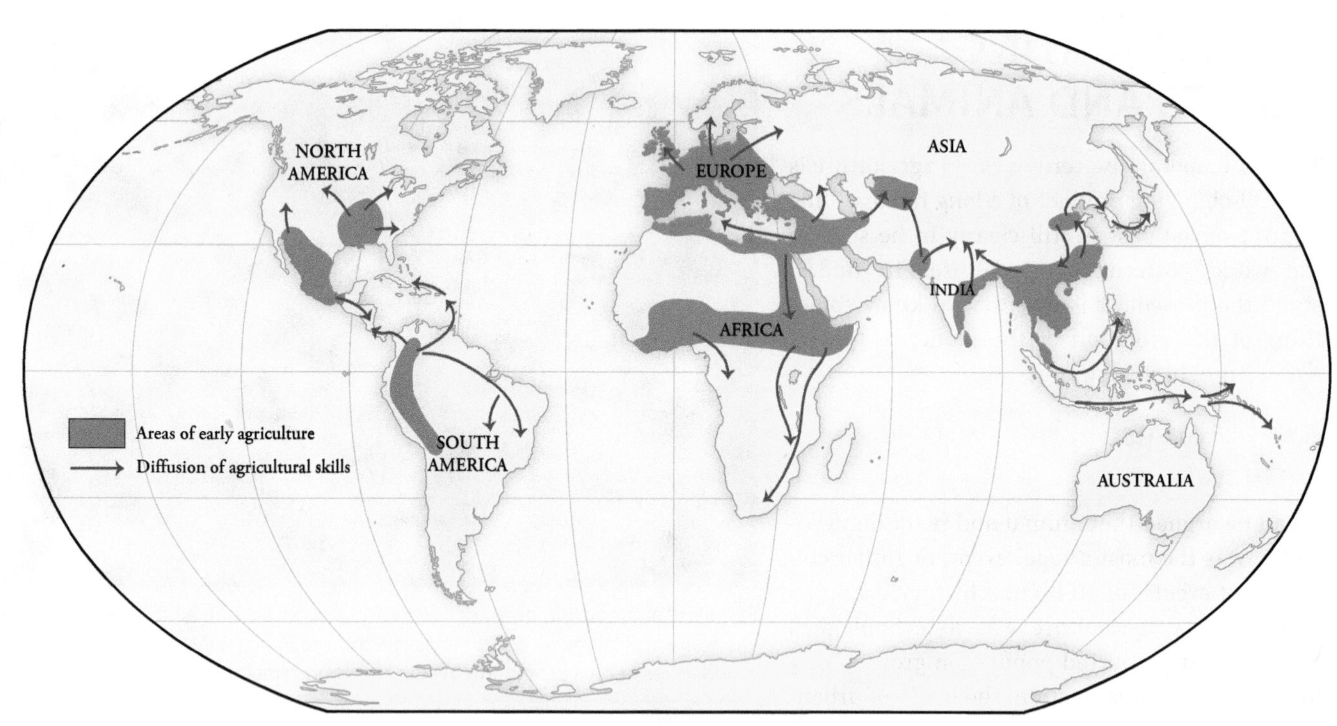

FIGURE 10.11 | Areas of agricultural domestication and early diffusion

Source: Stearns, P. N., M. Adas, S. B. Schwartz, and M. J. Gilbert. 2007. *World Civilizations: The Global Experience, Combined Volume, Atlas Edition, 5th.* New York: Pearson, 3. © 2008. Printed and electronically reproduced by permission of Pearson Education, Inc., New York, NY

environment. Artificial selection involves humans allowing certain plants and animals to survive and breed because they possess features judged desirable by humans, for example, plants with bigger seeds or animals that are less aggressive. The individual members of plant and animal species that humans favoured would reproduce and pass on the favoured characteristics, while the individual members with less desirable traits would be eliminated gradually.

Possible Causes of Domestication

Pleistocene
Geological time period, from about 1.5 million years ago to 10,000 years ago, characterized by a series of glacial advances and retreats; succeeded by the Holocene.

Animal species most suited to domestication were those already living in groups and having some form of hierarchical organization. Humans favoured animals that were valued for meat, milk, companionship, protection, and transport. But why domesticate at all?

Various explanations have been proposed. According to Sauer (1952), casual experimentation with plant and animal breeding probably began in a well-endowed environment that permitted a more sedentary way of life and an amount of leisure time. In his view, Southeast Asia, mainly the wooded hilly areas away from possible floods, was the most likely hearth area.

Perhaps the most satisfactory explanation combines a number of relevant demographic, environmental, and cultural variables. The key idea, not yet proven, is that either climate change or population pressure (or a combination of the two) might have prompted a search for new food supplies. This hypothesis is based on the idea that, although the process of domestication is not difficult, it can require a great deal of work and therefore is not especially attractive. Most societies did not need to increase food supplies because they had strategies to keep the population below carrying capacity. However, a change in climate at the end of the Pleistocene Era might have caused certain areas to become unusually rich in food resources, which would have made a more sedentary way of life possible and hence permitted an increase in population. Eventually, a larger population would have necessitated movement into marginal areas and competition for space, prompting the development of new strategies for food supply, specifically domestication.

Whatever the details, agriculture had become a major economic activity by about 500 BCE. Grigg (1974: 21–3) has identified five core areas and agricultural types evident by this date (Box 10.4).

Around the Globe

BOX 10.4 | Agricultural Core Areas, c. 500 BCE

1. Southwest Asia
   Basic crops: Wheat, barley, flax, lentils, peas, beans, vetch.
   Basic animals: Sheep, cattle, pigs, goats.
   Subtypes: Irrigated farming in the Nile Valley, Tigris and Euphrates, Turkestan, and Indus Valley; dry farming (without irrigation) was the basic Southwest Asian complex mix of agricultural activities; Mediterranean agriculture was a mix of cereals and tree crops such as figs, olives, and grapes; in Northern Europe, oats and rye were added to basic crops.
2. Southeast Asia
   Two types originated on the Southeast Asian mainland. Tropical vegeculture involved growing taro, greater yam, bananas, and coconuts, and pigs and poultry also were domesticated; this was a form of shifting agriculture. (Practised in tropical forest areas, this type of agriculture involves regular movement from one cut-and-burned area to another because of the rapidly declining soil fertility resulting from crop cultivation; also known as swidden agriculture, or slash-and-burn agriculture.) Wet rice cultivation gradually displaced the first type.
3. Northern China
   Based on local domesticates (e.g. pigs, foxtail millet, soy beans, and mulberry) and imported domesticates (e.g. wheat, barley, sheep, goats, and cattle).
4. Africa
   Two agricultural types, both shifting: tropical vegeculture (based on yams) and cereal cultivation (millets and sorghum).
5. America
   Two agricultural types, both shifting: root crops and corn/squash/beans complex. No livestock.

This listing must not disguise the fact that just three of the domesticated plants—wheat, rice, and corn—are of overwhelming importance in that, together, they have provided most of the calories that enabled populations to increase. Wheat was the oldest and most widespread of these.

Since about 500 BCE, links have been established between areas and types of agriculture. Of particular relevance are those agricultural changes prompted by European expansion from the fifteenth century onward.

THE EVOLUTION OF WORLD AGRICULTURAL LANDSCAPES

Contemporary agricultural landscapes are the outcome of a long evolutionary process. Since domestication began, plants and animals have diffused widely, far beyond the few areas where the wild varieties of the original species evolved. Wheat, for example, had diffused as far west as Ireland and as far east as China by about 2000 BCE, and this spread also meant the diffusion of related technologies and ways of life generally. One of the most important technologies was the plough, possibly first used about 4000 BCE. A major innovation took place in China in about 300 BCE with the invention of the horse collar, which allowed the animal to pull a greater weight.

Many agricultural landscapes emerged in response to European overseas movement and/or to the demands of new population concentrations. In Europe, for example, some agricultural crops and methods prior to 1500 were the result of diffusion from the Southwest Asian hearth areas. After 1500, two American crops, potatoes and corn, were widely adopted, and other agricultural changes—such as intensification, expansion of cultivated area, and increasing commercialization—reflected increased market size. Commercialization was facilitated by technical advances in transportation, particularly after the mid-nineteenth century. Agricultural commercialization at that time was merely one component of substantial growth and change in the larger world economy.

Five principal technological advances have transformed, or are transforming, agricultural landscapes around the world:

1. A second agricultural revolution associated with the onset of the Industrial Revolution in the eighteenth century.
2. The development of nitrogen fertilizers in the early twentieth century.
3. The "green revolution" that began in the mid-twentieth century.
4. The biotechnology revolution that began in the late twentieth century and that, despite much opposition, is proceeding apace.
5. The ongoing transition in some areas from ploughing the soil prior to planting to use of no-till strategies.

Following discussion of these advances, we briefly examine an unforeseen and potentially serious consequence of some of them, namely, lost animal and plant species.

A Second Agricultural Revolution: England after 1700

The most radical changes in agricultural activities since the beginning of domestication have occurred in the larger European world, helping to create what is today "the great gulf in productivity between the present agricultures of Asia, Africa and much of Latin America, and western Europe, North America and Australia" (Grigg, 1974: 53). In most cases, those changes did not begin until the seventeenth century and it was only in the mid-nineteenth century that commercial agriculture became established on a large scale and the present pattern of agricultural activity began to evolve. The link between industrial progress and agricultural progress seems clear.

Thus, a period of significant agricultural change—significant especially because of its links with other technological, social, and economic changes—began in England about 1700. Sometimes described as a second agricultural revolution, it involved a series of changes:

1. Development of new farming techniques, including introduction of fodder crops and new crop rotations.
2. Increases in crop output because of improvements in productivity.
3. Introduction of labour-saving machinery, most notably the 1701 invention in England of a seed drill that enabled much higher harvesting rates for a given number of seeds sown and, later in the nineteenth century, the tractor.
4. The ability to feed a growing population.

The first three of these represented changes in what Marx called the forces of production. But the new "revolution" also involved changes in the relations of production associated with the decline of the feudal system and the emergence of capitalism. Among the institutional changes was the imposition of "enclosure": the subdivision of large, open arable fields and large areas of pasture or wasteland into small fields divided by hedgerows, fences, or walls. Enclosure was part of a major social transformation from communal to private property rights. A second and related institutional change was the creation of farms that typically were rented from large landholders by capitalist farmers who employed labourers to work the land.

This second agricultural revolution—perhaps better described as an agrarian revolution—was one of the principal means by which England was able to begin supporting a growing population. "The transformation of output and land productivity enabled the country to break out of a 'Malthusian trap,' allowing the population to exceed the barrier of 5.5 million people for the first time. Rising labour productivity ensured that extra output could be produced with proportionately fewer workers, so making the industrial revolution possible" (Overton, 1996: 206).

Nitrogen Fertilizers

Perhaps the most important development after the second agricultural revolution was the introduction of nitrogen fertilizers. An essential nutrient for the cereal crops that have been staples for most people in most parts of the world since the beginnings of agriculture, nitrogen is added naturally to soil through rainfall. But rainfall alone does not add enough nitrogen, and it was soon discovered that growing a cereal crop on the same land year after year impoverished the soil and ultimately reduced yields. The favoured solution was to plant legumes along with the cereal crops to replenish the soil's nitrogen content: peas and lentils in the Middle East, beans and maize in the Americas, soy and mung in Asia, and peanuts in parts of sub-Saharan Africa. Other strategies to prevent soil impoverishment included crop rotation and leaving land fallow for a season.

By the nineteenth century, the demand for cereals was increasing so rapidly that concerted efforts were made to find new sources of nitrogen. One readily available source in much of Europe was horse manure. In Asia human waste

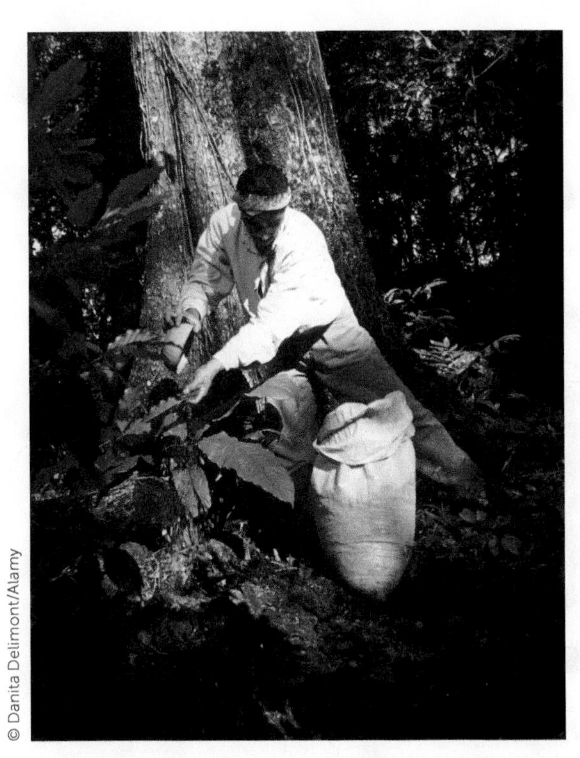

A Tzeltal Indian adds organic fertilizer to his shade-grown coffee in the Lacandon Jungle in Chiapas, Mexico.

was used. Another source was discovered about 1843—the island of Ichaboe off the coast of southwest Africa was covered in about 8 m (26 feet) of guano (seabird excreta), all of which was removed by 1850. Another source was discovered in 1850 off the coast of Peru, on a few arid Pacific islands with whole cliffs made of guano. Guano mining quickly proved a very profitable business there as well, and it is estimated that some 20 million tons were transported by ship from these islands to Europe between about 1850 and 1870. Once these sources were depleted, traders began exporting a fossil nitrate called caliche, found in desert areas in Chile, for use as a source of nitrogen. Even with the help of these organic fertilizers, however, crop yields were still too limited to permit human populations to grow beyond a maximum density of about five persons per hectare. One of the principal constraints on population growth during the nineteenth century was the lack of a reliable source of nitrogen sufficient to enable farmers to increase their cereal yields significantly.

This need was filled by Fritz Haber and Carl Bosch, two chemists who recognized that the vast ocean of air that surrounds us contains large quantities of nitrogen (Smil, 2001). Their solution was ammonia synthesis—a process that converts atmospheric nitrogen to ammonia, which is then used to produce synthetic nitrogen fertilizers. This "Haber–Bosch process" was discovered in the early twentieth century, and nitrogen fertilizer production began on a commercial scale in 1912. Nitrogen fertilizers were used widely by the 1950s and now produce about two million tons of ammonia each week, providing almost 100 per cent of the inorganic nitrogen used in agriculture. China is the single largest producer. In an important sense, humanity is dependent on nitrogen fertilizers. This technological breakthrough, more than any other, permitted world population to rise from 1.6 to 6.1 billion in the twentieth century. "In a very literal sense, without [this] invention, half the people alive today could not have been born" (Harman, 2004: 36).

At the same time, nitrogen fertilizers have been associated with environmental damage—including soil and water contamination, increasing soil acidity, and the release of nitrous oxide (a potent greenhouse gas) into the atmosphere—as well as increased risk for some cancers. Concerns such as these have prompted many to argue for a return to organic farming (Box 10.5).

The "Green Revolution"

Another major agricultural advance in the twentieth century was what came to be known as the green revolution—the rapid development of improved plant and animal strains and their introduction to the economies of the less developed world. As we have seen, such improvements had been ongoing in the Western world since the eighteenth century, as one aspect of larger economic and technological changes. Until the 1960s, however, the less developed world did not benefit from these changes, and there were many instances, especially in Asia, of widespread hunger, malnutrition, and dependence on food aid. The initial breakthrough came as a result of scientific work in the more developed world.

In the 1960s the Rockefeller and Ford foundations co-operated in the creation of an agricultural research system designed to transfer technologies to the less developed world. This system produced genetically improved strains of two key cereal crops, rice and wheat—often labelled high-yielding varieties (HYVs)—and eventually other crops such as beans, cassava, maize, millet, and sorghum were also improved. Although

Examining the Issues

BOX 10.5 | Organic Farming

Critics of contemporary conventional farming strategies argue that they should be replaced by more sustainable and ecologically more appropriate methods. Collectively, the alternative methods they propose differ from conventional farming in three ways (Atkins and Bowler, 2001: 68):

1. Alternative methods are less centralized: There are more farms, and marketing is local and regional rather than national and international.
2. Alternative methods emphasize the community rather than the individual: The focus is on co-operative activity, using labour rather than technology when appropriate, and giving due consideration to all costs, material and non-material, as well as moral values.
3. Alternative methods are less specialized: Farming is seen as a system of related activities rather than as a series of individual components.

In short, alternative farming involves a completely different value system, in which farming is regarded as a way of life and not simply a matter of producing a product for the market.

Organic farming is the best known of these methods. About 1 per cent of the world's agriculture is organic, mostly grassland and grazing; 15 per cent of the organic land is used for arable agriculture, mostly cereals. Although it remains a minor component of the agricultural sector in most countries, it is viewed favourably by many and is becoming increasingly popular. Organic agricultural land is highest in some European countries, with 19.7 per cent of all agricultural land in Austria, 15.6 per cent in Sweden, 12 per cent in Switzerland, and 11.5 per cent in the Czech Republic. Several other European countries have more than 5 per cent, Australia has 3 per cent, Canada has 1 per cent, and the United States and Japan have less than 1 per cent. In 2011 Canada had 3,713 organic farms (1.8 per cent of the total), most of which grow field crops, typically buckwheat,

Organic bell peppers grown in an Israeli greenhouse in the Jordan Valley.

the new strains tend to have lower protein content than their predecessors, they produce higher yields, respond well to fertilizer, are more disease- and pest-resistant, and require a shorter growing season. Besides genetic improvements, the green revolution involved expanded use of fertilizers, other chemical inputs, and irrigation. In many cases, the adoption of HYVs and other technologies quickly doubled crop yields—hence the movement's name.

The new strains and technologies allowed some farmers to grow enough not only to subsist but also to market surplus production, raising farm incomes and thus stimulating the rural non-farm economy as well. Increasing incomes and lower prices led to better nutrition, with higher calorie consumption and more diversified diets. These positive impacts have been most evident in Southeast Asia and the Indian subcontinent (Box 10.6).

Despite undoubted successes, the results have not always met expectations. As with efforts to introduce population control policies, full acceptance has sometimes been prevented or delayed by cultural and economic factors. Peasant farmers are usually conservative and reluctant to abandon traditional practices. To persuade them to accept new plant strains, governments have sometimes found it necessary to offer subsidies and guaranteed prices. Perhaps inevitably, the agricultural changes associated with the green revolution have had some unfortunate consequences, and criticisms are plentiful.

barley, and wheat; others grow fruits and vegetables. About 40 per cent of Canadian organic farms are in the prairies.

It appears that buyers of organic food, and also of organic non-food products, are motivated primarily by personal health considerations and concern for the environment. Organic farmers may spend somewhat less than regular farmers on purchased inputs such as pesticides and fertilizers, but their production levels are significantly lower; hence, their prices are higher. The farmers who pioneered organic production, and the consumers who supported them by purchasing their products, were motivated by real environmental and ethical concerns about conventional farming. As the demand for "healthier" food increases, however, it seems clear that a significant part of the organic sector is motivated primarily by profit. With increasing numbers of organic producers turning to more conventional marketing strategies—including supplying national or even international markets—organic agriculture is not always what consumers believe it to be. In effect, producers who adopt the "organic" label without the philosophy may be hijacking it.

Meanwhile, some observers are questioning whether foods produced using organic methods really are better for consumers than foods produced by more intensive methods. Some say that the organic sector simply offers a lifestyle package designed to make consumers feel good about paying higher prices for their food. However, a major 2014 scientific analysis showed that organic foods have more of the antioxidant compounds linked to better health and lower levels of toxic metals (Barański et al., 2014; Carrington and Arnett, 2014). It is noteworthy that yields under organic cultivation are almost 50 per cent less than under conventional strategies and, in a world of continuing population increases and therefore increased demand for food, organic methods might be an unaffordable luxury (Goodall, 2008). To produce the same amount of food we produce now organically would require several times more agricultural land than is currently used.

A recent development with some similarities to the original organic farming movement is fair trade. Still in its infancy, this grassroots social movement appeals directly to people's sense of justice. The idea is that consumers pay a guaranteed price plus a small premium to groups of small producers supplying commodity goods such as coffee, tea, chocolate, and fruit. Prior to the dramatic increases in food prices in 2007, supporters of this movement hoped that fair trade would counter one of the evils often attributed to globalization, namely, that the world price of many commodities had barely risen in 20 years, with the result that many small producers in the less developed world operated at a loss. However, a more significant concern might be that much of the markup on fair trade products goes to the retailer rather than the farmer.

One set of criticisms focuses on economic circumstances. New strains have been most readily adopted by farmers who are better off, who have some capital and relatively large holdings, with the result that, in some cases, improvements have led to the displacement of poor tenant farmers. Further, many of the poorest tenant farmers have been prevented from adopting improved strains because they cannot afford the required fertilizer and pesticides. This has led to increased inequality and exacerbated poverty in some regions. A related criticism is that new strains and techniques have been accompanied by unnecessary mechanization, which has reduced rural wages and increased unemployment.

A second set of criticisms focuses on undesirable environmental consequences. Excessive and sometimes inappropriate use of fertilizers and pesticides has resulted in water pollution, unwanted damage to insect and other wildlife populations, water shortages caused by increased irrigation, and some serious health problems.

Finally, a third type focuses on the broader social and political implications of the green revolution. There is a serious risk that imported knowledge, presumed to be expert, will be accepted at the expense of the local knowledge that has been acquired and successfully applied over centuries. This downgrading of local knowledge is one component of **neo-colonialism**, which in turn may be seen as one aspect of globalization. On balance, it is clear that the green revolution has helped to prevent possibly catastrophic food shortages.

neo-colonialism
Economic relationships of dominance and subordination between countries without equivalent political relationships; often develops after political colonialism ends and the former colony achieves independence but may also occur without prior political colonialism.

Around the Globe

BOX 10.6 | The Green Revolution in India

India depended on food aid for about 20 years after gaining independence in 1947, but green revolution technologies permitted steady increases in food production—although rapid population growth has offset some of the impact of these increases. You will recall from Chapter 5 that India has a population of over 1 billion and that it will shortly overtake China as the country with the world's largest population. Part of the success of the green revolution in India can be attributed to the presence of the Indian Agricultural Research Foundation (IARF), a government-funded organization that was established in 1905. With the adoption of the first new strain of rice in the late 1960s, funding to the IARF increased, and new technologies and a supportive infrastructure combined to put India at the forefront of green revolution advances.

The IARF is headquartered in Delhi and has nine regional research centres, a large library, and a modern mechanized farm. Since the late 1960s its principal goals have been to facilitate the spread and growth of green revolution technologies developed elsewhere while conducting its own research into developing higher-yielding and more disease-resistant varieties, enhancing genes, breeding for multiple cropping and intercropping, analyzing the links between photosynthesis and productivity, and improving the nutritional value of crops. This work has resulted in the development of 64 new crop varieties. In addition, the IARF has worked on related projects, including irrigation strategies, research on fertilizers and pesticides, and the use of manure.

Because India is a large country, diverse both physically and culturally, different strains and technologies are needed in different areas. For example, different physical settings need different environmentally friendly fertilizers. Overcoming cultural resistance is another problem, especially in areas of subsistence farming. Feeding refugees and displaced persons, such as those escaping from floods in Bangladesh and those who fled to India at the time of the 1979 Soviet invasion of Afghanistan, has been an additional challenge for Indian agriculture (Slatford and Fishpool, 2002).

However, it is also clear that—ironically—adoption of new crop strains sometimes increases the dependency of the less developed country, as use of new technology increases the need to import fuel, fertilizer, and pesticides and exposes the adopters to new risks, from environmental damage to rising oil prices. These problems notwithstanding, many observers see a need for additional green revolution strains and technologies in sub-Saharan Africa.

There is increasing evidence in Asia of what might be called a second green revolution, one focused on modifying existing seeds to specific and often difficult environments prone to floods, drought, or high salinity. This is a significant change as it may help to bring very poor people out of poverty.

Biotechnology: Another Agricultural Revolution

As a result of the green revolution, world grain production has increased spectacularly—about 45 per cent between 1970 and 1990. In the future, however, the most significant gains will likely be made through developments in the area of biotechnology. Tissue culturing and DNA sequencing are two of the methods that have helped to develop improved plant and animal varieties. Perhaps the most important—and controversial—aspect of biotechnology, however, is the alteration of the genetic composition of organisms, including food crops. Genetic modification is the process of changing the DNA of living things (plants, animals, or micro-organisms) in a way that does not occur in nature. The basic procedure for producing a genetically modified (GM) crop, developed in the early 1970s, involves introducing genes from other plants or micro-organisms, although recent advances make it possible to manufacture new genes by chemical means (synthetic biology). The insertion of new genes is intended to give the crop a favourable new characteristic. An early example was the creation of a frost-free tomato by adding genes from the flounder, a cold-water fish. Other crops have been modified to make them less vulnerable to pests, weeds, drought, salty soils, or diseases. In other cases, modification has focused on making a plant more nutritious.

The adoption of GM crops has been enormously successful, with the cultivated area increasing every year since they were first commercially available in 1995. By 2014 more than 170 million hectares (420 million acres) of land were growing GM crops (more than 10 per cent of global cropland), a huge increase from the 2000 figure of 28 million hectares (69 million acres). The leading GM crop-producing countries are the United States, Brazil, Argentina, India, and Canada. It is notable that 48 per cent of GM cropland is in less developed countries.

The principal GM crops are soybeans, corn, cotton, canola, and potatoes. Most of the area under GM crops is in the United States, with more than 40 per cent of the country's corn and 60 per cent of its cotton coming from genetically modified varieties. About 90 per cent of soybeans in Argentina come from a genetically modified variety that is tolerant to a specific herbicide, which allows for better weed control and higher yields. General Foods has made a caffeine-free coffee bean, while scientists in the UK have developed a rice strain that produces about 20 times more beta carotene (which the human body converts into vitamin A). Other GM products are regularly introduced. But despite all the promise, and all the successes, biotechnology brings much controversy (Box 10.7).

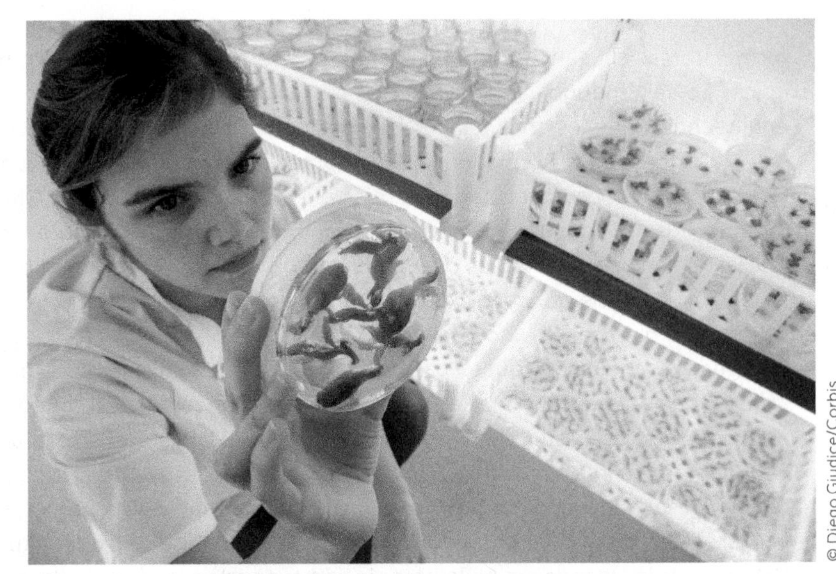
© Diego Giudice/Corbis

A scientist examines soy sprouts at the Institute of Agricultural Biotechnology or Rosario (INDEAR) in Argentina. The institute investigates and produces GMOs that will increase crop yield and be resistant to damage.

In the News

BOX 10.7 | For and Against Genetically Modified Crops

Public perception of genetically modified crops is rarely out of the news, and reports are often conflicting. According to a 2011 report from Friends of the Earth, dissatisfaction with GM crops is rising, with 61 per cent of EU citizens opposed and opposition increasingly evident in the United States, Brazil, India, and Uruguay. In the same year, the UK Food Standards Agency reported that opposition in the UK had declined to 22 per cent from 43 per cent in 2001. Clearly, the facts of this matter are ambiguous.

GM crops met with a mixed reception from the outset in Europe, with public opposition so significant that several countries banned already approved crops, actions that were opposed by the United States, Canada, and Argentina and that are being debated by the World Trade Organization. As of 2015 GM crops are relatively limited in Europe, although some countries grow a variety of GM corn, including Spain, France, Portugal, the Czech Republic, Germany, and Slovakia, while a GM potato has been approved and is being grown in Sweden and Germany. At present, the EU restricts the importation of many genetically modified products on the grounds that the consequences for consumers and the environment are uncertain. However, in 2015, the EU granted national governments increased power to decide whether to grow GM crops.

From the perspective of food production and the world food problem, a significant perspective given continued population growth, the greatest benefits of genetic modification are likely to come with the creation of crops resistant to viruses. Some observers also claim that foods will be produced that offer enhanced nutritional and possibly even medicinal qualities. Proponents argue that GM crops increase yields without decreasing biodiversity through

Continued

further forest clearance and decrease pesticide use and costs. So far, though, the main emphasis has been on engineering varieties that can resist specific herbicides (such as Monsanto's Roundup)—a development that is of less interest to consumers than to herbicide producers and farmers. It is not difficult to see why Monsanto might want to create plants resistant to its own herbicide.

But most of the concern expressed relates to possible risks to human health and the environment. Increasing numbers of interest groups, particularly environmental groups such as Friends of the Earth, are expressing concern about GM crops. Disturbingly, the charity ActionAid has warned farmers in Uganda that eating GM crops can cause cancer, despite the fact that the World Health Organization has ruled that they have no effects on human health. The GM projects in Uganda are philanthropic and supported by the Bill and Melinda Gates Foundation. Nevertheless, GM crops appear to represent everything that environmental groups dislike, at least partly because they are frequently promoted and marketed by companies such as Monsanto; much opposition focuses on the fact that a small number of chemical corporations (not only Monsanto but also Dow and Du Pont) dominate the GM industry. In January 2000, following a debate in Montreal, a global agreement was reached concerning safety rules for genetically modified products. This Biosafety Protocol to the UN Convention on Biodiversity includes rules designed to protect the environment from damage.

On balance, the evidence for health risks from GM crops does not seem to be significant, but several studies show gene flow from a modified plant to another plant that could possibly lead to some form of "superweed." It seems fair to say that scientists currently have insufficient understanding of this possible problem. Of course, it is also fair to say that many of the possible problems with GM crops are no different in principle from problems associated with conventional crops.

In Africa, where the need for increased food production is very real, only South Africa has shown substantial interest in modified crops, growing GM maize, cotton, and soybeans. Egypt and Burkina Faso grow GM maize and cotton, respectively, while Kenya, Tanzania, Uganda, Malawi, Mali, Zimbabwe, Nigeria, and Ghana are conducting field trials of GM maize, rice, wheat, sorghum, and cotton. Other countries, including Malawi, Mauritius, and Zimbabwe, have enacted biosafety laws, removing a major hindrance to GM adoption.

Although some commentators argue that the real problems of global food production are not to be solved by technological advances but by solutions addressing issues of spatial and human inequality (see Chapter 6), it may not be an exaggeration to suggest that developments in biotechnology could prove to be the most significant changes in agricultural production since the first animals and plants were domesticated. Some observers argue that the world faces a choice: either accept GM crops and thus help feed the world, especially through the introduction of drought- and disease-resistant varieties of such staple crops as rice and potatoes, or face the prospect of continuing famines in poor areas of the world. Recall, however, that discussions in Chapter 6 highlighted other reasons for famines (e.g. inefficient governments).

An important general point to bear in mind when debating the pros and cons of GM crops is that the humans who first domesticated plants and animals through selective breeding were causing genetic changes, meaning the genomes of species have been changed by humans for thousands of years through the process of artificial selection. The essential difference today is that modern genetic engineering is able to go much further much more quickly than conventional selective breeding. The technology is very different and the time factor hugely so, but the outcomes are similar.

A balanced assessment of this complex and widely covered topic needs to consider both the possible risks associated with GM production and the possible cost, in terms of feeding the world population, of not adopting this new technology.

No-Till: A Quiet Agricultural Revolution?

Ploughing, or tilling, soil is one of the oldest farming techniques. Turning the soil over either after a harvest or prior to planting is typically seen as beneficial because it helps remove weeds and residue from the previous crop as well as aerates the soil. But increasing evidence suggests that it may not be necessary in many places and, indeed, may be detrimental. When ploughing, farmers usually turn over between about 6 to 10 inches of soil, whereas no-till farming involves making a groove no more than 3 inches deep in which to plant seeds. Ploughing thus results in much more soil disturbance and can be a major cause of soil erosion.

No-till farming was the norm until the invention of the plough and the use of domesticated animals to pull ploughs. Ongoing improvements to plough design facilitated the expansion of agriculture globally, including in many grassland areas such as the North American prairies, where turning the sod required such technology. But does this mean that ploughing continues to be necessary? According to many critics, the answer is no, with no-till practices being feasible in many different physical environments. No-till farming is also suited to most crops, with the notable exceptions of wetland rice and root crops such as potatoes.

Huggins and Reganold (2008: 75) summarize six principal benefits of no-till agriculture and seven trade-offs. Weighing these up, they argue for no-till farming. The benefits are that it

- reduces soil erosion;
- conserves water;
- improves health of soil;
- lowers fuel and labour costs because of less tillage;
- reduces sediment and fertilizer pollution of nearby water bodies; and
- sequesters carbon.

The trade-offs are

- transition from conventional to no-till farming may be difficult because many other changes are needed;
- required equipment is costly;
- there is a heavier reliance on herbicides;
- there may be some unexpected changes in weeds and disease;
- more nitrogen fertilizer may be needed at first; and
- germination may be slower and there may be a reduction in yields.

TABLE 10.2 | Comparing tillage strategies

Number of Passes over a Field	No-Till	Conservation Tillage	Conventional Tillage
1	Apply herbicide	Till with chisel plow, burying up to 50 per cent of crop residue	Till with moldboard plow, burying up to 90 per cent of crop residue
2	Plant	Till with field cultivator	Till with disc to smooth the ground surface
3	Apply herbicide	Plant	Till with field cultivator to prepare the seedbed for planting
4	Harvest	Apply herbicide	Till with harrows to smooth seedbed
5		Till with row cultivator	Plant
6		Harvest	Apply herbicide
7			Till with row cultivator
8			Harvest

Source: Huggins, D. R., and J. P. Reganold. 2008. "No-Till: The Quiet Revolution." *Scientific American* 299 (5): 74. Reproduced with permission.

Table 10.2 describes what is involved in no-till farming, in conventional tillage farming, and in the intermediate strategy of conservation tillage (frequently called reduced tillage) farming in the specific case of corn–soybean crop rotation in the Corn Belt region. No-till practices, which require the fewest passes over a field, have spread across diverse soil types and agricultural production systems around the world over the last 30 years, although they are used on less than 10 per cent of global cropland. Most such farming is in the United States, Brazil, Argentina, Paraguay, Uruguay, Canada, and Australia. Adoption rates are much lower in Europe, Asia, and Africa. One reason that Europe lags is the lack of government policies promoting no-till, while in less developed countries the changes required are often too expensive to initiate. Most of the no-till agriculture in Canada is in the prairie region, where it has increased dramatically since the 1980s. Figure 10.12 shows the percentage of prairie land area under the three different tillage systems for each of the last four agricultural census years.

Huggins and Reganold (2008: 77) conclude that no-till, although "not a cure-all," is a sustainable strategy that is able to "deliver a host of benefits that are increasingly desirable in a world facing population growth, environmental degradation, rising energy costs and climate change."

Lost Animal and Plant Species

Together, the technological advances we've discussed have significantly improved the efficiency of the agricultural enterprise. But consider one unanticipated circumstance: the variety of food sources available has declined markedly since about 1900, meaning that global food supplies are

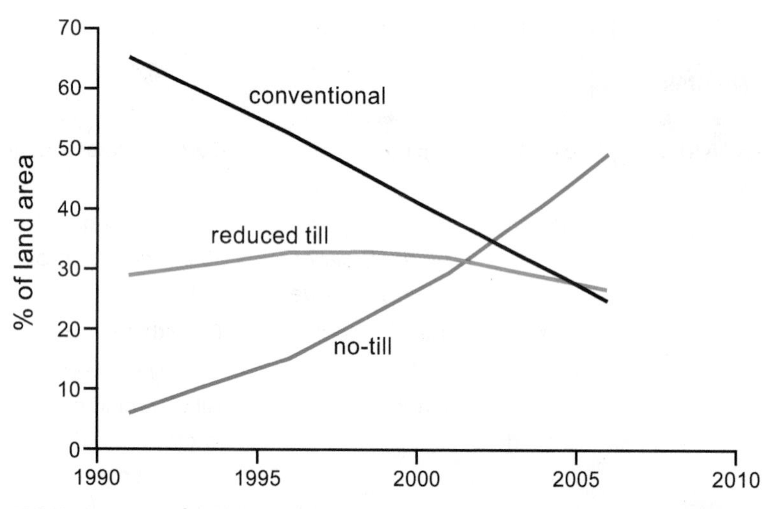

FIGURE 10.12 | Tillage system trends on the Canadian prairies

Source: Agriculture and Agri-Food Canada. 2009. "Flexibility of No Till and Reduced Till Systems Ensures Success in the Long Term 2009." www4.agr.gc.ca/AAFC-AAC/display-afficher.do?id=1219778199286&lang=eng.

now dependent on a relatively small number of animal and plant species compared to before that time.

The principal reasons for the decline are several of the technological advances explored in the previous sections, especially the increased use of hybrid seeds, fertilizers, pesticides, and crop rotations. These developments have combined to limit the genetic stores from which our crops were developed, effectively eliminating many of the traditional and well-adapted local varieties that had developed over the past 12,000 years. Historically, corn grown in a particular location was determined by how well it grew and what people wanted to do with the corn. In pre-industrial societies, natural and human selection produced types of corn with a genetic makeup different from other regions. The term *landrace* is used to identify these unique types. In a similar vein, heirloom varieties are those that were commercially available in the past but that fell out of favour with the development of hybrids.

landrace

A local variety of a domesticated animal or plant species that is well adapted to a particular physical and cultural environment.

The number of lost species, although unknown, is considerable. For example, it is estimated that 90 per cent of the traditional fruit and vegetable species in the United States no longer exist. In the less developed world, the green revolution was a principal contributor to this loss of species as farmers became dependent on imported high-yield species at the expense of landraces.

The loss of species resulting from technological advances designed to increase food supplies is a serious problem because some of our current species might succumb to disease or be affected by our changing climate. In short, our successful efforts using technology to increase food production have inadvertently increased the possibility of future food shortage problems. One solution adopted is the creation of seed banks to preserve as many remaining species as possible; there are about 1,400 such banks in the world.

WORLD AGRICULTURE TODAY: TYPES AND REGIONS

The noted American geographer Derwent Whittlesey identified nine major regional types of agriculture. Figure 10.13 shows the global distribution of these types as elaborated upon by Grigg (1974); it includes a twofold division of one of the types and identifies those areas with little or no agriculture. Even a cursory review of the figure highlights the relationships between agriculture and climate (Figure A2.7), between agriculture and generalized global environments (Figure A2.8), and between agriculture and population distribution and density (Figure 5.13). The figure also suggests the complex links between areas of production and areas of consumption.

Primitive Subsistence Agriculture

One of the earliest agricultural systems, primitive subsistence or shifting agriculture, is practised almost exclusively in tropical areas. It involves selecting a location, removing vegetation, and sowing crops on the cleared land. Land preparation is minimal, and little care is given to the crops. Typically, agricultural implements are limited and livestock are not normally part of the system, although fowl and pigs may be present. After a few years, the land is abandoned and a new location is sought. In this farming system, land is not owned or controlled by an individual or family but rather by some larger social unit such as the village or tribe. Shifting agriculture has traditionally been for subsistence purposes. Today, many such cultivators also produce a different crop for sale so that they have at least some market orientation.

Superficially, shifting agriculture appears to be wasteful and indicative of a low technology. Yet evidence suggests that it is a highly appropriate farming method. Not only is it an effective way of

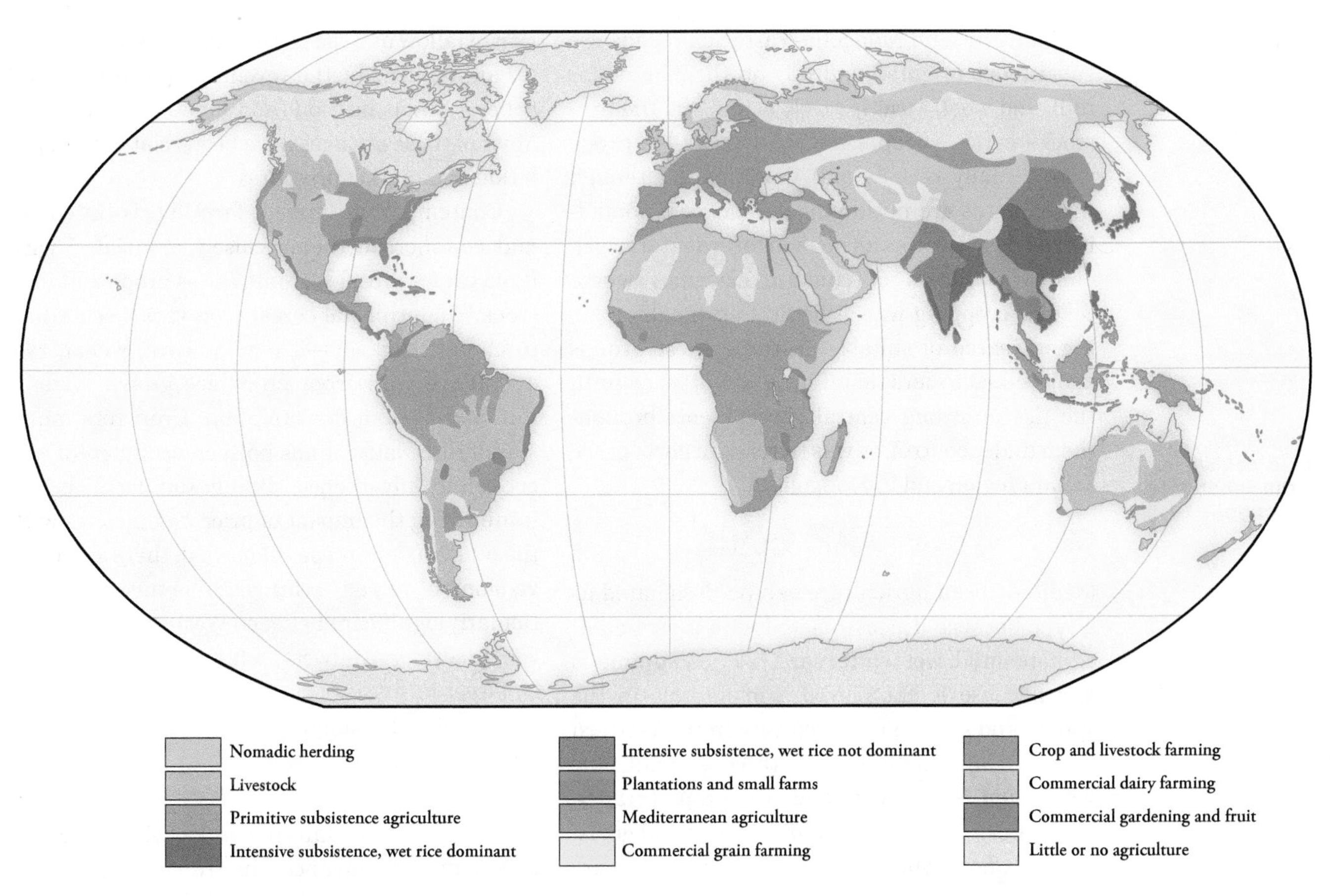

FIGURE 10.13 | World agricultural regions

Source: Grigg, D. B. 1974. *The Agricultural Systems of the World: An Evolutionary Approach*. New York: Cambridge University Press, 4. Copyright © Cambridge University Press 1974.

maintaining soil fertility in humid tropical areas and well suited to circumstances of low population density and ample land, but it also provides adequate returns for minimal capital and labour inputs. Shifting agriculture, then, can be explained by reference to a number of variables beyond economic considerations, including environment and population numbers.

Wet Rice Farming

Most of the rural population of East Asia is supported by wet rice farming, an intensive type of agriculture that requires only a small portion of the total land area—farms are small, perhaps only 1–2 ha (2.5–5 acres) and subdivided into fields—but large amounts of human labour. There are many versions of wet rice farming, but in all instances the crop is submerged under slowly moving water for much of the growing period. Wet rice farming is restricted to specific local environments, most notably flat land adjacent to rivers. Low walls delineate the fields and hold the water. Because this technology minimizes soil depletion, continuous cropping is often possible, producing multiple harvests each year. Cultivation techniques changed little until the green revolution of the 1960s, when some areas began using new rice varieties, fertilizers, and pesticides. There are close links between wet rice farming, high population densities, and flatland environments (wet rice fields on terraced hillsides are characteristic of South China and some areas of Southeast Asia). In some areas UN-sponsored projects train farmers to use an integrated crop management system designed to conserve biodiversity and to reduce dependence on chemicals (Rogers and Chew, 2005).

Pastoral Nomadism

Traditionally practised in the hot/dry and cold/dry areas of Africa and Asia, pastoral nomadism is declining as a result of new technologies and changing social and economic circumstances; however, it remains an important type of agriculture in the Atlantic Sahara region, Somalia, Iran,

Afghanistan, and Mongolia. Pastoral nomads are subsistence oriented and rely on their herds for milk and wool. Meat is rarely eaten, and the livestock—cattle, sheep, camels, or goats—is rarely sold. Usually one animal dominates, although mixed herds are not unknown. Pastoral nomadism likely evolved as an offshoot of sedentary agriculture in areas of climatic extremes where regular cropping was difficult. Continually moving in search of suitable pastures, these groups achieved considerable military importance until the rise of strong central governments brought them under control, as was the case in parts of the Sahara region and in Mongolia.

Pastoral Nomadism in the Sahara and Mongolia

Mediterranean Agriculture

Mediterranean agriculture can be designated as a type because it is associated with a particular climate (mild, wet winters and hot, dry summers) and because it has played a major role in the spread and growth of agriculture in the Western world. Traditionally, it has three components: wheat and barley, vine and tree crops (grapes, olives, figs), and grazing land for sheep and goats. Whereas both wheat cultivation and grazing are extensive land uses and low in productivity, vine and tree crops are intensive. Typically, all three activities are practised on all farms. Spatially, this type of agriculture evolved in the eastern Mediterranean. It had spread throughout the larger region by classical times and then was exported to environmentally suitable areas overseas, such as California, Chile, southern South Africa, and southern South Australia. Population growth and technological changes have led to several variations, with a general increase in irrigation and a decline in extensive wheat cultivation.

Mixed Farming

The four agricultural types discussed so far are generally subsistence in orientation (Mediterranean agriculture might be described as mixed subsistence/commercial); the remaining five are commercial. Mixed farming prevails throughout Europe, much of eastern North America, and other temperate areas of European overseas expansion. It is clearly a variant of the earliest farming in Southwest Asia, involving both crops and livestock. The transition to modern mixed farming involved a series of changes, such as the adoption of heavier ploughs and a three-field system (by the eleventh century), the reduction of fallow and the use of root crops and grasses for animal feed (*c.* the seventeenth century), and general intensification (mid-nineteenth century). Most of these changes have been related to population and market pressures.

Contemporary mixed farming is intensive and commercial (being closely associated with large urban areas) and integrates crops and livestock. The principal cereal crop varies according to climate and soil—it may be corn, wheat, rye, or barley—while root crops are grown for animal and human consumption. Crop rotation is standard because it has both environmental and economic advantages, aiding soil fertility and minimizing the impact of price changes. As with the other farming types identified, there are many versions of mixed farming. Differences are particularly significant between Western Europe and the American Midwest, which is characterized by larger farms, especially high productivity, and advanced technology.

Dairying

Specialization in dairying is closely related to urban market advances in transportation and to other technological changes that began in the nineteenth century. It is particularly associated with Europe and areas of European overseas expansion. Farms are relatively small and are capital intensive. The sources of the dairy products marketed in a given region often reflect distance from market, with those farms close to market specializing in fluid milk and those farther away producing butter, cheese, and processed milk. On the world scale, those dairy areas close to population concentrations—in Western Europe and North America—specialize in fluid milk, while relatively isolated dairy areas—New Zealand, for example—focus on less perishable products. Fluid milk producers are making increasing use of the feedlot system, in which cattle feed is purchased rather than grown on the farm. In such cases, dairying operations begin to look more like factory systems than the traditional image of an agricultural way of life. This development is most evident in North America.

Plantation Agriculture

A number of crops required for food and industrial uses can be efficiently produced in only tropical and subtropical areas. For this reason, as Europeans moved into the non-temperate world, they developed what became known as plantation agriculture. Plantations produce crops such as coffee, tea, oil-palm, cacao, coconuts, bananas, jute, sisal, hemp, rubber, tobacco, groundnuts, sugar cane, and cotton for export primarily to Europe and North America. This type of agriculture is extremely intensive, operating on a large scale using local labour (usually under European supervision) and producing only one crop on each plantation.

Because plantation agriculture evolved within the context of colonialism, it has profound social and economic implications and has been seen as exploiting both land and labour. However, the fault here lies with colonialism, not plantation agriculture itself. Because the plantation economy was and is so labour-intensive, the first plantations—established in the fifteenth century by the Portuguese in Brazil to produce sugar cane—frequently used slave labour imported from West Africa. Following the abolition of slavery, indentured labourers were brought from India and China (discussed in Chapter 8).

Designed to serve the needs of distant areas, plantation companies are often multinational. Many such companies are also characterized by what is known as vertical integration: the same company that produces the crop also refines, processes, packages, and sells the by-products. Thus, although political colonialism is largely gone, it has in many instances been replaced by economic colonialism.

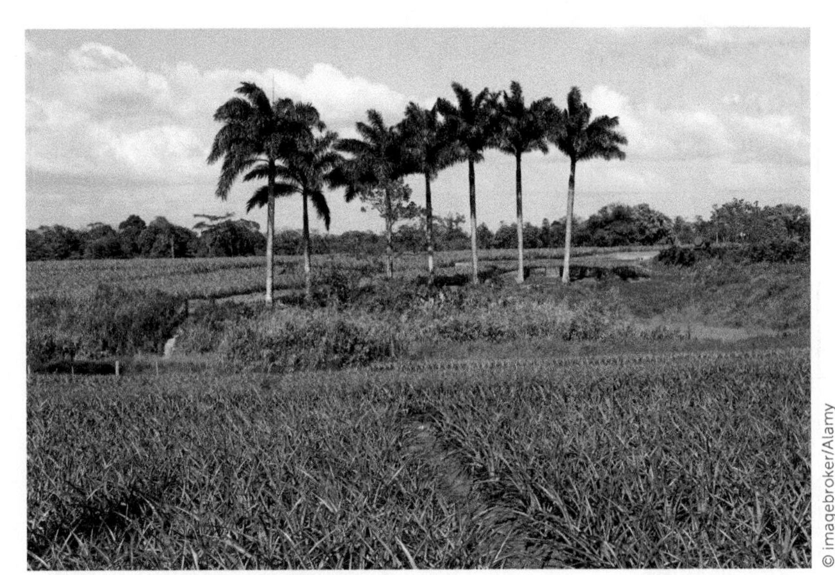

Pineapple plantation near Pital, Costa Rica.

Ranching

Commercial grazing—ranching—is generally limited to areas of European overseas expansion and is again closely related to the needs of urban populations. Cattle and sheep are the major ranch animals, since beef and wool are the products most in demand. Much of the European expansion into temperate overseas areas in North and South America, Australia, New Zealand, and South Africa was motivated by the desire to expand a grazing economy. Ranching is usually a large-scale operation because of the low productivity involved and hence is associated with areas of low population density. In some areas, the ranching economy has evolved into a ranching and livestock-fattening economy; the most notable example is the American Great Plains. Previously, cattle were shipped to the Corn Belt states for fattening, but the late twentieth-century proliferation of cattle feedlots within the Great Plains allows beef production to be vertically integrated in the region. This trend appears likely to continue; ranchers, especially in semi-arid areas of the temperate world, are at present experiencing difficult economic times.

Large-Scale Grain Production

Wheat is the dominant crop in large-scale grain production and is often grown for export. Major grain producers are the US, Canada, and Ukraine. Farms are large and highly mechanized. This type of production evolved in the nineteenth century to supply the growing urban markets of Western

A rancher rounds up cattle on the Dumbell Ranch in Alcova, Wyoming. The ranch covers approximately 115,000 acres and has 1,200 cattle.

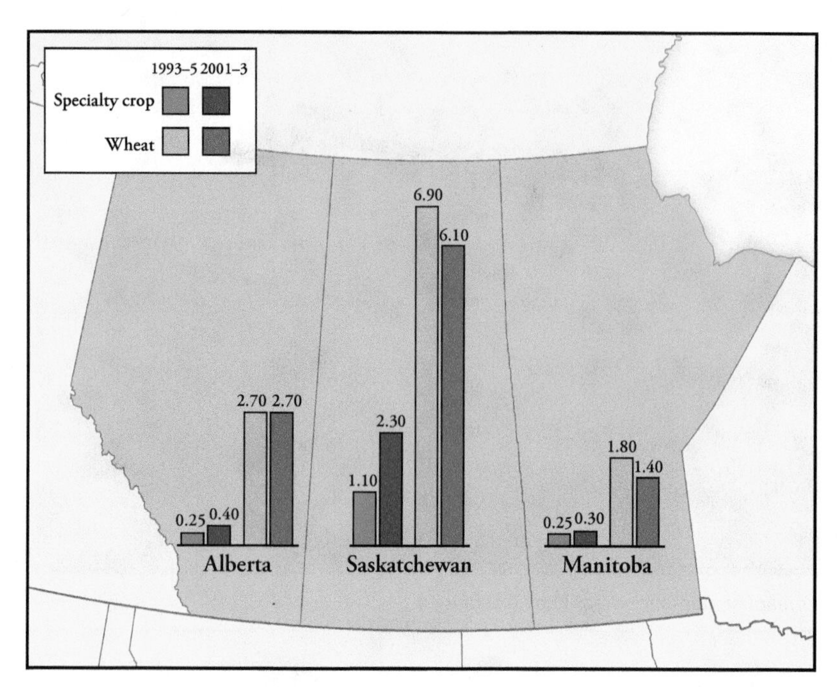

FIGURE 10.14 | Change in area devoted to wheat and specialty crops, Canadian prairies (millions of hectares)

Source: Fadellin, N., and M. J. Broadway. 2006. "Canada's Prairies: From Breadbasket to Feed Bunker and Hog Trough." *Geography* 91: 85. Reprinted by permission of the Geographical Association, www.geography.org.uk.

Europe and eastern North America, where in many cases it displaced a ranching economy. In some areas, such as Ontario in the mid-nineteenth century, wheat became the staple crop:

> A more classic case of a staple product would be difficult to imagine. More specialized in wheat production than the farmers of present-day Saskatchewan, Ontario farmers of the mid-nineteenth century exported at least four-fifths of their marketable surplus. Close to three-quarters of the cash income of Ontario farmers was derived from wheat and wheat and flour made up well over half of all exports from Ontario. (McCallum, 1980: 4)

Wheat and other grain staples were the engines of economic growth in other areas. Today, most of those areas are more diversified, and some rely more on mixed farming than on grain production. The transition from a predominantly wheat economy to a more diversified farming economy was evident on the Canadian prairies even in the 1980s, reflecting both global trends and national economic strategies (Carlyle, 1994; Seaborne, 2001).

restructuring
In a capitalist economy, changes in or between the various components of an economic system resulting from economic change.

Globally, as we saw in Chapter 3, world commerce is increasingly liberalized and competitive. Consequently, governments are being forced to reconsider their interventions into the agricultural sector. Particularly important for farmers on the Canadian prairies was the Crow rate, named for the Crow's Nest Pass Agreement of 1897, under which the Canadian Pacific Railway had guaranteed western grain farmers a special low rate "in perpetuity." In 1984 the terms of this agreement were changed and the rate became a direct federal subsidy, which was eliminated altogether in 1995. Also important for prairie grain agriculture was the 2011 federal government decision to strip the Canadian Wheat Board, founded after the Great Depression, of its monopoly over wheat and barley sales.

In recent decades non-grain crops have become increasingly important on the prairies, among them oil seeds, especially canola and flax, and such specialty crops as peas, lentils, mustard seed, and canary seed (Figure 10.14). The hog industry is also growing rapidly: in Saskatchewan hog numbers increased by almost 50 per cent between 1996 and 2001. Overall, traditional areas of large-scale grain production are undergoing significant changes as the need to adjust to changing global, national, and regional circumstances becomes evident. In turn, changes in production lead to a multitude of other changes (Box 10.8).

GLOBAL AGRICULTURAL RESTRUCTURING

For almost all human history, food was consumed in the same location where it was produced. Today, especially in the more developed world, producers and consumers are no longer in the same location. Agriculture is just one component of larger systems of production and consumption that are often affected by regulations with a global impact (e.g. trade agreements), the activities of transnational corporations, and various production, processing, and marketing networks. As noted in Box 10.9, these important changes are usefully addressed through a political economy approach, not least because they contribute to uneven development as capital moves into those places where profits will be highest.

Notwithstanding globalization processes, there are notable differences between the agricultural economies of the more and less developed worlds, especially concerning the use of capital. In the more developed world, much capital is

Examining the Issues

BOX 10.8 | Canadian Farmers: Fewer and Older

The trend towards fewer and older farmers in most parts of the more developed world is well established and is one aspect of larger economic changes related to the changing nature of employment (a topic discussed more generally in Chapter 13). In Canada the total number of farm operators continues to fall, declining by 10.1 per cent between 2006 and 2011. Although the decline in numbers of farmers has not reduced food production—with larger farms, those that remain are producing more than ever before—it has a significant impact on rural life and the health and perhaps even the survival of many small towns. Many farmers do not live on their farms but commute to their work from city residences.

The decline was neither new nor unexpected. On the prairies, rural populations grew rapidly until the 1930s, when the numbers began the drop that has continued into the twenty-first century. Meanwhile, as the farm landscape is gradually depopulated, the infrastructure of railway lines, grain elevators, small towns, schools, and so forth is being dismantled, and the large amounts of capital required to survive in an increasingly high-risk environment mean that more and more agricultural land is in the hands of corporations and huge farm conglomerates.

In short, it appears that the small family farm is a thing of the past. Most small farmers now require some off-farm income to support themselves, and the median age of farmers is steadily increasing across Canada; in Manitoba it rose from 46 to 52 from 1996 to 2006. To survive, some small farmers are moving into specialized areas—in Manitoba, some focus exclusively on raspberries or other soft fruits—or activities such as organic farming (see Box 10.5).

invested in machinery and farm inputs, which means less need for human labour. But in the less developed world, most notably in tropical regions as evident in the account of world agricultural regions, much less capital is invested and agriculture continues to be relatively labour intensive.

A process central to many of these changes in the agricultural industry (and other industries, as we will see in Chapter 13), is **restructuring**. In a capitalist economy, this term can refer to changes in either the type of capital invested in (for example, from human labour to machines) or the movement of capital resulting from changes in technology or labour relations. In agriculture, restructuring may involve spatial changes in agricultural activities, movements of capital from one level of the production process to another, or changes in the organization of production. Most simply expressed, restructuring can be seen as a process of ongoing adjustment to changing circumstances.

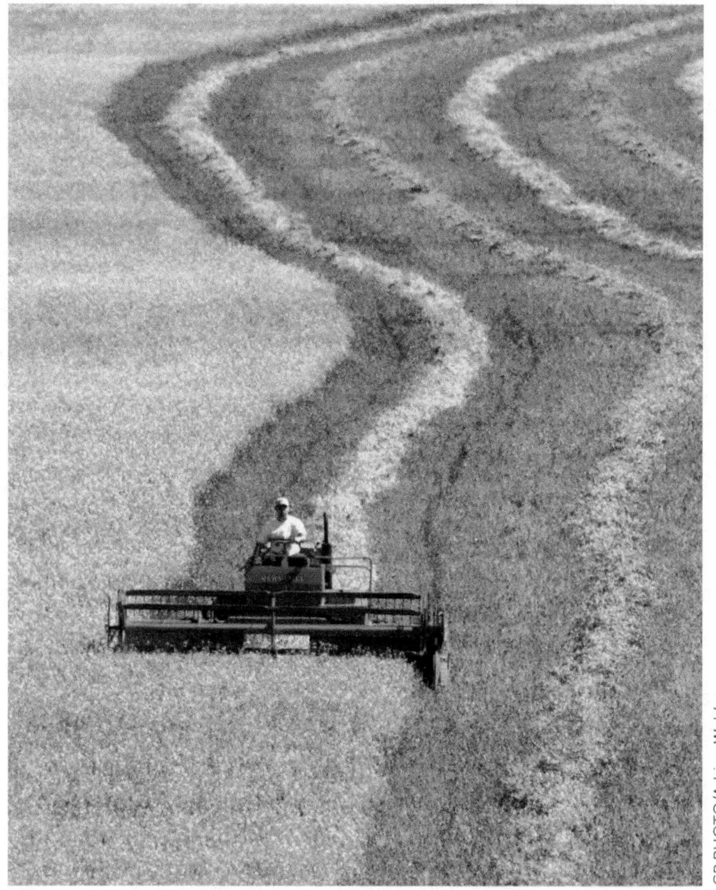

CP PHOTO/Adrian Wyld

Harvesting canola near Ponoka, Alberta (south of Edmonton), August 2002. A severe drought that summer reduced crop yields by two-thirds.

Agriculture in the World Economy

The relative importance of agriculture in an economy decreases with economic growth (Grigg, 1992). This relationship can be demonstrated in the following way:

Examining the Issues

BOX 10.9 | Marxist Political Economy

The Marxist perspective gained adherents when it became clear that neither the empiricist nor the positivist approach offered a meaningful account of the agricultural changes that began in the 1970s. Neither classical description nor location theory offered any conceptual insight into issues such as the industrialization of agriculture, the rise of agribusiness, state intervention, or the social impacts of technological change.

According to Marx, capitalism is contradictory, characterized by social tensions and conflicts. The essential argument is that a capitalist society is based on the circulation of capital—through production, exchange, and consumption—and this circulation leads to economic growth, or what is frequently called the ceaseless accumulation of capital. Because capital circulation involves the movement of money, goods, and labour, it creates (and continually recreates) geographies of production, consumption, and interaction. The requirement that circulation leads to economic growth implies that lack of growth, or decline, is untenable. For the capitalist system to work, labour must produce more value in the production process than it receives in the form of wages so that business owners can appropriate that "surplus value" as their profit. For Marx, the accumulation of capital by owners at the expense of workers inevitably leads to class conflict.

The capitalist system, as Marx describes it, has three key geographic implications: social and spatial divisions of labour (e.g. along class and gender lines); competition among owners for space, resources, and economic infrastructure; and, most generally, inherent instability (e.g. periods of depression and inflation). Marxist discourse also has two important advantages for the study of contemporary agricultural geography. It blurs the lines between agricultural and other economic activities, especially industry, and it integrates the agricultural experiences of the more and less developed worlds. A significant insight revealed by the Marxist perspective is that capitalist development in the broad agricultural and food system has increased dramatically as non-farm interests such as banks and the food-processing industry have gained increasing control of the agricultural production process.

In recent years, the Marxist-inspired political economy approach has provided a valuable perspective on agricultural change, particularly in the context of an increasingly globalized world and food system. The approach has been widely employed in studies of agriculture in the less developed world, especially as it has been merged with a more traditional ecological (humans and land) approach as discussed in Chapter 4, and can be summarized as follows:

1. At the regional level, human geographers attempt to integrate the multiple social and ecological relations involved in the organization of agriculture and land use.
2. This approach permits a focus on the politics of place, specifically regarding differences in power between groups and places (as discussed in a different context in Chapter 8).
3. It has clear implications for the design and implementation of rural development policies.
4. There is recognition of the need to involve local populations as well as governments and organizations in the planning process.
5. The interplay of social practices (agency) and political-economic conditions (structure) is often considered.
6. In many cases, the environmental impacts of agricultural activity can be addressed in political terms.

In this way, a political ecology perspective can add significant new dimensions to more traditional human ecological approaches. Examples include studies by Zimmerer (1991), which integrated regional political ecology concepts, structuration theory, a politics of place, and production ecology in an analysis of agricultural change in highland Peru; by Grossman (1993: 346), who used a political ecology perspective to "highlight not only the impact of political-economic relationships on resource-use patterns but also the significance of environmental variables and how their interaction with political-economic forces influences human–environment relations"; and by Young (1991), who showed that crop varieties imported into Bhutan were not well suited to the local environment.

The political ecology viewpoint also has much to say about the environmental impacts of agricultural activities (see Chapter 4), especially with regard to the unequal distribution of those impacts. For example, the hurricane that devastated parts of Central America in October 1998 did not affect all agricultural operations equally. Those farms using traditional methods suffered much less soil erosion and crop damage than did those using modern chemical-intensive methods. More generally, there is good evidence that, although modern agricultural methods are usually more productive than traditional methods, they are less environmentally friendly. The four most common types of environmental damage caused by agriculture are particularly dangerous for the less developed world: soil degradation, soil and water pollution, water scarcity, and reduction of genetic diversity.

1. For pre-industrial economies, data for the World Bank category of low-income countries suggest that about 30 per cent of their GDP is derived from agriculture, while about 60 per cent of their population is engaged in agriculture; similarly, the limited data available for contemporary industrial countries prior to the Industrial Revolution indicate that agriculture accounted for about 40 per cent of the GDP and about 70 per cent of the workforce.
2. For today's industrial economies, data for the World Bank category of high-income countries suggest that only 2.5 per cent of their GDP is derived from agriculture and that only 3.5 per cent of their population is actively engaged in agriculture.

These figures exaggerate the extent of the relative decline of agriculture because the data for contemporary industrial countries do not include food-processing and related activities, but the general picture is nevertheless clear: industrial and service activities have replaced agriculture as the most important sectors of national economies (Figure 10.15). Furthermore, agricultural incomes are lower on average than non-agricultural incomes, and this gap is constantly widening.

There are considerable spatial variations in the timing of the agricultural sector's relative decline. The decrease began in Britain, the home of the Industrial Revolution, where the agricultural percentage of the total labour force fell below 50 as early as the 1730s; in other European countries the decline to below 50 per cent had occurred by the 1840s. It has not yet occurred in much of Africa and Asia.

Changes to the Food Supply System

The food supply system is becoming increasingly complicated. Whereas the von Thünen theory focuses on horizontal relations on one level only

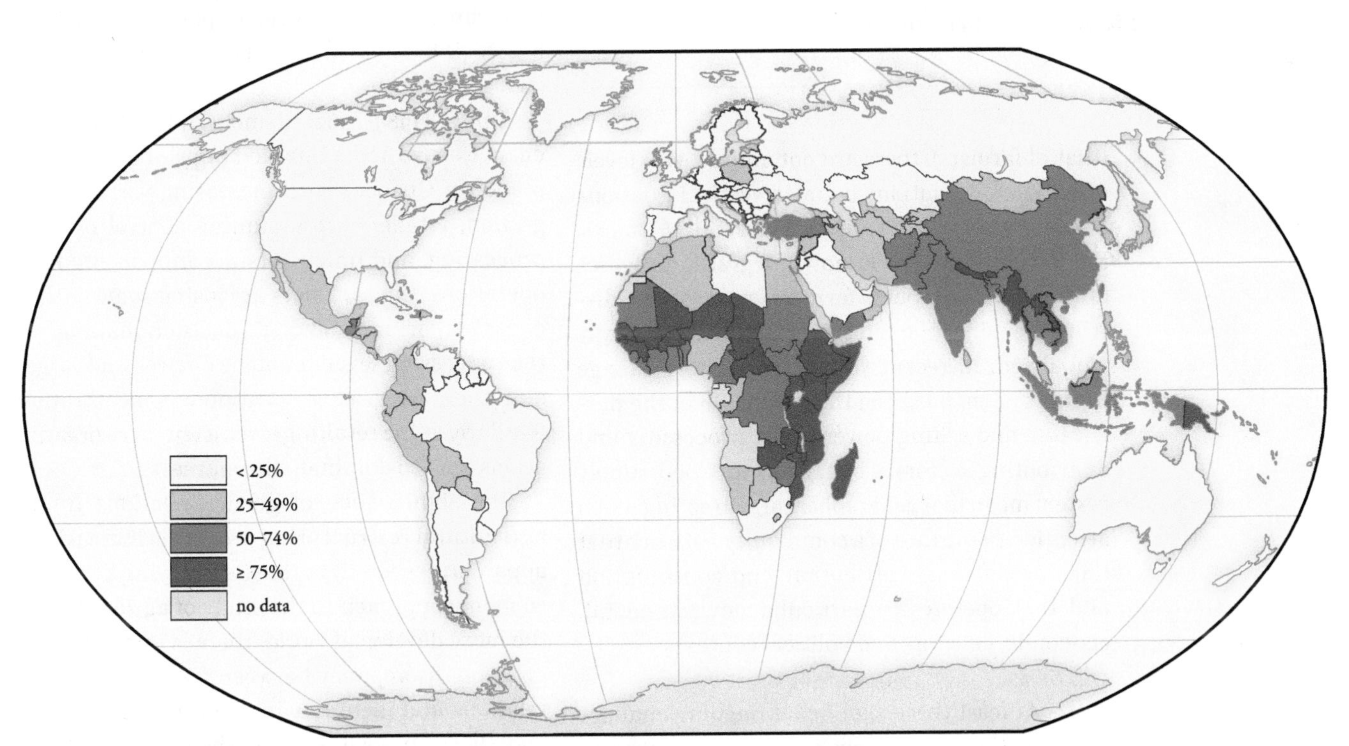

FIGURE 10.15 | Percentage of labour force in agriculture by country, 2012

This map provides a good indication of the importance of agricultural activity globally. The range is enormous. In many of the more developed countries, less than 25 per cent of the labour force works in agriculture; in both Canada and the United States less than 3 per cent do so. In some less developed countries, the proportion is more than 75 per cent; in Niger and Mali, more than 90 per cent of the people are agricultural workers. The correlation between development level and agriculture is far from perfect as several countries in Latin America have values below 25 per cent. The world average is 30.5 per cent. There are often significant differences between the genders: in Canada, 2 per cent of employed women and 4 per cent of employed men work in agriculture; in Bangladesh the corresponding percentages are 77 and 53.

Source: Updated from United Nations Development Program. 1996. *Human Development Report*. New York: Oxford University Press, tables 16 and 32.

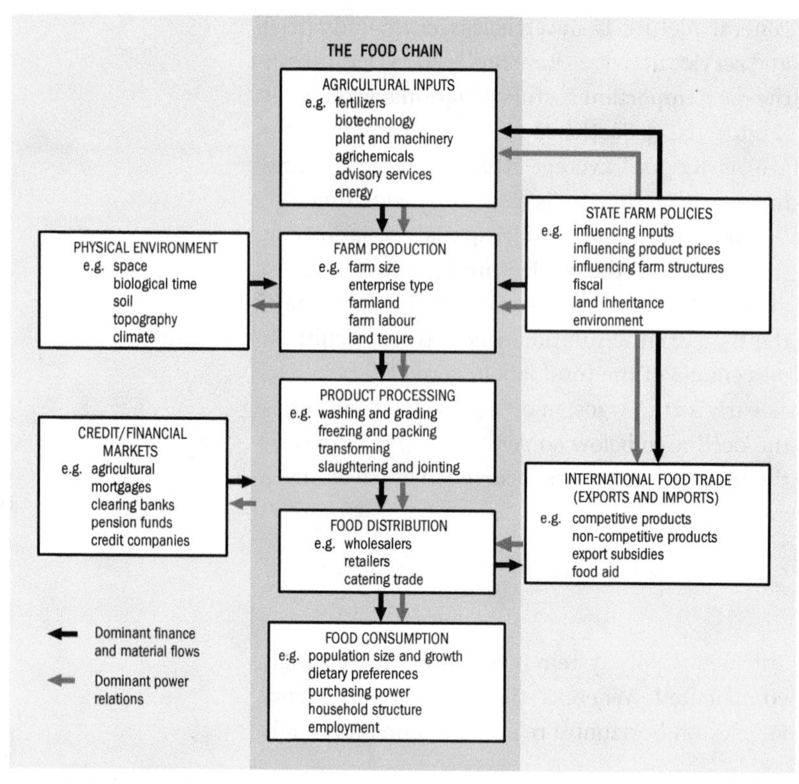

FIGURE 10.16 | The food supply system

Source: Pacione, M., ed. 1986. *Progress in Agricultural Geography*. Beckenham, UK: Croom Helm.

(that of farmers), there are not only several levels of horizontal relations but also vertical relations between different levels. As Figure 10.16 shows, the food supply system now integrates activities at five levels—inputs, farmers, processors, distributors, and consumers—and is increasingly globalized. In recent years, the greatest change in this system has been the expansion of the purchasing and selling power of the processing and distributing sectors. Changes to the food supply system mean that geographers are directing much attention to the idea of a commodity network that links production, distribution, and consumption and that operates in particular environmental, economic, cultural, and political contexts:

- At least three significant ongoing changes in the food supply system are related to globalization processes (Marsden, 1997).
- Globalization, such as corporate manufacturing of food, is accompanied by national-scale processes, for example, state policies governing product prices.
- Globalization, such as the spread of standardized food items, is accompanied by localization, for example, an increase in local organic production (with the outcome sometimes called glocalization).
- Increasingly, the non-farm sector of the food supply system is controlling what people are able to consume and eat; this sector includes consumer agencies and food safety organizations.

Agricultural change in the more developed world has gone through four distinct phases since World War II: mechanization, chemical farming, food manufacturing, and biotechnology. The 1950s brought increasing mechanization, which raised yields and reduced labour requirements, and the 1960s saw the advent of large-scale chemical farming with nitrogenous fertilizers, herbicides, fungicides, and pesticides. These first two stages modernized agriculture. The 1970s witnessed significant growth in the sector dealing with the manufacture of purchased agricultural inputs and in the processing and distributing sectors (Figure 10.16), while the 1980s brought biotechnological advances in plant varieties and livestock breeds. The last two stages industrialized agriculture.

Smith (1984: 362) demonstrates that these changes have had a "marked imprint on the landscape" in Quebec. With increasing vertical integration, retailers are assuming control of some processing, and processors are influencing farm operations; thus, farmers are losing some of their capacity for independent decision-making. At the processing level, plants are fewer and larger, and there is greater locational concentration. Similarly at the retailing level, a few large organizations tend to dominate the market.

The significance of gender relations in the agricultural restructuring process is increasingly apparent, especially to those employing a political economy approach to the study of agriculture. In the more developed world, there is interest in the growing discrepancy between traditional gender relations and identities in farming communities and those in society as a whole. The trend for women to leave farming is generally considered a reflection of the conservative gender relations associated with farming communities.

Industrializing Agriculture

For many human geographers, the most fascinating—most "geographical"—aspect of agriculture is that it is carried out at the interface

Focus on Geographers

Seeking Equity through the Study of Agricultural Geography | RAJU J. DAS

The majority of the world, mostly living outside of the few advanced countries, directly depends on agriculture for a living. Many countries still struggle to produce enough food, and many countries' economic development still depends on what happens in their agriculture. Globally, the crisis character of the world economy is expressed by famine and malnutrition in Africa and Asia, food shortages in parts of Eastern Europe, farmer debt and food overproduction in rich countries, and suicide of indebted farmers in India. Agricultural geography has shed light on many of these issues. Above everything else, agriculture signifies society's reciprocal interaction with nature to produce the most basic requirement for survival: food. After all, we need to eat before we can think or do anything else. This is a simple fact that many "cultural-turn geographers" tend to forget. Thus, agriculture—and therefore its study—is highly important. Some of the best-known geographers in the world, including David Harvey, Richard Peet, and Michael Watts, started their careers working on agricultural geography or continue to write within this field.

Traditionally, agricultural geography has studied how and why different kinds of agriculture occur in different countries/regions and with what effects on society and the environment. Changes in agricultural practices are reflected in changes in agricultural geography as a field. The attention is shifting from the geography of farming activities per se to the geography of global production of a system of food-related commodities, including food and the goods/services necessary to produce and distribute it.

Partly because many agricultural items cannot be produced economically and in sufficient amounts in the temperate lands and short growing seasons of advanced countries, human and land resources of poor countries such as India, Peru, and Kenya are being used to produce these items—high-value, non-traditional commodities such as shrimp and flowers—for export to richer countries. Among other things, diversion of land use from staple crops to high-value goods for export is contributing to food insecurity in those countries. Due to increasing costs and competitiveness nationally and internationally, many farmers are gradually losing their farms and only a small minority are able to make a profit. Millions of people, including economically viable peasants, are regularly forced, in spite of their valiant and sometimes bloody resistance, to surrender their land to domestic big business and foreign multinational corporations that use the land for industry, speculation, or large-scale farm activities. Further, in a context of market uncertainty and state withdrawal of support for farmers, many companies make farmers (especially those with a substantial amount of land) produce specific crops for a predetermined price. This process is known as contract farming. Companies (e.g. Pepsi) may provide credit and other services to them as part of the contract. The impact of capitalist globalization on various classes of agricultural populations needs to be studied in much greater detail and in varying contexts by agricultural geographers.

In Canada and other advanced countries, people consume food that comes from different parts of the world. When people buy flowers or shrimp, for example, they want to get the maximum number for their money. This objective makes them ignore the physically unsafe and economically low-wage conditions under which people work to produce these commodities. My own research, based on numerous interviews and field observations, has shown how women and men in India are forced by their desperate economic situation to work on shrimp farms as well as in shrimp-processing plants where the labour process is oppressive, super-exploitative, and dangerous to their health. Because their fingers are exposed to chemicals, they often cannot even eat food with their own hands.

In poor countries, agricultural labourers work long hours and have little employment security. Agricultural labourers are among the worst-paid workers in both poor and rich countries, just as they were 150 years ago in England. This is also the fate of agricultural labourers who come to Canada as temporary migrants from poorer countries, something Toronto's Robert Bridi (2010) has so eloquently written about. These foreign farm workers cannot form associations to go on a strike. They are unfree workers.

Many Canadians and others advocate ethical trade as a way of dealing with the problem of buying and consuming agricultural products produced under questionable conditions. Some advocate the removal of geographical ignorance about the labour processes in faraway places. Are these processes effective, long-term, global solutions to the problem? If not, why not, and what else is to be done to make sure that those who produce food, no matter where they live and work, have access to proper food and other needs?

Like any other form of production, agricultural production undergoes technological change, the effects of which

Continued

on ordinary people must be critically studied. An important conclusion from my research on technological change in agriculture, which is reported in *Geoforum* (2002), is that technology can have a positive physical effect (in that it may increase production of food-grain) but, in the absence of democratic, needs-based, non-exploitative social relations of control over land and labour process, technology alone cannot guarantee that everyone has enough food.

RAJU J. DAS is an associate professor in the Department of Geography at York University. His current work involves combining Marxist political economy and state and class theories to study the effects of capitalism, the capitalist state, and globalization on the conditions of peasants and rural labour.

References

Bridi, R.M. 2010. "Unveiling the Nature of the Labour Process in Tobacco Farming: Migrant Agricultural Workers in Southwestern Ontario," MA thesis, Department of Geography, York University.

Das, R.J. 2002. "The Green Revolution and Poverty: A Theoretical and Empirical Examination of the Relation between Technology and Society," *Geoforum* 33, 1: 55–72.

between humans and the physical environment. Unlike most other economic activities, agriculture is constrained by physical geography, and its success, as measured by productivity or return on investment, is beyond the farmer's control. But these physical constraints do not prevent the industrialization of agriculture. Rather, they guide that industrialization in particular directions.

As our discussion has shown, agriculture in the more developed world can be justifiably regarded as an industry, one in which the farmer is no longer the sole or even the primary decision-maker. According to Friedmann (1991: 65), "the distinction between agriculture and industry is no longer viable, and should be replaced by the conception of an agri-food sector central to capital accumulation in the world economy." The industrialization of agriculture has become especially evident since 1945. Both in the more developed world and increasingly in some parts of the less developed, agriculture has typically (but not always) become more intensive, more specialized, and more spatially concentrated.

Intensification, specialization, and spatial concentration are evident throughout much of the farming economy of the United States, for example, in the production of rice, sugar cane, tobacco, peanuts, and cotton in the South; corn, soybeans, pork, and beef in the Midwest; and beef in the West. The resulting increased production means that exports are of greater importance than previously.

Agriculture is one of the biggest industries in the world, employing about 1.3 billion people, although the number is decreasing. Agriculture produces about US$1.3 trillion worth of goods per year, and world output of food continues to increase. Nevertheless, farmers are always obliged to strive to be more competitive, more concerned with the environment, and more aware of changing consumer tastes. Some farmers are even entering the information age, for example, using GPSs (see Chapter 2) to monitor soil and moisture conditions on a two-hectare square of land.

State Intervention

Much contemporary agricultural activity can be understood only in the larger context of state agricultural policies and negotiated trading agreements. At first, states tried to respond to the inevitable fluctuations in production by regulating prices and marketing, with the ultimate goal of achieving some stability in agricultural production. In recent years, states have become increasingly involved in agriculture because improvements in productivity have led to the oversupply of domestic markets, reducing both product prices and farm income. Inadequate income for farmers is the principal reason behind government intervention in the form of income support.

In the more developed world, state policies aim to ensure the security of food supplies, price and income stability, protection of consumer interests, and regional development. Such policy objectives may be achieved through guaranteed prices, import controls, export subsidies, and various other amendments to the capitalist market system, such as marketing boards. Canada has price-support policies, income-support policies, and supply-management policies—all of which are subject to change as trading agreements become increasingly important. In the less developed world, most state policies are aimed at increasing productivity and achieving higher farm incomes and dietary standards.

As we saw in Chapter 3, international trading agreements, including the GATT (replaced by the WTO in 1995) and the Common Agricultural Policy (CAP) of the EU, have emphasized the need to reduce state support for agriculture, with a notable recent example being the 2005 reduction of subsidies paid to Europe's sugar farmers following demands from the WTO. The issue of state support for domestic agricultural production is very sensitive, especially in the EU, the world's leading importer and second largest exporter of food. Globally, the principal issue relates to the fact that state support means other countries, particularly those in the less developed world, are unable to compete in state-supported markets. Nationally, a concern with state support is that it needs to reconcile the consumer interests and farmer incomes.

Agribusiness

The term *agribusiness* refers to the activities of transnational corporations. Plantation agriculture was the first example of agribusiness, and even today most such businesses concentrate on production in less developed areas for markets in more developed areas. Table 10.3 suggests the extent to which agriculture in the less developed world is controlled by transnationals operating from and marketing their products in the more developed world.

Agribusiness has been widely criticized, most often on the grounds that, as Table 10.3 clearly shows, it is a form of economic colonialism, with little concern for the social, economic, and environmental consequences of activities in the producing area. More specifically, agribusiness frequently contributes to the creation of an agricultural economy vastly different from the local producing economy; benefits foreign investors and local elites rather than peasant groups; occupies prime agricultural land; imports labour that may lead to local ethnic conflict (see the discussion of indentured labour in Chapter 8); uses casual, part-time, and seasonal labour; and is reluctant to take responsibility for negative impacts on the physical environment.

TABLE 10.3 | Transnationals and crops in the less developed world

Crop	Companies	Observations
Cocoa	Cadbury-Schweppes Gill and Duffus Rowntree Nestlé	Control 60–80% of world cocoa sales
Tea	Brooke Bond Unilever Cadbury-Schweppes Allied Lyons Nestlé Standard Brands Kellogg Coca-Cola	Combined, hold about 90% of tea marketed in Western Europe and North America
Coffee	Nestlé General Foods	Control 20% of world market
Sugar	Tate and Lyle	Buys about 95% of cane sugar imported into the EU
Molasses	Tate and Lyle	Controls 40% of world trade
Tobacco	BAT R.J. Reynolds Philip Morris Imperial Group American Brands Rothmans	Combined, control over 90% of world leaf tobacco trade

Source: Adapted from Dixon, C. 1990. *Rural Development in the Third World*. London: Routledge, 18.

Global Trade in Food

Two causes of current agricultural change are the ever-increasing demand for food and the concentration of people in large urban centres. Given the differences in population growth rates between the more and less developed worlds, movements of food products from areas of surplus to areas of deficit (through both sales and aid) are bound to increase.

The global trade in food is a relatively recent phenomenon, associated with the Industrial Revolution and, in particular, technological advances in ocean transportation and refrigeration. Most of this trade involves exporting products from the less developed world, notably coffee, tea, and cocoa, to the more developed world.

At least half the world's people rely on one of three cereal crops: wheat, corn, or rice. In many cases these crops are produced for local consumption, but there is also substantial global trade in them. Most of the foods exported from more developed countries are cereals. Food exports are dominated by five countries—first and foremost the United States but also Canada, Argentina, France, and Australia—while distribution of the surpluses is concentrated in five private corporations. All five of these companies are international giants with diverse economic interests. The principal importer of grain in recent years has been the former USSR, but several Asian countries are increasing their imports, notably China, South Korea, Indonesia, the Philippines, and Japan.

Agricultural Reform in China

FOOD PRODUCTION, FOOD CONSUMPTION, AND IDENTITY

Much of this chapter has emphasized the political component of the traditionally economic study of agriculture. In this final section, we address the cultural component that has become such an important aspect of human geography in recent years. In the context of agriculture, the cultural turn (introduced in Chapter 8) is reflected by an interest not only in production but also in consumption. Historically—as geographers such as Vidal recognized in the early twentieth century—the kinds of food consumed in various parts of the world usually reflected the available resources. The food that people consumed represented one of the closest links between them and their environment, as well as one of the principal means by which groups of people were distinguished one from another. Despite the undeniable impact of interaction and globalization processes, this is still the case in many parts of the world (Figure 10.17). Box 10.10 considers the history of food in terms of consumption and the consequences of production practices (rather than those practices in themselves).

Food and Place Identity

Food consumption habits remain vital indicators of human identity, reflecting regional characteristics as well as individual and group tastes. Building on the ideas promoted by Sauer, much traditional cultural geography was concerned with regional and local foodways, particularly "ethnic preferences"—that is, the preferences of the specific group under study. By contrast, much contemporary cultural geography addresses the complex intertwining of consumption and identity, taking a distinct interest in the political

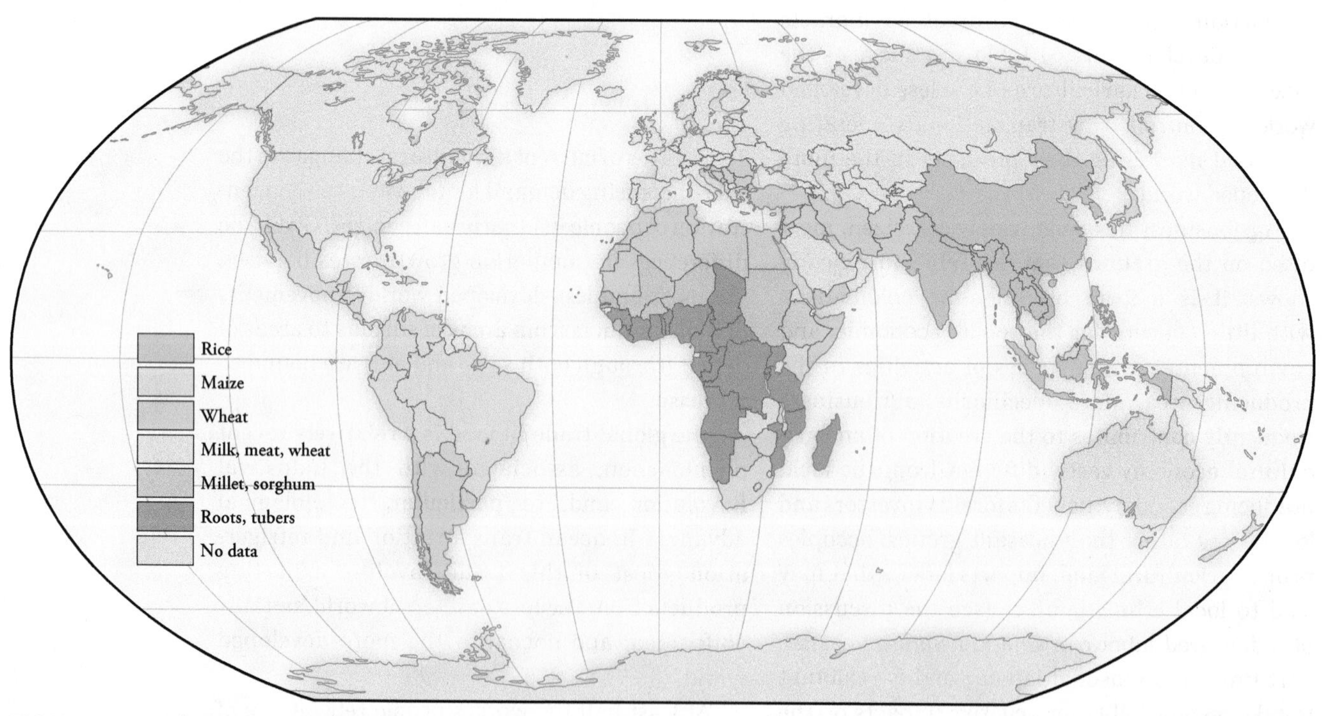

FIGURE 10.17 | Global dietary patterns

Examining the Issues

BOX 10.10 | You Are What, and Where, You Eat

Fernandez-Armesto (2001) identified eight key stages in the history of food, emphasizing the role played by consumption:

- Sometime before 150,000 years ago, humans began to use fire to cook food. This new technique for changing food—existing methods included burying, drying, masticating, and rotting—had significant social consequences, providing a new focus for human interaction.
- In the second stage, groups began to see food as representing more than just something to eat. Table rituals and dietary taboos began to appear in recognition of the other dimensions associated with food, particularly the moral and religious dimensions.
- In the third stage, food became associated with social status: people in positions of authority were able to eat not only more food than others but also different kinds.
- The fourth and fifth stages constitute what we earlier described as the first agricultural revolution: the domestication of plants and animals. These two stages—sometimes separate and sometimes related—represented a fundamental change, as humans began producing food, not merely collecting it. Plant domestication opened the way to new forms of social organization, greater population density, and the rise of urban centres.
- The sixth stage was a global movement of cultures in which food played several important parts. Food has been one of the most important motivations for both overland and overseas movement over the past 1,500 years. Demand for salt and spices played a major role in the development of transport routes and cultural contact. The wholesale movements of people over the following centuries involved widespread diffusion of animals and crops in all directions. Potatoes especially fuelled economic development, feeding the poor and underpinning the Industrial Revolution in Europe.
- The seventh stage is perhaps best known under the label McDonaldization, with globalization of food accompanied by homogenization. Globalized foods have been so successful because they have travelled the routes already laid out by American power, capital, and culture. But much more has been involved than the spread of American fast food to other parts of the world. In North America, for example, "ethnic" foods—from countries such as Italy, Mexico, China, India, and Thailand—have become popular.
- The eighth and most recent stage is the current trend towards fast food and the first real shift away from the social interaction that was initiated in the first stage. To the extent that fast food makes it unnecessary for members of a family to sit down to a meal together, it may well contribute to the fragmentation of the family. It is also possible that fast food is contributing to a general decline in health, not only because its nutritional value may be poor but also because, without the discipline and camaraderie associated with the shared table, the temptation to overeat can be hard to resist. The increase in obesity in North America seems to be closely related to the increasing popularity of fast food.

implications of consuming ethnic foods—foods associated with groups other than the one under study, both at home and elsewhere. Eating away from home is increasingly popular, especially in North America, where the remarkable variety of restaurants available, predominantly in urban areas, makes it possible not only to eat dozens of different sorts of food but also to experience something of dozens of different cultures. Some places to eat and drink have become what are known as third places, places other than home or work. Bars and coffee houses, for example, play important roles in many peoples' daily lives.

The popularity of ethnic food may well reflect a modern version of the historic association between food and social status. In Britain in 2001, the foreign secretary reported that the new national dish was neither fish and chips nor roast beef, but chicken tikka masala, a dish of Indian origin. Some commentators have suggested that this cross-cultural phenomenon perpetuates the former colonial relationship, although one could equally interpret it as representing a reversal of that relationship.

Specialty foods and drinks explicitly identified with a particular place (e.g. maple syrup as a distinctive product of eastern Canada) are increasing in popularity. Through advertising, frequently directed at potential tourists, a conscious attempt may be made to construct images of product quality and the related place. This development is perhaps best seen as an extension of the long-standing association between products such as wine and cheese and precise places, although there are many instances where the authenticity of the traditional product is challenged by new producers, as with Camembert cheese.

Production and Consumption of Cheeses

Camembert, like most other cheeses, was sold in only a few local markets until the mid-nineteenth century. But the extension of a rail link allowed Camembert producers in Normandy to sell their product in the much larger Paris market. Travel time was reduced from three days to just six hours, while the use of a new light round wooden box meant the cheese could travel without being damaged. To increase sales further, producers also worked towards a product that matched the expectations of consumers, with the most significant changes being pasteurization and the use of machines to ladle the cheese into moulds, which had become widespread by the 1950s. They began to produce a more standardized cheese in terms of appearance; by the 1970s the popular white Camembert product replaced earlier cheeses more varied in colour. In short, by the early twenty-first century Camembert cheese had been gradually stabilized and standardized in terms of taste and appearance, characteristics that in the past reflected local peoples and places. Cheese purists contend that traditional Camembert is creamier, has a sharper flavour, and has a different texture.

Small local producers continue to campaign against this standardization and won a landmark victory in 2008, when a French government committee decided that traditional methods of making Camembert, using raw milk and no machinery, had to be maintained. But there is now a row concerning what name can be used for the non-traditional versions. The traditional cheese is the only one legally entitled to be labelled Camembert de Normandie (Camembert of Normandy), but their industrial-scale competitors have adopted a very similar name, Camembert fabriqué en Normandie (Camembert made in Normandy). Although the latter name is geographically correct, traditional small-scale producers, who argue that mass-market industrial companies are intentionally misleading consumers with the marketing of their cheese, are suing over the legality of using it. The issue here is both cultural and economic, with traditional producers claiming that the distinctive identity of a place (the cheese is named after the village of Camembert) is threatened but also being concerned at their loss of market share (the industrial producers currently hold about 95 per cent of the market).

The importance of place labelling in food production and consumption is also evident in the case of Emmentaler cheese, often called Swiss cheese in North America. Since the twelfth century this cheese has been produced in the Emmentaler region of Switzerland and is an iconic Swiss food product. Because of its excellent reputation, falsely labelled inauthentic products have appeared in shops around the world. Estimates suggest that about 10 per cent of Emmentaler-labelled cheese available for purchase is not produced where or how it ought to be.

Darjeeling Tea

An interesting example of using a consumer preference to highlight regional political identity arose in 2008. Darjeeling tea—regularly described as the champagne of teas—is grown in Gorkhaland, in India's northern West Bengal state, an area where ethnic Gorkhas (or Gurkhas as they have often been called) are campaigning for a separate political unit (this is one of the several Indian cases of claims for linking ethnic identity and territory noted in Chapter 9). Estate owners in the region are being pressured into adding the phrase "The flavour of Gorkhaland" to packets of Darjeeling tea. The idea is that such labelling would highlight Gorkha ethnic and regional identity globally.

Changing Food Preferences

Globally, livestock contribute about 13 per cent of calories consumed and about 28 per cent of protein consumption (especially meat, milk, and eggs). It is possible to have a healthy diet without animal products, but even in small amounts they do provide nutritional benefits. Livestock protein is important for children and pregnant and lactating women, although consuming too much animal protein, specifically red meat, causes increased health risks.

The contribution of livestock products to diets has risen greatly in recent years and is expected to continue increasing. Livestock is the fastest-growing sector of the global agricultural economy and the FAO anticipates significant dietary changes in the less developed world by 2050. In 2000 cereals contributed 56 per cent of all calories consumed in less developed countries, but this amount is expected to decline to 46 per cent, while the calorific contribution of meat, dairy, and vegetable oils is expected to grow from 20 to 29 per cent. The example of China is compelling. In 1961, per capita meat consumption was 3.6 kg, but by 2002 it was 52.4. Today, about one-half of global pork consumption is in China.

Why are these changes occurring? As more and more of us move into cities and become more affluent, food consumption changes (the process of ongoing urbanization is discussed in the following chapter). In general, urbanites are wealthier and eat more food than poorer rural dwellers; they eat less fresh food but more processed food and more meat. They also are able to access food from a greater variety of sources than are rural dwellers. This shift in diets is having major environmental impacts as the livestock sector generates significant greenhouse gas emissions and is also a major source of land and water degradation. The one counter trend is that some established wealthy urbanites in more developed countries are consciously choosing to diversify their diet for health reasons, in particular by limiting (mainly red) meat consumption.

Losing and Wasting Food

A 2011 FAO report estimated that about one-third of all food produced globally is not consumed. This means that all the inputs, including land, water, human labour, finances, and energy sources are also wasted. Loss and waste occur throughout the agricultural supply chain, from farm to human consumption. FAO estimates suggest that per capita food loss or waste in more developed countries is about 280–300 kg/year (total per capita production is about 900 kg/year), while in less developed countries it is about 120–70 kg/year (total per capita production is about 460 kg/year).

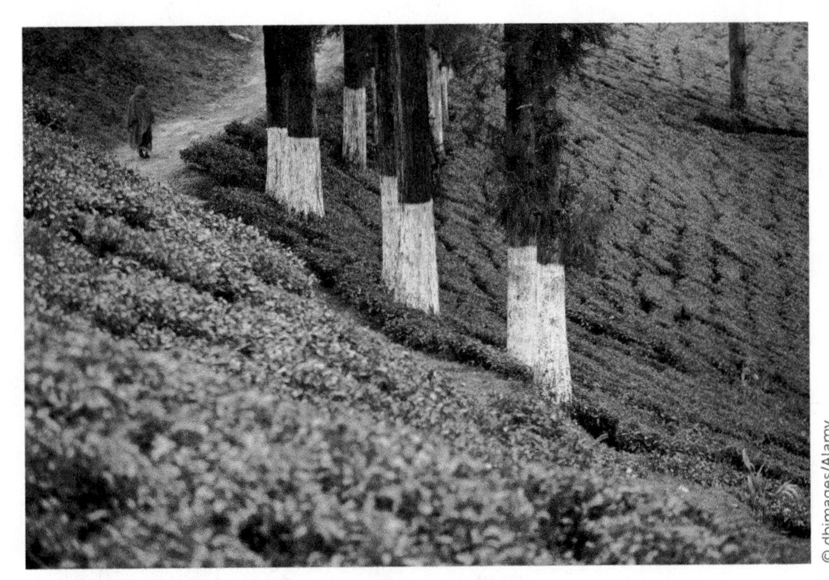

Happy Valley Tea Plantation in Darjeeling, India.

In less developed countries, food is lost rather than wasted. Most of this loss occurs in the production and transportation stages of the supply chain, occurring as it does on or close to the farm; very little is wasted at the consumer stage. For example, locusts, rats, and mice eat crops before harvesting or while stored prior to human consumption. Perishables, such as vegetables and milk, are liable to spoil while being transported. All these losses are avoidable in principle as they are related to problems of rural poverty, inadequate agricultural infrastructure, limited harvesting strategies, and inadequate storage and cooling facilities rather than to inappropriate human behaviour. Additional investment in the agricultural sector is needed to reduce these losses and thus increase the amount of food available for human consumption.

In more developed countries, the problem is one of food waste, occurring as it does principally at the consumption end of the supply chain. In per capita terms, this food waste exceeds the food loss in less developed countries. British and US studies suggest that at least 25 per cent of food purchased from shops is thrown away, mostly salad ingredients, while a similar amount of restaurant food is discarded. The problem here is avoidable as the cause is inappropriate human behaviour—people can afford to waste food, so they do. Some specific causes of waste are high consumer expectations of food quality, the practice of indicating a "sell by" date, and buffet-style restaurant meals. Of course, food waste would be lessened if more food items were purchased as frozen rather than fresh and, not insignificantly, a 2012 survey in Britain highlighted the fact that it was less expensive to buy frozen versions of many of the most popular food items, such as pizza, chicken, sausages, and vegetables.

CONCLUSION

Despite its age and some obvious weaknesses, von Thünen's work remains the fundamental answer to the question of agricultural location for at least two reasons. First, his decision to simplify the issue by excluding certain variables was a highly original contribution; this basic method has been used by many other location theorists, as we will see in Chapters 11 and 13. Second, the key explanatory variable is distance as measured in terms of transport cost. Human landscapes in general are clearly related to distance, regardless of how we choose to measure it—straight-line distance, transport-cost distance, time, or any other format. Relationships between distance and spatial patterns, such that some spatial regularity is evident, lie at the heart of much theory and analysis.

The technological advances in agriculture discussed in this chapter are impressive, having permitted a huge increase in population since about 1700. But an important (Malthusian) question remains. As described in Chapter 5, there is good reason to believe that the global population will increase from the current 7.3 billion to about 9.7 billion by 2050. Is it possible to produce enough food to feed this growing population?

Any attempt at a comprehensive answer must necessarily confront a number of interactive unknowns, all of which are discussed in this or earlier chapters. Principal issues discussed in this chapter relate to the decreasing variety of food available; changing consumption patterns, including an increasing preference for meat; food waste; development of new crop varieties; willingness to accept GM crops; and advances in farming strategies. Issues referred to earlier include diversion of grain to livestock and to biofuels, rising energy costs, the price of food, the impact of global warming, the availability of water, and competition for resources and related armed conflicts. Expressed more succinctly, the future of food supplies is uncertain.

But there are reasons to be optimistic. Although there seems little reason to believe that food supplies will increase through use of more land, irrigation, or fertilizer, other strategies may be highly effective. Most important, potentially, are advances in genetic technologies. While genetic modification is a matter of public concern, it is important to note that genetic selection permits faster and more exact breeding and will likely allow steadily increasing yields of staple crops. Continuing research into hybrid seed technologies, especially for corn and sorghum, promises increased yields in Latin America and Africa. Changes to livestock farming, involving abandonment of traditional outdoors feeding on farm waste and introduction of indoor confinement with controlled animal movement, diet, and health, will greatly increase yields. An important consideration of such changes is the severe compromise of animal welfare. Significant reduction in food loss and food waste, if these can be achieved, will also increase supplies considerably.

The message is clear: the best likelihood of producing enough food to feed 9.7 billion people is through a combination of increasing research, changing diets, and reducing food loss and waste.

Summary

Economic Geography

Much of our behaviour that creates human landscapes has strong economic motivations. Contemporary studies of agriculture, industry, and settlement take various approaches and may incorporate not only economic but also cultural and political content.

The Role of Theory

Location theory aims to explain geographic patterns. Theories are valuable because of their rigour and simplicity. Many of the theories used by geographers are normative, describing what the theorist thinks ought to be rather than what is. The concept of the economic operator assumes that the goal of any economic activity is the maximization of profit in the context of perfect knowledge and a perfect ability to use such knowledge in a rational fashion.

The Location of Agricultural Activities

Many factors can influence the location of agricultural activities. For some geographers, an ecological approach, considering all factors, is the most appropriate. Other geographers focus on particular factors, depending on the issue in question. Physical factors include climate, soils, and relief. As living things, animals and plants require appropriate physical environments, either natural or human-constructed. Cultural, social, and political factors also play a role. A farmer's religious and ethnic background affects his or her decisions, as do the policies of various levels of government. Economic factors include the basic principles of supply and demand.

Economic Rent

The concept of economic rent is central to agricultural location theory. Land uses compete for locations because it is not possible for all activities to occupy their economically optimal space. Instead, land is devoted to certain uses according to a spatial pattern that reflects their differing economic rents (itself a specific measure of rent-paying ability). Economic rent can be based on one or more variables; Ricardo based such values on land fertility, while von Thünen used distance from market.

Von Thünen Theory

In 1826 von Thünen, a German economist, published a landmark theory of agricultural location. A normative theory that holds all variables constant except distance from a central market, it generates two hypotheses: zones of land use develop around the market; and the intensity of each use decreases with increasing distance from the market. Von Thünen proposed particular contents of the zones for his time and place (early nineteenth-century north Germany). This theory and its variations have stimulated much geographic research.

Empirical Analyses

Studies of the relationships between land value and distance and between land use and distance can be conveniently subdivided into those with a historical focus, those in less developed areas of the world, and those on various scales ranging from the world to individual farms. Overall, these analyses confirm the value of the von Thünen theory as a basis for investigating agricultural patterns.

Agricultural Origins and Change

Agriculture originated in the domestication of plants and animals—a long process that began about 12,000 years ago, probably in several different areas, and gradually diffused. Pre-agricultural activities, such as hunting and gathering, are marginal today. Major technological changes include a series of advances in eighteenth-century England, the twentieth-century development of nitrogen fertilizers, and the green revolution in the less developed world. Most recently, changing the genetic composition of food crops may be the greatest change in agricultural production since domestication began.

Regions and Types

There are nine principal types of agriculture, each occupying a particular region or regions. Three types are classified as subsistence (shifting agriculture, wet rice farming, pastoral nomadism); a fourth type, Mediterranean agriculture, is mixed subsistence/commercial; and the remaining five are commercial (mixed farming, dairying, plantations, ranching, large-scale grain production). All but one type can be traced back thousands of years; the exception is plantation agriculture, which developed as a result of European overseas expansion. The modern versions of dairying, ranching, and large-scale grain production are very different from their antecedents.

Agriculture in the World Economy

Relative to other economic activities such as industry, agriculture has experienced a continuous decline in importance as measured by labour force and contributions to GDP.

Trade in Food

Food is an important component of world trade. Much food is sent as aid to poor countries. The former USSR and several countries in Asia are major importers of grain. The United States is the principal exporter.

Agriculture in the Less Developed World

China is currently experiencing major changes in the organization of agricultural activities as part of a transition from strictly communistic practice. In some other areas, such as Bhutan, green revolution technologies have caused environmental problems. Recently, human geographers have employed a political ecology approach, combining traditional human ecology with political economy to address specific problems such as land-use change, land-use conflict, and the involvement of women in agriculture.

Agriculture as an Industry

In many respects, agriculture functions as an industrial activity with several linked levels: input producers, farmers, processors, wholesalers and retailers, and consumers. National and international policies exert considerable influence on agricultural activities at all levels. Partly because of this organizational complexity, traditional von Thünen theory is inadequate to explain the present spatial distribution of agricultural activities.

Producing and Consuming Food

Historically, the kinds of food consumed in various parts of the world primarily reflected available resources. The contribution of livestock products to diets has increased greatly in recent years and is expected to continue increasing. Critically, much food produced is never consumed as huge amounts are lost or wasted. Much interest is focused on trends in consumption: the act of consuming is seen as a way of distinguishing oneself from others, and perhaps even of establishing superiority.

Links to Other Chapters

- **Globalization:** Chapter 2 (concepts); Chapter 3 (all of chapter); Chapter 4 (global environmental issues); Chapter 5 (population growth, fertility decline); Chapter 6 (refugees, disease, more and less developed worlds); Chapter 8 (popular culture); Chapter 9 (political futures); Chapter 11 (global cities); Chapter 13 (industrial restructuring)
- **Cultural and political considerations:** Chapter 7 (religion); Chapter 8 (ethnicity); Chapter 9 (role of the state)
- **Location and interaction theories; the economic operator concept; von Thünen theory:** Chapter 2 (positivism, concepts of space, location, and distance); Chapter 11 (central place theory); Chapter 13 (industrial location theory)
- **Correlation and regression analysis:** Chapter 2 (quantitative techniques)
- **Origins of agriculture:** Chapter 2 (environmental determinism, possibilism); Chapter 4 (energy and technology); Chapter 5 (population growth through time); Chapter 6 (origins of civilization)
- **Second agricultural revolution:** Chapter 2 (Marxist terminology); Chapters 4 and 13 (Industrial Revolution); Chapter 5 (Malthusian concepts)
- **Nine agricultural regions:** Chapter 2 (environmental determinism, possibilism)
- **Plantation agriculture; green revolution; agriculture in the less developed world:** Chapter 5 (fertility transition); Chapter 6 (population and food, world systems analysis); Chapter 8 (indentured labour)
- **Industrialization of agriculture; global trade in food:** Chapter 13 (restructuring)
- **Political ecology of agriculture:** Chapter 2 (Marxism); Chapter 4 (environmental ethics); Chapter 8 (gender)

Questions for Critical Thought

1. How has the subdiscipline of economic geography philosophically changed since the 1950s?
2. Summarize the general factors that affect where specific agricultural activities are located. How important are physical factors?
3. How important or relevant is von Thünen's theory in understanding patterns of agricultural production today?
4. What role did the Industrial Revolution play in the development of agriculture?
5. What agricultural advances have been made through the green revolution? What criticisms have been directed at the green revolution?
6. How has globalization affected world patterns of agricultural production in the late twentieth and early twenty-first centuries?
7. If the shelves of Canadian grocery stores can be stocked with food produced in other countries, why should we support Canadian farmers?

Suggested Readings

Visit the companion website for a list of suggested readings.

11 AN URBAN WORLD

This chapter, the first of two concerned with urban geography, focuses on the urban world, where more than half the world's population lives. Our discussion includes a substantial overview of ongoing urbanization processes at both the global and regional scales; we examine problems of definition, the different experiences of more and less developed countries, and the rise of megacities. An account of city origins builds on ideas about the rise of civilizations as discussed in Chapter 6, while one of pre-industrial and industrial cities relates to the rise of a capitalist mode of production. Central place theory—a highly influential contribution that asks and answers the question, "Why are urban centres located where they are?"—is described and evaluated in detail in the context of both urban systems and hierarchies of cities. The chapter concludes with a discussion of global cities, with a focus on their characteristics and their interconnectedness within an urban hierarchy. This section builds on the Chapter 3 concern with globalization, as global cities are seen as the command and control centres for the global economy.

Here are three points to consider as you read this chapter.

- How do ongoing urbanization experiences in more and less developed regions of the world differ?
- In what way is the origin and subsequent growth of cities connected to the development of agriculture?
- In what sense do global cities function as command centres in a globalizing world?

Motorists travel on a street in downtown Paris, one of the world's megacities. With most of the world's population now living in urban areas, topics such as defining urban areas in different parts of the world, the effect of technology on settlement patterns, and the development of global cities have become more significant.

Permanent settlements began to form with the introduction of agriculture. Currently, only a minority of human societies depend on hunting and gathering or agricultural activities that involve regular movement from place to place. Permanent settlements developed in association with specific economic activities, initially producing food for subsistence, later distributing surpluses, and eventually developing into trade and industrial centres. Principally agricultural settlements are usually called rural; those that are principally non-agricultural are called urban.

Urban geography, which focuses on cities and towns, has two main approaches. First, urban geographers investigate systems of cities by understanding the spatial distribution of cities and the complex patterns of interactions between them. Second, geographers study cities as systems, that is, the human geographic patterns and interactions within cities and the internal structure of cities. This chapter addresses the first approach; Chapter 12, the second.

A number of the concepts and empirical analyses presented in this chapter are relatively new. In particular, the discussion of global cities represents a marked departure from the urban geography studied just a few decades ago. Whereas the latter emphasized national urban systems and hierarchies, the former has an additional focus on world systems and hierarchies, which are emerging in response to and as a component of the ongoing processes of economic, cultural, and political globalization discussed in Chapter 3. Geographers are especially interested in the way new communication technologies liberate people and businesses from the need to be located near one another to work together.

urbanization
The spread and growth of cities; an increasing proportion of a population living in urban areas (cities and towns).

As urbanization unfolds, an increasing proportion of a country's population begins living in cities. This urbanization is matched by urban growth—the increase in the number of people actually living in particular places, which is typically facilitated by new housing construction on the periphery of the urban area. In Delhi, India, the new housing takes the form of high-rise residen. © Tristan Savatier.

AN URBANIZING WORLD

In 2007 the world's urban population exceeded the rural population for the first time in history. This change is a significant turning point both in terms of human settlement and in the relations between humans and land more generally. Although many of the world's major cities have long histories, the fundamental need to find or grow food meant that more people lived in rural rather than urban settings. Villages arose with the beginning of farming, approximately 12,000 years ago, but even then most people still needed to be close to their crops and animals. The first large permanent settlements began to develop about 3,500 years ago due to improvements in farming technologies and the expansion of trade networks, which freed some people from the need to produce food and allowed them to engage in other economic, political, or social activities.

For several thousands of years after the emergence of the first urban centres, the growth of urban populations was slow. It was not until the onset of the Industrial Revolution, when peasants left their rural homes to find work in factories, that the pace of urbanization picked up. The rapid growth of cities is a recent phenomenon that has followed different trajectories in the more and less developed parts of the world. These paths need to be examined in the context of recent increases in total world population and projected future increases, as discussed in Chapter 5.

The transition from a rural world to one of urban dominance has been very rapid and very recent. In 1800 urban dwellers were only 3 per cent of the world population. A century later, only 14 per cent lived in a town or city and just 16 cities in the world had more than a million people. But more than half the people in the world today are urban dwellers, and there are almost 500 cities with populations over a million. According to the UN, the urban share of the global population is expected to reach nearly two-thirds by 2050.

Defining Urban Areas

Intuitively, we all understand what an urban area is—a concentration of people living at a relatively high density and perhaps engaging in some forms of work and not others (i.e. manufacturing or office work, as opposed to farming or forestry). Further, urban areas are typically viewed in opposition to rural areas. We have a dichotomous system; whatever is urban is not rural and vice versa. The key is, "How do we more precisely define this dichotomy?"

To assess urban population trends over time, we need to ensure we are using consistent definitions. Fifty-four per cent of the world population in 2014 was defined by the UN as urban. However, this level of precision relies on definitions that differ between countries and therefore creates a degree of uncertainty. Table 11.1 shows some of these definitions. If a country changes its definition of ***urban***—for example, by raising or lowering the minimum number of residents required for a settlement to qualify for urban status—the percentages of urban and rural dwellers in that country will change accordingly. If that country has an extremely large population, the shift may affect the urban–rural proportions for the world as a whole. China is an example of this challenge.

Defining Urban Areas in China

China has more urban residents than any other country in the world. However, the exact number is largely unclear because the country has probably the most complex and confusing set of guidelines for identifying whether a place is urban. The trouble stems from the fact that a Chinese city, or municipality (known as a *shi*), is regarded as the basic administrative unit for urban areas. However, most Chinese cities are situated within a sometimes very large and highly populated rural area (usually agricultural) and consist of an urbanized core (i.e. a highly built-up and densely settled area) that is surrounded by a scattering of towns, suburbs, and other cities.

TABLE 11.1 | Some definitions of urban centres

France	Communes containing an agglomeration of more than 2,000 inhabitants living in contiguous houses or with no more than 200 metres between houses; communes of which the major part of the population is part of a multi-communal agglomeration of this nature.
Portugal	Agglomerations of 10,000 or more inhabitants.
Norway	Localities of 200 or more inhabitants.
Israel	All settlements of more than 2,000 inhabitants, except those where at least one-third of the heads of households, participating in the civilian labour force, earn their living from agriculture.
Canada	Places of 1,000 or more inhabitants having a population density of 400 or more per km^2.
United States	Places of 2,500 or more inhabitants with a population density of at least 500 persons per square mile.
Botswana	Agglomerations of 5,000 or more inhabitants and where 75 per cent of the economic activity is non-agricultural.
Ethiopia	Localities of 2,000 or more inhabitants.
Mexico	Localities of 2,500 or more inhabitants.
Argentina	Populated centres with 2,000 or more inhabitants.
Japan	Cities (*shi*) having 50,000 or more inhabitants with 60 per cent or more of the houses located in the main built-up areas and 60 per cent or more of the population (including their dependants) engaged in manufacturing, trade, or other urban types of business. Alternatively, a *shi* having urban facilities and conditions as defined by the prefectural order is considered urban.
India	All places with a municipality, corporation, cantonment board, or notified town area committee, etc. A place satisfying the following three criteria simultaneously: a minimum population of 5,000; at least 75 per cent of the male working population engaged in non-agricultural pursuits; and a population density of at least 400 per km^2.

Note: This table confirms that there is no simple, agreed-upon definition of *urban centre* based on population size and emphasizes the danger of comparing data for different countries.

Sources: Carter, H. 1995. *The Study of Urban Geography*, 4th edn. London: Arnold, 10–12; www.iiasa.ac.at/Research/LUC/ChinaFood/data/urban/urban_8.htm; http://72.14.207.104/search?q=cache:mDsAr85KWvEJ:www.censusindia.net/2001housing/metadata.pdf+.

Chongqing, which has rapidly ascended the list of the world's largest cities (Box 11.1), is an example. The areal extent of its administrative area is about the size of Austria, and the population is about the same as Canada. About two-thirds of the population work as farmers. Chongqing is not a city, or even a metropolitan area, but a region of 32 million people, only a fifth of whom actually live in the city; the rest live in the surrounding towns and rural areas. As a result, Chongqing is not the largest Chinese city but is more accurately ranked as the seventh largest urban population (Chan, 2009).

Defining Urban Areas in Canada

As noted earlier and shown in Table 11.1, defining the dichotomous nature of urban and rural areas can be difficult and is often culturally specific. It can be even more challenging to define these different settlement types when we recognize that they fall along a continuum from highly urbanized at one end to widely dispersed (rural) at the other.

city
A legally incorporated self-governing unit; an inhabited place of greater size, population, or importance than a town or village.

Canada began to try to address this issue in 2011. Since 1971 the government has defined urban areas as any place with a population in excess of 1,000 people, living at a population density of 400 persons per square kilometre or greater. Recognizing that an urban area of a thousand people is very different from one of five million, Statistics Canada—the official statistical gathering organization in Canada—has made some changes. It will no longer use the term *urban area* but will consider places as population centres, although the basic definition will remain the same. Population centres will be categorized into one of three types, based on population size:

- Small population centre: Population between 1,000 and 29,999
- Medium population centre: Population between 30,000 and 99,999
- Large population centre: Population greater than 100,000

Conceptualizing Cities

City and *urban area* are often used interchangeably. Should they be? Do they mean the same thing? Technically, the answer is no. A **city** is

In the News

BOX 11.1 | Listing the "20 Largest Cities"

Compiling a list of the world's 20 largest cities based on population is not as straightforward as one might expect. While most scholars agree that Tokyo is the largest city in the world, this idea is not universally accepted. One list published in 2003 had New York as the largest; a 2005 issue of *Time* stated that Chongqing is the largest; and BBC news reported in 2012 that Tokyo, Seoul, Chongqing, or Shanghai might be considered the largest. Why the uncertainty?

The principal reason is that a specific urban centre can be defined in more than one way, with the result that population counts for that centre can often vary widely. Three commonly used definitions are the municipal, or legal, definition; the urban agglomeration, or physical city definition; and the sphere of influence definition. By way of illustration, consider the case of Toronto. In 2006 Toronto proper (the municipality) covered about 640 km^2 and had a population of 2.6 million; the census metropolitan area (CMA; see Box 11.2) covered about 5,500 km^2 and had a population of 5.1 million; and the Greater Toronto Area (the sphere of influence) covered more than 7,000 km^2 and had a population of 5.6 million. As such, how do we definitively state Toronto's actual size?

Returning to the world's largest cities, Tokyo's urban population can range from about 8 million if only the central city is included to more than 40 million if the entire National Capital Region is counted; this continuous urban region includes Yokohama and more than 80 other urban areas. As noted in the main text, it appears that many people in Chongqing are not urbanites but are agricultural workers living rural lives in a mostly rural setting.

No two lists of the world's largest cities are likely to be identical. A recent analysis of 8 lists of the world's top 20 cities found that, together, they included a total of 30 such cities. To remain reasonably consistent, most of the data used in this chapter are derived from the United Nations Department of Economic and Social Affairs, Population Division. This body favours the urban agglomeration definition, as it is perhaps the most meaningful and readily understood method of thinking about urban areas. That said, some countries (including China) do not provide data in this way.

a legally defined entity, a municipality. Cities are incorporated places, typically bigger than a town or village, with fixed boundaries and a locally elected government (i.e. a city council) that makes decisions about taxes and the provision of essential services. In many cases, cities include both densely settled areas along with lower density suburbs. But in other cases, especially in the United States, many suburban areas lie outside the city's boundaries and are thus governed separately; the result is often the duplication of services and unplanned and uncoordinated suburban growth. Nevertheless, when most people refer to New York City, they do not mean the incorporated city of New York but the entire place, including the main five boroughs and its suburban reaches in New Jersey, Pennsylvania, and Connecticut. This latter conceptualization is either the urban area or the metropolitan area.

Urban areas, as previously outlined, include built-up areas of various population sizes, from about 200 people (as defined in many northern European countries) to tens of thousands of people (as defined in Japan). As in Canada, urban areas can be villages, towns, cities, suburbs, or a mix of cities and suburbs. One simple way of thinking about an urban area is to recall what you can see when you fly over a place at night; the extent of the illuminated part can be equated with its urbanized area. Typically, the urban area is more spatially expansive than the city proper, as the built-up area usually spills across the municipal boundaries. But, as with China, the city's (*shi*'s) jurisdictional boundaries can sometimes be larger than the built-up urban area.

A third conceptualization, known as the metropolitan area, includes urban areas (which by definition include incorporated municipalities) and non-urban (rural) areas that are functionally connected to the city. In some cases, people live in one urban area but commute to another to work or shop; if these urban areas are deemed functionally connected in such a way, the entire zone—including the urban areas of two or more municipalities—and the intervening rural area may be considered a metropolitan area. For example, people commute to Manhattan from places such as New Haven (Connecticut), Newark (New Jersey), and Allentown (Pennsylvania), all of which are urban areas on their own but are functionally connected to New York as well. Therefore, these suburbs are part of metropolitan New York.

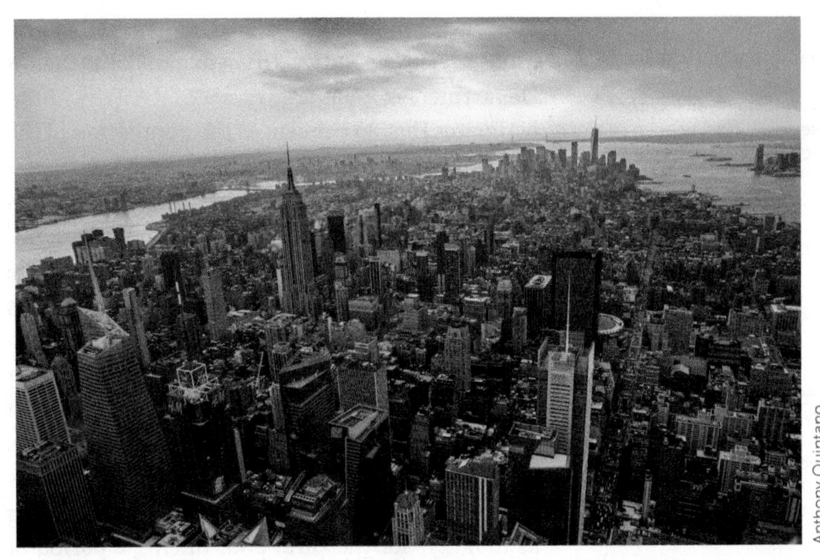

Anthony Quintano

A view of midtown Manhattan, the most densely populated of New York City's five boroughs. The city is also the workplace of over two million people from the entire metropolitan area.

Urbanization in More and Less Developed Regions of the World

In the more developed world, the Industrial Revolution provided the impetus for the rapid growth in the number and size of urban centres. This change was partly caused by the increase in migration of people from rural to urban areas and the rapid total population increase during the second and third stages of the demographic transition (see Chapter 5). Growth involved an expansion of the urban area, the creation of suburbs,

suburb
A residential or mixed (residential and employment) use area on the periphery of the city, typically displaying some degree of homogeneity in terms of economic, social-cultural, and/or built form.

Andreas Praefcke

Low density suburban development, such as this one in Dallas, Texas, has become the norm in most North American cities since the end of the Second World War.

urban area
The spatial extent of the built-up area surrounding and including an incorporated municipality, such as a city; typically assessed by its population size and/or population density and/or nature of residents' employment.

metropolitan area
A region comprising two or more functionally connected urban areas and the less densely populated (or built-up) areas in between; examples include metropolitan New York and the Greater Toronto Area.

urban sprawl
The largely unplanned expansion of an urban area into rural areas.

and, increasingly, **urban sprawl** (whereby rural areas are enveloped by urban expansion). The UN Department of Economic and Social Affairs (2015) estimates that 80 per cent of the population of the more developed world was urban and only 20 per cent rural in 2014. That same year, urban dwellers accounted for 82 per cent of the Canadian population and 81 per cent of the US population. The urban proportion in the more developed world is expected to increase to 85 per cent by 2050.

In the less developed world, the urbanization process was and is very different. Rapid urban growth started later, in many cases as recently as the mid-twentieth century, and the process was even more dramatic, with urban centres acting as magnets for the rural poor during a period of exponential population growth in these regions. In fact, there is a close relationship between the rapid urbanization of the less developed world and rural poverty. In some cases, cities grew through rural-to-urban migration; in other cases, rural settlements were (and continue to be) transformed into urban centres through natural increases in population. Both processes often involve the addition of informal (squatter) settlements at the margins of the urban area. By 2014, 48 per cent of the population of the less developed world was urban and 52 per cent rural. The former is expected to increase dramatically to 63 per cent by 2050 (UN DESA, 2015). The social, environmental, and economic implications of this dramatic urban growth in many parts of the less developed world are discussed towards the end of Chapter 12.

Figure 11.1 shows the relative numbers of rural and urban dwellers in the world from 1950 to 2050. The rapid convergence of the two lines between 1950 and 2007 is clear, as is the subsequent rapid divergence, with the urban population reaching a projected 66 per cent by 2050. Figure 11.2 distinguishes between the more and less developed worlds through the same time period, showing that the greatest increases have and will occur in the less developed world's urban centres. As the figure demonstrates, the number of urban dwellers in the less developed world is expected to exceed the number of rural dwellers in the years to come.

Table 11.2 provides data on the average annual rate of change for urban, rural, and total populations over three periods. As these data show, the rates of change are declining in all cases. Especially notable are

- the relatively high projected growth rate for the urban population of the less developed world;
- the negative projected growth rate for the rural population in both the more and less developed parts of the world; and
- the world's projected urban population growth rate, which is almost twice the projected total population growth rate for the 2015 to 2050 period.

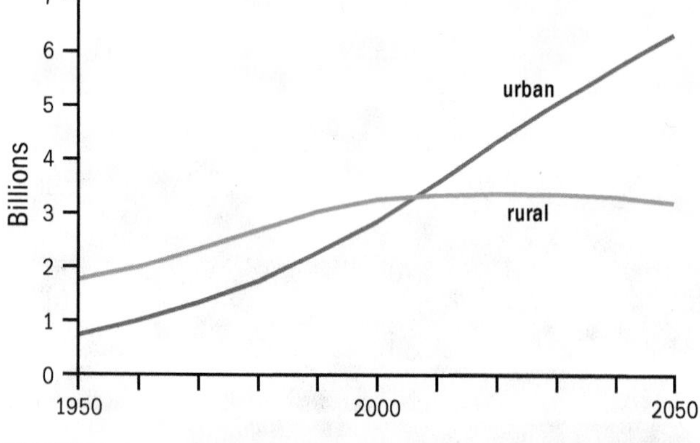

FIGURE 11.1 | World rural and urban populations, 1950–2050

Source: Adapted from United Nations Department of Economic and Social Affairs, Population Division. 2015. *World Urbanization Prospects: The 2014 Revision*. New York: United Nations. http://esa.un.org/unpd/wup/.

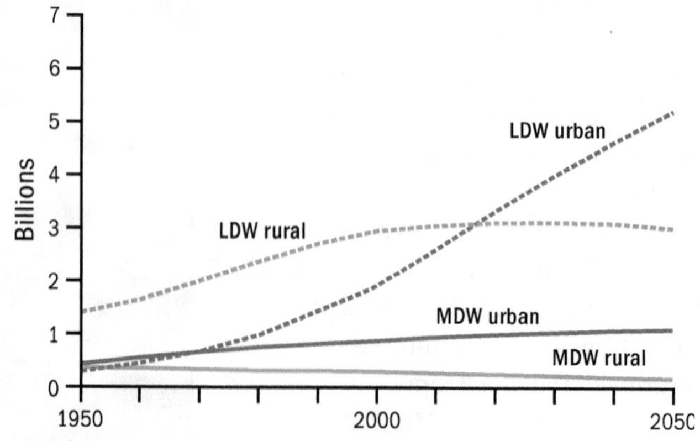

FIGURE 11.2 | Rural and urban populations of the more developed and less developed regions of the world, 1950–2050

Source: Adapted from United Nations Department of Economic and Social Affairs, Population Division. 2015. *World Urbanization Prospects: The 2014 Revision*. New York: United Nations. http://esa.un.org/unpd/wup/.

TABLE 11.2 | Population growth rates (total, urban, rural): 1950–2000, 2000–2015, and 2015–2030

	Annual Rate of Change (%)		
	1950–2000	2000–2015	2015–2030
Total Population			
World	1.77	1.19	0.76
More developed world	0.77	0.36	0.10
Less developed world	2.12	1.38	0.88
Urban Population			
World	2.68	2.17	1.35
More developed world	1.38	0.72	0.35
Less developed world	3.75	2.74	1.61
Rural Population			
World	1.22	0.19	−0.13
More developed world	−0.36	−0.79	−1.05
Less developed world	1.48	0.29	−0.07

Source: Adapted from United Nations Department of Economic and Social Affairs, Population Division. 2015. *World Population Prospects: The 2014 Revision, Data Tables and Highlights*. New York: United Nations.

With reference to the last point, the projected annual world urban growth rate of 1.35 per cent means that the urban population will double in about 52 years.

Other Regional Variations

The differences between values in the more and less developed regions of the world show how misleading using world averages can be. But there are also some notable regional differences even within these two categories (Figure 11.3). In the less developed world, Latin America and the Caribbean were already highly urbanized in 2011, with about 80 per cent urban population compared to 40 per cent in Africa and 45 per cent in Asia. All

Already among the world's largest cities, Lagos, Nigeria, is also one of the fastest growing. The city struggles to provide sufficient housing, transportation infrastructure, and other services quickly enough to meet the demand.

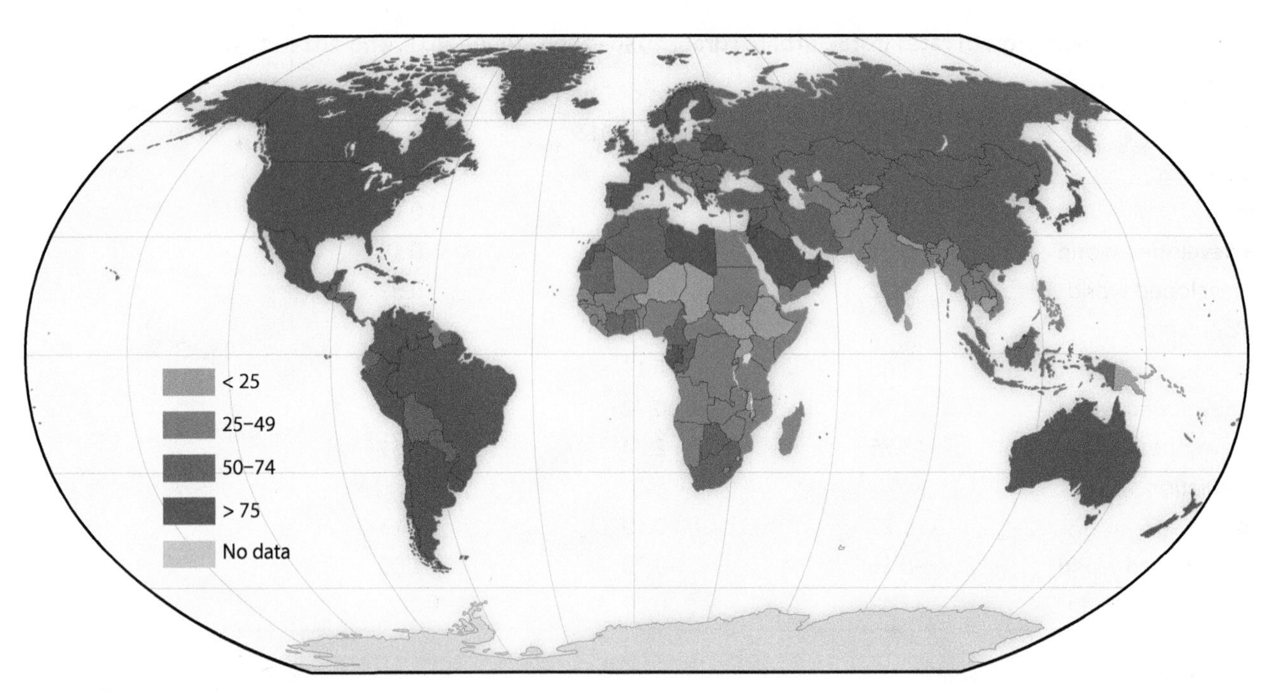

FIGURE 11.3 | Percentage of urban population by country, 2014

Source: United Nations Department of Economic and Social Affairs, Population Division. 2015. *World Urbanization Prospects: The 2014 Revision*. New York: United Nations. http://esa.un.org/unpd/wup/.

these percentages will increase by 2050, with Latin America and the Caribbean reaching 86 per cent, Africa 56 per cent, and Asia 64 per cent. Much of the expected growth in the Asian urban population will come from its two largest countries, China and India, which are still relatively non-urban; the proportion of the population living in urban areas is 55 per cent in China and only 31 per cent in India.

In the more developed world, the European urban population is projected to increase between 2011 and 2050 from 73 to 82 per cent, while in North America the projected increase is from 81 to 87 per cent. The situation of urban settlement in Canada is discussed in more detail in Box 11.2. Given that countries of the more developed world are already highly urbanized, there is little opportunity for continued urbanization. However, urban growth can still occur, as cities can continue to get larger via immigration and natural increases in population. But once practically everyone lives in cities, not much can occur other than deurbanization—a decrease in the proportion of the population living in cities. The future of urbanization in the more developed world is one of the twenty-first century's challenging questions.

Around the Globe

BOX 11.2 | Changing Settlement in Canada

As in most of the world, urbanization continues in Canada, albeit rather slowly. Changes in total, urban, and rural population between 1950 and 2000 and projected through 2015 and 2050 are shown in Table 11.3 and graphed in Figure 11.4. From 60.9 per cent in 1950, Canada's urban population for 2050 is projected to reach 87.6 per cent. In 2015, it is 81.8 per cent.

The urban centres increasing most in population are those located relatively close to the US border, those attracting migrants both from elsewhere in Canada and from overseas, and those with economies based on manufacturing or services. Newfoundland and Labrador is the only region in Canada experiencing a declining urban population. A notable feature of Canadian urbanization

TABLE 11.3 | Canada: Total, urban, and rural population (thousands), 1950–2050

Year	Total Population	Urban Population	Percentage Urban	Rural Population	Percentage Rural
1950	13,737	8,372	60.9	5,365	39.1
2000	30,697	24,398	79.5	6,300	20.5
2015	35,871	29,353	81.8	6,519	18.2
2030	40,617	34,304	84.5	6,313	15.5
2050	45,228	39,616	87.6	5,611	12.4

Source: Adapted from United Nations Department of Economic and Social Affairs, Population Division. 2015. *World Urbanization Prospects: The 2014 Revision*. New York: United Nations. http://esa.un.org/unpd/wup/.

is the continuing concentration of urban population in four areas:

- the Golden Horseshoe region, extending from Oshawa through Toronto to St Catharines in southern Ontario (an area with about 22 per cent of the Canadian population)
- Montreal and the adjacent region
- the Lower Mainland of British Columbia
- the Calgary–Edmonton corridor

Another useful way to describe urbanization in Canada is with reference to CMAs. A CMA is a very large urban area (the urban core) together with the adjacent urban and rural areas (the urban and rural fringes) that have a high degree of social and economic integration with the core. To be defined as a CMA, the urban core population must be at least 100,000 based on the previous census (Statistics Canada, 1999: 183).

In 2011 about 69 per cent of Canada's population lived in the 33 CMAs (Table 11.4 and Figure 11.5). Between 2006 and 2011 only the CMAs as a group have grown at a rate above the national average—7.4 per cent compared with 5.9 per cent. The CMA with the strongest growth rate is Calgary (12.6 per cent, 2006–11), and there is also substantial growth in Edmonton, Saskatoon, Kelowna, Moncton, Vancouver, Toronto, and Ottawa-Gatineau. In some CMAs, such as Greater Sudbury, Saguenay, and St Catharines–Niagara, the population is relatively stable or declining slightly. A phenomenon known as the **donut effect**, where population in the central area is growing more slowly than in the surrounding area, is evident in some CMAs, most notably Regina and Saskatoon.

donut effect
A popular but colloquial term that refers to a pronounced difference in the growth rates between a core city (slow growth or no growth) and its surrounding areas (faster growth), a pattern that resembles the North American fried confection; usually characterized by people moving out of the core or inner-suburbs of a city and moving into newer peripheral suburbs.

Canada's rural population is relatively stable and thus is a declining proportion of the total population. The rural population was 18.2 per cent in 2015. This comparatively small figure is a result of decades of rural-to-urban migration and continued urban sprawl.

TABLE 11.4 | Canada: Census metropolitan area populations (thousands), 2011

Toronto, ON	5,583.1
Montreal, QC	3,824.2
Vancouver, BC	2,313.3
Ottawa–Gatineau, ON–QC	1,236.3
Calgary, AB	1,214.8
Edmonton, AB	1,159.9
Quebec, QC	765.7
Winnipeg, MB	730.0
Hamilton, ON	721.1
Kitchener–Cambridge–Waterloo, ON	477.1
London, ON	474.8
St Catharines–Niagara, ON	392.2
Halifax, NS	390.3
Oshawa, ON	356.2
Victoria, BC	344.6
Windsor, ON	319.2
Saskatoon, SK	260.6
Regina, SK	210.6
Sherbrooke, QC	201.9
St John's, NL	197.0
Barrie, ON	187.0
Kelowna, BC	179.8
Abbotsford–Mission, BC	170.2
Greater Sudbury/Grand Sudbury, ON	160.8
Kingston, ON	159.6
Saguenay, QC	157.8
Trois-Rivières, QC	151.8
Guelph, ON	141.1
Moncton, NB	138.7
Brantford, ON	135.5
Saint John, NB	127.8
Thunder Bay, ON	121.6
Peterborough, ON	119.0

Source: Statistics Canada. 2015. "Population and Dwelling Counts, for Census Metropolitan Areas, 2011 and 2006 Censuses." www12.statcan.gc.ca/census-recensement/2011/dp-pd/hlt-fst/pd-pl/Table-Tableau.cfm?LANG=Eng&T=205&S=3&RPP=50.

Continued

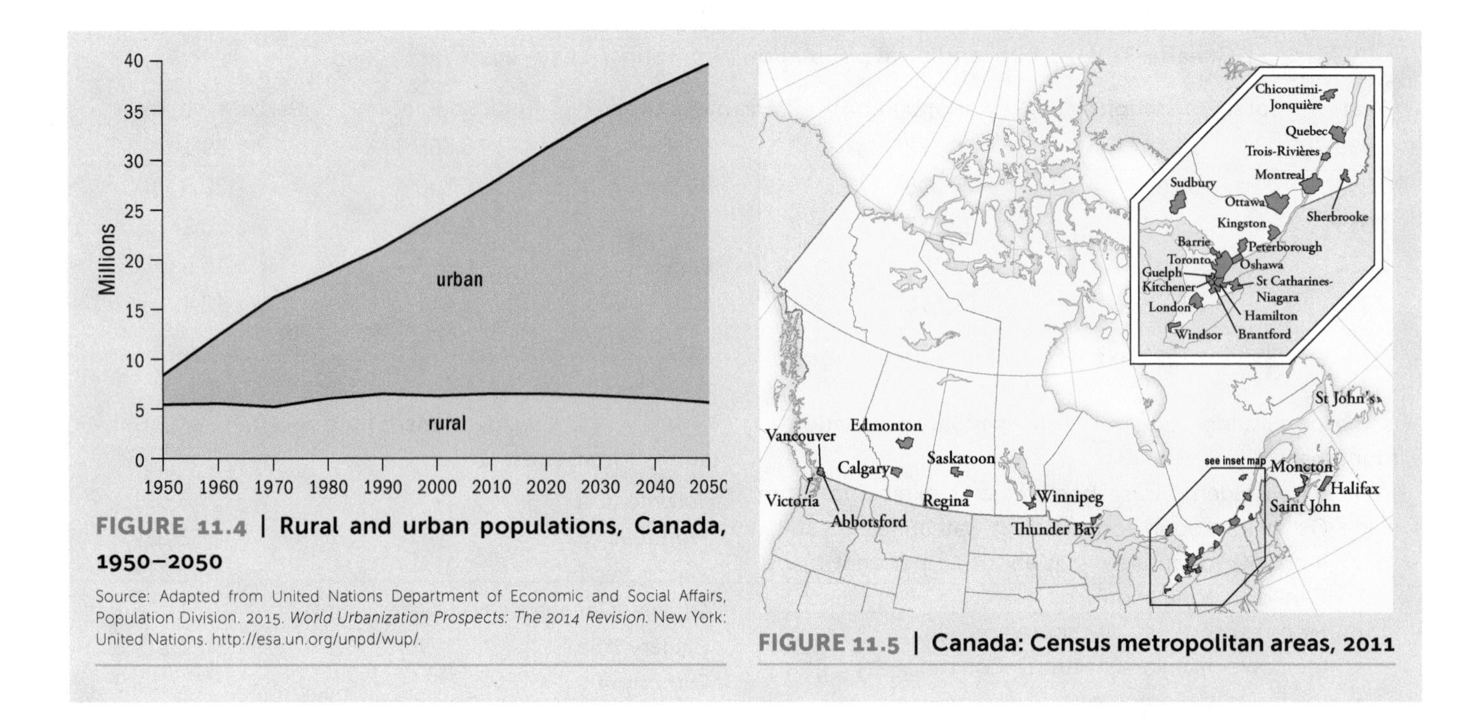

FIGURE 11.4 | Rural and urban populations, Canada, 1950–2050

Source: Adapted from United Nations Department of Economic and Social Affairs, Population Division. 2015. *World Urbanization Prospects: The 2014 Revision*. New York: United Nations. http://esa.un.org/unpd/wup/.

FIGURE 11.5 | Canada: Census metropolitan areas, 2011

Megacities and Megaregions

The urban population is increasing rapidly, both in absolute numbers and in relation to the rural population. This growth is both concentrated in a handful of very large cities and spread across many smaller ones. While the populations of **megacities** are increasing and new ones appear regularly, most additional urban population growth is actually occurring in smaller cities, some with fewer than 500,000 people. Despite the shock value of considering the consequences of cities with 10 million people or more, the more typical urban experience is living in smaller cities, as approximately half of all urban dwellers currently live in cities with less than 500,000 people.

megacity
Metropolitan areas with populations of more than 10 million.

The growing number of megacities, and their increasing size, is highlighted by the following facts:

- In 1950 there were just two megacities: New York with 12.3 million people and Tokyo with 11.3.
- By 1975 there were four megacities: Tokyo with 26.6 million people, New York with 15.9, Shanghai with 11.4, and Mexico City with 10.7.
- By 2014 there were 28 megacities; that number is expected to rise to 40 by 2030, with most of the new megacities located in the less developed world.

Table 11.5 provides data on the number and size of megacities in 2014 and projections for 2030, and Figure 11.6 maps the two patterns (but recall the cautionary comments about city size made in Box 11.1). The vast majority of megacities are located in less developed countries. In 2014 only 6 of the 28—Tokyo, New York, London,

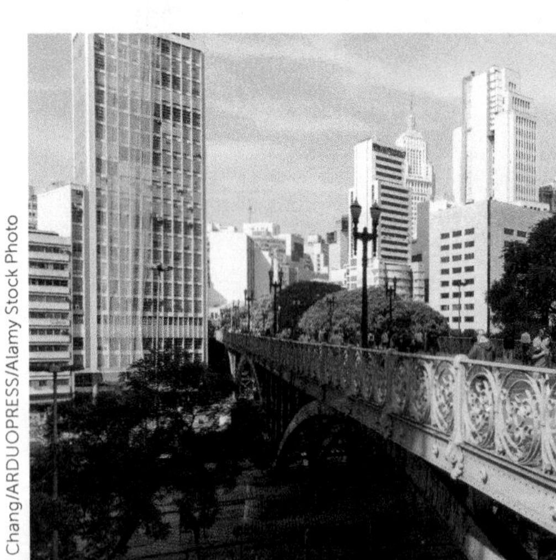

People walk across the Viaduto Santa Ifigenia in São Paulo, Brazil, one of the world's megacities.

Los Angeles, Kinki MMA (Osaka-Kobe), and Paris—were located in the more developed world. The basic pattern is projected to change significantly by 2030, with 12 additional cities added to the list (Table 11.5). Perhaps the most notable change between 2014 and 2030 is the substantial growth anticipated for the African cities of Lagos and Kinshasa and the South Asian cities of Delhi, Dhaka, and Karachi. Each is projected to grow by more than 8 million over the next 15 years, which is almost equivalent to adding a megacity onto an already established one. At the other end of the spectrum, Tokyo is projected to remain the world's largest city in 2030, but its population will likely stay the same or fall slightly between now and then. This pattern of relatively slow growth, if not decline, is common across many megacities in the more developed world, including New York, Los Angeles, London, and Paris.

Shanghai, China, is one of the world's largest urban areas and is becoming one of a handful of top-tier global cities. The district of Pudong, shown here, features many of the city's famous buildings, including the Oriental Pearl Tower (far left).

TABLE 11.5 | Cities with more than 10 million people, 2014 and 2030

2014			2030		
Rank	**City**	**Population (millions)**	**Rank**	**City**	**Population (millions)**
1	Tokyo	37.8	1	Tokyo	37.2
2	Delhi	25.0	2	Delhi	36.1
3	Shanghai	23.0	3	Shanghai	30.7
4	Mexico City	20.8	4	Mumbai (Bombay)	27.8
5	São Paulo	20.8	5	Beijing	27.7
6	Mumbai (Bombay)	20.7	6	Dhaka	27.4
7	Kinki MMA (Osake-Kobe)	20.1	7	Karachi	24.8
8	Beijing	19.5	8	Al-Qahirah (Cairo)	24.5
9	New York–Newark	18.6	9	Lagos	24.2
10	Al-Qahirah (Cairo)	18.4	10	Mexico City	23.9
11	Dhaka	17.0	11	São Paulo	23.4
12	Karachi	16.1	12	Kinshasa	20.0
13	Buenos Aries	15.0	13	Kinki MMA (Osake-Kobe)	20.0
14	Kolkata (Calcutta)	14.8	14	New York–Newark	19.9
15	Istanbul	13.9	15	Kolkata (Calcutta)	19.1
16	Chongqing	12.9	16	Guangzhou-Guangdong	17.6
17	Rio de Janeiro	12.8	17	Chongqing	17.4
18	Manila	12.7	18	Buenos Aires	17.0
19	Lagos	12.6	19	Manila	16.8
20	Los Angeles–Long Beach–Santa Ana	12.3	20	Istanbul	16.7

Continued

TABLE 11.5 | *Continued*

2014			2030		
Rank	**City**	**Population (millions)**	**Rank**	**City**	**Population (millions)**
21	Moscow	12.1	21	Bangalore	14.8
22	Guangzhou-Guangdong	11.8	22	Tianjin	14.7
23	Kinshasa	11.1	23	Rio de Janeiro	14.2
24	Tianjin	10.9	24	Chennai	13.9
25	Paris	10.8	25	Jakarta	13.8
26	Shenzhen	10.7	26	Los Angeles–Long Beach–Santa Ana	13.2
27	London	10.2	27	Lahore	13.0
28	Jakarta	10.2	28	Hyderabad	12.8
			29	Shenzhen	12.7
			30	Moscow	12.2
			31	Lima	12.2
			32	Bogotá	11.9
			33	Paris	11.8
			34	Johannesburg	11.6
			35	Bangkok	11.5
			36	London	11.5
			37	Ahmadabad	10.5
			38	Luanda	10.4
			39	Ho Chi Minh	10.2
			40	Chengdu	10.1

Source: United Nations Department of Economic and Social Affairs, Population Division. 2015. *World Urbanization Prospects: The 2014 Revision*. New York: United Nations. http://esa.un.org/unpd/wup/FinalReport/WUP2014-Report.pdf.

Another feature of changing urbanization patterns is the possible emergence of mega-urban regions. In China, the Hong Kong–Shenzhen–Guangzhou region is home to over 120 million people (again, recall the cautionary comments about city size in Box 11.1). In West Africa a 600 km (373 mile) urban corridor runs through Nigeria, Benin, Togo, and Ghana.

Approximately 10 per cent of the world's urban population lives in megacities, and they are of interest to geographers for two reasons. First, three of these places—Tokyo, London, and New York—play significant roles in controlling the global economy (a topic addressed later in this chapter). Second, many of the other megacities are seen as representative of the global urban future because they are located in less developed countries and beset with problems of inadequate infrastructure, congestion, pollution, crime, and poverty (discussed in Chapter 12).

THE ORIGINS AND GROWTH OF CITIES

Cities are both a cause and a consequence of larger economic and social transformations, from the agricultural revolution and its associated urban revolution to the Industrial Revolution and the second urban revolution. The agricultural revolution inspired a cultural transformation that culminated in the rise of new forms of

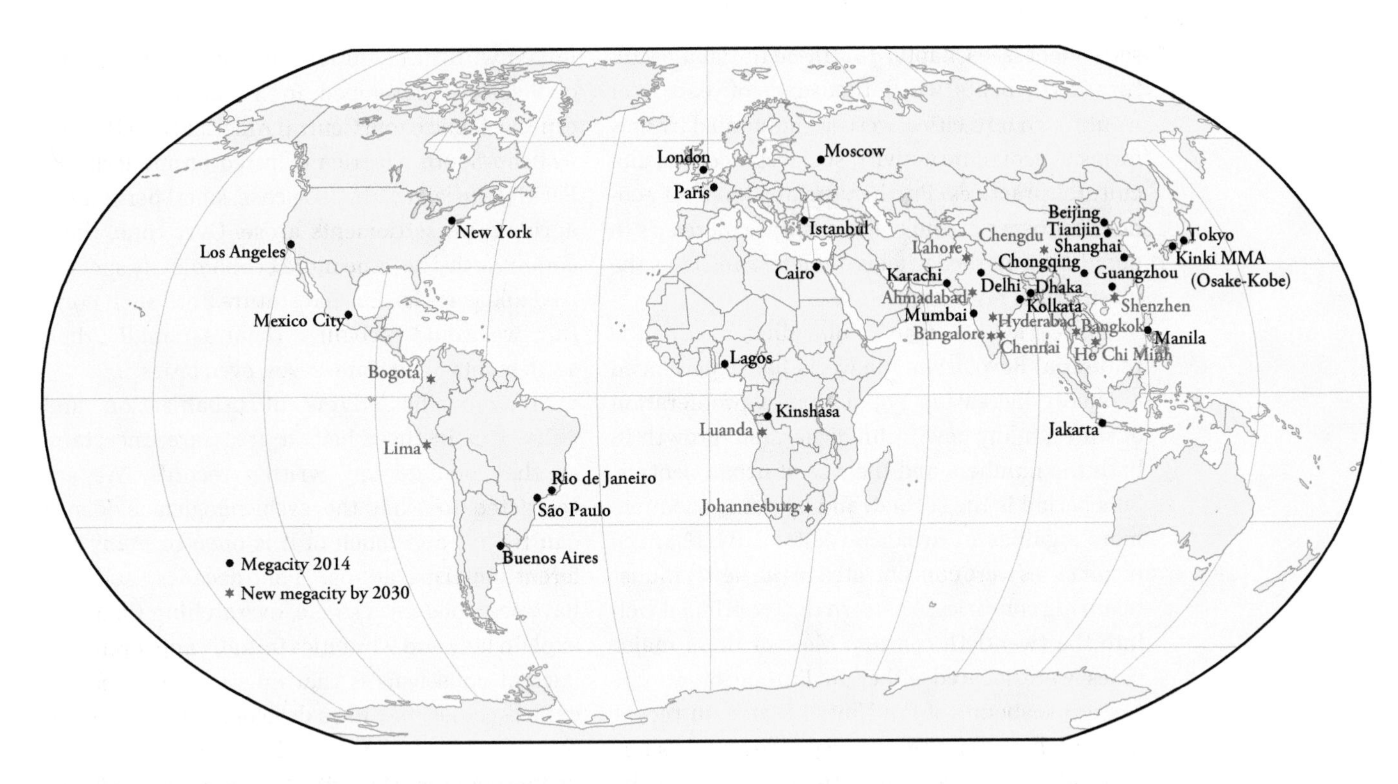

FIGURE 11.6 | Megacities, 2014 and 2030

Source: United Nations Department of Economic and Social Affairs, Population Division. 2015. World Urbanization Prospects: The 2014 Revision. New York: United Nations. http://esa.un.org/unpd/wup/FinalReport/WUP2014-Report.pdf.

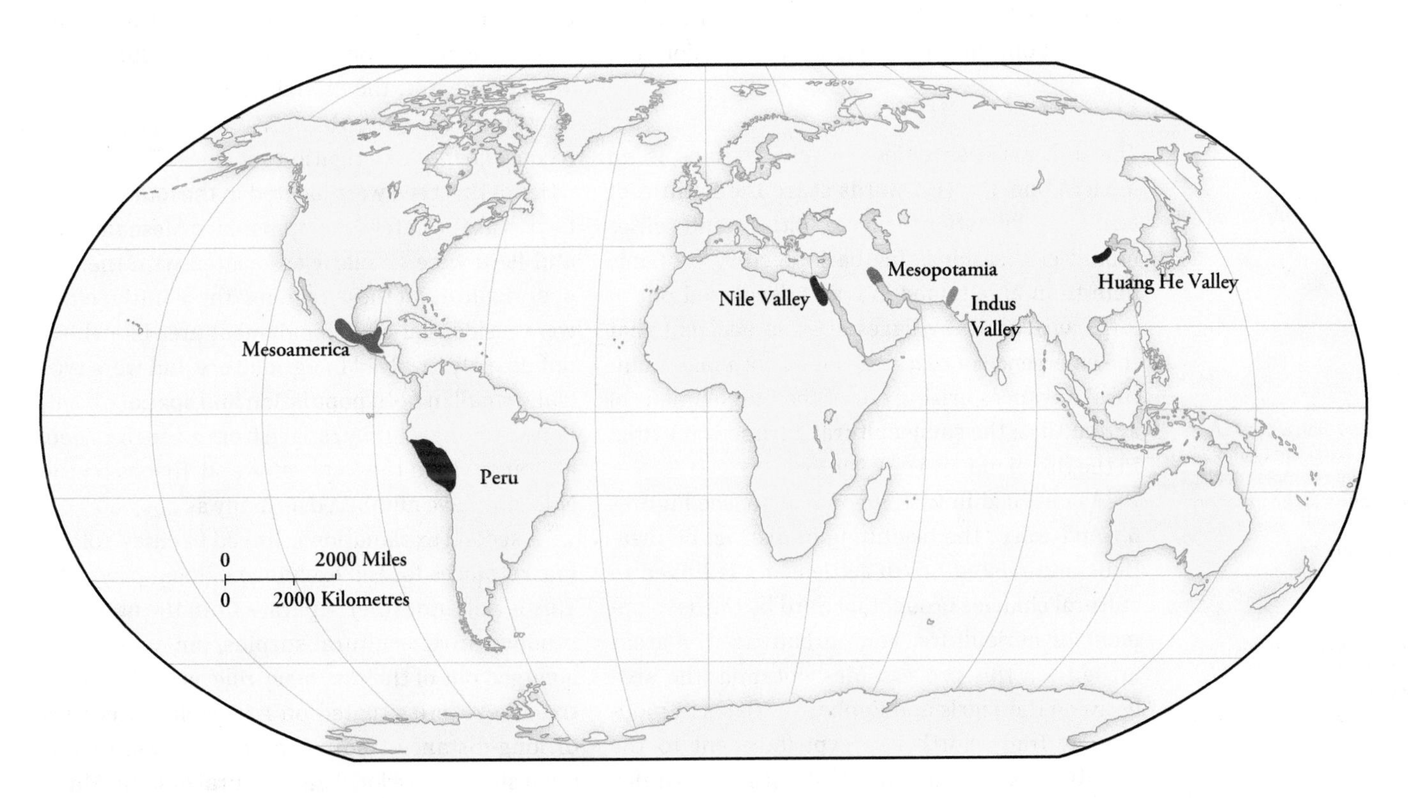

FIGURE 11.7 | The six urban hearths

settlement (see Chapter 10). These initial agricultural settlements, which thousands of years later would turn into cities, were accompanied by new forms of economic activity, social structures, and cultural practices. Further, greater levels of economic prosperity and commensurate increases in population marked what is often referred to as the (first) urban revolution.

Since about 1750, coinciding with the Industrial Revolution, the world has experienced a rapidly increasing population, a proliferation of states, many new technologies, and growth in both the numbers and the size of urban centres. This period is the second, and by many accounts more significant, urban revolution. By 1850 the major cities were concentrated in the newly industrializing countries, a pattern that continued well into the twentieth century. Most of these major cities were located either in Europe or on the eastern seaboard of the United States. In recent decades, however, the majority of the world's largest cities are located in former colonies in the less developed world. Before 1950 cities in the less developed world were typically transportation centres or colonial government centres that formally excluded most of the native population. With independence, these have become population magnets, offering employment and wealth to some but only minimal benefits to the majority.

Urban Origins

The link between cities and civilizations is an ancient one; the two words share the Latin root *civitas*. Civilizations create cities, and cities mould civilizations. The earliest cities probably date from about 3500 BCE and developed out of large agricultural villages. It is no accident that the emergence of cities coincided with major cultural advances arising out of the development of agriculture; the same cultural advances resulted in the rise of modern civilizations.

As outlined in Chapters 6 and 10 (see Figures 6.1 and 10.11), the beginning of distinct civilizations and related urban settlements is linked to cultural changes brought forward by the development of agriculture. Four urban **hearth** areas arose from this process: Mesopotamia (the area between the Tigris and Euphrates Rivers in modern-day Iraq), northern Egypt (adjacent to the Nile River), the Indus River Valley (in modern-day Pakistan), and the Huang (Yellow) River Valley (in modern-day China). Later, but not apparently linked with these initial areas, additional urban civilizations developed in Mesoamerica (modern-day Mexico and Central America) and Pacific-Andean South America (centred on modern-day Peru) (Figure 11.7). In each case, small permanent agricultural settlements arose. Over time, these communities became more economically specialized and grew in size and stature until such point that we would recognize them as small urban settlements or, in some cases, even cities.

The specific origins of urbanization and **urbanism** in these hearth areas are uncertain, as they predate any written records. We are restricted to what the archaeological evidence can tell us, and much of it is open to many different interpretations. For decades scholars have searched for a single, overarching theory to explain how and why cities initially appeared. The current consensus is that no single explanation suffices; cities may have developed in one area in response to a set of stimuli and out of necessity in another area. Most cities of the world likely originated in connection to one or more of the following four ways.

First, we know that cities were initially established in agricultural regions and that city life did not become possible in any of the initial hearth areas until the progress of agriculture freed some group members from the need to produce food. To some degree, the first cities reflected the production of an **agricultural surplus**, possibly as a consequence of irrigation schemes. The earliest cities of this type were located in the four original hearth areas. Later developments in Mesoamerica and Peru were similarly associated with the rise of agriculture in these regions. These initial cities were residential and employment areas for those not directly involved in agriculture and were typically small in both population and spatial extent. Populations generally ranged from 2,000 to 20,000, although Ur on the Euphrates and Thebes on the Nile may have numbered as many as 200,000.

A second explanation is linked to cities' role as marketplaces for the exchange of local products. This is not altogether separate from the ability to produce an agricultural surplus, but some cities emerged out of their primary role as trading centres; they were situated on navigable waterways or long-distance trading routes, and were without a substantive local agricultural system. Many southern Asian cities along the Silk Road—a network of trading routes connecting Europe and the

urbanism
The urban way of life; associated with a declining sense of community and increasingly complex social and economic organization as a result of increasing size, density, and heterogeneity.

agricultural surplus
Agricultural production that exceeds the sustenance needs of the producer and is sold/exchanged with others.

hearth
The area where a particular cultural trait originates.

Middle East in the west and China and Southeast Asia in the east—began in exactly this way. In other words, they were rest stops or places of river crossings for traders' caravans.

Third, cities may have started as defensive or administrative centres. Early cities, from Mesopotamia to China, had defensive fortifications and walls, indicative of societies concerned about protecting themselves and their property. This defensive imperative might have been a major driving force in the creation of cities in the first place—people choosing to cluster together for collective protection behind walls and inside fortresses. Further, the Greeks built cities in their colonial territories throughout the Mediterranean as centres from which to exercise control over their empire(s). In many cases, they planned cities before settling in them, using a grid pattern of streets with a central marketplace and places for political meetings and military preparedness. The Romans, who knew that urbanization was the key to controlling their conquered areas, followed this pattern as their own empire expanded. Many modern-day North African, Middle Eastern, and European cities can trace their origins to either Greek or Roman times.

Fourth, some cities arose as ceremonial centres for religious activity. Mesopotamian, Mesoamerican, Chinese, and Egyptian cities all contained major ceremonial elements, from temples to pyramids and ziggurats, signifying that they were of potential importance for urbanization. Early Chinese cities also had ceremonial dimensions; urban locations were selected using geomancy, an art of divination through signs derived from the earth. Once a promising site was located, the city was laid out in geometric fashion. Chinese cities were square in shape, reflecting two fundamental beliefs: the earth was square and humans should be part of nature rather than dominate it.

Pre-Industrial Cities

The pre-industrial period dates from the emergence of cities in *circa* 3500 BCE to the Industrial Revolution in the eighteenth century CE. The term *pre-industrial* may be misleading in that it suggests that all cities were the same before the Industrial Revolution; in fact, there were notable distinctions between cities in different parts of the world and at different times, reflecting a wealth of cultural variations. Nevertheless, it is a useful term as it emphasizes the significance of the changes that accompanied industrialization.

The Ziggurat of Ur in modern-day Iraq was the centrepiece of the city's temple complex, which acted as both an administrative centre and ceremonial shrine. Ziggurats are a common feature of Mesopotamian cities.

Before the Industrial Revolution, most cities were concerned with marketing, commercial activities, and craft industries and served as religious and administrative centres. As new features in the human landscape, cities developed a new way of life—a new social system and a more diverse division of labour. Political, cultural, and social changes followed the economic change from pre-agricultural to agricultural societies. Rural and urban settlements were complementary, the former producing a food surplus and the latter performing a series of new functions. From the beginning, then, cities have been functionally different from but connected to surrounding rural areas.

While it is useful to consider this entire pre-industrial period as a single entity, it is similarly important to address some of the regional and temporal variations that existed within this lengthy period of human history. The following discussion is divided into six sections addressing urban growth and development in the urban hearth areas, Greece and the eastern Mediterranean, the Roman Empire, China, the Islamic world, and pre-Industrial Europe.

The Urban Hearths

The urban revolution occurred at somewhat different times for each of the urban hearth areas. One unanswerable question is whether the emergence of cities in each area was a result of relatively simultaneous and independent origin or diffusion from one locale to another. The fact that the process occurred in much the same way in each of the hearth areas suggests the latter. But the lack of known interactions between people in pre-urban Mesopotamia and the Nile Valley, to say nothing of China or Mesoamerica, suggests that diffusion was unlikely. Among the six urban hearth areas, the first visible signs of urban development were as follows:

- Mesopotamia: Ur, Babylon (*c.* 3500 BCE)
- Nile Valley: Memphis, Thebes (*c.* 3200 BCE)
- Indus Valley: Harappa, Mohenjo-Daro (*c.* 2200 BCE)
- Huang Valley: Erlitou, Sanxingdui (*c.* 1500 BCE)
- Pacific-Andes (Peru): Aspero, Caral (*c.* 900 BCE)
- Mesoamerica: San Lorenzo, La Venta (*c.* 200 BCE)

These areas share a link between agriculture and subsequent urban development. As noted earlier, this process was not instantaneous; in the case of Mesopotamia, it took approximately 6,000 or 7,000 years for the first agricultural settlements to develop into the first cities. The "discovery" of agriculture via the domestication of plants and animals about 12,000 years ago presented the opportunity for permanent settlement. Some people abandoned their nomadic ways and became farmers, with an attachment to place. Over time, farmers sought protection, places to exchange their surplus produce, and places to worship gods or pay homage to their leaders. Over thousands of years these initial settlements grew into cities as greater and greater agricultural surpluses were assured through greater use of technology (i.e. irrigation, further plant and animal domestication, development of the plough and other tools, etc.). With a surplus assured, individuals could be freed to engage in non-agricultural activities, such as specialized craft industries (i.e. making tools and other commercial goods), trade, administration, and military/defence. All these non-agricultural economic and social activities became the focus of urban life, a tradition that continues in its most basic form in cities around the world.

While some regional differences exist between the basic forms of cities in these urban hearth areas, many similarities exist as well. Most cities had ceremonial features and areas (i.e. the pyramids and ziggurats), which were the most elaborate and distinctive features of the city. All cities had marketplaces for the exchange of goods from local, regional, and (in some cases) long-distance locales. Cities in many of these hearth areas had protective walls and fortresses, with the cities of the Nile Valley being a notable exception.

In modern industrial cities, different areas serve different functions. By contrast, homes, workshops, markets, and other functions in these urban hearth cities were located in a relatively haphazard fashion. There was, however, evidence of coordinated planning within many early cities, as the residential areas of the elite were often separated from the rest; ceremonial features were constructed at great effort; protective walls and fortresses were built in a coordinated way; and some places, such as the cities of the Indus Valley, had centralized drinking and wastewater systems.

Greece and the Eastern Mediterranean

Urbanization, and the urban ways of life, almost certainly diffused via interactions through trade from Mesopotamia and the Nile Valley to the eastern Mediterranean. Over a thousand-year period, beginning about 1500 BCE, ancient Greece became the most highly urbanized civilization, with more than 500 cities and towns on the mainland and on islands throughout the eastern Mediterranean. A few urban places, such as Athens, Sparta, Corinth, and Thebes, were sizable. The largest, Athens, probably reached a population of 250,000 during the height of its power and supremacy.

Aside from their philosophical, political, and cultural contributions to modern Western civilization, the Greeks also made several contributions to urban form. Every Greek city had an **acropolis** on which they built their religious buildings, such as the Parthenon in Athens. Using the highest point in the city for such ceremonial purposes is a direct translation of many urban hearth regions' practice of employing architectural techniques to create inspiringly tall ceremonial buildings. The Greeks also improved upon the public spaces of Mesopotamian cities. Because public space was an urban priority in ancient Greek cities, ample land was set aside for the public meetings, military preparations, and (later) marketplace functions of the **agora**.

acropolis
The fortified religious centre of cities in ancient Greece; the literal translation is "highest point in the city."

agora
The centre of ancient Greek civic life; the place where public meetings, trials of justice, social interaction, and commercial exchange took place.

The Roman Empire

When the Romans succeeded the Greeks as having the dominant empire of the Mediterranean world, they borrowed much of their predecessor's urban expertise. As a result, the Roman Empire (*c.* 500 BCE–500 CE) became even more urbanized than that of the Greeks. The Romans developed a hierarchical network of towns and cities scattered along the Mediterranean coasts and extending inland into North Africa, Southwest Asia (including Mesopotamia), and Europe. This network was linked via an elaborate system of roads, many of which became the basis for modern highways in Europe. At the pinnacle of this urban hierarchy was the city of Rome. In about 200 CE, the empire's peak, the city had a population in excess of one million, a threshold that would not be surpassed anywhere in the world until the industrial era. Like the Greeks, the Romans used urban settlements as a means to control territory and population. They established cities across their territorial empire, many of which remain, including Vienna (Austria), Paris and Lyon (France), Cologne and Bonn (Germany), London and Manchester (England), and Seville (Spain).

In terms of urban form, the Romans borrowed the Greek grid-patterned streets (where feasible) and the ceremonial and civic aspects of urban life. The ceremonial and religious functions of the acropolis and the civic, commercial, and administrative functions of the agora were combined into a single entity known as the **forum**. The forum became the centre of Roman life, and most cities had one. Roman cities also included other easily recognizable urban features, such as stadiums, amphitheatres, monumental buildings, spacious streets, welcoming public spaces, and an elaborate system of aqueducts, fountains, and baths. While many of these features are still visible as ruins in many former Roman cities, aspects of the common person's life are less visible. We have much less sense of what life was like for the majority of the urban population, to say nothing of the urban poor and especially the many slaves, upon whose effort the splendour of the Roman urban empire was built. Housing for the majority of the population in Roman society was likely cramped, squalid, and unsafe.

China and Eastern Asia

As population movement and trade declined in Europe after the collapse of the Roman Empire in the fifth century CE, the process of urbanization faltered and did not resume for some 600 years. But urbanization in China and elsewhere continued. In Eastern Asia, for example, Chinese ideas about urban growth diffused out of the Huang Valley to other parts of the country, as well as the Korean peninsula, Japan, and the Southeast Asian peninsula. Emerging at this time were the Korean city of Seoul and the Japanese cities of Tokyo and Kyoto.

China's largest city during European urban stagnation in the Middle Ages was Chang'an (now Xi'an), the eastern terminus of the Silk Road. Chang'an was a heterogeneous city, as population from all parts of the Silk Road made their way east. Other major Chinese cities of this period include the modern cities of Beijing, Kaifeng, Shanghai, Jinling (now Nanjing), and Hangzhou, all connected via the Grand Canal, which was built in the fifth century BCE as a way to stimulate urban and economic development of eastern China. In the late thirteenth century CE, the European merchant traveller Marco Polo characterized Hangzhou as the finest and most splendid city in the world because of its markets, heterogeneous population, and cultural vibrancy.

The Islamic World

In much the same way that Chinese and other Eastern Asian cities flourished during the period of European urban stagnation, the cities of the Islamic world (loosely grouped together here) grew in size and stature. Cities such as Constantinople (now Istanbul and once the eastern Roman capital), Baghdad (Iraq), Medina (Saudi Arabia), and Timbuktu (Mali) became the leading Muslim cities of the world. The Islamic Golden Age, roughly from 800 to 1500 CE, saw much of the world's advances in science, mathematics, technology, and urban architecture originate in Islamic cities.

forum
The centre of Roman civic, commercial, administrative, and ceremonial life; combined the functions of the ancient Greek acropolis and agora.

Pre-Industrial Europe

The resumption of European urbanization in the eleventh century was fuelled by increasing commercial activities, political power gains by political units (later to be recognized as countries), population increases, and agricultural technology advancements. Old cities were revived and new ones were established throughout Europe, especially in areas well located for trade, such as on the Baltic and North Seas and the Mediterranean coasts. Urbanization was encouraged in the seventeenth century by the development of mercantilism (see Chapter 5). Following this approach, European countries established colonial empires through exploration, territorial expansion, and economic exploitation, all of which was aided by the development of cities as **entrepots**.

entrepot
A city, usually a port, that functions as an intermediary for trade and transshipment and that exports raw materials and manufactured goods.

Pre-industrial cities throughout the world, then, followed the development of agriculture and were typically small and compact in form. They soon took on distinctive functions and characters. The typical city functioned both as support for an agricultural population and as a local or regional trading centre. Beyond economic functions, however, pre-industrial cities had important roles to play as ceremonial locations, centres of governance for large empires, and places for social and cultural innovation and change. All this would change dramatically with the advent of industrialization.

The Urban Revolution and Industrial Cities

The Industrial Revolution brought a series of changes that altered urban landscapes, creating an urban revolution. The principal economic changes were related to a series of technological advances, particularly in the areas of agriculture and manufacturing, which led to the appearance of the industrial city.

Like the transformative nature of the agricultural revolution of about 12,000 years ago, scientific discoveries in the 1700s led to a second revolution, which further increased the amount of food produced per unit of labour. Increased agricultural productivity resulted in an oversupply of labour in rural areas. In search of employment, many rural folk migrated to cities. Coinciding with these changes in the agricultural sector, and in many ways the driving force behind them, were changes occurring within the city in terms of manufacturing and industry. New sources of energy had been harnessed, which could power great machines. These new machines spurred new and much greater scaled industrial enterprises, which required labour. This rural to urban migration of millions of people over the late eighteenth and nineteenth centuries resulted in the urban revolution of the industrial era.

Not all mercantile cities turned into industrial ones. Industrial city location was determined primarily by the proximity of available raw materials for the industrial processes. New cities emerged adjacent to coal fields such as in the Midlands of England, the Ruhr Valley in Northwestern Europe, and in Appalachia in the United States. After 1760 the steam engine became available and old industrial activities, especially textile and metal production, were transformed by steam-powered machinery. The importance of coal for power and transportation resulted in major new concentrations of activity, machinery, capital, and labour in the new industrial cities. In the nineteenth century, steam-powered engines on iron railroads facilitated the development of major transportation networks that allowed industrial cities to be linked for the efficient flow of goods and people.

Urbanization has been one of the key phenomena of the industrial age, involving the spatial movement of large numbers of people and major changes in social and economic life. Some of this change was positive, but much of it made lives worse. A great deal of industrial city growth occurred in an unregulated way and in an uncontrolled environment. Consequently, the haphazard placement of factories, homes, shops, transportation routes, etc., made for a confusing and unpleasant place. Factories became the hallmark of the industrial city, with their smokestacks belching black smoke as an apparent sign of "progress." But these factories, or entire districts of them, were located proximate to the housing of many of the rural migrants who worked in them. This housing was often of very poor quality, overcrowded, and heavily polluted: factory smoke choked the air, effluent polluted the local water supply, garbage and refuse lay uncollected in open spaces and on the streets, and human sewage was dumped into the city drains because centralized sewerage systems could not be built fast enough to accommodate the speed of urban growth. These housing areas, which became known as urban slums, were as much a hallmark of the industrial city as the factory smokestacks.

At the same time, cities constructed great monuments of the industrial age, including elaborate railroad stations, bridges, industrial exhibition halls, government buildings, and (by the late nineteenth century) skyscrapers. This construction was made possible by the technological advances of the industrial area, including the making of steel and the harnessing of new forms of energy—first steam, then electricity. The industrial city also changed as a result of innovations in not only railroads but also forms of intra-urban transportation technology, such as the electric streetcar and, later, automobiles, buses, and trucks via the internal combustion engine. These innovations changed the urban form of the highly centralized city to the current decentralized cities.

Urban growth proceeded apace in the new industrial areas—in Britain from about 1750 onward and then in Western Europe, the United States and Canada, parts of Southern Europe, Russia, and Japan by the end of the nineteenth century. The rapidity of this growth is illustrated by the United States, where the urban population increased from approximately 5 per cent in 1800 to 50 per cent by the 1920s and more than 80 per cent today. As illustrated earlier, much of the more developed world has undergone this same transformation over the past two centuries, and the less developed world is currently doing so even more quickly, making the world truly urban.

THE LOCATION OF CITIES

Why are urban centres located where they are? This is a quintessentially geographical question that geographers frequently ask. The answer is not simple. Many factors contribute to the location of any particular city. Among the most influential are physical geographic considerations, such as topography, and the period of initial settlement (i.e. when the community was first established). Geographers typically understand a city's location with respect to its site and its situation. A city's site refers to its physical location and the characteristics of that location, such as being at the head of a bay or estuary, at a strategic river crossing, etc. The city's situation refers to its location relative to others. In other words, we often think of how accessible a city is, which is a direct reference to its situation. Remote, isolated, and inaccessible cities are situationally poor; cities that are well connected to others are situationally advantaged. We can examine the urban growth of two places and conclude that the greater success of one is linked to its site and situational characteristics. In general, cities are located in accessible areas, typically along a coast or at the confluence of two rivers, etc., and not typically in mountainous, desert, or tundra regions. Because interconnectedness is important, they usually are not isolated from one another.

The period of initial development is also a central component of understanding the location of cities. For example, cities of the pre-industrial period had very different locational needs than did cities of the industrial age. While few modern cities are founded from scratch in this era of globalization, the relative importance of each is changing (rising or falling) as a result of its location and function in the global economy.

In medieval times, urban defensibility was of paramount importance; therefore, cities were built on hilltops or in sheltered harbours and with impenetrable walls and fortresses. At other times, being on a regional trade route was essential for urban growth, and so cities located at river crossing points or within a single day's travel of another city were more likely to flourish than cities in less accessible or strategic locations. During the industrial era, different factors became important for urban success: access to raw materials such as coal, metals, cotton, etc.; access to a large pool of manufacturing labour; and access to a marketplace for the manufactured goods. In the present era of increased globalization, with the costs of global transportation at an all-time low, production can occur almost anywhere. Cities with economies based on resource extraction or manufacturing have declined in relative importance and been replaced by cities where urban amenities are of greater import.

A quick consideration of the population shift in American cities over the last half century or so, from the northeast and Midwest (an area known as the rust belt) versus those in the southeast and southwest (the sun belt) illustrates this point. As Table 11.6 shows, the annual urban growth rate for the entire country between 1950 and 2010 was 1.5 per cent. Much of this growth was driven by cities in the south, which averaged annual population increases of 12.0 per cent. For the select rust belt cities, the growth rate was less than 0.2 per cent. The amenities and associated lifestyle opportunities, mostly related to climate, represent primary differences between these cities.

To answer our initial question about the location of cities, we need to understand that past cities of

TABLE 11.6 | Change in population in rust belt vs sun belt cities in the United States, 1950–2030

	1950	1970	1990	2010	1950–2010 (%)	2030[1]
Buffalo	899	1,084	955	935	0.07	999
Cincinnati	881	1,202	1,335	1,628	1.41	1,916
Cleveland	1,392	1,954	1,680	1,781	0.47	1,948
Detroit	2,769	3,966	3,703	3,730	0.58	3,886
Pittsburgh	1,539	1,845	1,681	1,733	0.21	1,884
Rust belt cities (selected)					**0.19**	
Atlanta	513	1,182	2,184	4,544	13.10	6,140
Dallas-Fort Worth	866	2,025	3.219	5,149	8.24	6,683
Miami	622	2,141	3,969	5,518	13.12	6,554
Phoenix	221	874	2,025	3,649	25.85	4,808
San Diego	440	1,209	2,356	2,964	9.56	3,522
Sun belt cities (selected)					**12.00**	
All US cities					**1.52**	

Note:
1. All population figures are in thousands; the 2030 figures are projected.
Source: Adapted from United Nations Department of Economic and Social Affairs, Population Division. 2015. *World Urbanization Prospects: The 2014 Revision*. New York: United Nations. http://esa.un.org/unpd/wup/.

importance may or may not hold this status today. While the characteristics of a city's site may remain constant (though global climate change may have something to say about that), the situational characteristics can change dramatically. A strategically located and thriving community in one time period can become an isolated backwater when global economic and political circumstances change. The opposite can apply to places that were relatively unheard of in the past but have become megacities. The contrasting American rust belt and sun belt are examples.

central place theory
A theory to explain the spatial distribution of urban centres with respect to their size and function.

URBAN SYSTEMS AND HIERARCHIES

Cities function together as an economic, political, cultural, and environmental system. We outlined the idea of a system—a set of interrelated components or objects linked to form a unified whole—in Chapter 4. In an urban system, the whole can be seen as a region, a country, or even the entire world, and cities comprise the interrelated components or objects. One way of examining the nature of the relationship between cities is through their function, in the sense that each city contributes to the urban system by providing people with access to a specific set of goods and services. Others consider how the distribution of cities relates to their population, through what we call the rank-size distribution and urban primacy. We will consider both approaches in this section.

Central Place Theory

Central place theory attempts to explain why cities are located where they are and how they work in relation to one another in an urban system. The theory is associated with the German geographer Walter Christaller (1966), who observed that southern Germany had a large number of very small communities, each providing a narrow range of goods and services to the local population, and only a handful of larger communities, each providing a much greater range of goods and services. This observation fuelled Christaller's interest in developing a theoretical explanation for the distribution of different sized communities across space, as connected to their economic functions.

In Chapter 10 we outlined the agricultural land-use theory developed by von Thünen. One of this theory's most important contributions to geography is the concept of the isolated state, a grossly simplified ideal situation to observe the effects of a particular variable (in von Thünen's case, distance to market as expressed by transportation costs). Christaller used a similar model to theoretically observe the role played by distance in the location of urban centres. In this isolated state, the topography is flat and uniform, the rural population is distributed evenly across the landscape, and people are able to travel in all directions without constraints of roads or other transportation routes. While these conditions are overly simplistic and unrealistic, they allowed Christaller to isolate the spatial phenomena of interest—the distribution of central places.

The urban centre of Winterset, Iowa, performs a series of functions for the surrounding rural population.

According to Christaller, **central places** are understood as economic or market centres, where people come to acquire goods and **consumer services**. These central places can be cities, towns, or even tiny villages and hamlets. For the purposes of this theoretical understanding of the urban system, the number of people residing in these central places is irrelevant. Of central importance, however, is the economic function of the community vis-à-vis the goods and services they provide to people in the surrounding area. Not surprisingly, hamlets and villages provide a limited number of goods and services whereas towns, cities, and megacities provide incrementally more.

The area surrounding a central place is its **hinterland**. People come from the hinterland to acquire the goods and services that the centre provides. Imagine a hamlet, of which there are thousands in Canada, that has a single main intersection (with or without a stop light). The hamlet likely has a gas station on one corner, a general store or a restaurant on another, a church or school on a third, and perhaps a post office or a bank on the final one. This small centre provides these goods and services to a local population; people travel from the surrounding area to get gas, buy some milk and a newspaper, post a letter, and attend school or church. While it is difficult to ascertain exactly how big the hinterland is for this hypothetical community, we can imagine that, given the isolated state conditions of the landscape as described by Von Thünen and Christaller, we could draw a circle around the hamlet that encapsulates its hinterland. We have many such hamlets in an urban system, each competing with one another for customers and each with its own hinterland. Because most people prefer to frequent the nearest provider of goods and services (assuming they are of equal quality to those farther away), everyone would travel to their nearest centre. However, towards the edges of the hinterland, individuals begin to have a choice of which centre to visit.

Within a central place system, the ideal shape of each hinterland is a circle, representing equal accessibility to the central place from all directions. However, as Figure 11.8 shows, a central place system comprised of a set of circles means that either some areas are not served by any centre or some areas are served by two; according to economic logic, both are inefficient solutions. To completely cover an area without any overlap, the ideal shape is the hexagon. The nearest geometric figure to a circle, the hexagon has the greatest number of sides and provides total coverage without duplication.

Each centre, no matter how big or small, provides more than one good or service; the bigger the centre, the more goods and services it provides. Therefore, the size of each centre's hinterland depends on which goods and services it provides and how many. In addition, the hinterland's size is

FIGURE 11.8 | Theoretical hinterlands (or market areas) for central places

central place
An urban centre that provides goods and services for the surrounding population; may take the form of a hamlet, village, town, city, or megacity.

consumer services
Services that are provided primarily for individual consumers, such as retail, hospitality, food, leisure, health care, education, and social welfare; represents approximately 50 per cent of employment in most countries of the more developed world.

hinterland
The market area surrounding a central place; the spatial area from which the providers of goods and services in a central place draw their customers.

range
The maximum distance that people are prepared to travel to obtain a particular good or service.

threshold
The minimum number of people (market size) required to support the existence of a particular economic function.

influenced by two components of each good and service: range and threshold.

Each good or service that a centre provides has a **range**, the maximum distance that people are willing to travel to acquire a certain good or service. For example, people are willing to travel only a short distance to acquire everyday items such as coffee, milk, and gas or frequent basic services such as getting a haircut or dropping clothes at a dry cleaner; Christaller labelled these goods and services low-order. On the other hand, people are willing to travel much further for high-order goods and services, such as a car, an appointment with a medical specialist, a sporting event, or a live theatrical performance.

Every good or service also needs a minimum number of potential customers to draw from in order to make a profit (or at least stay in business). This is called the **threshold**. Most people need to visit the grocery store, get gas for their car, or get a haircut fairly regularly; hence, the threshold for these low-order goods and services tends to be quite small. In other words, a relatively small population will provide enough customers to make the provision of these low-order goods and services potentially profitable. At the same time, how often does one need to visit a cardiologist, car dealership, or international airport? Not very often. The threshold is quite large; providers of these high-order goods and services need to draw from a very large population to make their business viable.

According to the central place theory, then, small centres (such as our hypothetical hamlet) provide low-order goods and services that cater to the local population and for which people do not have to travel very far. Larger places, such as towns, have a mix of these same low-order goods and services (because people in towns still need to buy milk, get their hair cut, and visit the post office), as well as a selection of some higher-order ones, such as law offices, department stores, and high schools, for which people would be willing to travel a greater distance. Cities, and even our largest megacities, provide all the same low-order and high-order goods and services as provided by smaller places, but they also offer some high-order goods and services that are not provided elsewhere, including international airports, professional sporting events, luxury shops, and exclusive restaurants. Because of the exclusivity of these items, people are willing to travel quite long distances to get them (in other words, their range is large). But providers of these goods and services also need to have a large population from which to draw potential customers (their threshold is big).

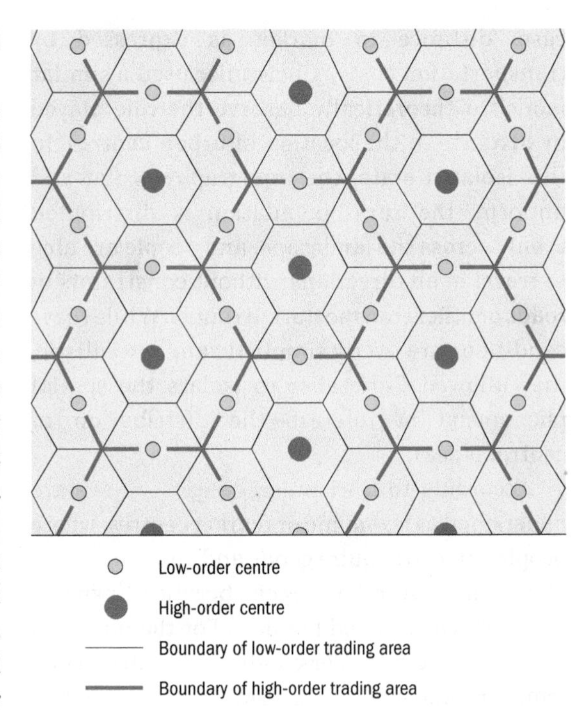

FIGURE 11.9 | A simplified (two-order) central place system

This pattern creates a pyramid-shaped hierarchy of centres of various sizes and functions; there are far more hamlets and villages than there are towns and more towns than cities. Christaller identified seven such levels to this hierarchy and showed that the highest-order settlements (the largest) will be fewest in number, while the lowest-order (the smallest) will be the most numerous. The spatial regularity of these centres is not disturbed by these number differences because, for any given level, settlements are equidistant from one another based on the hexagonal patterns of hinterlands described earlier. Figure 11.9 shows a simplified two-order situation. Thus, central place theory suggests both a hierarchy of centres and a nested hierarchy of hinterlands. For an illustration of this hierarchy with respect to hockey teams, see Box 11.3.

Urban Hierarchies

The urban hierarchy exists to a certain degree in all urban systems. Canada has 147 urban centres with a population of at least 10,000 and nearly 750 population centres with between 1,000 and

Examining the Issues

BOX 11.3 | Application of Central Places to Hockey Associations and Teams in Canada

Canada has thousands of youth house league hockey associations. Most small communities and neighbourhoods in the country have a local arena where these teams play. In some areas, these places are truly neighbourhood or community hubs; neighbours interact with one another while their kids practise and play hockey.

In many parts of the country, small rural communities or urban neighbourhoods band together to create an umbrella association from which a representative (or rep) hockey team is formed from the most talented players. This rep team travels and plays other rep teams from nearby towns and cities. There are hundreds of such teams, at all age levels, in Canada.

Some small cities—including Kingston, Oshawa, Sherbrooke, Halifax, Saskatoon, Red Deer, and Victoria—have a junior hockey team as part of the Canadian Hockey League. These quasi-professional teams (the players do not get paid) play throughout their respective regions of the country. Canada has dozens of these teams, and their players represent the cream of the crop for their age.

Finally, seven of Canada's largest cities (Montreal, Ottawa, Toronto, Winnipeg, Calgary, Edmonton, and Vancouver) have professional hockey teams. Several other cities, including Quebec City and Hamilton, would like to have one. These teams, along with others in the United States, employ the world's best hockey players.

This example illustrates the principles of the central place theory. The range for taking your kids to house league hockey practices and games is much smaller (perhaps a few kilometres) than the range for an NHL hockey team (perhaps 100 km). Similarly, the threshold is much smaller for a house league association (perhaps a few thousand families) when compared to a rep team (perhaps 10,000 families) and an NHL team (perhaps a million people).

10,000 people. From these numbers, we can begin to see the Canadian urban hierarchy, in which Toronto is at the pinnacle and some 750 small population centres are at the base. The Canadian urban system displays a remarkably consistent pyramid-shaped hierarchy, with an increasing number of cities as population size declines (Table 11.7).

Of Canada's nearly 35 million people, nearly half live in its six largest cities (i.e. those in excess of a million). Approximately the same number of people live in Canada's other 884 urban centres, which vary in size from just less than a million to a thousand. This concentration is only increasing, as the six largest cities saw an average population growth of nearly 10 per cent between 2006 and 2011, double the rate for Canada as a whole. The leaders of this growth in percentage terms were Calgary and Edmonton, but Toronto's growth was the largest in absolute terms with nearly half a million people added. Most of Canada's rapidly growing places are in the west, with Alberta leading the way, and most declining in population are in eastern Canada, especially the Atlantic provinces. Further, most places demonstrating strong population growth are large cities, but most seeing a drop are smaller urban centres (with the exception of Windsor and Thunder Bay, two relatively large cities with recently declining populations). While there are notable areas of greater and lesser urban growth, the dominant pattern across Canada is characterized by slow growth.

This hierarchy is based solely on population. But as we discussed with reference to central place theory, there is more to an urban hierarchy than simply the number of people living in each centre. The central place's economic and political function and its cultural importance in the urban system are also relevant. Toronto is Canada's economic centre by almost any measure and is, as discussed later in the chapter, the most visible Canadian example of a global city. A few examples of Toronto's role at the pinnacle of Canada's urban hierarchy include the following:

- It is home to more corporate head offices than anywhere else in Canada.
- It is the financial and insurance capital of the country.

- It is the telecommunications and media heart of the country.
- It is the primary destination of more immigrants than anywhere else, resulting in it having the most culturally diverse population.

Canada's next two largest cities, Montreal and Vancouver, are similarly important to the global economy, and both are primarily regional economic and cultural centres. Connected to Christaller's view of central places, these and the other cities towards the top of the urban hierarchy provide a greater range of goods and services to a much larger spatial area than do the many smaller towns and cities at the base of the hierarchy.

rank-size distribution
A descriptive regularity among cities in an urban system; the numerical relationship between city size and rank in an urban system; sometimes referred to as the rank-size rule.

The Rank-Size Distribution

Closely linked to central place theory, though separate in origin, the **rank-size distribution** (or rank-size rule) establishes a simple numerical-size relationship between cities in an urban system. The relationship is specified as follows:

$$P_x = P_1/R$$

where

P_x = population of city x

P_1 = population of the largest city (the first ranked city)

R = population rank of city x

In an urban system conforming to the rank-size distribution, a city's population is inversely proportional to its population rank. In general, the nth largest city will have a population 1/n the size of the largest city—the second largest city in the system will have half the population of the largest city, the third largest city will have one-third the population of the largest, and so on.

The principle of the rank-size distribution generally applies to many urban systems around the world. Cities in many of the world's most economically developed countries follow this distribution remarkably well. Perhaps the most widely studied examples include the urban systems of the United States, Germany, and Japan, which all generally follow the rule. Table 11.8 shows that the rank-size distribution also applies reasonably well to Canada, at least for the country's largest cities. Amongst this group, the largest ones tend to be somewhat

TABLE 11.7 | The hierarchical Canadian urban system, 2011

Urban Order	Population Range	Number of Population Centres	Select Places	Total Population (thousands)
1	> 4,000,000	1	Toronto	5,583
2	2,000,000–3,999,999	2	Montreal, Vancouver	6,138
3	1,000,000–1,999,999	3	Ottawa, Calgary, Edmonton	3,611
4	500,000–999,999	5	Quebec City, Winnipeg, Hamilton, Kitchener–Waterloo, London	3,169
5	200,000–499,999	9	Halifax, Victoria, Saskatoon, Regina, St John's	2,673
6	100,000–199,999	13	Kelowna, Kingston, Trois-Rivières, Guelph, Saint John	1,951
7	30,000–99,999	54	Lethbridge, Chilliwack, Belleville, Shawinigan, Cape Breton–Sydney	2,927
8	10,000–29,999	60	Okotoks, Truro, Whitehorse, Collingwood, Summerside	1,834
9	1,000–9,999	743	Taber, Napanee, Revelstoke, Yarmouth, Lac Megantic	2,311
10	< 1,000	N/A	N/A	6,329

Sources: Adapted from Statistics Canada. 2015. "Population and Dwelling Counts, for Population Centres, 2011 and 2006 Censuses." https://www12.statcan.gc.ca/census-recensement/2011/dp-pd/hlt-fst/pd-pl/Table-Tableau.cfm?LANG=Eng&T=801&S=51&O=A; Statistics Canada. 2015. "Population and Dwelling Counts, for Census Metropolitan Areas and Census Agglomerations, 2011 and 2006 Censuses." http://www12.statcan.gc.ca/census-recensement/2011/dp-pd/hlt-fst/pd-pl/Table-Tableau.cfm?T=201&S=3&O=D&RPP=150.

TABLE 11.8 | Canada's urban system and the rank-size distribution, 2014

City	Population (thousands), 2014	Population Rank	Rank-Size Distribution Predicted Population (thousands)	Difference between Actual and Predicted Population (thousands)
Toronto	5,901	1	5,901	0
Montreal	3,948	2	2,951	+998
Vancouver	2,447	3	1,967	+480
Ottawa–Gatineau	1,309	4	1,475	-166
Calgary	1,306	5	1,180	+126
Edmonton	1,246	6	984	+263
Quebec	797	7	843	-46
Winnipeg	753	8	738	+16
Hamilton	739	9	656	+84
Kitchener–Waterloo	494	10	590	-96
London	486	11	536	-51
St Catharines–Niagara	402	12	492	-90
Halifax	393	13	454	-61
Oshawa	373	14	422	-48
Victoria	354	15	393	-39

Source: Adapted from United Nations Department of Economic and Social Affairs, Population Division. 2015. *World Urbanization Prospects: The 2014 Revision*. New York: United Nations. http://esa.un.org/unpd/wup/.

bigger than the rank-size distribution predicted, while the smaller ones are a bit less populous.

Thus, the rank-size distribution does not predict a given city's population with certainty, but it does provide a useful description of the nature of the population distribution. Rank-size city distributions tend to occur in large urban systems, in countries with a long history of urbanization, and in areas that are economically and politically stable and complex.

Urban Primacy

Urban systems that do not adhere to the rank-size distribution tend to follow an urban primacy distribution, where the largest city is named the **primate city**. A primate city is disproportionately large, such that it is more than twice the size of the system's next largest city. Population size, economic output, political clout, and cultural importance are significant characteristics of the primate city.

Primate cities are often the national (or regional) capital, and they dominate the urban system in terms of influence. Because of their size and economic, political, and cultural importance, primate cities become magnets for rural and urban migrants, who are drawn to the city for its greater employment and housing opportunities. This process makes the primate city even more disproportionate to others in the urban system.

While primate cities can be found around the world, they tend to be in small urban systems (with a small number of cities and/or a small total population) and those with a short history of urbanization. Countries of the less developed world, especially those that were former European colonies, are more likely to have primate cities. The development of primacy is a result of the colonial power establishing a single dominant centre of control, leaving the rest of the country to languish. As Table 11.9 shows, not all countries that follow the urban primacy distribution are from the less developed world.

primate city
The largest city in an urban system, usually the capital, which dominates its political, economic, and social life; a city that is more than twice the size of the next largest city in the system.

A specific variant of urban primacy is the binary urban system, which is dominated by two cities. Typically, one city is the national (or regional or former colonial) capital, while the other is a more recent economic centre, often a port or strategic hub. Here are a few examples, from different areas of the world:

- Spain: Barcelona (port) and Madrid (capital)
- Brazil: Rio de Janeiro (port and former capital) and São Paulo (modern financial centre)
- China: Beijing (capital) and Shanghai (port and financial and economic centre)

TABLE 11.9 | Urban systems of the more and less developed worlds with primate cities, 2014

City	Population (thousands)	City	Population (thousands)	City	Population (thousands)
More Developed World Examples					
United Kingdom		**France**		**Republic of Korea**	
London	10,189	Paris	10,764	Seoul	9,775
Manchester	2,624	Lyon	1,597	Busan	3,237
Birmingham	2,497	Marseille	1,595	Incheon	2,659
Less Developed World Examples					
Mexico		**Argentina**		**Thailand**	
Mexico City	20,843	Buenos Aires	15,024	Bangkok	9,098
Guadalajara	4,766	Cordoba	1,504	Samut Prakan	1,652
Monterrey	4,435	Rosario	1,317	Udon Thani	501
Less Developed World Examples (continued)					
Indonesia		**Nigeria**		**Democratic Republic of the Congo**	
Jakarta	10,176	Lagos	12,614	Kinshasa	11,116
Surabaya	2,834	Kano	3,508	Lubumbashi	1,936
Bandung	2,513	Ibadan	3,085	Mbuji-Mayi	1,919

Source: Adapted from United Nations Department of Economic and Social Affairs, Population Division. 2015. *World Urbanization Prospects: The 2014 Revision*. New York: United Nations. http://esa.un.org/unpd/wup/.

business services
Services that are provided primarily for other businesses, including financial, administrative, and professional activities such as accounting, advertising, banking, consulting, insurance, law, and marketing.

global city
A city that is an important node in the global economy; a dominant city in the global urban hierarchy; sometimes referred to as a world city.

A country is a typical way of thinking about an urban system. But we can also think about the entire world as a particular form of an urban system. In the next section, we look at global cities and consider how they are organized into a global hierarchy.

GLOBAL CITIES

Every urban centre provides some kind of consumer services (e.g. retail, hospitality, leisure, education). As illustrated by the central place theory, small centres provide fewer services than do larger centres. The distribution of **business services** is different, as not every central place provides the same range and number of business services. These services cluster in a small number of places, called **global cities** (or world cities), that are closely connected to the global economy.

Throughout the various stages of human history, different cities have emerged as the largest or most important places in the world, from Babylon to Rome, Constantinople (Istanbul), and Beijing. In the current era of globalization, the task of identifying the most dominant cities has become more prominent. Hall (1966) was the first person to identify and think about cities in this way, and Friedmann (1986, 2002), Sassen (1991, 2002), Beaverstock et al. (1999), Taylor (2004), Abrahamson (2004), and Huang et al. (2007) also conducted important empirical and conceptual analyses. Since the 1980s recognition of the global hierarchy's importance has grown, with global cities becoming recognized as the key command and control centres of the globalizing world economy.

Any discussion of a global hierarchy starts with the cities at or near the top. These cities have significant populations and are considered the control centres of the global economy. Not surprisingly, they are disproportionately located in more developed countries, as these are the world economy's core areas. Most rankings identify three cities at the top of the hierarchy: London in Western Europe, New York in North America, and Tokyo in Pacific Asia. In some accounts, Paris is identified as well. The idea that there are only two to four global cities in the world oversimplifies the concept that only these cities influence the global economy.

Defining Global Cities

A global city can be identified in several ways. Size certainly matters. While New York, London, Tokyo, and Paris are all megacities, being one does not equate to being a global city. Geographers generally agree that larger populations are not the only qualification for global-city status. The factors that contribute to this position are primarily

Pawel Libera/Getty Images

One of three or four cities at the top of the global city hierarchy, London is also a megacity, a national capital, a major financial centre, and home to several supranational organizations.

economic and include both management activities and business services, but the importance of political, cultural, and environmental aspects is increasingly being recognized.

Economic Characteristics

The economic attributes most commonly cited as characteristics of a global city have three main dimensions. First, global cities contain the headquarters of large corporations, many of which are transnational companies. Decision-makers of, for example, transnational manufacturing companies, retail giants, or natural resource extraction companies cluster in global cities to decide what kinds of things to make, how much to sell them for, and how to market them to consumers. As a result, these cities can influence global patterns of trade. Further, high-skilled professional staff in offices in global cities perform the administrative support (e.g. human resources, accounting, and consulting work). The urban locational choices made by transnational companies (see Chapter 2) reflect a general shift in many national economies from manufacturing to service; hence, older industrial cities have suffered as new service centres have emerged.

Second, global cities are key centres for financial institutions, such as banking, insurance, specialized investment organizations, and stock exchanges. Through these institutions, global cities are able to manage and coordinate global economic power, thereby serving as the control centres for capital in the new international division of labour, in which transnational companies have their headquarters in one place and the related manufacturing activities in another (often the less developed world, where labour costs are low). In their capacity as home to transnational company headquarters, financial institutions, and stock markets, global cities mediate between the world economy and specific nation-states.

Third, business services such as accounting, advertising, marketing, and law gather in global cities. Through these business services, large corporations and financial institutions get advice and support.

Political Characteristics

Politically, the relationship between global cities and levels of government is complex. Global cities are typically home to national and/or subnational governments and/or **supranational organizations** such as the United Nations, the European Union, and the World Bank. These cities also tend to receive support from governments as they strive to compete with other global cities, as evidenced by the national government support that

supranational organization
A multinational grouping of independent states, where power is delegated to an authority by member governments.

London received for its successful bid to stage the 2012 Olympic Games or that Brazil gave to its cities to host the 2014 FIFA World Cup. In some cases, a city plays a distinct political role. Consider the unusual example of The Hague. Although not the official capital city of the Netherlands, this city is a political centre and has achieved significant stature as home to several international organizations concerned with law, peace, and justice, including the International Court of Justice and the International Criminal Court. Similarly, New York is the site of the UN, even though it is not even the capital of New York state.

Cultural Characteristics

Global cities are also cultural centres, implicated in and responding to cultural globalization trends. As part of this involvement, global cities are usually **gateway cities**. Therefore, they are commonly culturally heterogeneous, serving as home to diverse ethnic identities, linguistic groups, and cultural traditions.

gateway city
A city that is a key point of entry to a major geographic region or country for goods or people, often via an international airport, container shipping port, or major rail centre; a city in which several different cultural traditions are absorbed and assimilated.

Global cities are also major players in such cultural activities as film, television, media, publishing, and sport. They are likely to be sites of various spectacles and events, such as world fairs, theme parks, and major sporting events. They are also home to such major cultural institutions as theatres, museums, and art galleries.

Necessarily, a global city is a transportation and communications centre, with efficient intra-urban movement and various inter-regional and international links. This aspect of global cities is evident in the emerging geography of the Internet, which is widely dispersed in its use but favours some places over others and has a commercial concentration in global cities (Zook, 2005). E-commerce and e-business are encouraging further growth of and concentration of both consumer and business services in global cities.

Environmental Characteristics

Several less tangible aspects of global cities are accounted for in what researchers call environmental characteristics. In this case, environment includes not only the physical environment but also the social. Levels of pollution and cleanliness are important, as are issues such as traffic congestion, safety, security, and freedom of expression. Global cities tend to be at the forefront of making cities more "livable."

The Hierarchy of Global Cities

Researchers compile indices of the various components just discussed to "rank" cities according to their global-city status. A number of rankings exist, including several by the Globalization and World Cities (GaWC) research network in the UK, the Hong Kong-based International Cities ranking (Huang et al., 2007), the Global Power City Index by the Institute for Urban Strategies in Tokyo, and the American-based Global Cities Index. These rankings are important for understanding the characteristics of global cities, their major functions, and their relative position in the global hierarchy. And while researchers generally agree on which cities are in the upper tiers of this hierarchy, they disagree about the precise position of various cities because each ranking system uses a slightly different set of indicators for evaluating each city and for weighting these factors.

The most commonly cited example of such a hierarchical ranking system is that developed by geographers in the GaWC group. Revised regularly, the GaWC rankings assess global cities by a mix of economic, political, and cultural indicators. The 2012 rankings group the global cities into three broad categories—alpha, beta, and gamma—each of which is further subdivided into three or four smaller groups. Alpha cities are the most important nodes of the global economic system and facilitate linkages between major economic regions (such as North America, Europe, and Pacific Asia) and the global economy. Beta and gamma cities represent the main connections with the global economy and more moderate and less significant economic regions, respectively. The result is the identification of 182 world cities grouped into these three levels: 45 at the alpha level, 78 at the beta, and 59 at the gamma level.

Figure 11.10 shows the world's alpha cities, divided into four categories. Not surprisingly, many of the 45 highest ranking cities are located in more developed world countries on the leading edge of economic globalization. The two leading global cities identified in this analysis are London and New York, with Paris, Tokyo, and six others in the second ranked group. The alpha world cities are evenly distributed across the

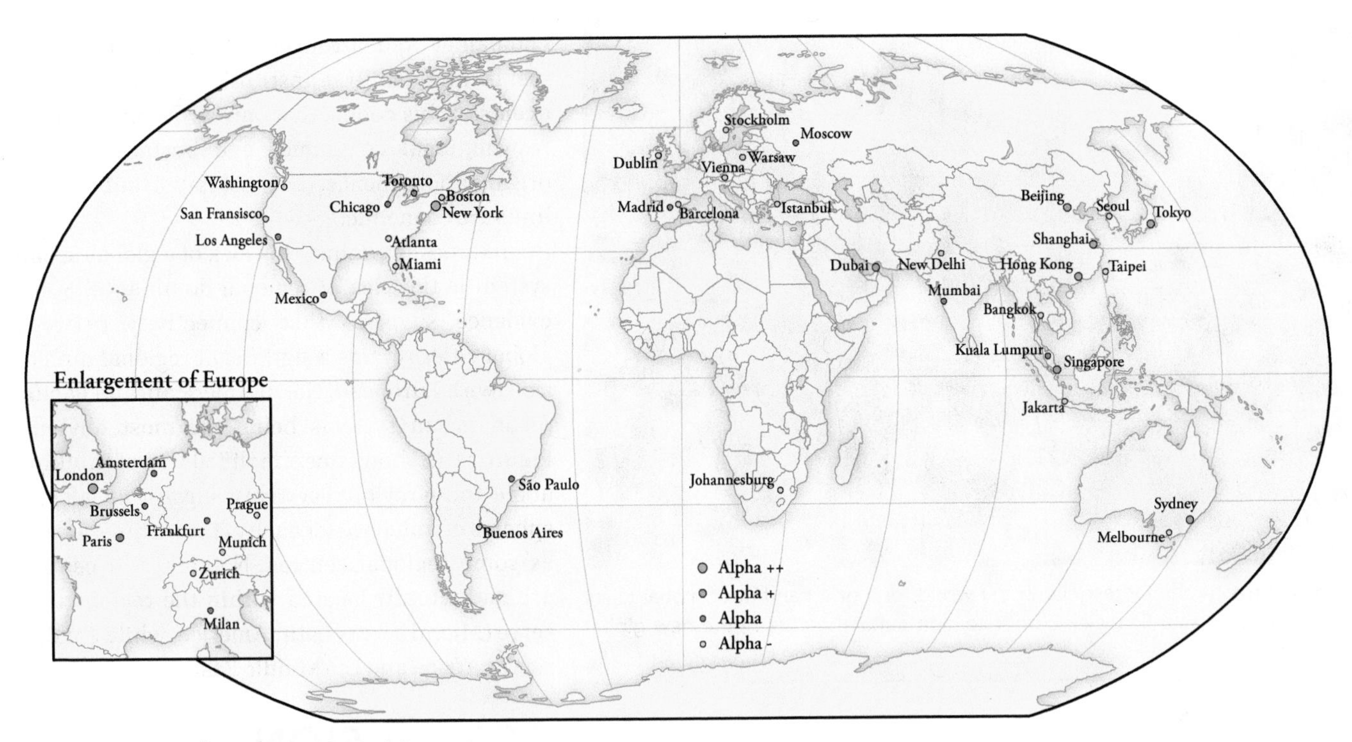

FIGURE 11.10 | Alpha global cities, 2012

Source: GaWC. 2012. *The World According to GaWC, 2012*. www.lboro.ac.uk/gawc/world2012t.html.

three globalization arenas of Western Europe (13 cities), North America (9 cities), and Pacific Asia (10 cities) and represent the main command and control centres of the global economy. Toronto is the only Canadian city among the alpha group (Montreal, Vancouver, and Calgary are beta cities). Only three cities in Latin America (São Paulo, Buenos Aires, and Mexico City), one city in sub-Saharan Africa (Johannesburg), and one city in Southwest Asia (Dubai) are among those at the pinnacle of the global cities ranking. Each of these cities represents the fundamental linkages between emerging market economies and the world economy and, in many cases, are simultaneously primate cities.

Figure 11.11 outlines the result of a second comprehensive analysis, which ranks cities using 26 specific indicators across these dimensions:

- *Business activity:* Number of major global corporate headquarters, the value of the city's capital markets, the flow of goods through its ports and airports, etc.
- *Human capital:* Proportion of population that is foreign-born, quality of post-secondary institutions, proportion of population with post-secondary education, etc.

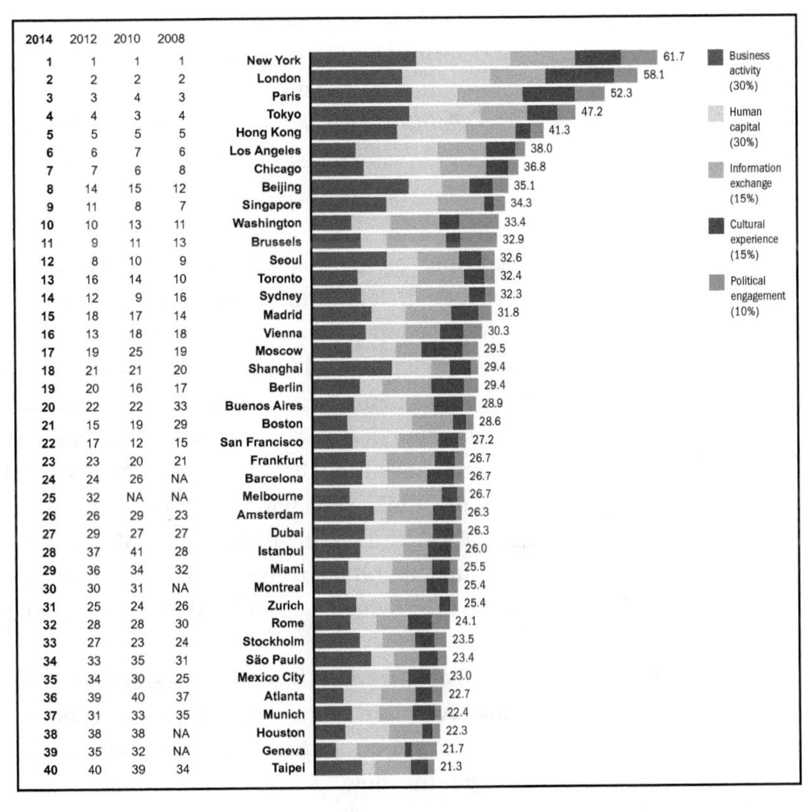

FIGURE 11.11 | The Global Cities Index, 2014

Source: Adapted from Kearney, A. T. 2014. "2014 Global Cities Index and Emerging Cities Outlook: Global Cities, Present and Future." https://www.atkearney.com/documents/10192/4461492/Global+Cities+Present+and+Future-GCI+2014.pdf/3628fd7d-70be-41bf-99d6-4c8eaf984cd5.

Rich Legg/iStockphoto

Tokyo is currently the largest city in the world, one of a handful of global cities based on its role as a financial, corporate, and cultural centre in the new global economy.

- *Information exchange:* Internet penetration rates, number of international news bureaus, freedom of expression, etc.
- *Cultural experience:* Major sporting events hosted, number of museums and performing-arts venues, number of international tourists, etc.
- *Political engagement:* Number of embassies and consulates, international organizations, political conferences, etc.

According to this ranking, the top four cities are, as expected, New York, London, Paris, and Tokyo. Among the top 40, the vast majority are from the more developed world, including 9 American, 2 Canadian, 15 European, 1 Asian, and 2 Australian. Of the remainder, all are in newly industrializing countries, including the three leading commercial cities of China, Singapore, and Seoul; the three Latin American urban powerhouses (Mexico City, São Paulo, and Buenos Aires); and the two rising southwestern Asian commercial hubs, Dubai and Istanbul.

Global cities can thus be understood as occupying levels in a hierarchy, but they are not isolated one from another. They are nodes linked through the flow of capital, knowledge, information, commodities, economic activities, and people (especially people as high-value labour), which form an urban system. Global cities are also connected through the activities of transnational companies and arrangements between governments and cultural institutions. The principal agents of this connectivity are firms, particularly transnational companies and business service organizations; hence, the most important of these links are economic.

Juxtaposed against this idea of a global-urban system is the idea of regional dominance. Some evidence suggests that connectivity between global cities retains a significant regional dimension, with European, North American, and Pacific Asian urban systems being the most obvious. Figure 11.12 shows one simplified way of thinking about these regional systems, suggesting general spheres of influence for each of these three, as well as some regional centres. Notably, two centres are not actually located within the regions they serve (Miami serves Latin America, while London serves Africa and the Middle East).

CONCLUSION

Recall four facts discussed earlier in this chapter. As of 2014, approximately 80 per cent of people in the more developed world and 48 per cent of people in the less developed world live in urban centres. Both percentages are increasing; the rise in the latter is especially rapid. The population of the world's megacities (more than 10 million people) is growing, and new megacities are appearing. Most of the projected additional urban population in the coming years will go to other cities, including some with fewer than 500,000 people today. These changes are significant, but they are not the only ones occurring on the global scale.

Now consider the possibility that cities are the solution, not the problem. From the outset, cities have been associated with progress (what we often call civilization) and have served as centres of the human experience. The fact that we are living in an age of unprecedented urban growth suggests that new versions of the human experience are emerging. China is the most dramatic example of growth—it is estimated that 50,000 skyscrapers will be built there during the next 50 years—but rapidly growing cities are the norm in most less developed countries while urban numbers continue to creep upwards in more developed countries. The increasing number of megacities is just one indicator of this ongoing transformation.

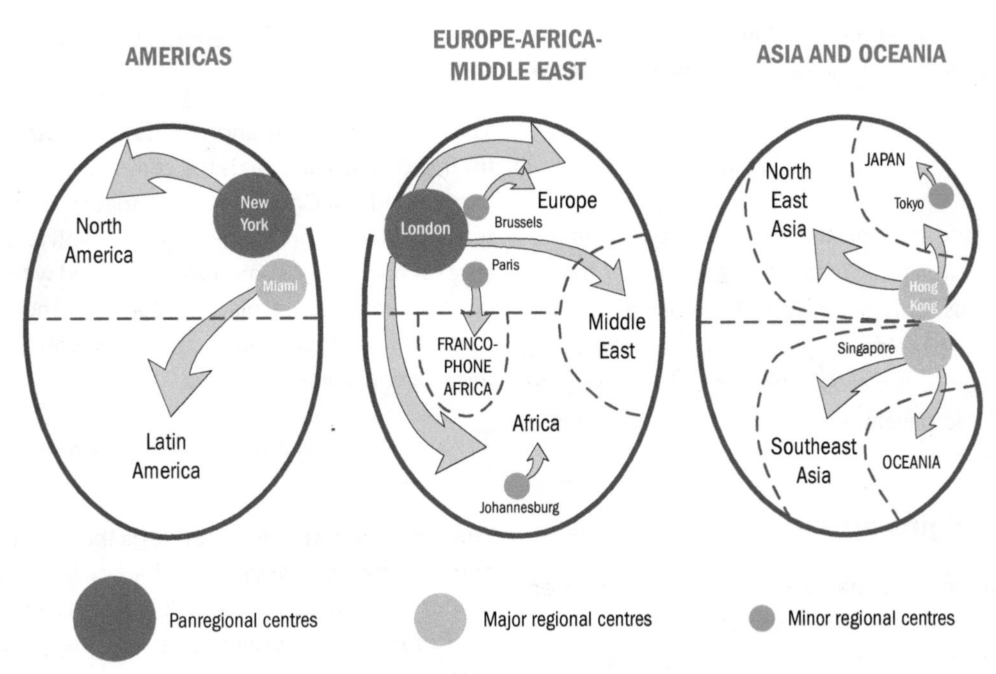

FIGURE 11.12 | Global cities and spheres of influence
This diagram is derived from a detailed analysis of information about the location of business headquarters, subsidiary offices, and capital flows that identifies 55 global cities of varying degrees of importance and questions many of the general assumptions about global cities. The patterns shown in this figure are quite different from some of the complex hierarchies described in other studies.

Humans are social animals. The urban experience may fulfill a desire for regular and varied human contact that is necessarily lacking in rural areas. Cities provide opportunities for social and economic interaction, innovation, and creativity that are not possible in rural areas. It is through their cities that poor countries make connections with overseas markets, thus encouraging further economic investment. Cities are part of the solution to national economic concerns and to global warming for the simple reason that dense concentrations of population are the most efficient way to provide people with basic services such as health and education and to supply people with clean water, sanitation, and energy.

Summary

An Urbanizing World

The first permanent settlements appeared with the domestication of plants and animals approximately 12,000 years ago; the first cities emerged several thousand years later. Dramatic urban growth was not initiated until the beginnings of industrialization around 1750 CE. Globally, humans crossed the line from predominantly rural to predominantly urban in 2007, though significant differences exist between the more and less developed worlds, with the former already experiencing high levels of urbanization. Today, much of the growth in cities occurs in the less developed world.

Defining Urban Areas

Two points are relevant here. First, the precise way in which urban and rural dwellers are defined varies by country. Any changes in these definitions necessarily alter the numbers of each group in a country and, if the country has a large population, has notable impacts on global statistics. Second, the spatial extent of many cities can be measured in several ways, meaning that there are often different statements regarding how many people live in a city.

Megacities and Megaregions

About one-half of the world's urban population lives in cities of less than 500,000. Of the 28 megacities, only 6 are from the more developed world. The projected growth of megacities and megaregions is almost wholly expected to occur in the less developed world, especially in Africa and South Asia.

The Origins and Growth of Cities

Permanent settlements were initially associated with economic changes, specifically the production of an agricultural surplus; they also reflect social needs. The earliest cities developed about 3500 BCE as successors to large agricultural villages. Cities established before about 1750 can be conveniently labelled pre-industrial: they functioned primarily as marketing and commercial centres, although ceremonial and administrative functions were also important. Urbanization was stimulated by the mercantilist philosophy that flourished in Europe from the beginning of European overseas expansion until the emergence of capitalism. With the rise of industrialization, cities increased in number and size at a rapid pace. The locations of the industrial cities that appeared at this time were often determined by the location of the resources they depended on. For their inhabitants, ways of life in these cities represented a marked departure from earlier experiences.

Urban Systems and Hierarchies

Many attempts have been made to explain why cities are located where they are. Traditionally, attention was focused on specific cities, but it is increasingly shifting to systems of cities.

Central Place Theory and Urban Hierarchies

The most influential and persuasive explanation for the location of cities is the central place theory, expounded by Christaller. This theory, conceptually similar to von Thünen's work, involves a series of simplifying assumptions combined with four key concepts (central place, hinterland, range, and threshold) and generates principles concerning the size and spacing of cities.

The Rank-Size Distribution and Urban Primacy

The rank-size distribution suggests that, in a given country, the population size of a city is inversely proportional to that city's rank, where the largest city is rank 1. The urban system of many countries, including Canada and the United States, fits this general descriptive pattern. Many countries that do not conform to the rank-size distribution fit what is known as urban primacy, where a single primate city dominates the national urban system. The United Kingdom (London), France (Paris), Mexico (Mexico City), Thailand (Bangkok), and others fit this pattern.

Global Cities

Since the 1980s a hierarchy of cities in a global urban system has become increasingly evident. Initially, Tokyo, London, New York, and possibly Paris were recognized as global cities, but it is now usual to recognize many more and to divide them into categories. Most global cities are in more developed regions of the world. In general terms, they are understood in the larger context of globalization processes. Economically, global cities are typically home to transnational companies, financial institutions, and business services. Politically, these cities are usually home to one or more levels of government and supranational organizations. Culturally, they are important gateway centres and are frequently the sites of world fairs and major sporting events.

Links to Other Chapters

- **Globalization and global cities:** Chapter 2 (concepts); Chapter 3 (all of chapter); Chapter 4 (global environmental issues); Chapter 5 (population growth, fertility decline); Chapter 6 (refugees, disease, more and less developed worlds); Chapter 8 (popular culture); Chapter 9 (political futures); Chapter 10 (agricultural restructuring); Chapter 13 (industrial restructuring)
- **Urbanization:** Chapter 5 (population growth); Chapter 6 (more and less developed worlds)
- **Central place theory:** Chapter 10 (agricultural location theory); Chapter 13 (industrial location theory)
- **Global cities:** Chapter 3 (economic globalization)

Questions for Critical Thought

1. Compare urbanization (level and rate) in less developed and more developed nations. In what ways are urban problems in less developed nations different from those in more developed nations?
2. What are the implications of the proliferation of cities with more than 10 million people (see Figure 11.6)?
3. Discuss the relevance of Christaller's central place theory to our understanding of contemporary urban systems.
4. Why are New York, Tokyo, and London classed as global cities? What would it take for a city like Toronto to become a true global city?
5. We have recently reached the point where half the world's population is classed as urban. What are the implications of the continued rapid growth of the urban population on a global scale?

Suggested Readings

Visit the companion website for a list of suggested readings.

12 THE CITY AND THE URBAN FORM

Cities around the world are different, yet they display some fundamental similarities, at least partly because they are shaped by comparable processes. This chapter begins with a consideration of the internal structure of the city and the distribution of urban land uses, which typically relate to distance and direction away from the city centre. A few conceptual formulations were developed to help understand the internal structure of cities, especially those in North America.

We then examine the various constituents of cities, namely the places where people live and work and how they get around. The bulk of urban land is allocated to places for people to live, and this chapter explores residential patterns (neighbourhoods) and processes (gentrification and segregation), as well as the continued spatial spread of residential developments beyond the traditional urban boundaries into the suburban areas. All cities have residential areas and areas dedicated to different forms of employment, from manufacturing to office work, consumer services such as retailing, and public spaces dedicated to recreational activities. Transportation and communication networks connect the places that people live to the places they work. The organization of land uses in cities is, to some degree, planned in accordance with changing political, social, and economic objectives.

The chapter concludes by discussing one of the main ways that cities in the less developed world differ from those in the more developed, namely the presence of informal settlements and slums.

Here are three points to consider as you read this chapter.

- Should housing be provided by the state as a social service, or is housing a commodity to be bought and sold?
- What are some of the social inequalities evident in cities, and how might they be modified?
- What role do local (municipal) and subnational (provincial or state) governments play in the planning of cities? Should this role be strengthened or de-emphasized?

A streetcar passes condominiums and the CN Tower in downtown Toronto. Examining urban structures and land use, particularly in terms of housing and business development, reveals a great deal about urban populations and the social inequalities within them.

4068
UNION STATION
STOP

In this chapter, we focus on three related trends: the evolving notion of urban residential space; the spread of population and other urban functions beyond city areas towards the suburbs; and the allocation of urban space to its many uses and ways the land uses have changed over time. The urban environments that people live in today are often very different from those of years gone by.

EXPLAINING URBAN FORM

We typically refer to a city's most visually distinctive area as the downtown, which geographers define more precisely as the **central business district (CBD)**. The CBD is the city's social and cultural hub, economic engine, and political nerve centre. Business, personal, and public services are usually all drawn to the area because of its high level of accessibility. The CBD of many large urban cities contains office skyscrapers, municipal and other forms of governance, and retail businesses, as well as entertainment and recreation centres such as sports stadiums, museums, theatres, and associated activities like hotels and restaurants. Historically, we would also have found manufacturing activities located adjacent to major transportation facilities such as a port or railway line, but they have largely moved to the city's periphery. At one time, many people lived in the CBD, but most moved to the suburbs over the course of the nineteenth and twentieth centuries.

central business district
(CBD) The social, cultural, commercial, and political centre of the city; usually characterized by high-rise office and residential towers, key municipal government buildings, and civic amenities.

urban structure
The arrangement of land uses in cities; related to urban morphology.

The diversity of the CBD is one of its most characteristic elements. It is a key area for many aspects of modern society, but it is also ever-changing. The CBD's boundaries expand as the city grows, and the dominant functions change as certain land uses move to the city's periphery. Recent efforts of urban revitalization, combined with demographic and economic changes, have drawn people back to the city centre to live in medium- and high-density residential units such as apartments and condominiums.

In addition to the CBD, cities contain many other characteristic elements, each of which is addressed in this chapter. For example, cities have residential areas, areas of employment, and more recent areas of mixed-use development on the periphery. Not surprisingly, there is diversity amongst the land-use types, as not all residential or industrial areas are alike. Some inner-city residential areas are older and may be undergoing revitalization, while other areas are more suburban and cater to the upper or middle class.

One of the most defining elements of cities over the twentieth century was an increasing focus on their suburbs. These places have been a part of cities since their inception but are increasingly important because they house the majority of the city's population and employ many of them too. In the decades since the end of World War II, there has been a steady decentralization of residents and businesses as they occupy suburban areas.

Urban Structure

Urban geographers are interested in the ways that cities are constructed and experienced by those who live and work in them. Part of how people experience cities is framed by the city's **urban structure**, the generalized arrangement of land uses within the urban area. Urban land uses are broadly categorized into six types: residential, commercial, industrial, institutional, recreational, and transportation (including roads, parking lots, public transit, railways, and airports). There is significant variation within each of these broad land-use types. For example, low-density suburban residential areas consist primarily of single-family homes; while medium-density have town homes and low-rise apartments and high-density have towers of rental apartments or condominiums. The allocation of these different land uses across the urban landscape is a central concern for urban geographers.

Every city is unique, with its specific economic, cultural, and historic characteristics. At the same time, there are broad similarities in how they are all structured. North American cities display similar land-use patterns. The patterns in Europe or Latin America are analogous but different from those in North American cities.

An important concern in a city's geography is where and in what numbers people live, work, shop, engage in recreational activities, etc. Interestingly, the distribution of population densities tends to be remarkably consistent for all urban areas, bearing little relation to the internal structure. In particular, urban population densities typically decline as a negative exponential function of distance from the centre of the city (see Figure 2.5). In other words, population density is highest in the centre and declines towards the urban periphery. Generalizations of urban land use can be conceived similarly, operating on the same principles as von Thünen theory, with different land uses having different economic rent lines (see Chapter 10). Based on this premise, the value of land also characteristically decreases with increasing distance from the centre (Figure 12.1). Thus, land uses with the greatest values of economic rent at the city centre will have steep rent lines and occupy land adjacent to the centre. Theoretically, then, a city is made up of a series of concentric zones. A typical arrangement might have businesses at the centre with industrial and then residential uses at increasing distances from the city centre (see Figure 10.4).

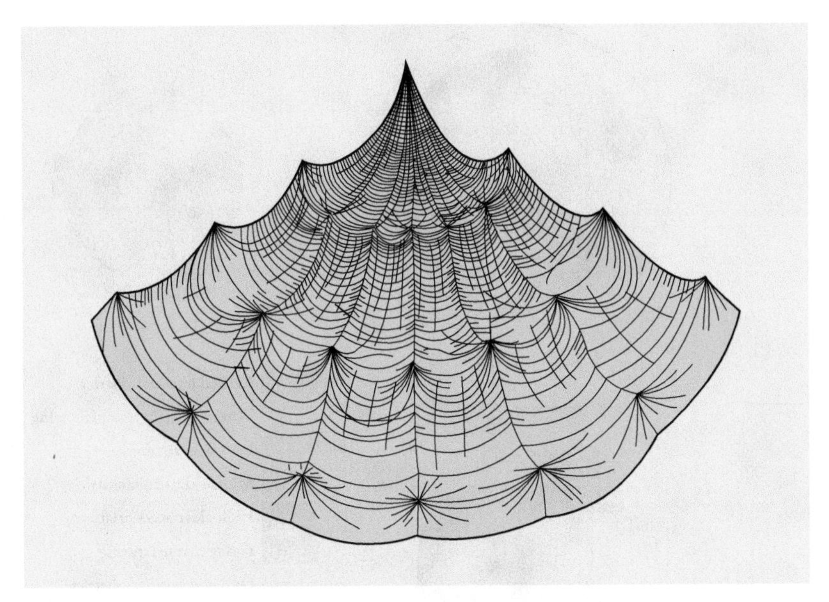

FIGURE 12.1 | Urban land values
This classic "circus tent" schematic diagram is often used by geographers to provide a generalized representation of urban land values. The peaks in the diagram are areas of high value, and the intervening areas are lower value. High-value areas are commercial or industrial locations on transport routes; low-value areas are residential or vacant.

Modelling the North American City

To understand the modelling of North American cities, we turn to the Chicago School, an informal group of scholars from the early twentieth century who were mainly concentrated at the University of Chicago but included others at neighbouring institutions. Many of them were sociologists, but other members were geographers, economists, architects, anthropologists, etc. The Chicago School developed three generalized descriptions of twentieth-century North American cities: the concentric zone, sector, and multiple nuclei models (Figure 12.2). These "classic" models were based on observations of Chicago and other American cities over the course of the twentieth century.

Concentric Zone Model

In 1925 Canadian-born sociologist Ernest Burgess developed the concentric zone model, which focuses primarily on residential land uses. In his study of Chicago, Burgess observed a spatial relationship between a household's socio-economic status (mostly income) and distance from the central business district: the greater the distance, the greater the wealth and, usually, the better the housing quality. The latter involved a trade-off in terms of length and cost of the commute to work, as most employment areas were located within or adjacent to the CBD.

Based on this general spatial association, Burgess identified five zones. The first is the CBD, which is typically subdivided into areas such as financial, commercial (retail), entertainment, etc. Immediately surrounding the CBD is an area Burgess termed the zone in transition. Much of the blue-collar (industrial) employment occurs in this area, as factories and related rail yards and warehouses are located here. Because this zone is the least desirable for residential use, urban land values are normally lowest. Consequently, the residential areas are of poor quality and home to the city's poorest residents, many of whom are recent immigrants. At the time Burgess developed

Downtown Vancouver is home to some of the highest urban property values in Canada.

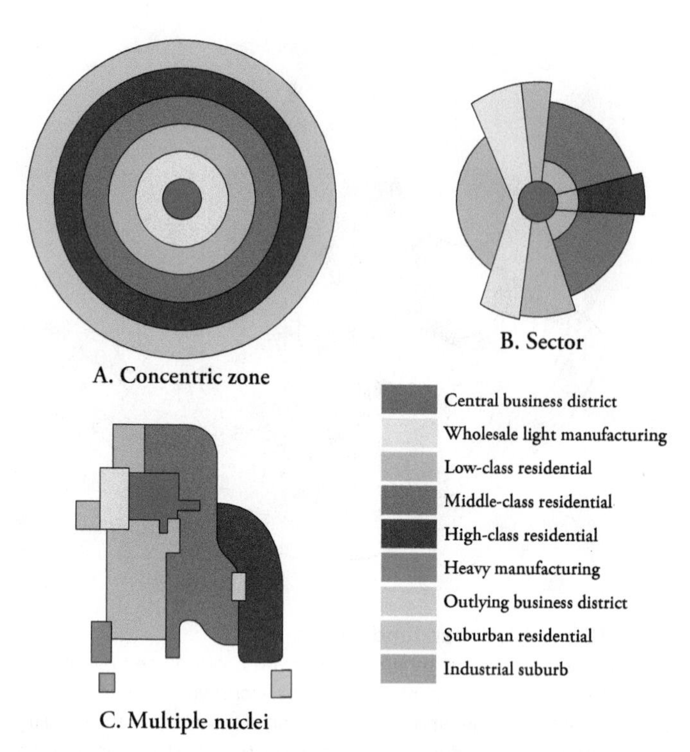

FIGURE 12.2 | Three classic models of the internal structure of urban areas

this model, the zone in transition was seen as an immigrant receiving area—the area where new immigrants would first settle, in highly segregated neighbourhoods, while they worked to establish themselves.

The remaining three zones are all residential, becoming increasingly affluent with increasing distance from the CBD. The third zone contains older homes occupied by working-class families, many of whom are second-generation immigrants who had moved out of the zone in transition. For these people, most of whom work in the nearby factories, commuting to work is relatively straightforward and inexpensive. The fourth zone contains the newer, more spacious, and more expensive homes of the middle-class, who can afford the cost of a longer commute to the CBD. The fifth, and most distant zone, is comprised of a number of satellite residential suburbs at the end of commuter rail lines, which are home to the city's wealthiest residents living in spacious homes in almost rural-like settings.

As a descriptive device, the concentric zone model applies best to North American cities and proved especially useful in understanding the context of changing residential character. Specifically, urban growth is understood as a process of expansion and invasion (or the conversion of one land-use type into another) as each inner zone expands into its neighbouring outer one. Over time, a sequence of invasions leads to a succession of land uses.

While simplistic, this model was highly influential and remains valuable as a general descriptive device. It is important to appreciate that the model describes an ideal situation given certain assumptions and does not describe reality. The explicit identification of land-use types and zones remains an important contribution to our understanding of the geography of the city. However, as the automobile became ubiquitous in the latter half of the twentieth century, the notion of land-use conversion became almost redundant, as new suburban land could be more easily and inexpensively developed.

Sector Model

As a result of an extensive empirical analysis of urban land market data from across the United States, the economist Homer Hoyt (1939) developed the sector model of urban land use as a modification of the concentric zone. He believed that Burgess's conceptualization had merit but was insufficient to fully explain land-use patterns. Hoyt surmised that direction from the CBD, as predicated by the increased accessibility of certain areas via transportation corridors, was also important. In other words, distance and direction from the CBD were key determinants of land usage.

According to the sector model, the city expands not just in concentric rings but in a series of sectors or wedges. Transportation corridors (e.g. railroad lines, public transit routes, or commercial main streets), which emerge around infrastructure, are responsible for creating these sectors within the broadly concentric pattern described by Burgess. Once a given land use is established in a specific sector of the city, growth occurs outward in a wedge-like fashion. For example, once a high-status residential neighbourhood is established, the most expensive new housing is built on the outer edge of that district, farther away from the city centre. Similarly, new industrial development, such as new factories, are built along the railroad lines that already service existing factories in previously established industrial sectors.

Multiple Nuclei Model

The third classic model, developed by geographers Chauncey Harris and Edward Ullman in 1945, is the multiple nuclei model. According to Harris and Ullman, while cities may initially develop around a single CBD, additional nodes (or nuclei) emerge within the urban area over time, around which particular activities take place. This differs from the concentric zone and sector models, which are premised on a single nucleus, the downtown or CBD.

Additional nuclei may include airports, suburban business centres, or universities, among others. Their influence on the structure of urban land uses is explained by what Harris and Ullman identify as the forces of attraction and repulsion. We can similarly think of these as centripetal and centrifugal forces, respectively. According to this idea, certain land uses attract other, complementary, uses and simultaneously exclude others. For example, industrial land uses tend to attract more industry and working-class residential land uses yet repel high-status residential uses at the same time. As such, there are two key contributions of this model. Cities are more complex than assumed in the two earlier models, although the basic premises are still valid. Existing land uses within the city are understood to play a pivotal role in attracting or repelling new land uses in adjacent areas.

This model proved extremely attractive to geographers and influenced much subsequent research, as outlined later in this chapter. It anticipated many of the late twentieth-century changes that cities experienced, in particular the decline of CBDs and the growth of suburban business and retailing centres.

Modelling the European City

Geographers have applied the three classic urban structure models to the European context, with varying success. For example, the sector model has been applied to explain the urban structure of cities such as Paris, where industrial land uses exist in the northern and southeastern areas and high-status residential areas exist in a southwestern sector and near the royal palaces.

However, in contrast to the North American city, the central core of many European cities is characterized by narrow streets, a compact and dense urban form, and an economically and culturally vibrant downtown facilitated by public squares, plazas, and local marketplaces. Notable examples include Bruges, Venice, London, and Moscow.

Furthermore, European cities tend to have a higher density of people living near the city centre but a much lower vertical profile of the built form than their North American counterparts. Modern steel and glass skyscrapers dominate the skylines of Toronto, Chicago, and New York; London, Paris, and Rome have notably fewer tall buildings, though this is beginning to change. In terms of spatial extent, European cities are compact, with a greater emphasis on public transportation, more trains and light-rail lines, walkable districts, etc. North American cities are characterized by wide multi-lane automobile thoroughfares and comparably more limited public transit lines.

While remarkably dense, the skyline of central Paris shows only limited vertical reach, as Parisians and the city's urban planners generally dislike the look of skyscrapers.

Given the cultural diversity and varying ages of European cities, no single urban structure model can be applied to all the continent's cities, but notable regional and cultural variations exist in the urban form. As such, the urban structure of ancient cities like Athens differs greatly from that of medieval cities like Prague, industrial cities like Manchester, or commercial centres like London or Paris. Despite these regional and historical variations, White (1984) tried to construct a composite model of the continental West European city (Figure 12.3). This model begins with a historic core that remains central to commercial, civic, and social life and that is ringed with residential areas undergoing change through gentrification and/or urban decay. Suburban residential areas, representing the interwar and postwar housing booms, contain a broader range of households than their American counterparts, most notably with a sizable minority of social housing. Distinct dormitory (commuting) villages and mixed middle-class and high-status households sit on the urban periphery.

FIGURE 12.3 | The West European city

Source: White, P. 1984. *The West European City.* New York and London: Longman.

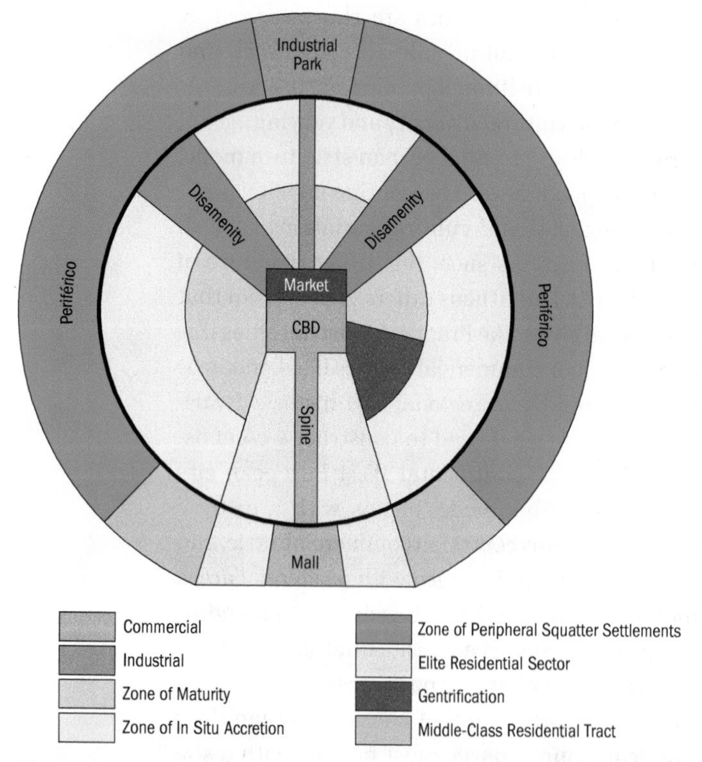

FIGURE 12.4 | Modelling the Latin American city

Source: Ford, L. 1996. "A New and Improved Model of Latin American City Structure." *Geographical Review* 83: 438. Copyright © 1996 The American Geographical Society.

Modelling the Latin American City

Prior to colonial European settlement, the region of Central America reaching from central Mexico to Honduras and the South American Andean and Pacific coast regions extending from Ecuador to Bolivia had important urban centres. Europeans transformed many of these areas and built many others. Several of the new colonial cities were planned to include a central plaza with streets radiating outward in a rectangular grid pattern. Wealthy Europeans lived close to the city centre while the city outskirts attracted locals seeking employment. Latin American cities have expanded and diversified dramatically since colonial times, and attempts to model internal structure reflect both colonial foundations and subsequent developments.

Figure 12.4 shows Larry R. Ford's (1996) model of the Latin American city, which is centred on the CBD and a large market plaza. Other land uses in the city vary according to distance and direction from the centre. A commercial spine, considered an extension of the CBD, is a grand ceremonial boulevard with adjacent elite-quality residences running from the city centre to the suburbs. In general, residential quality declines with increasing distance from the centre; the zone of maturity (high-quality residential) is closest to the centre, followed by a mixed residential, transitional area (termed *in situ accretion*) and wedged-shaped sectors of low-quality, disadvantaged homes (termed *disamenity* as a result of the presence of hazardous or noxious features such as railway tracks, open sewers, etc.). Surrounding all these areas is an outer ring of the city, defined by informal squatter settlements.

In general, the urban structure of the Latin American city as defined by Ford (1996) is similar in form to the sector model, as it incorporates concentric zones and wedge-shaped sectors surrounding a CBD.

Modelling the Sub-Saharan African City

The simple models we have discussed so far are not easily developed elsewhere in the world. It is particularly difficult to generalize African and Asian cities as there are numerous regional traditions and varied colonial imprints. O'Connor (1983) noted several different types of African cities, including the following:

- Indigenous: pre-colonial cities such as Addis Ababa
- Islamic: located mostly in the Saharan region (North Africa), such as Timbuktu
- Colonial: located throughout much of the extensive area colonized by Europeans, often as capitals, mining towns, or trading centres, such as Kinshasa
- Dual: comprising at least two of the previous types, such as Khartoum-Omdurman

These different types are not readily generalized into a single model. As urbanization proceeds, this simple classification actually becomes less helpful. Nevertheless, the internal structure of large African cities displays variations according to distance and direction, as do most cities throughout the world. There is usually a commercial centre, comprised of a former colonial business district as well as a traditional and sometimes informal marketplace. Encircling the commercial centre are sectors of residential neighbourhoods, usually differentiated by ethnicity, and manufacturing clusters. Like European and Latin American cities, high-quality housing tends to be located close to the centre. An outer zone of informal squatter settlements is a common characteristic of cities in much of the less developed world.

Squatter shacks outside the township of Soweto, south of Johannesburg, South Africa.

Modelling the Asian City

There are numerous regional urban traditions and a varied colonial imprint in Asia. Regardless of location, date, and reason for foundation, most Asian cities have an internal structure that reflects the principles of both distance and direction as outlined previously. Five of the many examples of Asian city structure are described here.

Most cities in Southwest Asia and North Africa display a dominant Islamic influence, being most obviously characterized by a major central mosque, a historic kasbah (a fortified citadel), a central market or bazaar (suq), and an irregular and narrow street pattern. As urban growth occurs, suburban mosques and bazaars serving the local suburban community emerge, creating separate nuclei. Many Islamic cities reluctantly reflect the influence of globalization and the emergence of Westernized districts with hotels, skyscrapers, office buildings, and increasingly exclusive residential districts.

South Asia has two principal city types—indigenous and colonial. Many indigenous cities developed along important trade routes and are centred on a bazaar that often offers a wide variety of retail outlets and accommodations. Close to the bazaar are high-quality residences, and housing quality generally declines with increasing distance. There are also enclaves scattered throughout the city for minority ethnic groups and people of lower caste (social status).

Most of the major colonial cities in South Asia are coastal and in strategic political or economic locations. As shown in Figure 12.5, the centre is a historic military fort with adjacent open space. Several sectors lie beyond the open space, including an administrative area, a business district, a housing area for the indigenous population, and an area of European settlement. Commercial nuclei are centred on bazaars, and the outer fringe of the city contains a series of residential areas. Today, most of these cities are surrounded by vast informal squatter settlements.

Figure 12.6 illustrates the generalized internal structure of the Southeast Asian city, which has experienced a variety of cultural influences, including Indian, Chinese, Arab, and European. Ethnic diversity dominates this model. There is a port zone, as well as a series of other zones and sectors. The overall zonal pattern is one of declining housing quality with increasing distance, while the sectors include commercial zones distinguished according to ethnicity. The outer zones include both formal suburbs and areas of informal squatter settlement.

© Kazuyoshi Nomachi/Corbis

The Great Mosque is located at the centre of Mecca, Saudi Arabia.

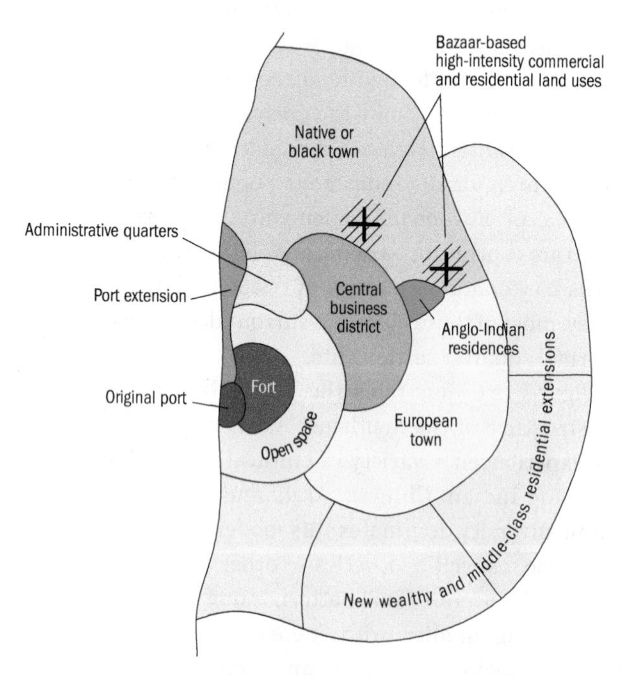

FIGURE 12.5 | Modelling the Asian colonial city

Source: Brun, S., and J. Williams. 1993. *Cities of the World*, 2nd edn. New York: Harper and Row, 360.

China is home to many of the world's largest and fastest growing cities. Recent and rapid growth, as well as regional diversity in the country's urban system, has led to few comprehensive generalizations of the internal structure of Chinese cities. However, several characterizations can be made. First, many Chinese cities are characterized by a high population density that does not translate into a compact urban form; many of the most widely recognized Chinese cities are sprawling megacities.

Second, most cities are polycentric, meaning they have multiple nuclei rather than a singular downtown. Each nucleus tends to mirror the main central district, in that it comprises a mix of commercial and residential buildings. These structures can be a mix of modern and traditional architectural styles, as well as high- and low-rise.

Third, the unique influence of a top-down communist, and more recently market socialist, urban planning approach has led to a somewhat distinct urban form. The central core of many Chinese

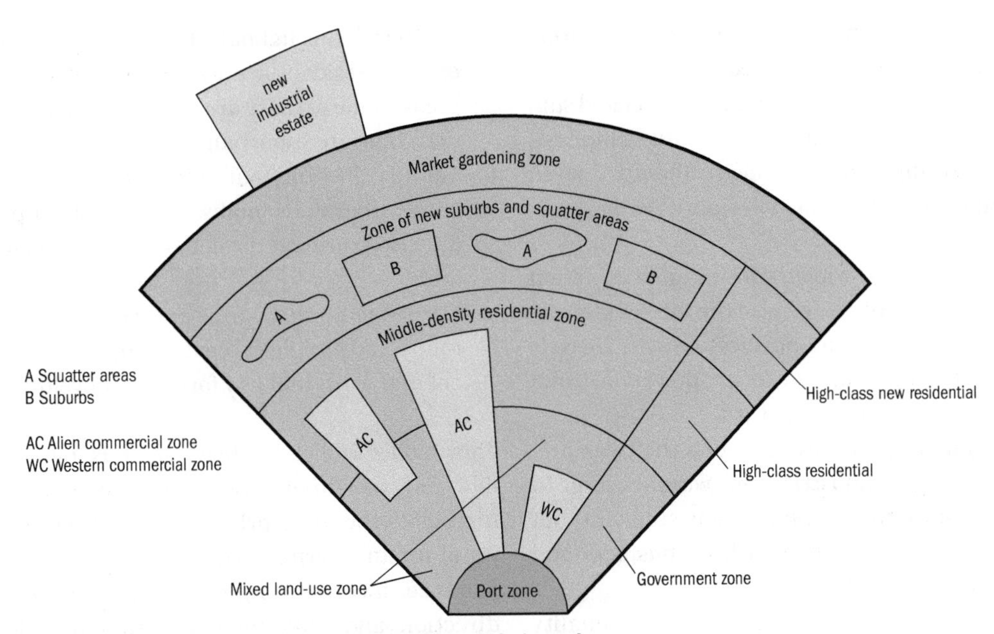

FIGURE 12.6 | Modelling the Southeast Asian city

Source: McGee, T. 1967. *The Southeast Asian City*. New York: Praeger, 128.

cities, Beijing in particular, is reserved for public squares (Tiananmen Square), wide boulevards, and many monumental administrative and civic buildings. As a consequence, the level of development intensity is lower here than elsewhere in the city; this situation contrasts sharply with cities in North America and elsewhere. A push towards a market economy and an embracement of globalization in the latter part of the twentieth century resulted in tremendous growth in manufacturing and industrial activity in many Chinese cities. Much of this development occurs on the periphery of the city, in special industrial sectors.

Rethinking Models of the City

As noted, the concentric zone model focuses on distance, the sector on distance and direction, and the multiple nuclei on distance, direction, and multiple centres. Similarly, the models developed for other regions of the world are closely related to these classic types, with the basic focus on distance, direction, and nuclei. All these models simplify a complex reality; none of them purports to be universally correct. Although they are valuable as descriptions against which real-world structures can be assessed, they do not necessarily explain the internal structure of current urban areas. There is a general failure to answer (or even to ask) questions about processes, and the causes of certain spatial configurations. Further, the three classic models are essentially economic, largely ignoring social, cultural, and political variables.

A useful revision of the classic models is that by White (1987), who suggested the general form of an early twenty-first-century city in the more developed world. Although this model was created in response to changes affecting the city's internal structure—changes such as deindustrialization

A remnant of the French colonial administration in Vietnam, the Hanoi Opera House is modelled on European architectural styles. Hanoi was of great political importance to the French, functioning as the capital of French Indochina.

and the rise of a service economy, decentralization of retailing activities, increased government intervention, increased automobile use, and related suburban expansion—it maintains the earlier emphasis on distance, direction, and nuclei. There are seven components to White's model (Figure 12.7):

1. The core area, which continues to function as the heart of the city, is the site of government, financial, and business offices. There is less retailing than previously. This zone grows upward rather than outward.
2. Surrounding the core is a zone that was previously light industrial and warehousing. In some cities, this second zone is stagnant, but in others it has benefited from business and residential investment.
3. Most cities include several areas of low-quality housing, characterized by poverty and often occupied by minority ethnic groups. Most of these areas are adjacent to the second zone, although some are located elsewhere.
4. Much of the rest of the city consists of middle-class residences, frequently divided into relatively distinct neighbourhoods. Although this is the largest area of the city, it is not continuous; it is interrupted here and there by the three remaining components.
5. Scattered throughout the middle-class area are elite residential enclaves. As with the poor areas, some of these are close to the city centre and some are suburban.
6. Various institutional and business centres, such as hospitals, malls, and industrial parks, are also scattered throughout the middle-class area.
7. Finally, most cities grow outward along major roads, and peripheral centres largely independent of the original city form may develop.

neighbourhood
A part of the city that displays some internal homogeneity regarding type of housing, may be characterized by a relatively uniform income level and/or ethnic identity, and usually reflects certain shared social values.

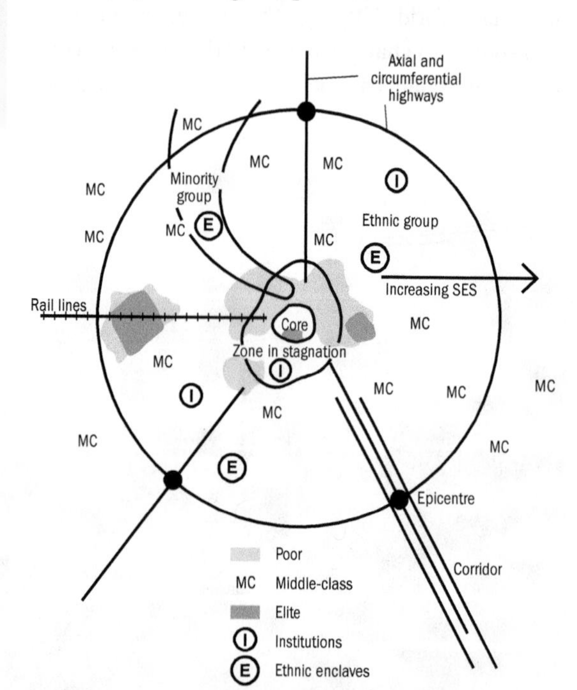

FIGURE 12.7 | White's model of the twenty-first-century city

Source: White, Michael J. 1987. *American Neighborhoods and Residential Differentiation*. New York: Russell Sage Foundation, 237. © 1987 Russell Sage Foundation.

This model is helpful because it is more flexible than the three classic ones. Our discussion highlights the unsurprising fact that most models of urban internal structure rely primarily on distance, incorporating other variables such as direction and multiple nuclei to more closely approximate reality.

HOUSING AND NEIGHBOURHOODS

Shelter is a basic human need, and housing is the key feature of the urban social and economic landscape. For most of us, understanding a city includes recognizing distinct housing areas, or **neighbourhoods**, with some relatively privileged areas in terms of housing quality and related lifestyle and others that are more deprived. In many respects, the residents of a large detached suburban home with multiple garages and a substantial landscaped lot live in a different world than do the residents of a small dilapidated inner-city home in a crowded, impoverished, and crime-ridden area. Housing and the areas in which it is located mirror and enhance larger social inequalities. One useful way to introduce the idea of the city as social space is to identify four overarching social trends that have the potential to affect the social geography of the city (Box 12.1).

Housing Markets

In more developed countries, there are two ways of thinking about housing. It is primarily viewed as a commodity; housing is most people's largest expenditure and is often also a major capital investment. This perspective encourages the creation of socially variable housing areas. Secondarily, it is viewed as an entitlement, a universal right, regardless of ability to pay. There is a fundamental ideological distinction between

Examining the Issues

BOX 12.1 | Social Trends and the Social Geography of the City

Any focus on the social geography of the city needs to take into account a number of overarching processes. Following Murdie and Teixeira (2006), we note four ongoing processes that apply in the Canadian context.

Economic restructuring, as it involves a relative decline in manufacturing employment and an increase in service employment, is characteristic of many cities in post-industrial societies. These trends have been accelerated in Canada through trade liberalization agreements such as NAFTA. The significance of this restructuring is that many of the new service employment opportunities are low paid, part-time, and temporary and that low-paid, service-sector employment has relocated from the city centre to the suburbs. Together, these changes led to increased social inequalities inside the city and a spatial mismatch between where many low-income people live (in the city centre) and where suitable jobs are available (increasingly in the suburbs).

Changes in age structure and family and household formation, such as a decrease in the percentage of younger age groups and an increase in the percentage of the elderly population, are also typical of cities in post-industrial societies. In Canada age structures are spatially variable because of internal migration. For example, Victoria receives many elderly immigrants whereas Calgary receives job-seekers more likely to have young families.

Other important changes in household formation include decreasing household size as fertility declines and more diversified family structures. Both contribute to changes in the city's social geography, particularly those involving neighbourhood change. Some older industrial cities are experiencing significant demographic change as deaths outnumber births, a circumstance with implications, especially for health care and educational facilities. A specific example is Pittsburgh, where obstetrics wards were converted to acute care while public school enrolment fell from 70,000 in the late 1980s to 30,000 in 2008.

Increased internationalization, with growing proportions of immigrants coming from Asia, Africa, the Middle East, the Caribbean, and Latin America and decreasing proportions from traditional European source areas, is another trend in many countries. Two major social geographic impacts on Canadian cities are those of diversifying the cultural mosaic and sharpening differences within many immigrant communities, with some newcomers financially well-off and others possessing limited resources. In many cities, there may be few contacts between these two groups within any given ethnic identity.

Retrenchment of the welfare state, especially as it involves reductions in subsidized housing construction and welfare payments, has occurred in several provinces with relatively right-wing governments, significantly affecting opportunities for low-income Canadians, including many recent immigrants.

these two ideas. In the simplest terms, those with a capitalist perspective believe that housing is a consumer good (like clothing or electronics) to be bought and sold without any state intervention. Those with a socialist perspective believe that housing ought to be provided by the state as a social service (like education or health care). In most countries, attitudes lie somewhere between those two scenarios.

The United States has minimal state intervention in this matter, which equates to little public housing and relatively few people receiving financial support to buy or rent a home. Most other more developed countries, including Canada and those in Europe, work to ensure a basic level of housing for all people. Public housing dominates in Singapore and Hong Kong, where the state is the principal provider of housing. Of course, political perspectives change, and the details of residential location and therefore of social areas inside cities are closely related to shifting political attitudes. The former communist countries of Eastern Europe aimed, in principle, for total state intervention and a complete lack of private ownership, but housing policies in these countries are now more in line with the rest of Europe.

Housing as a commodity is appropriately analyzed in the context of a market, a locus where buyers and sellers interact to exchange services or goods. But housing is different from most other commodities. Most notably, with just a few exceptions, housing cannot be moved from one place to another and its location is a critical determinant of value. The value of housing is affected by its spatial relationship with other urban land uses, such as industry, and with the social characteristics of the area within which it is located. The value is also affected by many specific attributes of the house, including lot size, whether or not it is detached, number of levels, amount of living space, and state of repair both outside and inside.

Our ability to purchase many goods is determined essentially by personal financial circumstances, availability of the good, and awareness of options. But the housing market is much more complex. Like any other market, the housing market responds to supply-and-demand considerations, but this simple economic logic is disturbed by the market comprising many different participants who often display different political perspectives. Important constraints on individual residential location decisions are imposed especially by urban planners (described later in this chapter), speculators who hold land in anticipation of making a profit, developers and builders who regularly collaborate to influence neighbourhood character and housing type, and real-estate agents, lawyers, and financiers (sometimes called gatekeepers), who frequently work (separately or collaboratively) to encourage sales in some areas of the city and not in others. These components of and interventions into the market have many and varied effects.

For example, when prospective purchasers or renters seek information about available homes, some relevant information may intentionally be withheld. The result is that they lack possession of all the facts needed to make an informed decision. In this way, they can be directed to certain areas of the city and denied access to others. When sales support and financial assistance are required, some people may be treated differently than others. Such spatially discriminatory lending behaviours, commonly based on skin colour, were routine in the past and, although illegal today, continue to occur.

redlining
A spatially discriminatory practice, favoured by financial institutions, that identified parts of the city regarded as high risk in terms of loans for property purchase and home improvement; affected areas were typically outlined in red on maps.

A historically common practice in the United States is **redlining**—so named because the affected areas were often marked in red on maps used by financial institutions. Redlining denies capital (mortgage money for purchases and loans for home improvement) to certain areas of the city, especially those with high proportions of renters, cultural minorities, and low-income households. These denials are based on the perception that the areas are likely to experience decline in property values over time. Such a practice can be understood as a logical consequence of a capitalist market, where lenders are concerned about property values and the prospects for loan repayment, and as a response to the fact that real-estate agents and lenders tend to be conservative and opposed to spatial social change. It is also an illegal practice that discriminates between groups. Attempts by governments to prevent redlining are very difficult to enforce.

The most obvious effects of these illegal interventions into the housing market are to create and reinforce segregated neighbourhoods and to make some properties in some affected areas difficult to sell and therefore liable to fall into disrepair and be taken over by slumlords. A related consequence is to encourage home-building in suburban areas that are likely to develop as socially, economically, and ethnically homogeneous neighbourhoods.

The practice of redlining was comparatively less common in Canada, partially as a result of the development of the Central (now Canada) Mortgage and Housing Corporation (CMHC). The CMHC was established in the 1940s to grant and, most importantly, insure mortgages as a way to make home ownership a reality for a greater range of the population. As a result of this program, home ownership rates are generally higher in the central parts of Canadian cities than they are in American ones (Harris, 2015).

It has recently been common practice for banking and other mainstream financial institutions to withdraw from many inner-city areas, with their void being filled by payday loan, cheque-cashing, and other institutions charging high interest rates. This is effectively a form of redlining as it reduces the feasibility of people buying, selling, or renovating their homes.

Residential Mobility

Like any other migration decision, the choice to move residences within a city can be conceptualized as being prompted by a mix of the push and pull factors that reflect spatial inequalities (see Chapter 5). Put simply, if they are able, people move to improve their quality of life. Push factors include having too little living space and other perceived inadequacies in the design and quality of the home; these deficiencies frequently relate to the age of the home. Additional push factors relate to neighbourhood characteristics such as the availability of recreational facilities, good-quality schooling, ethnic composition, and likelihood of criminal activity. A third set involves accessibility, including distance from employment opportunities and quality of transport services. Another consideration for many people is the desire to move out of rented accommodation and into a private home. Pull factors can be generally viewed as the reverse of push factors, for example, the prospect of more living space, shorter commute times, better neighbourhood amenities, and so on.

These push and pull factors are understood differently by different individuals and families with varying expectations of what a home in the city can offer them. For example, some prefer the peace, quiet, and privacy afforded by a single-family home in the suburbs, but others prefer the excitement and opportunities available through a condominium or apartment in the downtown core. While individuals may be influenced by a host of push factors, they may be financially unable to relocate. Furthermore, these factors also change through time, with residential relocation often occurring at significant moments during the life cycle (Table 12.1).

While residential mobility refers to individuals' decisions to change their place of residence within the city, **filtering** is the idea that, through time, a housing unit is occupied sequentially by people from steadily changing income groups. The sequence is usually one of downward filtering, with income level decreasing because housing units typically decline in quality as they age and the neighbourhood becomes less desirable. One outcome of this process may be housing abandonment, a major problem in many older industrial cities, that may contribute to urban financial problems because of unpaid property taxes. In some cases, the reverse scenario applies and upward filtering occurs (discussed in the section on gentrification).

The filtering concept can be applied at the level of the individual housing unit and at the neighbourhood level. Some factors that underlie neighbourhood decline, or alternatively revitalization, are outlined in Table 12.2. As the number of revitalization factors increases, there is a higher demand for property and a greater likelihood that the properties will be maintained and improved. As the number of factors contributing

TABLE 12.1 | Selected life-cycle events that can cause residential relocation (chronological)

Completion of secondary education
Completion of tertiary education
Completion of occupational training
Marriage
Separation or divorce
Birth of first child
Birth of last child
First child reaches secondary-school age
Last child leaves home
Retirement
Death of a spouse

Source: Adapted from Pacione, M. 2005. *Urban Geography: A Global Perspective*, 2nd edn. New York: Routledge, 204.

filtering
The idea that, through time, housing units typically experience a transition from being occupied by members of one income group to members of a different income group; downward filtering is more usual than upward filtering.

TABLE 12.2 | Factors underlying neighbourhood decline or revitalization

Revitalization Factors	Decline Factors
High-income households	Low-income households
New buildings with good design or old buildings with good design or historic interest	Old buildings with poor design and no historic interest
Distant from neighbourhoods of very low income	Close to neighbourhoods of very low income or to those shifting to low-income occupancy
In a city gaining (or not losing) population	In a city rapidly losing population
High owner occupancy	Low owner occupancy
Small rental units with owners living on premises	Large rental apartments with absentee owners
Close to strong institutions or desirable amenities, such as a university, a lakefront, or downtown	Far from strong institutions and desirable amenities
Strong, active community organizations	No strong community organization
Low vacancy rates in homes and rental apartments	High vacancy rates in homes and rental apartments
Low turnover and transiency among residents	High turnover and transiency among residents
Little vehicle traffic, especially trucks, on residential streets	Heavy vehicle traffic, especially trucks, on residential streets
Low crime and vandalism	High crime and vandalism

Source: Downs, A. 1981. *Neighborhoods and Urban Development*. Washington, DC: Brookings Institution, 66.

AP Photo/Bebeto Matthews

A billboard above a local hardware store on Malcolm X Boulevard in New York City's Harlem neighbourhood advertises new condos for sale, from $400,000 to $4 million, just a block away from Marcus Garvey Park (named after the Jamaican black nationalist and Rastafari prophet). In a clash of cultures, gentrification continues to affect Harlem—drummers, who have been playing African drums in the park since the 1960s, were forced to relocate after nearby condo residents called police with noise complaints.

gentrification
A process of inner-city urban neighbourhood social change resulting from the in-movement of higher-income groups; originates from *gentry*, a term referring to people of high social standing and immediately below those of noble birth.

to neighbourhood decline increases, there is a reduced demand for property and a greater likelihood of property neglect and abandonment.

Because housing is often their biggest personal capital investment, people may aim to earn income by buying and then selling their home at a profit. Even if profit accumulation is not the overriding concern, most hope that their home will retain or increase its value over time. In some cases, properties decline in value such that the homeowners owe more on their mortgage than the home is worth, a term referred to as negative equity. These homeowners are much less likely to be able to sell their homes than people whose properties have increased in value since the time of purchase. For most, such considerations will always play an important role in any decision to relocate.

Decisions made by individuals and families about where to live in the city reflect existing spatial social variations and then reinforce or change those variations. In other words, these decisions contribute to the ever-changing social mosaic of the urban landscape, including the formation of relatively distinct neighbourhoods.

Gentrification

Some areas of the city—usually those in the central core, where houses and neighbourhoods have experienced many cycles of downward filtering—reach a point where people perceive them as investment opportunities. In many cases, middle-class individuals or families see the advantages of these highly accessible neighbourhoods, namely, that living in the city is more convenient for people who work there and it offers less tangible quality-of-life benefits. Instead of demolishing the existing housing stock, they choose to renovate. This process of people moving in and transforming formerly derelict or low-quality housing areas into higher status neighbourhoods is known as **gentrification**. The redevelopment and revitalization, through renovation and other investments, increases the value of houses and land, which attracts other individuals to do the same.

Most cities, whether large and small, in the more developed world have areas that are being gentrified to some degree (Ley, 1996; Fraser, 2004). Gentrification involves both housing upgrades and changes in neighbourhood character and identity, frequently with trendy new coffee houses, bookstores, antique shops, and specialty services replacing working-class homes and independent services such as small grocery stores and small owner-operated neighbourhood restaurants.

Like many changes, gentrification has both positive and negative aspects. It represents a benefit for middle-class people who seek the character and architectural details that houses in these areas possess, as well as the convenience of living close to the city centre. For the middle-class, it also represents an opportunity to exploit the differences between the actual and potential value of land. For cities, the built environment is upgraded, and local governments receive increased revenue from property taxes.

On the downside, gentrification for the poor, especially renters, represents a threat because rising house prices and property taxes may result in their displacement. Gentrification often contributes to greater social polarization, which can lead to social conflict as some inner-city residents actively resist gentrification. The Grandview–Woodlands area of east Vancouver is an inner-city neighbourhood that has bucked the trend and avoided decline and gentrification. The area has retained its distinct character and affordable housing, possibly because residents have wanted to preserve its diversity and identity (Ventimiglia, 2007).

Residential Segregation

Most cities have residential areas that can be clearly distinguished on the basis of income, class, ethnicity, religion, or some other economic or cultural variable. The process creating this spatial separation of distinct subgroups with the larger population is termed **segregation**. These areas commonly include not only the subgroup members' homes but also the social and cultural institutions, such as places of worship, schools, and businesses, that are an important part of social life. As a result, these segregated areas maintain a spatial separation to varying degrees, one that is apart from (as opposed to a part of) the larger city.

Historically, especially in Europe prior to the Industrial Revolution, the most distinctive urban residential district was the Jewish district, usually labelled the ghetto. During the Industrial Revolution, class divisions became more evident and were expressed in spatial terms. By the beginning of the nineteenth century, British commentators were acutely aware of the division between the working-class districts close to the factories and the middle-class districts located elsewhere, especially on the outskirts of the urban area.

Distinct residential areas also emerged in large immigrant-receiving North American cities in the nineteenth century, with divisions based on ethnicity and class. Once an ethnic group was large enough, it settled as a group, usually in an inexpensive area close to employment opportunities. In the United States, African Americans who moved from the rural south to the urban northeast and Midwest settled in highly segregated neighbourhoods.

Elsewhere in the world, segregation can be based on religion, as in Belfast, Northern Ireland, and many cities of the Middle East. Cities in colonial areas usually included separate districts for Europeans, other immigrant groups, and the local population. Even before the formal institutionalization of apartheid in South Africa, cities were clearly divided on explicitly ethnic lines. In other cases, residential variation may reflect lifestyle preferences.

To a large extent, economic forces play a significant role in residential segregation, with most cities having a number of residential areas distinguished principally by income, wealth, and social class (i.e. the ability of individuals to pay for homes in the area). All cities have working-class, middle-class, and exclusive neighbourhoods that exclude particular individuals based on their inability to afford property. In many cases, this simplistic economic segregation and broader social segregation overlap significantly, as members of certain cultural groups, and primarily new immigrants, have lower incomes. Their segregation may therefore be based on economic circumstances, social characteristics, or a combination. There is no doubt that society and space are irrevocably entwined and that most large cities include segregated neighbourhoods as part of the larger social spatial mosaic.

Social Segregation

Consideration of social segregation—that which is based on social or cultural characteristics, such as ethnicity, language, religion, lifestyle, etc.—begins with an understanding of population similarities and differences. Two main groups are important in these discussions about social segregation: the **charter population** and **minority populations (groups)**. Social segregation assesses the spatial distribution of minority groups as compared to the charter, or dominant, population. Minority groups that live in close proximity to one another are considered segregated, whereas those minority groups that are dispersed among the city's neighbourhoods are considered non-segregated.

Two primary forms of social segregation exist. First, congregation is the result of (mostly) voluntary processes. The choice to live in a highly segregated neighbourhood has several advantages for minority groups. Most important, the congregation of minority groups helps to maintain and preserve their cultural and religious practices through daily interactions with others of a similar background. Furthermore, as a result of a strong sense of cohesion, this process allows for the mutual support of individuals who share a particular way of life via the availability of culturally

segregation
The spatial separation of population subgroups within the wider urban population.

charter population
The dominant or majority cultural group in an urban area; the host community.

minority population (or groups)
A population subgroup that is seen, or that views itself, as somehow different from the general (charter) population; this difference is normally expressed by ethnicity, language, religion, nationality, sexual orientation, lifestyle, or even income (as in the case of the homeless or the extremely wealthy).

specific institutions, businesses, and social networks. Congregation can also aid in minimizing social conflict between different cultural groups by helping to ensure they remain apart. Finally, congregation can help minority groups to develop a local political power base by facilitating electoral representation.

Second, when social segregation results from either overt or sometimes subtle forms of discrimination, we have involuntary segregation. This type can be the result of discrimination in the housing and/or labour market. Residential gate-keepers, such as real-estate agents, developers, and financial institutions (via redlining), have been accused of purposely excluding individuals of particular minority groups from certain residential neighbourhoods.

visible minority
A member of a minority group whose minority status is based wholly on the colour of his or her skin; the Canadian government recognizes anyone that is neither white nor Aboriginal as a visible minority.

cultural minority
A member of a minority group whose minority status is based on factors other than skin colour, such as language, religion, lifestyle, ethnic origin, etc.

It is not always clear, and nor does it necessarily matter, which form of segregation occurs. For example, it can be argued that some of the many North American Chinatowns are the result of both congregation and discrimination resulting in involuntary segregation. Whatever the cause, the outcome is the same—a highly segregated neighbourhood.

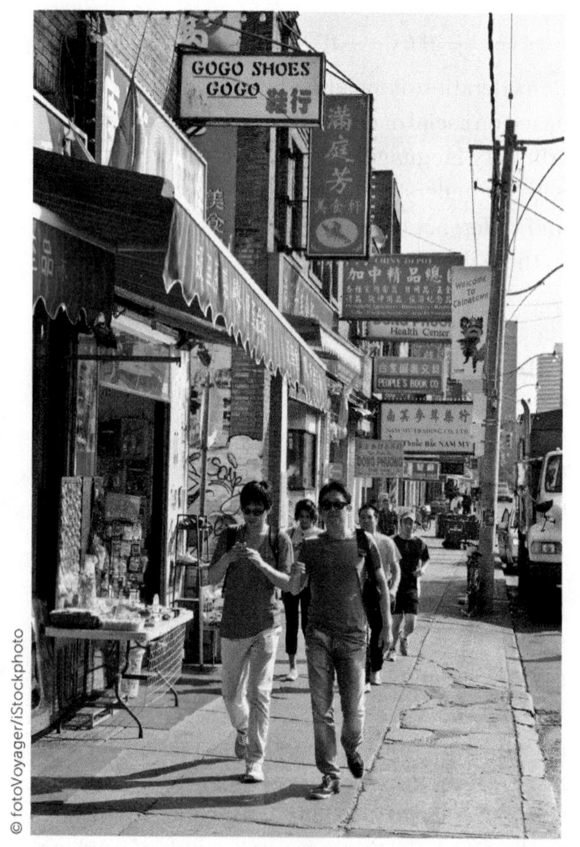

The location of Toronto's Chinatown—in and around the city's long-time garment manufacturing district—is typical of many such highly segregated ethnic enclaves.

Another spatial outcome of social segregation is the creation of social places. In addition to the homes that members of segregated minority groups reside in, these neighbourhoods contain businesses and social and cultural institutions that cater to the minority group. A visitor to any North American Chinatown or Little Italy is struck by the visible imprinting of the cultural group on the urban landscape. Stores, restaurants, places of worship, community centres, street signs, and property decorations are some of the many visible signs of this social segregation.

The level of voluntary or involuntary segregation varies in degree and form, from place to place, and between different minority groups. **Visible minorities**, for example, African Americans in US cities, tend to be more highly segregated than **cultural minorities**—those that do not look different but perhaps speak a different language or practise a different religion. Several measures have been developed to quantify the level of segregation. The most commonly used is the Index of Dissimilarity, which estimates the proportion of the minority group's population that would have to move from the neighbourhood in order for that group's spatial distribution to equal the charter population. A second measure of segregation is the Isolation Index, which quantifies the proportion of a minority group's daily interactions with other members of the group as opposed to those from others.

Segregation may take one of three spatial types—a ghetto, an enclave, or a colony—all of which are formed through congregation, discrimination, or a combination of the two. Enclaves are spatial concentrations of minority groups that use them to protect their economic, social, or cultural practices. The best examples of these long-lasting areas are the Jewish districts, or settlement areas, in many eastern American or European cities.

Ghettos, in contrast, are the product of long-standing discrimination. In this case, a dominant society imposed the segregation on a minority group as a means of separating or limiting it and treating it as inferior on the basis of racial or ethnic differences. Examples include the segregation of African Americans and Hispanic Americans in many cities in the United States.

Finally, colonies result from discrimination, congregation, or a combination and are less persistent in the urban landscape than enclaves or ghettos.

Colonies form when the differences between the minority and the charter groups are relatively small and the segregation of the former is a temporary step in assimilating into society. Examples include the immigrant reception areas in many North American cities in the 1920s and 1930s.

Assimilation, Acculturation, and Multiculturalism

A related set of concepts is assimilation and acculturation (see Chapter 8). Assimilation is the process through which an ethnic minority group is absorbed into the larger society, thereby losing its unique cultural identity. Acculturation involves a similar absorption into the mainstream culture, but the minority group retains certain aspects of its identity. Both processes are acutely tied to social segregation since groups that are more highly segregated, whether by choice or discrimination, will take longer to be absorbed into the mainstream culture (if they ever are). Despite the sometimes difficult transition to a new culture, most immigrant groups eventually assimilate or acculturate.

Assimilation is the overarching goal of American immigration policy, in which the prevailing view is of society as a "melting pot"—individual immigrants are added and become part of American culture over time. In Canada immigrant ideals and practices are framed in the context of multiculturalism, which is more closely aligned with acculturation. Multiculturalism became official Canadian policy in 1971 and a core value of Canadian identity in the decades since. The intent of this policy is to nurture the increasingly pluralistic character of the country while recognizing the importance of equality for all Canadians. The ideal, but not necessarily realistic, outcome is a society in which immigrants and cultural groups acculturate through the explicit assistance of the federal government and its immigrant settlement agencies (Teixeira, Li, and Kobayashi, 2012). However, the reality for most newcomers to Canada is likely not all that different from the experience of those arriving in the United States.

SUBURBS AND SPRAWL

Suburbs are as old as cities (Bourne, 1996: 166). Historically, urban areas were concentrated around a central area—perhaps a bridging point across a river, a confluence of two rivers, or a market—and gradually expanded outward in the form of residential suburbs as population increased. Beyond the urban periphery, most land was in agricultural use. As noted earlier, the favoured residences were most often those close to the city centre, with poorer people relegated to the more distant and therefore less accessible homes.

In the more developed world especially, this long-standing spatial organization has been transformed in two general ways. Beginning in the late nineteenth century, increasing numbers of people opted for homes outside the established urban area, a location decision made possible by improved public transportation and, as the twentieth century progressed, increased private automobile ownership. In recent decades the globalization processes that have contributed to the world city phenomenon have also worked to transform suburbs and to initiate the growth of new cities beyond established urban areas. These two transformations are described in this section.

Suburbanization

Suburban growth was mostly in response to changes taking place inside the city, including the establishment of factories and rising population densities. The favoured locations outside the city were directionally biased along transportation routes, and new streetcar suburbs appeared around many large cities in Europe and North America during the second half of the nineteenth century. This change was dramatically affected by the introduction of the automobile, which allowed access to the large areas of land that lay between the streetcar routes and meant that distance from the centre of the urban area became the key variable rather than direction. Most of these new suburbs comprised housing and basic services designed for the working and/or middle class and represented a compromise between the need to live close to employment either in the central city or industrial districts and the desire to be close to open spaces and rural areas.

A form of decentralization, suburbanization proceeded rapidly throughout the twentieth century. It was mainly prevalent where land was readily available, planning regulations were weak, populations were wealthy and could afford large homes, and levels of physical mobility were high. In many countries suburbanization has become

Focus on Geographers

Human Geography and the Housing Experiences of Immigrants | CARLOS TEIXEIRA

My research interests in settlement patterns and urban geography are in part a product of my biography: I am a "new" Canadian who emigrated from the Azores islands three decades ago. Like many immigrants, I first settled in an ethnic neighbourhood, specifically the Quartier Portugais/Little Portugal in Montreal. I later moved to Toronto, where I chose to reside in Little Portugal, which was very close to both Little Italy and colourful Kensington Market—for more than a century, a major reception area for immigrants from all over the world. Living in such rich multicultural environments greatly shaped the way I perceived the impact of immigrant groups on our cities and their neighbourhoods, from an early research focus on the formation and mobility of the residential Portuguese communities in Montreal and Toronto to my current interest in different immigrant groups and research topics ranging from population and migration issues to housing and ethnic entrepreneurship. My research interests today include urban and social geography, with an emphasis on housing and ethnic entrepreneurship and the social structure of North American cities. In the last five years, I have concentrated my research in two major projects—one in Kelowna (a mid-sized city in British Columbia's Okanagan Valley) and the other in Richmond and Surrey (outer suburbs of Vancouver)—that examined the housing experiences and coping strategies of recent immigrants.

While immigrants continue to arrive in Canada's traditional metropolitan gateway areas, recent data from the Canadian census has sparked significant interest in immigrant dispersal to new destinations outside major urban centres, especially the suburbs and, more recently, small/mid-sized cities. Rapid population growth and concentration of immigrants and minorities in these new destinations have led to an increasing demand for affordable housing. The scarcity of research on the housing experiences of immigrants and minorities in mid-sized cities and the outer suburbs of major metropolitan areas prevents a full understanding of why certain immigrant groups are more successful than others in locating appropriate housing in a suitable or comfortable neighbourhood and of the factors that facilitate or prevent this phenomenon in mid-sized cities and the outer suburbs of major Canadian cities.

The evidence from my research indicates that new immigrants face numerous difficulties (e.g. high rents, overcrowding, poor-quality housing, discrimination) in the rental housing market. Most immigrants that I interviewed spent 50 per cent of their monthly household income on housing, putting them at risk of homelessness. These findings suggest that the housing crisis affecting Kelowna, Surrey, and Richmond—a limited supply of affordable rental housing and high living costs for many new immigrants—makes these three cities unique and challenging for immigrant settlement. Funding from all levels of government to stimulate the creation of both for-profit and non-profit housing is urgently needed. The shortage of appropriate housing services and programs is also a major gap in the settlement in these regions.

Canadian geographers like me have tried to find solutions to the numerous problems that immigrants face in Canada's rental and homeownership markets. I strongly believe that, as geographers, we need to continue doing the best research possible, but our work should also make a difference, that is, to change the "real world" and to influence public policy. In the new era of globalization, sharing our research with our communities and the media also becomes a priority and a means of spreading the word about the value of geography and what we do as geographers!

CARLOS TEIXEIRA is professor in the Department of Geography at the University of British Columbia—Okanagan. In 2005 he received the Order of Infante D. Henrique from the Portuguese government, one of the highest awards a Portuguese citizen residing overseas can receive for work in service of the Portuguese diaspora.

the dominant urban experience for most people, producing a landscape of conformity both visually and behaviourally, with a heavy emphasis on consumption.

The characteristic landscape feature (particularly in North America, where there was often much space available) was a single-family house on a large lot. The appeal of the suburb proved undeniable, and large areas of agricultural land were replaced with standardized residences. A symbiotic relationship existed between suburban growth and related sprawl and the automobile: the automobile made it possible to build suburbs, and it was impossible to live in the suburbs without having an automobile. Travelling to work, shopping, and visiting friends all required use of a car.

Many twentieth-century suburbs, specifically those in Europe, were designed for the working class. Incorporating large tracts of public housing, they have frequently become areas of urban deprivation, social anonymity, residential congestion, and pollution. The fact that public transportation systems have not been adequately modernized in many cities adds yet another set of difficulties to working-class suburban life.

In North America especially, spatial growth, population increase, and diversification of suburbs are quite remarkable. With regard to population, consider that there were fewer people in US suburbs than in either central cities or rural areas in 1960, but by 1970 suburban numbers exceeded both central city and rural numbers, and by 2000 suburban exceeded central city and rural combined. Between 1990 and 2006 the population of the city of Chicago increased by 50,000 people, but that of suburban Chicago increased by more than a million. Current North American suburbs are ethnically varied and include many different economic activities.

Urban Sprawl

In many accounts, the distinction between suburbanization and urban sprawl (defined in Chapter 11) is unclear, although the latter term clearly evokes a negative assessment that suggests a lack of urban planning. Sprawl is often taken to imply not just residential development but also the supporting commercial landscape, with the strip mall and retail power centre as the leading components. These malls prioritize the automobile with their linear extension, digital LED signs designed to inform and advertise, and extensive parking lots. Sprawl and consumption go hand in hand. The single most obvious development affecting retailing, industry, and residential areas since the end of World War II is the ubiquity of urban sprawl. Sprawl also implies residential spread without infilling, thus resulting in low population densities, frequent discontinuities in land use, and the creation of numerous single-use locations that are not clearly linked to other nearby land uses.

Commentators are divided as to the merits of these new landscapes. Some write about blight and soulless postmodern places while others consider these automobile-inspired landscapes as legitimate new architectural forms. Sprawl is regularly blamed for destroying farmland and increasing commuting times (which in turn may contribute to the deterioration of family life), for increased pollution from automobiles, and for a physically inactive population. Regardless of such issues, North Americans expressed their opinions with their residential decisions—by 1990 about half the population lived in suburbs.

Sprawl can lead to the formation of **conurbations** that are perhaps better understood as city regions. The Greater Toronto Area and Hamilton (GTAH) and Megalopolis (or BOSNYWASH), which extends from north of Boston through New York, Philadelphia, Baltimore, and Washington, DC, and south to Richmond, Virginia, are two North American examples of conurbations. Each is comprised of many distinct urban areas, but the growth of suburban areas has led them to develop into extremely large contiguous urban–suburban regions. Such regions can grow at the expense of the inner city, leading to reductions in the urban tax base, underutilization of urban infrastructure, and lower property values inside the traditional city.

conurbation
A continuously built-up area formed by the coalescing of several expanding cities that were originally separate.

edge city
A centre of office and retail activities located on the edge of a large urban centre.

Post-Suburbia

Many commentators see both ongoing suburbanization and a new process of post-suburbanization, involving movement of people and activities to locations beyond the suburbs, as the future of suburban life. But what does this mean in terms of how people live and work?

Edge cities are completely new urban centres, often located on the periphery of a city (Garreau, 1988). These new and rapidly growing centres are post-industrial in character, having limited connections with industry; instead of offering manufacturing jobs, they offer work in the professions, management, administration, and skilled labour. They include office buildings and public institutions but few factories. As suburban areas have matured in recent years, many have acquired features formerly associated with city centres, such as high-rise office and residential towers, to become post-suburban locations. Markham (Ontario), Surrey (British Columbia), and Longeuil (Quebec) are edge cities.

Edge city development is not always easy to distinguish from more general changes in suburban areas. These and other post-suburban developments are one component of the increasing decentralization of offices and factories, the decline of manufacturing employment, and the processes of economic restructuring associated with larger globalization processes (discussed later in this chapter). They involve a continuing exodus of financial and producer services from older downtowns. An excellent example of an edge city is Tyson's Corner, Virginia, just outside of Washington, DC. It was a rural area in the mid-1960s but now includes the seventh largest concentration of retail space in the United States and is one of 17 edge cities close to Washington (Bourne, 1996: 173).

gated community
A high-status residential subdivision or community with access limited to residents and other authorized people such as domestic workers, tradespeople, and visitors; often surrounded by a perimeter wall, fence, or buffer zone such as a golf course.

Edge cities are all over the United States, frequently dominating the urban landscape. Many of the fastest-growing urban regions—such as those centred on Las Vegas, Denver, Phoenix, Dallas, Austin, Houston, and Atlanta—include numerous edge cities. Edge city growth is usually accompanied by a corresponding weakening central city area, but it may also occur around a viable downtown. They are less prominent in some other more developed countries. Growth outside many European and Asian cities is severely constrained by space limitations, so solutions to problems of congestion are being sought (Box 12.2).

Edge city is not the only term geographers use to describe these new urban landscapes. Some choose to talk about peri-metropolitan growth, polynucleation, spillover cities, suburban activity centres, boomburbs, superburbia, technoburbs, galactic cities, dispersed cities, exploding cities, and more. This proliferation of labels suggests the complexity of the unfolding urban landscape and the difficulties in describing it. In the case of the United States, the new landscape of distinct and often separate urban areas has appeared in conjunction with a major shift in population and employment from the northeast and Midwest to the south and southwest. This shift in population has occurred alongside the emergence of **gated communities**. Many social commentators view this as a regrettable trend as gated communities are explicitly landscapes of exclusion. In many cases, edge cities exclude the financially and socially disadvantaged population.

But in other cases, edge cities are seen as hugely variable places. They are landscapes of excess and want, of prestigious designer homes and trailer parks. They are home not only to the rich and powerful but also to marginalized social and ethnic groups. They are organized in some places and disorganized in others. In many gateway cities, such as Vancouver and Sydney (Australia), new immigrants (mostly from Asia) are attracted to outlying suburban areas perhaps because of job opportunities and, in the Canadian case, because of the availability of subsidized housing.

© Steve Skjold/Alamy Stock Photo

Gated communities such as this one in Florida are popular in several parts of the United States. Palm Springs, California, has the highest number in the country and likely the world.

INEQUALITY AND POVERTY

Cities are a collection of unequal places. In many countries, poverty has increasingly become concentrated in cities and in the older central area within cities, partly because of the disadvantaged people who generally live there.

Cities throughout the more developed world typically experienced a prolonged period of substantial growth, beginning with a process of concentration that created the industrial cities of the nineteenth century and followed by a process of decentralization that created the conurbations and urban sprawl of the twentieth century. The rapid expansion of urban industrial areas in the nineteenth century resulted in high-density,

In the News

BOX 12.2 | Making the Most of Space in Modern Cities

How might cities of the future deal with constraints on urban space, and what might they look like? Most envisage, or hope, our cities will be more sustainable and livable. This box discusses possible urban futures with reference to Tokyo and Dubai, two large modern cities.

Tokyo is a crowded city with a constant demand for more space. Initially, this demand expressed itself in urban sprawl, but more recently it has involved internal reorganization. The population of about 37 million (as of 2013) in the larger urban region is making ingenious use of the urban area by building upward, downward, and into the sea. Building up is a relatively recent alternative for a city prone to earth tremors. In addition to skyscrapers, there are elevated highways and railways and even multi-storey golf driving ranges. Building down is also popular; some buildings use at least four below-ground levels for retail and office purposes. Perhaps most dramatically, Tokyo is reclaiming land from the sea to permit more expansion. It has even been suggested that the city might extend out onto a floating platform in Tokyo Bay.

The largest city in the United Arab Emirates, Dubai is building a series of islands in the Persian Gulf. There are three massive palm-shaped islands and a cluster of 300 islets that are designed to appear from the air as a map of the world. The intent is that these islands, when complete, will house about a million people. These islands are aimed at wealthy people and are not a response to issues of congestion in the city. The first residents occupied homes in 2007, but whether these ambitious projects will be completed as planned is in some doubt. Dubai is also home to the world's tallest building, Burj Khalifa (completed in 2010). Plans were announced in 2008 for the building of the world's first "moving" skyscraper, the Dynamic Tower (or Da Vinci Tower), an 80-storey building with each level designed to move independently such that the building will constantly change shape. As of early 2015, construction of this project has not begun.

© Jorge Ferrari/epa/Corbis

Locals visit a model of The World Islands at the Arabian Travel Market in Dubai, United Arab Emirates (UAE). The World Islands is an artificial archipelago of sand islands created through the dredging of sand from the Persian Gulf, which are in the process of being developed into luxury residential communities.

poorly serviced housing areas of poor quality. These areas contrast markedly both with the neighbourhoods of pre-industrial cities and with rural villages, where the separation of the rich and poor was not nearly so marked. In 1845 the slums of Nottingham, England, were described as follows:

> Nowhere else shall we find so large a mass of inhabitants crowded into courts, alleys, and lanes as in Nottingham, and those, too, of the worst possible construction. Here they are so clustered upon each other; court within court, yard within yard, and lane within lane, in a manner to defy description. Some parts of Nottingham [are] so very bad as hardly to be surpassed in misery by anything to be found within the entire range of our manufacturing cities. (Hoskins, 1955: 218)

These nineteenth-century slums were cleared as part of the twentieth century urban redevelopment programs and were replaced by equally poor forms of low-cost, often high-rise, housing. Both types of residential area have been associated not only with poverty and deprivation but also with high levels of crime, vandalism, and substance abuse. Worsening social circumstances for many residents are evident in the increasing number of homeless people and the rise in crime that is commonly related to organized gang activity and ethnic tensions. Many urban areas experience a number of new problems related to economic restructuring, which exacerbates the existing social and economic inequality.

The Urban Poor

Although *ghetto* and *slum* are not synonyms, deprivation is frequently associated with distinct ethnic groups, and the current spatial concentration of poverty often shows a close relationship to ethnic neighbourhoods. Aboriginal populations are especially vulnerable in Canada, as are African-American populations in the United States. But poverty knows no ethnic boundaries.

Why are some poor people deprived of such basic needs as shelter? Why are many people both undernourished and malnourished? We can divide the responses to these important questions into four categories:

cycle of poverty
The idea that poverty and deprivation are transmitted intergenerationally, reflecting home background and spatial variations in opportunities.

1. Some commentators have proposed that a culture of poverty typically leads to a **cycle of poverty**. This argument centres on some people's inability to cope effectively in the larger world such that poverty becomes their normal circumstance. Once in place, this culture of poverty is perpetuated, with poverty, deprivation, and slum living becoming the usual way of life and limiting the prospects for social and spatial integration with the larger urban area. This influential argument has been severely critiqued because of the emphasis placed on people's inability to cope and on deficient child-rearing practices. In short, this argument can be interpreted as blaming the poor for being poor.
2. An extension of this logic focuses on personal characteristics that limit the ability of individuals to cope and care for themselves, characteristics such as physical or mental illness, deviant behaviour (criminal activity or substance abuse), personal disaffection with prevailing lifestyles, and lack of education and employment skills.
3. Others have argued for an emphasis on structural class conflicts and claimed that the capitalist system necessarily concentrates power and wealth in the hands of a few and requires a majority population who are relatively dispossessed both materially and socially. Extensions of this logic stress the consequences of three factors: economic restructuring, such as the recent deindustrialization that results in both increasing numbers of poorly paid low-level service jobs and increased levels of unemployment; gentrification, which makes some homes inaccessible to poor people; and the general failure of both public and private sectors to build low-cost housing.
4. Another explanation, as suggested in the account of residential mobility, emphasizes the roles played by those who have the power to control access to urban resources, such as real-estate agents and mortgage companies. This institutional argument centres on the fact that material and knowledge resources are unevenly distributed and on the resulting uneven relations between the people in positions of power and authority and the disadvantaged.

Each category assists in our understanding the reasons for poverty and in identifying possible solutions. Nevertheless, it cannot be denied that the issues are far from being understood and resolved; most current evidence suggests that urban problems such as poverty, inadequate housing, and homelessness are increasing. A common response on the part of rich residents is to widen the gap between themselves and the poor; the differences between the predatory landscape of a run-down inner-city neighbourhood and the architecture of a suburban gated community speak directly to the failure of many large cities to provide fair and just communities for all residents.

Recent research suggests that cities, and the countries in which they are located, can suffer serious social and health problems precisely because there are such inequalities. Compared to places whose populations are relatively equal, places with a significant income and lifestyle gap between rich and poor have higher incidences of mental illness, substance abuse, obesity, and criminal activity; lower educational standards; and a shorter life expectancy.

Poverty and Urban Life

In some large American cities, it is possible to identify "welfare neighbourhoods," where most people rely on welfare payments and these funds are the basis for the local economy (DeVerteuil, 2005a). These neighbourhoods typically lose health, education, and financial and basic retailing services, including inexpensive supermarkets.

Although members of all social groups are susceptible to poverty, compelling evidence links it with ethnic minority identity, employment status, level of education, health, age, and gender issues. For most cities, the transition to the current post-industrial economy has resulted in major losses of manufacturing employment opportunities, and the most deindustrialized areas have received no new investment. The principal exceptions to this negative trend are the global cities discussed in Chapter 11. Elsewhere, the new employment opportunities in the high-technology, service, and tourist industries are rarely open to former members of the industrial labour force, who frequently lack the needed educational qualifications. Increasing levels of unemployment lead to problems of social service supply and, often, increasing social tension and criminal activity, as well as personal difficulties related to poverty.

Poverty and health problems also go hand in hand as inadequate housing or homelessness increases the risk of hypothermia during cold periods and of tuberculosis, bronchitis, and skin infections. In countries that lack an effective, free health-care system, these illnesses may go untreated. Substance abuse, as one avenue out of despair, is also a major health problem in poor areas of the city. A 2011 UN report on drugs highlighted several British cities (notably Manchester, Liverpool, and Birmingham) as containing "no go" areas that are effectively controlled by drug gangs.

With reference to age and gender, single mothers and older women living alone are vulnerable, although in many countries the gender-based divisions are being reduced through subsidized daycare and related measures. Most current research on these matters stresses that gender, class, ethnicity, and sexuality are frequently interwoven, and analyses rarely focus on just one (Ray and Rose, 2000: 518).

Poverty is also closely related to criminal activity, with poor people regularly overrepresented as both perpetrators and victims. There is a close relationship between unemployment and criminal behaviour but not a causal relationship. It is important to recognize that criminal activity is related to inequality and social exclusion (Box 12.3); further, poor groups within society, especially visible minorities and others readily identifiable by authorities, are often targeted for harassment and arrest.

Homelessness

One of the functions of a city is to provide its people with shelter. There are many different forms of accommodation in the typical city, from single-family homes, apartments, and condominiums to rooming houses and group homes. But not all residents can afford a place of their own, and some rely on institutional settings or public spaces instead. Homelessness is often precipitated by a specific circumstance, such as loss of employment, eviction, or domestic violence. People without a home gradually become labelled as homeless and even self-identify as such, a situation that increases the likelihood of remaining homeless. Being homeless does not necessarily mean sleeping in the open or on the street (Box 12.4). Four typical degrees of homelessness are

- *Rooflessness:* Sleeping "rough" (i.e. in the open air) is the most visible sign of being homeless.
- *Houselessness:* This term applies to people who routinely sleep in shelters.
- *Living in insecure housing:* This circumstance arises when permanent housing is unavailable, when people are obliged to share with others, or when a person's or family's housing is likely to be lost if the rent is raised or payments cannot be made.
- *Living in inadequate accommodation:* Some housing is of such poor quality—overcrowded or otherwise unfit for habitation—that it cannot be considered adequate shelter. This is the situation of many Aboriginal people living on reserves in the northern reaches of some Canadian provinces.

Consider the following disturbing example of the spatial juxtaposition of privilege and deprivation. New Haven, Connecticut, is the home of Yale University, one of the wealthiest and most prestigious schools in the United States. In 2002 a news report described a tent city adjacent to the university. Established after authorities closed down an overflow homeless shelter, the tent city was quickly cleared by police. The contrasts are frightening. Proportionally, Connecticut has more millionaires than any other American state, and yet New Haven has an infant mortality rate comparable to that of Malaysia. It has suffered a devastating number

Around the Globe

BOX 12.3 | Eight Mile Road

Detroit's Eight Mile Road is familiar to many, particularly after the release of *8 Mile*, a semi-biographical film of the rapper Eminem. The road, so named because it is located eight miles from the Detroit River (Figure 12.8) is the actual and symbolic dividing line between city and suburbs, separating blacks and whites, poor and rich, powerless and powerful, underprivileged and privileged. Eight Mile Road is a major eight-lane highway, frequented by prostitutes and drug dealers and with mixed land uses including pawn shops, topless bars, sex shops, and cheap motels. In 2006 the rapper Proof (a close friend of Eminem and the best man at his wedding) was murdered outside the road's Triple C bar.

Previously a vibrant industrial and working-class residential area, Detroit has suffered from two circumstances. Riots in 1967 prompted many white residents to flee to the suburbs, taking their jobs and money with them. Industrial restructuring (discussed in Chapter 13), especially the long, drawn-out decline of the automobile industry, has led to huge job losses. The urban landscape to the south of Eight Mile Road, the city proper, lost 25 per cent of its population between 2000 and 2010, an unprecedented demographic collapse (with the exception of New Orleans following Hurricane Katrina). There are 60,000 empty houses, and the drastically reduced tax base means declining services. This area is more than 80 per cent African American, and estimates suggest that there are about 13,000 homeless people. The landscape includes scenes of severe deprivation with burned-out and abandoned buildings. Criminal activity is rampant.

The urban landscape to the north, Oakland County, is more than 80 per cent white and is a more typical representation of suburban America, comprising pleasant neighbourhoods of painted homes and parks with names such as Hazel Park, Oak Park, and Pleasant Ridge, as well as coffee shops and attractive office blocks. The difference between city and suburb, north and south of Eight Mile Road, could not be more marked. People live separate lives in these separate communities.

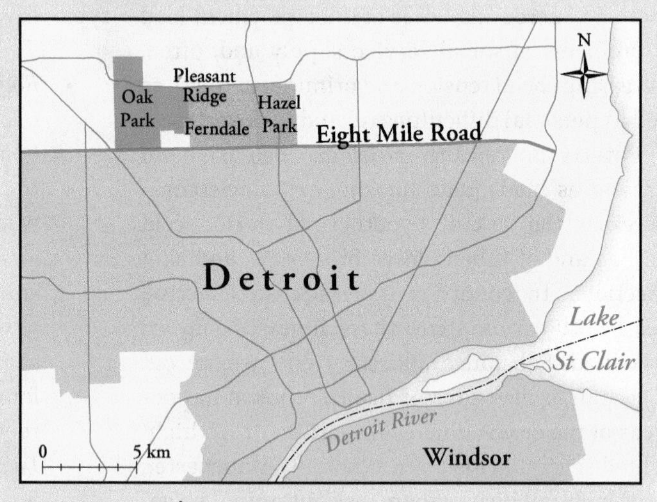

FIGURE 12.8 | Locating Eight Mile Road

of deaths associated with AIDS, and a 2002 report found that almost 70 per cent of the city's children were without health insurance. Critics suggest that New Haven is a metaphor for the United States in the early twenty-first century. Most circumstances of homelessness are not quite as dramatic as this example, although this serious social problem is experienced by increasing numbers of people.

How many people are homeless? There is no easy answer to this question, both because there is no precise definition of *homelessness* and because estimates frequently reflect vested social and political interests. For the United States in the 1980s, social advocates estimated several million while government officials estimated 250,000–350,000; the 1991 Canadian census attempted to count the number of homeless people by reference to the use of soup kitchens, but the figure was assumed to be inaccurate and was thus never released. A recent study counted 200,000 Canadians accessing homeless emergency shelters or sleeping outside in a given year, but this figure undercounts the true level of homelessness as it does not account for those staying with friends or relatives or somehow not coming into contact with shelter staff (Gaetz et al., 2013). Further, a recent survey suggests that nearly 1.5 million Canadians, approximately 5 per cent of the total population, had experienced homelessness or insecure housing in the previous five years (Gaetz et al., 2013).

But, as Collins (2010) demonstrates, homelessness need not be a problem. State housing and cultural factors, especially the prevailing willingness of friends and relatives to provide shelter for otherwise homeless people, have resulted in homelessness being a minor issue in New Zealand. Having few homeless people means that there is no need for shelters or soup kitchens, which have become so much a part of the urban landscape in Canadian and American cities. Although few data are available on the financial cost of homelessness to the various levels of government and to society more generally, it is substantial because of the pressures placed on emergency medical

Around the Globe

BOX 12.4 | The Homeless Experience

What does it mean to be homeless? In-depth studies that interviewed homeless people in Calgary (Peressini and McDonald, 2000) and in Los Angeles (DeVerteuil, 2003, 2005b) offer some valuable and disturbing insights.

The Calgary study suggests that the typical homeless person is single, male, and poor. Such persons are undereducated and unemployed, but these circumstances are not of their choosing. Many work on a casual or temporary basis. They invest much time and energy into seeking employment, contrary to stereotypical perceptions of the homeless as lazy and unwilling to work. Also contrary to some claims, they do not choose to separate themselves from larger society or constitute a deviant subculture. Many have friends and families and are homeless despite their best efforts to be and do otherwise. When asked about their personal situations, some respondents understood homelessness as being without a home, but others understood it as being without a source of income or as being unemployed. Reported reasons for being homeless included lack of money, unemployment, and larger economic circumstances. Many homeless respondents rarely slept on the street, with shelters being their principal means of accommodation.

The Los Angeles study focused on 25 single, homeless women. Interviews uncovered the sequence and variety of residential patterns, the circumstances of movement from one setting to another, and respondents' perceptions of the various settings. Results showed that the women lacked employment and other income, that at least some used mobility as a coping mechanism, that sometimes they were not merely passive users of institutions but reinterpreted settings in ways that affirmed their identities, and that stability was not necessarily indicative of coping. A 50-year-old Caucasian woman, Ellie, cycled through 10 settings during a three-year period (Figure 12.9). Much of this was intentional as she often remained at shelters as long as permitted, kept storage units in both Los Angeles and San Francisco, and withheld information about her circumstances from friends.

Other city residents tend to encounter homelessness on a routine but casual basis. It is common for some homeless people to seek financial support at traffic intersections and downtown traffic arteries and to seek refuge in or adjacent to public buildings. Sadly, most people go about their lives with little awareness of the homeless and, certainly, little understanding of how the homeless live their lives.

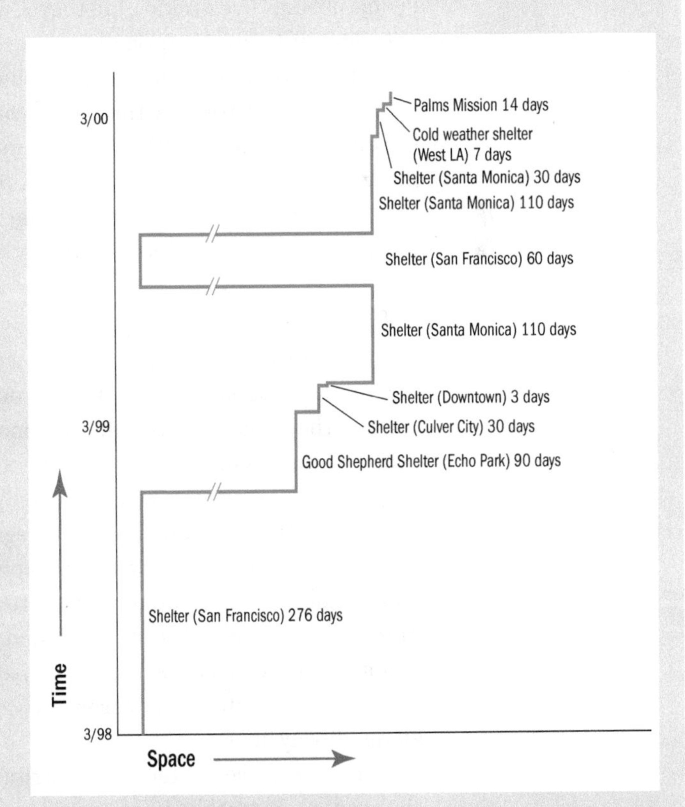

FIGURE 12.9 | Space–time prism for Ellie, March 1998 to March 2000

Source: De Verteuil, G. 2003. "Household Mobility, Institutional Settings, and the New Poverty Management." *Environment and Planning A* 35(2): 369. Pion Limited, London.

facilities, police services and the justice system, social service agencies, and charitable organizations. One UK estimate is that homeless people have a life expectancy of only 47, more than 30 years less than the norm, mostly because of drug and alcohol abuse. DeVerteuil (2003: 378) is surely correct to query: "Might it not be more fiscally sound for governments simply to provide affordable housing?" The case of New Zealand provides an answer to this question. Closer to home, the Canadian Housing & Renewal Association provides the following telling figures: a psychiatric in-patient bed costs $665 per day; jail equals $143 per day; provision of a shelter bed costs $69 per day; and supportive/social housing would cost $25–$31 per day.

CITIES AS CENTRES OF PRODUCTION AND CONSUMPTION

People not only live in cities, but they also work and consume in them. Historically, people moved to cities from rural areas in search of jobs and, once employed and settled in the city, their everyday lives played out in the urban environment. Cities, then, are also centres of employment, production, and consumption. What are the main forms of urban employment, and where do they occur?

Manufacturing

As noted in Chapter 11, urban growth, urbanization, and industrialization go hand in hand. During the Industrial Revolution, many established urban centres grew rapidly and many new urban centres were established. By the early twentieth century, manufacturing regions came into being and many of their cities concentrated on just one or a few related manufacturing activities. Inside these cities, it was usual for many industrial activities to be located adjacent to the central business district, near waterways, and/or along railway lines.

However, these patterns of distribution have changed considerably in recent decades. Most countries in the more developed world have experienced some loss of manufacturing activity due to a shift to newly industrializing countries such as Mexico and China. Inside the more developed world city, the central industrial locations have lost manufacturing activities while outer suburban areas have gained. This movement from the city centre towards the urban periphery is part of larger economic restructuring processes and is associated with processes that have reduced the friction of distance, such as innovations in transportation and logistics. The specifics of these transformations are discussed in greater detail in Chapter 13.

To some degree, manufacturing activities in the inner city have been replaced by various service activities, such as business services (i.e. law, finance, advertising, publishing, insurance, etc.), public services (i.e. education, health care, government, etc.), and consumer services (i.e. retail, food services, hospitality, etc.).

Offices and Professional/ Business Services

In a similar way, the geography of office locations has changed over the past few decades in large urban cities across North America. As Canada's leading business centre, Toronto provides an excellent case study for the migration of offices from the central business core to the suburbs. In the latter, office space is more cost-effective and businesses are able to attract and retain workers who reside in these areas.

The Canadian Urban Institute's (2011) study of the changing geography of Toronto's office space noted that, in the late 1970s, over 60 per cent of its capacity was concentrated in the downtown core. Offices of engineering, publishing, financial, and legal firms, as well as the head offices of manufacturing companies, were all housed in this area. Today, only 20 per cent of the city's office capacity is in the downtown core, with primarily only the financial, insurance, and real-estate firms remaining. The bulk of the other businesses have moved to suburban office parks in employment nodes and edge cities surrounding the older city. This general decentralizing trend is seen in many other large cities across Canada and the United States. Yet the office spaces of downtowns everywhere remain the most expensive and most prestigious.

Retailing and Consumer Services

Retailing is a vital component of national and urban economies, a significant user of urban land, a key factor in the image that residents have of their cities, and a major leisure activity for urban residents. Retailing is a consumer service, as are entertainment and recreation.

Traditionally, retailing has been located on streets in the central areas of cities. The location of department stores in suburban areas, which began in the United States, marked a major transformation of retail landscapes. In 1947 a suburban shopping centre comprising two department stores, several smaller stores, and a large car park opened in Los Angeles. The building of enclosed, climate-controlled shopping malls in suburban areas soon followed, with the first being Southdale in Minneapolis. There are currently thousands of enclosed malls in the United States and around the world.

Since about the 1970s, urban retailing activity has undergone a number of organizational and related locational changes in response to broader post-industrial trends. There are three principal organizational changes. First, the number of independent retailers has declined sharply as a consequence of competition between large business enterprises. Second, the largest retailers, of both goods and services, operate as transnational firms with stores in more than one country; major examples include Walmart (American), Ikea (Swedish) and Marks & Spencer (British). Third, new communications technology has permitted the creation and rapid growth of major online retailers, such as Amazon (American) and Alibaba (Chinese), that have no physical urban retail presence but utilize major regional distribution hubs (suburban warehouses) and efficient delivery services to deliver goods to consumers. Even traditional "brick and mortar" retailers have recognized the opportunities of online retailing and have mimicked the large online companies. Ironically, Amazon, the world's largest online retailer, opened its first physical store in Seattle in November 2015.

The most evident recent locational change is a further decentralization of shopping centres, many of which are located in out-of-town areas accessible only by automobile. These retail areas, often called power retail developments, include many new retailers. Such centres are controversial in many areas because they may compete with small rural retailers, pave over prime arable land, replace attractive green space, bring significant increases in traffic, and require large parking areas. These locational changes challenge the viability of older retailing areas, especially in downtowns, as well as the older suburban enclosed shopping malls, many of which were already declining in popularity as suburbs spread further outward and demographic circumstances changed.

"Big box" stores such as The Home Depot, Costco, and Best Buy are ubiquitous in urban areas.

TRANSPORTATION AND COMMUNICATION

As noted during the discussion of trade theories in Chapter 3, location and interaction are opposite sides of the same coin. Cities occupy locations separated from one another by distance yet need to interact with each other. Inside a city, spatial phenomena such as homes, workplaces, and sites of consumption similarly are located apart but need to interact. Both interurban and intra-urban interactions require transportation and communication services. We are concerned here with how transportation and communication facilitate and influence movement inside cities.

Cities are concentrations of people and activities; to function effectively, they must allow people, goods, and information to move around without constraint. Even a cursory consideration of urban centres illustrates the important role of transportation. Roads, rail, (sometimes) air, and telecommunications services are key to understanding how cities function. Most people have to travel to earn, or spend, their money.

People in cities move around as part of their everyday routine. They commute between home and their school or workplace several days each week; they travel elsewhere in the city to shopping centres or places of entertainment and to visit friends and relatives. They are also in frequent electronic contact with others, and this movement of ideas and information using telephones, computers, and mobile phones is increasingly important both economically and socially. Goods also move into, out of, and around the city as part of everyday economic transactions.

The importance of transportation and movement becomes clear when a city is without transportation services for a period of time, perhaps due to weather conditions or labour unrest. In these cases, people might not be able to go to work or purchase food, and emergency services may have difficulty responding to urgent medical needs. Even in situations where only a portion of a city transport network or a mode of transportation is unavailable due to localized flooding, a damaged bridge, or large highway accident, lives and activities can be seriously disrupted.

Accessibility and Mobility

Two closely related concepts—accessibility and mobility—help illustrate the valuable role of urban transportation. As stated in Chapter 2, accessibility refers to the ease at which it is possible to reach a certain location from other locations. For example, a home in the suburbs is farther from many workplaces and from the financial, legal, and medical services located in the city centre than an apartment closer to the downtown core is. Consequently, these places and services are considered less accessible from the suburban locations than from the downtown. Accessibility needs to use an appropriate distance concept, for example, physical distance, time distance, or cost distance.

In general, **mobility** has increased with technological improvements in transportation and with rising incomes, which has meant that physical distance has become less of a constraining factor. Speaking more geographically, the friction of distance is reduced. In many cities, widespread automobile ownership increases personal mobility. But mobility is closely related to economic circumstances, with low-income earners more dependent on public transit systems and generally less able to move from place to place than high-income earners. Transportation issues cannot be considered apart from larger questions of social and economic well-being in the city.

mobility
The ability to move from one location to another.

back-office activities
Repetitive office operations, usually clerical in nature and performed using telecommunications, that can be located anywhere in or out of the city, including relatively low-rent areas.

front-office activities
Skilled occupations requiring an educated, well-paid workforce; because image and face-to-face contact with others is important, these activities favour prestige locations in major office buildings in city centres.

Role of Information Technologies (IT)

Accessibility should not be thought of merely in terms of measures of physical distance because information technologies (IT) provide people with virtual accessibility to an increasing number of activities, such as online shopping, home entertainment, and even dating. Fundamental changes are occurring in both personal and business communications, and these changes affect the movement of people inside the city.

For many people and businesses, IT advances mean that more and more of their daily routine can be conducted from their home or business without the need to move around the city. Electronic commerce (e-commerce), such as online banking, telemedicine, and online shopping, has radically changed the commercial landscape of the city. Most commentators acknowledge the rise of e-commerce but do not go so far as to suggest that more traditional shopping behaviour will disappear entirely. This is partly because, for some, shopping is more than simply making purchases; it is also a social, and often recreational, activity.

IT advances enable more people to work or study from home (telecommute), thus removing the daily need to travel to other areas in the city. Businesses are tending to subcontract more work, which means working from a home office becomes more feasible for some workers. IT advances also permit many business activities, including **back-office activities**, to locate far away from established office location areas. This decentralization reduces movement into and out of the city centre. However, in the case of **front-office activities**, IT intensifies existing concentration in established office areas, and this situation usually increases movement into and out of the city centre.

Overall, the consequences of these changes on the lives of people and the functioning of businesses is not clear. However, a reduction in the number of travel trips is likely, and future transport planning needs to anticipate and accommodate these ongoing changes.

Urban Transportation and Land Use

Transportation is a major user of urban land (i.e. roads and parking lots) and, as such, it affects and responds to changes in land use. The transportation system, including roads and the availability of public transit, affects accessibility. In turn, accessibility is one factor considered when businesses and other land users make location decisions that affect the pattern of urban land use. Along with the transportation system, this pattern affects the activity structures of people and businesses contributing to the city's overall travel pattern.

Much of the empirical work on the relationship between transportation and land use has looked at the impact of highway construction. Land values close to the highway are higher than those farther away because highways enhance accessibility. If accessibility improves in an already established area, few land-use changes are expected to occur; however, if accessibility improves in an undeveloped area, significant land-use change is likely to follow. Recent research on public transit, such as light rail transit (LRT), suggests similar outcomes. For example, the locations of LRT stations has a positive impact on urban investment and development as residents and businesses opt for places with greater levels of accessibility.

PLANNING THE CITY

Every day, businesses make decisions about where to invest capital, for example, in a new coffee shop, clothing store, or laundromat. Individuals also make decisions daily about where to spend money or shop or how to get from one place to another. Although these choices are constrained in various ways, they are not planned. They are, however, important in shaping the city's social and economic geography: each decision has a small impact, but the cumulative effect contributes greatly to the city's urban geography. That said, cities are not left to develop solely through the decisions of individuals and the operations of the free market.

Cities are planned as part of larger urban policies designed to manage them, for example, to ensure the provision of essential services such as water supply, policing, and transportation networks. Further, cities are planned with constraints, such as zoning regulations that dictate the locations of various land uses. Such regulations are one component of a larger urban planning exercise intended to influence the location of land uses inside the city and the spatial spread and temporal growth of the city. It is inconceivable to think of a modern-day city that could develop and effectively function without some planning activity.

Urban planning provides the broad outlines, or the template of the city that is moulded into a particular and ever-changing urban landscape by the actions of those who work, live, and interact within the city.

Ancient Origins

The first permanent settlements and some of the earliest large cities did not just emerge; they were at least partially created, as evidenced by the rectangular grid street layout and central squares that must have been designed. These layouts have order and regularity and required a strong central authority that asserted itself on and through the urban landscape. The grid system was applied in about 2500 BCE in Mohenjo Daro, located on the Indus River in what is now Pakistan. It was popularized by Hippodamus of Miletus, a fifth-century BCE Greek architect who planned several settlements, including the harbour town of Peiraeus at Athens and the new city of Rhodes. The grid system was also widely used throughout the Roman Empire and medieval Europe. It has been, and continues to be, frequently employed and is regarded as a normal feature of most planned cities.

Many later cities were also planned, at least in part. Prior to the rapid growth of cities during the Industrial Revolution, most European cities show evidence of the expression of authority on the urban landscape, notably in the orientation of streets and the location of public places such as squares and market areas. Europeans brought these ideas to Canada, as seen in the layouts of Vancouver, Quebec City, Guelph, and Charlottetown (Figure 12.10).

It was not unusual for the central areas of large cities to be designed in a monumental style, with wide and long avenues and intentional focal points at intersections where statues or other imposing structures were located. Examples of such design exist from all areas of the world, such as Giza (Egypt), Teotihuacan (Mexico), Babylon

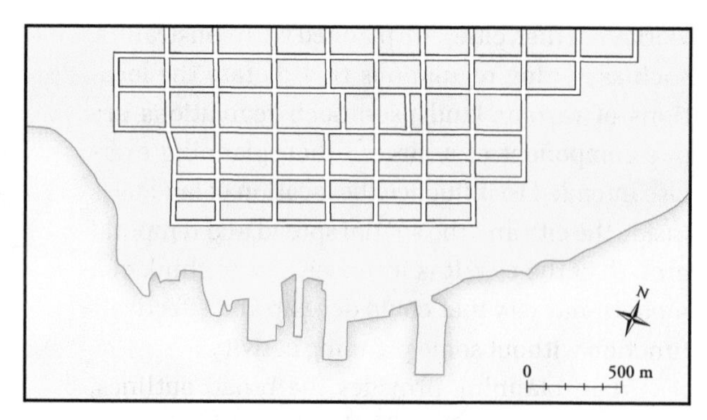

FIGURE 12.10 | Street layout in Charlottetown, 1768

Source: Grant, J. 2006. "Shaped by Planning: The Canadian City Through Time." In *Canadian Cities in Transition: Local Through Global Perspectives*, 3rd edn, edited by T. Bunting and P. Filion, 321. Toronto: Oxford University Press.

(Iraq), and Beijing (China). As with some of the early grid pattern cities, the motivation for this type of planning was primarily to express and defend authority in landscape, not to improve the quality of life for all residents.

Improving the Industrial City

While some cities included at least some degree of planning, the majority of land use in almost all cities was largely unregulated and uncontrolled and, as a result, cities often grew haphazardly in response to specific local circumstances. This situation changed in the first half of the nineteenth century with the onset of rapid urbanization associated with industrialization. As urbanization proceeded, many cities in Europe and North America had major sanitation problems that could be solved only through the planning and building of citywide sewage and water-supply systems. Early industrial cities were dirty and overcrowded, inviting diseases such as typhoid, typhus, and cholera. Modern urban planning arose from the need for these essential improvements. While the initial focus of this planning was on improving health, planners also focused attention on related concerns, such as building better housing and providing public open space.

garden city
A planned settlement designed to combine the advantages of urban and rural living; an urban centre emphasizing spaciousness and quality of life.

green belt
A planned area of open, partially rural, land surrounding an urban area; an area where urban development is restricted.

Planning Cities and Suburbs

By the late nineteenth century there was an interest in improving existing urban landscapes and in creating entirely new urban places.

Beautiful Cities

The City Beautiful movement was one of two major planning directions proposed in reaction to the pollution and crowding of industrial cities. This movement advocated reconstructing cities in a more aesthetically pleasing fashion with a focus on redesigning the parts of the urban landscape that governments had control over, including street layouts, public buildings, and open spaces. The most substantive outcome of this movement was the comprehensive integrated plan for Chicago, which proposed a series of ring and radial roads, street widening, an integrated public transportation system, a uniform height for public buildings, and extensive open spaces. The plan was not fully implemented, but many components were approved and put in place.

The movement aroused much interest in other countries, including Canada, where city beautification groups arose across the country and plans were drawn up for many cities, including Toronto, Ottawa, Calgary, and Edmonton. In retrospect, the City Beautiful movement was a brief effort to impose a social order on urban areas in an attempt to control the often chaotic growth of large cities.

Garden Cities and Suburbs

The horror of urban life in the industrial era inspired a second urban planning movement, the **garden city** movement. The idea of garden cities was to blend city and rural (natural) areas through the design of completely new urban areas. According to this movement, each new city was to be built according to a master plan that aimed to provide a spacious and high-quality environment for working and living. There were concentric patterns of land use, wide streets, low-density housing, public open spaces, and a **green belt**. Figure 12.11 shows a planned garden city and its agricultural belt, as proposed by the movement's major proponent, Ebenezer Howard.

Inspired by the garden city concept, the American architects Charles Stein and Henry Wright created the garden suburb of Radburn, New Jersey, in the 1920s. This innovative plan was designed to be safe for children, with pedestrian and vehicle traffic kept apart by means of a road network designed for vehicles only and a separate network of internal pedestrian pathways. The suburb comprises superblocks sharing an area of common parkland that can be thought of as the heart of the community. Innovatively, the houses are turned around such that the back doors face the roadway, and front doors face the parkland and pathways. Although the scale of the original plan was never fulfilled, with only 667 homes built, Radburn is a highly desirable place to live, judging by home sales. The Radburn plan has not been widely copied, but one successful Canadian suburb designed in this way is described in Box 12.5.

Cities of Towers

While other planners advocated garden suburbs and low-rise dwellings, Le Corbusier, a Swiss-French architect, was planning large, dramatic-looking urban centres with massive multi-storey buildings. He advocated increasing both open space and population density, through use of high-rise buildings separated by parkland, and clusters of buildings linked by road or rail. In the 1920s these were revolutionary ideas. Such planned cities were to have an organized spatial structure with residences located by social class. In some cases, people were to live and work in the same building. Only two cities have been broadly built along these lines: Chandigarh, the capital of the Punjab, India, and Brasilia, the capital of Brazil. However, Le Corbusier's influence is seen in many cities, especially the high-rise apartment buildings that were part of the North American urban renewal efforts of the 1950s and 1960s, and in the suburban apartment complexes so prominent in Toronto's Scarborough, East York, and North York suburbs.

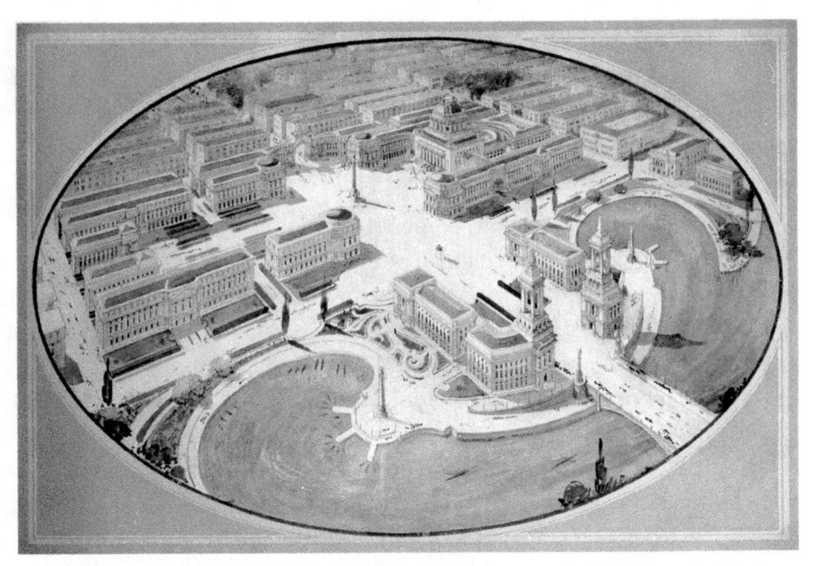

"The Civic Centre of Calgary as It May Appear Many Years Hence": This frontispiece of a plan by Thomas Mawson shows the influence of the City Beautiful movement, which favoured impressive malls with sweeping views.

Canadian Architectural Archives, University of Calgary, Mawson plan for the civic centre (Thomas Mawson fonds Acc. 59A/79.15, MAW C25.p1).

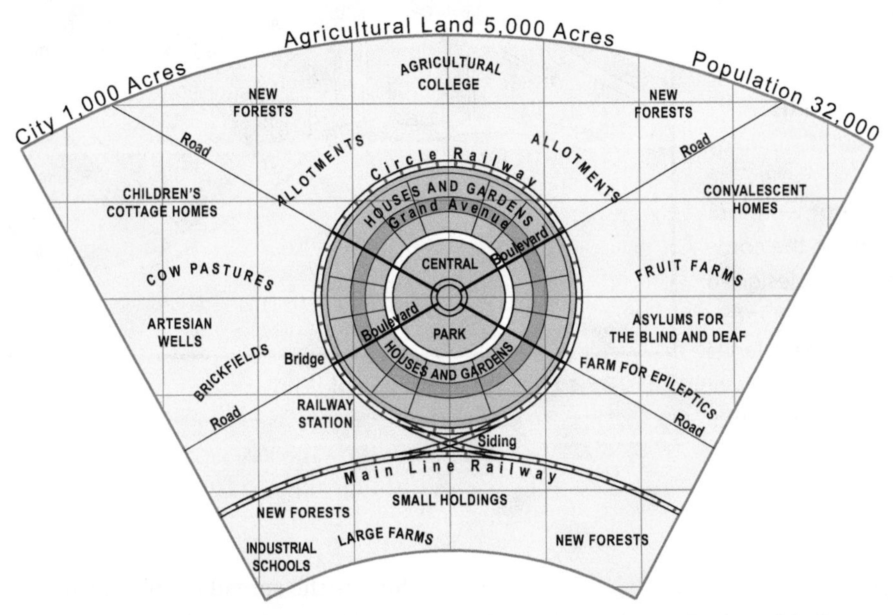

FIGURE 12.11 | Ebenezer Howard's garden city and its agricultural belt

This diagram shows the entire 6,000 acres. Note that radial routes divide the city into sectors, a rail line encircles the city creating concentric zones, and social institutions create nuclei. Howard also produced plans showing the layout of land uses inside the city and plans showing the pattern of several of these cities.

Source: Cherry, G. 1988. *Cities and Plans: The Shaping of Urban Britain in the Nineteenth and Twentieth Centuries*. New York: Arnold, 66–7. Reproduced by permission of Taylor & Francis.

Around the Globe

BOX 12.5 | Wildwood Park Community in Winnipeg

Wildwood Park, a residential area in south Winnipeg, is one of two Canadian communities that are at least partly patterned on the Radburn model (the other is Kitimat, British Columbia). Hubert Bird, a local developer and builder, established Wildwood in the late 1940s. The neighbourhood comprises 286 homes organized around 10 horseshoe-shaped bays (Figure 12.12). The front of the houses are located on the outside of each looped lane looking onto a community park and the back looks onto a vehicular lane. Houses on the inside of each looped lane have the front looking onto a pedestrian walkway that provides direct access to the park. Like Radburn, the motivation is to separate people and vehicles and eliminate through traffic. In Wildwood, all homes are single-family dwellings. Residents have a strong sense of community, and property values indicate that homes in the community are considered desirable.

The reversed-design concept is not without problems, critics, and controversy. Reversing the traditionally public and private faces of homes has consequences for both front and back landscapes. One issue is that living in a reversed house results in social contacts with neighbours being centred on back doors rather than front, a circumstance that is different to that experienced in other suburbs. Also, visitors to the community encounter difficulty when first trying to locate a specific home. The reversed-design communities did not work out as the planners intended. In Wildwood the communal parkland at the front of homes—designed to be the place for movement and social interaction—is not well used. Instead, the back lanes fulfill both of these roles. These lanes are so well used that they are often extremely congested. Increased car ownership adds to this congestion and has changed the landscape markedly, with additional garage construction frequently creating a barrier between home and lane.

It is also clear that some current residents question the design's original logic and that unanticipated changes are taking place. For example, a controversy erupted in 2008, when one resident received planning permission to build a 300-square-foot deck on the front of a house despite the local councillor and some residents expressing strenuous objections to this loss of shared green space. The failure of the reversed-design community to function as planners intended is one of the reasons that many planners advocate streetscapes as the principal public space.

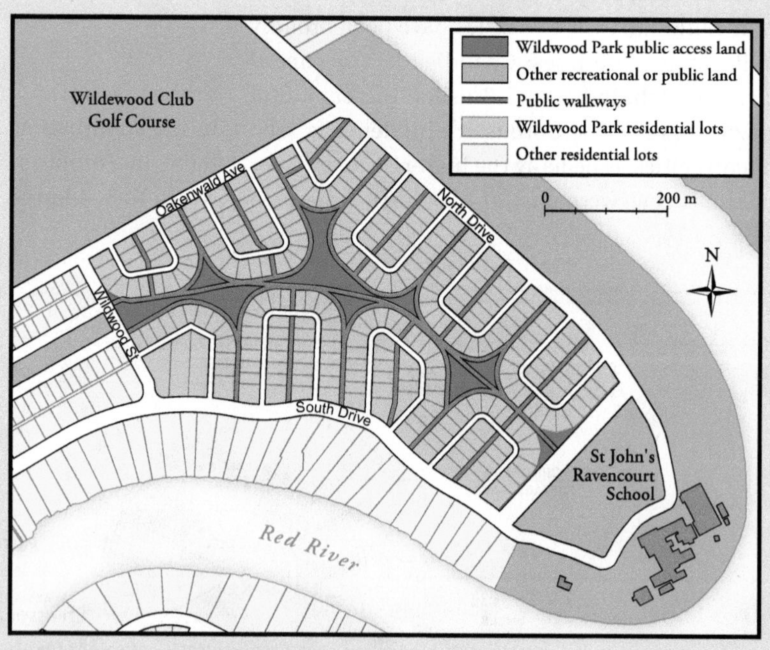

FIGURE 12.12 | The layout of Wildwood

Source: Gillmor, D. 2005. "Wildwood Childhood." *Canadian Geographic* July–August: 57. Used with permission of Canadian Geographic.

New Urbanism

The notion of building ideal communities was resurrected in the 1980s in the form of what has been called New Urbanism. This movement was developed in response to the predominant residential trend of twentieth-century urbanism, which is the spread of suburbs on the periphery of a city. For many planners, this suburban sprawl is seen as a regrettable trend because it consumes large tracts of land and creates residential areas that often lack focus and any sense of community. New Urbanism was influenced by the experiences of some European urban centres that have maintained a concentrated population without excessive suburbanization.

New Urbanism is focused on building communities based on pedestrian movement and mass transit rather than private vehicles. Businesses are intended to be readily accessible by foot or public transit. Streets are intentionally narrow, in an effort to reduce traffic speed and make the communities friendlier and safer for all residents. Homes should have front porches close to the street to encourage social interaction. Furthermore, each neighbourhood is to have an evident centre, and like the Radburn garden suburb concept, parking lots and garages are located at the rear of homes and businesses.

Examples of this type of planned community are Seaside, Florida, Mackenzie Towne in Calgary, and Cornell in Markham, Ontario. The intent of New Urbanism is to plan the city and to advocate and sell an image of what urban life ought to be. Urban places are being mythologized, and urban dwellers are actively encouraged to behave in particular ways.

Planning in Practice

Urban planning is not only focused on individual planners and their often utopian aspirations, but it is also a function of government intervention into the lives of individuals and businesses. Planning is a pragmatic activity intended to influence both the form and function of the city and to make it a more livable environment.

Since World War II, planning in many countries has centred on physical design, urban renewal, and suburban expansion. Planning is typically undertaken in the context of legislation from a senior government that outlines constraints for what can and cannot be done and usually involves some public participation. In Canada planning is well established, with plans prepared and

The planned community of Seaside, Florida, is an early example of New Urbanism. It was established in the early 1980s in a conscious effort to create a modern version of the idealized old-fashioned American town.

implemented by all local governments. A notable planned community is the Don Mills suburb of Toronto. Inspired by the garden city concept, Don Mills is characterized by low-density development and clear separation of land uses. This modernist approach to suburban expansion was promoted at the federal level and has been imitated across Canada. Two generally acknowledged drawbacks of this type of expansion are the related decline of city centres and the inevitable dependence on automobiles.

Zoning

zoning
One of the most important tools of the urban planner; controls the use of land and the buildings constructed there.

Zoning is one of the major tools of urban planning. It functions to regulate land use and has a significant impact on the social and economic geography of the city. The roots of zoning are connected to the nineteenth century industrial city, when incompatible land uses were often found adjacent to one another, for example, new factories built adjacent to a residential area. The first zoning regulations were not introduced until the 1920s; three decades later, they had become the principal tool used by urban planners to regulate urban land use.

In most current cities, urban plans include a classification of the urban area into distinct zones, for example, several distinct classes of residential, commercial, and industrial land uses. The fundamental goal of zoning is to ensure that those land uses judged to be incompatible are not located adjacent to each other.

The positive intent of zoning regulations is clear, but one criticism is that they lead to inflexibility and, for some critics, a boring urban landscape. In some American cities, one consequence of zoning is that areas of the city may become accessible only to specific income groups.

The Inner City as Cultural Capital

Many cities, first in Europe and North America and now throughout the world, have initiated a process of planned redevelopment of city centres, focusing on cultural capital such as historic preservation and the promotion of the arts. This use of culture as an engine of economic growth often involves plans to build museums, galleries, theatres, bookshops, and cafés, as well as preserve buildings and areas of historical and cultural value. For example, the EU initiated a competition in 1985 to select an annual City of Culture, which has encouraged cities to plan along these lines.

Urban Transportation Planning

One of the most challenging aspects of planning today involves the urban transportation system. Several levels of government are usually responsible for planning and financing transportation systems, a process that is complicated by the fact that new routes and modes are needed and older routes and modes require upgrading or become redundant. Many cities face significant planning and financial challenges related to capital investment in roads, highways, and public transit infrastructure (e.g. subways) as cities grow and land uses change.

In many cities, there are ongoing debates about the respective roles of public transit and private automobile use; methods of limiting private automobile use seem to dominate these discussions. In general, public transit systems are well developed in European and some Asian cities, which at least partly reflects the limited space available for urban expansion, the cities' congested character, and pollution concerns. London and Singapore, amongst other cities, limit and charge for private automobile access to the city centre. Copenhagen's and Amsterdam's city centres are primarily home to public transit and to bicycles rather than cars. In contrast, most North American cities are designed with automobile travel in mind.

A related debate concerns the merits of adding new, or improving existing, highways in a city rather than investing in public transit upgrades. Opponents of such highway expansions contend that it is bad public policy because more (or better) highways increase demand rather than relieve congestion or improve movement. This is simple supply-and-demand logic—if a new highway enhances accessibility to a workplace or mall, people are more likely to use that highway. Costs are also an issue here. Many public transit systems, especially subways and light rail systems, are expensive to build in comparison to some highway projects.

The question of who pays for transportation services is also regularly debated. The principle of public provision of transportation services and infrastructure is widely accepted but not universally applied. In many countries, some components have been privatized—the London underground rail system has operated as a partnership since 2003, with the infrastructure maintained by private companies but the underground owned and operated by a public company.

It costs money for people to move around the city. Most high-income people accept this expense as a necessary and valued expenditure, but the cost of movement may be excessive for low-income people. In most cities, transportation planning is a technical activity, usually based on quantitative data concerned with matters such as engineering problems, traffic flows, and predictions of use with little concern for social justice. Debates about private versus public transportation and about financing have social justice implications, but they are rarely front and centre.

Transportation systems are intended to improve accessibility, not only for suburban commuters travelling between home and downtown but also for inner-city residents who need to travel to work, shops, hospitals, and recreational sites. Highway improvements typically benefit suburban developers and their high-income residents. But lower-income suburban residents, particularly those without access to an automobile, become effectively stranded or face long commutes on generally inefficient public transit. In addition to promoting urban sprawl, highways and private automobile ownership contribute to congestion in city centres, air pollution problems, and the incidence of accidents. Not surprisingly, then, highway development and private automobile use are viewed unfavorably by many people concerned with the creation and maintenance of sustainable and socially just urban environments.

CITIES OF THE LESS DEVELOPED WORLD

Much of this chapter has focused on the cities in the more developed world. While cities of the less developed world also have residential, industrial, and commercial areas; extensive transportation networks; and, to a lesser extent, are planned, important distinctions exist between the cities of the two worlds. In this section, we outline two ways that many of the cities of the less developed world are different from those in the more developed: the prevalence of peripheral informal (squatter) settlements and the existence of inner-city slums.

What is life like in a less developed world city? As addressed in Chapter 6, citizens of the less developed world face a litany of challenges that few in the more developed world have to deal with. Some of the aspects that make less developed world cities less livable than those in the more developed world include health (premature death due to infectious disease and higher rates of infant and material mortality); housing (insufficient supply of shelter and often insecure tenure); employment (generally poorly paid, frequently unsafe and insecure); food and water (often expensive, of poor quality, and/or of insufficient quantity); and education (unavailable to many, resulting in high rates of illiteracy). All these factors result in cities that make it difficult for many to live their lives. However, not all cities in the less developed world are like this, nor are all areas prone to such problems. Portions of many cities resemble the typical more developed world's middle-class neighbourhoods.

Notwithstanding dire circumstances, residents note that their best hope for a reasonable quality of life comes from living in the city in safe and affordable housing, not in rural areas. As suggested, most cities in the less developed world perform the same urban functions as cities in the more developed but generally less effectively. Thus, they provide homes, places to work, and places to consume. The challenges of daily life are real and may prove very difficult to resolve, but many residents work extremely hard to cope with these problems and to provide their children with a better life. The struggles experienced by (especially) the urban poor are not of their own making but the result of larger economic and social circumstances that need to be addressed at a root level.

Colonial Origins and Recent Growth

We have indicated throughout this book that many parts of the less developed world have a colonial heritage. Many cities in Africa and Latin America, and some in Asia, were essentially creations of European powers developed to serve European needs, a fact that in many cases is reflected in their colonial/pre-independence names, their architecture, their locations, and their internal structure. In most instances, the pre-independence name has been changed for powerful symbolic reasons, although colonial architecture has generally remained. With regard to internal structure, it was common for colonial powers to implement some degree of formal spatial segregation on the basis of ethnic identity, a situation that reached its extreme in South Africa before 1994 (discussed in Chapter 8).

It is important to recognize these European-created cities as centres of exploitation: their functions were primarily administrative and military, and they controlled the export of valuable primary products. They failed to generate local growth, and little of the income generated benefited local populations. As a consequence of the importation of indentured labour (see Chapter 8), many of these cities became culturally pluralistic, and some have experienced cultural conflict as a result. Because the colonial powers failed to initiate local industrial development, the rapidly growing cities of the post-independence period have had no pre-existing industrial base to build on. For many of these cities, a number of the issues and problems discussed in this chapter originated in the colonial phase.

informal settlement

A concentration of temporary dwellings, neither owned nor rented, at the city's periphery; related to rural-to-urban migration, especially in less developed countries; sometimes referred to as squatter settlement or shanty town.

slum

A heavily populated informal settlement, usually located within the urban core, and characterized by poverty, substandard housing, crime, and a lack of sanitation, water, electricity, or other basic services; common in less developed world cities today and in more developed world cities in the nineteenth and early twentieth centuries.

As we saw in Chapter 11, rapid city growth in the less developed world is a relatively recent phenomenon. In 1950 only 17 per cent of the population in the less developed world lived in cities; today the percentage is about 50. This growth is the result of large movements of people from rural to urban areas, movements driven by changes to the demand and supply of employment opportunities in rural and urban areas and by a (not always accurate) perception that the quality of life would be better in the city.

AP Photo/Bullit Marquez

One squatters' community north of Manila, Philippines, is located in a cemetery. Residents live in dwellings built on top of above-ground tombs, representative of the disamenity zones that tend to attract informal housing.

One common feature in rural-to-urban migration has been a tendency to gravitate towards a primate city. This tendency has been particularly strong in less developed countries, where the population typically gravitates towards the capital. It is common for primate cities in the less developed world to have between 20 and 30 per cent of the country's total population; in many cases these cities are true megacities. It can certainly be argued that these exceptionally large cities are too large, especially given the economic circumstances of the country. (See Chapter 11 for more on primate cities and megacities.)

Informal Settlements and Slums

Many cities in the less developed world are economically and socially vibrant. Tall buildings with impressive architecture are home to major corporations, indicating that they are command centres of national or regional economies, and several of these cities are also increasingly involved in the global economy (Box 12.6). Unfortunately, some of these cities have areas that suffer from overcrowding, crime, poverty, disease, limited provision of services, traffic congestion, unemployment, damaged environments, and ethnic conflicts. In all these situations, the poor and underprivileged suffer most. In essence, cities in the less developed world face intense pressure to accommodate additional population but are generally incapable of responding adequately.

It must be noted, however, that the slums of British and North American industrial cities in the mid-nineteenth century were just as dire as those in many cities in the less developed world today. Also, people moved to them for essentially the same reason: the perceived difficulties in the slums were outweighed by the perceived opportunities.

The Growth of Informal Settlements and Slums

Much of the rapid urban growth in the less developed world has involved the emergence and spread of **informal settlements**, areas of uncontrolled expansion on the periphery of a city, made up of low-quality, poorly serviced temporary dwellings. In addition to these informal settlements, many cities include long-standing **slum** areas located throughout the city but usually towards the centre. While their geographic location may be different, these two types of urban

Around the Globe

BOX 12.6 | Globalization and the Less Developed World City

When discussing cities in the less developed world, it is easy to emphasize the problems and fail to credit the other changes. Globalization processes are occurring, with both negative and positive impacts.

The often unfortunate social consequences of economic restructuring as a part of globalization are not evident in only more developed world cities. Changing job markets, combined with reductions in public expenditures and rising prices, are evident around the world and frequently contribute to a loss of social stability. But globalization processes have also facilitated the growth of new and vibrant business centres in many cities. A detailed analysis of Accra (Ghana) and Mumbai (India) indicates the links between globalization and the spatial structure of urban corporate geography (Grant and Nijman, 2002). The urban structure of both cities (and many others) have changed through the years with shifting economic demands, and both have entered a global phase of restructuring. This stage has transformed aspects of these cities' geographies through an increased foreign presence since the early 1980s.

Financial and producer services are a rapidly growing sector of the economies in both cities, contributing to their overall economic health. These activities are spatially concentrated, and new centres of corporate control have arisen. Figure 12.14 maps the urban economic geographies during the colonial phase, and Figure 12.15 maps these geographies during the global phase. The most notable spatial development is that both cities have three separate business districts (one global, one national, and one local) that tie in with the larger economy. It isn't clear whether the companies are under domestic or foreign ownership; however, both types are integrated in the global economy.

The two cities have also undergone similar structural changes. Especially among foreign companies, the fastest-growing aspects of the urban economies are financial and producer services. In Accra, the emphasis in this sector is on communication, real estate, advertising, and consulting. These are important in Mumbai, as are banking and finance.

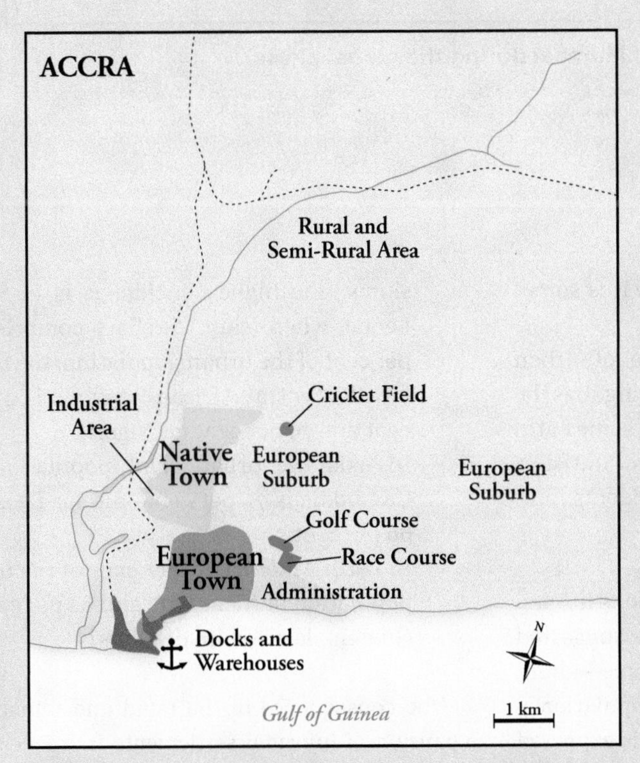

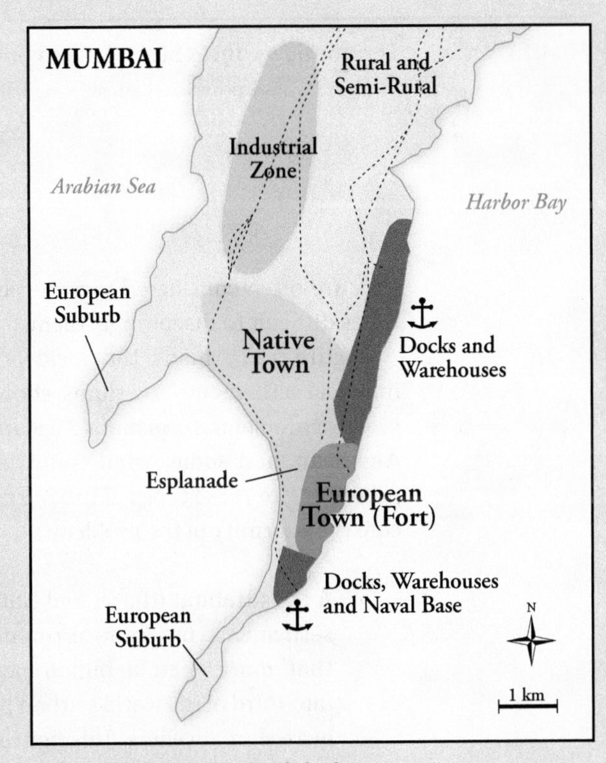

FIGURE 12.13 | Economic geographies of Accra and Mumbai during the colonial phase

Source: Grant, R., and J. Nijman. 2002. "Globalization and the Corporate Geography of Cities in the Less-Developed World." *Annals, Association of American Geographers* 92: 325, Figure 1 (MW/RAAG/P1890). Published by Blackwell Publishing Ltd. Used with permission of Taylor & Francis Ltd., www.informaworld.com.

Continued

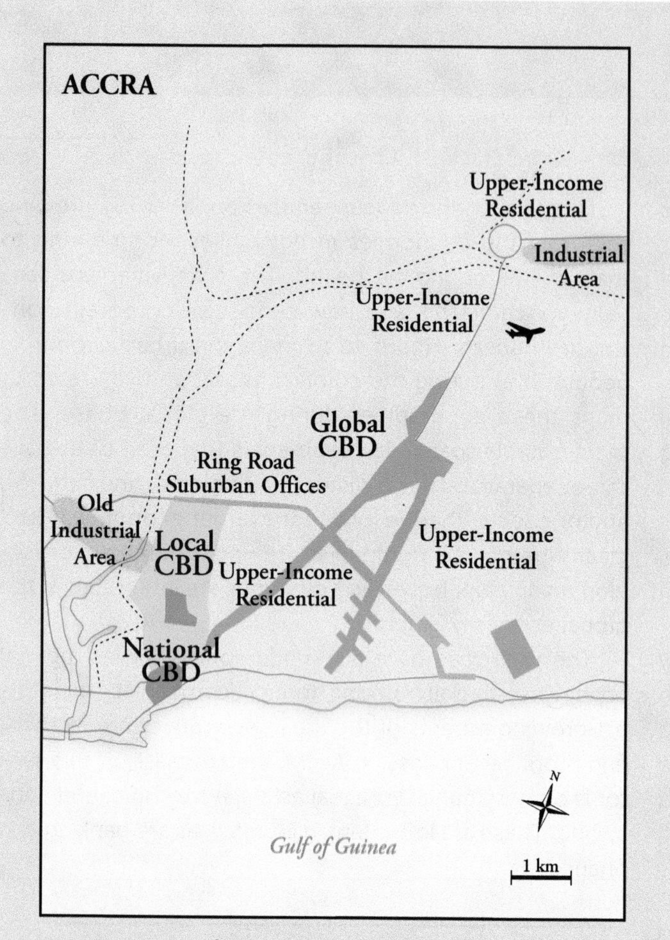

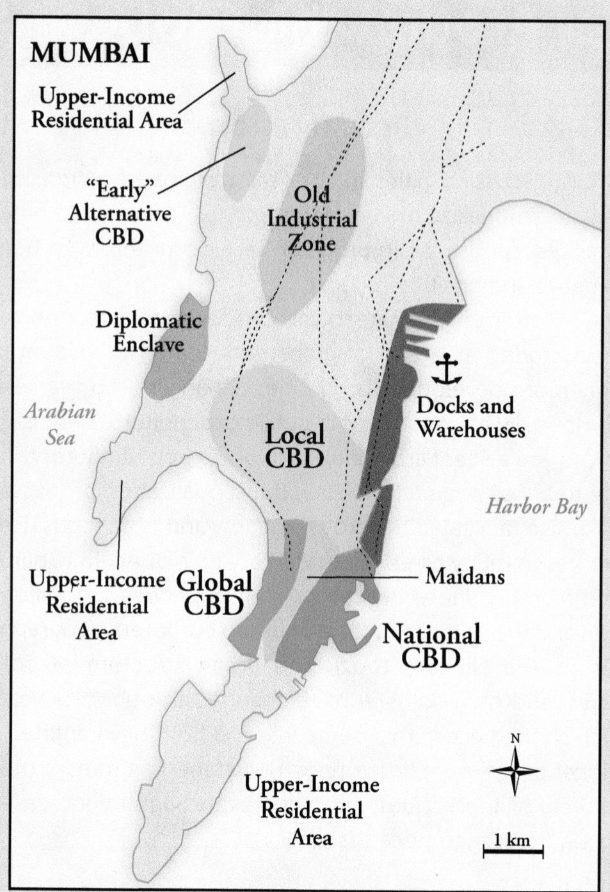

FIGURE 12.14 | Economic geographies of Accra and Mumbai during the global phase

Source: Grant, R., and J. Nijman. 2002. "Globalization and the Corporate Geography of Cities in the Less-Developed World." *Annals, Association of American Geographers* 92: 325, Figure 11 (MW/RAAG/P1890). Published by Blackwell Publishing Ltd. Used with permission of Taylor & Francis Ltd., www.informaworld.com.

area are often considered together as it is sometimes difficult to disentangle them.

Figure 12.15 maps the incidence of urban informal settlements and slums, showing that the greatest problems are in many African, some Latin American, and some Asian countries. Statistics pertaining to these types of urban residents indicate the enormity of the problem:

- A 2005 Habitat (the United Nations human settlements program) estimate suggested that more than a billion people—about one-third of the world's urban population—were slum dwellers. This figure is expected to reach 1.4 billion by 2020 and 3.5 billion by 2050 unless significant action is taken.
- About 70 per cent of the urban population in Africa live in informal settlements and slums. The highest incidence is in Sierra Leone, where slum dwellers comprise 96 per cent of the urban population; the figure for the Central African Republic is 92 per cent and 79 per cent for Nigeria.
- In Asia the urban slum population for Cambodia is 72 per cent, and for Laos it is 66 per cent.
- In Latin America, 81 per cent of the urban population in Nicaragua and 62 per cent in Guatemala are slum dwellers.

One consequence of the rapid and unplanned expansion of informal settlements is frequently a chaotic, uncoordinated system of governance. For example, Bangkok's city core is governed by a single authority, while the larger metropolitan area is divided into about 2,000 small areas, each with its own

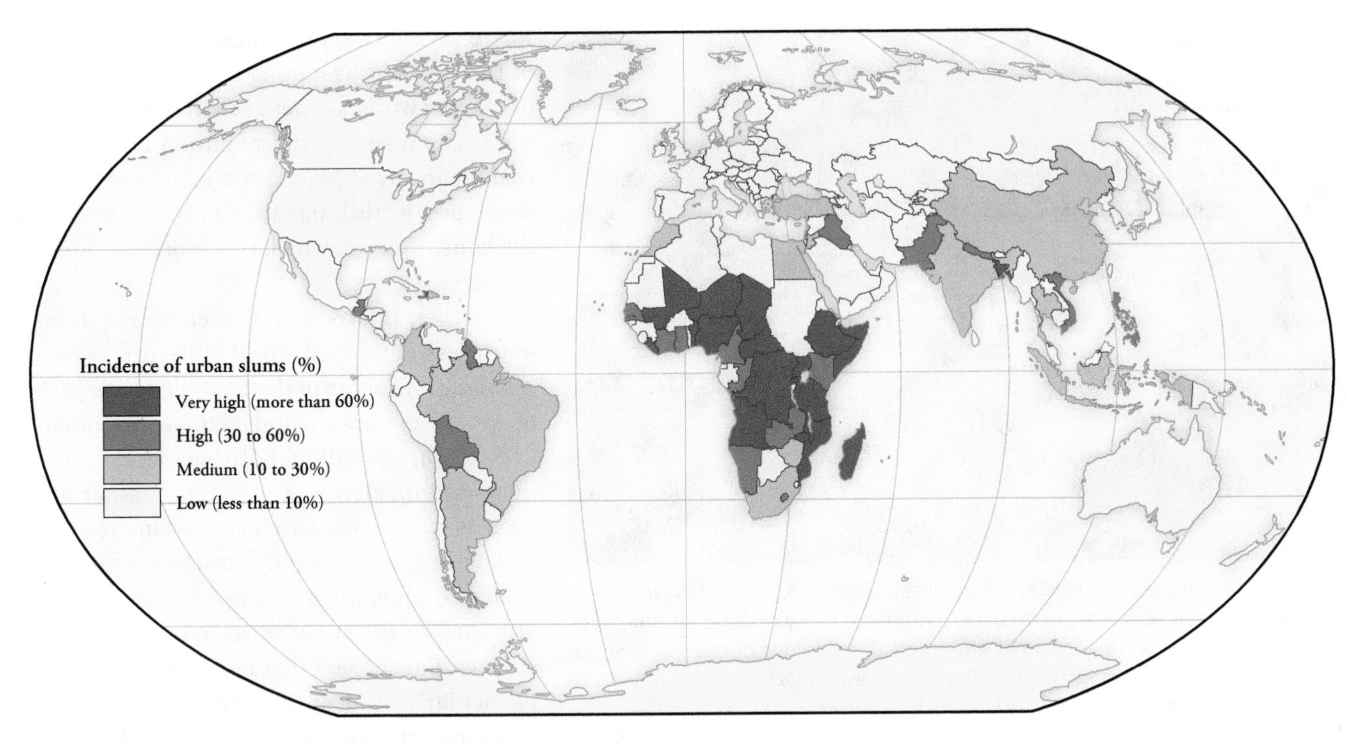

FIGURE 12.15 | Incidence of urban slums as a percentage of urban population, 2009

Source: Adapted from United Nations Human Settlement Program (UN-Habitat). 2013. *State of the World's Cities, 2012–2013: Prosperity of Cities*. Routledge: New York.

local government. In many other cities, the informal settlements on the periphery have no political representation at all. It is easy to appreciate just how difficult it is to build transportation lines and provide essential services under these circumstances.

Employment in Informal Settlements and Slums

New migrants to cities are drawn towards the city by the prospect of employment. Many do not obtain permanent full-time employment but subsist by becoming involved in the **informal sector**. People working as part of this sector may be involved in retail distribution by selling goods such as fresh water, food, newspapers, or jewellery, or they may provide services such as car-washing, laundries, gambling, or prostitution. Others earn money by making items or growing food for sale in small garden areas, and others beg and scavenge to survive.

The labour market is commonly segmented on the basis of gender, with women performing the least attractive labour as measured by the quality of the work environment and remuneration. Child labour is also a common circumstance, including working in the home, the informal sector, and waged labour. Children are especially valued by some employers because of their willingness to work long hours for little pay in often unsafe and unhealthy environments, such as mining and brick-making.

Much of the available employment is low-paying and sometimes unsafe, but such work is usually regarded by the worker as temporary. For migrants, the attraction to cities is not the informal settlement or the possibility of informal sector employment but the perception that the city offers better schooling, better medical services, a better water supply, and the prospect of permanent employment. As informal settlements grow, enormous pressure is put on urban governments to provide the needed basic services, but many fear that extending these urban services will attract even more migrants. This situation may be unfortunate for the migrants, because informal settlements are usually decrepit and lack basic amenities, but these areas can be important reception areas, providing a transition between rural and urban ways of life.

informal sector
A part of a national economy involved in productive paid labour but without any formal recognition, governmental control, or remuneration.

Not only do the residents of *favelas* (informal settlements) in Rio de Janeiro, Brazil, face poverty, crime, and insecure housing tenure, but many of these communities also exist on unstable hillslopes that are prone to landslides during the tropical country's wet season. Rocinho (pictured) is the largest *favela* in Rio.

Health Issues

Those living in slums have serious health problems, many caused by unsafe water and inadequate sanitation. For example, Bangui in the Central African Republic has a current population of about 700,000 but a sewage system built for a population of 26,000 and that has never been extended or improved. Cairo treats less than half of its sewage; the remainder ends up in the Nile or in local lakes. Not surprisingly, acute respiratory infections and diarrhea are major causes of death in many urban slums and informal settlements, specifically among young children, while malaria, typhoid, and cholera outbreaks remain major problems in some cities. Air pollution is often at a dangerous level in many of these housing areas because of the proximity of industrial activities (a result of a lack of urban planning and zoning) and minimal emissions control.

Environmental conditions in these informal settlements and slums, and the cities they are a part of, has been labelled the brown agenda (Cohen, 1993). This term is an explicit acknowledgement of the links between urban and economic growth and environmental deterioration. There are two principal concerns: conventional environmental health problems, such as limited land for housing and lack of services, and problems related to rapid industrialization, such as waste disposal and pollution. The significance of this twofold agenda is that cities in the less developed world are being challenged to address the waste disposal and pollution issues (which remain unsolved even by many cities in the more developed world), but they have not yet solved the housing crisis, which is of more fundamental concern.

Further, informal settlements and slums in many less developed world cities are especially vulnerable to some of the possible consequences of global environmental change discussed in Chapter 4, including both sea-level rise and increased incidence of extreme weather events such as hurricanes. Often located in areas prone to flooding, lacking infrastructure and protection, and unplanned and poorly managed (if at all), these areas suffer most in disaster circumstances. But the fact that more and more people, particularly in less developed countries, are choosing an urban way of life clearly suggests that it is perceived as preferable to living in rural areas (Box 12.7).

CONCLUSION

Throughout the discussions in this chapter, it is evident that cities are highly complex organizational and institutional forms that serve as centres for social interaction, cultural activities, economic development, and technological change. Further, shifting economic, cultural, social, political, and technological circumstances combine to create changing urban areas and urban experiences. Not surprisingly, then, our discussions of urban areas and the urban way of life are conceptually diverse, reflecting a wide range of economic and social perspectives. For example, geographers have used simple models to describe and generalize about the internal structure of cities, as well as examine the changing social and economic geography of the city. Perhaps the most suggestive indication that cities are simultaneously both hopeful and hopeless places is in Box 12.7. Much current thought argues that the slums of less developed world cities are gateways to the good life, but as the account of selected cities suggests, they are also fraught with serious economic, social, and environmental problems.

In the News

BOX 12.7 | Slum Areas as Gateways to Prosperity

Notwithstanding the often dire circumstances encountered by migrants to cities in less developed countries, newcomers feel their best hope for a reasonable quality of life is inside the city, not in a rural area. For example, Chapter 5 noted that fertility rates typically decline with increasing industrialization and urbanization. In most less developed countries, a clear distinction exists between urban areas with relatively low fertility and rural areas with relatively high fertility.

Scholars and mainstream media accept that squatter settlements have advantages over rural areas, especially for women who find themselves freed from the constraints of a patriarchal system and with previously unheard of opportunities for playing a role outside the home. Urban informal settlements and slums in less developed countries provide low-cost housing and facilitate the creation of social support networks not possible in rural areas. Evidence suggests that poverty in rural areas leads to children dying from starvation, while poverty in urban areas leads to children selling goods on the street. People rarely die of hunger inside a city.

Some scholars contend that what is sometimes called the arrival city, the place where rural migrants first experience an urban life, is an engine of social change and a spur to social mobility. Poor people enter these places, take advantage of cheap rent, make crucial social contacts, accumulate capital, and pass through them, emerging as true urban dwellers. In this sense, informal settlements are crucial places connecting rural areas and cities.

Given this positive understanding of informal settlements and slums, along with the fact that ongoing growth seems inevitable, how can life in these areas best be improved? Three suggestions are offered.

1. Most countries have moved away from negative policies such as forced eviction, neglect, and involuntary resettlement. Urban authorities are making efforts to formally integrate informal settlements into the city proper. For example, beginning in 1994, a plan was developed for Rio de Janeiro that anticipated the integration of approximately 600 *favelas* (the Brazilian term for slum) into the city, providing assistance for people to improve their housing. Although there have been many setbacks, this plan appears to be improving conditions in many areas. Some argue that this type of plan rationalizes and perpetuates poverty while relieving governments of responsibility for the welfare of people.
2. International aid could be redirected from rural to urban areas. Western governments and international agencies have long focused their attention on rural development, neglecting urban areas. Rural areas need aid, but the reality is that the need for flush toilets is far greater in a congested city than in a relatively sparsely populated rural area.
3. It is not appropriate to simply impose change from outside (recall the Chapter 10 account of the green revolution, for example). Improvements in urban areas need to actively involve local residents as participants. If the people living in informal settlements had some security of tenure, they would likely be more willing to work to improve their homes.

Summary

Explaining Urban Form

Cities are comprised of several distinct areas and land uses. The most obvious area of the city is the downtown, or what geographers call the central business district (CBD). The CBD is the focus of economic, residential, political, and civic life, with high-rise office and residential towers, city hall and other government buildings, as well as museums, theatres, restaurants, and hotels. Cities also have a diverse array of residential, commercial, and industrial areas, divided between the central city and the suburbs.

Urban Structure

Three classic models of urban structure describe the concentric zones, sectors, and multiple nuclei of cities in North America. Cities in different regions of the world typically display internal structures similar in basic principles but often different in detail.

Housing and Neighbourhoods

In more developed countries, there are two different ways of thinking about housing: as a commodity or as a universal right. Usually, the reality lies somewhere between the two. Housing as a commodity—something that is bought, sold, and rented—is appropriately analyzed in the context of a market. This market does not work in the same way for all urban residents. For example, redlining denies capital to certain areas based on the perception that the areas will decline in property values over time. Filtering is the idea that, through time, a housing unit or a neighbourhood is occupied sequentially by people from steadily changing income groups.

Gentrification

The process of people moving in and transforming a formerly derelict or low-quality housing area is known as gentrification. It is sometimes explained in terms of the rent gap hypothesis, which argues that there is a discrepancy between the current value of land and the potential value. Alternatively, it is explained in terms of larger social changes. Gentrification frequently benefits middle-income groups but poses problems for low-income groups, who become displaced and more spatially marginalized.

Residential Segregation

Most urban areas contain residential districts distinguished on the basis of income, class, ethnicity, religion, or some other economic or cultural variable. Distinct residential areas emerged in large immigrant-receiving North American cities in the nineteenth century, with spatial differentiation based on ethnicity and class. Explanations for the creation and maintenance of ethnic neighbourhoods have focused on economic and cultural forces. In general, two processes are at work: the internal cohesion of the group and the desire of non-group members to resist spatial expansion of the group, the result of which is congregation and involuntary segregation, respectively.

Suburbs and Sprawl

Cities have always contained suburbs, but suburbanization came to dominate urban areas of the more developed world beginning in the late nineteenth century, as increasing numbers of people opted for homes outside the established urban area, a location decision made possible by improved transportation and increased automobile ownership. Suburbanization is a form of decentralization and has involved a reorientation of the city, with people and businesses locating in the suburbs and often a corresponding decline of traditional city centres.

Inequality and Poverty

Cities are unequal places. In many countries, poverty has increasingly become concentrated in cities and, within cities, concentrated in the older central areas. Although members of all social groups are susceptible to poverty, there is compelling evidence of links with ethnic minority identity, level of education, employment status, health, age, and gender.

Homelessness

A principal problem in many large cities is homelessness. Four versions of homelessness are rooflessness, houselessness, insecure housing, and inadequate accommodation. Homelessness is frequently precipitated by a specific circumstance such as loss of employment, eviction, or domestic violence. Once homeless, people may gradually become labelled as homeless and even self-identify as being homeless, which hinders opportunities to change circumstances.

Cities as Centres of Production and Consumption

Manufacturing, along with services of various forms (including professional/business services and consumer services such as retailing), are the key components of the urban economy and changing geographies. Most cities have experienced some significant decentralization of manufacturing activity away from central areas towards outer suburban areas. Office employment, as well as urban retailing, has undergone similar organizational and related locational changes in response to broader post-industrial trends, including a decline in the number of independent retailers as a consequence of competition from large multinational enterprises and innovations in communication technology that have permitted the development of largely online retail enterprises. The most evident locational change is a decentralization of shopping centres, many of which are now located in out-of-town areas accessible only by automobile.

Transportation and Communication

To function effectively, people, goods, and information need to move between and within cities without significant constraints. Movement is a part of everyday routine. The role played by information technologies is of particular importance, and changing accessibility and mobility are affecting many personal and business location decisions. There are often debates concerning the respective roles of public transit and private automobile use and the implications for larger issues of social and environmental justice.

Urban Planning

Cities are both planned and unplanned. The earliest evidence of town planning is the use of a rectangular grid street layout followed by the building of monumental structures. In the nineteenth century, the need for planning to correct serious urban health and overcrowding problems initiated the building of sewage systems and other improvements. Some planners proposed ambitious new urban centres, as in the City Beautiful and garden city movements. Notable modernist planners, such as Le Corbusier, saw few of their plans come to fruition, but their ideas are part of every city, for example, in the form of high-rise apartment blocks. More recently, the New Urbanism movement has proved a popular approach. One of the most important tools of urban planners today is zoning.

Cities in the Less Developed World

The colonial heritage of many cities in the less developed world left them ill-prepared to cope with the population explosions that have occurred since about 1950. Many of these cities are experiencing social and environmental strains far more severe than those affecting cities in the more developed world. One common feature in rural-to-urban migration has been a tendency to gravitate towards a single well-known, or primate, city. Regardless of location, many of these cities suffer from overcrowding, crime, poverty, disease, limited provision of services, traffic congestion, unemployment, damaged environments, and social conflicts.

Informal Settlements and Slums

Informal settlements are areas of uncontrolled expansion on the periphery of a city, made up of low-quality, poorly serviced temporary dwellings. Slums are similar in nature, but tend to be located adjacent to the central core of the city. The incidence of slums and informal settlements is highly correlated with the rate of urbanization. One hopeful sign in recent years is that many countries have moved away from negative policies such as forced eviction, neglect, and involuntary resettlement as strategies to solve informal settlement problems.

Links to Other Chapters

- **Urban structure:** Chapter 2 (concepts of space, location, and distance); Chapter 5 (migration); Chapter 10 (agricultural location theory); Chapter 13 (industrial location theory)
- **Housing and neighbourhoods:** Chapter 7 (cultural identity, language, religion); Chapter 8 (cultural identity, ethnicity)
- **Inequality and poverty:** Chapter 8 (feminism, crime, health)
- **Cities as centres of production and consumption:** Chapter 3 (economic globalization); Chapter 13 (industrial change)
- **Transportation and communication:** Chapter 3 (concepts of distance and space, transportation)

- **Urban planning:** Chapter 9 (the role of the state)
- **Cities of the less developed world:** Chapter 4 (human impacts); Chapter 5 (fertility and mortality, history of population growth, explaining population growth); Chapter 6 (less developed world, world systems analysis); Chapter 9 (colonial activity and the colonial experience)

Questions for Critical Thought

1. Despite sharing certain similarities in terms of their form, cities differ from one part of the world to another. Why do these differences exist? What role does culture play in explaining these differences?
2. What factors have formed the social geography of present-day North American cities?
3. What are the causes and consequences of gentrification? Explain why you would (or would not) support government incentives designed to encourage gentrification.
4. Why is it so difficult to control urban sprawl? What is the single most important factor underlying sprawl? How can/should urban sprawl be managed or controlled?
5. Suppose that you had the resources needed to instantly create enough units of housing to shelter all those who were homeless. Would the problem of homelessness then be "solved"?
6. How and why have the retail and industrial landscapes of North American cities changed in the post–World War II era?
7. Do you think that the ideals of New Urbanism can be successfully implemented in our cities today? Why or why not?
8. How would you compare living in cities (i.e. the "urban experience") in more developed nations to that in less developed nations? In other words, how and why are these urban experiences different?

Suggested Readings

Visit the companion website for a list of suggested readings.

13 GEOGRAPHIES OF ENERGY AND INDUSTRY

This chapter begins by asking why industries are located where they are. The answer once again takes the form of a neo-classical economic theory that displays all the advantages and disadvantages of that approach; the conceptual similarities with both agricultural (Chapter 10) and urban (Chapter 11) location theories are clear.

The origins and evolution of modern industrial landscapes are discussed with emphasis on the many changes associated with the period of the Industrial Revolution. We next turn to a discussion of fossil fuel energy sources: oil, coal, and natural gas. Coal was the key energy source that fuelled the Industrial Revolution, and all three are significant industrial inputs today. There are particular emphases on oil production, movement, and consumption and on the increasing role played by natural gas. This is followed by a description of major world industrial regions, the dramatic rise of Japan and the newly industrializing countries, and the current emergence of China as an industrial country. Next, our discussion of the industrial restructuring that has been apparent since the 1970s (and was introduced in previous chapters) emphasizes the growth and increasing concentration of service activities—a process that requires further consideration of the transition from Fordism to post-Fordism and the deindustrialization evident in the more developed world, as well as a variety of theoretical perspectives, especially forms of Marxism.

Here are three points to consider as you read this chapter.

- Do you think that the many changes associated with the period of the Industrial Revolution are fundamental to an understanding of global human geography?
- There are two divergent attitudes towards non-renewable energy resources. On the one hand, they are an essential input in our everyday lives; on the other hand, we talk about our addiction to fossil fuels. Is energy friend or enemy?
- What might be the global implications of the continuing rise of China as both producer and consumer of industrial goods?

An oil worker views a tar sands mining plant in Alberta. Extracting the vast amount of oil available in the tar sands—estimated to be up to two trillion barrels—helps satisfy the world demand for the fossil fuel but raises environmental concerns, such as the destruction of land and the large carbon footprint.

primary activities
Economic activities concerned directly with the collection and utilization of natural resources.

secondary activities
Economic activities that process, transform, fabricate, or assemble raw materials derived from primary activities; also activities that reassemble, refinish, or package manufactured goods.

tertiary activities
Economic activities involving the sale or exchange of goods and services; includes distributive trades, such as wholesaling and retailing, and also personal services.

quaternary activities
Economic activities concerned with handling or processing knowledge and information; typically involve a high level of skill; highly specialized.

In the study of spatial variations and changes in economic activity, human geographers have traditionally distinguished three levels of economic activity. **Primary activities** such as farming, fishing, forestry, and mining occur where appropriate resources are available. **Secondary activities** convert primary products into other items of greater value. This process is called manufacturing ("making by hand") and requires three things: a labour force, an energy supply, and a market. Most manufacturing is done in factories, and the result is a uniform product. **Tertiary activities** are those involved in moving, selling, and trading the goods produced at the first two levels, as well as activities such as professional and financial services. It is also helpful to identify a fourth level: **quaternary activities** specialize in assembling, transmitting, and processing information and controlling other business enterprises. They include professional and intellectual services, such as those provided by management consultants and educators. Quaternary-sector employment is the favoured growth area in the post-industrial city discussed in Chapter 12. The distinction between tertiary and quaternary is not always made, as both involve services; a useful characterization is to note that tertiary activities focus on goods whereas quaternary activities focus on people and information.

Like any classification, this one is not ideal, and the divisions between the four levels are not absolute; but it is still useful. For our purposes, it is convenient to consider all four kinds of economic activity here, with two exceptions: transport-related activities and farming (discussed in Chapters 3 and 10, respectively).

THE INDUSTRIAL LOCATION PROBLEM

Locational questions and theories are central issues for geography, and they are just as important for industry as for agricultural and settlement issues. Once again, theories need to be considered in the context of particular social and economic systems, especially when we realize that industries strive to minimize costs, including labour costs, and maximize profits. Many different types of economic organization play roles in the industrial process. Even the household plays a role in production and reproduction, although that role is less important today than it was in the past; firms, which make up the commercial sector, are a second type of economic organization. Industrial location theory explains why firms locate their factories where they do (Box 13.1). Some firms are owned by one person; others are partnerships, co-operatives, or corporations. Corporations play the most important role in a country's economy.

Early Location Theory

For economists such as Adam Smith, J.S. Mill, and David Ricardo, industrial location was related to the location of agricultural food surpluses that could be used to feed industrial workers. Probably more important was Smith's realization that, under the new doctrine of capitalism, the central consideration in location decisions was purely financial. This incentive remains paramount. In some form, it has been at the heart of most industrial location theory since the writings of Smith. The only major exception to this generalization is Marx, who saw industrial location as one of many issues best explained in terms of political inequalities. The classic industrial location theory, however, is least-cost theory.

Least-Cost Theory

The first attempt to develop a general theory of the location of industry was that of the German economist Alfred Weber (1868–1958), brother of the sociologist Max Weber. Alfred Weber's work, published in 1909 and translated into English in 1929 (Friedrich, 1929), is another example of a theory that does not purport to summarize reality. A normative model, it aims to prescribe where industrial activities ought to be located. Weber follows the von Thünen tradition of setting up a number of simplifying assumptions:

1. Some raw materials are ubiquitous; that is, they are found everywhere. Examples suggested by Weber are water, air, and sand.
2. Most raw materials are localized; that is, they are found only in certain locations. Sources of energy fall into this category.
3. Labour is available only in specific locations; it is not mobile.
4. Markets are fixed locations, not continuous areas.
5. The cost of transporting raw material, energy, or the finished product is a direct function of weight and distance. Thus, the greater the weight or distance, the greater the cost.

Examining the Issues

BOX 13.1 | Factors Related to Industrial Location

Traditionally, a number of factors have been considered in the typical decision process to locate an industry:

- Various measures of *distance* have been very important. Distance from sources of raw material has become less important as a result of improvements in transport and changing industrial circumstances, but it used to be a major consideration because such materials are unevenly distributed and vary in quality, quantity, cost of production, and perishability. Similarly, distance from an energy supply is declining in importance.

  During the nineteenth century, it was essential to locate factories near a source of coal, but since then the emergence of new energy sources, such as electricity, has freed industry from this locational constraint. Distance from market is another factor. Like the other distance variables, it tends to decrease in relevance with improvements in transportation.
- Availability of *labour* also shows spatial variations in quality, quantity, and cost. Although theoretically perfectly mobile, labour is frequently limited to specific locations for social reasons or because of political restrictions on movement.
- *Capital*, in the form of tangible assets (such as machinery) or intangible assets (such as money), can be a key locational consideration; capital may be mobile.
- *Transport* was traditionally a main factor, linked to several of the variables already noted.

Each of these factors plays a role in the location decision, although the extent to which each is significant may vary according to the production process. In some cases, for example, capital might be substituted for labour. When an industrialist has many options, the number of feasible locations usually is increased.

Also important for understanding industrial location are factors related to the nature of the product, as well as various internal and external economies. With regard to internal economies, each manufacturing activity has an optimum size and benefits may be gained from either horizontal integration (enlarging facilities) or vertical integration (involvement in related activities at one location). External economies often result when similar industries are located in close proximity.

Finally, industrial location decisions frequently reflect various human and institutional considerations. The most important human factor is uncertainty. Decision-makers do not have adequate information about all the relevant factors. They may wish to minimize their costs, but they are unlikely to be able to do so. We have already distinguished between optimizing and satisficing behaviour, and this distinction is crucial in the current context. Location decisions, then, often reflect ambiguity, "educated guesses," and chance factors. The central institutional factor is the role played by *government*. Conflicts regularly arise between economic logic and what governments see as socially desirable.

6. Perfect economic competition exists. This means that the industry consists of many buyers and sellers, and no single participant can affect product price.
7. Industrialists are economic operators interested in minimizing costs and maximizing sales.
8. It is assumed that both physical geography (climate and relief) and human geography (cultural and political systems) are uniform.

These assumptions are clearly unrealistic, but they are necessary for a least-cost location format. Given these assumptions, industrialists will locate industries at least-cost locations in response to four general factors: transport and labour (interregional factors) and agglomeration and deglomeration (intraregional factors).

Transport Costs

Weber's first concern was transport cost. To find the point of least transport cost, two factors must be assessed: distance and the weight being transported. According to Weber, the point of least transport cost is the location where the combined weight movements involved in assembling finished products from their sources and distributing finished products to markets are at the minimum.

material index

An index devised by Weber and used in industrial location theory to show the extent to which the least-cost location for a particular industrial firm will be either material- or market-oriented.

isotims

In Weberian least-cost industrial location theory, lines of equal transport costs around material sources and markets.

isodapanes

In Weberian least-cost industrial-location theory, lines of equal additional transport cost drawn around the point of minimum transport cost.

Weber also introduced the **material index**, the weight of localized material inputs divided by the finished product weight. If an industry has a material index greater than one, the point of least transport cost, and hence the appropriate location, will be nearer the localized material sources. If the material index is less than one, the point will be nearer to the market. The historical importance of orientation to material resources has declined because fewer industries use heavy, bulky materials and because transportation has improved.

In general, Weber's predictions are correct. M.J. Webber (1984: 57) gives the example of the soft drink industry, where 1 ounce of syrup, a ubiquitous material (water), and a 4-ounce can combine to produce 12 ounces of soft drink and a 4-ounce can. The material index is (1 + 4)/(12 + 4) = 0.31. Therefore, the soft drink industry is typically market oriented. By contrast, industries such as the smelting of primary metals remain predominantly material oriented, as do many agricultural processing industries, because the cost of materials still forms a substantial share of the total cost.

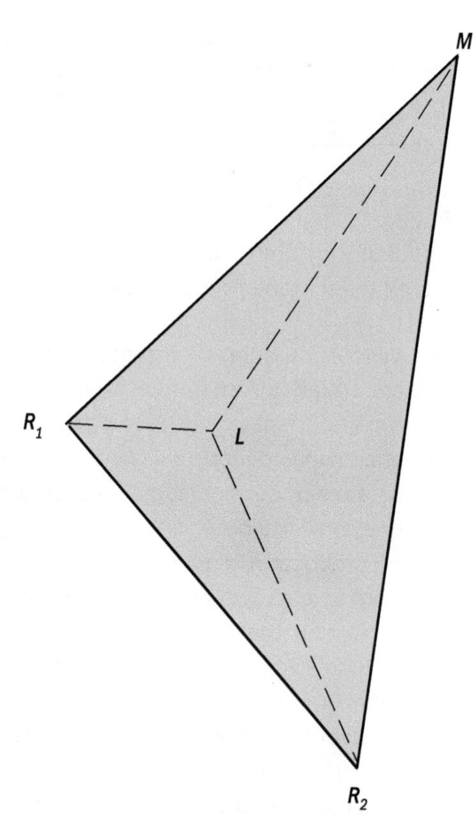

FIGURE 13.1 | A locational triangle
The position of the least-cost location (*L*) will depend on the relative attractiveness of each of R_1, R_2, and *M*.

To help explain location decisions, Weber used a graphic approach. A simple version of this procedure is shown in Figure 13.1. There are two raw material sources (R_1 and R_2) and a market (M), and the industry locates according to the relative attraction of each. If only ubiquities are used, the locational figure is reduced to one point, the market. If only one raw material is used and loses no weight in the production process, the industry can locate at the material source, at the market, or anywhere on a straight line between the two. The locational figure is needed only in those cases where weight-losing materials are used, but this is the most common type of manufacturing. Thus, the locational figure will usually have more than three sides, generating a complex problem. Geographers usually solve such locational problems algebraically rather than by referring to diagrams.

Labour Costs

Weber also acknowledged the importance of labour costs. Such costs represent a first distortion of the basic situations already described and can be approached by mapping the spatial pattern of transport costs and then comparing it with that of the relevant labour costs.

To achieve this comparison, Weber introduced the concepts of **isotims** (lines of equal transport costs around material sources and markets) and **isodapanes** (lines of equal additional transport costs drawn around the point of minimum transport cost). Maps of isotims and isodapanes are examples of cost surfaces. Figure 13.2 shows isotims around two material sources and one market, while Figure 13.3 superimposes isodapanes on the isotims. Weber identified a critical isodapane, outside of which industrial location would not occur—industries can be attracted to low-labour-cost locations only if they lie inside the critical isodapane. A low-labour-cost location will attract industrial activity only if the savings in labour cost exceed the additional transportation costs of moving the raw materials to, and the finished products from, the low-cost labour site.

Agglomeration and Deglomeration

Transport and labour are interregional considerations. Agglomeration and deglomeration are intraregional considerations that, like labour, can potentially cause deviation from the location with the least transport cost. Agglomeration economies result from locating a production facility

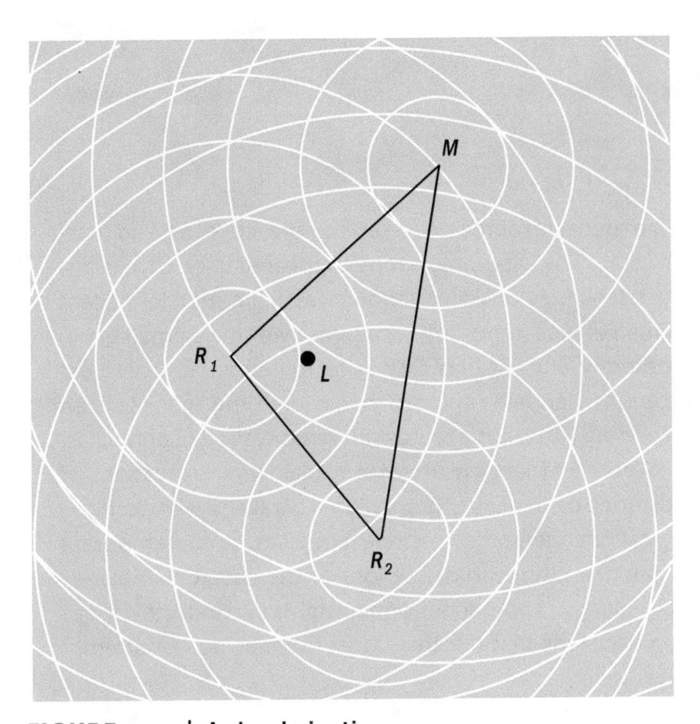

FIGURE 13.2 | A simple isotim map
The diagram shows isotims around the two raw-material locations (R_1 and R_2) and the market (*M*). *L* is the least-cost location.

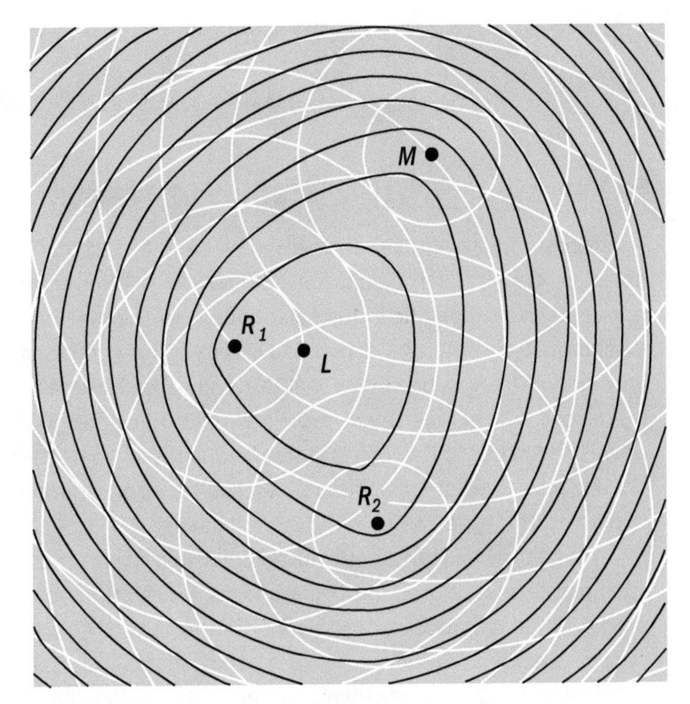

FIGURE 13.3 | An isodapane map
Superimposing isodapanes on an isotim map indicates how far from *L* an industry may locate to take advantage of a particular low-cost labour location.

close to similar industrial plants, allowing plants to share equipment and services, and generating large market areas that aid the circulation of capital, commodities, labour, and information. Deglomeration economies are the reverse of agglomeration economies and result from location away from congested and high-rent areas.

Least-Cost Theory: An Evaluation

Weber's theory is normative and thus does not attempt to describe or summarize reality. The major criticisms are directed at its simplifying assumptions. Markets, for example, are not the simple fixed points that Weber chose to assume. Labour markets are characterized by discontinuities associated with age, gender, ethnicity, and skill. Nevertheless, Weber's basic logic is simple and sound (Box 13.2).

Hoover (1948) continued Weber's focus on transport costs and added substantially to it. The Weberian locational figure changes when transport costs are not directly proportional to distance. The cost per unit of distance is less for long hauls than for short hauls (Figure 13.4), and a step structure is often used (Figure 13.5).

Market-Area Analysis

According to proponents of market-area analysis, Weber's fundamental error is his assumption that industrialists seek the lowest-cost location. For the market-area analyst, industrialists are profit maximizers. Least-cost theory is a form of variable cost analysis (concerned with spatial variations in production costs), while the market-area argument is a form of variable revenue analysis (concerned with spatial variations in revenue).

In market-area analysis, the location that will result in the greatest profit can be determined by identifying production costs at various locations and then taking into account the size of the market area that each location is able to control. The argument is that industries will attempt to monopolize as many consumers as possible—they seek a spatial monopoly and, in doing so, base their location decisions on those of their rivals.

spatial monopoly
The situation in which a single producer sells the entire output of a particular industrial good or service in a given area.

This argument about seeking a spatial monopoly is relevant, but if least-cost theory tends to de-emphasize demand, market-area analysis tends to de-emphasize other factors. The most important market-area theorist is August Lösch, who (like Christaller) determined that the ideal

Examining the Issues

BOX 13.2 | Testing Weberian Theory

Least-cost industrial location theory has been tested many times. A classic example is the application of Weberian concepts to the specific industrial locations of the Mexican steel industry in the 1950s (Kennelly, 1968). The choice of an example dating back over half a century is deliberate; least-cost theory was much more relevant then than it is now. The choice of the iron and steel industry is also deliberate because least-cost theory can still be appropriately applied to this type of heavy industrial activity.

In 1950 the Mexican steel industry consisted of two main steel plants and a number of other plants. Production covered a range of products and accounted for about 50 per cent of Mexican requirements. Location of plants, particularly the major ones, was closely related to the availability of raw material resources, such as iron ore, coke, oil, and market scrap. The location of labour appeared to be of minimal importance, as unskilled labour was quite mobile. Access to market was important, however, with major markets in Monterrey and Mexico City. After determining least-cost locations, Kennelly reported that, overall, the 1950 Mexican steel industry was in accord with basic Weberian principles.

Weberian theory is useful in explaining the locations of industries of this type, but its logic weakens with technological change and is less relevant to most contemporary location decisions than to those of the past. Other (especially Marxist) approaches offer insights into industrial location decisions that the traditional theories do not attempt to consider.

market area is hexagonal. Like least-cost theory, market-area analysis is a normative approach.

Behavioural Approaches

Normative approaches allow us to identify rational or best locations and understand aspects of actual locations through comparison. But is such an approach satisfactory? Many geographers think not; they prefer to focus on explicit analyses of why things are where they are rather than on abstract analyses of where they ought to be, given certain assumptions. Normative approaches consider why things are where they are only in an indirect fashion. A set of approaches known as behavioural attempt to correct this situation.

In brief, behavioural approaches centre on individuals' subjective views. The concept of the

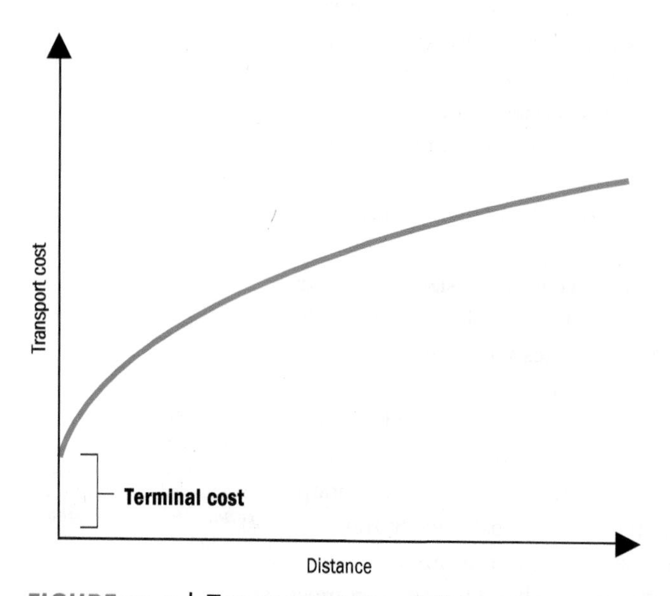

FIGURE 13.4 | Transport cost and distance
Note that the cost does not begin at $0 because there is a basic terminal charge.

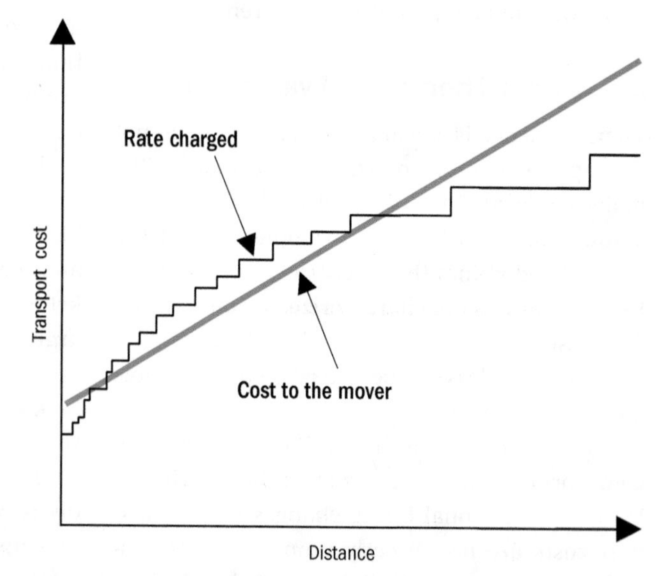

FIGURE 13.5 | Stepped transport costs
The rate charged here does not relate directly to distance because of the rate structures used by the movers.

economic operator is rejected and that of satisficing behaviour is paramount. A simple example demonstrates the relevance of this approach. The beginning industrialist does not really debate where to locate; the selected location is the existing location of the industrialist (or the nearest feasible location). Oxford, England, became the home of Morris, a major automobile manufacturer, for this non-least-cost and non-market-area reason. The location selected was the home area of the industrialist William Morris. Had that home area been a remote Scottish island, some other location would have been selected. Nonetheless, specific locations are frequently chosen for subjective reasons.

Contemporary industrial firms, especially large corporations, are complex decision-making entities. Location decisions made by small, often single-owner, firms are less likely to be economically rational than those made by large firms, but other considerations compete with least-cost and maximum-profit logic for all firms. It is also clear that decision-makers do not typically have access to all relevant information and, furthermore, that such information is subject to various interpretations. Geographers are well aware of the less than perfect mental maps held by decision-makers. Although behavioural issues are not as easily theorized or quantified as the more directly economic issues, they are always factors to be considered in any analysis of industrial location.

THE INDUSTRIAL REVOLUTION

Before the Industrial Revolution, industrial activity was important but limited. Every major civilization was involved in a few basic industrial activities such as bread-making, brick-making, pottery, and cloth manufacture. These activities were located in the principal urban centres and close to raw materials. In addition, each household was an industrial organization that produced clothing, furniture, and shelter. All industries involved minimal capital and equipment, were structured simply, and were small in size and output. Energy sources (such as wood) and raw materials (such as agricultural products and sand) were relatively ubiquitous, not localized. Given these characteristics, industries could be dispersed; thus, there was little evidence of an industrial landscape. Transport systems were limited and markets for products were local rather than regional or national. Everything changed, however, as a result of the developments that we call the Industrial Revolution (Box 13.3).

Examining the Issues

BOX 13.3 | The Period of the Industrial Revolution

Earlier in this book, we referred to the Industrial Revolution in contexts without any direct connection to industry. The reason is that the revolution involved much more than industrial change. These changes, which originated in Europe around the middle of the eighteenth century and continued until roughly the middle of the nineteenth, have collectively contributed much to our contemporary world.

Industrial Revolution is the name given to a series of technical changes that involved the large-scale use of new energy sources, particularly coal, through inanimate converters. After the first successful steam engine was built in 1712, other new machines were able to use steam to improve industrial output. The introduction of new machines led to the construction of factories close to the necessary energy sources, raw materials, or transport routes. The resulting concentrations of factories promoted urban growth, including the emergence of distinct industrial landscapes and working-class residential areas, since workers needed to live near their places of employment. Industrial cities replaced the earlier pre-industrial forms of settlement. As we saw in Chapter 3, new transport links were essential for moving raw materials to the factories and finished products to the markets; in England they initially took the form of improved roads, then canals and, most critically, railways. Agriculture became increasingly mechanized, and there was a demand for new raw materials,

Continued

especially wool and cotton, from overseas areas. Overall, the percentage of the total labour force engaged in agriculture declined rapidly.

Technical changes were accompanied by significant changes in demographic characteristics. As described in the demographic transition model (see Chapter 5), the first phase of the Industrial Revolution involved a dramatic reduction in death rates, followed by reductions in birth rates. The interval between these two drops was a period of rapid population growth: the world population was estimated at 500 million in 1650 and 1.6 billion by 1900. Industrialization and the associated population increases contributed to new areas of high population density, rural-to-urban migration, rapid urban growth, and considerable European movement overseas to the new colonial territories.

A major social and economic change was the collapse of feudal societies and the rise of capitalism. As labour was transformed into a commodity to be sold, the producer became separated from the means of production. Politically, this period witnessed the rise of nationalism and the emergence of the modern nation-state, as well as the expansion of several countries around the globe to create empires. Although these empires were short-lived, they contributed enormously to the contemporary political map and to the division between more and less developed worlds. The nation-state, as we have already seen, remains one of the most potent forces in the contemporary world. The rise of modernism was linked to the Industrial Revolution through a common emphasis on the practicality and desirability of scientific knowledge. The transition to an industrial way of life was also a shift from tradition to modernity, from *Gemeinschaft* to **Gesellschaft**. Even though new post-industrial forces have emerged since about 1970, our world is unmistakably a product of the Industrial Revolution.

Gesellschaft
A term introduced by Tönnies; a form of human association based on rationality and depersonalization; assumed to be characteristic of urban dwellers.

There were also important advances in the organization of academic knowledge during this period; the social sciences are related to these various changes. Geography was one of these disciplines. Major contributions to the advancement of knowledge included the works of Marx and Darwin.

Origins

Between about the mid-eighteenth and mid-nineteenth centuries, the industrial geography of England was dramatically transformed. In fact, England was the first industrial area. Elsewhere in Western Europe, the transformation did not begin until the nineteenth century, but it was accomplished over shorter periods of time.

Why was England, and not China or Germany, first? Why did it occur in the eighteenth century, not earlier or later? A conventional answer is that the Industrial Revolution was triggered by the rapid onset of new and stable political, legal, and economic institutions throughout much of seventeenth-century Europe—including the rule of law, secure property rights, and social mobility—with England the first country to build on these changes. An alternative elitist answer, proposed by Clark (2007), is that England had a particularly long history of settlement, stability, and security and that this history created the necessary cultural preconditions for industrialization. Clark contends that these preconditions are hard work, rationality, frugality, and education. A form of natural selection took place as rich and successful people—who worked hard and were rational, frugal, and well educated—propagated themselves, and the poor died out. Such an idea, while certainly contentious, might be described as a pattern of differential fertility that favours the rich. According to this argument, then, the greater prevalence of these traits by about 1700 made the Industrial Revolution possible in England.

In general, the term *Industrial Revolution* is apt, for it was a period of rapid and cumulative change unlike anything that had occurred previously: machines replaced many hands, and inanimate energy sources were efficiently harnessed. However, the term may be misleading to the extent that it suggests that the process of industrialization took the same form in every case. In truth, the process was uneven, and other countries did not simply imitate the English example. Belgium followed it most closely but emphasized the metallurgical industries far more than the textile industries.

The most important component was the rise of large-scale factory production. Developed out of earlier small establishments and domestic household activities, factories were both much larger and more mechanized; they also required more capital. Because they relied on new, localized energy sources, they tended to agglomerate. The key energy source was coal, and the first truly industrial landscapes arose in the coalfield areas of England. But these developments were only one part of the Industrial Revolution.

The late eighteenth century witnessed the rise of a new economic system that emphasized individual success and profits, namely, capitalism. The new capitalists seized the opportunities provided by a whole series of technological developments in the metals and clothing industries: a furnace that burned coal to smelt iron (first used in 1709); a spinning jenny for multiple-thread spinning (1767); a steam engine used as an energy source (1769); and an energy loom for weaving (1785). These are only a few examples of the new inventions, most of which were developed in England.

Recall our definition and discussion of energy as the capacity to do work, and the more successfully we can use forms of energy other than our own, the more successfully we can achieve our work goals (see Chapter 4). Also, recall our definition of technology as the ability to convert energy into forms that are useful to us. The fact that the first agglomerations developed on the coalfields of northern and central England and in the traditional areas of cloth manufacture in northern England reflected the importance of coal for energy and the focus on the clothing industries. These industrial agglomerations caused rapid increases in urban populations as workers moved closer to places of employment.

The basic logic of the Weberian least-cost theory is helpful in understanding these developments. Using localized energy sources meant that industries were located at those sources and, if possible, at the sources of raw materials. The need to reduce costs meant that the Industrial Revolution was paralleled by a transport revolution.

To clarify some of these generalizations, let us take a closer look at examples of nineteenth-century industrialization, especially the early British textile and iron and steel industries.

Early Industrial Geography

Analyses of these early industrial landscapes include traditional descriptive accounts based on Weberian theory and other theoretical accounts that acknowledge the role played by economic power and organizations.

Textiles

Before the Industrial Revolution, the textile industries were specialized and spatially concentrated in three areas: southwest, east-central, and northern England. The manufacturing process had remained unchanged from the fourteenth century to the late eighteenth century, when various inventions that were part of the larger Industrial Revolution prompted radical changes. The textile industry's principal technological breakthroughs occurred in villages and small towns in northern England, as mechanized factories were constructed near suitable sources of water. Of the two major textile industries, cotton was the first to experience change since most of the new procedures were more easily applied to cotton than to wool. In both industries, the main changes followed the same pattern: a decline in home production (the "cottage industry") and a rise in factory production; use of steam engines; rapid expansion in northern England, reflecting the availability of coal; and associated declines in the other two areas.

In the cotton industry, raw material came initially from the East Indies, but supplies were irregular and a new source had to be found. The Caribbean and southern United States were suitable source areas, but the task of separating cotton fibres from seeds was extremely labour-intensive—even with a slave economy, supplies were limited. The invention of the cotton gin in 1793 allowed American plantations to flourish and English factories to be adequately supplied. The cotton mills of northern England became so dependent on US imports that mill workers were laid off when supplies were reduced during the American Civil War. In the case of wool, the new area of supply was southeastern Australia.

Iron and Steel

Prior to the Industrial Revolution, Britain had two principal centres of iron production—one in central England and one in southern England. Both used local ores and relied on wood (charcoal) as their energy source. The iron industry expanded rapidly after coal was substituted for charcoal from 1709 onward and an efficient steam engine was developed. As iron production increased, new uses for it were found: the first cast-iron bridge was built in 1779 in central England, and most new industrial machinery was constructed of iron.

By the early nineteenth century, new iron areas dominated as technological advances continued to favour coalfield locations such as south Wales and north-central England. An important new factor emerged after 1825, when the first steam-powered train was operated in northeast England. The railway boom reached its height in the 1840s and gradually allowed some movement away from coalfields. Railways, together with the discovery of new iron ore sources, allowed a new industrial centre to develop in northeast England. Other major mid-nineteenth-century iron and steel areas were located in the English Midlands, south Wales, and central Scotland.

Industrial Landscapes

The Industrial Revolution and related rapid population and urban growth combined to create new landscapes. Factory towns arose, especially in northern England, and migrants from the south and rural areas poured into the new employment centres. In environmental and social terms, the results were often disastrous.

Before the onset of industrial change, the landscape was agricultural and predominantly rural. Even before the introduction of steam energy, however, some industrial production was shifting from the household to factories and the landscape was becoming more congested. Probably the first real factory in Britain was a silk mill in north-central England that was completed in 1722; five to six storeys high, it employed 300 people and used water energy. This example shows that coal was not essential for industrialization, but it certainly accelerated the process of change. The factory was followed by many others as the century progressed, but they produced neither smoke nor dirt. Steam energy resulted in major landscape change. More factories meant more workers crowded into small areas, and the use of coal quickly polluted the landscape. Land was expensive, and the areas of small, terraced houses where workers lived quickly deteriorated into slums. These industrial cities were subject to heavy smoke pollution—a situation that, in many locations, was not addressed until the 1960s.

Diffusion of Industrialization

After about 1825 the technological advances developed in Britain diffused rapidly to mainland Europe (particularly Belgium, Germany, and France), to North America, and to Russia. Britain did not welcome the spread of its industrial innovations and unsuccessfully tried to keep some advances secret. In the United States, Pennsylvania and Ohio became the early industrial leaders because of their high-quality coal sources, while in the Russian Empire, Ukraine became the driving force when coal was discovered there. Japan began to industrialize after establishing cultural contact with the United States and rejecting feudalism in 1854.

In Europe the Ruhr region offered not only coal and small, local iron ore deposits but also an excellent location for water transport. As the nineteenth century progressed, the demand for labour became so great that immigrants moved to the Ruhr from elsewhere in Europe, giving the region an unusually diverse ethnic character.

FOSSIL FUEL SOURCES OF ENERGY

It is difficult to exaggerate the importance of energy as an international resource system, an economic consideration affecting the activities of the largest companies, and a political consideration affecting the foreign policy of many governments. The more successfully we can utilize energy sources, the more easily we can fulfill our needs and wants. We become aware of energy sources and acquire the ability to use them through the development of new technology. To put energy use in context, consider that human "progress" has always been tied to increasing energy use, from the use of fire to agricultural domestication and then to the Industrial Revolution and the use of fossil fuels. Since the onset of the Industrial Revolution most energy has come from three fossil fuels—first coal and later oil and natural gas—and this discussion focuses on these sources. (Because both production and consumption of fossil fuels can have negative impacts on environment, Chapter 4 includes a discussion of alternative renewable sources.)

Coal was the leading source of energy from the beginning of the Industrial Revolution until it was replaced by oil in the 1960s. By the early 1970s oil provided almost 50 per cent of the world's energy. Since then, its use has declined to 32.9 per cent because of a combination of high prices, occasionally erratic supplies, more efficient use, and increasing reliance on other energy sources.

However, oil continues to be the chief source of industrial energy, and the geographies of oil reserves, production, movement, and consumption are important aspects of contemporary life. The other fossil fuel energy sources are coal, which contributes 30.1 per cent of the world's energy, and natural gas, contributing 23.7 per cent. Thus, fossil fuels contribute 87 per cent of energy consumption worldwide. The remaining 13 per cent is made up of various renewables, mainly hydroelectricity and nuclear energy.

Predicting future energy use is fraught with difficulty, but one estimate is that global consumption of the three fossil fuels—oil, coal, and natural gas—will be about 27 per cent each by 2030, meaning that renewables will increase to 19 per cent (British Petroleum, 2011). Both energy production and consumption are spatially variable. Countries in the more developed world generally use more energy per capita than do countries in the less developed world. China, with its large population and growing industrial economy, consumes 21 per cent of the global total. It surpassed the US as the world's largest energy user in 2010.

CP/Andrew Vaughan

The oil rig *Rowan Gorilla III* is towed into Halifax harbour from the Cohasset oil field for a scheduled refit. Nova Scotia's three producing oil fields—Cohasset, Panuke, and Balmoral—are about 250 kilometres southeast of Halifax.

Oil

The global geography of oil production, movement, and consumption is a fundamental component of the larger global economy and is also intimately tied up with global politics. There are ongoing worries about price, remaining reserves, political stability of major producing states (as this affects supply), activities of transnational oil companies, corruption in producing countries, and a host of environmental issues relating to production, transportation, and use.

With reference to oil, Pasqualetti (2011: 972) writes: "Its discovery, development, and sale have for about 150 years brought wealth to a few, convenience to some, and avarice to many." Oil is rarely out of the news for the simple reason that much of the world runs on this fuel: it is a global commodity that is in high demand and the price at any given time has a significant impact on the global economy. As is typically the case with commodities, oil prices are determined by supply and demand considerations but with the added complexity that the price is established through binding market agreements to purchase oil at a predefined price on a predefined date in the future (i.e. futures contracts).

The most important player identified in any account of global energy is the Organization of Petroleum Exporting Countries (OPEC), a cartel founded in 1960 by Iran, Iraq, Kuwait, Saudi Arabia, and Venezuela and later joined by Algeria, Gabon, Indonesia, Nigeria, Qatar, and the United Arab Emirates (UAE). As the major supplier of oil, OPEC's principal purpose is to set levels of production in order to affect the price of oil. Therefore, the organization is able to wield enormous influence on the economies of more developed countries—a price increase in 1973, for example, was a leading cause of a widespread recession. In the first years of the twenty-first century, OPEC worked to keep the price of oil around US$22–$28 per barrel, but a series of economic uncertainties, political events, and natural disasters combined to increase the price dramatically to more than $150 by 2008, after which prices fell as part of the larger global recession that began in that year. In 2014–15 the price of oil fell about 40 per cent because of increased supplies from non-OPEC countries, notably the United States. At present, much uncertainty exists concerning whether OPEC will be able to retain the power to control oil prices.

In 1960 the major sources of oil were the United States, Russia, and Venezuela; only about 15 per cent came from the Middle East. Today, this area dominates oil production, with 32.2

Around the Globe

BOX 13.4 | Oil Reserves and Oil Production

The importance of oil for individual countries and within the global economy can hardly be overestimated. Oil drives the global economy and is central to the economies of many countries, including Canada. It is hugely important for countries such as Kazakhstan, which ranks eighteenth in the world in production (1.8 million barrels per day) and twelfth in known reserves (30.0 billion barrels having been discovered). These reserves (and those in Azerbaijan) are more accessible since the 2005 completion of a pipeline linking Baku on the Caspian Sea with the Turkish Mediterranean port of Ceyhan. For Europe, this pipeline is a strategically significant source of oil outside the Middle East and Russia.

In other countries, oil is a potential lifeline for survival and a cause of bitter conflict. Nigeria is the largest African producer, at 2.3 million barrels a day, and has the eleventh largest proven reserves, at 37.1 billion barrels, in the world; yet the World Bank estimates that 80 per cent of oil wealth is held by just 1 per cent of the population. This gross inequality, as well as the environmental damage caused by production, has contributed to ethnic tensions and violence. The Ogoni people of the Niger Delta, agriculturalists and fishers whose lands and livelihoods have largely been destroyed by the oil exploration and extraction—led by Shell Oil and the national government—organized in the early 1990s to protest their losses. In 1995 Ken Saro-Wiwa (the leading Ogoni activist, environmentalist, and noted author) and eight other leaders of the resistance movement were executed by the state. Sudan/South Sudan is not a major oil producer, ranking thirty-first in the world, but oil is of great importance for these impoverished countries. Physical geography, ethnicity, and religion have been at play in these two areas in Africa. Much of their oil reserves are located in the predominantly Christian southern parts of the countries, while Muslim majorities, with more political power, are located to the north. Thus, the human geography of oil is a complex story.

Table 13.2 lists the top 12 countries in oil production and oil reserves. Of special note is the extent to which reserves and production do not mesh. Venezuela, for example, has the largest proven reserves (298.3 billion barrels) and is eleventh in production, at least partly because much of these reserves are in oil sands, where production is more costly and environmentally damaging. Three of the top oil producers—Russia, the US, and China—are major economic powers but rank considerably lower in proven reserves, and Mexico (which, like Canada, works to feed the American appetite for oil) ranks tenth in oil production but stands at eighteenth in known reserves. On the other hand, Libya has the ninth highest total proven reserves but ranks twentieth in production; Libyan production has declined notably since the Arab Spring and the continuing political turmoil.

TABLE 13.2 | Top 12 countries in oil production and oil reserves, 2013

Country	Production (Million Barrels/Day)	Country	Proven Reserves (Billion Barrels)
Saudi Arabia	11.5	Venezuela	298.3
Russian Federation	10.8	Saudi Arabia	265.9
United States	10.0	Canada	174.3
China	4.2	Iran	157.0
Canada	3.9	Iraq	150.0
United Arab Emirates	3.7	Kuwait	101.5
Iran	3.6	United Arab Emirates	97.8
Kuwait	3.1	Russian Federation	93.0
Iraq	3.1	Libya	48.5
Mexico	2.9	United States	44.2
Venezuela	2.6	Nigeria	37.1
Nigeria	2.3	Kazakhstan	30.0

Note: China has 18.1 billion barrels of proven reserves, and Mexico has 11.1 billion. Libya produces 1.0 million barrels, and Kazakhstan produces 1.8 million barrels.
Source: Adapted from BP *Statistical Review of World Energy 2014*.

per cent of the global total. As detailed in Box 13.4, a few countries tend to dominate world production.

As shown in Table 13.1, the United States is the world's leading consumer of oil, with about 21.9 per cent of global consumption, followed by China and Japan. In general, most of the principal consuming countries are likely to become increasingly dependent on imports. In short, there is a growing gap between the places where oil is produced and where it is consumed. For this fundamental geographic reason, there is much movement of oil from producing to consuming countries. Figure 13.6 shows the major patterns of trade in oil.

TABLE 13.1 | Principal oil-consuming countries, 2013

Country	Share of Total (%)
United States	19.9
China (including Hong Kong)	12.1
Japan	5.0
India	4.2
Russian Federation	3.7
Saudi Arabia	3.2
Brazil	3.2
Germany	2.7
South Korea	2.6
Canada	2.5
Iran	2.2
Mexico	2.1
France	1.9
Indonesia	1.8
United Kingdom	1.7

Source: BP *Statistical Review of World Energy 2014*.

The Future of Oil

One fundamental question is: What will happen when the oil runs out? Note that the data in Box 13.4 cannot answer this question. Proven oil reserves in the Middle East and North Africa increased by more than 80 per cent since 1973, and other reserves continue to be discovered. Such new discoveries and, eventually, exploitation could see a significant upward bump in the coming years as the Arctic Ocean becomes more accessible, a consequence of climate change and ice melt. Nevertheless, our current dependence on oil cannot be sustained. Notwithstanding the recent fall in price, the ever-increasing demand both in the more developed world and in industrializing countries (especially China) will likely

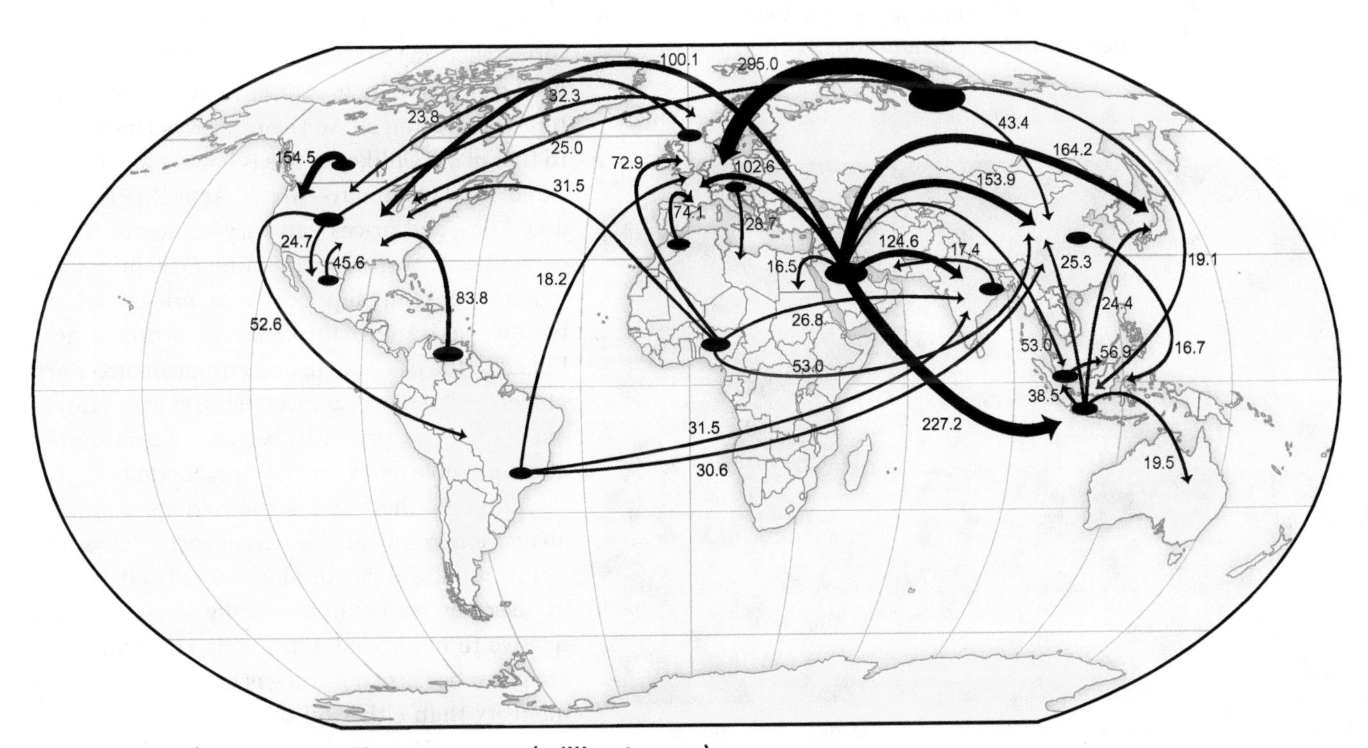

FIGURE 13.6 | Major oil trade movements (million tonnes), 2014

Source: Updated from BP *Statistical Review of World Energy 2011*. www.bp.com/content/dam/bp-country/de_de/PDFs/brochures/statistical_review_of_world_energy_full_report_2011.pdf.

force prices to rise, as occurred prior to the 2008 recession, and the supply will eventually be unable to meet demand in an economically feasible way. Nobody knows when this will occur because, along with new reserves, new technologies are being developed to make other known reserves, such as the Alberta oil sands, profitable to exploit (some of the environmental impacts of this activity are discussed in Chapter 4). Despite the uncertainties, the question is pressing. As already noted, the United States is the world's largest importer of oil, and about one-third of its imports come from OPEC countries. As a result of shale oil production, however, the United States will likely become the world's largest oil producer by 2020, decreasing its reliance on OPEC countries.

A few small countries are anticipating significant wealth from oil in the foreseeable future. The 169,000 people living on the islands of São Tomé and Principe off the coast of West Africa live in poverty, but the government reached an agreement with Nigeria in 2005 to explore for oil in a jointly owned maritime area. Substantial reserves have been discovered, and it is possible that the country will not follow the route of some other African countries where oil revenues have not benefited the majority of people. However, this determination is complicated by the fact that power is constitutionally divided between a prime minister and a president. East Timor, which became independent in 2002, is often described as the world's poorest country, but it is also anticipating income from oil following a 2006 agreement with Australia to share revenues from oil and gas in the Timor Sea. More generally, a 2006 agreement between Exxon Mobil and Indonesia's state oil firm, Pertamina, has led to the development of the large Cepu oil field on Java and is likely to encourage other big oil companies to invest more in oil and gas throughout Southeast Asia.

As noted in Box 13.4, the populations of some countries with large oil reserves have not benefited significantly from this resource. Nigeria's coastal producing areas are severely damaged both environmentally and socially—the Niger Delta is home to warring groups instead of marine life. Oil-producing areas of the Amazon basin in Ecuador are polluted and impoverished. While these are exceptionally negative cases, much evidence suggests that oil states are commonly dysfunctional, at least partly because a reliance on oil seems to discourage countries from investing in other areas of the economy.

CP/AP Photo/Bruce Stanley

The Tin Fouye Tabankort gas processing plant is the high-tech showpiece of Algeria's natural gas industry.

Natural Gas

When gas was first found in the context of oil extraction, it was not considered useful because of the difficulties involved in transporting it to market, and it was either burned or vented into the atmosphere. Such processes continue in those oil production areas that lack local markets or the infrastructure to transport the fuel. Gas was first used commercially in the 1960s as an alternative to home heating oil, so the price was closely tied to that of oil. Unlike oil, gas is bought and sold in three markets today—North America, Europe, and Asia—and prices can vary markedly. In the deregulated North American market, prices are much lower than oil prices; Asian prices are high because gas is traded through long-term contracts linked to the price of oil; and European prices are moderate. Like oil, however, natural gas is rarely out of the news, primarily because of controversy about possible new sources of production.

Although there are some serious concerns about environmental consequences of some forms of natural gas production, it is often seen by users as an environmentally attractive fuel relative to oil or coal. Carbon dioxide emissions are lower because natural gas has a lower carbon intensity than either oil or coal and thus burns more cleanly. It is used particularly in the electric power and industrial sectors.

Reserves and Production

Most of the natural gas deposits known and exploited to date are those found in association with oil, with the gas either mixed with the oil or floating on top. Like oil, these conventional sources are located using seismic methods and are easily tapped by standard vertical wells.

Table 13.3 lists the seven countries with the largest known reserves of natural gas. With the exception of Turkmenistan, these countries are also important sources of oil; other countries with more than 2 per cent of global production are Venezuela (3.0 per cent), Nigeria (2.7 per cent), and Algeria (2.4 per cent) (Canada has 1.1 per cent). Table 13.4 lists the 11 countries with the highest production levels; no other country has more than 2 per cent of production.

Despite several decades of extraction, the known reserves of natural gas have been rising in many countries since the 1990s. This increase is a result of new technologies allowing for the development of unconventional resources, particularly natural gas in shale formations. Two new technologies are those of horizontal well drilling and hydraulic fracturing (often called fracking). Extracting gas from shale is not simple: the gas is deep underground and requires a vertical well and then a horizontal one that passes through the shale deposit, permitting access to more of the shale gas reservoir. Hydraulic fracturing technology involves injecting very large volumes of water mixed with sand and fluid chemicals into the well at high pressure to fracture the rock and release the gas. This process is controversial, especially outside the United States, which has fewer environmental/geological requirements. In most other countries with shale gas deposits, there are concerns about groundwater pollution, methane leakage, and seismic activity. Extracting shale gas also requires many wells to be drilled, and bringing in the needed water is a substantial transportation project. In densely populated countries many people will live close to the wells and be affected.

In the United States shale provided 40 per cent of the extracted natural gas in 2013, a dramatic increase from 5 per cent in 2007. There are significant shale gas resources in several of the other major consuming countries, including China and some European countries. Poland has substantial deposits and extraction is beginning. Both France and South Africa have deposits but are assessing environmental implications of hydraulic fracturing. In the United Kingdom a huge field has been discovered in northwest England. There are also prospects for exploiting other unconventional sources, such as tight gas that is found in sandstones, coal bed methane, and methane hydrates.

TABLE 13.3 | Natural gas: Proven reserves, 2013

Country	Proven Reserves (% of Global Total)
Iran	18.2
Russian Federation	16.8
Qatar	13.3
Turkmenistan	9.4
United States	5.0
Saudi Arabia	4.4
United Arab Emirates	3.3

Source: *BP Statistical Review of World Energy 2014*.

TABLE 13.4 | Natural gas production, 2013

Country	Production (% of Global Total)
United States	20.6
Russian Federation	17.9
Iran	4.9
Qatar	4.7
Canada	4.6
China	3.5
Norway	3.2
Saudi Arabia	3.0
Algeria	2.3
Indonesia	2.0
Netherlands	2.0

Source: *BP Statistical Review of World Energy 2014*.

Movement and Consumption

Global consumption of natural gas is increasing more rapidly than is consumption of either oil or coal. Table 13.5 shows current consumption data.

The ongoing global increases in consumption are related to improved distribution of gas. There are two main ways of moving this fuel: directly through pipelines or transportation in liquid form (LNG). In the United States an extensive network of pipelines transports natural gas from production

TABLE 13.5 | Natural gas consumption, 2013

Country	Consumption (% of Global Total)
United States	22.2
Russian Federation	12.3
Iran	4.8
China	4.8
Japan	3.5
Canada	3.1
Saudi Arabia	3.1
Germany	2.5
Mexico	2.5

Source: *BP Statistical Review of World Energy 2014*.

areas to consuming areas, but because production is less than consumption, gas is also imported through pipelines from Canada and as LNG from overseas.

LNG trade has increased dramatically in recent years and will continue to rise as more LNG terminals are built. More countries have LNG terminals, and China is constructing six new ones that will double the amount of LNG that can be imported. But pipeline flows remain crucial in North America and Europe, where much gas from Russia is transported by pipeline to other European countries.

Coal

Coal continues to be a major source of energy. Table 13.6 ranks countries by proven reserves and includes data on production and consumption. Nine countries account for about 90 per cent of proven reserves; six of them account for more than 80 per cent of production. The major producers generally are also the major consumers, and China's contribution to both production and consumption is especially striking.

TABLE 13.6 | Coal: Proven reserves, production, and consumption, 2013

Country	Proven Reserves (% of Global Total)	Production (% of Global Total)	Consumption (% of Global Total)
United States	26.6	12.9	11.9
Russia	17.6	4.3	2.4
China	12.8	47.4	50.3
Australia	8.6	6.9	1.2
India	6.8	5.9	8.5
Germany	4.5	1.1	2.1
Ukraine	3.8	1.2	1.1
Kazakhstan	3.8	1.5	0.9
South Africa	3.4	3.7	2.3

Source: *BP Statistical Review of World Energy 2014*.

Reserves and Production

Coal is a major global resource. Depending on the geology of the deposit, coal is removed either by surface opencast (strip) mining or by underground mining. Globally, underground mining is the more important, accounting for about 60 per cent of production, although surface mining dominates in both Australia and the United States (Box 13.5). There are several types of coal, depending on how far along the deposit is in the process of coal formation, or coalification. The first stage is peat, then brown coal or lignite, then sub-bituminous and bituminous coals, and finally anthracite. The further along in the process, the higher the rank of the coal. Different types have different uses. Lower-ranking coals are used mostly to generate power and in various industries, while higher-ranked coals are used for these purposes and also for iron and steel and as smokeless fuel.

Perhaps the major factors affecting coal production are the need to consider safety and environmental issues, particularly in more developed countries that have relatively stringent requirements. Extracting coal (especially underground coal) is a hazardous activity, and loss of life is all too common. Historically, coal production has been a leading cause of environmental damage. Major concerns relate to the release of methane, the presence of various waste products, the effects on groundwater, and impacts on visible landscape.

Movement and Consumption

Coal is both bulky and heavy, meaning movement adds significantly to the price. This factor was key in Weber's theory of industrial location, discussed earlier. There are two principal regional coal markets, Western Europe (primarily the United Kingdom, Germany, and Spain) and East Asia (mainly Japan and South Korea). Australia is a major supplier of the Asian market.

The mode of transportation used to move coal depends on the distance to be travelled. Short distance movements are usually by truck, train, or barge, while ships are used for overseas movement.

Examining the Issues

BOX 13.5 | The World's Biggest Coalmine

The North Antelope Rochelle Mine is a surface mine located in the Powder River basin of Wyoming, a state that has overtaken both West Virginia and Kentucky as the biggest producer of coal in the United States. Originally two separate mines—the North Antelope Mine opened in 1983 and the Rochelle Mine in 1985—the two combined in 1999. Operated by Peabody Energy, the mine is often described as the biggest in the world; it covers about 260 sq km (100 sq miles) and has vast and easily accessible deposits of low sulphur coal. "No other coal seam on the planet is so big, so close to the surface, and so cheap to mine" (Graasch and The Daily Climate, 2013).

With about 2.4 billion tonnes of reserves and a current annual production of more than 100 million tonnes, the mine will not be depleted for approximately another 25–30 years. (Peabody is looking to acquire additional land in Wyoming that contains an estimated one billion tonnes of coal.) About 21 long freight trains leave the mine each day, carrying coal to power plants across the country; about 40 per cent of US electricity comes from coal, most of which is from Powder River. The mine also exports some of its coal to China and India, where demand is high. Peabody has three large mines in Australia that also satisfy some of this Asian demand.

Peabody received an Excellence in Surface Mining Award from the Wyoming Department of Environmental Quality in 2011 and an Excellence in Surface Coal Mining Reclamation honour from the U.S. Department of the Interior in 2012. But the North Antelope Rochelle Mine, and coal production more generally, has many critics. Local ranchers argue that there are health impacts for both sheep and cattle, aquifers are draining away, and grazing land is lost. Environmental groups bemoan coal burning's impact on global warming. As of 2015, there is increasing pressure from environmental groups, with several lawsuits filed against the Environmental Protection Agency. Not surprisingly, officials at Peabody routinely deny that any possible global warming is caused by humans (Goldberg, 2014).

The industry has many different suppliers; hence, the market for coal is competitive. Because much coal is consumed in the country of production (Table 13.6), the international trade in coal is relatively limited, amounting to only about 16 per cent of all coal consumption.

It is not only production of coal that has major negative environmental impacts but also consumption. Chinese consumption of coal is the single greatest source of carbon dioxide emissions.

Energy Supplies in Britain

Looking at the experience of Britain is instructive, as it was the home of the Industrial Revolution. The revolution relied heavily on coal found in several British locations, especially in southern Wales, northern England, and western Scotland. As these domestic coal sources were depleted, imports became necessary, but they proved very expensive given the significant price increase for transport. As early as 1945, Britain turned to oil, with the first imports coming from the Middle East (where the British had colonial interests). By the 1960s cost and politically related supply uncertainties prompted a search for oil and gas in the North Sea. Gas was discovered in 1965 and oil in 1969, and both were soon determined to be available in commercial quantities. By 1980 production was equal to domestic demand, with 36 oilfields and 25 gas fields developed. These supply areas are now being depleted.

Most recently, the discovery of substantial shale gas fields offers the opportunity for substantial domestic energy production, but this potential might not be fulfilled if the environmental impacts of hydraulic fracturing are judged to be too damaging to justify extraction. So what does the future hold? The answer is likely to depend on the extent to which declines in the North Sea supply might be countered by extraction of gas from shale.

WORLD INDUSTRIAL GEOGRAPHY

The global manufacturing system is dominated by three regions: North America, that is, the United States and Canada; Europe, particularly

Germany, Italy, the United Kingdom, and western Russia; and Pacific Asia, especially Japan, South Korea, and eastern China. These three global regions, identified in more general terms in Chapter 3 (especially Figure 3.10), control the export and import of manufactured goods. Many industries are part of this global manufacturing system through trading activities. Other industries are transnational, owning or controlling production in more than one country. Transnational corporations and other large firms tend to be less committed to specific industrial locations than small firms because of the spatial separation of production from organization (Dicken, 1992).

Regarding individual countries, the United States continues to be the world's leading producer of manufactured goods by far, followed by China, Japan, and Germany. Industrial activity in all these countries except China is framed within a capitalist social and economic system. Thus, decisions that affect the geography of industry and employment opportunities are made by industrial firms rather than governments. While small firms must cope with larger economic changes, large firms often have the power to influence their economic environment. Nevertheless, all firms must take international trends into account. In some cases, they may find their home markets threatened by imports; in others, they may rely heavily on export markets.

Industrial firms are still concerned with costs and profits; consequently, the basic logic of least-cost theory and market-area analysis, as previously discussed, continues to be relevant in all countries. While there is no doubt that many location (and other industrial) decisions are not optimal, this may not be crucial in the long term. Therefore, firms in capitalist economies continue to make location decisions on the basis of the products, production technology, labour costs, sources of raw material and energy, capital availability, markets, land costs, and, in some instances, environmental restrictions—all considerations implicitly or explicitly contained within the traditional theories. That said, making location decisions on the basis of two major industrial inputs, energy and raw materials, involves much uncertainty. As the discussion of energy stressed, new areas of production and even different types of product frequently become available and these necessarily affect the cost of this input. This is also the case with some major raw material inputs, as suggested by Box 13.6.

In the News

BOX 13.6 | Mongolia—A New Resource Frontier

The present country of Mongolia does not comprise all the traditional homeland of the ethnic Mongolian people, and there are more ethnic Mongolians living in the Inner Mongolia Autonomous Region in China than in Mongolia itself. Located between China and Russia, most of Mongolia is grassland, semi-desert, desert, or grassy steppe, with mountains in the west and southwest. With just 2.9 million people, the population density is only 2 per sq km. The CDR is low at 6 while the relatively high CBR of 28 looks likely to decline as the economy grows.

And grow it will. Mongolia is experiencing the first stages of an anticipated massive resource boom and is undergoing a dramatic economic and social transformation. Since the early 1990s this former Soviet satellite has gradually made the transition from communism to democracy, but political change was not accompanied by economic growth. In the 1990s poverty was prevalent and rationing of basic items was necessary.

However, the country has significant deposits of coal, gold, silver, copper, fluorspar, tungsten, and uranium. The global resources industry has been aware of these huge mineral deposits for some years, but exploration and production are only now commencing on a large scale as investors judge that the political climate is favourable following the 2009 election of a pro-business government. The key move that attracted investors was the decision to remove a windfall tax on copper and gold profits. Soon after that decision was announced, Rio Tinto and Ivanhoe Mines (a Canadian-listed company) agreed to develop the US$5 billion Oyu Tolgoi deposits. Located in the south Gobi region, these are the world's largest undeveloped copper and gold deposits. The IMF has estimated that this project might account for 30 per cent of the Mongolian economy by 2021. Another large project is the Tevan Tolgoi coal mine, also in the south Gobi region, which is being developed by both local and foreign firms. Most of the coal mined is transported 600 km to the important railway hub of Baotou in China and then moved on to major steel producing areas.

Other megaprojects look destined to follow these examples, potentially tripling the national economy by 2020. But economic growth is not the whole story. It seems possible that, in this semi-arid country, mining will consume large quantities of precious water, damage valuable grassland, and affect the migration routes of animals. Much depends on what environmental policies are implemented. Also important for local people is whether some of the income from mining is used for education, health, and public housing. That more and more mines will open seems certain. What is less assured is the impact on environment and on the everyday lives of local people.

Industrial Regions in More Developed Countries

As Figure 13.7 shows, there are five principal centres of industrial activity in the world and a large number of secondary centres. Four of the five principal centres are in the more developed world—eastern North America, Western Europe, western Russia and Ukraine, and Japan. The fifth region is in China, a country of the less developed world. Notably, the Pearl River Delta in southern China (discussed later) is the world's most dynamic industrial region. Together, these five regions account for over 80 per cent of global industrial production.

Eastern North America

This part of the world is a highly productive industrial region. Following European settlement, which began in the seventeenth century, it developed an industrial geography based on close ties to Europe, the proximity of raw materials such as coal, iron ore, and limestone, availability of labour, rapid local urbanization and market growth, easy movement of materials and finished goods along the natural waterways, and the building of canals and railways. There are numerous local industrial areas within this region, including southern Ontario, southern Quebec, southern New England, the Mohawk Valley, the southern Lake Erie shore, and the western Great Lakes. Box 13.7 outlines the industrial geography of Canada.

Western Europe

This region is also highly productive and, like the North American region, consists of numerous local industrial areas, the most important of which are central and northern Britain, the Ruhr and mid-Rhine valleys in Germany, and northern Italy. Although Britain had the initial advantage of the earliest industrial start, by the mid-twentieth century it was having difficulty keeping up with more recent industrial developments. Much the same is true of major European coalfield areas. However, the Ruhr Valley, which contains the most important coalfield in Europe, developed a large iron and steel industry and has been much more successful in coping with industrial changes. Rapid diversification in northern Italy after World War II transformed a textile region into a textile, engineering, chemical, and iron and steel region using local gas and imported oil.

Western Russia and Ukraine

Until the dramatic political upheavals of the late 1980s and early 1990s, industrial location and production in the former USSR and the Eastern European countries were determined by central planning agencies. Recent changes have not significantly altered the spatial pattern of industrial activities. In the former USSR there are five major industrial areas, four of which are now in Russia.

Those industrial centres based in Moscow and Ukraine were established in the nineteenth century. The Moscow area is market oriented and began its industrial development specializing in textiles—today it is diversified, with a wide variety of metal and chemical industries using oil and gas. Industry in Ukraine is centred on a coalfield but also has access to supplies of iron, manganese, salt, and gas; Ukraine is a major iron and steel and chemical industrial area. The other three regions were developed by

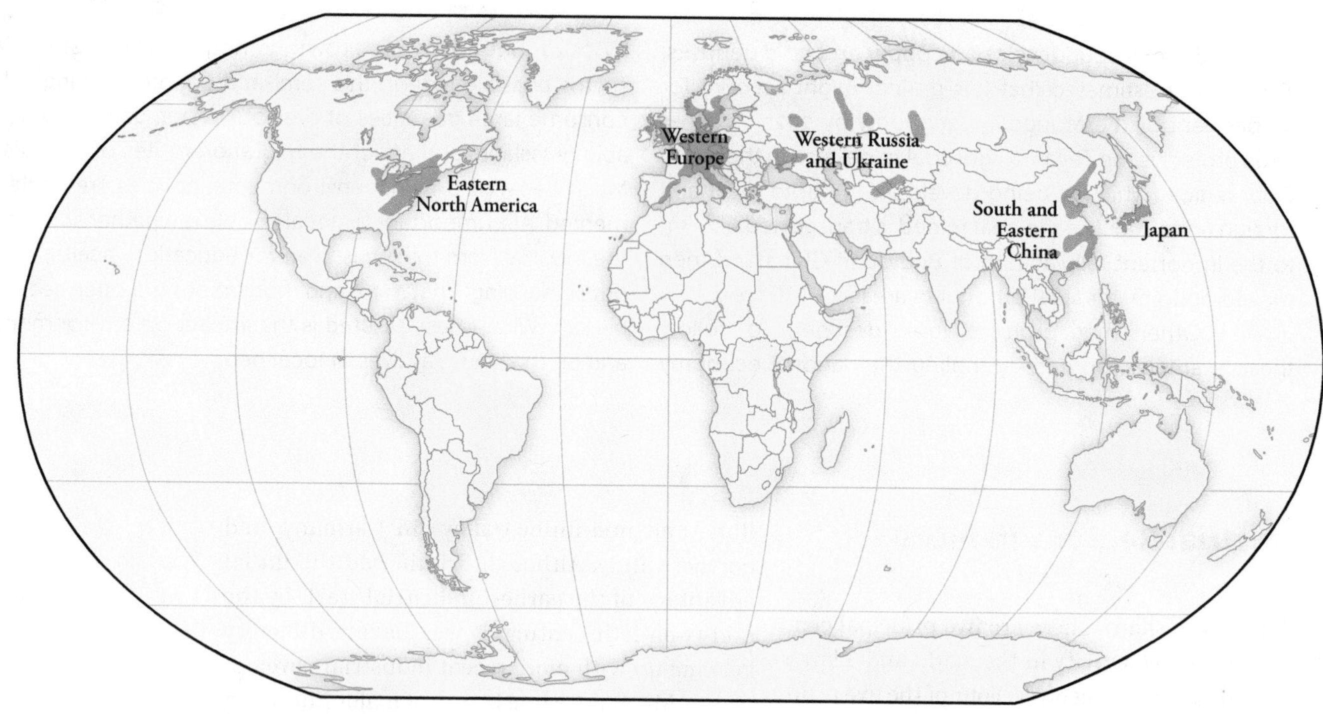

FIGURE 13.7 | Major world industrial regions

the USSR government after the 1917 revolution and are located in southern Russia. The Volga area to the east of Ukraine has local oil and gas sources and was a relatively secure location during World War II. Its major industries are machinery, chemicals, and food processing. Farther east, the Urals area is both a source of many raw materials, especially minerals, and a major producer of iron and steel and chemicals. Industrial growth in the Urals was promoted by the former USSR government because of its great distance from the western frontier. Farthest east is the Kuznetsk area, which is similarly endowed with minerals as well as a major coalfield.

Industry in Japan

Japan consists of four large islands and several thousand smaller islands stretching over an area roughly 2,500 km (1,500 miles) from north to south. Although cultural contact with Europe and North America did not begin until after 1854, with the end of feudalism, by 1939 Japan had expanded territorially and developed into an industrial power comparable to those in the West. Industrialization continued after World War II despite the loss of Korea and considerable bomb damage. During the 1950s and 1960s Japan excelled in heavy industry, particularly shipbuilding, but by the late 1960s the emphasis had shifted to automobiles and electronic products, and Japan has most recently focused on computers and biotechnology. During each of these three post–World War II phases, Japan has been a world industrial power.

Japan has had remarkable success despite the fact that it is relatively small in size and, because it has few industrial raw materials of its own, is forced to import almost everything it needs. Further, it is also a substantial distance from major world markets. But Japan has overcome these obstacles by making use of its large population.

The principal factors behind these extraordinary industrial achievements are low labour costs, high levels of productivity, an emphasis on technical education, minimal defence expenditures, aid from the United States (motivated by the perception that Japan served as a bulwark against Chinese communism), and a distinctive industrial structure in which small specialized firms are linked to corporate giants such as Nissan and Sony. This structure facilitates rapid acceptance of various new technologies. Another aspect of this success story is Japanese business's considerable investment overseas. Many Japanese companies operate in North America and Europe

Around the Globe

BOX 13.7 | Industry in Canada

Plentiful natural resources mean that Canada has a well-developed primary (resource extraction) industrial sector. Its immense geographic extent and (consequently) dispersed national market pose a challenge to many firms because of high transportation costs. The primary sector is scattered throughout Canada, while the secondary sector is concentrated in the Great Lakes–St Lawrence Lowlands in southern Ontario and southern Quebec. Proximity to the United States has resulted in a manufacturing sector that includes many branch plants of American parent companies.

Following the initial in-movement of Europeans, a series of staple economies developed: fish, fur, timber, wheat, and minerals. By the mid-nineteenth century Canada was a major exporter of primary (resource) products and an importer of secondary (manufactured) products. Manufacturing in Canada at that time was mostly limited to the processing of agricultural and other primary products for both domestic and export markets. Following Confederation in 1867, the Canadian government pursued two policies with direct relevance to the infant industrial geography: subsidizing railway building and introducing tariff protection for manufactured products. The immediate beneficiaries were the established areas of the Great Lakes–St Lawrence Lowlands, and a factory system was soon established. Montreal and Toronto in particular became industrial centres.

The capital needed to develop manufacturing industry initially came from Britain, but as this investment dropped off, a branch-plant economy controlled by Americans developed. American firms were thus able to bypass the protective tariffs in place, while Canada benefited by the introduction of financial and other capital and technology. After World War II, the US became less globally competitive (especially compared to Japan), the branch-plant economy became less and less attractive for industrialists from the US as Canada became increasingly nationalistic.

The Canadian manufacturing industry is highly regionalized, and many geographers see the country in terms of heartland and hinterland, or core and periphery. The hinterland produces the resources on which the manufacturing industries of the heartland depend. Toronto and Montreal have fabrication economies, while all other areas have resource-transforming economies. The two types of economy face different sets of problems. The industrial economies of Toronto and Montreal have suffered because of imports, mainly from Japan and other Asian countries. The economy of the rest of Canada, which is largely dependent on raw materials, is subject to fluctuations in world price and demand over which Canadians have no control. At the national level, because Canada has many resources, economies facing problems are usually buffered by the successes of other sectors. At the local level, however, many areas are subject to booms and busts. Single-resource towns, of which there are many, are the most vulnerable (Britton, 1996).

especially, and Japanese banks play an important role in international finance.

Since the late 1990s Japan has shown signs of a substantial economic downturn with a declining growth rate, increasing numbers of bankruptcies, corporate indebtedness, and rising unemployment. A major concern is competition for production, with the loss of many manufacturing jobs to China (where production costs are lower) and competition for high-technology production from South Korea.

Newly Industrializing Countries

Japan is no longer alone. Its story has encouraged many imitators in other Asian countries. During the 1970s the economies of South Korea, Taiwan, Hong Kong, and Singapore all accelerated rapidly, and those of Malaysia, Thailand, Indonesia, and the Philippines followed. These eight countries are the leaders among what became known as the newly industrializing countries, or NICs (others include Brazil, Mexico, Greece, Spain, Portugal, and, most recently, Vietnam). It is notable, however, that several of these economies suffered during the financial crisis of 1997–8 and that all shared in the global recession that began in 2008 and resulted in a shrinking of the world economy in 2009.

Industries in the Pacific Rim countries follow the Japanese example of low labour costs and high productivity. All these countries have high growth rates and are experiencing the same labour shifts

© Brett Gundlock/Corbis

A man works on a Volkswagen in the company's Puebla, Mexico, factory. This plant is one of the biggest in North America; vehicles manufactured here are shipped all over the world.

from agriculture to industry that much of Europe experienced in the nineteenth century. The most successful of the NICs is South Korea. Until 1950 South Korea was a poor, less developed country characterized by subsistence rice production; at present, it is an industrial giant that began with heavy industry, then focused on automobiles, and now specializes in high-technology products. Its experience is the Japanese transformation repeated over a period of less than 40 years. Current evidence suggests that these trends will continue. One motivation for regional free trade agreements in recent years and for continuing European integration has been awareness of the need to respond to the dramatic industrial growth in much of Asia.

To attract transnational corporations, several NICs—including South Korea, Singapore, Taiwan, Hong Kong, China, the Philippines, and Mexico—and some other less developed countries have set up export-processing zones (EPZs), manufacturing areas that export both raw materials and finished products (Figure 13.8). There are more than 3,500 such zones in about 130 countries, employing more than 66 million people, about 40 million of whom are in China. Industries are attracted to these zones for three general reasons: inexpensive land, buildings, energy, water, and transport; a range of financial concessions in such areas as import and export duties; and low workplace health and safety standards and an inexpensive labour force made up largely of young women, who are regarded as less likely than men to be disruptive and more willing to accept low wages and difficult working conditions. There are few advantages other than waged employment for the processing country, and even this advantage varies depending on larger international economic circumstances. The parallel between these export-processing zones and the agricultural plantations discussed in Chapter 10 is compelling.

There are close links between these export-processing zones and high-technology companies. In Mexico, EPZs known as maquiladoras are set up within easy reach of the high-tech suppliers in such areas as the Santa Clara Valley in California (Silicon Valley) and the Dallas–Fort Worth area in Texas (Silicon Prairie). The high-tech materials, such as silicon chips, are manufactured in the United States and transported to the maquiladoras along the US–Mexico border for the simple but labour-intensive process of assembly; then the finished product is transported to the US for sale.

Industry in China

It is sometimes too easy for people in the Western world to forget that China has historically been a major player in the world economy. It is estimated that it produced about 25 per cent of total world output about 2,000 years ago and about 33 per cent as recently as 1820. These numbers highlight the fact that China's poor economic performance for much of the past 200 years, as it failed to follow the European and North American examples, has been atypical. In 1950, for example, China was producing only 5 per cent of total world output.

But China is making up for lost time and growing rapidly in what might be called a second "industrial revolution." After several decades of rapid growth, the country is now best described as the workshop of the world. The port of Qingdao, for example, did not exist in 2004 but is one of the biggest container ports in the world (see Chapter 3 for a brief account of recent developments in the Chinese transport network). Have you ever wondered where the majority, about 60 per cent, of ornaments, artificial snow, Santa hats, and other Christmas paraphernalia come from? The answer is factories in Yiwu, a city located about 300 km south of Shanghai. These same factories also make novelty items for Easter and Valentine's Day. More generally, it is widely expected that China will soon become the world's leading manufacturing country, although specific predictions are very difficult in light of larger global economic changes.

Liberalization of China's communist economy began in 1978 with the opening up of the country to international trade and foreign investment. Since the 1980s both the agricultural and industrial economies have undergone reform and modernization. These important changes were initiated during the 18-year leadership of Deng Xiaoping, who effectively determined that economic growth was more important than continued class struggle.

During the 1950s industrial development was based on a combination of large technology-intensive, state-funded factories and small labour-intensive, locally organized units (Morrish, 1994, 1997). The large state-funded units made up about 100,000 of the total of about 8 million units and produced 55 per cent of the national industrial output. Despite this high percentage, the strategy proved unsuccessful because the state-owned

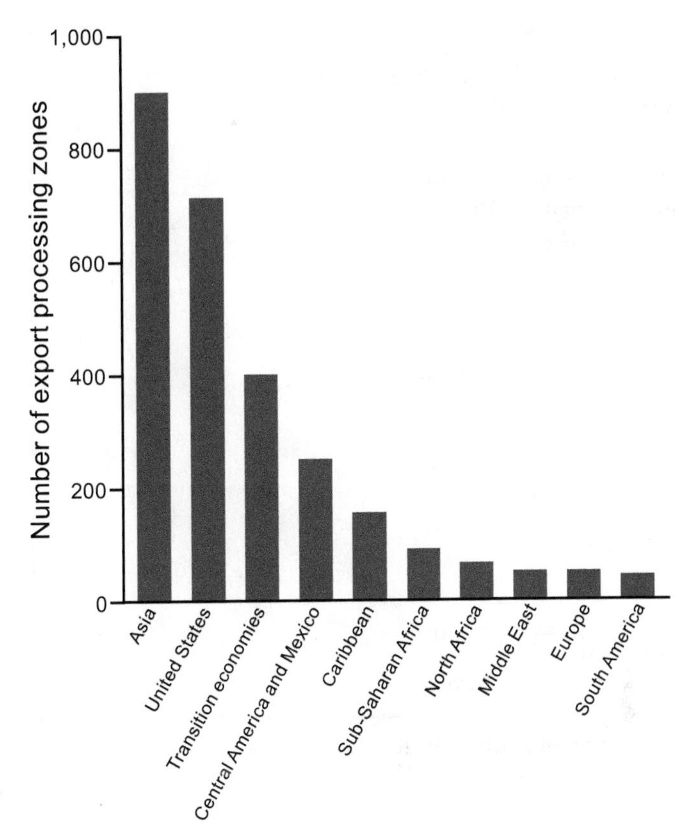

FIGURE 13.8 | Export-processing zones

Source: Adapted from Boyenge, J.-P. S. 2007. *ILO Database on Export Processing Zones (Revised)*. Geneva: International Labour Office, Sectoral Activities Programme, Working Paper, 2.

factories were expensive to operate and highly inefficient. Shortages of consumer goods were commonplace.

A new regional development policy was introduced in 1980 with the establishment of several special economic zones (similar to EPZs) that enjoy low taxes and other financial concessions to attract foreign investment. A reform program was initiated in 1984 that involved less state control, decreased subsidies, and increased response to market forces and that encouraged local collective and private industrial enterprises. One result was that China's industrial production outstripped that of South Korea, increasing at 12 per cent per annum between 1980 and 1990. In 1994 the state relaxed its control even further and, to encourage collective and private enterprise, introduced a new program requiring urban industrial enterprises to respond to market forces directly, without state involvement.

The special economic zones have played a major role in the recent expansion of Chinese industrial activity (Figure 13.9). Four zones were established

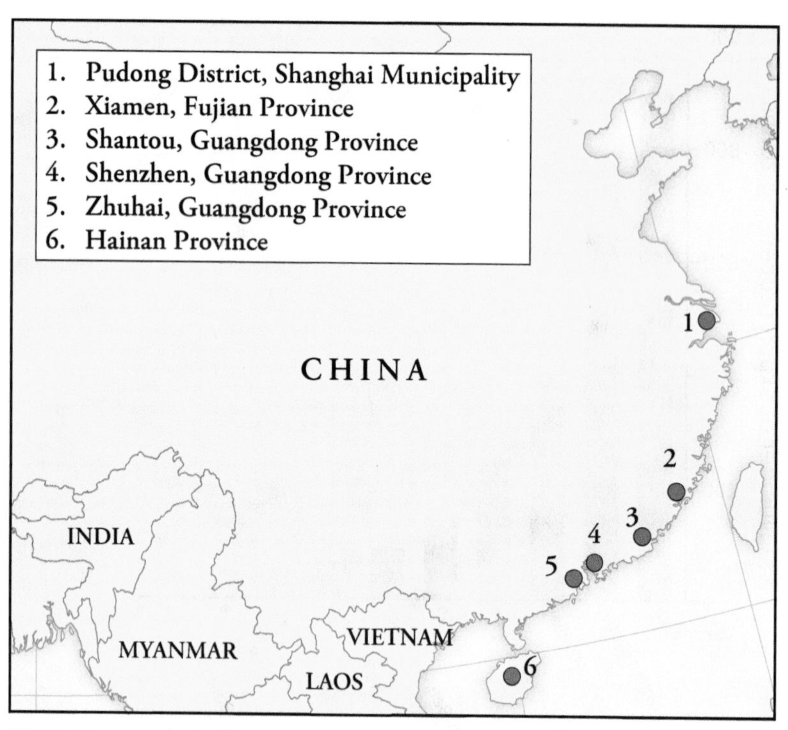

FIGURE 13.9 | Special economic zones in China

in 1980 at Shenzhen, Zhuhai, Guangzhou (all located in Guangdong province north of Hong Kong), and Xiamen (in neighbouring Fujian province). These zones have grown dramatically with the introduction of "labour-intensive factories that manufacture everything from computer keyboards and dishwashers to leather coats and Mighty Morphin Power Rangers" (Edwards, 1997: 12). Most of the foreign investment has come from Hong Kong, attracted by cheap labour and land. Together, the three zones north of Hong Kong, all located in the Pearl River Delta, form what many consider the world's most dynamic industrial region, sufficiently large that it is affecting global trading patterns and investment flows. This delta area attracts about one-fourth of China's foreign direct investment and generates about one-third of its exports. In a sense, it is the early twenty-first-century equivalent of mid-nineteenth-century industrial northern England.

In 1984, 14 locations were designated "open coastal cities," free to trade outside China, and an

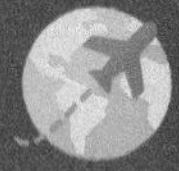

Around the Globe

BOX 13.8 | China: From Struggling Peasant Economy to Industrial Giant

Napoleon famously stated: "Let China sleep, for when she awakes, she will shake the world." China has awoken. Since 1978 its GDP has increased by an average of just under 10 per cent each year, about three times more than the increase for the United States (excepting in recent years because of the 2008 global recession).

China's economy is enormous, with the second largest GDP in the world. Only the United States has a larger GDP, but China is likely to move into first place before 2025. In addition to dominating production, China is increasingly important as a trading nation and as a consumer.

The world's biggest net recipient of foreign investment, China manufactures 60 per cent of the world's bicycles, 40 per cent of the world's socks, and 50 per cent of the world's shoes. It is also by far the largest garment exporter, with almost 50 per cent of the global total. Wages are usually low: garment workers are paid about US40 cents an hour—less than a third of what their counterparts in Mexico receive. China is the favoured location for transnational firms to contract out manufacturing employment. But this situation is changing. Inflation is high and wages are rising. One response is for Chinese manufacturers to move their factories further inland, where workers accept lower wages.

In recent decades the United States has been the only significant mass market in the world, but that is changing with the rise of China and the emergence of a new and huge mass market. China is more than just a supplier of cheap products for the rich countries of the world; it is also a rival purchaser of resources and products. Until recently, the country was self-sufficient in most primary products, but it is now the most important player in global commodity markets. It is the world's largest consumer of grain, meat, coal, and steel. China is even importing rice because so much land is being taken over for housing, factories, and shopping malls. There are several reasons for rising oil prices, but Chinese demand is certainly a significant part of the explanation. The EU, the United States, and Japan are China's three most important trading partners.

China is also increasingly a major global consumer: with about one-fifth of the world's population, it consumes half of its cement, one-third of its steel, and one-quarter of its aluminum. This rise of a new consuming China creates a problem for the United States, not only because many jobs are lost to this growing economy, but also because American influence weakens as Chinese trade with other countries increases. It seems clear that Chinese influence around the world generally will increase, not least in the context of trade, investment, and other economic activities.

China is heavily involved in other countries that are able to supply needed raw materials, especially Australia, DR of Congo, Angola, Sudan, South Sudan, Mongolia, and Burma. Consider, for example, that Chinese firms are currently building or renewing transport infrastructure in DR of Congo of about US$12 billion in exchange for rights to copper mining. Chinese investment in the oil industry in Sudan (prior to 2011) allowed that country to ignore Western concerns and sanctions about the conflict in the Darfur region. At present, about 10 per cent of China's oil imports are from Sudan and South Sudan. The rise of China also has impacts on the Canadian economy, principally because of the related commodities boom and resulting surge in investment, but also because China is increasing trade with and investment in Canada. By mid-2012 Chinese oil sands investments over the previous two years had topped $13 billion; in July of that year, the China National Offshore Oil Corporation (CNOOC) made a takeover bid of $15.1 billion for Nexen, a Canadian company with extensive oil sands assets. This bid was approved in 2013.

Chinese industrial growth has major consequences for the environment. Most of the power plants burn coal, which causes acid rain and pollution with impacts on agriculture and health. In many parts of China there is already insufficient water for both agriculture and industry. The clearest examples of regional development policies involve the planned economies in socialist and communist states. By definition, these states do not rely on the capitalistic dynamics of private ownership and entrepreneurship. Rather, they depend on the rationalization of industry and the implementation of comprehensive planning. The majority of such states have been in the less developed world. In China regional policies have included efforts to industrialize large cities (as during the Great Leap Forward of 1958–9, when perhaps 20 million rural residents moved into cities) and to limit the growth of large cities. The seventh five-year plan in China (1986–90), which focused on uneven development and rapid urbanization, was a direct reflection of the fact that earlier plans had largely failed to address two issues: income differences between urban and rural areas and the domination of industry by urban areas located on the country's coast. The constants in Chinese planning are state direction, collective ownership of the means of production, and control over population movements—all features rare in capitalist countries. The fact that Chinese policies have prompted some significant spatial inequalities in economic development and wealth is not a concern. Deng Xiaoping, the political leader of China for more than two decades before his death in 1997, once remarked that it was perfectly in order for some parts of China to become rich before other parts; in fact, this was the way it had to be (Freeberne, 1993: 420).

China is indeed a major global producer and consumer and is playing an increasingly important role in the global economy with each passing year. All this growth is occurring in a country that continues to deny basic democratic rights to its citizens, a fact that must not be forgotten.

additional three larger open zones were created one year later. Another special economic zone, Hainan Island, was established in 1988. By the early 1990s it was obvious that the coastal economy was growing too quickly, and the Chinese government began to focus attention on some inland areas. As discussed in Box 13.8, China's growth has been so dramatic not only in production but also in consumption that it is a cause of some concern to other nations and producers.

Industry in India

The example of India highlights many of the general points noted earlier. Following independence in 1947, the country was a producer of agricultural products, but today it has a much more diversified industrial structure. Industrial landscapes have been created, and the government has attempted to reduce regional disparities. Since 1951 India has used a series of five-year plans to guide development. Initially, the emphasis was on heavy industry, but later plans have stressed self-reliance and social justice. Given that its industrial transformation did not begin until 1951, India has achieved remarkable success; the main reasons are its substantial market, available resources, adequate labour, and relatively sound government planning. India is also experiencing the benefits of outsourcing as both North American and European countries relocate employment. Of note is the movement of call centres for many major businesses, with the United Kingdom losing about

33,000 jobs by the end of 2003. A well-known feature of the Indian economy is the vibrant software and information technology industry, especially in Bangalore, yet this sector accounts for only 3 per cent of India's GDP.

By 2014 India ranked tenth in the world in terms of GDP. Indian industrial employment and output will likely continue to grow in the immediate future; many observers expect it to emulate China as it also has a very large population. But this must be some years in the future because India's GDP is about half that of China, while its exports of goods and services are much less. Also, there is some recent evidence of slowing economic growth and movement of economic activity elsewhere. Most notably, because of various political squabbles and bureaucratic issues, Coal India (a state-owned mining monopoly) is planning to spend billions on mines outside India, notwithstanding the abundance of reserves at home.

GLOBALIZATION AND INDUSTRIAL GEOGRAPHIES

Globalization processes play a significant role in contemporary industrial geography, affecting location, organization, and activity. Although these processes have already been discussed, specifically in Chapter 3, it is important to review two essential issues related to industrial change in both the more and the less developed worlds. Industrial restructuring amounts to a global shift in industrial investment and activity. This process was well underway by the 1980s, along with the other classic component of globalization—a decline in the friction of distance that is a consequence of new communication technologies. Together these two processes have produced the new and dynamic industrial geography. Related to the two globalization processes is the transition from Fordism to post-Fordism.

Fordism to Post-Fordism

To understand the current post-Fordist phase of industrial activity in the more developed world, it is important to understand the Fordism that preceded it. The term *Fordism* refers to the methods of mass production first introduced by Henry Ford in 1920s America, particularly fragmentation of production—best illustrated by the use of assembly lines that reduced labour time and related costs. In sharp contrast to most nineteenth-century industrial activity, the Fordist system gave workers the necessary income and leisure time to become consumers of the many new mass-produced goods. Economic policies based on the theories of the British economist John Maynard Keynes were widely implemented, resulting in rising living standards throughout the 1950s and 1960s. These two decades of sustained economic growth witnessed the rise of the first transnationals, among them Ford, based in the United States (automobiles), Nestlé, based in Switzerland (foodstuffs), and Imperial Chemicals, based in the United Kingdom (chemicals). These early transnationals were limited in their ability to invest outside their home countries, but they were able to take advantage of technological advances in transportation.

This economic boom slowed down in the early 1970s for two principal reasons. First was the termination of the provision in the Bretton Woods system (see Chapter 3) that required the United States to convert overseas holdings of US dollars to gold at a fixed rate. When the US ended this practice because of the expenses incurred in the Cold War and the resulting budget deficit, other currencies fluctuated, resulting in price fluctuations and some business losses. The second reason was the 1973 OPEC decision to raise the price of oil—a commodity essential to industrial production.

One important outcome of the recession of the early 1970s was industrial restructuring in the more developed world. Basically, this restructuring involved deindustrialization, especially in such traditional industrial activities as textiles and shipbuilding and in automobile manufacturing, and corresponding reindustrialization as firms established branch plants overseas, either in NICs or in less developed countries.

Industrial Restructuring

The transition to post-Fordism entails major changes in industrial geographies, resulting essentially from technological advances and globalization processes. Three technological changes are most significant:

1. Production technologies, such as electronically controlled assembly lines and automated tools, are increasing the separability and flexibility of the production process.

2. Transaction technologies, such as computer-based, just-in-time inventory control systems, also increase locational and organizational flexibility.
3. Circulation technologies, such as satellites and fibre optic networks, facilitate the exchange of information and increase market size.

Together, these three changes represent a transition to flexible accumulation, making it easier for companies to take advantage of spatial variations in land and labour costs and to serve larger markets. The subsequent industrial restructuring takes three principal forms:

1. The relationship between corporate capital and labour is changing as machines replace people, manufacturing industry declines, and transnationals seek locations with low labour costs (as noted in the account of export-processing zones).
2. Both the state and the public sector are playing new roles with the shift away from collective consumption in areas such as education and health care to joint public–private projects and deregulation.
3. There is a new division of labour at various spatial scales as the new technologies allow corporations to respond rapidly to variations in labour costs (as in the account of export-processing zones, as well as discussions in Chapter 12).

Deindustrialization and reindustrialization are spatial trends caused by economic restructuring, specifically shifts in investment between the manufacturing and service sectors or, to put it another way, shifts between production and consumption. Both can be seen as components of the transition to a post-industrial society. A simple way to conceptualize industrial change is to recognize four stages:

1. Infancy: Initial primary activities and domestic manufacturing.
2. Growth: The beginnings of a factory system.
3. Maturity: Full-scale development of manufacturing and related infrastructure.
4. Old age: Decline and inappropriate industrial activity resulting in a depressed economic region.

Deindustrialization is a reduction in manufacturing in more developed countries that is usually most easily measured by reference to employment data. It is most evident in the older industrial areas and in such industries as iron and steel, textiles, and shipbuilding. Many deindustrialized areas are depressed as a result of outdated industrial infrastructures. It is not so common in high-tech industries such as pharmaceuticals and electronics. The major social consequence is unacceptably high unemployment.

Deindustrialization is usually seen as a negative development. Because the decline takes place in areas where industry used to be highly concentrated, it initiates a larger economic and social decline. The fact that it tends to occur rapidly, often during periods of economic recession, makes regional and local adjustment all the more difficult. Box 13.9 outlines the particular circumstances evident in the UK.

Reindustrialization at least partially counters industrial decline. This process may take various forms. First, there is an increasing tendency for small and/or new firms to be more competitive. This is most likely to occur outside the traditional industrial areas and reflects the information exchange made possible by electronic means. Second, high-tech industrial activities, especially microelectronics, are expanding rapidly in output, if not also in employment. Such industries are locating in environmentally attractive areas where skilled workers choose to live. A third aspect of reindustrialization is the expanding service industry. Rising incomes and changing lifestyles are bringing significant growth in tourism and recreation industries, with accompanying impacts on environmentally attractive areas. Other service industries, particularly banking and information services, are also expanding in major urban centres.

In the more developed world, the new industrial landscape is quite different from the nineteenth-century version. Not only is the landscape visually different with the decline of smoke-producing heavy industry, but both the work experience and the location of the landscape have also changed. The need for coherent regional policies seems obvious.

Information Technologies and Location

As the preceding discussion implies, decision-making by industrial firms, concerning location specifically, often differs markedly from the

flexible accumulation
Industrial technologies, labour practices, relations between firms, and consumption patterns that are increasingly flexible.

collective consumption
The use of services produced and managed on a collective basis.

deindustrialization
Loss of manufacturing activity and related employment in a traditional manufacturing region in the more developed world.

reindustrialization
The development of new industrial activity in a region that has earlier experienced substantial loss of traditional industrial activity.

Around the Globe

BOX 13.9 | The Deindustrialization Revolution in the United Kingdom

The role of the UK as the home of the Industrial Revolution was detailed earlier in this chapter. Major industrial areas developed in locations where coal was readily available and transport by water and/or rail was feasible. In the mid-twentieth century the UK was an industrial giant, with manufacturing producing about 33 per cent of the national GDP and employing 40 per cent of the workforce. Today, manufacturing accounts for only about 10 per cent of GDP and employs only 8 per cent of the workforce. In its place, service-sector activities have become increasingly important. The basic explanation for this dramatic transformation is that globalization processes, including the rise of transnationals and improvements in communication technologies, involved much of the manufacturing in the more developed world being transferred to less developed countries—especially China. This same comment should apply to other established industrial countries, such as Germany and France, but they have not experienced comparable levels of deindustrialization. Why has the UK suffered more than other countries?

Although there is no simple answer to this question, we can identify two factors specific to the UK. First, there was a sense of complacency in the 1950s. Both Germany and Japan were defeated powers, and the UK continued to control a major world empire that provided raw materials and opportunities for export of finished products. Major UK industries such as automobile, railway, aircraft, and shipbuilding felt no need to modernize plants or improve labour practices.

Second, more so than most other established industrial countries, the UK has experienced erratic and largely ineffective government policies. While it was widely acknowledged that manufacturing was in crisis and that a transition to services was needed, the policies introduced to facilitate this shift typically focused on short-term issues rather than long-term goals. Although some companies were nationalized, the more popular strategy was to expect companies to compete effectively in the globalizing world. Most did not. Some companies have been sold to foreign firms, as was the case with leading chocolate maker Cadbury, sold to Kraft in 2010. Iconic manufacturers such as Dorman Long (the iron and steel company that built Sydney Harbour Bridge), Swan Hunter (shipbuilders), and automobile companies such as Rover have disappeared or are of much decreased significance. Not so in France, where Assault (aircraft) and Peugeot and Renault (automobiles) continue to thrive. Similarly, German firms such as Mercedes (automobiles) and Myles (domestic appliances) remain household names.

The typical UK deindustrialized area is located on a coalfield and was overly reliant on heavy industry. Formerly heavy industrial areas have been designated as special development areas, development areas, and intermediate areas—designations that change according to political circumstances and that have had little real impact. Unemployment is high and out-migration is normal. Government intervention now takes the form of development controls, financial incentives, and creation of industrial estates.

model proposed by Weber. Increasing emphasis on technology and decreasing emphasis on materials and localized energy sources mean that, for many industries, transport cost is no longer the main criterion. Many contemporary firms are trading off between two types of cost: the labour and land costs considered by Weber and a whole set of new costs associated with exchanging information between firms.

A major debate in studies of industrial activity, especially location decisions, concerns the implications of labour (and, to a lesser extent, land) costs on the one hand and the costs of information exchange on the other. For many industrial activities, information can be rapidly exchanged at low cost by electronic means, but there is a danger in such impersonal exchanges if the information is lacking in clarity. In principle, firms that make location decisions on the assumption that they will be able to exchange information successfully are able to seek out locations with low labour costs. The result is a decentralized (deliberated) industrial pattern. A prime example is the successful expansion of many high-tech Japanese firms to other countries. In situations where firms lack confidence in their ability to exchange unambiguous information, they will prefer to locate in close proximity (to agglomerate) in order to facilitate personal, face-to-face exchange of information.

Acknowledging the contemporary relevance of information exchange introduces two possible patterns of industrial location. Decentralization occurs if long-distance electronic information exchange is feasible, but centralization occurs if the information is lacking in clarity and requires personal contact to be effective. Decentralization is typically associated with firms (frequently transnationals) that mass-produce a standard product and thus do not need to worry about ambiguities in the exchange of information, while centralization occurs when functionally related firms need to exchange information on a person-to-person basis because their products are not identical. Much evidence suggests that firms concentrating spatially achieve a competitive advantage. In the case of the automotive tool, die, and mould industry in southwestern Ontario, a collection of small, specialized firms co-operate closely and have access to a pool of skilled labour, thriving on social networks and on flows of knowledge related to local labour mobility (Holmes, Rutherford, and Fitzgibbon, 2005). Another example is the mining supply and service industry concentrated in Sudbury, which similarly depends on a pool of skilled labour (Robinson, 2005).

Service Industries

It is generally assumed that, as an economy progresses, it undergoes a transition from primary (extractive) to secondary (manufacturing) to tertiary (service) activities. Figure 13.10 provides a simple illustration of the relative importance, in terms of employment, of the different sectors of industrial activity. Some geographers contend that, especially in the more developed world, the emerging post-industrial society is increasingly service-oriented. This may be so, but service industries have existed for a very long time.

According to Daniels (1985: 1), a service "is probably most easily expressed as the exchange of a commodity, which may either be marketable or provided by public agencies, and which often does not have a tangible form." Service industries, including transportation, utilities, insurance, real estate, education, health, and government, are crucial components of any economy. The service industry grew along with manufacturing during the Industrial Revolution, but it has achieved its most rapid expansion since World War II. Since 1960 employment in services has increased globally from about 20 per cent to over 40 per cent (Figure 13.11). In the less developed world, retailing and distribution are among the dominant service activities, while in the more developed world more specialized services such as banking and advertising are also important.

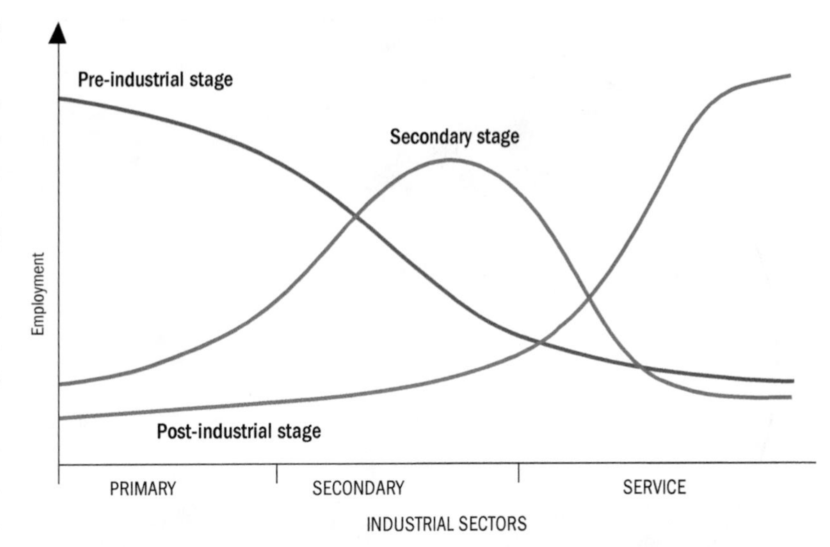

FIGURE 13.10 | Economic growth and employment distribution
A simplified diagram of the relationship between stages of economic growth and distribution of employment in the three principal economic sectors.

The global variations in employment in industry and in services are indicated in Figures 13.12 and 13.13. It is useful to compare the two figures with Figure 10.15, which maps the percentage of the labour force employed in agriculture. The

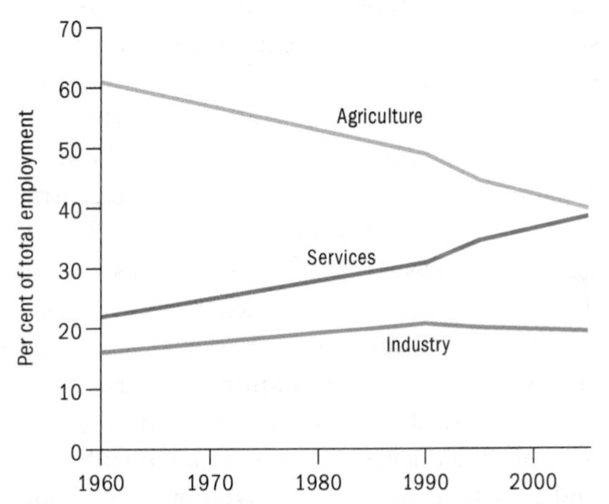

FIGURE 13.11 | The changing structure of world employment

Source: Based on: International Monetary Fund. 2006. *Finance and Development: A Quarterly Magazine* 43 (1); United Nations Development Program. 1996. *Human Development Report*. New York: Oxford University Press, Table 16.

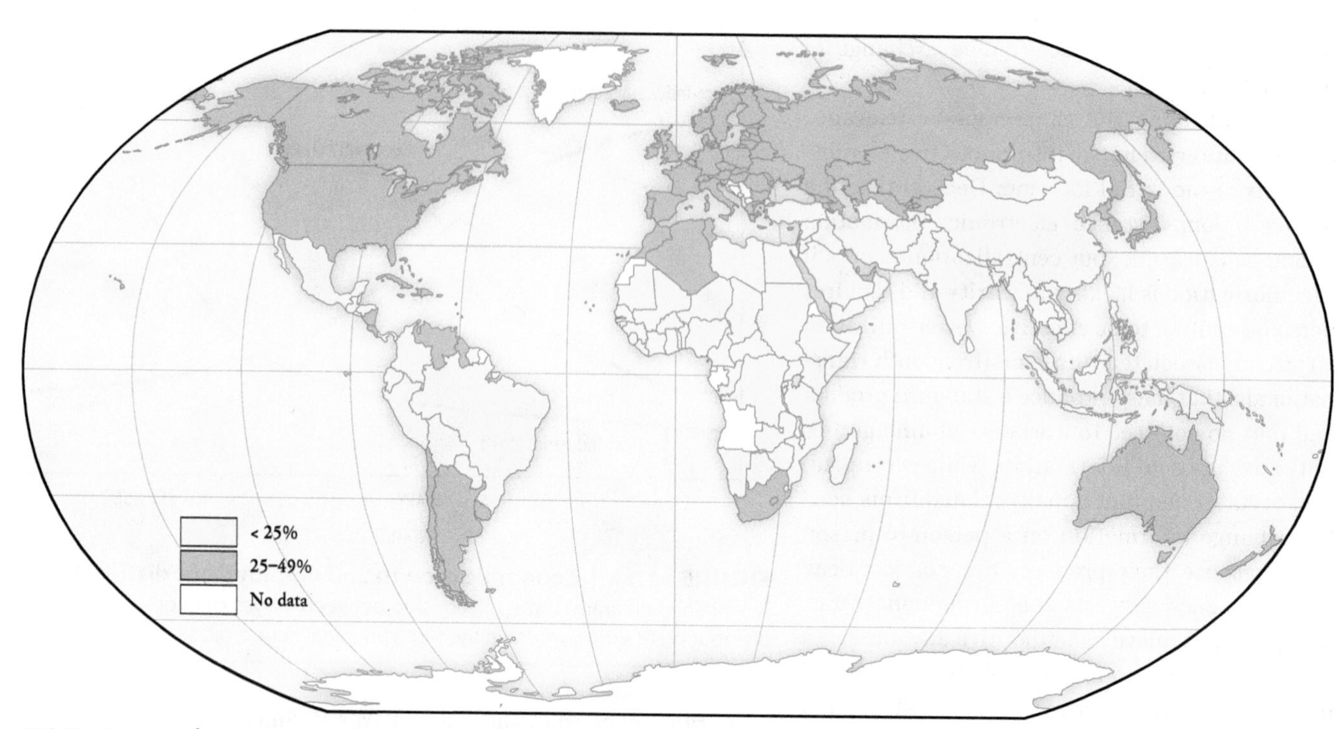

FIGURE 13.12 | Percentage of labour force in industry by country, 2010

Figure 10.15 showed not only enormous variations between countries but also a basic distinction between the more developed and less developed worlds. Although this map of percentages in the industrial labour force shows less variety, it indicates a distinction between the more developed and less developed worlds. In general, the percentages are between 25 and 49 in the more developed world and under 25 in the less developed world. The world average is 20 per cent. There are significant gender differences: in Canada 11 per cent of employed women and 33 per cent of employed men work in industry; in Bangladesh the corresponding figures are 9 and 11.

Source: Updated from United Nations Development Program. 1996. *Human Development Report*. New York: Oxford University Press, Tables 16 and 32.

figures provide some compelling additional evidence of the differences between the more and less developed worlds. Some summary data for selected transition economies are presented in Table 13.7. This table shows the growing role of services in these economies at the turn of the century, a result of increasing local demand and also of offshore outsourcing from other countries.

Geographic attempts to explain the location of service industries have focused on the central place model described in Chapter 11. This work has made it clear that service locations can be determined by considering such standard causal variables as transport costs, market location, and economies of scale. In addition, services have an especially strong tendency towards agglomeration. Their locations depend on population, and they tend to cluster. Furthermore, location decisions in the service industry appear to be particularly vulnerable to behavioural variables such as the availability of information and the interpretation and use of information. A consideration of behavioural variables helps to explain the clustering tendencies. Information diffusion takes place most rapidly and effectively in local networks. Some argue that recent changes in the technology of information diffusion have led not only to the clustering of certain services but also to spatial changes in the roles played by urban centres. Others, however, contend that new information technologies are leading to increased decentralization, possibly even (in some cases) home-based service work. Regardless of where service industries locate, there is little doubt that they are able to contribute significantly to increases in productivity and living standards. In Canada, for example, service-sector growth is a crucial component of larger economic prosperity (Gruel and Walker, 1989).

A geography of service industries must also consider the availability of those services that are central to human well-being, among them health care and educational facilities. In fact, any study of service industries is a virtual microcosm of contemporary geography because services are linked to so many other geographic topics and

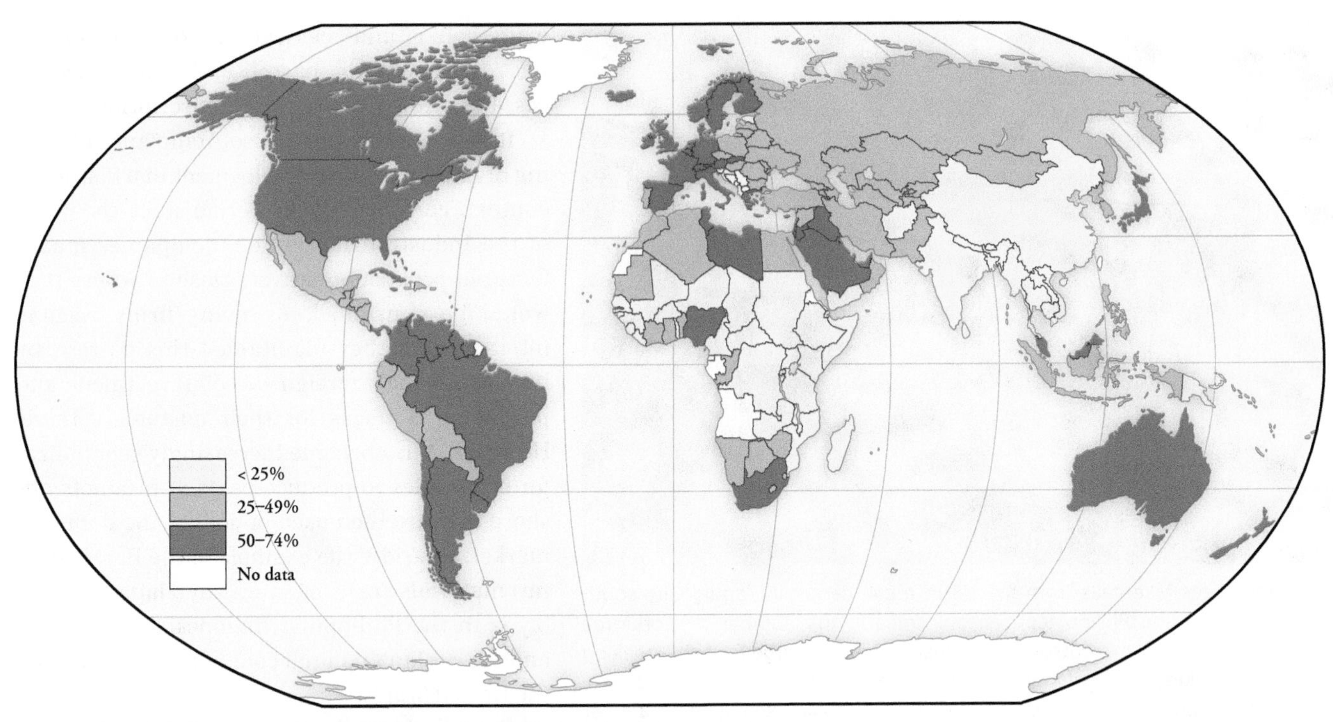

FIGURE 13.13 | Percentage of labour force in services by country, 2010

This map shows the highest values (between 50 and 74 per cent) occurring in the more developed world and a few other countries (especially in Latin America). In general, the less developed world has lower proportions of service workers, often below 25 per cent. The world average is 31 per cent. There are significant gender differences: in Canada 87 per cent of employed women and 64 per cent of employed men work in services; in Bangladesh the corresponding figures are 12 and 30.

Source: Updated from United Nations Development Program. 1996. *Human Development Report*. New York: Oxford University Press, Tables 16 and 32.

because they are so important in contemporary economies and societies.

Outsourcing

Outsourcing refers to the situation where a company hands work that was previously completed in-house to other firms. If this involves work being outsourced to other countries in order to take advantage of inexpensive labour, it is a type of **offshoring**, adding to the internationalization of employment. Offshore outsourcing was initially viewed by many in North America and Europe as one indicator of a globalizing economy's flexibility but is now viewed more typically in terms of the negative impact of job losses, as many firms have an increasing percentage of employees in countries such as India, China, and Mexico, where inexpensive labour is plentiful.

Service activities generally are increasingly outsourced offshore as companies are able to hand over work to specialist service suppliers located elsewhere. This is occurring largely

outsourcing
Paying an outside firm to handle functions previously handled inside the company (or government) with the intent to save money or improve quality; a term often used interchangeably (and incorrectly) with *offshoring*; much outsourcing involves offshoring jobs from more developed countries such that they become much lower-paid jobs in less developed countries.

offshoring
The transferring by a company of production or service provision to another country.

TABLE 13.7 | Employment by sector, 1990 and 2001, selected transition economies

	1990			2001		
Country	Agriculture	Industry	Services	Agriculture	Industry	Services
Bulgaria	18.5	44.2	37.3	26.3	27.6	46.0
Estonia	21.0	36.8	41.8	6.9	33.0	60.1
Poland	25.2	37.0	35.8	19.1	30.5	50.4
Romania	29.1	43.5	27.4	42.3	26.2	31.5
Russia	13.9	40.2	45.6	11.8	29.4	58.8

Source: Adapted from International Labour Organization. 2004. *Global Employment Trends*. Geneva: ILO, 34.

Filipino employees take calls from the US in a call centre in Taguig City south of Manila. In 2011 the Philippines overtook India in the number of workforce employed by the call centre outsourcing industry due to its large English-speaking population. Revenues from outsourcing call centres, health insurance processing, animation development, and software programming totalled $9 billion, or 4.5 per cent of the Philippine GDP.

because of the improved communications associated with the globalizing economy. It is possible that Russia, some Eastern European countries, China, and especially India will gain more and more business employment with the services provided being consumed in more developed world countries. However, some commentators see a new trend developing. As wages increase in less developed countries, the benefits of outsourcing will be less evident and jobs may begin to return to North America and Europe.

The discussion of industry in India noted the importance of offshore outsourcing to industrial growth in that country, and Table 13.8 shows recent increases in information technology and business processing outsourcing (BPO) employment. BPO is the outsourcing of back-office and front-office functions typically performed by white-collar and clerical workers. Examples include accounting, payroll and human resources, and medical coding and transcription.

It is helpful to view outsourcing and offshoring of manufacturing employment in a historical context. For much of the period since the onset of the Industrial Revolution, companies manufactured products relatively close to where they would be consumed. As many firms became international, they maintained this pattern by increasing their foreign direct investment and producing overseas for their customers there. However, it has become increasingly economical for companies to produce parts or products offshore that are then used or sold in the domestic market. Consider the example of the Barbie doll: raw materials are from Taiwan and Japan; assembly is in the Philippines, Indonesia, and China; and the design and final coat of paint come from the United States.

Overall, India is the favoured location for offshoring service employment while China is favoured for offshoring manufacturing employment, although China seems likely to gain more of the former. An important debate centres on the question of who benefits from offshore outsourcing. It is correct to note that employment opportunities are created in countries where they are needed, but this necessarily involves job losses in the more developed countries.

Industry and Society

The industrial geography discussed so far has said little about the role played by social matters. Our concerns have been essentially spatial and economic—a fair reflection of the dominant geographic traditions. As we have seen, however, human geographers are increasingly turning their attention to issues related to social circumstances. Since about 1970 many researchers have realized that much industrial geography can be understood best by reference to social theory. They argue that industrial activity does not take place in a politically and socially neutral setting, according to the principles of neo-classical economics. They claim that we must turn to contemporary social theory to gain real insights into the geography of industrial capitalism. Unfortunately for beginners, the range of that theory is considerable, but most of it consists of variations on the Marxist themes discussed in Chapter 10. Massey (1984), for instance, produced a Marxist theoretical analysis

TABLE 13.8 | Information technology and business process outsourcing employment in India (thousands of jobs)

	2002	2003	2006	2009	2012
Information technology	106	160	379	1,004	2,717
Business process outsourcing	170	205	285	479	972

Note: Data for 2006, 2009, and 2012 are projections.
Source: "Survey: A World of Work." 2004. *The Economist*, 13 November, 10.

Focus on Geographers

Historical Industrial Geography: Population, Health, and the Cape Breton Sydney Coalfield | NATALIE C. LUDLOW

As a multidisciplinary researcher who focuses on demography and epidemiology, I am always intrigued by the impact of industrialization on human society. The longitudinal recurring human problems of urbanization (disparity, inequality, and segregation) become visible through a historical lens. Through the study of the history of industrial regions, cities, and towns, questions relating to population, health, environment, economy, and geography can be examined. Historically within Canada, two locations were deemed hubs of industrialization: Hamilton, Ontario, and the Sydney Coalfield of Cape Breton, Nova Scotia. The historical coal and steel towns that developed along the latter present an interesting case of early Canadian industrialization.

Coal mining along the Sydney Coalfield dates to the early 1700s. It was often small-scale in nature and, depending on the period, either government or independently owned. Between the 1890s and 1910s, however, Cape Breton Island saw immense economic growth, the result of expanded underground mining, industrial agglomeration, and spatial monopoly of two competing industrial giants: Dominion Coal Company (DOMCO) and the General Mining Association (which became the Nova Scotia Iron and Steel Company around 1900). DOMCO and its sister firm, Dominion Iron and Steel Company (DISCO), owned the largest colliery, at Glace Bay, and turned the village of Sydney into a bustling steel city.

Cape Breton, far removed from the industrial heartland of Canada, does not present an ideal industrial location, but the confinement of geological fixity of a material (e.g. coal) encourages migration to isolated locations. Fort McMurray, Alberta, is a current example. Such isolation can present corporate advantage. The industrial location problem of isolation and distance from the market was offset by the fundamental Marxist dilemma of capitalism: labour as a commodity and labour surplus exert control over workers. Isolation and company agglomeration and monopoly gave rise to a stratified socio-economic and socio-political pattern—the company town—along the Sydney Coalfield. Company towns, prominent throughout the Industrial Revolution, are labour-intensive single-industry towns where industry governs municipal and industrial affairs, exerting authority over community members. Along the Sydney Coalfield, DISCO built the first hospital in Sydney near the steel plant; the company store in Glace Bay increased the probability of incurring debt to DOMCO; and company houses and boarding houses were a visible characteristic of the built environment.

In these early years of industry, both Sydney and Glace Bay suffered under the company town structure, as reflected in historical annual reports of health and industry and in provincial death records. Many homes lacked public health infrastructure (water supply and sewage disposal), causing health problems such as typhoid fever, gastroenteritis, cholera infantum, and infantile diarrhea. Occupational health and safety was minimal. Skill was not an asset in the hiring process, likely keeping labour costs low and unintentional workplace hazards high. Expenditures were directed towards improving and modernizing the industry and transportation system. Labour recruitment from Europe, Canada, and Newfoundland created such an influx of workers in the 1900s that they were considered a dime a dozen, frequently living in overcrowded, unpleasant conditions. An overabundance of workers maintained the corporate exploitation of labour, as any worker who attempted to fight for better wages, hours, or safety could find himself easily replaced or, worse, industry-blacklisted across the province. Thus, industry along the Sydney Coalfield significantly affected the demographic and epidemiologic landscape.

By 1911 the demographic profile of Sydney and Glace Bay revealed a predominantly single young male population, exemplified through the construction of population pyramids, sex ratios, and marital statuses for each community. For example, working-age sex ratios reveal the single-industry landscape of Sydney and Glace Bay (139.3 and 126.5 working age males per 100 working-age females, respectively). From an epidemiological standpoint, tuberculosis (an infectious disease) made a significant mark on mortality patterns around 1911 in Sydney and Glace Bay, as it did in other high-density, overcrowded urban locations during this time. Accidental deaths, however, were the number-one cause of death in both communities, are statistically related to the occupational environment, and are representative of Canada's early twentieth-century industrial landscape.

An industrial geographic perspective can illuminate the history of a community, its people, development, health, environment, and economy. As a researcher interested in population health and industry, I often use a variety of

Continued

archival resources (e.g. censuses, annual reports, death registry, city directory, maps, and photographs) to understand the complex spatiotemporal relationships of population, health, and industrialization. Examining the geography and history of industry is integral to comprehending human demographic and epidemiologic patterns, especially in a country with a large economic interest in natural resources (such as Canada). In addition, such research enhances knowledge of past, present, and future affairs and issues relating to industry in Canada and abroad.

NATALIE C. LUDLOW has an MA in anthropology and is a doctoral candidate in the Department of Geography and Planning at the University of Saskatchewan. Her multidisciplinary approach to research focuses on urban health ecology, urban development, and industrialization.

of the changing industrial geography of the United Kingdom, while Graham et al. (1988) outlined a framework based on Marxism for investigating structural change in industries. Such industrial geographies represent significant departures from the usual Weberian economic theory; instead of emphasizing distance and cost, they give priority to relating spatial and social issues by reference to such topics as class and gender as they have evolved in a capitalist framework.

For example, the fact that transnationals actively seek low-wage locations for manufacturing activities has attracted charges of labour exploitation. High-profile companies such as Nike have been especially subject to criticism. There is also evidence that some transnationals locate plants in countries where the costs of production are kept low not only because the wages are low but also because there are fewer requirements concerning matters such as pollution control.

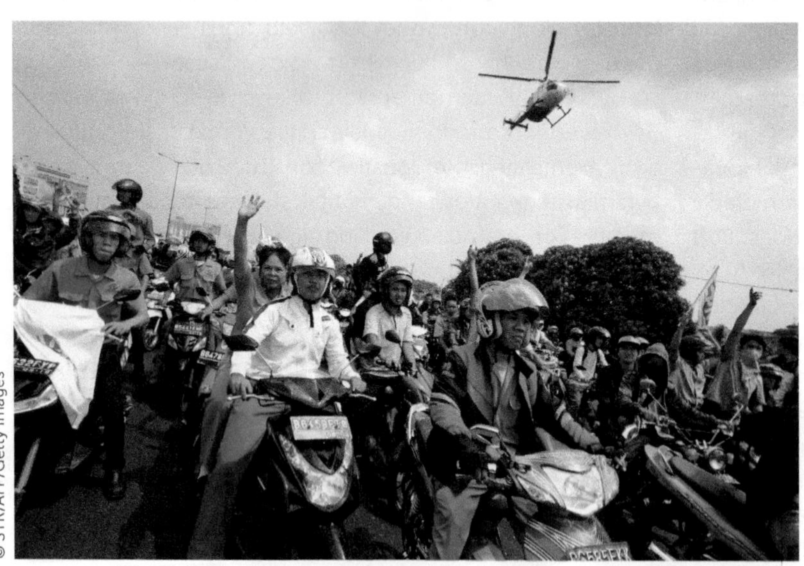

© STR/AFP/Getty Images

Twenty-thousand Indonesian workers occupy Jakarta's main toll road in Bekasi to demand a minimum wage increase while a police helicopter flies overhead. For years big-name manufacturers such as Honda, Samsung, and Nike have quietly harvested profits from factories in Indonesia, where wages are even lower than in China or India. Indonesian workers are hitting back with a wave of industrial strikes, demanding a bigger chunk of the profits in one of Southeast Asia's fastest growing economies.

It is now commonplace to argue that no hypothesis about industrial (or other) locations can be complete unless it is placed in some appropriate social, political, and institutional context. Massey (1984) argued that places vary in many respects, from workplace social relations and the division of labour to local political circumstances, and each of these must be considered in any analysis of industrial location and the changing distribution of employment. Similarly, a focus on workplace social relations often requires consideration of gender differences.

Changing Local Labour Markets

As we saw in Box 13.1, industrial geography traditionally treated labour as either a location factor or a commodity. Reacting against these interpretations, Marxist theory focused on the spatial division of labour—the idea that regional development and employment distribution are spatial expressions of the labour process, subject to control and manipulation. More recently, industrial geographers have focused on the geographies of labour markets, particularly as these are affected by the transition from Fordism to post-Fordism. Table 13.9 summarizes the differences between these two phases.

Gendered Employment

As noted in Chapter 8, a majority of women of working age in most of the more developed world are in the labour force. This marks a significant change. In Canada, for example, female participation rates

TABLE 13.9 | Labour Markets: From Fordism to Post-Fordism

Factor	Fordist Labour Markets	Post-Fordist Labour Markets
Labour process	Mass production involving large workforces within firms. Productivity achieved by division of labour into detailed standardized tasks.	Flexible, specialized mass production. Productivity achieved by division of labour within firms into core and peripheral categories, with functional flexibility of core workers and numerical flexibility of peripheral workers.
Employment	Substantial proportion in manufacturing. Predominantly male. Most jobs full-time. High degree of job security. Generally full employment.	Small proportion in manufacturing, most in services. Increased female participation in labour force. Many jobs part-time and temporary. Increasing job insecurity. High unemployment.
Wages	Most set by collective bargaining (by industry-wide or nationwide unions). Income inequalities stable or falling.	De-collectivization and localization of wage determination. Increasing polarization between high-paid full-time workers and low-paid part-time workers.
Labour relations	Highly formalized and confrontational. Labour organized into strong unions. Collective strike action by workers common and protected in law.	Much more individualistic and co-operative. Unions are in decline and solidaristic labour cultures in retreat.
Labour market regulation	Macroeconomic demand management to maintain full employment. Welfare-oriented. Benefits unconditional and funded by progressive taxation system. Extensive employment protection and workplace regulation.	Abandonment of full employment policies. Shift from welfare to workfare and training. Benefits increasingly conditional. Deregulation of employment and workplaces to promote labour flexibility.
Spatial features	Relatively self-contained local labour markets. Distinct local employment structures and labour cultures. Typified by centres of mass production industry. Local disparities in unemployment are minimal.	Local labour markets are still important, but segmentation is greater within than between them. Typified by new local centres of flexible production and service activity. Substantial local disparities in unemployment and growth of localized concentrations of social exclusion.

Source: Adapted from Martin, R. L. 2000. "Local Labour Markets: Their Nature, Performance, and Regulation." In *Oxford Handbook of Economic Geography*, edited by G. L. Clark, M. P. Feldman, and M. S. Gertler, 459, Table 23.1. New York: Oxford University Press.

in the labour force more than doubled between 1951 and 1981.Yet for the most part the type of work done by females has not changed: a majority still are employed in clerical, service, and low-skill jobs. Why is this so? One reason is discrimination by employers. In an account of the steel industry in Hamilton, Pollard (1989) cited evidence of sexist hiring practices by the principal steel company, Stelco. Between 1961 and 1978 Stelco received between 10,000 and 30,000 applications from women for production jobs—none of whom was hired. Over the same period, about 33,000 men were hired for production jobs. Following considerable publicity and intervention by the union and the Human Rights Commission, some women were hired. But the evidence suggests that the underlying social problem is much larger. While on the job, women were made to feel uncomfortable not only by men in the workplace but also by members of the community, especially wives of male employees. Larger social traditions are at stake in any attempt to change the gendered division of labour.

UNEVEN DEVELOPMENT IN MORE DEVELOPED COUNTRIES

Problems of and prospects for economic growth in the less developed countries, including considerations of the role of physical geography, the idea of developmental stages, and the claims made by new economic geography were discussed in the larger context of global inequalities in Chapter 6. Although they may be thought of as undesirable, it seems clear that spatial variations in economic development and quality of life are normal outcomes of any process of change, such as industrialization of an economy. Central or core areas will usually experience innovations before outer or peripheral areas, and the uneven distribution of resources prompts spatial variations. Unfortunately, in many parts of the world these natural spatial variations have resulted in differences that society considers inappropriate. Hence, most countries have regional development policies. With these points in mind, three ideas about the causes of uneven development have been discussed primarily in the context of more developed countries.

Explaining Uneven Development

First, there is the staples theory of economic growth. A staple is a primary industrial product that can be extracted at low cost and for which there is a market demand. Originally developed by Canadian historians, the staples theory appears particularly applicable to areas of European overseas expansion. Staple success means economic growth both in the areas of extraction and in the export centres; as such, staple production has direct impacts on regional growth. Other scholars have built models describing this process. Pred (1966), for example, focused on the staple's multiplier effects.

Second, a general approach to regional differences in economic growth, proposed by Friedmann (1972), is based on the core–periphery concept. A core region is a dominant urban area with potential for further growth. Peripheries include areas of old established settlement characterized by stagnant, perhaps declining, economies, some of which may be former staple production areas; these are called downward transition areas. There are two other types of peripheral regions: upward transition regions are linked to cores and continue to be important resource areas, and resource frontier regions are peripheral new settlement areas. "Peripheral regions can be identified by their relations of dependency to a core area" (Friedmann, 1972: 93).

Third, it can be argued that growth does not occur everywhere at the same time but manifests itself in points of growth. Thus, a new area of resource exploitation will induce economic growth but not necessarily at the source. Urban centres often serve as growth poles, as Calgary has for the Alberta oil industry. Although a substantial literature has developed around the idea of growth poles, the concept has rarely been employed effectively.

© JENNY

People working on the assembly line at Huajian shoe factory in Dukem, Ethiopia. Huajian is one of six Chinese factories operating in the Chinese-built Eastern Industry Zone—Ethiopia's first industrial park—which the government hopes will attract private foreign investment and boost the country's manufacturing and export sector. As China's own economy matures, companies such as Huajian are moving to Africa to seek affordable labour. With a population of 82 million, Ethiopia is Africa's second most populous nation and is rich in natural resources such as leather and cotton, a major attraction for investors.

The Canadian Example

As noted in the context of deindustrialization in the United Kingdom, governments have long intervened in the processes of economic change. In Canada companies were granted monopolies, the progress of settlement was dictated, and tariff barriers were installed all before 1900. In the twentieth century, however, such intervention began to play a major role in redressing spatial imbalances. Typically, regional development policies have been designed to encourage growth in depressed, usually peripheral, areas.

A number of regions in Canada qualify as relatively underdeveloped. Some would contend that almost all of Canada outside the Great Lakes–St Lawrence Lowlands qualifies for this description. In the case of the Atlantic provinces—Newfoundland, New Brunswick, Nova Scotia, and Prince Edward Island—economic and related social problems stem from a peripheral location some 1,600 km (1,000 miles) from the major centres, a poor resource base, national tariff policies, and increasingly centralized transport and production systems. Inequality is evidenced by low wages and high unemployment. The Canadian government has attempted to address these issues by stimulating growth through low-interest loans to farmers and others and tax subsidies for industrialists. Some results have been achieved, but the basic problems have not been solved. The principal goals are to foster a self-sustaining entrepreneurial climate, more successful medium and small business enterprises, lasting employment opportunities, and an expanding competitive economy. These are not simple goals in a globalizing world that tends to reinforce the dominance of already successful regions rather than spread investment and opportunity to peripheral locations.

In the eyes of many Canadians, the Canadian Shield and the North—which together make up some 75 per cent of Canada—are frontiers that produce primary products for export that benefit primarily the Canadian core area; in many respects, then, these resource extraction areas are part of the less developed world. Others, notably Aboriginal populations, see these areas as a homeland. These very different images are not easy to reconcile (see Bone, 2011).

The European Example

The European Union is a second example of a political unit that contains core and peripheral regions and, as a result, regional inequities. As of 2015 there are 28 EU members, with Croatia the most recent addition. Moves towards enlargement and integration have been taking place within a context of significant regional disparities. The number of peripheral regions with low incomes and high unemployment rates increases with each expansion. Several countries that joined the EU in the 1980s, especially Greece (1981), Spain (1986), and Portugal (1986), brought serious regional development problems, while the subsequent addition of many East European countries has further increased the breadth of regional variations. There are, therefore, very strong arguments for a European regional policy. To date, however, the principal policy has been simply to provide funds to depressed regions. The fact that an effective regional policy is needed became strikingly evident in 2011 as several countries struggled economically, including Greece, Portugal, and Ireland.

CONCLUSION

The organization of this textbook has resulted in the separation of some topics that could be discussed together. Perhaps most notably, much of what is included in the Chapter 4 account of human impacts might be placed elsewhere. For example, the environmental impacts of energy production and consumption and of industrial activity are covered in Chapter 4. Hence, there is only incidental reference to environmental issues in this chapter. This is not to suggest that environmental concerns do not matter—far from it. Major players in both energy and industrial sectors are unable to operate without having to consider the environmental impacts of their activities.

Today more than ever, there are locational implications of the environmental consequences of industrial activity. In many countries, particularly democratic ones, industries face new rules designed to control pollution by regulating the quantities of waste produced and the disposal of waste products. In principle, environmental considerations add a new cost factor that may affect both location and production decisions. Interestingly, in a study of the effects of the National Environmental Policy Act in the United States, Stafford (1985) concluded that the policy would not lead to major locational changes. As regulations become increasingly more rigorous, however, locational effects will likely become apparent.

But there is much more to this matter of environmental impacts. Some mainstream commentators are rethinking the whole question of what we understand by the idea of growth. As noted in Chapter 6, thinkers on the ideological left have long thought along these lines, but their ideas are becoming more widely acknowledged. There is increasing acceptance of the idea that the prevailing GDP model of growth that focuses on production and consumption serves both to encourage overconsumption and to devalue the environment. What seems to be needed is an understanding of growth that considers well-being and that places value on critical environmental assets such as clean water and clean air. In a similar vein, some critics are quite sensibly revisiting the conventional division of the world into more and less developed countries, a division that at heart has economic growth as its base. Such a thought-provoking shift in view has increasingly become central to the study of human geography.

Summary

Levels of Production

Traditionally, three levels are recognized: primary, which is extractive; secondary, which constitutes manufacturing; and tertiary, which is service. A fourth (quaternary) level involves the transfer of information (i.e. communications).

Levels of Economic Organization

Two levels are recognized. Households produce and reproduce; as economic organizations they are of decreasing importance. Firms make up the commercial sector; manufacturing firms operate in factories.

Factors Related to Location

Industrial location theory explains why factories are located where they are. Traditionally, the relevant factors have included transport; distance from raw material sources, energy supplies, and the market; availability of labour and capital; the nature of the industrial product; internal and external economies; entrepreneurial uncertainty; and governmental considerations.

Least-Cost Theory

The typical location decision aims to maximize profits by minimizing costs or maximizing sales (or both). Least-cost theory, developed by Weber and published in 1909, is a normative theory that identifies where industries ought to be located. Following a series of simplifying assumptions, Weber concluded that industries locate at least-cost sites determined by transport costs, labour costs, and agglomeration–deglomeration benefits. Solving locational problems required defining concepts such as material index, locational figure, isodapane, critical isodapane, and isotim. Other researchers, especially Hoover, have improved Weber's formulation of the transport-cost variable.

Market-Area Analysis

Market-area analysis, the second major theoretical approach, centres on profit maximization rather than cost minimization. Lösch is the principal theorist.

Behavioural Approaches

Behavioural approaches to the industrial location problem focus on the subjective views of the people involved and the common preference for satisficing rather than optimizing behaviour. Behavioural issues are important, albeit more difficult to quantify or theorize than strictly economic considerations. Contemporary industrial firms are, typically, complex decision-making entities.

Industrial Revolution

Prior to the Industrial Revolution, industrial activity was domestic, small in scale, and dispersed. A revolution in industrial activity and related landscapes occurred between 1760 and 1860, beginning in Britain. Factories replaced households as production sites, mechanization increased, and localized energy sources were used. All these changes were related to the rise of capitalism, which emphasized individual initiative and profits. New industrial landscapes appeared on the coalfields of Britain and were accompanied by city growth, rural-to-urban migration, and a series of transport innovations. Inner-city slum landscapes developed in many of the new agglomerations on coalfields and in the traditional textile areas of northern England.

Energy

Coal was the main fuel during the Industrial Revolution but was replaced by oil in the 1960s. OPEC is a major player in international affairs because of the importance of oil in the contemporary world. This role may be declining, especially because of increased American output. Today, three fossil fuels—oil, coal, and natural gas—are important sources of energy for industrial activities.

World Industry

Most industrial activity continues to take place in a capitalist economic and social framework. The global manufacturing system is dominated by North America, Western Europe, and the Asian Pacific region. Multinational corporations are increasing in importance.

Canadian Industry

The dominant characteristics of Canada's industrial geography include abundant natural resources, a well-developed extraction industry, a branch-plant economy reflecting the country's proximity to the United States, a dispersed national market, a dispersed primary sector, and highly concentrated secondary and tertiary sectors.

Newly Industrializing Countries

Japan was the first of the NICs and has been followed by other Asian countries (notably South Korea), some Southern European countries, Brazil, and Mexico. Export-processing zones using inexpensive and relatively unskilled labour have been established in some countries (for example, the maquiladoras in Mexico).

Industry in the Less Developed World

Industrial growth is highly desired by many less developed countries to bolster weak economies, provide employment, demonstrate economic independence, encourage urbanization, help create a better economic infrastructure, and reduce dependence on overseas markets for their primary products. But industrialization is difficult to achieve, not least because of colonialism's legacy. Many countries have experienced their greatest industrial successes in import substitution. India has thrived in this regard since 1951, while China opened the doors to international trade and foreign investment and also reduced state control of industries during the 1980s. The most dynamic industrial region in the world is the Pearl River Delta in south China, just north of Hong Kong.

Industrial Restructuring

Technological changes in production, transaction, and circulation are currently prompting some major changes in the global economy and in national and regional industrial geographies. Offshore outsourcing of both manufacturing and service employment is increasingly important.

Information Exchange

Industrial location decisions used to be made largely on the basis of transport costs, but today the costs of information exchange and labour are much more important. Efficient electronic transfer of reliable information allows firms to decentralize by seeking out areas of low labour cost. Where information exchange requires personal contact, location patterns tend to be centralized.

Service Industries

As part of the process of economic restructuring and the transition from Fordism to post-Fordism, industrial societies are moving to a post-industrial service stage. Because the service industry is so diverse, a wide range of factors may determine locations. Central place theory offers the most important set of explanatory concepts, but behavioural approaches also seem relevant. New information technologies are undoubtedly affecting the location of services.

Some Social Issues

Industrial geography, like other branches of economic geography, has largely neglected social issues. Now that industrial geography is becoming more concerned with issues such as the gender division of labour, it is increasingly turning to various social theories.

Spatial Inequalities

Correcting spatial inequalities is seen as a government responsibility. Governments in capitalist countries frequently implement policies to improve the economic status of regions that are less developed, either because they are peripheral or because their industrial infrastructure is outdated. The most obvious instances of planning the spatial economy take place in socialist countries. Economic restructuring continues to contribute to uneven development.

Links to Other Chapters

- **Globalization:** Chapter 2 (concepts); Chapter 3 (all of chapter); Chapter 4 (global environmental issues); Chapter 5 (population growth, fertility decline); Chapter 6 (refugees, disease, more and less developed worlds); Chapter 8 (popular culture); Chapter 9 (political futures); Chapter 10 (agricultural restructuring); Chapter 11 (global cities)
- **Industrial location theory:** Chapter 2 (positivism); Chapter 10 (agricultural location); Chapter 11 (settlement location)
- **Industrial Revolution:** All chapters
- **Energy:** Chapter 4 (renewable energy sources)
- **Differences in industrial geography of more and less developed worlds:** Chapter 6 (global inequalities); Chapter 10 (differences in agriculture); Chapter 12 (differences in settlement)
- **Industry in Japan; NICs; industrial geography of China:** Chapters 3, 10, and 11 (transnationals)
- **Industrial restructuring (especially in the service industry):** Chapter 11 (industry and services in cities, global cities); Chapter 12 (cities as centres of production and consumption)
- **Industry and society:** Chapter 2 (Marxism); Chapter 8 (gender)
- **Environmental considerations in industry:** Chapter 4 (environmental issues)
- **Uneven development:** Chapter 3 (trade and trade blocs); Chapter 6 (developing the less developed world); Chapter 11 (core and periphery)

Questions for Critical Thought

1. How has the "industrial location problem" changed from the late nineteenth century/early twentieth century to today? What (if any) locational considerations have remained in place over the past century or so?
2. How relevant/useful is Weber's least-cost theory to understanding industrial location today?
3. What were the effects of the Industrial Revolution on cities and agriculture in the eighteenth and nineteenth centuries?
4. How has the global pattern of industrial production been affected by globalization?
5. To what extent will the expansion of industry in developing nations enhance social and economic well-being in those nations?
6. What are the advantages and disadvantages of promoting tourism as a stimulus for economic growth and development?
7. Why does "uneven development" exist? Is it possible (and if so, under what circumstances) that all nations could be more or less equally developed?

Suggested Readings

Visit the companion website for a list of suggested readings.

CONCLUSION

WHERE NEXT?

We began this book by identifying three recurring themes: relations between humans and land, regional studies, and spatial analysis. By now you should have a clearer idea of why these themes are so central to human geography.

The introduction also pointed out that one goal of human geography, and of this book, is to help us understand our changing world as it is today and how it came to be so. Human geography is a discipline of tremendous breadth and diversity, with especially close links to the natural sciences (particularly physical geography), and a healthy academic pedigree. A human geographic perspective—a geographical imagination—is an invaluable asset. The editor of *The Times* of London once wrote of school geography:

> It seemed a discipline that took a child out of the classroom, into the street and the park and said: "Look what's here! How did it come to be here? What is it made of?" On such empirical enquiry rests all learning. Geography offered such empiricism in the most comprehensible and immediate form. It also offered the basis for argument, for disagreement, for controversy, for the spirit of dialectic that makes for true understanding, not rote learning. (Jenkins, 1992: 193–2)

If it is not possible to answer the question that titles this final chapter, we must at least attempt to comment on changes both in our subject matter—human behaviour as it affects the earth's surface—and in our methods of studying it.

CHANGING HUMAN GEOGRAPHIES

Human geography is an exciting field. If the first chapters of this book suggested that human geographers must look back nostalgically to the golden days when every journey meant new facts to be discovered, catalogued, and understood, you now probably realize

People walk across the High Line, a rail-to-trails park in New York City's Chelsea neighbourhood. Begun in 2006 and completed in 2014, the High Line receives approximately five million visitors every year. The success of this repurposed section of disused railway has led to the revitalization of the neighbourhood, real estate development in the surrounding areas, and plans for similar urban parks in cities across the United States and the world.

that the end of the exploration/invasion era did not by any means signify an end to the discovery of new facts. Humans are constantly changing landscapes in myriad ways—culturally, socially, politically, and economically. There may not be a wealth of new facts over the horizon, waiting to be discovered, but the facts within the horizon are constantly changing, as is our understanding of them. Indeed, the facts are changing at an ever-increasing pace because there are so many of us to cause change, because technology is always improving, and because we live in a globalizing world that is shrinking as a result of advances in communications and other forms of movement. Five examples of changing human landscapes—the human geography of the future—will clarify these generalizations.

Population

Human geography is about human beings, and the single most important change related to us is probably the continuing increase in our numbers. As we saw in Chapter 5, the rate of natural increase in the world population peaked in the late 1980s and is declining (the fertility transition), but the total number of people continues to increase from the current 7.3 billion. The United Nations estimates 8.4 billion by the year 2025, 9.7 billion by 2050, and a stable population of just over 10 billion by 2200. Most growth takes place in the less developed world; many countries in the more developed world have natural increase rates close to zero. The composition of the world population is also changing as the numbers of elderly people increase, and important social transformations are underway as gender roles and family patterns change.

With population growth below replacement levels, population aging, and corresponding declines in the labour force, immigration is likely to become an increasingly important—and divisive—issue in the more developed world. Meanwhile, the shift will continue away from industrial activity and towards "knowledge" or "information" as the economic engine of society. Since knowledge travels freely, without regard for physical borders, the prospects for upward social mobility through higher education will increase, but so will the competition that individuals face.

As recently as about 1750, most people lived in rural areas and derived their living from agriculture. Many of the different human group identities apparent today developed in these conditions of relative isolation from others—before the massive changes initiated by the Industrial Revolution. These differences of language, religion, and ethnicity have left a legacy of cultural and social diversity that continues to shape the contemporary world.

Urbanization

The percentage of the world's population living in urbanized areas reached 50 per cent in 2007 and continues to increase; the number of very large cities will also continue to rise, especially in the less developed world. These ongoing increases will bring further changes in the urban way of life, and if the existing problems of housing availability and basic infrastructure evident in urban areas of the less developed world are any indication, many of these changes will be in an undesirable direction. In many more developed countries, cities continue to grow outward and upward, often in response to a new set of postmodern processes. For example, edge cities are now commonplace in areas where sufficient space is available. Such cities are centres of consumption, not production.

Human Impacts

More people mean more demands both on resources and on agricultural and industrial outputs. Energy use per capita is much higher in the more developed world than it is in the less developed. Necessarily, then, energy demands in the latter are increasing as expectations rise and industrialization proceeds. The basic conclusion is unavoidable: the impacts on our fragile home, as discussed in Chapter 4, will also increase.

Although it appears that awareness of human impacts is encouraging a shift, at least in the more developed countries, from an anthropocentric to an ecocentric view of the world—an idea discussed in the context of sustainable development in Chapter 4—there is still a long way to go. A recent report in *Science*, based on extensive and intensive research on global fisheries, indicated that, without major change in current practices, there will be little sustainable seafood or fish by mid-century. As one of the authors noted, without real change in how ocean species are managed and harvested, "We'll be eating sea squid soup and jellyfish pie" because all the recognizable seafood, except for that from aquaculture, will be gone (Calamai, 2006). To put the issue in slightly different terms, many human geographers challenge the basic premise on which industrial societies have been established: the idea that progress depends on conquering nature and increasing production. This idea, centred on what has been called the "treadmill of production" (Schnaiberg and Gould, 1994: v), can be seen as the cause of both environmental degradation and inequalities in the social well-being of different groups. Problems of the global environment and social justice are not separate issues.

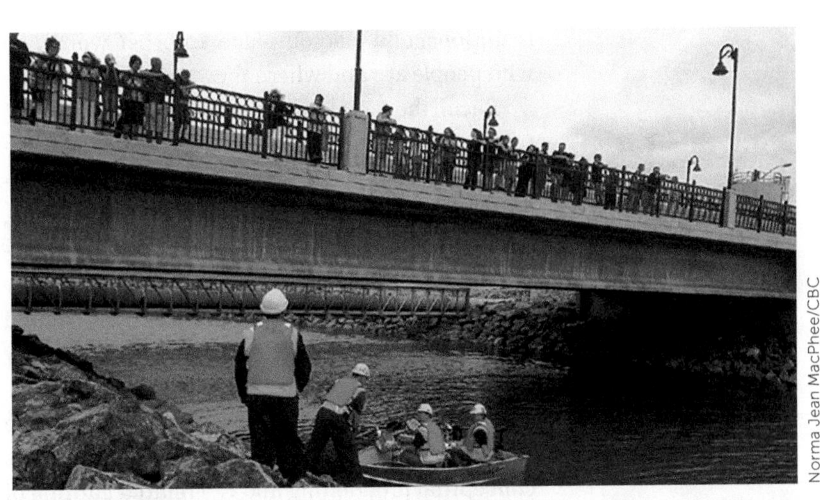

Norma Jean MacPhee/CBC

The Ferry Street Bridge crosses the former Sydney Tar Ponds—one of Canada's worst toxic waste sites, in Cape Breton, Nova Scotia. Run-off from coke ovens had filled the Muggah Creek with coal-based sludge. A massive cleanup of the area began in 2007 and was completed in 2013.

Political Worlds

Every day, the media report on political events of geographic interest. Perhaps most notably, the rise of the Islamic State and other fundamentalist groups such as Boko Haram is impacting landscapes throughout much of the Middle East and North Africa. Also, the political map of the world changes: in some cases, countries choose to sacrifice some of their national independence as they integrate with others, with the most advanced example of this trend being the European Union; in other cases, areas within countries assert their independence and strive for nationhood, as occurred with the creation of the new state of South Sudan.

The trend towards political disintegration was one factor behind the terrible conflict unleashed by the political collapse of the former Yugoslavia in the early 1990s, as different linguistic, religious, and ethnic groups appealed to perceived ethnic identities and historical injustices, and was seen more recently in Sri Lanka when the minority Tamils fought unsuccessfully for autonomy from the majority Sinhalese. It seems highly probable that there will be further assertions of autonomy by groups who perceive themselves as different from the majority groups in their states.

Certainly, geography aids in any attempt to explain political conflict. For example, to understand the 2003 Iraq War and its aftermath, we need to know something about at least five thoroughly geographic subjects: the way the states in the area were created; the spatial distribution of various ethnic groups (Shiite Muslims in the south, Sunni Muslims in the centre, and Kurds in the north); the importance of oil and natural gas to the economies both of producing countries and of their customers in the more developed world and the location of known deposits in Iraq (in the north and south but not in the centre); the relations of dependency that exist between more and less developed worlds; and the geopolitical perception of the United States following the 2001 terrorist attacks.

Sameness and Difference

The uniqueness of human beings resides in our biological attributes and in our ability to adapt culturally. We humans have managed to avoid extinction because of our culture—essentially, our ability to develop ideas from experience and to act on the basis of them. Once established, however, culture can be an impediment to change. The fact that most people are most comfortable with familiar patterns goes a long way towards explaining our desire to identify with place in a world where, for many of us, places are constantly being created and destroyed. In *A Question of Place*, Johnston (1991) argues that geography needs to study what he calls milieux, places within which ways of life are constructed and reconstructed (in short, places in their entirety).

An important foundational idea in any discussion of human identity is the fact that humans have the capacity to identify and name different groups, with the most fundamental identification and naming involving "us" and "them." This distinction is typically based on what is loosely called culture, including language, religion, and traditions, and also on place; in other words, on who people are and where they are.

Recall the social constructionist argument that rejects the essentialist claim that a specific identity or group membership is fixed and unchangeable, arguing instead that identities are fluid, contested, and negotiated. From this perspective, self, identity, community, and social reality all are creations of the human mind and should not be regarded as objective entities in some way separate from ourselves. This idea informs numerous geographic studies of difference; one of the great achievements of human geography is this conceptual broadening and the related gaining of insights into the lives and experiences of people. One of the most notable aspects of the concern with geographies of identity is the realization that dominant societies construct identities of others and also construct the landscapes in which others might be obliged to live. In doing so, they construct landscapes that take the characteristics of the dominant group for granted.

Notwithstanding the increasing evidence of cultural difference based on such considerations as language, religion, ethnicity, gender, sexuality, physical well-being, age, and place, many countries believe in the existence of and need for an essentially unchanging national culture. But is there, or ought there to be, any such thing? Are there any cultural identities, national or otherwise, that have some inalienable right to exist and to remain relatively unchanged? Agreeing with the legitimacy of static, or relatively static, national identities raises numerous questions concerning who has the right to define that identity. Some of these questions were considered in this book, especially in Chapters 7, 8, and 9.

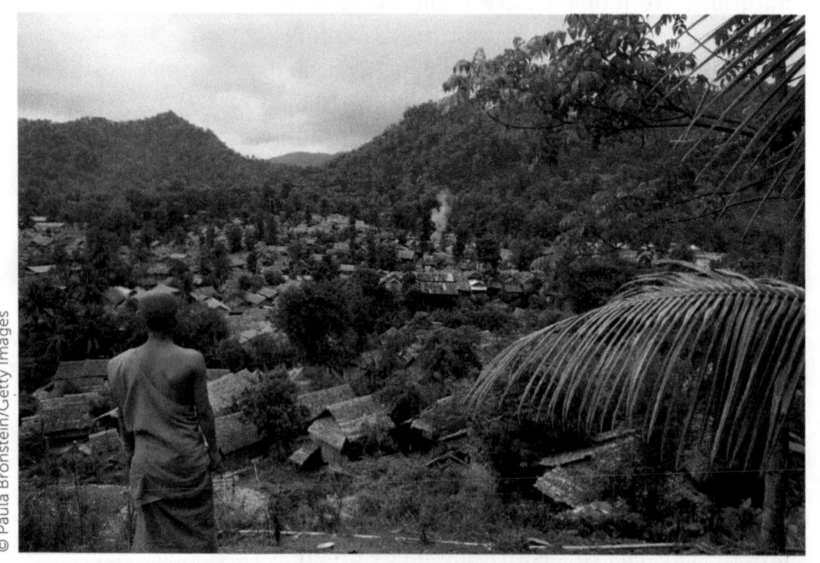

A Burmese monk looks out from a viewpoint at the Mae La refugee camp in Tak province, Thailand. The camp is situated along the Burma–Thailand border and was home to around 50,000 refugees in 2012. Mae La is the largest of nine camps along the Thai border, where Burmese refugees have lived in a stateless limbo for many years.

A CHANGING DISCIPLINE

Chapters 1 and 2 outlined the evolution of human geography and identified key philosophies, concepts, and techniques of analysis. One purpose of these accounts was to demonstrate how our discipline is always changing, largely in response to the changing needs of society. Human geography, like other academic disciplines, exists to serve society. There is, then, every reason to believe that the discipline described in this book will be a different discipline tomorrow. Attempting to predict the details of such change is hazardous, but we may at least identify six of what seem to be the most pressing societal demands on human geography.

Space and Place

The need for an academic discipline focusing on space and place seems self-evident. Space is important because we need to understand location, distances between locations, and movements between locations if we are to make sense of the world. For this reason, the single most important geographic contribution to human life is the map. Maps allow us to understand and explain why things are where they are and to express the closely related concepts of space, location, region, and distance. Because maps are so important, geographers have taken advantage of recent technological advances to develop the more sophisticated tool of geographic information systems (GIS).

Place is important because understanding the meaning that people assign to locations is crucial if we are to comprehend their attitudes and behaviours. All people have an innate sense of place, but human geographers are able to refine this innate sense into a knowledge that facilitates understanding of the changing human world. Our recurring discussions of globalization, regional character, and local change have continually reinforced these points.

Integrating Human and Physical Geographies

As discussed in Chapter 1, geography occupies "a very puzzling position within the traditional organization of knowledge. . . . It is neither a purely natural nor a purely social science" (Haggett, 1990: 9).

- For some, geography is a single, coherent discipline that integrates physical and human geography. For example, Stoddart (1987: 330) wrote that "Human geography as an exclusively social science loses its distinctive identity" and "outside a more general framework physical geography loses its coherence."
- For others, disciplinary coherence is less clear. Johnston (2005: 9) noted: "For most outsiders, an encounter with the discipline of geography may suggest that it studies everything, from global environmental change at one extreme to the minutiae of body-space at the other. It spans the physical, environmental, and social sciences, and reaches into the humanities too." Johnston queried whether geography "is now coming apart at the seams."

Perhaps the most helpful way to conceive of geography is to acknowledge that it is characterized by diversity and fragmentation but that, at the same time, the core ideas—what are called recurring themes in the introduction to this book—are always evident (Johnston, 2005; see Box C.1).

Examining the Issues

BOX C.1 | Moving Beyond the Introductory Level: Be Prepared to Think Again

As you worked through your human geography course, with this textbook as one guide, you may have noticed that human geography is not quite as neatly structured as you imagined it to be. In this respect, the discipline is no different from any other social science. Contrary to first appearances, it does not consist of a set of tidy packages, each with content clearly distinguished from all the others. This textbook may be divided into chapters, each neatly labelled to suggest the contents, but such packaging and labelling are necessary simplifications of a very complex reality.

Simplification and categorization are essential to enable understanding of complexity. But one danger of working within the framework suggested by this or any other book is that some very important tensions are lost. In particular, human geography includes several important cases of one type of categorization—**binary thinking**. Most cases of binary thinking are attempts to handle complexity, but they also have some important consequences for understanding our world and ourselves.

> **binary thinking**
> A simple form of "A or B" categorizing that, especially when applied to humans, often leads to "A or not A" thinking, where the categorization goes beyond merely acknowledging difference to include value judgments, so that, for example, male/female becomes male/not male and white/black becomes white/not white; can lead to racist, sexist, homophobic, and xenophobic thinking and behaviour.

This is especially so because some binaries go beyond simply dividing into two, as in A or B, but imply a power relationship between the two categories, as in A or not A.

Perhaps the single most important binary in human geography is that of thinking about people in terms of self and others or, more bluntly put, of "us" and "them." This is of the A or not A type. This binary is encouraged specifically by our membership in a state—*we* are members of a certain state, but *they* are not—and refers to one aspect of our identity. In common with other A or not A ways of thinking about identity, this one is not simply about difference but also about status. Members of a state often see themselves as in some way superior to those who are not members of that state and hence this identity binary contributes to conflict. Similar comments can be made about binaries such as white-skinned/dark-skinned (white or not-white), man/woman (man or not man), and heterosexual/homosexual (heterosexual or not heterosexual).

Another influential human geographic binary concerns the concepts of culture/economy. These concepts are typically assigned to different boxes—recall that Marx packaged economy in the infrastructure and culture in the superstructure. But this binary disguises the evident fact that many places, such as skyscrapers, and many events, such as Christmas, are both cultural and economic and

Continued

cannot be understood with reference to just one of the two concepts.

In *Spaces of Geographical Thought* (2005), edited by Cloke and Johnston, additional binaries noted include agency/structure (the tension between individuals and the societies of which they are a part), space/place (an evident tension between positivist and humanist positions), nature/culture (an abiding tension in Chapter 4), and local/global (central to the discussion in Chapter 3). We can add others that appear throughout this book: relevant/esoteric; image/reality; control/freedom; inclusion/exclusion; and science/art.

Human geography has many examples of binary thinking. Perhaps the time has come to think instead about building bridges.

It has been claimed that physical geographers are from Mars and human geographers are from Venus (see Viles, 2005). But even if these metaphors contain truth, relationships have always existed between the human and physical worlds, at all spatial scales and at all times since human life first appeared. It is true that much contemporary human geography does not make any explicit reference to physical geography. Yet human beings have always had to take physical geography into account in their shaping of the human world, and human activities have had an ever-increasing impact on the physical world. In order to grasp these fundamental relationships, a basic understanding of the physical world is essential.

To say that human activities are closely related to physical environments is not to validate environmental determinism. Rather, it is to acknowledge a central and pervasive relationship. Humans make decisions, as individuals and as groups, but they make them with a multitude of physical factors in mind. Success in adaptation—that is, coping—means adapting to both the physical environment and relevant human variables.

In their book on geography and social theory, Dear and Wolch (1989: 3) make the following statement:

> The journey along Mulholland Drive, atop the Hollywood Hills, provides one of the world's great urban vistas. To the south lies the Los Angeles basin, a glittering carpet. To the north, the San Fernando Valley (still part of the City of Los Angeles) unfolds in an equivalent mass of freeways, office towers, and residential subdivisions. There is probably no other place in North America where such an overpowering expression of the human impact on landscape can be witnessed. And yet, the physical landscape cannot be denied. Even in this region of almost 12 million people, the landscape still contains and molds the city.

A Synthesizing Discipline

Notwithstanding debates about what geography may or may not be, the discipline is important to society because it offers the best framework for investigating human impacts on global and local environments. Most ecological research benefits immeasurably from some understanding of the links between human and physical geography. The nineteenth-century American geographer and congressman George Perkins Marsh wrote in 1864:

> the earth is fast becoming an unfit home for its noblest inhabitant, and another era of equal human crime and human improvidence, and of like duration . . . would reduce it to such a condition of impoverished productiveness, of shattered surface, of climatic excess, as to threaten the depravation, barbarism, and perhaps even extinction of the species. (Marsh, 1965: 43)

Focused, like Marsh, on the relationship between humans and land, geographers continue to play an important role in both environmental analysis and environmental education.

Handling Data

Human and physical geography serve society through the development and application of remote sensing and electronic data processing. In an increasingly complex world with ever-larger data sets, remote sensing and global positioning systems (GPS) allow us to acquire data rapidly and repeatedly, while electronic data processing permits the conversion of data into maps and facilitates data analysis. Perhaps the best-known recent advance is the development of GIS—integrated computer systems for the input, storage, analysis, and output of spatial data.

These new ways of handling data are much more than mere extensions of earlier procedures. In an important sense, they are changing the nature of geographic information and the role that such information plays in society. A traditional map is a single product manufactured for users, whereas a GIS can be manipulated by the user. Data can be added or removed depending on specific interests, scales can be changed (the user can zoom in on a certain area, for example), and the data can be updated as new information becomes available. Remember, new tools prompt new thoughts.

Understanding and Solving Problems

Human geography facilitates understanding of people and place by explaining how different peoples see and organize space differently. Hence, geographers are acutely aware that there are no simple or universal solutions to problems. Effective changes can be made only following full analysis of spatial variations. Sensitive to spatial variations and the significance of place, human geographers know that solutions to a problem in one area may not be solutions to similar problems in another area. This awareness of the need for sensitivity is evident, albeit in different ways, in all our major philosophical traditions—empiricist, positivist, humanist, Marxist, and postmodernist. Understanding space and place is important in itself and because it allows us to propose solutions to social and environmental problems. This focus is particularly evident in the vital subdiscipline of cultural geography, introduced in Chapters 6 and 7. This progressive subdiscipline is increasingly concerned with highlighting regional and global inequalities and working towards solutions.

Values

What underlying values direct human geographic research? The general answer is that human geographic values are socially specific, reflecting the values that legitimate the societies in which we live. As we saw in Chapter 2, the positivist school of geography purported to be value-free whereas Marxist and humanistic geography contributed the important insight that values are socially constructed and therefore subject to change. This general answer does not address precise questions about values, such as the relative merits of environmental preservation on the one hand and economic gain on the other hand (Buttimer, 1974). As discussed on several occasions in this text, the views held on such matters tend to reflect larger ideological preferences.

BEING A HUMAN GEOGRAPHER: WHERE WE BEGAN

By now the advantages of training in human geography should be clear; the benefits of being human geographers follow logically. Training in human geography allows us to understand our world, and being human geographers allows us to put that training into practice. Most students of human geography do not formally become human geographers but find employment in a wide range of professions. Training in human geography encourages understanding of the importance of spatial scale, global and local alike. It emphasizes interactions between humans and land as well as emotional attachments to place and develops an appreciation of the major global changes in population numbers, culture, political identity, and land use. This understanding is combined with an enviable set of skills that includes not only traditional quantitative and qualitative research procedures but also the distinctive tools of cartography, remote sensing, and GIS. It is not surprising that human geographers find employment in such diverse arenas as teaching, business, industry, government, consulting, urban and regional planning, conservation and historical preservation, and libraries and archives. Most important, any training in human geography provides lifelong benefits that we might summarize under the heading of "geographical imagination."

C. Wright Mills (1959) introduced the important concept of the sociological imagination to describe the value of seeing the connections between one's own experiences and problems and the larger social world. A related and no less valuable perspective has developed over the years in the field of human geography. What is the geographical imagination? Simply put, it is appreciation of the relevance of space and place to all aspects of human endeavour. Peattie (1940: 33) recognized this nearly three-quarters of a century ago, when he wrote that geography "explains ways of living in all their myriad diversity." More recently, Gregory (1994) offered a detailed commentary on geographical imagination as sensitivity to space and place in the effort to understand ways of life.

Geographical imagination is independent of philosophical preference:

- From an essentially positivistic viewpoint, Morrill (1987: 535) observed, "If there is not a convincing theory of why and how humans create places and imbue them with meaning, then it is time to develop that theory."
- From an essentially humanistic viewpoint, Prince (1961: 231) asserted, "Good geographical description demands not only respect for truth, but also inspiration and direction from a creative imagination. . . . It is the imagination that gives [the facts] meaning and purpose through the exercise of judgement and insight." This is similar to what Bunkše (2004: 12) refers to as a "geographic sensibility."
- Finally, from a Marxist viewpoint, Harvey (1973: 24) observed, "This imagination enables the individual to recognize the role of space and place in his own biography, to relate to the spaces he sees around him, and to recognize how transactions between individuals and between organizations are affected by the space that separates them."

Three different geographers, three different theoretical perspectives, one fundamental conclusion. As Daniels (2011: 186) has recently asserted: "In keeping with the exploratory tradition of geography, it is worth affirming the importance of the geographical imagination, as a matter of both practical wisdom and scholarly reflection, and not least for its pleasure and enchantment, for people's love of learning about the world and their place in it."

The concept of geographical imagination is an appropriate note on which to conclude this introductory text. I hope that, as students of human geography, you have gained an appreciation and a sensitivity to the importance of space and place in understanding the world that is our home.

Summary

Changing Human Landscapes

Although the days of discovering new worlds are over, human geography is by no means fixed or static. As our world is constantly changing, old facts disappear and new ones appear. Among the most important areas where changes are occurring are population numbers, urban growth, and resource use. Political, economic, and cultural circumstances also change at the global scale, the precise consequences of which remain uncertain. Yet one clear impact is increasing pressure on tribal peoples, who are disappearing at an alarming rate.

Changing Human Geography

Human geography is an academic discipline that changes in response to the changing requirements of the society of which it is a part. Space and place are constantly being reinforced as our key concepts; our links with physical geography are a major strength; new technical approaches offer ever-improving ways to handle data; and our increasing understanding of issues is leading to more and better solutions.

Being A Human Geographer

Develop and use your geographical imagination.

Suggested Readings

Visit the companion website for a list of suggested readings.

APPENDIX 1

ON THE WEB

This appendix provides useful links to online resources that expand on topics covered in the text. The list also appears on this book's companion website.

Introduction

Central Intelligence Agency: The World Factbook

https://www.cia.gov/library/publications/the-world-factbook/index.html

Updated every two weeks, this publication is a major source of basic information, with maps and country data.

Geography in the News

www.rgs.org/OurWork/Schools/School+Members+Area/School+Members+Area

This site is maintained by the Royal Geographical Society with the Institute of British Geographers. It is an up-to-date source for all things geographical and a good source of quality information.

The Atlas of Canada

www.nrcan.gc.ca/earth-sciences/geography/atlas-canada

The atlas is an excellent resource for learning about the geography of Canada, both physical and human.

Chapter 1

Canadian Association of Geographers

www.cag-acg.ca/en

This site offers useful information on geography in Canada. The association is intended primarily for professional geographers and those whose employment is closely related to their geographic education.

Canadian Geographic

www.cangeo.ca

This is the website for *Canadian Geographic* magazine, an informative source that deals with a wide range of geographic issues for Canada.

National Geographic

www.nationalgeographic.com

Undoubtedly the best-known geography magazine, *National Geographic* is popular yet often of scholarly significance. The magazine and website are very informative and well written, with inspiring photography.

Chapter 2

Worldmapper

www.worldmapper.org

This site includes a quite wonderful and thought-provoking collection of maps on a wide range of topics. It employs an innovative technique: the maps are equal-area cartograms, also known as density-equalizing maps. The cartogram resizes each territory according to the variable being mapped.

Environmental Systems Research Institute

www.esri.com

ESRI created ArcView and ArcInfo, two of the most widely used GIS software packages. The company's site includes useful general information on GIS technologies.

Chapter 3

European Union

europa.eu/index_en.htm

The EU's website presents news and statistics on its member countries.

International Forum on Globalization

www.ifg.org

The IFG brings together leading activists concerned with the consequences of globalization processes.

World Trade Organization

www.wto.org/index.htm

The WTO's site contains a wealth of information on the organization, as well as on trade and related issues.

Chapter 4

World Resources Institute

www.wri.org

This site includes information and regional analyses on the full range of environmental issues, including climate change, biodiversity, ecosystem changes, and resource use and abuse. (The information is also useful for Chapter 10.)

Earthwatch

www.un.org/earthwatch

Earthwatch is a broad UN initiative aimed at coordinating, harmonizing, and encouraging environmental observation activities among all UN agencies. It covers all major issues and world regions.

Environment Canada

www.ec.gc.ca

Environment Canada offers a comprehensive overview of the Canadian environment and of environmental issues.

Chapter 5

Population Reference Bureau

www.prb.org

The most fundamental and reliable of many websites dealing with population matters, the private, non-profit, US-based Population Reference Bureau publishes an annual *Population Data Sheet* that provides basic demographic data for most countries in the world. This site includes such data and commentaries on specific issues.

Daily Population Updates

www.census.gov/cgi-bin/ipc/popclockw

This US Census Bureau website provides the estimated total world population on a daily basis.

Canadian Population Data

www.statcan.gc.ca/start-debut-eng.html

Statistics Canada maintains this comprehensive site, which includes recent census data.

Chapter 6

The UN Refugees Agency

www.unhcr.org

The Office of the United Nations High Commissioner for Refugees maintains this regularly updated site that provides a wealth of information and commentary.

UNDP

www.undp.org

The home page of the United Nations Development Programme includes links to past volumes of the annual *Human Development Report* and other useful publications.

Millennium Development Goals

www.worldbank.org/mdgs

This site offers comprehensive and regularly updated World Bank information about the Millennium Development Goals.

Chapter 7

Landscapes as World Heritage

whc.unesco.org/en/culturallandscape

This UNESCO website details World Heritage Sites designated as cultural landscapes. These areas are viewed as expressions of the long and intimate relationships between humans and the environments they occupy.

World Languages

www.ethnologue.com

This site is a comprehensive coverage of language families and all known languages of the world.

Religious Adherents

www.adherents.com/Religions_By_Adherents.html

This site features data on major world religions, with detailed discussions.

Chapter 8

The Myth of Race

www.pbs.org/race/000_General/000_00-Home.htm

An excellent source of reasoned information about the idea of race, this site is a companion to the PBS television series *RACE—The Power of an Illusion*.

Gender

www.globalfundforwomen.org

The Global Fund for Women promotes women's economic security, health, education, and leadership. The organization discusses issues, highlights success stories, and encourages involvement.

Tourism

www.wttc.org

The World Travel & Tourism Council's website includes press releases, news, reports, and statistical data on global tourism from an industry, pro-tourism perspective.

Chapter 9

Global Issues

www.globalissues.org/issue/65/geopolitics

This part of the Global Issues website includes a wide variety of articles on various geopolitical issues.

Democracy

www.freedomhouse.org

Freedom House reports on democracy and freedom globally, from an American perspective.

Chapter 10

Food and Agriculture Organization

www.fao.org

This UN agency offers relevant facts and figures and topical discussions on all matters related to agricultural activity and food supply. (The site is also very useful for Chapters 4 and 6.)

Agricultural and Environmental Education

www.world-agriculture.com

The intent of this website is to promote agricultural lifestyles that respect the natural world and facilitate human relations.

Organic Farming

ec.europa.eu/agriculture/organic/home_en

This European Commission website features much useful discussion of organic farming.

Chapter 11

UN Habitat

http://unhabitat.org

The UN Human Settlements program is mandated to promote socially and environmentally sustainable urban centres.

Statistics Canada

www.statcan.gc.ca/daily-quotidien/120208/dq120208a-eng.htm

This page contains information on population numbers and has links to specific urban data.

Globalization and World Cities Research and Network (GaWC)

www.lboro.ac.uk/gawc/index.html

This valuable source for researching global cities provides much data and discussion.

Chapter 12

Canada Mortgage and Housing Corporation

www.cmhc-schl.gc.ca/en

The CMHC website has information about housing and related issues in Canada.

Improving the Living Environment

www.bestpractices.org

This UN site contains over 1,100 proven solutions, from more than 120 countries, to the common social, economic, and environmental problems of an urbanizing world.

Homelessness in Canada

www.homelesshub.ca/SOHC2014

The Homeless Hub offers resources, news, and statistics on homelessness in Canada.

UN-Habitat

urbandata.unhabitat.org

This page from the UN-Habitat site features statistics and resources about the world's cities, prospects, and problems.

Chapter 13

Modern History Sourcebook

www.fordham.edu/halsall/mod/modsbook14.html

Fordham University's website includes a detailed overview of the Industrial Revolution.

International Energy Authority

www.iea.org

The International Energy Authority, established in November 1974, has gained recognition as one of the world's most authoritative sources for energy statistics. Its massive annual studies of oil, natural gas, coal, and electricity are indispensable tools for scholars as well as energy policy-makers and companies involved in the energy field.

Canada

ic.gc.ca

Industry Canada's site mostly covers news items and governmental matters.

Conclusion

Facing the Future
www.facingthefuture.org/default.aspx

This site discusses some global problems, focusing on scarce resources, poverty, conflict, and the environment.

Now is also a good time to revisit the sites listed for the introduction.

APPENDIX 2

GLOBAL PHYSICAL GEOGRAPHY

This appendix provides an overview of planet earth as a habitable environment for humans and an account of 10 principal global environment regions.

A HABITABLE PLANET

The earth is a habitable planet for humans—the only such planet known to us. By *habitable*, we mean that physical processes and physical environments on this planet permitted the emergence and subsequent spread and growth of human life. Indeed, only a series of quite remarkable coincidences have made earth habitable for humans. The planet's distance from the sun, its axial tilt, its 24-hour rotation, and its "minor" variations in surface relief all combine to spread the sun's energy relatively evenly. The evolution of the earth required the presence of water to form oceans, an average surface temperature of 15°C (59°F), and the presence of a particular kind of atmosphere. Interestingly, the first life on earth evolved in water; the subsequent movement of life onto land converted the initial carbon dioxide atmosphere into an oxygen-rich one. This atmospheric transformation included the creation of the ozone layer (ozone is a form of oxygen) that prevents the penetration of ultraviolet rays that would sterilize the land surface. The oxygen-rich atmosphere also allowed the emergence of animal life.

From our earthbound perspective, the planet's surface is tremendously varied. The atmosphere is constantly changing; there are mountains as high as 8,854 m (29,048 feet) above sea level (Mount Everest), land areas 408 m (1,338 feet) below sea level (the shores of the Dead Sea), and known ocean depths of 10,927 m (35,849 feet) below sea level (south of the Mariana Islands in the Pacific Ocean). The planet frequently experiences dramatic changes—volcanic eruptions, earthquakes, tidal waves. On another level, all these variations and changes are insignificant. An accurate hand-sized model of the earth feels as smooth as a peach.

The earth is approximately spherical in shape, with an equatorial circumference of 40,067 km (24,897 miles) and a polar circumference of 40,000 km (24,855 miles). This shape is the result of gravity, the force that is also responsible for the tendency of the earth's surface area to be layered: a dense substance, rock, is at the bottom; a less dense substance, water, is often above the rock; and the least dense substance, air, is above the other two. Such an arrangement is crucial for human life, which has access to all three essential substances. The sphere moves in two ways. First, it rotates on its axis; one rotation defines a day. Second, it orbits the sun; one revolution defines a year.

Rotation

Rotation imposes a diurnal cycle on much of the life on earth. The end points of the axis of rotation are defined as the poles, north and south. Except for the poles, all locations on the earth's surface move as the earth rotates, and one circuit of the earth is a line of latitude. Because the earth is a sphere, the longest line of latitude lies midway between the poles: this is the equator. Lines from one pole to the other are lines of longitude. Unlike lines of latitude, these are not parallel, but they converge at the poles and are farthest apart at the equator. Lines of latitude are measured in degrees, with the equator defined as 0° and the poles therefore as 90°N and 90°S; lines of longitude are similarly measured in degrees, in this case from an arbitrary 0° line known as the prime meridian, which runs through the Greenwich Observatory in

London, England. The geographic grid commonly used to describe locations is based on latitude and longitude.

Revolution

Revolution produces seasonal climatic changes and variations in length of daylight. Figure A2.1 illustrates the four seasonal positions of the earth relative to the sun and introduces the terms *solstice* and *equinox*. The summer solstice—that is, when the hemisphere reaches maximum inclination towards the sun so that it is directly overhead at latitude 23.5°—occurs on 21 or 22 June in the northern hemisphere and on 21 or 22 December in the southern hemisphere. The winter solstice occurs on 21 or 22 December in the northern hemisphere and on 21 or 22 June in the southern hemisphere. On this date, the hemisphere reaches maximum inclination away from the sun. The two equinoxes, vernal and autumnal (spring and fall), are the halfway points between the solstices. On both occasions the axial inclination is at exactly 90° with respect to the sun.

solstice
Occurs twice each year when the sun is vertically overhead at the farthest distance from the equator, once 23.5°N and once 23.5°S of the equator; for example, when at 23.5°N, there is maximum daylight (longest day) in the northern hemisphere and minimum daylight (shortest day) in the southern hemisphere.

equinox
Occurs twice each year, in spring and fall, when the sun is vertically overhead at the equator; on this day the periods of daylight and darkness are both 12 hours long all over the earth.

A great deal of human activity is related to these global physical factors. Diurnal, seasonal, and annual cycles are basic to much social and economic organization. The durations of the day and year are defined by earth movements, while the month is defined by lunar movement around the earth. The hour, by contrast, is an arbitrary human division, as is the use of the equinoxes and solstices as the beginnings of seasons.

The Earth's Crust

The crust or outer layer of the earth is composed of rocks—aggregates of minerals—that have continued forming and changing for at least three billion years. The various rock types are usually classified as sedimentary (those that were formed at low temperatures on the earth's surface); igneous (those that solidified from a molten state); or metamorphic (those that have changed under high temperature and pressure).

One of the most important features of the earth's surface is that it is part land (29 per cent) and part water (71 per cent). The present distribution is the result of a long and ongoing process of movement by the lithosphere—the earth's crust and the uppermost mantle below the crust. The lithosphere is divided into six major and several minor plates that move relative to one another as a result of movements in the liquid mantle deep inside the earth. The continents quite literally float on this liquid mantle, with some four-fifths of their mass below the surface. The idea that the continents also move was suggested as long ago as the seventeenth century, when the English philosopher Francis Bacon (1561–1626) observed that

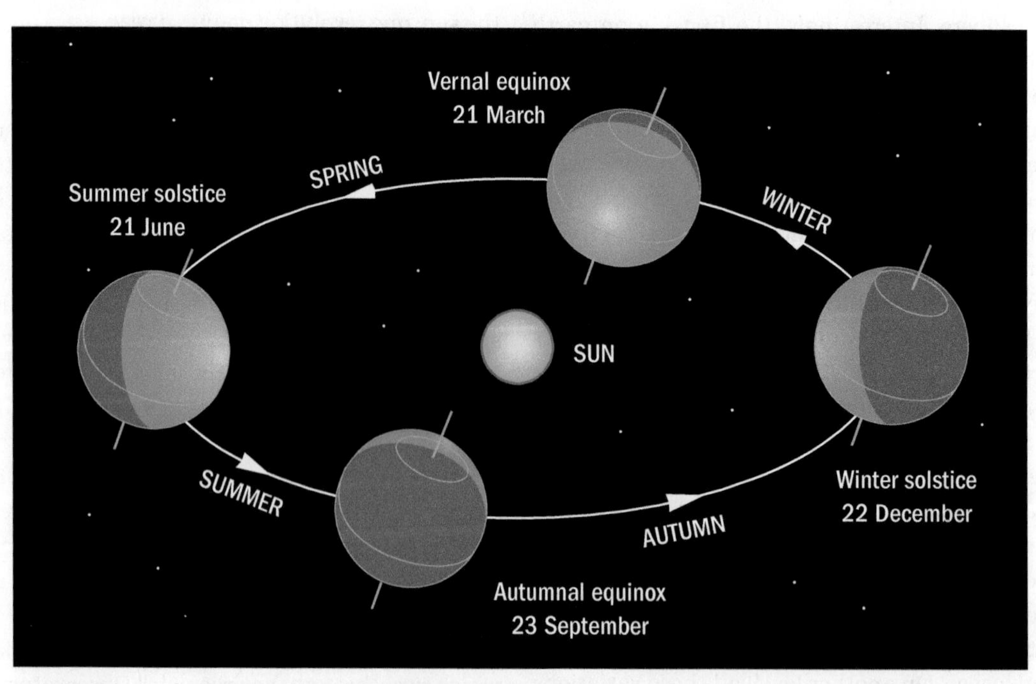

FIGURE A2.1 | The revolution of the earth around the sun

the Atlantic coasts of South America and Africa fitted together like pieces of a jigsaw puzzle. Early scientific evidence was gathered by Alfred Wegener (1880–1930), a German meteorologist, but this idea of moving continents, known as continental drift, did not achieve scientific credibility until the 1960s. Figure A2.2 illustrates the positions of the continents since outward movements began roughly 225 million years ago (MYA). At the present time, the Atlantic Ocean is widening about 2 cm (0.75 inches) each year, Australia and Africa are moving northward, and parts of Africa and western North America may separate from their respective continents.

A second consequence of moving continents is the creation of mountain chains such as the Himalayas. When continents collide—as India and Asia did, perhaps 150 MYA—mountain chains are thrust upwards. The Appalachians are much older, resulting from a collision of North America and Africa perhaps 350 MYA.

Earthquakes are a third consequence; they occur when parts of the crust move over other parts as a result of movements in the liquid mantle. Once stress exceeds rock strength, an earthquake occurs. Such movements typically take place along fault lines (such as the San Andreas fault in California), which are lines of crustal weakness. Volcanoes also occur along lines of crustal weakness, often plate boundaries. A volcanic eruption involves liquid from the mantle (magma) being forced upwards and onto the earth's surface.

As already noted, approximately two-thirds of the earth's crust is covered by water. The physical landscape beneath is constantly changing, while the water itself is also involved in a series of movements. Waves are caused by wind; tides result from the gravitational pull of the moon (and, to a

continental drift
The idea that the present continents were originally connected as one or two land masses that separated and drifted apart.

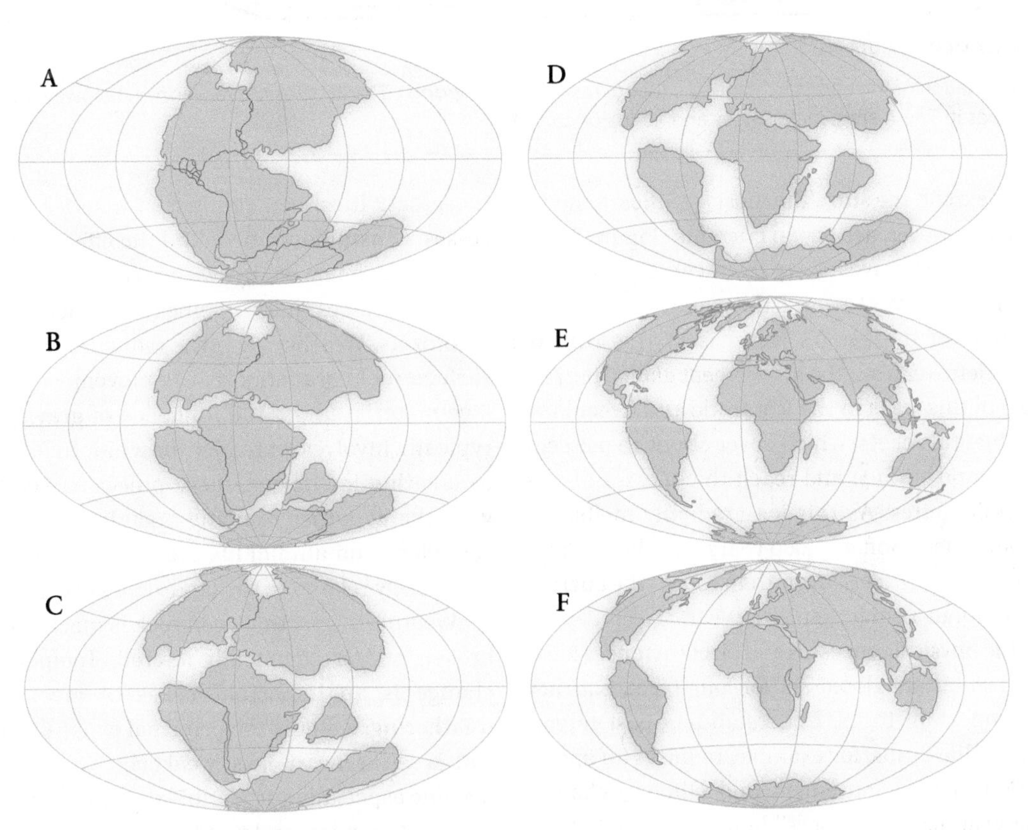

FIGURE A2.2 | Moving continents

In (A), beginning 222 MYA, there is one land mass, a super-continent called Pangaea. By 180 MYA (B), Pangaea is breaking up; Laurasia is moving north, and the southern land mass, Gondwanaland, is also breaking away. In (C), 135 MYA, the process continues, with the beginnings of the North and South Atlantic oceans. In (D), 65 MYA, Madagascar has separated from Africa and India is moving towards Asia, but Australia remains linked to Antarctica. Map (E) shows the present distribution of land and sea, and (F) shows a likely scenario 50 million years in the future: Australia continues to move north; East Africa is shifting farther east, California (west of the San Andreas fault) has separated from North America, and the Mediterranean Sea is shrinking as Africa moves north.

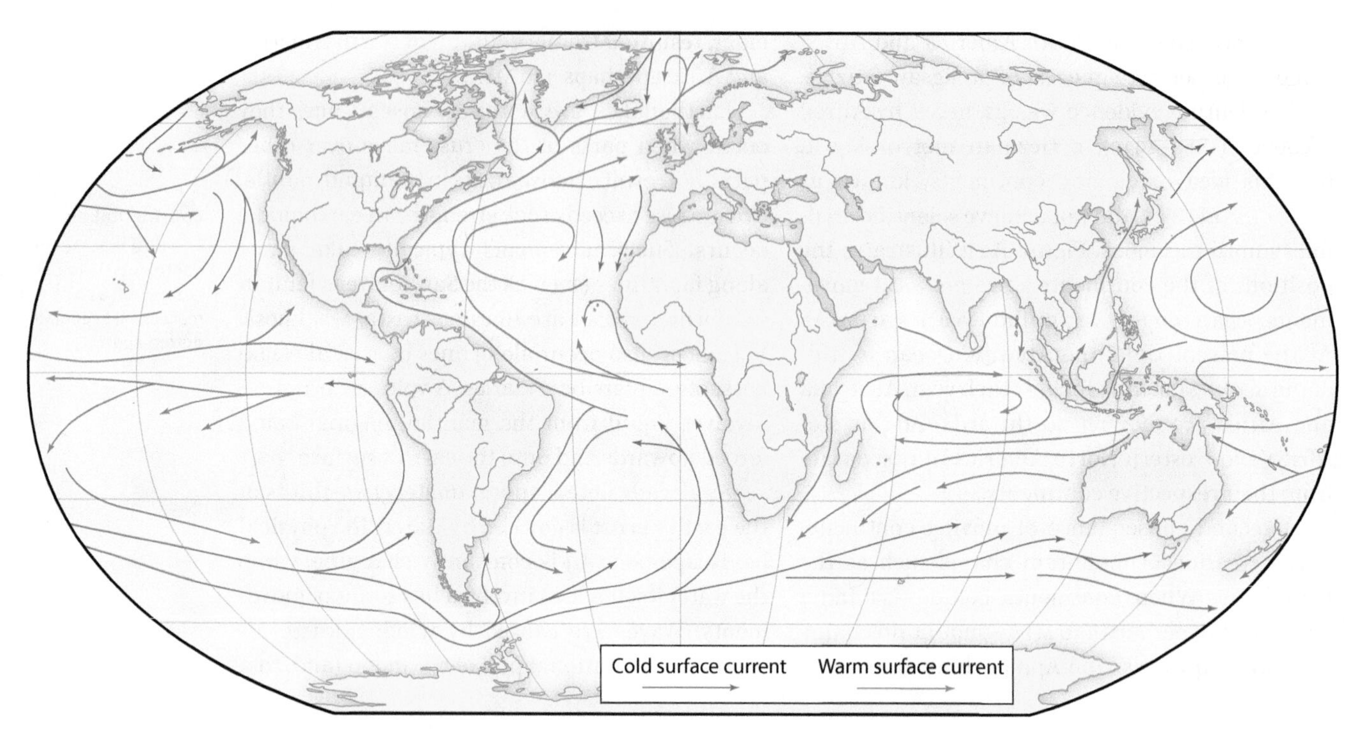

FIGURE A2.3 | The major ocean currents

The prevailing winds and the rotation of the earth combine to cause gyres (large whirlpools). These gyres move warm water away from the equator and cold water towards it. There are 5 major gyres and 30 major currents.

lesser extent, the sun). The most significant movements, however, are ocean currents (Figure A2.3). Warm water moves away from the equator and cold water moves towards the equator.

Some of the earth's water is in the form of ice—a leftover from the most recent glacial period. Most of this ice is in the Antarctic and Greenland continental sheets, which cover about 10 per cent of the land surface and reach thicknesses of several kilometres. As this ice gradually melts, the sea level rises. Some 20,000 years ago, the sea level was 100 m (328 feet) lower than today; the current rise is about 1 cm (0.39 inch) a year.

The physical processes and circumstances identified so far are global and/or long-term, but they are important to us. Understanding past migrations of life forms, for example, requires an appreciation of land movement and climatic change. One example of the possible long-term effects of a physical process—exacerbated by human activity—is the melting of all ice, which would result in a sea-level rise of 60 m (197 feet). Such a rise would flood many of the largest cities in the world.

The Physical Landscape

The physical landscape in which human activities take place (and which is continually modified by those activities) is a product of a series of processes. Crustal movement and the effects already noted are followed by weathering and gradational processes that combine to produce specific landforms. Weathering is the reduction in size of surface rock; gradation is the movement of surface material under the influence of gravity and typically involves water, ice, or wind. In addition to creating landforms, these processes combine with the activities of living organisms to create soil. Plant and animal life are, in a circular fashion, closely related to the landscapes created.

Weathering occurs when rock is mechanically broken and/or chemically altered. Temperature change is an important cause of mechanical weathering; water is the principal cause of chemical weathering. Rocks weakened in this way become especially susceptible to gradational processes. Water, ice, and wind can erode, move, and deposit material. Many of the earth's distinctive physical landscapes are clear evidence of these processes at work.

Water is probably the most important cause of gradation. A continuous transfer of water from sea to air to land to sea is taking place. River erosion creates V-shaped valleys and waterfalls, while river deposition creates flood plains and deltas. Seawater also affects land surfaces, and coastlines undergo constant change; cliffs are the result of erosion, beaches the result of deposition. Other parts of the earth's surface show the effects of ice and snow. During the most recent glacial period, ice covered an area approximately three times greater than that covered today, including Northern Europe and Canada. Areas formerly covered by ice show distinctive evidence both of erosion (e.g. U-shaped valleys) and of deposition (e.g. the low hills called drumlins). Wind erosion results either from wind-blown materials colliding with stable materials (as in sandblasting) or, more often, the wind's removal of loose particles from a surface. Sand dunes are among the best-known products of wind deposition of material. Each of these processes affects parts of the earth's surface at all times. Physical landscapes, then, are subject to continuous modification—almost fine-tuning. In principle, the eventual outcome, without crustal movement or climatic change, would be an almost featureless physical landscape.

Our physical environment is much more than the distribution of landforms. Soils, vegetation, and climate are additional components of a physical environment. Figures A2.4, A2.5, A2.6, and A2.7 provide general outlines of global landforms, soils, vegetation, and climate, respectively. Figure A2.4 maps the major cordilleran *belts*—the backbones of the continents. The slope and ruggedness of local landforms in particular are major considerations in human decisions about the location of economic activities, such as agriculture, industry, and transport.

Soil types are distributed in a systematic fashion (Figure A2.5). There are close general links both to climate and to natural vegetation; these are two of the most important determinants of soil type. Soils are important for their agricultural potential and for their capacity to support buildings. Much of the European migration to the New World was closely tied to the availability of soils suitable for commercial agriculture; in the late nineteenth century, many settlers found suitable grassland soils in Canada, Argentina, and Australia.

Figure A2.6 maps the global distribution of natural vegetation. The four principal types are grassland, forest, desert scrub, and tundra. Much

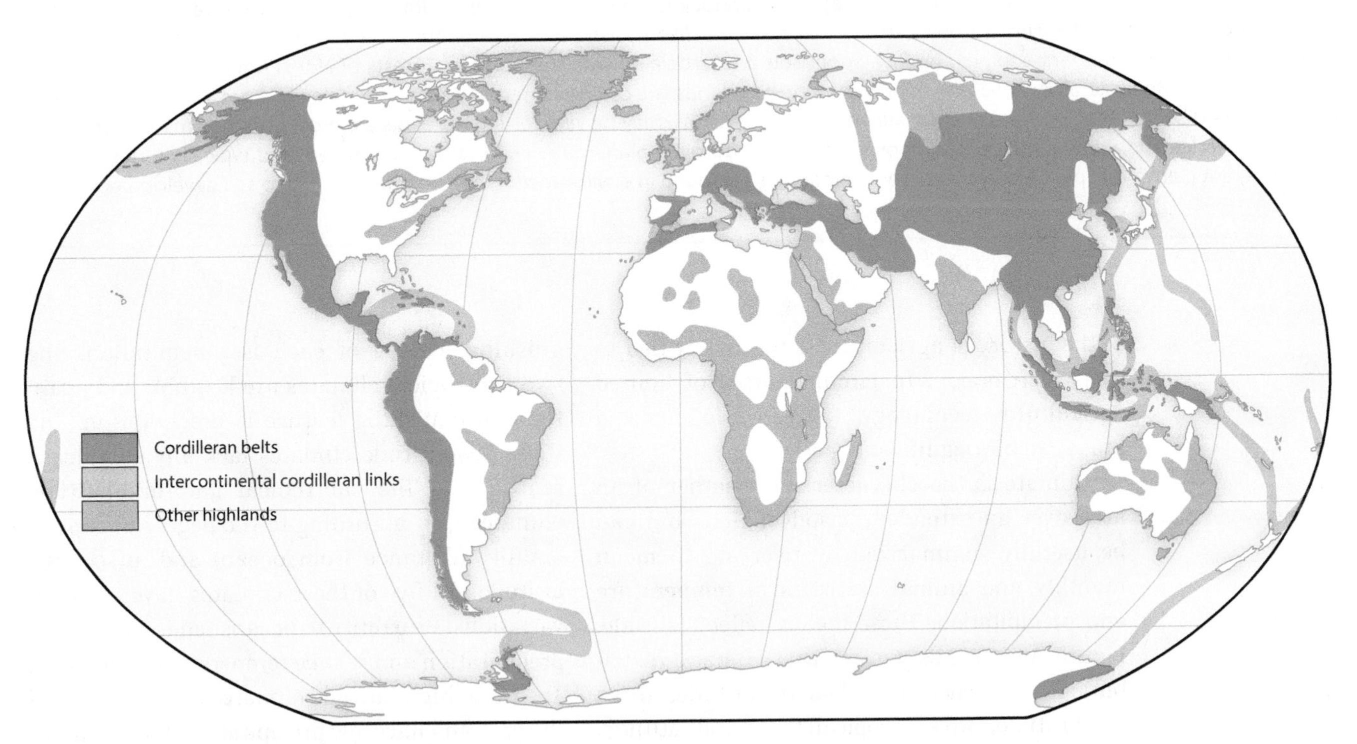

FIGURE A2.4 | Global cordilleran belts

The backbones of the continents, these cordilleran belts also extend through ocean areas.

Source: Adapted from Scott, R. C. 1989. *Physical Geography*. St Paul, MN: West Publishing, 295.

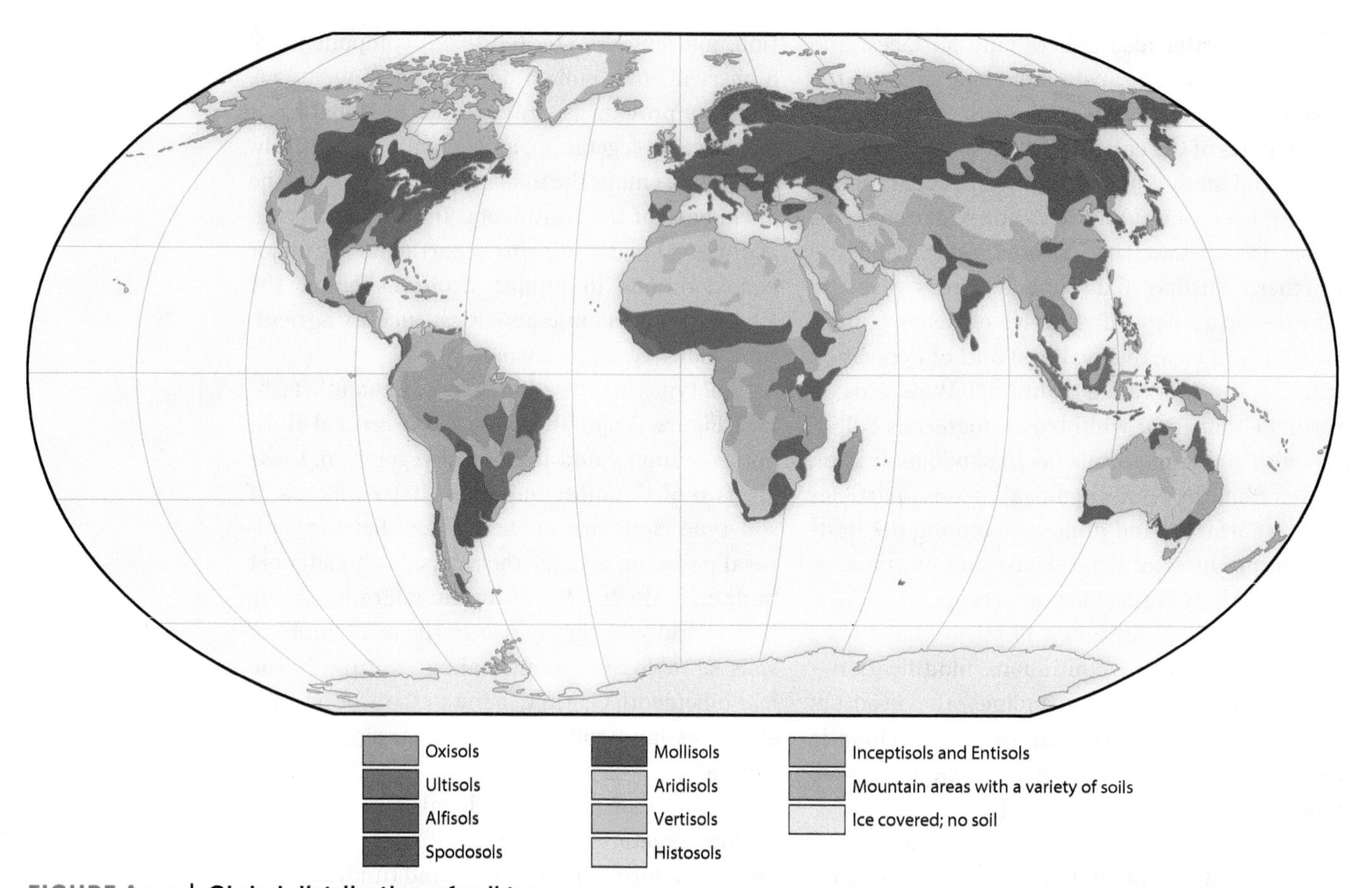

FIGURE A2.5 | Global distribution of soil types

Oxisols are typical of tropical rain forests and are deep soils but not inherently fertile: if the natural vegetation is removed, the soil is soon impoverished. Ultisols, in the humid tropics, are low in fertility. Alfisols, dispersed throughout low- and mid-latitude forest/grassland transition areas, are of moderate to high fertility. Spodosols are associated with the northern forests of North America and Eurasia; low fertility is typical. Mollisols, associated with middle-latitude grasslands, are often very fertile. Aridisols, located in areas of low precipitation, lack water. Vertisols, limited to areas of alternate wet and dry seasons, may be very fertile. Histosols are most common in northern areas of North America and Eurasia; they are composed of partially decayed plant material, usually waterlogged, and typically fertile for only water-tolerant plants. Inceptisols and entisols are immature soils found in environments that are poorly suited to soil development.

Source: Adapted from Scott, R. C. 1989. *Physical Geography.* St Paul, MN: West Publishing, 276–7.

of the history of agriculture is intimately tied to their distribution. In circumstances of limited agricultural technology, for example, forested areas can be a significant barrier.

Climate is the characteristic weather of an area over an extended period of time and can be usefully summarized by referring to mean monthly and annual statistics of temperature and precipitation. These means reflect latitude. Figure A2.7 makes it clear that climatic distributions are generally linked to latitude. Three low-latitude, two subtropical, three mid-latitude, and three high-latitude climates are mapped, along with one highland climate. Low-latitude climates are characteristically hot; the distinguishing feature of each is precipitation. The two subtropical climates are less hot, and again the distinguishing feature is precipitation. The three low-latitude climates lack any significant seasonality. The subtropical and mid-latitude climates vary according to the degree of continentality (distance from ocean) and, of course, latitude. All five of these climates have seasonal variations. Proximity to oceans tends to increase precipitation and reduce temperature variations. The three high-latitude climates are cold most of the year and have low precipitation. Finally, highland climates display many local variations, since temperature and precipitation are affected by altitude in addition to latitude.

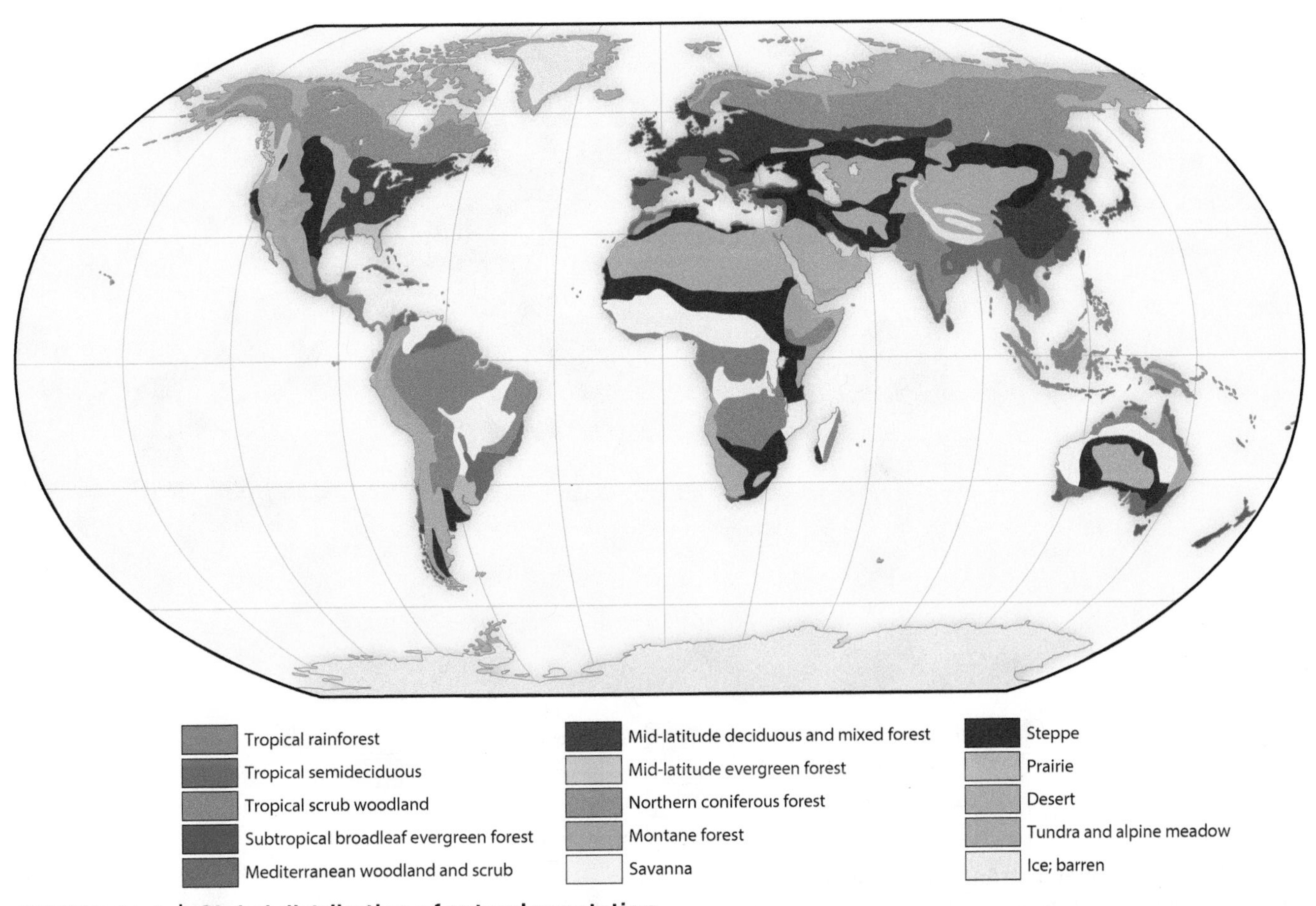

FIGURE A2.6 | Global distribution of natural vegetation
Differences in forest types are related to climate. Forests become less dense and have fewer species as they move northward or southward from the equator. Deciduous trees dominate the middle latitudes, evergreens the higher latitudes. Savanna, steppe, and prairie are composed chiefly of grasses. Desert vegetation is closely linked with an arid climate. Tundra vegetation is the most cold-resistant.

Source: Adapted from Scott, R. C. 1989. *Physical Geography*. St Paul, MN: West Publishing, 232–3.

GLOBAL ENVIRONMENTS

The preceding section highlighted a wide variety of physical processes and environments. These processes often have an intimate relationship to human activities but do not in any sense determine them. Although we need to be aware of physical factors at all levels, we must not assume any causal relationships. As this book demonstrates, we humans make the decisions—albeit with many physical factors in mind! This section presents a brief summary of the major global environments (Figure A2.8), integrating much of the material above in a more applied human context. The notion of global environmental regions is useful here. Indeed, those regions are closely correlated with climatic regions in particular. Humans have favoured certain environments and have made changes to all areas they have settled in.

Tropical Rain Forests

These areas are located close to the equator and experience high rainfall—typically exceeding 2,000 mm (78 inches) per annum—and high

Guenter Guni/iStockphoto

Evergreen cloud forest on the slopes of Mt Ruwenzori on the border of Uganda and DR of Congo. Mt Ruwenzori (5,114 metres) is the third highest mountain in Africa.

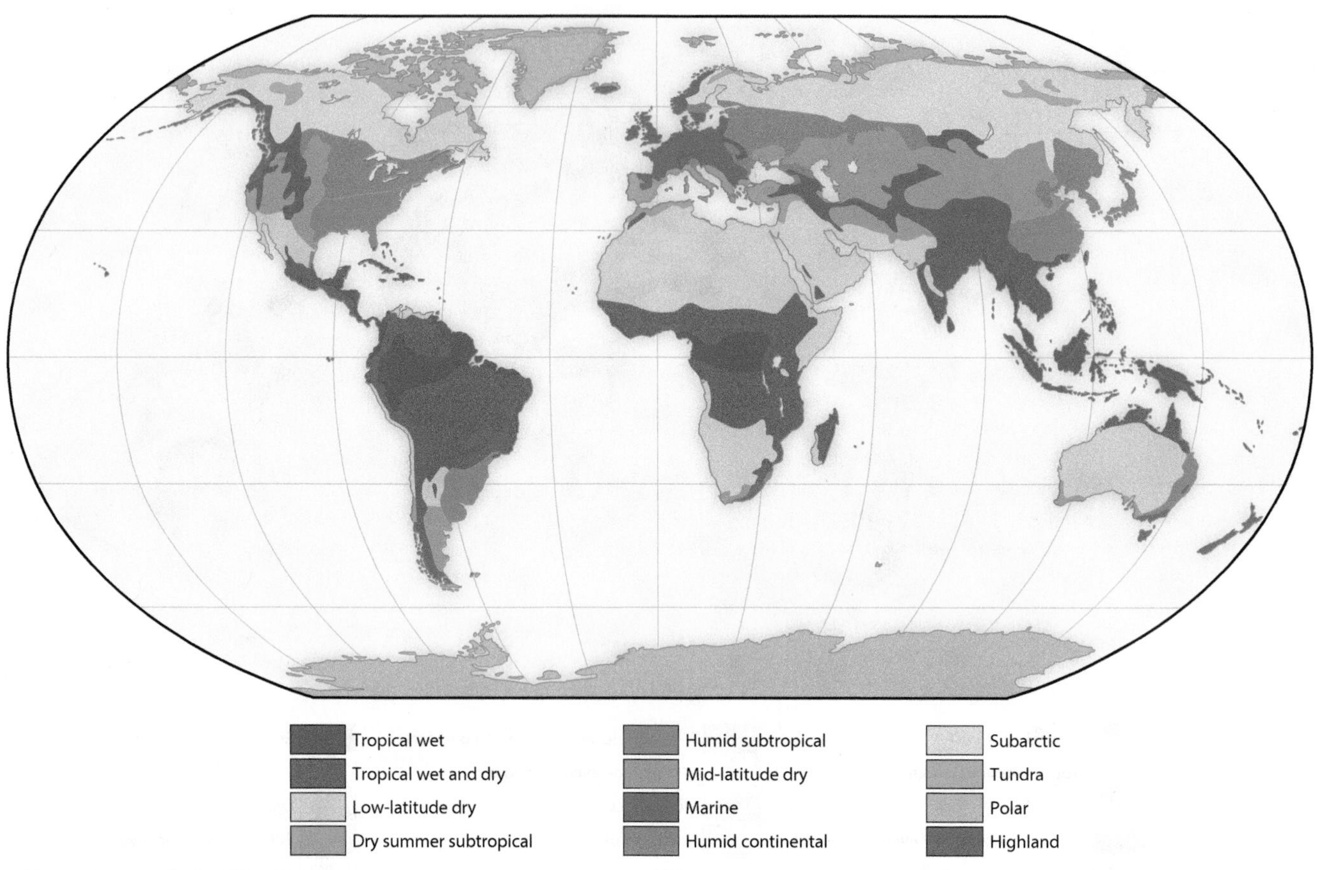

FIGURE A2.7 | Global distribution of climate

These types of climate, distinguished on the basis of moisture and temperature, are derived from the Köppen classification. Tropical wet, tropical wet and dry, and low-latitude dry climates may be grouped as low-latitude climates; dry summer subtropical, humid subtropical, mid-latitude dry, marine, humid, and continental climates as mid-latitude climates; and subarctic, tundra, and polar climates as high-latitude climates.

Source: Adapted from Scott, R. C. 1989. *Physical Geography*. St Paul, MN: West Publishing, 158–9.

temperatures—24°C to 30°C (75°F to 86°F). They lack seasonal variation. Broad-leaf trees dominate, and the vegetation is dense and highly varied. Nevertheless, once the vegetation is removed, the soil is characteristically shallow and easily eroded. Tropical rain forests have not become major agricultural regions. Among the economically valuable products of rain forests are mahogany and rubber, as well as numerous starchy food plants, such as manioc, yams, and bananas.

Flooded fields around Inle Lake in the monsoon area of Burma.

Monsoon Areas

Monsoon areas are characterized by marked seasonality, especially an extended dry season and heavy rainfall. Rainfall may be as high as 2,000 cm (780 inches) in some areas. The typical natural vegetation is deciduous forest, but all monsoon areas have been changed dramatically as a result of human activity. Agriculture has been practised for a long time, with rice as the basic crop in many such areas. Population densities are usually high.

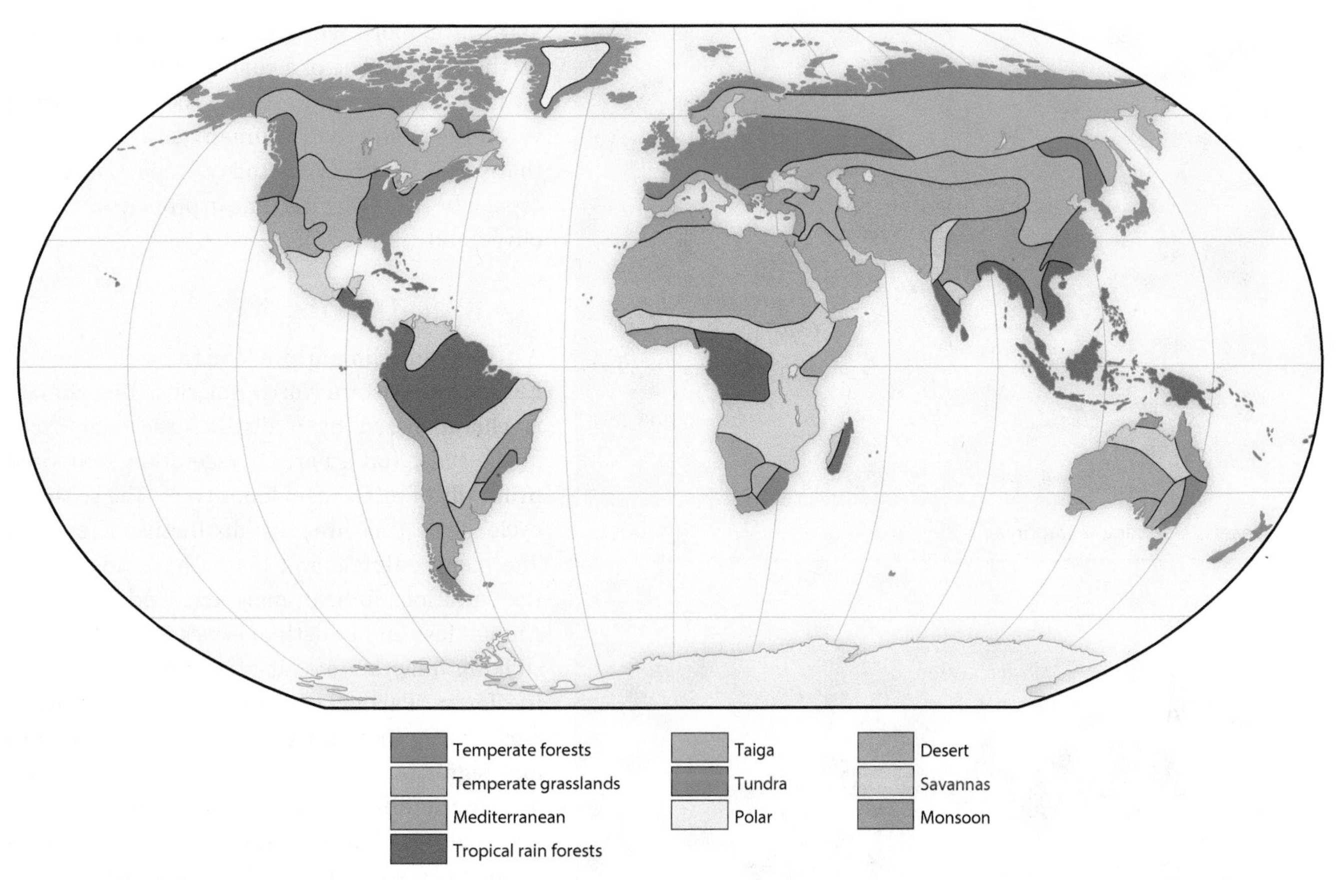

FIGURE A2.8 | Generalized global environments

Climate is clearly a key consideration in many human decisions, but weather is also important. Weather can have a major impact on agricultural yields. Atypical or violent weather can cause floods or droughts, and hurricanes—vast tropical storms—can cause serious damage to populated land areas.

Savannas

Savannas are located to the north and south of the tropical rain forests. They have a distinctive climate: a very wet season and a very dry season. The natural vegetation is a combination of trees and grassland. Africa has the most extensive savanna areas, which are rich in animal species. To date, most savannas are relatively undeveloped economically, but they offer great promise for increases in animal and crop yields. Important grain crops include sorghum in Africa and wheat in Asia.

Deserts

A desert area has low and infrequent rainfall, perhaps as low as 100 mm (4 inches) per annum and perhaps no rain for 10 years. Areas are typically defined as arid if they receive less than 255 mm (10 inches) of rain per annum and as semi-arid if they receive less than 380 mm (15 inches). There are hot, cold, sandy, and rocky deserts. In all deserts, water is a crucial consideration in human activities. Irrigation and nomadic pastoralism are two human responses to desert environments.

Mediterranean Areas

There are only a few Mediterranean environments. They are characterized by long, hot, dry summers

Savanna, Naukluft, Namibia.

Xebeche/iStockphoto

Desert, Argentina (Patagonia).

akrp/iStockphoto

Mediterranean area, Provence-Alpes-Côte d'Azur, France.

Вадим Орлов/iStockphoto

Temperate grassland, rural Russia.

and warm, moist winters. The typical location is on the western side of a continent in a temperate latitude. Evergreen forests are the usual natural vegetation, although human activity means that scrub land prevails today. Important crops, especially in the Mediterranean proper, are wheat, olives, and grapes.

Temperate Forests

A temperate climate predominates in much of Europe and eastern North America. This climate is characterized by a distinct seasonal cycle, fertile soils, and a natural vegetation composed primarily of mixed deciduous trees. The seasonal cycle gives this area its distinctive character. The many animals and plants have adapted to the variations in heat, moisture, food, and light. Annual loss and growth of leaves is an especially striking adaptation that affects all other forest life forms. Particularly since the beginnings of agriculture, these areas have been attractive to and significantly altered by humans, and they are now dominated by agriculture, industry, and urbanization. These changes began in Europe about 7,000 years ago, in Asia about 6,000 years ago, and in North America about 1,000 years ago. Indeed, the removal of forest cover has been so dramatic that today there is a general recognition of the need to preserve and even expand the remaining forests.

Temperate Grasslands

North American prairies, Russian steppes, South African veld, and Argentine pampas are four of the large temperate grasslands. Located in continental interiors, they have insufficient rainfall for forest cover and tend to show great annual temperature variations. Such grasslands were once occupied by particular large migrating herbivores such as the bison (North America); antelopes, horses, and asses (Eurasia); and pampas deer (South America). After remaining almost devoid of humans until the late seventeenth century, these grasslands became the granaries of the world in the twentieth century. Their soils are typically suited to grains; wheat became dominant in most areas.

Boreal Forests

Boreal forests, often called taiga, are located only in the northern hemisphere—in North America and Eurasia. Their winters are long and cold, their summers brief and warm, and precipitation is limited. The natural vegetation is evergreen forest. Agriculture is very limited, and the major economic importance of taiga consists in forest products and minerals.

Tundra

Located north of the boreal forest, this area experiences long, cold winters; temperatures average −5°C (23°F) and there are few frost-free days. Perennially frozen ground—permafrost—is characteristic and presents a severe challenge to human activities. Only a few plants and animals have adapted to this environment because the land has been exposed for only about 8,000 years, since the last retreat of ice caps. There has been little time for soil to form.

Polar Areas

Polar conditions prevail in both hemispheres and are best represented by the Greenland and Antarctic ice sheets. Antarctica is a continental land mass under an ice sheet that reaches thicknesses of 4,000 m (more than 13,000 feet). These areas lack vegetation and soil and have proven hostile to human settlement and activity. Indigenous inhabitants of these areas lived in close relationship to the physical environment, subsisting on hunting and fishing. In recent years non-indigenous inhabitants have greatly increased in number, particularly because of the discovery of oil.

These brief descriptions of 10 global environments make no attempt to describe all areas. Nor do all areas within a particular environment have

Dougall Photography/iStockphoto

Boreal forest, Prince Albert National Park, Saskatchewan.

kalimf/iStockphoto

Temperate mixed deciduous forest of lower Great Lakes–St Lawrence Valley region.

Philip Dearden

Tundra, Herschel Island, Yukon.

Philip Dearden

Polar region, Ellesmere Island.

uniform characteristics. Regions merge into one another; the tropical rain forest merges into savanna and then into desert, for example. Rarely are there clear, sharp divisions. As a result, many areas are difficult to classify. This outline emphasizes that the earth comprises many environments with which humans have to cope. All these environments have been affected by human activities, and all need to be considered in any attempt to comprehend our human use of the earth.

Suggested Readings

de Blij, H. J., P. O. Muller, R. S. Williams Jr, C. T. Conrad, and P. Long. 2009. *Physical Geography: The Global Environment*, 2nd Canadian edn. Toronto: Oxford University Press.

A detailed, comprehensive text on physical geography; very readable, well illustrated, and student-friendly throughout.

Trenhaile, A. S. 2009. *Geomorphology: A Canadian Perspective*, 4th edn. Toronto: Oxford University Press.

A textbook presenting a systematic explanation of Canada's landforms.

On the Web

Canada

www.nrcan.gc.ca

The home page of Natural Resources Canada, on which you can find the Geological Survey of Canada, offering a wide variety of information on physical geography and links to relevant provincial and territorial websites.

United States

www.usgs.gov

The US Geological Survey site includes clear overviews of physical geographic landscapes; offers a good account of the application of GIS technologies.

GLOSSARY

accessibility A variable quality of a location, expressing the ease with which it may be reached from other locations.

acculturation The process by which an ethnic group is absorbed into a larger society while retaining aspects of distinct identity.

acid rain The deposition on the earth's surface of sulphuric and nitric acids formed in the atmosphere as a result of fossil fuel and biomass burning; causes significant damage to vegetation, lakes, wildlife, and built environments.

acropolis The fortified religious centre of cities in ancient Greece; the literal translation is "highest point in the city."

adaptation The process by which humans adjust to a particular set of circumstances; changes in behaviour that reduce conflict with the environment.

agglomeration The spatial grouping of humans or human activities to minimize the distances between them.

agora The centre of ancient Greek civic life; the place where public meetings, trials of justice, social interaction, and commercial exchange took place.

agricultural revolution The slow transition, beginning about 12,000 years ago, from foraging to food production through plant and animal domestication.

agricultural surplus Agricultural production that exceeds the sustenance needs of the producer and is sold/exchanged with others.

alienation The circumstance in which a person is indifferent to or estranged from nature or the means of production.

anarchism A political philosophy that rejects the state and argues that social order is possible without a state.

animism A general name for beliefs that attribute a spirit or soul to natural phenomena and inanimate objects.

anthropocentric Regarding humans as the central fact of the world; stressing the centrality of humans to the detriment of the rest of the world.

apartheid The South African policy by which four groups of people, as defined by the authorities, were spatially separated between 1948 and 1994.

areal differentiation From Hartshorne, a synonym for *regional geography*.

assimilation The process by which an ethnic group is absorbed into a larger society and loses its own identity.

authority The power or right, usually mutually recognized, to require and receive the obedience of others.

back-office activities Repetitive office operations, usually clerical in nature and performed using telecommunications, that can be located anywhere in or out of the city, including relatively low-rent areas.

binary thinking A simple form of "A or B" categorizing that, especially when applied to humans, often leads to "A or not A" thinking, where the categorization goes beyond merely acknowledging difference to include value judgments, so that, for example, male/female becomes male/not male and white/black becomes white/not white; can lead to racist, sexist, homophobic, and xenophobic thinking and behaviour.

biomass The mass of biological material present in an area, including both living and dead plant material.

business services Services that are provided primarily for other businesses, including financial, administrative, and professional activities such as accounting, advertising, banking, consulting, insurance, law, and marketing.

capitalism A social and economic system for the production of goods and services based on private enterprise.

carrying capacity The maximum population that can be supported by a given set of resources and a given level of technology.

cartography The conception, production, dissemination, and study of maps.

caste A social rank, based solely on birth, to which an individual belongs for life and that limits interaction with members of other castes.

catastrophists Those who argue that population increases and continuing environmental deterioration are leading to a nightmarish future of food shortages, disease, and conflict.

ceiling rent The maximum rent that a potential land user can be charged for use of a given piece of land.

census The periodic collection and compilation of demographic and other data relating to all individuals in a given country at a particular time.

central business district (CBD) The social, cultural, commercial, and political centre of the city; usually characterized by high-rise office and residential towers, key municipal government buildings, and civic amenities.

central place An urban centre that provides goods and services for the surrounding population; may take the form of a hamlet, village, town, city, or megacity.

central place theory A theory to explain the spatial distribution of urban centres with respect to their size and function.

centrifugal forces In political geography, forces that make it difficult to bind an area together as an effective state; in urban geography, forces that favour the decentralization of urban land uses.

centripetal forces In political geography, forces that pull an area together as one unit to create a relatively stable state; in urban geography, forces that favour the concentration of urban land uses in a central area.

chain migration A process of movement from one location to another through time sustained by social links of kinship or friendship; often results in distinct areas of ethnic settlement in rural or urban areas.

charter population The dominant or majority cultural group in an urban area; the host community.

chorology A Greek term revived by nineteenth-century German geographers as a synonym for *regional geography*.

choropleth map A thematic map using colour (or shading) to indicate density of a particular phenomenon in a given area.

city A legally incorporated self-governing unit; an inhabited place of greater size, population, or importance than a town or village.

civilization A contested term because it can be understood to mean that some groups are civilized while others are not; traditionally understood to refer to a culture with agriculture and cities, food and labour surpluses, labour specialization, social stratification, and state organization.

class A large group of individuals of similar social status, income, and culture.
Cold War The period of confrontation without direct military conflict between Western (led by the US) and communist (led by the USSR) powers that began shortly after the end of World War II and lasted until the early 1990s.
collective consumption The use of services produced and managed on a collective basis.
colonialism The policy of a state or people seeking to establish and maintain authority over another state or people.
commercial agriculture An agricultural system in which the production is primarily for sale.
competitive capitalism The first of three phases of capitalism, beginning in the early eighteenth century; characterized by free-market competition and laissez-faire economic development.
conservation A general term referring to any form of environmental protection, including preservation.
constructionism The school of thought according to which all our conceptual underpinnings (e.g. ideas about identity) are socially constructed and therefore contingent and dynamic, not given or absolute.
consumer services Services that are provided primarily for individual consumers, such as retail, hospitality, food, leisure, health care, education, and social welfare; represents approximately 50 per cent of employment in most countries of the more developed world.
contextualism Broadly, the idea that it is necessary to take into account the specific context within which any research is conducted.
conurbation A continuously built-up area formed by the coalescing of several expanding cities that were originally separate.
core–periphery The concept that states are often unequally divided between powerful cores and dependent peripheries.
cornucopians Those who argue that advances in science and technology will continue to create resources sufficient to support the growing world population.
creole A pidgin language that assumes the status of a mother tongue for a group.
critical geography A collection of ideas and practices concerned with challenging inequalities, as these are evident in landscape.
cultural adaptation Changes in technology, organization, and ideology that permit sound relationships to develop between humans and their physical environment.
cultural minority A member of a minority group whose minority status is based on factors other than skin colour, such as language, religion, lifestyle, ethnic origin, etc.
cultural regions Areas in which there is a degree of homogeneity in cultural characteristics; areas with similar landscapes.
culture A complex term that typically refers to the way of life of a society's members; also usefully understood as referring to our ability first to analyze and then to change the physical environments that we encounter.
cycle of poverty The idea that poverty and deprivation are transmitted intergenerationally, reflecting home background and spatial variations in opportunities.
deconstruction A method of critical interpretation applied to texts, including landscapes, that aims to show how an author's or reader's multiple positioning (in terms of class, gender, and so on) affects the creation or reading of the text.
deglomeration The spatial separation of humans or human activities so as to maximize the distances between them.
deindustrialization Loss of manufacturing activity and related employment in a traditional manufacturing region in the more developed world.
democracy A form of government involving free and fair elections, openness and accountability, civil and political rights, and the rule of law.
demographic transition The historical shift of birth and death rates from high to low levels in a population; mortality declines before fertility, resulting in substantial population increase during the transition phase.
demography The study of human populations.
density A measure of the number of geographic facts (for example, people) per unit area.
dependence In political contexts, a relationship in which one state or people is dependent on, and therefore dominated by, another state or people.
dependency theory A theory that centres on the relationship between dependence and underdevelopment.
desertification The process by which an area of land becomes a desert; typically involves the impoverishment of an ecosystem because of climate change and/or human impact.
development A term that should be handled with caution because it has often been used in an ethnocentric fashion; typically understood to refer to a process of becoming larger, more mature, and better organized; often measured by economic criteria.
developmentalism Analysis of cultural and economic change that treats each country or region of the world separately in an evolutionary manner; assumes that all areas are autonomous and proceed through the same series of stages.
devolution A process of transferring power from central to regional or local levels of government.
dictatorship An oppressive, anti-democratic form of government in which the leader is often backed by the military.
diffusion The spread of any phenomenon over space and its growth through time.
discourse A system of ideas or knowledge that serves as the context through which new facts and ideas are understood.
disorganized capitalism The most recent form of capitalism, characterized by disorganization and industrial restructuring.
distance The spatial dimension of separation; a fundamental concept in spatial analysis.
distance decay The declining intensity of any pattern or process with increasing distance from a given location.
distribution The pattern of geographic facts (for example, people) within an area.
domestication The process of making plants and/or animals more useful to humans through selective breeding.
donut effect A popular but colloquial term that refers to a pronounced difference in the growth rates between a core city (slow growth or no growth) and its surrounding areas (faster growth), a pattern that resembles the North American fried confection; usually characterized by people moving out of the core or inner-suburbs of a city and moving into newer peripheral suburbs.
doubling time The number of years required for the population of an area to double its present size, given the current rate of population growth.
ecocentric Emphasizing the value of all parts of an ecosystem rather than, for example, placing humans at the centre, as in an anthropocentric emphasis.
ecology The study of relationships between organisms and their environments.

economic operator A model of human behaviour in which each individual is assumed to be completely rational; economic operators maximize returns and minimize costs.
economic rent The surplus income that accrues to a unit of land above the minimum income needed to bring a unit of new land into production at the margins of production.
ecosystem An ecological system; comprises a set of interacting and interdependent organisms and their physical, chemical, and biological environment.
ecotourism Tourism that is environmentally friendly and allows participants to experience a distinctive ecosystem.
edge city A centre of office and retail activities located on the edge of a large urban centre.
effect (or friction) of distance A measure of the restraining effect of distance on human movement.
empiricism A philosophy of science based on the belief that all knowledge results from experience and therefore gives priority to factual observations over theoretical statements.
energy The capacity of a physical system for doing work.
entrepot A city, usually a port, that functions as an intermediary for trade and transshipment and that exports raw materials and manufactured goods.
environmental determinism The view that human activities are controlled by the physical environment.
essentialism Belief in the existence of fixed unchanging properties; attribution of "essential" characteristics to groups.
ethnic group A group whose members perceive themselves as different from others because of a common ancestry and shared culture.
ethnocentrism A form of prejudice or stereotyping that presumes that one's own culture is normal and natural and that all other cultures are inferior.
ethnography The study and description of social groups based on researcher involvement and first-hand observation in the field; a qualitative rather than quantitative approach.
existentialism A philosophy that sees humans as responsible for making their own natures; it stresses personal freedom, decision-making, and commitment in a world without absolute values outside individuals' personal preferences.
exonym A name given to people (or a place) by a group other than the people to which the name refers (or who are not native to the territory within which the place is situated).
fecundity A biological term; the ability of a woman or man to produce a live child; refers to potential rather than actual number of live births.
federalism A form of government in which power and authority are divided between central and regional governments.
feminism The movement for and advocacy of equal rights for women and men and a commitment to improve the position of women in society.
fertility Generally, all aspects of human reproduction that lead to live births; also used specifically to refer to the actual number of live births produced by a woman.
feudalism A social and economic system prevalent in Europe prior to the Industrial Revolution; land was owned by the monarch, controlled by lords, and worked by peasants who were bound to the land and subject to the lords' authority.
fieldwork A means of data collection; includes both qualitative (for example, observation) and quantitative (for example, questionnaire) methods.
filtering The idea that, through time, housing units typically experience a transition from being occupied by members of one income group to members of a different income group; downward filtering is more usual than upward filtering.
first effective settlement A concept based on the likely importance of the initial occupancy of an area in determining later landscapes.
flexible accumulation Industrial technologies, labour practices, relations between firms, and consumption patterns that are increasingly flexible.
Fordism A group of industrial and broader social practices introduced by Henry Ford, including the mass-production assembly line, higher wages, and shorter working hours.
foreign direct investment (FDI) Direct investment by a government or multinational corporation in another country, often in the form of a manufacturing plant.
formal (or uniform) region A region identified as such because of the presence of some particular characteristic(s).
forum The centre of Roman civic, commercial, administrative, and ceremonial life; combined the functions of the ancient Greek acropolis and agora.
front-office activities Skilled occupations requiring an educated, well-paid workforce; because image and face-to-face contact with others is important, these activities favour prestige locations in major office buildings in city centres.
functional (or nodal) region A region that comprises a series of linked locations.
garden city A planned settlement designed to combine the advantages of urban and rural living; an urban centre emphasizing spaciousness and quality of life.
gated community A high-status residential subdivision or community with access limited to residents and other authorized people such as domestic workers, tradespeople, and visitors; often surrounded by a perimeter wall, fence, or buffer zone such as a golf course.
gateway city A city that is a key point of entry to a major geographic region or country for goods or people, often via an international airport, container shipping port, or major rail centre; a city in which several different cultural traditions are absorbed and assimilated.
Gemeinschaft A term introduced by Tönnies; a form of human association based on loyalty, informality, and personal contact; assumed to be characteristic of traditional village communities.
gender The social aspect of the relations between the sexes.
genocide An organized, systematic effort to destroy a group defined in ethnic terms; usually the targeted group is seen as living in the "wrong place."
gentrification A process of inner-city urban neighbourhood social change resulting from the in-movement of higher-income groups; originates from *gentry*, a term referring to people of high social standing and immediately below those of noble birth.
geographic information system (GIS) A computer-based tool that combines the storage, display, analysis, and mapping of spatially referenced data.
***géographie Vidalienne* (or *la tradition Vidalienne*)** French school of geography initiated by Paul Vidal de la Blache at the end of the nineteenth century and still influential today, focusing on the study of human-made (cultural) landscapes.
geopolitics The study of the importance of space in understanding international relations.

geopolitik The study of states as organisms that choose to expand in territory in order to fulfill their "destinies" as nation-states.
gerrymandering The realignment of electoral boundaries to benefit a particular political party.
Gesellschaft A term introduced by Tönnies; a form of human association based on rationality and depersonalization; assumed to be characteristic of urban dwellers.
ghetto A residential district in an urban area with a concentration of a particular ethnic group.
global city A city that is an important node in the global economy; a dominant city in the global urban hierarchy; sometimes referred to as a world city.
globalization A complex combination of economic, political, and cultural changes that have long been evident but that have accelerated markedly since about 1980, bringing about a seemingly ever-increasing connectedness of both people and places.
green belt A planned area of open, partially rural, land surrounding an urban area; an area where urban development is restricted.
gross domestic product (GDP) A monetary measure of the market value of goods and services produced by a country over a given time period (usually one year); provides a better indication of domestic production than GNP.
gross national income (GNI) or gross national product (GNP) A monetary measure of the market value of goods and services produced by a country, plus net income from abroad, over a given period (usually one year).
hearth The area where a particular cultural trait originates.
heartland theory A geopolitical theory of world power based on the assumption that the land-based state controlling the Eurasian heartland held the key to world domination.
hegemony A social condition in which members of a society interpret their interests in terms of the world view of a dominant group.
hinterland The market area surrounding a central place; the spatial area from which the providers of goods and services in a central place draw their customers.
historical materialism An approach associated with Marxism that explains social change by reference to historical changes in social and material relations.
Holocene Literally, "wholly recent"; the post-glacial period that began 10,000 years ago and was preceded by the Pleistocene.
homeland A cultural region especially closely associated with a particular cultural group; the term usually suggests a strong emotional attachment to place.
humanism A philosophy centred on such aspects of human life as value, quality, meaning, and significance.
hypothesis In positivist philosophy, a general statement deduced from theory but not yet verified.
iconography The description and interpretation of visual images, including landscape, in order to uncover their symbolic meanings; the identity of a region as expressed through symbols.
idealism A humanistic philosophy according to which human actions can be understood only by reference to the thought behind them.
idiographic Concerned with the unique and particular.
image The perception of reality held by an individual or group.
Industrial Revolution The process that converted a fundamentally rural society into an industrial society, beginning in England around 1750; primarily a technological revolution associated with new energy sources.
informal sector A part of a national economy involved in productive paid labour but without any formal recognition, governmental control, or remuneration.
informal settlement A concentration of temporary dwellings, neither owned nor rented, at the city's periphery; related to rural-to-urban migration, especially in less developed countries; sometimes referred to as squatter settlement or shanty town.
innovations Introduction of new inventions or ideas, especially ones that lead to change in human behaviour or production processes.
interaction The relationship or linkage between locations.
international division of labour The current tendency for high-wage and high-skill employment opportunities, often in the service sector, to be located in the more developed world, while low-wage and low-skill employment opportunities, often in the industrial sector, are located in the less developed world.
irredentism The view held by one country that a minority living in an adjacent country rightfully belongs to the first country.
isochrones Lines on a map of equal travel time from a given starting point. One example of an isoline, which generally allows map readers to infer change with distance and to estimate specific values at any location on the map.
isodapanes In Weberian least-cost industrial-location theory, lines of equal additional transport cost drawn around the point of minimum transport cost.
isopleth map A map using lines to connect locations of equal data value.
isotims In Weberian least-cost industrial location theory, lines of equal transport costs around material sources and markets.
landrace A local variety of a domesticated animal or plant species that is well adapted to a particular physical and cultural environment.
landscape A major concern of geographic study; the characteristics of a particular area especially as created through human activity.
landscape school American school of geography initiated by Carl Sauer in the 1920s and still influential today; an alternative to environmental determinism, focusing on human-made (cultural) landscapes.
Landschaftskunde A German term, introduced in the late nineteenth century and best translated as "landscape science"; refers to geography as the study of the landscapes of particular regions.
latitude Angular distance on the surface of the earth, measured in degrees, minutes, and seconds, north and south of the equator (which is the line of 0° latitude); lines of constant latitude are called parallels.
law In positivist philosophy, a hypothesis that has been proven correct and is taken to be universally true; once formulated, laws can be used to construct theories.
less developed world All countries not classified as "more developed" (see *more developed world*); countries characterized by a low standard of living.
life cycle The process of change experienced by individuals over their lifespans; often divided into stages (such as childhood, adolescence, adulthood, old age), each of which is associated with particular forms of behaviour.
limits to growth The argument that both world population and world economy may collapse because available world resources are inadequate.

lingua franca An existing language used as a common means of communication between different language groups.
lithosphere The outer layer of rock on earth; includes crust and upper mantle.
locale The setting or context for social interaction; a term that has become popular in human geography as an alternative to *place*.
location A specific part of the earth's surface; an area where something is situated.
location theory A body of theories explaining the distribution of economic activities.
longitude Angular distance on the surface of the earth, measured in degrees, minutes, and seconds, east and west of the prime meridian (the line of 0° longitude that runs through Greenwich, England); lines of constant longitude are called meridians.
malapportionment A form of gerrymandering, involving the creation of electoral districts of varying population sizes so that one party will benefit.
malnutrition A condition caused by a diet lacking some food necessary for health.
Maoism The revolutionary thought and practice of Mao Zedong (1893–1976); based on protracted revolution to achieve power and socialist policies after power is achieved.
Marxism The body of social and political theory developed by Karl Marx, in which mode of production is the key to understanding society and class struggle is the key to historical change.
material index An index devised by Weber and used in industrial location theory to show the extent to which the least-cost location for a particular industrial firm will be either material- or market-oriented.
megacity Metropolitan areas with populations of more than 10 million.
mental map The individual psychological representation of space.
mercantilism A school of economic thought dominant in Europe in the seventeenth and early eighteenth centuries that argued for the involvement of the state in economic life so as to increase national wealth and power.
metropolitan area A region comprising two or more functionally connected urban areas and the less densely populated (or built-up) areas in between; examples include metropolitan New York and the Greater Toronto Area.
minority language A language spoken by a minority group in a state in which the majority of the population speaks another language; may or may not be an official language.
minority population (or groups) A population subgroup that is seen, or that views itself, as somehow different from the general (charter) population; this difference is normally expressed by ethnicity, language, religion, nationality, sexual orientation, lifestyle, or even income (as in the case of the homeless or the extremely wealthy).
mobility The ability to move from one location to another.
model An idealized and structured representation of the real world.
mode of production The organized social relations through which a human society organizes productive activity.
modernism A view that assumes the existence of a reality characterized by structure, order, pattern, and causality.
monarchy The institution of rule over a state by the hereditary head of a family; monarchists are those who favour this system.
more developed world (According to a United Nations classification) Europe, North America, Australia, Japan, and New Zealand; countries characterized by a high standard of living.
mortality Deaths as a component of population change.
multiculturalism A policy that endorses the right of ethnic groups to remain distinct rather than to be assimilated into a dominant society.
multilingual state A state in which the population includes at least one linguistic minority.
nation A group of people sharing a common culture and an attachment to some territory; a term difficult to define objectively.
nationalism The political expression of nationhood or aspiring nationhood; reflects a consciousness of belonging to a nation.
nation-state A political unit that contains one principal national group that gives it its identity and defines its territory.
nativism Intense opposition to an internal minority on the grounds that the minority is foreign.
neighbourhood A part of the city that displays some internal homogeneity regarding type of housing, may be characterized by a relatively uniform income level and/or ethnic identity, and usually reflects certain shared social values.
neo-colonialism Economic relationships of dominance and subordination between countries without equivalent political relationships; often develops after political colonialism ends and the former colony achieves independence but may also occur without prior political colonialism.
nomothetic Concerned with the universal and the general.
normative Focusing on what ought to be rather than what actually is; in normative theory, the aim is to seek what is rational or optimal according to some given criteria.
nuptiality The extent to which a population marries.
offshoring The transferring by a company of production or service provision to another country.
oligarchy Rule by an elite group of people, typically the wealthy.
organized capitalism The second phase of capitalism, beginning after World War II; increased growth of major corporations and increased state involvement in the economy.
Orientalism Western views of the Orient, implying a view of the periphery from the centre; closely associated with post-colonial theory, especially the work of Edward Said.
Other Subordinate group as seen by and contrasted to dominant groups; implies both difference and inferiority.
outsourcing Paying an outside firm to handle functions previously handled inside the company (or government) with the intent to save money or improve quality; a term often used interchangeably (and incorrectly) with *offshoring*; much outsourcing involves offshoring jobs from more developed countries such that they become much lower-paid jobs in less developed countries.
ozone layer Layer in the atmosphere 16–40 km (10–25 miles) above the earth that absorbs dangerous ultraviolet solar radiation; ozone is a gas composed of molecules consisting of three atoms of oxygen (O_3).
pandemic A term used to designate diseases with very wide distribution (a whole country, or even the world); epidemic diseases have more limited distribution.
Paris Club An ideologically neo-liberal grouping of financial officials from 19 of the biggest developed economies—15 European countries plus the US, Canada, Japan, and Australia—loosely formed in 1956 and more formally structured in the 1970s, which provides financial services and organizes debt restructuring, debt relief, and debt cancellation for indebted countries and their creditors.
participant observation A qualitative method in which the researcher is directly involved with the subjects in question.
patriarchy A social system in which men dominate, oppress, and exploit women.

perception The process by which humans acquire information about physical and social environments.
phenomenology A humanistic philosophy based on the ways in which humans experience everyday life and imbue activities with meaning.
phenotype Any physical or chemical trait that can be observed or measured.
physiological density Population per unit of cultivable land.
pidgin A new language designed to serve the purposes of commerce between different language groups; typically has a limited vocabulary.
place Location; in humanistic geography, place has acquired a particular meaning as a context for human action that is rich in human significance and meaning.
placelessness Homogeneous and standardized landscapes that lack local variety and character.
place utility A measure of the satisfaction an individual derives from a location relative to his or her goals.
Pleistocene Geological time period, from about 1.5 million years ago to 10,000 years ago, characterized by a series of glacial advances and retreats; succeeded by the Holocene.
pollution The release of substances that degrade air, land, or water into the environment.
population aging A process in which the proportion of elderly people in a population increases and the proportion of younger people decreases, resulting in increased median age of the population.
population momentum The tendency for population growth to continue beyond the time that replacement-level fertility has been reached because of the relatively high number of people in the child-bearing years.
population pyramid A diagrammatic representation of the age and sex composition of a population; by convention, the younger ages are at the bottom, males are on the left, and females on the right.
positionality The ideological preference and the identity of the researcher as they relate to the subjects of the research.
positivism A philosophy that contends that science is able to deal with only empirical questions (those with factual content), that scientific observations are repeatable, and that science progresses through the construction of theories and derivation of laws.
possibilism The view that the environment does not determine either human history or present conditions; rather, humans pursue a course of action that they select from among a number of possibilities.
post-Fordism A group of industrial and broader social practices evident in industrial countries since about 1970; involves more flexible production methods than those associated with Fordism.
postmodernism A movement in philosophy, social science, and the arts; based on the idea that reality cannot be studied objectively and that multiple interpretations are possible.
power The capacity to affect outcomes; more specifically, to dominate others by means of violence, force, manipulation, or authority.
pragmatism A humanistic philosophy that focuses on the construction of meaning through the practical activities of humans.
primary activities Economic activities concerned directly with the collection and utilization of natural resources.
primate city The largest city in an urban system, usually the capital, which dominates its political, economic, and social life; a city that is more than twice the size of the next largest city in the system.
principle of least effort Considered a guiding principle in human activities; for human geographers, refers to minimizing distances and related movements.
producer services Activities that offer a wide range of services to multinational and other companies that need to respond quickly to changing circumstances, including banking, insurance, marketing, accountancy, advertising, legal matters, consultancy, and innovation services; in recent years, the fastest-growing sector of national economies in most of the more developed countries.
projection Any procedure employed to represent positions of all or a part of the earth's spherical (three-dimensional) surface onto a flat (two-dimensional) surface.
public goods Goods that are freely available to all or that are provided (equally or unequally) to citizens by the state.
qualitative methods A set of tools used to collect and analyze data in order to subjectively understand the phenomena being studied; the methods include passive observation, participation, and active intervention.
quantitative methods A set of tools used to collect and analyze data to achieve a statistical description and scientific explanation of the phenomena being studied; the methods include sampling, models, and statistical testing.
quaternary activities Economic activities concerned with handling or processing knowledge and information; typically involve a high level of skill; highly specialized.
queer theory Ideas developed in gay and lesbian studies and concerned with oppressed sexualities in terms of both social rights and cultural politics.
questionnaire A structured and ordered set of questions designed to collect unambiguous and unbiased data.
race A subspecies; a physically distinguishable population within a species.
racism A particular form of prejudice that attributes characteristics of superiority or inferiority to a group of people who share some physically inherited characteristics.
random sampling Sampling such that every part of the study area or every item in the data set has an equal chance of being selected and is independent of all other parts or items.
range The maximum distance that people are prepared to travel to obtain a particular good or service.
rank-size distribution A descriptive regularity among cities in an urban system; the numerical relationship between city size and rank in an urban system; sometimes referred to as the rank-size rule.
raster A method used in GIS to represent spatial data; divides the area into numerous small cells and pixels and describes the content of each cell.
rational choice theory The theory that social life can be explained by models of rational individual action; an extension of the economic operator concept to other areas of human life.
recycling The reuse of material and energy resources.
redlining A spatially discriminatory practice, favoured by financial institutions, that identified parts of the city regarded as high risk in terms of loans for property purchase and home improvement; affected areas were typically outlined in red on maps.
region A part of the earth's surface that displays internal homogeneity and is relatively distinct from surrounding areas according to some criterion or criteria; regions are intellectual creations.
regionalization A special kind of classification in which locations on the earth's surface are assigned to various regions, which must be contiguous spatial units.

reindustrialization The development of new industrial activity in a region that has earlier experienced substantial loss of traditional industrial activity.
remote sensing A variety of techniques used for acquiring and recording data from points that are not in contact with the phenomena of interest.
renewable resources Resources that regenerate naturally to provide a new supply within a human lifespan.
replacement-level fertility The level of fertility at which a couple has only enough children to replace themselves.
representation A depiction of the world, acknowledging the impossibility to be exact as all such depictions are affected by the researcher's identity.
restructuring In a capitalist economy, changes in or between the various components of an economic system resulting from economic change.
rimland theory A geopolitical theory of world power based on the assumption that the state controlling the area surrounding the Eurasian heartland held the key to world domination.
sacred space A landscape particularly esteemed by an individual or a group, usually (but not necessarily) for religious reasons.
sampling The selection of a subset from a defined population of individuals to acquire data representative of that larger population.
satisficing behaviour A model of human behaviour that rejects the rationality assumptions of the economic operator model; assumes that the objective is to reach a level of acceptable satisfaction.
scale The resolution level(s) used in any human geographic research; most characteristically refers to the size of the area studied but also to the time period covered and the number of people investigated.
scientific method The various steps taken in a science to obtain knowledge; a phrase most commonly associated with a positivist philosophy.
secondary activities Economic activities that process, transform, fabricate, or assemble raw materials derived from primary activities; also activities that reassemble, refinish, or package manufactured goods.
segregation The spatial separation of population subgroups within the wider urban population.
sense of place The deep attachments that humans have to specific locations, such as home, and to particularly distinctive locations.
sexism Attitudes or beliefs that serve to justify sexual inequalities by incorrectly attributing or denying certain capacities either to women or to men.
sex ratio The number of males per 100 females in a population.
sexuality In some feminist and psychoanalytic theory, interpreted as a cultural construct rather than as a biological given; aligned with power and control.
simulation Representation of a real-world process in an abstract form for purposes of experimentation.
site The location of a geographic fact with reference to the immediate local environment.
situatedness An idea that rejects notions of researcher authority and impartiality—knowledge is not neutral and cannot be acquired in some detached and disembodied manner but is partial and located somewhere.
situation The location of a geographic fact with reference to the broad spatial system of which it is a part.
slavery Labour that is controlled through compulsion and is not remunerated; in Marxist terminology, one particular mode of production.
slum A heavily populated informal settlement, usually located within the urban core, and characterized by poverty, substandard housing, crime, and a lack of sanitation, water, electricity, or other basic services; common in less developed world cities today and in more developed world cities in the nineteenth and early twentieth centuries.
socialism A social and economic system that involves common ownership of the means of production and distribution.
society The interrelationships that connect individuals as members of a culture.
sovereignty Supreme authority over the territory and population of a state, vested in its government; the most basic right of a state understood as a political community.
space A real extent; used in both absolute (objective) and relative (perceptual) forms.
spatial monopoly The situation in which a single producer sells the entire output of a particular industrial good or service in a given area.
spatial preferences Individual (sometimes group) evaluation of the relative attractiveness of different locations.
species A group of organisms able to produce fertile offspring among themselves but not with any other group.
spectacle A term referring to places and events that are carefully constructed for the purposes of mass leisure and consumption.
squatter settlements A shanty town; a concentration of temporary dwellings, neither owned nor rented, at the city's edge; related to rural-to-urban migration, especially in less developed countries.
state An area with defined and internationally acknowledged boundaries; a political unit.
state apparatus The institutions and organizations through which the state exercises its power.
stock resources Minerals and land that take a long time to form and hence, from a human perspective, are fixed in supply.
subsistence agriculture An agricultural system in which the production is not primarily for sale but is consumed by the farmer's household.
suburb A residential or mixed (residential and employment) use area on the periphery of the city, typically displaying some degree of homogeneity in terms of economic, social-cultural, and/or built form.
superstructure A Marxist concept that refers to the political, legal, and social systems of a society.
supranational organization A multinational grouping of independent states, where power is delegated to an authority by member governments.
sustainable development A term popularized by the 1987 report of the World Commission on Environment and Development; refers to economic development that sustains the natural environment for future generations.
system A set of interrelated components or objects linked together to form a unified whole.
tariff A tax or customs duty on imports from other countries.
technology The ability to convert energy into forms useful to humans.

telecommute Work at home, using a computer to complete tasks and to communicate with others.
teleology The doctrine that everything in the world has been designed by God; also refers to the study of purposiveness in the world and to a recurring theme in history, such as progress or class conflict.
tertiary activities Economic activities involving the sale or exchange of goods and services; includes distributive trades, such as wholesaling and retailing, and also personal services.
text A term that originally referred to the written or printed page but that has broadened to include such products of culture as maps and landscape; postmodernists recognize that there may be any number of realities, depending on how a text is read.
theory In positivist philosophy, an interconnected set of statements, often called assumptions or axioms, that deductively generates testable hypotheses.
threshold The minimum number of people (market size) required to support the existence of a particular economic function.
time–space convergence A decrease in the friction of distance between locations as a result of improvements in transportation and communication technologies.
topography A Greek term, revived by nineteenth-century German geographers to refer to regional descriptions of local areas.
toponym Place name; evidence provided by place names can be crucial in a historical study of movement and settlement if other sources of information are unavailable.
topophilia The affective ties that humans have with particular places; literally, love of place.
topophobia The feelings of dislike, anxiety, fear, or suffering associated with a particular landscape.
transnationals Large business organizations that operate in two or more countries.
undernutrition Diet inadequate to sustain normal activity.
urban area The spatial extent of the built-up area surrounding and including an incorporated municipality, such as a city; typically assessed by its population size and/or population density and/or nature of residents' employment.
urbanism The urban way of life; associated with a declining sense of community and increasingly complex social and economic organization as a result of increasing size, density, and heterogeneity.
urbanization The spread and growth of cities; an increasing proportion of a population living in urban areas (cities and towns).
urban sprawl The largely unplanned expansion of an urban area into rural areas.
urban structure The arrangement of land uses in cities; related to urban morphology.
vector A method used in GIS to represent spatial data; describes the data as a collection of points, lines, and areas and describes the location of each of these.
vernacular region A region identified on the basis of the perceptions held by people inside and outside the region.
verstehen A research method, associated primarily with phenomenology, in which the researcher adopts the perspective of the individual or group under investigation; a German term best translated as "empathetic understanding."
visible minority A member of a minority group whose minority status is based wholly on the colour of his or her skin; the Canadian government recognizes anyone that is neither white nor Aboriginal as a visible minority.
welfare geography An approach to human geography that maps and explains social and spatial variations.
well-being The degree to which the needs and wants of a society are satisfied.
world systems theory A body of ideas that suggests a division of the world into a core, semi-periphery, and periphery, stressing that the periphery is dependent on the core; has numerous implications for an understanding of the less developed world.
zoning One of the most important tools of the urban planner; controls the use of land and the buildings constructed there.

REFERENCES

INTRODUCTION

Fairgrieve, J. 1926. *Geography in School*. London: University of London Press.

Gritzner, C. F. 2002. "What Is Where, Why There, and Why Care?" *Journal of Geography* 101: 38–40.

Lewis, P. F. 2002. *Careers in Geography*. Washington, DC: Association of American Geographers.

CHAPTER 1

Chiasson, P. 2006. *The Island of Seven Cities: Where the Chinese Settled When They Discovered North America*. Toronto: Random House.

Dohrs, F. E., and L. M. Sommers, eds. 1967. *Introduction to Geography: Selected Readings*. New York: Crowell.

Febvre, L. 1925. *A Geographical Introduction to History*. London: Routledge & Kegan Paul.

Hart, J. F. 1982. "The Highest Form of the Geographer's Art." *Annals, Association of American Geographers* 72: 1–29.

Hartshorne, R. 1939. *The Nature of Geography: A Critical Survey of Current Thought in the Light of the Past*. Lancaster, PA: Association of American Geographers.

Heffernan, M. 2003. "Histories of Geography." In *Key Concepts in Geography*, edited by S. L. Holloway, S. P. Rice, and G. Valentine, 3–22. London: Sage.

Huntington, E. 1927. *The Human Habitat*. New York: Van Nostrand.

Lewthwaite, G. R. 1966. "Environmentalism and Determinism: A Search for Clarification." *Annals, Association of American Geographers* 56: 1–23.

May, J. A. 1970. *Kant's Concept of Geography and Its Relation to Recent Geographical Thought*. Toronto: University of Toronto Press.

Menzies, G. 2002. *1421: The Year China Discovered the World*. London: Bantam Press.

Nisbet, R. 1980. *History of the Idea of Progress*. New York: Basic Books.

Semple, E. 1911. *Influences of Geographic Environment*. New York: Henry Holt.

Spate, O. H. K. 1952. "Toynbee and Huntington: A Study in Determinism." *Geographical Journal* 118: 406–28.

Wright, J. K. 1926. "A Plea for the History of Geography." *Isis* 8: 477–91.

———. 1947. "Terra Incognitae: The Place of Imagination in Geography." *Annals, Association of American Geographers* 37: 1–15.

CHAPTER 2

Dear, M. J. 1988. "The Postmodern Challenge: Reconstructing Human Geography." *Transactions, Institute of British Geographers* (new series) 13: 262–74.

Hamnett, C. 2003. "Editorial. Contemporary Human Geography: Fiddling While Rome Burns?" *Geoforum* 34: 1–3.

Holland, P., ed. 1991. "Qualitative Resources in Geography." *New Zealand Journal of Geography* 92: 1–28.

Minshull, R. 1967. *Regional Geography: Theory and Practice*. London: Hutchinson.

Monmonier, M. 1991. *How to Lie with Maps*. Chicago: University of Chicago Press.

Peet, R. 1989. "World Capitalism and the Destruction of Regional Cultures." In R. J. Johnston and P. J. Taylor, eds, *A World in Crisis: Geographical Perspectives*, 2nd edn, 175–99. Oxford: Blackwell.

Relph, E. 1976. *Place and Placelessness*. London: Pion.

Saarinen, T. F. 1974. "Environmental Perception." In I. R. Manners and M. W. Mikesell, eds, *Perspectives on Environment*, 252–89. Washington, DC: Association of American Geographers, Commission on College Geography Publication 13.

Tobler, W. 1970. "A Computer Movie." *Economic Geography* 46: 234–40.

Tuan, Yi-Fu. 1979. *Landscapes of Fear*. Oxford: Blackwell.

———. 1982. *Segmented Worlds and Self: Group Life and Individual Consciousness*. Minneapolis: University of Minnesota Press.

———. 1983. "Geographical Theory: Queries from a Cultural Geographer." *Geographical Analysis* 15: 69–72.

———. 1984. *Dominance and Affection*. New Haven: Yale University Press.

Whittlesey, D. 1954. "The Regional Concept and the Regional Method." In P. E. James and C. F. James, eds, *American Geography: Inventory and Prospect*, 19–68. Syracuse, NY: Syracuse University Press.

Yeates, M. H. 1968. *An Introduction to Quantitative Analysis in Economic Geography*. Toronto: McGraw-Hill.

CHAPTER 3

Amnesty International. 2003. "Human Rights on the Line: The Baku-Tbilisi-Ceyhan Pipeline Project." London: Amnesty International. www.amnesty.org.uk/sites/default/files/baku_line_0.pdf.

Blainey, G. 1968. *The Tyranny of Distance*. London: Macmillan.

Cleary, M., and R. Bedford. 1993. "Globalisation and the New Regionalism: Some Implications for New Zealand Trade." *New Zealand Journal of Geography* 20: 19–22.

Conkling, E. C., and M. H. Yeates. 1976. *Man's Economic Environment*. Toronto: McGraw-Hill.

Dicken, P. 1992. *Global Shift: The Internationalization of Economic Activity*. London: Paul Chapman.

Friedman, T. L. 2005. *The World Is Flat: A Brief History of the Twenty-First Century*. New York: Farrar, Straus, and Giroux.

Ghemawat, P. 2011. *World 3.0: Global Prosperity and How to Achieve It*. Boston: Harvard Business Press Books.

———, and S. A. Altman. 2013. *Depth Index of Globalization and the Big Shift to Emerging Economies*. www.ghemawat.com/Dig/Files/Depth_Index_of_Globalization_2013_(Full_Report).pdf.

Globe and Mail, The. 2010. "How Are You Affected by Canada's Digital Divide?" 6 April, online discussion.

Halliday, Josh. 2011. "London Riots: How Blackberry Messenger Played a Key Role." *The Guardian*, 8 August. www.theguardian.com/media/2011/aug/08/london-riots-facebook-twitter-blackberry.

Harvey, D. W. 1990. "Between Space and Time: Reflections on the Geographical Imagination." *Annals, Association of American Geographers* 80: 418–34.

Hoad, D. 2002. "The World Trade Organisation, Corporate Interests, and Global Opposition: Seattle and After." *Geography* 87: 148–54.

Janelle, D. G. 1968. "Central Place Development in a Time-Space Framework." *Professional Geographer* 20: 5–10.

———. 1969. "Spatial Reorganization: A Model and Concept." *Annals, Association of American Geographers* 59: 348–64.

Johnston, R. J. 1986. "Individual Freedom and the World Economy." In R. J. Johnston and P. J. Taylor, eds, *A World in Crisis: Geographical Perspectives*, 173–95. Oxford: Blackwell.

Keynes, J. M. 1926. *The End of Laissez-Faire*. London: Hogarth Press.

Levinson, M. 2006. *The Box: How the Shipping Container Made the World Smaller and the World Economy Bigger*. Princeton, NJ: Princeton University Press.

Massey, D. 2002. "Globalization: What Does It Mean for Geography?" *Geography* 87: 293–6.

Norberg, J. 2001. *In Defence of Global Capitalism*. Stockholm: Timbro.

Ohmae, K. 1993. "The Rise of the Region State." *Foreign Affairs* 76 (2): 78–87.

Rigg, J. 2001. "Is Globalization Good?" *Geography Review* 14 (4): 36–7.

Robinson, M., L. Jane, E. Duggan, and S. Law. 2011. "Vancouverites Fight Back against Rioters through Social Media." *The Vancouver Sun*, 17 June.

Saul, J. R. 2005. *The Collapse of Globalism and the Reinvention of the World*. New York: Overlook Press.

Swiss Economic Institute. 2011 (6 August). *KOF Index of Globalization*. globalization.kof.ethz.ch/static/pdf/method_2011.pdf.

Ullman, E. L. 1956. "The Role of Transportation and the Bases for Interaction." In W. L. Thomas Jr, ed., *Man's Role in Changing the Face of the Earth*, 862–80. Chicago: University of Chicago Press.

Warman, M. 2011. "London Riots: Police Could Get Power over Social Media." *The Daily Telegraph*, 11 August. www.telegraph.co.uk/technology/social-media/8695798/London-riots-police-could-get-powers-over-social-media.html.

CHAPTER 4

Agnew, C. 1990. "Green Belt around the Sahara." *Geographical Magazine* 62 (4): 26–30.

Anderson, D. 2000. "Abrupt Climatic Change" *Geography Review* 13 (1): 2–6.

Bahn, P., and J. Flenley. 1992. *Easter Island: Earth Island*. New York: Thames and Hudson.

Carson, R. 1962. *Silent Spring*. Boston: Houghton Mifflin.

Connor, S. 2014. "Climate Change 'Final Warning' as IPCC Report Pushes for Fossil Fuel Phase-Out by 2100." *The Independent*, 2 November. www.independent.co.uk/environment/climate-change/scientific-evidence-proves-climate-change-is-man-made-un-experts-conclude-9833748.html.

Diamond, J. 2005. *Collapse: How Societies Choose to Fail or Succeed*. New York: Viking Penguin.

Dregne, H. E. 1977. "Desertification of Arid Lands." *Economic Geography* 53: 322–31.

Environment Canada. 1991. *The State of Canada's Environment*. Ottawa: Supply and Services Canada.

Friedman, T. L. 2008. *Hot, Flat, and Crowded: Why We Need a Green Revolution—And How It Can Renew America*. New York: Farrar, Straus and Giroux.

Gale, R. J. P. 1992. "Environment and Economy: The Policy Models of Development." *Environment and Behavior* 24: 723–37.

Glance, N. S., and B. A. Huberman. 1994. "The Dynamics of Social Dilemmas." *Scientific American* 270 (3): 76–81.

Goodall, C. 2008. *Ten Technologies to Save the Planet*. London: Profile Books.

Goudie, A. 1981. *The Human Impact: Man's Role in Environmental Change*. Oxford: Blackwell.

Grove, R. H. 1992. "Origins of Western Environmentalism." *Scientific American* 267 (1): 42–7.

Hardin, G. 1968. "The Tragedy of the Commons." *Science* 162: 1243–8.

Hunt, T., and C. Lipo. 2011. *The Statues That Walked: Unraveling the Mystery of Easter Island*. New York: Free Press

Intergovernmental Panel on Climate Change (IPCC). 2011. *IPCC Special Report on Renewable Energy Sources and Climate Change Mitigation*. Cambridge and New York: Cambridge University Press. http://srren.ipcc-wg3.de/report/IPCC_SRREN_Full_Report.pdf.

——. 2014. *Climate Change 2014: Impacts, Adaptation, and Vulnerability*. New York: Cambridge University Press.

Johnston, R. J. 1992. "Laws, States and Superstates: International Law and the Environment." *Applied Geography* 12: 211–28.

——and P. J. Taylor. 1986. "Introduction: A World in Crisis?" In R. J. Johnston and P. J. Taylor, eds, *A World in Crisis: Geographical Perspectives*, 1–11. Oxford: Blackwell.

Kaplan, R. D. 1994. "The Coming Anarchy." *Atlantic Monthly* 273 (2): 44–76.

——. 1996. *The Ends of the Earth: A Journey at the Dawn of the 21st Century*. New York: Random House.

Katz, E., A. Light, and D. Rothenberg, eds. 2000. *Beneath the Surface: Critical Essays in Deep Ecology*. Cambridge, MA: MIT Press.

Kelly, K. 1974. "The Changing Attitudes of Farmers to Forest in Nineteenth Century Ontario." *Ontario Geography* 8: 67–77.

McKibben, B. 2006. "The Coming Meltdown." *The New York Review of Books* 53 (1): 16–18.

Marsh, G. P. (1864) 1965. *Man and Nature, or Physical Geography as Modified by Human Action*. Edited by D. Lowenthal. Cambridge, MA: Harvard University Press.

Ponting, C. 1991. *A Green History of the World*. New York: St Martin's Press.

Powell, J. M. 1976. *Environmental Management in Australia, 1788–1814*. New York: Oxford University Press.

Rautiainen, A., I. Wernick, P. E. Waggoner, J. H. Ausubel, and P. E. Kauppi. 2011. "A National and International Analysis of Changing Forest Density." *PLoS One* 6 (5). www.ncbi.nlm.nih.gov/pmc/articles/PMC3089630/?tool=pubmed.

Repetto, R. 1990. "Deforestation in the Tropics." *Scientific American* 260: 36–42.

Ruckelshaus, W. N. 1989. "Toward a Sustainable World." *Scientific American* 261 (3): 166–75.

Ruddiman, W. F. 2005. *Plows, Plagues, and Petroleum: How Humans Took Control of Climate*. Princeton, NJ: Princeton University Press.

Simon, J., and H. Kahn, eds. 1984. *The Resourceful Earth*. Oxford: Blackwell.

Smil, V. 1987. *Energy, Food and Environment*. New York: Oxford University Press.

——. 1993. *Global Ecology: Environmental Changes and Social Flexibility*. New York: Routledge.

——. 2005. "The Next 50 Years: Fatal Discontinuities." *Population and Development Review* 31 (2): 201–36.

Thomas, W. L., W. L. Thomas, Jr, C. Ortwin Sauer, and L. Mumford, eds. 1956. *Man's Role in Changing the Face of the Earth*. Chicago: University of Chicago Press.

Tuan, Y. F. 1971. *Man and Nature*. Washington, DC: Association of American Geographers, Commission on College Geography, Resource Paper No. 10.

Victor, D. G. 2011. *Global Warming Gridlock: Creating More Effective Strategies for Protecting the Planet*. New York: Cambridge University Press.

Wilkinson, H. R. 1963. *Man and the Natural Environment*. Hull, UK: University of Hull, Department of Geography, Occasional Papers in Geography No. 1.

World Bank. 2013. *World Development Report 2014: Risk and Opportunity—Managing Risk for Development*. Washington, DC: World Bank. http://econ.worldbank.org/WBSITE/EXTERNAL/EXTDEC/EXTRESEARCH/EXTWDRS/EXTNWDR2013/0,,content MDK:23459971~pagePK:8261309~piPK:8258028~theSitePK:8258025,00.html.

World Commission on Environment and Development. 1987. *Our Common Future*. Oxford: Oxford University Press.

World Wildlife Federation. 2014. *Living Planet Report 2014*. Gland, Switzerland: WWF. www.livingplanetindex.org/projects?main_page_project=LivingPlanetReport&home_flag=1.

CHAPTER 5

Boserup, E. 1965. *The Conditions of Agricultural Change*. London: Allen and Unwin.

Demeny, P. 1974. "The Populations of the Underdeveloped Countries." *Scientific American* 231: 148–59.

Dwyer, D. J. 1987. "New Population Policies in Malaysia and Singapore." *Geography* 72: 248–50.

Economist, The. 1993. "Eastern Germany: Living and Dying in a Barren Land." 331 (23): 54.

Ehrlich, P. 1968. *The Population Bomb*. New York: Ballantine Books.

Grigg, D. 1977. "E.G. Ravenstein and the 'Laws' of Migration." *Journal of Historical Geography* 3: 41–54.

Halfacree, K., and P. J. Boyle. 1993. "The Challenge Facing Migration Research: The Case for a Biographical Approach." *Progress in Human Geography* 17: 333–48.

Hall, R. 1993. "Europe's Changing Population." *Geography* 78: 3–15.

Jowett, J. 1993. "China's Population: 1,133,709,738 and Still Counting." *Geography* 78: 401–19.

King, R. 1993. "Italy Reaches Zero Population Growth." *Geography* 78: 63–9.

Lee, E. S. 1966. "A Theory of Migration." *Demography* 3 (1): 47–57.

Meadows, D. H., D. L. Meadows, J. Randers, and W. W. Behrens III. 1972. *The Limits to Growth*. New York: Universe Books.

Moon, B. 1995. "Paradigms in Migration Research: Exploring 'Moorings' as a Schema." *Progress in Human Geography* 19: 504–24.

Nelson, F. 2006. "Where Have All the Babies Gone?" *The Spectator*, 4 March, 24.

Petersen, W. 1958. "A General Typology of Migration." *American Sociological Review* 23: 256–65.

Peterson, P. G. 1999. "Gray Dawn: The Global Aging Crisis." *Foreign Affairs* 78: 42–55.

Ravenstein, E. G. 1876. "Census of the British Isles, 1871; Birthplaces and Migration." *Geographical Magazine* 3: 173–7, 201–06, 229–33.

——. 1885. "The Laws of Migration." *Journal of the Statistical Society* 48: 167–227.

——. 1889. "The Laws of Migration." *Journal of the Statistical Society* 52: 214–301.

Ridker, R. G., and E. W. Cecelski. 1979. "Resources, Environment and Population: The Nature of Future Limits." *Population Bulletin* 34 (3): 3–4.

Robey, B., S. O. Rutstein, and L. Morris. 1993. "The Fertility Decline in Developing Countries." *Scientific American* 269 (6): 60–7.

Taylor, P. J. 1989. "The Error of Developmentalism in Human Geography." In D. Gregory and R. Walford, eds, *Horizons in Human Geography*, 303–19. London: Macmillan.

United Nations. 2002. *World Population Aging: 1950–2050*. New York: United Nations, Department of Economic and Social Affairs, Population Division.

Wolpert, J. 1965. "Behavioural Aspects of the Decision to Migrate." *Papers of the Regional Science Association* 15: 159–69.

Zelinsky, W. 1971. "The Hypothesis of the Mobility Transition." *Geographical Review* 61: 219–49.

CHAPTER 6

Barberis, M. 1994. "Haiti." *Population Today* 22 (1): 7.

Brandt, W. 1980. *North–South: A Programme for Survival*. London: Pan.

Chew, S. C., and R. A. Denemark, eds. 1996. *The Underdevelopment of Development: Essays in Honor of Andre Gunder Frank*. Thousand Oaks, CA: Sage.

Collier, P. 2007. *The Bottom Billion: Why the Poorest Countries Are Failing and What Can Be Done About It*. New York: Oxford University Press.

Conway, G. 2008. "Presidential Address: The Food Crisis." *Geographical Journal* 174: 269–73.

Daniel, M. L. 2000. "The Demographic Impact of HIV/AIDS in Sub-Saharan Africa." *Geography* 85: 46–55.

Davis, M. 2001. *Late Victorian Holocausts: El Niño, Famines, and the Making of the Third World*. New York: Verso.

Degg, M. 1992. "Natural Disasters: Recent Trends and Future Prospects." *Geography* 77: 198–209.

Diamond, J. 1997. *Guns, Germs and Steel: A Short History of Everybody for the Last 13,000 Years*. London: Random House.

Dowden, R. 2008. *Africa: Altered States, Ordinary Miracles*. London: Portobello Books.

Economist, The. 2002. "The Next Wave." 19 October: 75–6.

———. 2011 (26 February). *The 9 Billion-People Question: A Special Report on Feeding the World*. www.economist.com/sites/default/files/special-reports-pdfs/18205243.pdf.

Gee, M. 1994. "Apocalypse Deferred." *The Globe and Mail*, 9 April, D1, D3.

Ignatieff, M. 1996. "Review of *The Ends of the Earth*, by R.D. Kaplan." *The New York Times Book Review*, 31 March.

Juma, C. 2010. *The New Harvest: Agricultural Innovation in Africa*. New York: Oxford University Press.

Kaufmann, D., A. Kraay, and M. Mastruzzi. 2006. *Governance Matters V: Aggregate and Individual Governance Indicators for 1996–2005*. Washington, DC: World Bank.

Kaplan, R. D. 1994. "The Coming Anarchy." *Atlantic Monthly* 273 (2): 44–76.

———. 1996. *The Ends of the Earth: A Journey at the Dawn of the 21st Century*. New York: Random House.

Mahmud, A. 1989. "Grameen Bank Bangladesh: A Workable Solution." *Geographical Magazine* 61 (10): 14–16.

Ó Gráda, C. 2009. *Famine: A Short History*. Princeton, NJ: Princeton University Press.

Rigg, J., A. Bebbington, K. V. Gough, D. F. Bryceson, J. Agergaard, N. Fold, and C. Tacoli. 2009. "The World Development Report 2009 'Reshapes Economic Geography': Geographical Reflections." *Area* 34: 128–36.

Rostow, W. W. 1960. *The Stages of Economic Growth*. Cambridge: Cambridge University Press.

Sen, A. 1981. *Poverty and Famines: An Essay on Entitlement and Deprivation*. New York: Oxford University Press.

Smil, V. 2000. *Feeding the World: A Challenge for the Twenty-First Century*. Cambridge, MA: MIT Press.

Sowden, C. 1993. "Debt Swaps—For or Against Development." *Geographical Magazine* 25 (12): 56–9.

Todd, H. 1996. *Women at the Center: Grameen Bank Borrowers after One Decade*. Boulder: Westview Press.

Toolis, K. 2000. "While the World Looks Away." *The Guardian Weekend*, 2 December.

Wallerstein, I. 1979. *The Capitalist World Economy*. Cambridge: Cambridge University Press.

World Bank. 1998. *Global Development Finance, 1998: Analysis and Summary Tables*. Washington, DC: World Bank.

———. 2009. *World Development Report: Reshaping Economic Geography*. Washington, DC: World Bank.

Young, L. 1996. "World Hunger: A Framework for Analysis." *Geography* 81: 97–110.

CHAPTER 7

Biswas, L. 1984. "Evolution of Hindu Temples in Calcutta." *Journal of Cultural Geography* 4: 73–84.

Cartwright, D. 1988. "Linguistic Territorialization: Is Canada Approaching the Belgian Model?" *Journal of Cultural Geography* 8: 115–34.

Church News. 1979. *Special Edition: The Era of Mormon Colonization*. Salt Lake City: Deseret News (26 May).

Francaviglia, R. V. 1978. *The Mormon Landscape*. New York: AMS Press.

Freeman, D. B. 1985. "The Importance of Being First: Preemption by Early Adopters of Farming Innovations in Kenya." *Annals, Association of American Geographers* 75: 17–28.

Gale, D. T., and P. M. Koroscil. 1977. "Doukhobor Settlements: Experiments in Idealism." *Canadian Ethnic Studies* 9: 53–71.

Garreau, J. 1981. *The Nine Nations of North America*. Boston: Houghton Mifflin.

Hägerstrand, T. 1951. "Migration and the Growth of Culture Regions." In *Lund Studies in Geography, Series B, no. 3*. Lund, Sweden: Gleerup.

———. 1967. *Innovation Diffusion as a Spatial Process*. Translated by A. Pred. Chicago: University of Chicago Press.

Hale, R. F. 1984. "Vernacular Regions of America." *Journal of Cultural Geography* 5: 131–40.

Hoernig, H., and M. Walton-Roberts. 2006. "Immigration and Urban Change: National, Regional, and Local Perspectives." In *Canadian Cities in Transition: Local Through Global Perspectives*, 3rd edn, edited by T. Bunting and P. Filion, 408–18. Toronto: Oxford University Press.

James, P. E. 1964. *One World Divided*, 2nd edn. Toronto: Xerox College Publishing.

Jordan, T. G. 1988. *The European Culture Area: A Systematic Geography*, 2nd edn. New York: Harper and Row.

———, and M. Kaups. 1989. *The American Backwoods Frontier: An Ethnic and Ecological Interpretation*. Baltimore: Johns Hopkins University Press.

Kearns, K. C. 1974. "Resuscitation of the Irish Gaeltacht." *Geographical Review* 64: 82–110.

Kniffen, F. 1951. "The American Covered Bridge." *Geographical Review* 41: 114–23.

Lewis, B. 2003. *The Crisis of Islam: Holy War and Unholy Terror*. New York: Weidenfeld and Nicholson.

McCrum, R., W. Cran, and R. MacNeil, eds. 1986. *The Story of English*. London: BBC.

Mackay, J. R. 1958. "The Interactance Hypothesis and Boundaries in Canada: A Preliminary Study." *Canadian Geographer* 3 (11): 1–8.

McWhorter, R. 2002. *The Power of Babel: A Natural History of Language*. New York: W.H. Freeman.

Meinig, D. W. 1965. "The Mormon Culture Region: Strategies and Patterns in the Geography of the American West, 1847–1964." *Annals, Association of American Geographers* 55: 191–220.

Nostrand, R. L., and L. E. Estaville Jr. 1993. "Introduction: The Homeland Concept." *Journal of Cultural Geography* 13 (2): 1–4.

Park, C. 1994. *Sacred Worlds: An Introduction to Geography and Religion*. New York: Routledge.

Pyle, G. F. 1969. "The Diffusion of Cholera in the United States in the Nineteenth Century." *Geographical Analysis* 1: 59–75.

Raitz, K. B. 1973. "Ethnicity and the Diffusion and Distribution of Cigar Tobacco Production in Wisconsin and Ohio." *Tijdschrifte voor Economische en Sociale Geografie* 64: 293–306.

Russell, R. J., and F. B. Kniffen. 1951. *Culture Worlds*. New York: Macmillan.

Simpson-Housley, P. 1978. "Hutterian Religious Ideology, Environmental Perception, and Attitudes towards Agriculture." *Journal of Geography* 77: 145–8.

Sommers, B. J. 2008. *The Geography of Wine: How Landscapes, Cultures, Terroir, and the Weather Make a Good Drop*. New York: Plume.

Spencer, J. E., and R. J. Horvath. 1963. "How Does an Agricultural Region Originate?" *Annals, Association of American Geographers* 53: 74–92.

Toynbee, A. J. 1935–61. *A Study of History*, 12 vols. New York: Oxford University Press.

Trépanier, C. 1991. "The Cajunization of French Louisiana: Forging a Regional Identity." *Geographical Journal* 157: 161–71.

Zelinsky, W. 1973. *The Cultural Geography of the United States*. Englewood Cliffs, NJ: Prentice-Hall.

———. 1980. "North America's Vernacular Regions." *Annals, Association of American Geographers* 70: 1–16.

CHAPTER 8

Alvarez, A. 2001. *Governments, Citizens and Genocide: A Comparative and Interdisciplinary Approach*. Bloomington: Indiana University Press.

American Anthropological Association. 1998. "Statement on Race." www.aaanet.org/stmts/racepp.htm.

Bone, R. M. 1992. *The Geography of the Canadian North*. Toronto: Oxford University Press.

Chalk, F., and K. Jonassohn. 1990. *The History and Sociology of Genocide*. New Haven: Yale University Press.

Chouinard, V. 1997. "Guest Editorial. Making Space for Disabling Differences: Challenging Ableist Geographies." *Environment and Planning D: Society and Space* 15: 379–87.

Doyal, L., and I. Gough. 1991. *A Theory of Human Need*. London: Macmillan.

Economist, The. 2006. "Happiness (and How to Measure It)." 19 December. www.economist.com/node/8450035.

Evans, R. 1989. "Consigned to the Shadows." *Geographical Magazine* 61 (12): 23–5.

Eyles, J., and W. Peace. 1990. "Signs and Symbols in Hamilton: An Iconology of Steeltown." *Geografiska Annaler* 72B: 73–88.

Fennell, D. A. 1999. *Ecotourism*. London: Routledge.

Gaye, A., J. Klugman, M. Kovacevic, S. Twigg, and E. Zambrano. 2010. *Measuring Key Disparities in Human Development: Gender Inequality Index*. Human Development Research Paper 2010/46. New York: UNDP.

Giddens, A. 1984. *The Constitution of Society*. Cambridge: Polity Press.

Gould, P., and R. White. 1986. *Mental Maps*, 2nd edn. Boston: Allen and Unwin.

Gould, S. J. 1981. *The Mismeasure of Man*. New York: Norton.

———. 1985. "Human Equality Is a Contingent Fact of History." In *The Flamingo's Smile*, edited by S. J. Gould, 185–98. New York: Norton.

———. 1987. "Bushes All the Way Down." *Natural History* 96 (6): 12–19.

Harding, K. 2007. "It's Just Too Late in Nunavut." *The Globe and Mail*, 12 January.

Hopkins, J. S. P. 1990. "West Edmonton Mall: Landscape of Myths and Elsewhereness." *Canadian Geographer* 34: 2–17.

Jackson, E. L., and D. B. Johnson. 1991a. "Geographic Implications of Mega-Malls, with Special Reference to West Edmonton Mall." *Canadian Geographer* 35: 226–32.

——— and ———, eds. 1991b. "The West Edmonton Mall and Mega Malls." *Canadian Geographer* 35: 226–305.

Jackson, P. 1989. *Maps of Meaning: An Introduction to Cultural Geography*. London: Unwin Hyman.

Jakle, J. A. 1985. *The Tourist: Travel in Twentieth-Century North America*. Lincoln: University of Nebraska Press.

James, P. E. 1964. *One World Divided*, 2nd edn. Toronto: Xerox College Publishing.

Jayne, M. 2006. "Cultural Geography, Consumption and the City." *Geography* 91: 34–42.

Johnston, L. 1997. "Queen(s') Street or Ponsonby Poofters? Embodied Hero Parade Sites." *New Zealand Geographer* 53 (2): 29–33.

Kennedy, K. A. R. 1976. *Human Variation in Space and Time*. Dubuque, IA: Brown.

Kobayashi, A. 1993. "Multiculturalism: Representing a Canadian Institution." In *Place/Culture/Representation*, edited by J. Duncan and D. Ley, 205–31. London: Routledge.

Ley, D., and K. Olds. 1988. "Landscape as Spectacle: World's Fairs and the Culture of Heroic Consumption." *Environment and Planning D: Society and Space* 6: 191–212.

McKittrick, K., and L. Peake. 2005. "What Difference Does Difference Make to Geography?" In *Questioning Geography*, edited by N. Castree, A. Rogers, and D. Sherman, 39–54. New York: Blackwell.

McQuillan, A. 1993. "Historical Geography and Ethnic Communities in North America." *Progress in Human Geography* 17: 355–66.

Monk, J. 1992. "Gender in the Landscape: Expressions of Power and Meaning." In *Inventing Places: Studies in Cultural Geography*, edited by K. Anderson and F. Gale, 123–38. Melbourne: Longman Cheshire.

Montague, A., ed. 1964. *The Concept of Race*. New York: Collier.

Nelson, L., and J. Seager. 2005. "Introduction." In *A Companion to Feminist Geography*, edited by L. Nelson and J. Seager, 1–11. Malden, MA: Blackwell.

O'Hare, G., and H. Barrett. 1993. "The Fall and Rise of the Sri Lankan Tourist Industry." *Geography* 78: 438–42.

Osborne, B. S. 1988. "The Iconography of Nationhood in Canadian Art." In *The Iconography of Landscape: Essays on the Symbolic Representation, Design and Use of Past Environments*, edited by D. Cosgrove and S. Daniels, 162–78. New York: Cambridge University Press.

Pain, R. 1992. "Space, Sexual Violence and Social Control: Integrating Geographical and Feminist Analyses of Women's Fear of Crime." *Progress in Human Geography* 15: 415–31.

Pawson, E., and G. Banks. 1993. "Rape and Fear in a New Zealand City." *Area* 25: 55–63.

Pratt, G. 2004. "Feminist Geographies: Spatialising Feminist Politics." In *Envisioning Human Geographies*, edited by P. Cloke, P. Crang, and M. Goodwin, 128–145. London: Arnold.

Raitz, K. B. 1979. "Themes in the Cultural Geography of European Ethnic Groups in the United States." *Geographical Review* 69: 79–94.

Robinson, M. 1999. "Cultural Conflicts in Tourism: Inevitability and Inequality." In *Tourism and Cultural Conflicts*, edited by M. Robinson and P. Boniface, 1–32. New York: CABI.

Rooney, J. F., Jr. 1974. *A Geography of American Sport*. Reading, MA: Addison-Wesley.

Rose, G. 1994. *Feminism and Geography: The Limits of Geographical Knowledge*. Minneapolis: University of Minnesota Press.

Shields, R. 1989. "Social Spatialisation and the Built Environment: The Example of West Edmonton Mall." *Environment and Planning D: Society and Space* 7: 147–64.

Smith, D. M. 1973. *A Geography of Social Well-Being in the United States*. New York: McGraw-Hill.

Smith, S. J. 1987. "Fear of Crime: Beyond a Geography of Deviance." *Progress in Human Geography* 11: 1–23.

Thomas, B., and D. Dorling. 2007. *Identity in Britain: A Cradle-to-Grave Atlas*. Bristol, UK: Policy Press.

Wheat, S. 1994. "Taming Tourism." *Geographical Magazine* 67: 16–19.

CHAPTER 9

Anderson. B. 1983. *Imagined Communities: Reflections on the Origins and Spread of Nationalism*. London: Verso.

Bunge, W. 1988. *Nuclear War Atlas*. Oxford: Blackwell.

Carroll, W. K. 1992. *Organizing Dissent: Contemporary Social Movements in Theory and Practice, Studies in the Politics of Counter-Hegemony*. Toronto: Garamond Press.

Clark, G. L., and M. J. Dear. 1984. *State Apparatus: Structures and Language of Legitimacy*. Boston: Allen and Unwin.

Cohen, S. B. 1991. "Global Geopolitical Change in the Post-Cold War Era." *Annals, Association of American Geographers* 81: 551–80.

———. 2003. *Geopolitics of the World System*. Lanham, MD: Rowman & Littlefield.

de Blij, H. 1992. "Political Geography of the Post-Cold War." *Professional Geographer* 44: 16–19.

East, W. G., and J. R. V. Prescott. 1975. *Our Fragmented World: Introduction to Political Geography*. London: Macmillan.

Elsom, D. 1985. "Climatological Effects of a Nuclear Exchange: A Review." In *The Geography of Peace and War*, edited by D. Pepper and A. Jenkins, 126–47. Oxford: Blackwell.

Evans, R. 1991. "Legacy of Woe." *Geographical Magazine* 63 (6): 34–8.

Fukuyama, F. 1992. *The End of History and the Last Man*. New York: Free Press.

Hartshorne, R. 1950. "The Functional Approach in Political Geography." *Annals, Association of American Geographers* 40: 95–130.

Hirsch, P. 1993. "The Socialist Developing World in the 1990s." *Geography Review* 7 (2): 35–7.

Huntington, S. P. 1993. "The Clash of Civilizations." *Foreign Affairs* 72 (3): 22–49.

———. 1996. *The Clash of Civilizations and the Remaking of World Order*. New York: Simon & Schuster.

Johnston, R. J. 1982. *Geography and the State: An Essay in Political Geography*. New York: St Martin's Press.

———. 1985. *The Geography of English Politics*. London: Croom Helm.
———. 1993. "Tackling Global Environmental Problems." *Geography Review* 6 (2): 27–30.
Klare, M. T. 2001. "The New Geography of Conflict." *Foreign Affairs* 80: 49–61.
Kohr, L. 1957. *The Breakdown of Nations*. Swansea: Christopher Davies.
Krebheil, E. 1916. "Geographic Influences in British Elections." *Geographical Review* 2: 419–32.
Lemon, A. 1996. "Lesotho and the New South Africa: The Question of Incorporation." *Geographical Journal* 162: 263–72.
Mackinder, H. J. 1919. *Democratic Ideals and Reality*. New York: Henry Holt.
Messick, D. M., and D. M. Mackie. 1989. "Intergroup Relations." *Annual Review of Psychology* 40: 45–81.
O'Loughlin, J. 1992. "Ten Scenarios for a 'New World Order.'" *Professional Geographer* 44: 22–8.
Openshaw, S., P. Steadman, and O. Greene. 1983. *Doomsday: Britain after Nuclear Attack*. Oxford: Blackwell.
Rokkan, S. 1980. "Territories, Centres and Peripheries." In *Centre and Periphery*, edited by J. Gottman, 163–204. London: Sage.
Royle, S. 1991. "St Helena: A Geographical Summary." *Geography* 76: 266–8.
Times, The (London). 2006. "Comment: Till the Votes Do Them Part." 20 May.
Wilkinson, P. 1971. *Social Movement*. London: Macmillan.
Wood, W. B. 2001. "Geographic Aspects of Genocide: A Comparison of Bosnia and Rwanda." *Transactions, Institute of British Geographers* (new series) 26: 57–75.
Zurick, D. 1999. "Lands of Conflict in South Asia." *Focus* 45 (3): 33–7.

CHAPTER 10

Atkins, P., and I. Bowler. 2001. *Food in Society: Economy, Culture, Geography*. London: Arnold.
Barański, M., D. Srednicka-Tober, N. Volakakis, C. Seal, R. Sanderson, G. B. Stewart, C. Benrook, et al. 2014. "Higher Antioxidant and Lower Cadmium Concentrations and Lower Incidence of Pesticide Residues in Organically Grown Crops: A Systematic Literature Review and Meta-Analyses." *British Journal of Nutrition* 112: 794–811.
Barnes, T. J. 2001. "Retheorizing Economic Geography: From the Quantitative Revolution to the 'Turn.'" *Annals, Association of American Geographers* 91: 546–65.
Carlyle, W. J. 1994. "Rural Population in the Canadian Prairies." *Great Plains Research* 4: 65–87.
Carrington, D., and G. Arnett. 2014. "Clear Differences between Organic and Non-Organic Food, Study Finds." *The Guardian*, 11 July.
Chisholm, M. 1962. *Rural Settlement and Land Use: An Essay in Location*. London: Hutchinson.
Conzen, M. P. 1971. *Frontier Farming in an Urban Shadow*. Madison: State Historical Society of Wisconsin.
Fernandez-Armesto, F. 2001. *Food: A History*. London: Macmillan.
Friedmann, H. 1991. "Changes in the International Division of Labour: Agri-Food Complexes and Export Agriculture." In *Towards a New Political Economy of Agriculture*, edited by W. Friedland, L. Busch, F. H. Buttel, and A. Rudy, 65–93. Boulder: Westview Press.
Goodall, C. 2008. *Ten Technologies to Save the Planet*. London: Profile Books.
Griffin, E. 1973. "Testing the Von Thünen Theory in Uruguay." *Geographical Review* 63: 500–16.
Grigg, D. B. 1974. *The Agricultural Systems of the World: An Evolutionary Approach*. New York: Cambridge University Press.
———. 1992. "Agriculture in the World Economy: An Historical Geography of Decline." *Geography* 77: 210–22.
Grossman, L. S. 1993. "The Political Ecology of Banana Exports and Local Food Production in St Vincent, Eastern Caribbean." *Annals, Association of American Geographers* 83: 347–67.
Hall, P. G., ed. 1966. *Von Thünen's Isolated State*. Oxford: Pergamon.
Harman, O. 2004. "The Essential Element" *New Republic* (30 August): 31–7.
Horvath, R. J. 1969. "Von Thünen's Isolated State and the Area around Addis Ababa, Ethiopia." *Annals, Association of American Geographers* 59: 308–23.
Huggins, D. R., and J. P. Reganold. 2008. "No-Till: The Quiet Revolution." *Scientific American* 299, 5: 70–7.
Johnson, H. B. 1962. "A Note on Thünen's Circles." *Annals, Association of American Geographers* 52: 213–20.
Jonasson, O. 1925. "Agricultural Regions of Europe." *Economic Geography* 1: 277–314.
King, L. J. 1969. *Statistical Analysis in Geography*. Englewood Cliffs, NJ: Prentice-Hall.
Leaman, J. H., and E. C. Conkling. 1975. "Transport Change and Agricultural Specialization." *Annals, Association of American Geographers* 65: 425–37.
McCallum, J. 1980. *Unequal Beginnings: Agriculture and Economic Development in Quebec and Ontario until 1870*. Toronto: University of Toronto Press.
Marsden, T. K. 1997. "Creating Space for Food: The Distinctiveness of Recent Agrarian Development." In *Globalising Food: Agrarian Questions and Global Restructuring*, edited by D. Goodman and M.J. Watts, 169–91. New York: Routledge.
Muller, P. O. 1973. "Trend Surfaces of American Agricultural Patterns: A Macro-Thünen Analysis." *Economic Geography* 49: 228–42.
Norton, W., and E. C. Conkling. 1974. "Land Use Theory and the Pioneering Economy." *Geografiska Annaler* 56B: 44–56.
Overton, M. 1996. *Agricultural Revolution in England: The Transformation of the Agrarian Economy, 1500–1850*. Cambridge: Cambridge University Press.
Peet, J. R. 1969. "The Spatial Expansion of Commercial Agriculture in the Nineteenth Century: A Von Thünen Interpretation." *Economic Geography* 45: 283–301.
Rogers, A., and F. Chew. 2005. "Ducks for Bugs: Rice Farming in Malaysia." *Geography Review* 18 (5): 34–7.
Sauer, C. O. 1952. *Agricultural Origins and Dispersals*. New York: American Geographical Society.
Schlebecker, J. T. 1960. "The World Metropolis and the History of American Agriculture." *Journal of Economic History* 20: 187–208.
Seaborne, A. 2001. "Crop Diversification in Canada's Breadbasket: Land Use Changes in Saskatchewan's Agriculture." *Geography* 86: 151–8.
Sinclair, R. 1967. "Von Thünen and Urban Sprawl." *Annals, Association of American Geographers* 57: 72–87.
Slatford, R., and I. Fishpool. 2002. "India and the Green Revolution." *Geography Review* 15 (2): 2–5.
Smil, V. 2001. *Enriching the Earth: Fritz Haber, Carl Bosch and the Transformation of World Food Production*. Cambridge, MA: MIT Press.
Smith, W. 1984. "The 'Vortex' Model and the Changing Agricultural Landscape of Quebec." *Canadian Geographer* 28: 358–72.
Wolpert, J. 1964. "The Decision Process in Spatial Context." *Annals, Association of American Geographers* 54: 537–58.
Young, L. J. 1991. "Agricultural Changes in Bhutan: Some Environmental Questions." *Geographical Journal* 157 (2): 172–8.
Zimmerer, K. S. 1991. "Wetland Production and Smallholder Persistence: Agriculture Change in a Highland Peruvian Region." *Annals, Association of American Geographers* 81: 443–63.

CHAPTER 11

Abrahamson, M. 2004. *Global Cities*. New York: Oxford University Press.
Beaverstock, J. V., R. G. Smith, and P. J. Taylor. 1999. "A Roster of World Cities." *Cities* 16: 6.
Chan, K. W. 2009. "Urbanization in China: What Is the True Urban Population of China? Which Is the Largest City in China?" faculty.washington.edu/kwchan/Chan-urban.pdf.
Christaller, W. 1966. *Central Places in Southern Germany*. Translated by C. W. Baskin. Englewood Cliffs, NJ: Prentice-Hall.
Friedmann, J. 1986. "The World City Hypothesis." *Development & Change* 17: 69–83.
———. 2002. "Intercity Networks in a Globalizing Era." In *Global City-Regions*, edited by A. Scott, 119–36. New York: Oxford University Press.
Hall, P. 1966. *The World Cities*. London: Heinemann.
———, H. Gracey, R. Drewett, and R. Thomas. 1973. *The Containment of Urban England*. London: Allen and Unwin.
Huang, Yefany, Yee Leung, and Jianfa Shen. 2007. "Cities and Globalization: An International Cities Perspective." *Urban Geography* 28: 3.
Sassen, S. 1991. *The Global City: New York, London, Tokyo*. Princeton, NJ: Princeton University Press.
———, ed. 2002. *Global Networks—Linked Cities*. New York: Routledge.

Statistics Canada. 1999. *1996 Census Dictionary* (Catalogue no.92-351-UIE). Ottawa: Minister of Industry.

Taylor, P. 2004. *World City Network: A Global Urban Analysis*. New York: Routledge.

United Nations, Department of Economic and Social Affairs (UN DESA), Population Division. 2015. *World Urbanization Prospects: The 2014 Revision*. New York: United Nations. http://esa.un.org/unpd/wup/.

Zook, M. A. 2005. *The Geography of the Internet Industry*. New York: Blackwell.

CHAPTER 12

Bourne, L. S. 1996. "Reinventing the Suburbs: Old Myths and New Realities." *Progress in Planning* 46: 163–84.

Canadian Urban Institute. 2011. *The New Geography of Office Locations and the Consequences of Business as Usual in the Greater Toronto Area*. Report prepared for the Toronto Office Coalition, 18 March.

Cohen, M. 1993. "Megacities and the Environment," *Finance and Development* 30, 2: 44–7.

Collins, D. 2010. "Homelessness in Canada and New Zealand: A Comparative Perspective on Numbers and Policy Responses," *Urban Geography* 31: 932–52.

DeVerteuil, G. 2003. "Household Mobility, Institutional Settings, and the New Poverty Management." *Environment and Planning A* 35: 3661–79.

———. 2005a. "Welfare Neighborhoods: Anatomy of a Concept." *Journal of Poverty* 9: 23–41.

———. 2005b. "The Relationship between Government Assistance and Housing Outcomes among Extremely Low-Income Individuals: A Qualitative Inquiry in Los Angeles." *Housing Studies* 20: 383–99.

Ford, L. 1996. "A New and Improved Model of Latin American City Structure." *Geographical Review* 86: 437–40.

Fraser, J. C. 2004. "Beyond Gentrification: Mobilizing Communities and Claiming Space," *Urban Geography* 25: 437–57.

Gaetz, S., J. Donaldson, T. Richter, and T. Gulliver. 2013. *The State of Homelessness in Canada, 2013*. Toronto: Canadian Homelessness Research Network Press.

Garreau, J. 1988. *Edge City: Life on the New Frontier*. Toronto: Doubleday.

Grant, R., and J. Nijman. 2002. "Globalization and the Corporate Geography of Cities in the Less-Developed World," *Annals, Association of American Geographers* 92: 320–40.

Harris, C. D., and E. L. Ullman. 1945. "The Nature of Cities." *Annals, American Academy of Political and Social Science* 37: 7–17.

Harris, R. 2015. "Housing: Dreams and Nightmares." In *Canadian Cities in Transition: Perspectives for an Urban Age*, edited by P. Filion, M. Moos, T. Vinodrai, and R. Walker, 325–42. Toronto: Oxford University Press.

Hoskins, W.G. 1955. *The Making of the English Landscape*. London: Hodder and Stoughton.

Hoyt, H. 1939. *The Structure and Growth of Residential Neighbourhoods in American Cities*. Washington, DC: Federal Housing Administration.

Ley, D. 1996. *The New Middle Class and the Remaking of the Central City*. Oxford: Oxford University Press.

Murdie, R. A., and C. Teixeira. 2006. "Urban Social Space." In *Canadian Cities in Transition: Local Through Global Perspectives*, edited by T. Bunting and P. Filion, 3rd edn, 154–70. Toronto: Oxford University Press.

O'Connor, A. 1983. *The African City*. London: Hutchinson.

Peressini, T., and L. McDonald. 2000. "Urban Homelessness in Canada," in T. Bunting and P. Filion, eds, *Canadian Cities in Transition: The Twenty-First Century*, 2nd edn. Toronto: Oxford University Press, 525–43.

Ray, B., and D. Rose. 2000. "Cities of the Everyday: Socio-Spatial Perspectives on Gender, Difference, and Diversity." In *Canadian Cities in Transition: The Twenty-First Century*, 2nd edn, edited by T. Bunting and P. Filion, 502–24. Toronto: Oxford University Press.

Teixeira, C., W. Li, and A. Kobayashi, eds. 2012. *Immigrant Geographies of North American Cities*. Toronto: Oxford University Press.

Ventimiglia, A. 2007. "The Inside Story: Curbing Gentrification." *Canadian Geographic* 127 (3): 9–12.

White, M. 1987. *American Neighborhoods and Residential Differentiation*. New York: Russell Sage Foundation.

White, P. 1984. *The West European City*. New York and London: Longman.

CHAPTER 13

Bone, R. M. 2011. *The Canadian North: Issues and Challenges*, 5th edn. Toronto: Oxford University Press.

British Petroleum. 2011. *BP Energy Outlook 2030*. www.bp.com/genericarticle.do?categoryId=2012968&contentId=7066695.

Britton, J. 1996. *Canada and the Global Economy: The Geography of Structural and Technological Change*. Montreal and Kingston: McGill-Queen's University Press.

Clark, G. 2007. *A Farewell to Alms: A Brief Economic History of the World*. Princeton, NJ: Princeton University Press.

Daniels, P. W. 1985. *Service Industries: A Geographical Appraisal*. New York: Methuen.

Dicken, P. 1992. *Global Shift: The Internationalization of Productive Activity*. London: Chapman.

Edwards, M. 1997. "China's Gold Coast." *National Geographic* 191 (3): 2–31.

Freeberne, M. 1993. "The Northeast Asia Regional Development Area: Land of Metal, Wood, Water, Fire and Earth." *Geography* 78: 420–32.

Friedmann, J. 1972. "A General Theory of Polarized Development." In *Growth Centers in Regional Economic Development*, edited by N. M. Hansen, 82–102. New York: Free Press.

Friedrich, C., trans. 1929. *Alfred Weber's Theory of the Location of Industries*. Cambridge, MA: Harvard University Press.

Goldberg, S. 2014, November 10th. "The Real Story of US Coal: Inside the World's Biggest Coalmine." *The Guardian*, 10 November.

Graasch, G., and The Daily Climate. 2013 (9 December). "Powder River Basin Coal on the Move." *Scientific American*.

Graham, J., K. Gibson, R. Horvath, and D. M. Shakow. 1988. "Restructuring in U.S. Manufacturing: The Decline of Monopoly Capitalism." *Annals, Association of American Geographers* 78: 473–90.

Holmes, J., T. Rutherford, and S. Fitzgibbon. 2005. "Innovation in the Automotive Tool, Die and Mould Industry: A Case Study of the Windsor–Essex Region." In *Global Networks and Local Linkages: The Paradox of Cluster Development in an Open Economy*, edited by D. A. Wolfe and M. Lucas, 119–53. Montreal and Kingston: McGill-Queen's University Press.

Hoover, E. M. 1948. *The Location of Economic Activity*. Toronto: McGraw-Hill.

Kennelly, R. A. 1968. "The Location of the Mexican Steel Industry." In *Readings in Economic Geography: The Location of Economic Activity*, edited by R. H. T. Smith, E. J. Taafe, and L. J. King, 126–57. Chicago: Rand McNally.

Massey, D. 1984. *Spatial Divisions of Labour: Social Structures and the Geography of Production*. New York: Methuen.

Morrish, M. 1994. "China Takes the Road to Market." *Geographical Magazine* 67 (4): 43–5.

———. 1997. "The Living Geography of China." *Geography* 82: 3–16.

Pasqualetti, M. J. 2011. "The Geography of Energy and the Wealth of the World." *Annals, Association of American Geographers* 101: 971–80.

Pollard, J. S. 1989. "Gender and Manufacturing Employment: The Case of Hamilton." *Area* 21: 377–84.

Pred, A. 1966. *The Spatial Dynamics of U.S. Urban-Industrial Growth, 1800–1914*. Cambridge, MA: MIT Press.

Robinson, D. 2005. "Sudbury's Mining Supply and Service Industry: From a Cluster 'In Itself' to a Cluster 'For Itself.'" In *Global Networks and Local Linkages: The Paradox of Cluster Development in an Open Economy*, edited by D. A. Wolfe and M. Lucas, 155–76. Montreal and Kingston: McGill-Queen's University Press.

Stafford, H. A. 1985. "Environmental Protection and Industrial Location." *Annals, Association of American Geographers* 75: 227–40.

Webber, M. J. 1984. *Industrial Location*. Beverly Hills: Sage.

CONCLUSION

Bunkše, E. V. 2004. *Geography and the Art of Life*. Baltimore: Johns Hopkins University Press.

Buttimer, A. 1974. *Values in Geography*. Washington, DC: Association of American Geographers, Commission on College Geography, Resource Paper No. 24.

Calamai, P. 2006. "Seafood Species Face Extinction." *Toronto Star*, 3 November, A3.

Cloke, P., and R. J. Johnston, eds. 2005. *Spaces of Geographical Thought*. Thousand Oaks, CA: Sage.

Daniels, S. 2011. "Geographical Imagination." *Transactions of the Institute of British Geographers* 36: 182–7.

Dear, M., and J. Wolch. 1989. "How Territory Shapes Social Life." In *The Power of Geography: How Territory Shapes Social Life*, edited by J. Wolch and M. Dear, 3–18. Boston: Unwin Hyman.

Gregory, D. 1994. *Geographical Imaginations*. Oxford: Blackwell.

Haggett, P. 1990. *The Geographer's Art*. Oxford: Blackwell.

Harvey, D. W. 1973. *Social Justice and the City*. London: Arnold.

Jenkins, S. 1992. "Four Cheers for Geography." *Geography* 77: 193–7.

Johnston, R. J. 1991. *A Question of Place: Exploring the Practice of Human Geography*. Oxford: Blackwell.

———. 2005. "Geography—Coming Apart at the Seams?" In *Questioning Geography*, edited by N. Castree, A. Rogers, and D. Sherman, 9–25. New York: Blackwell.

Marsh, G. P. (1864) 1965. *Man and Nature; or, Physical Geography as Modified by Human Action*. Cambridge, MA: Belknap Press.

Mills, C. Wright. 1959. *The Sociological Imagination*. New York: Oxford University Press.

Morrill, R. L. 1987. "A Theoretical Imperative." *Annals, Association of American Geographers* 77: 535–41.

Peattie, R. 1940. *Geography as Human Destiny*. Port Washington, NY: Kennikat Press.

Prince, H. C. 1961. "The Geographical Imagination." *Landscape* 11 (2): 22–5.

Schnaiberg, A., and K. A. Gould. 1994. *Environment and Society: The Enduring Conflict*. New York: St Martin's Press.

Stoddart, D. R. 1987. "To Claim the High Ground: Geography for the End of the Century." *Transactions, Institute of British Geographers* (new series) 12: 327–36.

Viles, H. 2005. "A Divided Discipline?" in *Questioning Geography*, edited by N. Castree, A. Rogers, and D. Sherman, 26–38. New York: Blackwell.

INDEX

Page numbers in italics indicate selected maps and captions.

Abkhazia, 317
Aboriginal peoples, 142, 285, 305; in Canada, 8, *233*, 282, 284, 311, 443, 503; sustainability and, 107, 127–8
abortion, 137–8, 144; selective, 147, 148
accessibility, 38
Accra, Ghana, 457, *457–8*
acculturation, 268, 437
acid rain, 118–19, 491
acropolis, 402, 403
ActionAid, 189
adaptation: cultural, 149–51, 224–5; environmental, 125
Afghanistan, 331, 332; refugees and, 192, 193; US-led invasion of, xxvi, 247, 329, 333, 336
Africa, 66, 167, 175, 204–5, 278, 295; agriculture in, 71, 275, 357, 358, 367–8; cities in, 426–7, 455, 457–8; civil wars in, 190, 307, 314, 330; colonialism and, 204, 207, 306, 312–14, 426–7, 455–6; disease in, 198–202; environmental issues in, 111, 118, 122, 123–4; food shortages in, 183–91; future of, 190–1, 208–9; governance/corruption in, 189–91; human origins in, 161, 261, 263; nations and states in, 204, 207, 303, 311–14; population of, 136, 140, 142–3, 156–7, 160; refugees of, 192–3, 195–6
African Development Bank, 205
African Union (AU), 71, 313, 323
Afrikaners, 265–6, 279
age: fertility and, 134–5, 136, 138, 142; median, 148–9; migration and, 166; mortality and, 140; urban poverty and, 443
age structure, 140, 146–8, *148*, *149*, *151*, 431
agglomeration, 38, 40, *41*, 469, 470–1, 475, 494–5, 496; urban, 389, 390
agora, 402, 403
agrarian revolution, 358
agribusiness, 377
agricultural revolution, 99, 355–6, 379; second, 358
agricultural surplus, 400
agriculture, 342–84; alternative, 360; ancient core areas of, 356, 357; biotechnology and, 345, 358, 362–4, 374; chemical, 359, 360–2; cities and, 400, 437, 439, 450; climate and, 356, 366, 368; commercial, 185–6, 347, 352, 353–4, 357, 358, 368–70; contract, 375; criticisms of, 360–2; dairy, 368; domestication and, 99, 355–7; early, 174–5; environment and, 107–12, 115–17, 119–20, 361–2, 364, 372; gender and, 273–4, 275, 374; global restructuring of, 370–8; government policy and, 69, 345, 346, 374, 376–7; industrialization of, 358, 374, 376; Industrial Revolution and, 358, 373, 377, 379, 473–4; investment in, 370–1, 381; landscape of, 344–6, 357–66; land use and, 348–54; large-scale grain, 369–70; location theory and, 344–50; Mediterranean, 357, 367; migration and, 162, 164, 167, 404; mixed, 368; no-till, 364–5; organic, 188, 360–1, 371, 520; pastoral nomadic, 367–8; peasant, 185–6, 360; physical factors and, 344–5; plantation, 180, 266–7, 368–9, 377; population and, 150–1, 156–8; primitive subsistence, 366–7; protectionism and, 71, 81–2, 89; ranching, 369; regional types of, 357, 366–70; settlement and, 388, 398, 400, 401–4; shifting, 357, 366–7; subsistence, 156–7, 182, 347, 353–4, 366–8; transportation and, 62–3, 349–50, 351, 352, 357, 368; water and, 117; wet rice, 357, 367; *see also* food
aid, 175, 190–1, 207, 208; food, 179, 188–90, 195–6, 197
AIDS, 19, 149, 186, 198, 199–202, 206, 207
air: human impacts on, 105, 116
air transportation, 66
Alberta oil sands, 115, 480, 491
alienation, 89
Alpine rail tunnels, 65, 68
Amazon (online retailer), 447
Amazon River basin, 86, 101, 480
American Anthropological Association, 264
Amish, 249, 251
Amnesty International, 86, 180
anarchism, 323
animals: domestication of, 99, 112–13, 114, 355–7, 379, 402; as food, 150, 248–9, 345, 355–6, 380–1; human impacts on, 112–15; lost species of, 113–15, 365–6
animism, 243
Antarctica, 10, 126, ozone hole over, 105, 116
Anthropocene (human) epoch, 96, 114
anthropocentric world view, 107
anthropogeography, 13, 14
anthropology, 51
anti-natalist policies, 145–6
apartheid, 80, 265–6, 332, 334, 435
Apian, Peter, 11
Arab Spring, 330, 331, 334, 336, 338; social media and, 74
Aral Sea, 117, 118
ARC/INFO (software), 48
Arctic, 86; climate change and, 121, 123, 124, 333, 479; ecotourism in, 291–2; sovereignty over, 333–4
areal differentiation, 17, 35
Aristotle, 5, 16, 323
Asia, 34, 76, 80, *81*; cities in, 427–9, 455, 458; immigration from, 267–8; industry in, 72, 486–92
Asia–Pacific Economic Co-operation (APEC), 323
assimilation, 268–9, 437
Association of Southeast Asian Nations (ASEAN), 68, *69*, 323, 338
asylum seekers, 192
atlases: early, 9, 11, 13; modern, 282, 334, 518
Attawapiskat, ON, 284
attractions, tourist, 290, 292–4
Australia, 15, 65, 84, 89, 241, 284; animals' impact in, 113, *113*, 114; distance in, 62; elections in, 325; immigration to, 267
Australopithecus, 6
authority, 260
auto industry, 444, 473, 488, 492, 495
automobiles, 437, 439, 447, 448, 454–5

back-office activities, 448, 498
Balfour Declaration, 309
Bangkok, 458–9
Bangladesh, 184, 206, 336; creation of, 319; flooding in, 120, 196; Grameen Bank of, 112, 204; population of, 160; refugees in, 193
barriers/walls: states and, 309, 338
Basque people, 237, 310, 332
BAT industries, 72, 377
Begin, Menachem, 332
behavioural approaches: to industrial location, 472–3; to migration, 165
Beijing, 65, 66, 429, 450
Belgium, 236, 237, 315
Belize, 292
Benedict XVI, 201
Berlin Wall, 309, 338
Bhopal, India, environmental disaster in, 105
"Bible belt," of US, 37, 219
"big box" stores, *447*
binary thinking, 513–14
bin Laden, Osama, 247, 332
biodiversity: loss of, 113–15, 122
biofuels, 100, 103–4, 120, 188
biogeochemical cycles, 96, 97–8
biomass, 99
Biosafety Protocol, 364
biosphere, 97
biotechnology, 345, 358, 362–4, 374
Bird, Hubert, 452
Blackberry Messenger (BBM), 75
Blainey, Geoffrey, 62
blogs, 74
Boko Haram, 511; Nigerian parents' protest against, *75*
Boserup, Ester, 156–7
Bosnia-Herzegovina, 316
Botswana, 136, 140, 201; gender in, 273–4; good governance in, 189, 190
boundaries, 308–9; as colonial constructs, 204, 207, 309, 312–14; religion and, 245, 315, *316*
brain, human: evolution of, 4, 6
Brandt Report, 177
Branson, MO, 294
Brasilia, 451

Brazil, 66, 71, 89, 487; biofuels in, 103; environment in, 101, 105, 110–11, 117; population of, 148, *149*
Bretton Woods system, 79–80, 88
bridges, 66
Bridi, Robert, 375
Britain, 14, 176, 222, 240, 281, 283, 312, 334, 405; agrarian revolution in, 358; Canada and, 270, 305, 487; coalfields of, 404, 475–6, 494; deindustrialization in, 494; energy in, 103, 483; environment and, 103, 107, 122, 123; food consumption in, 379–80, 381, 382; geography in, 14; indentured labour and, 266–7, 321; Industrial Revolution and, 373, 435, 475–6, 485; industrial slums of, 404, 441, 456; migration research in, 163, 164; monarchy of, 323; Palestine and, 309, 332; politics/voting in, 325, 326–8; popular culture in, 287, 290; riots in, 74–5; segregation in, 284, 285, 435; separatist movements in, 314–15; as state, 303; transport systems in, 63–4, *64*; urban problems in, 443
British Commonwealth, 307, 323
British Empire, 34, *304*, 304–5, 306, 318
British Petroleum (BP), 86, 104
Brundtland Commission, 105, 127
Buddhism, 88, 241, 243, 244–5
Buffon, Georges-Louis Leclerc, comte de, 104
Bulgaria, 69
Burgess, Ernest, 423–4
Burma, 244–5, 267; government of, 197, 240, 335; refugees and, 192, *512*
business processing outsourcing (BPO), 498
business services, 412, 413, 414, 416, 446

Cajuns, 220–1
Calgary, 431, 450, *451*, 453; homelessness in, 445
caliche (nitrogen source), 359
Cameron, David, 75
Canada, 62, 63, 71, 73, 84; Aboriginal peoples in, *233*, 282, 284, 311, 443, 503; agriculture and, 344, 346, 351, 360–1, 365, 370, 371, 376–7; Arctic and, 123, 291–2, 333–4; changing settlement patterns in, 394–6; cities in, 431, 432, 437, 438, 440; city planning in, 449–54; environmental issues in, 99, 105, 106, 115, 122, 123, 127–8; environmental management in, 107, 110, 118, 126–7, 291; geography in, 15; homelessness in, 444–6, 520; iconography of, 259; immigration to, 267, 268, 269–71; industry in, 487, 491, 499–501, 502–3; multiculturalism in, 84, 269–71, 437; oil/gas industries in, 86–7, 115, 480, 482, 491; politics in, 326, *328*; population in, 139, 148, *151*, 159; regionalization of, 217–18, *219*; tourism in, 290, 291–2; urban centres/areas in, 389, 390; urban hierarchies in, 408–11; urbanization in, 394–6, *396*; *see also* Quebec
Canadian Housing & Renewal Association, 446
Canadian Pacific Railway, 370
Canadian Urban Institute, 446
Canadian Wheat Board, 370
canals, 63, 64, 66; impact of, *65*
cane toad, 114
Cape Breton Island, NS, 8, 499–500
capital: Canadian industry and, 487; factories and, 475; industrial location and, 469, 471; movement of, 60, 72, 76, 78, 79, 80
capitalism, 88–9, 177, 182, 282, 284; agriculture and, 358, 372; assessment of, 88; competitive, 88; culture and, 30, 31, 223; democracy and, 89, 303, 323, 326–8, 338; developmental stages of, 182; disorganized, 88; employment and, 266; environment and, 105, 125–6, 223; global inequalities and, 175–6, 185–6; housing and, 430–2; Industrial Revolution and, 473–4, 475; industry and, 468, 484, 498–501; Marxism on, 29–30, 31, 36, 44–5, 82, 88, 89; nation-states and, 303, 323–5, 326–8, 338, 474; modernism and, 32; organized, 88; population and, 139, 156; poverty and, 442; socialist/communist states and, 80, 85, 324–5, 484, 491
carbon dioxide, 104, 111, 119–20, 121, 123, *125*, 292
Caribbean Community (CARICOM), 71
carrying capacity, 145, 154
Carson, Rachel, 105
cartography, 46–8; computer-assisted, 48; *see also* maps
caste system, 244
Castro, Fidel, 324
catastrophists, 124, 154–5, 190, 208–9
cattle ranching, 111, 369
Celtic languages, 239
census, 159
Census Metropolitan Areas (CMAs), 395
Centers for Disease Control and Prevention, 202
Central African Customs and Economic Union (UDEAC), 71
Central American Common Market (CACM), 71
central business district (CBD), 422; in various models, 423–6
centralization, 308, 495
central place, 407
central place theory, 406–8, 409, 496
centrifugal/centripetal forces, 308
Chad, 190, 331, 333
Chamberlain, Houston Stewart, 264
Chang Chi'en, 8
Channel Tunnel, 65, 68
Charlottetown, 449, *450*
charter population, 435
cheese, 222, 350, 368, 380
Chernobyl nuclear disaster, 102, 105
Chiasson, Paul, 8
Chicago, 423, 425, 439, 450
Chicago School, 423; North American city models of, 423–5
chimpanzees, 264
China, 75, 202, 208, *231*, 267, 276, 334, 381; agriculture in, 115, 120, 150, 357, 359, 367; capitalism in, 80, 85, 324, 484, 491; cities in, 389–90, 391, 394, 398, 403, 411, 416, 428–9, 450; conflicts/tensions of, 321, 329; disease in, 198, 202; early civilization/urbanism in, 160, 306, 309, 400, 401–2, 403; early world exploration by, 8, 175; economic rise of, 34, 80, 100, 176, 338, 490–1; energy and, 100, 477, 478, 482; environment in, 100–1, *101*, 115, 117, 118, 120, 122, 491; famines in, 153, 184, 185; forced labour of, 267, 305, 369; geography in, 4, 7–8, 9, 11, 13, 19; industry in, 446, 484, 485, 487, 488, 489–91, 492, *502*; language in, 232, 235, 237; outsourcing in, 497–8; population in, 34, 136, 137, 138, 143, 146, 147, 148, 152, 160, 184, 188, 324, 362; religion in, 243, 244–5, 250; as socialist state, 324; tourism in, 295–7; transportation in, 64, 65–6; Uighurs and, 318
China National Offshore Oil Corporation (CNOOC), 491
Chinese Exclusion Act (US), 267
chlorofluorocarbons (CFCs), 105–6, 116, 119
cholera, 228–9, 450, 460
Chongqing, China, 390
chorology, 13, 14, 17
Christaller, Walter, 406–8
Christian Aid, 203
Christianity, 88, 217, 235, 241, 243, 248–9; churches of, 249, 250; environment and, 105; European dividing line of, 245, 315, *316*; Islam and, 246, 247–8, 316
Churchill, MB, 64, 291–2
cities, 388–416, 421–61; apartheid, 435; "arrival," 461; beautiful, 450; as central places, 406–8, 409; civilization and, 400; classic models of, 423–5, 429–30; coastal, 427; colonial, 400, 426–7, 455–6, 457; conceptualization of, 390–1; "of Culture," 454; dual, 427; edge, 439–40, 446; form/structure of, 422–30; garden, 450–1, 453, 454; gateway, 414, 440; global, 412–16; governments of, 391, 422, 430, 434–5, 458–9; hierarchies of, 409–12, 414–16; high-rise buildings and, 441, 451; indigenous, 427; industrial, 402, 404–5, 413, 431, 433; inner, 432, 434–5, 439, 446, 454; Islamic, 403, 427; "largest," 390; in less developed world, 392–4, 396–7, 455–61; location of, 405–6; mega-, 396–8, 428, 456; in more developed world, 396–7, 412, 414, 416; origins/growth of, 398–405; planning of, 449–55; post-industrial, 431, 439–40; poverty in, 440–6; pre-industrial, 401–4, 441; primate, 411–12, 415, 456; production/consumption in, 446–7; social geography of, 430, 431; space constraints in, 441; transportation/communication in, 447–9; vs urban areas, 390–1; *see also* colonial cities; global cities; informal settlements and slums; inner city; settlements; urbanization
City Beautiful movement, 450, *451*
civilizations: cities and, 400; "clash of," xxvi–xxvii, 247–8, 333, 336–8; development of, 174–5; religions and, 215, 241, 243; Toynbee on, 215; *see also* hearths
civil wars, 270, 325, 329–30, 331, 336; in Africa, 190, 307, 314, 330; drug trafficking and, 330; poverty and, 187; refugees and, 191–4, 330; resource competition and, 330, 333; in Sri Lanka, 192, 321; in Syria, 191, 310, 330; in US, 326
class, 29, 30, 241, 258; landscapes and, 283–5, 426, 430; politics/voting and, 314, 324, 326–8
Clayoquot Sound, BC, 127–8
climate: agriculture and, 356, 366, 368; global distribution of, *8*; human impacts on, 119–24
climate change, 119–24; cities and, 406, 417; debate over, 123–4; human-induced, 111, 119–20, 123; ice melting/rising sea levels and, 120, 121, 122, 123–4, IPCC report on, 100, 121–2, 124; politics and, 100, 105–6, 122; renewable energy and, 100, 103–4; tourism and, 291–2; uncertainty over, 100, 109, 110–11, 113, 120, 121–2, 123–6
Cloke, Paul, and R.J. Johnston, 514

Club of Rome, 153–4, 208
coal, 477, 480, 482–3, 485, 492; cities and, 404; Industrial Revolution and, 473, 475–6, 494
coalfields: Britain, 404, 475–6, 494; Mongolia, 484–5; Ruhr Valley, 404, 476, 485; Sydney, NS, 499; Ukraine/Russia, 485, 486; US, 404, 483
Coal India, 492
Cold War, 338; end of, 83, 328, 331, 334–5
Colombia, 203, 331, 333
colonial cities, 400, 435, 455–6; as entrepots, 404; exploitation in, 456; primacy of, 411–12; as tourist attractions, 294, *295*; world examples of, *426*, 426–7, *428*, *429*, 457, *457*
colonialism, 20, 21, 50, 79, 175–7, 245, 328; agriculture and, 180, 266–7, 361, 368–9, 377; constructionist views of, 260–1, 279; economic, 176, 204, 266, 369, 377, 456; exploration and, 304–7, 404; food consumption and, 379–80; food shortages and, 185–6; independence from, 84, 306–7; language and, 232, 237, 240; migration and, 162, 266–8, 303, 474; neo-, 361; place-names and, 240, 455; racism and, 264, 265–8, 279; tourism and, 290, 292–4; *see also entry above;* Europeans, as overseas explorers/colonizers
Columbus, Christopher, 8, 10
commodity flows, 67, 68
commodity network, 374
Common Agricultural Policy, 377
Common Market for Eastern and Southern Africa (COMESA), 71
common markets, 68
Commonwealth of Independent States (CIS), 71, 317
communism, 328; collapse of, 338
company towns, 499
concentric zone models: agricultural, 348–54, 373–4, 382, 406–7; urban, 423–4, *424*
condominiums, 422, 433, *434*
conflicts, 302–3, 328–9; in Africa, 190; categories of, 329; civil wars as, 329–30; in former Yugoslavia, 194, 315–16; Israeli–Palestinian, 261, 309–10; refugees and, 186, 191–4; religion and, 246–8, 261, 314–16, 333, 336–8, 511; resource competition and, 330, 333–4; separatist, 314–15; South Asian, 180, 318–21; tourism and, 296; *see also* war; *specific nations and conflicts*
Confucianism, 88, 244, 245
Congo, Belgian, 307, 329; *see also* Democratic Republic of Congo
conservation, 125, 127
constructionism, 32, 258, 260–1, 512
consumer services, 407, 412, 446–7
consumption: cities and, 259, 437–8, 439, 446–7; collective, 493; energy, 477, 481–3; food, 378–82; geography of, 288; tourism as, 290, 294
container shipping, 66, 489
contextualism, 261
continental drift, *3*
continents: shape of, 174–5, *175*; "southern," 10
contraception, 137, 138, 146, 156
conurbations, 439
convergence, time-space, 61–2
Cook, James, 8, 9, 10
coral reefs, 96
Corbett, Jon, 44–5
cordilleran belts, global, *5*
core–domain–sphere model, of cultural adaptation, 225, 226
core–periphery models: of economic growth, 502–3; of nations, 314, 487; of world systems theory, 175–6, *177*, 185–6
cornucopians, 124, 154–5, 190, 208–9
corporations: *see* transnational corporations
correlation and regression analysis, 353
corruption: aid funding and, 189–90
Council for Mutual Economic Assistance (Comecon), 323
countries, xxix; *see also* nation-states; states
Cousin, Victor, 16
creole languages, 237–8
crime, 282, 283; in cities, 430, 441–2, 444, 456
Croatia, 69, 316, 503
crop theory, 349–50, *350*
"Crow rate," 370
crude birth rate (CBR), 134–5, *138*
crude death rate (CDR), 139–40, *141*
Cuba, 324, 329; US and, *324*
cultural adaptation, 149–51, 224–5, 511
cultural geography: *see* culture; language; religion; *see also* difference; identity
cultural landscapes, 222–9; adaptation and, 224–5; diffusion and, 225–9
cultural regions: core–domain–sphere model of, 225; formal, 214–18, *216*; vernacular, 218–22
"cultural turn," 258–61, 280
culture, 214–29; agriculture and, 345, 378; "City of," 454; collapse of, 98; conflict and, 214, 336–8; diffusion of, 225–9; ethnicity and, 268–71; folk, 286–7; food and, 378–80; future of, 511–12; gender and, 30–2, 271–6; of global cities, 414; globalization and, 60, 80–3, 89–90, 287–9, 379; of inner city, 454; landscapes of, 222–9; local, 61, 82–3; Marxism and, 30, 31, 223, 227–9, 258, 271, 282, 513–14; material, 225; micro-, 279–81; migration and, 161, 162–3; nation-state and, 80, 82, 302–3, 311, 314–15; non-material, 82, 258; plurality of, 258; popular, 82, 287–9, 290; population density and, 160; population growth and, 149–51, 224–5; regions of, 214–22; *see also* identity; symbolic approach, to culture
currency, North American, 70
customs union, 68, 71
cycle of poverty, 442
cyclones, 197–8
Cyprus, 69
Czechoslovakia, 84

Dalai Lama, *245*
Dalrymple, Alexander, 10
dams, 100–1, *101*
Darwin, Charles, 13, 16, 113, 264
Das, Raju J., 375–6
data collection and analysis techniques, 46–53; cartography, 46–8; geographic information systems, 48–9, 51, 512, 514–15; positivism and, 28, 52; qualitative methods, 51–3; quantitative methods, 52–3; remote sensing, 49–51
Davis, William Morris, 14, 16, 17
Davis Inlet, NL, 284
D'Azeglio, Massimo, 236
Dear, Michael J., and Jennifer Wolch, 514
death, causes of: disease, 198–202, 206; natural disasters, 196, 197; poor nutrition, 135, 183–4, 195, 206; pregnancy/childbirth, 207; *see also* mortality
debt crisis, 186, 202–3
decentralization, 308; business/retail, 422, 430, 440, 446–7, 448; industrial, 494–5, 496; residential, 422, 437–40
decolonization, 34, 306–7, 310, 329, 334
deconstruction, 33
deforestation, 98, 107–11; agriculture and, 104, 108, 109, 111; "great reversal" of, 109–10
deglomeration, 38, 40, *41*, 469, 470–1
deindustrialization, 492, 493, 494, 502
democracy, 323; in Africa, 189, *189*, 190; capitalism and, 89, 303, 323, 326–8, 338; development aid and, 190–1; food shortages and, 186–7, 188, 207; globalization and, 84; peace and, 334–6; religion and, 248; world status of, *335*, 335–6
Democratic Republic of Congo (DRC), 74, 111, 160, 199, 202, 333, 397, 491; conflicts in, 140, 190, 193, 307, 329
demographic transition theory, 157–8, 182, 474
demography, 134; sample data of, xxxi; *see also* population
demonstrations: *see* protest, public
Deng Xiaoping, 489, 491
Denmark, 69, 187, 248, 303, 323, 333–4
density: forest, 110; land use/residential, 422, 440–1, 450, 454; *see also entry below*
density, population, 5, 12, 159–60, *160*; in cities, 422, 425, 428, 451; physiological, 159
department stores, 447
dependence, 175
dependency theory, 175–6, 204, 304
depopulation, rural, 371
Depth Index of Globalization (DIG), 78, *78*, *79*
desertification, 111–12
deserts, 111
determinism, environmental, 15, 16, 17, 104, 174
Detroit, 444
development: as concept, 42–5; defining, 178, 181; human, 182–3, *183*, 184; inequalities and, 174–83, 184; MDGs and, 206–7; measuring, 181–3; population and, 135, 136, 138–9, 144; sustainable, 126–9; uneven, 501–3
developmentalism, 164, 181–2
devolution, 314–15
Diamond, Jared, 98, 175
Dias, Bartolomeu, 10
Diaspora, 309
dictatorship, 323, 335
diet, 98, 117, 287, 360, 368, 377, 380–1; global patterns of, *378*, 378–82; religion and, 248–9, 345; *see also* food
difference, 256–99; "cultural turn" and, 258–61; focus on, 259–61; geographic scales of, 261; inclusions/exclusions and, 262; landscape and, 279–81; myth of race and, 262–8; popular culture and, 287–9; power relations and, 260–1; theories of, 260–1; *see also* identity; identity/difference
diffusion, 42; cultural, 225–9; of disease, 42, 228–9; early adopters and, 228–9, *230*; linguistic, 234–5, *235*; of religions, *242*, 243–6; S-shaped curve of, 42, *42*, 227, *230*

digital divide, 62, 72–4
disabled community, 280
disasters, 194, 196–8; as "natural," 194
discourse, 5, 45, 260–1
discrimination: housing and, 432, 436
disease: Aboriginal peoples and, 282, 284; diffusion of, 42, 228–9; in less developed world, 198–202; population aging and, 152; *see also specific diseases*
Disney theme parks, 223–4, 288, 294
distance: agriculture and, 348–54; cities and, 422–9; as concept, 37–40; decay of, 38, *38*; distribution and, 37–8; economic, 39–40; friction of, 38, 60, 61–2, 64, 66–7; geography as "discipline in," 61–2; historians on, 62; industrial location and, 469, 471, *472*; physical, 38, *39*, *40*; time, 38–9, *39*, *40*; *see also* globalization
distribution, *37*, 37–8; language, 232–3, *233*, 241; population, *160*, 161; religion, 241, *242*, 243
Djibouti, 193, 195
domestication: of animals, 99, 112–13, 114, 355–7, 379, 402; causes of, 356; centres of, 355, *356*; consequences of, 113, *113*; of plants, 99, 100, 104, 109, 355–6, 379, 402
Don Mills, ON, 454
donut effect, 395
doubling time, 143–4, 157
Doukhobors, 251
Drake, Francis, 10
droughts, 34, 117, 118, 183, 184; food aid and, 188–9, 194; in Horn of Africa, *185*, 195; in Sahel region, 105, 111, 122
drug production/trafficking, 267; conflict and, 330, 331, *331*
drug users: AIDS and, 200, 202; in cities, 443, 444, 446; slang of, 281
Dubai, 441
Dubai Mall, 287–8, 294
Durkheim, Émile, 16
"Dust Bowl," of Depression era, 116

early adopters, 228–9, *230*
earthquakes, 181, 196–7
Earth Resources Technology Satellite (ERTS; later Landsat), 49–50
Easter Island, 98
Eastern Europe, 308, 323, 324, 334, 335, 336; heartland theory and, 307, *308*, 338; NATO and, 338; religious dividing line in, 245, 315, *316*
Eastern Orthodox Church, 245, 315, *316*
East Timor, 480
Ebola, 19, *21*, 202, 207, 331
ECO Canada, 126
ecocentric world view, 107
ecocide, 98
ecology, 97, 129, 514; "deep," 107; political, 372
e-commerce, 447, 448
Economic Community of West African States (ECOWAS), 71, 323
economic operator, 349, 469, 472–3
economic rent, 347–50, 353; calculation of, 348
economic system, global, 78–80; energy and, 476–83; geo-economic regions of, 80, *81*, 338; industry and, 468–73, 483–501; *see also* globalization
ecosphere (global ecosystem), 97
ecosystems, 97; animal impacts on, 114; chemical cycling/energy flows in, *97*, 97–8; human impacts on, 96, 98, 104–7; preservation of, 107, 126, 127–8; *see also* human impacts
ecotourism, 291–2
education: fertility and, 136, 139, 146; in less developed world, 182, 189, 201, 203, 204, 206
effect (or friction) of distance, 38, 60, 61–2, 64, 66–7
Egypt, 64, 310; Arab Spring and, 74, 336, *336*
Ehrlich, Paul, 154, 208
Eight Mile Road, Detroit, 444
elections, 325–8; bias in, 326–7; legitimacy of, 325–6
electronic waste (e-waste), *87*
elites: landscapes and, 283–5, 426, 430
Ellie (homeless woman in California), 445
Eminem, 444
empires, 304–7, 315–18; *see also* colonialism; *specific empires*
empiricism, 26, 35
employment: cities and, 431, 439–40, 446–7, 468; environmental, 126–7; outsourcing and, 491–2, 496, 497–8; postwar changes in, 495; in service industries, 495–7; in slums, 459; by transnationals, 72, 85, 488, 490, 493, 500; *see also* labour; workforce
enclosures, 358
"end of history" thesis, 334–5
energy: fossil fuel, 476–83; Industrial Revolution and, 473–6; industry and, 483–501; renewable, 99–104, 477; technology and, 98–9
Engels, Friedrich, 29, *30*
English language, 231, 232, 233–5, 237–8
entitlements, 186
entrepot, 404
environment, 96–129; agriculture and, 107–12, 115–17, 119–20, 361–2, 364, 372; careers in, 126–7; in cities, 450–1, 476; disasters of, 102–3, 105, 118, 119; energy/technology and, 98–9; ethics and, 104–7; fossil fuels and, 96, 115, 119, 478, 481, 482, 483; globalization and, 86–7; global perspective on, 96–104; human impacts on, 107–24, *125*; issues of, 106–7; migration and, 161, 163, 167; politics of, 105–6; renewable energy and, 99–104; responses to issues of, 124–6; in slums, 460; state power and, 106, 119, 324–5; sustainability and, 126–9; *see also* environments, global; human impacts
environmental determinism, 15, 16, 104, 174
Environmental Protection Agency (EPA), 105, 483
environments, global, *9*
epidemics, 150, 153
equinox, vernal and autumnal, 2
Eratosthenes, 5–6; map of, *5*
Eritrea, 179, 189, 195; *see also* Ethiopia
erosion, soil, 98, 111, 112, 115–16
Esperanto, 238
essentialism, 32, 258
Esteva, Gustavo, 181
ETA (Basque separatist group), 332
ethanol, 103
ethics, environmental, 104–7
Ethiopia, 160, 162, 179, 184, 198; case study of, 179; conflict in, 179, 189, 190, 193, 195; famine in, 185, 195; industrial park in, *502*; land use in, 353–4; safe water in, 206
ethnic cleansing, 316, 329
ethnic diversity, 270–1
ethnic groups, 268; migration by, 268; neighbourhoods/enclaves of, 268, 285, 435–6, 442
ethnicity, 268–71; assimilation/acculturation and, 268–71, 437; multiculturalism and, 269–71, 437; poverty and, 430, 442, 443; of states, 312–22; as term, 268
ethnic regions: in Africa, 311–14, *313*; in Europe, 314–18, *322*; in South Asia, 318–19, *320*, 321
ethnocentrism, 52
ethnography, 51
euro, 70
Europe: cities in, 425, *426*, 439, 454; as cultural region, 217; early maps of, 5–6, *5–6*, *9*; early travellers to, 8–9; environment in, 102, 103, 109–10, 119, 120; Industrial Revolution in, 473–6; industry in, 476, 485–6, 494; medieval geographic decline in, 6–7; migration from, 62, 161–2, *162*, *166*, 167–8; nation-states in, 303, 314–18; population in, 139, 140, 143, 152, 156, 159–60, 188; regionalism in, 314–18, *322*; *see also entries below*
Europeans, as overseas explorers/colonizers, 7, 8, 10–11, 176, 304–7; boundaries set by, 204, 207, 303, 309, 312–14; cities built by, 426–9, 455–6; diseases carried by, 98; environmental destruction by, 104–5, 107, 109, 113, *113*, 114; imperialism and, 304–7, 315–18; imperial territory of, *304*, *306*; indentured labour used by, 167, 266–7, 305, 321, 369, 456; industrialization difficulties and, 204, 456; *see also* colonialism
European Space Agency, 51
European Union, 34, 68, 71, 82, 89, 413, 518; agriculture and, 69, 345, 363, 377, 520; as grouping of states, 310, 312, 321–2, 323, 338; history of, 69; illegal migration to, 168; uneven development in, 503
evolution, human, 262, 263
existentialism, 27
exonyms, 240
exoticism, 292–4, 295
exploitation, 29, 329–30, 456
exploration: colonialism and, 304–7; early travel and, 7, 8, 9, 10–11; *see also* colonialism; Europeans, as overseas explorers/colonizers
export-centred economies, 86
export-processing zones (EPZs), 488, 489
extinction, of species, 111, 113–15
Exxon Valdez oil spill, 119

Facebook, 74–5
factories, 468, 469; Chinese, 489–90, *502*; early, *40*, 473–4, 475, 476
fair trade movement, 361
family: geography and, 223–4; size of, 135, 136, 139, 158
famine and food shortages, 183–90; explanations of, 184–8; food aid and, 188–90
fascism, 323
favelas, *460*, 461
fear: geography of, 282, 283, 285, 443
fecundity, 134–5, 136–7
federalism, 308; in Africa, 313
feedbacks, 97

feedlot system, 368, 369–70
feminism, 30–2, 35, 53; geography of, 271–6; types of, 31; *see also* gender
fertility, 134–9; factors affecting, 135–8, 324; measures of, 134–5; population aging and, 148–9, 152; population growth and, 155–8; replacement-level, 135; variations in, 138
fertility transition theory, 135–6, 153, 510
fertilizers, 156, 344–5, 360, 361–2; nitrogen, 145, 345, 358–9, 365, 374
feudalism, 29, 266, 474, 486
fieldwork, 51
filtering, 433–4
Finland, xxxi, 69, 72, 74, 101
fire, 96, 149–50; for cooking, 99, 379; environment and, 108–9
first effective settlement, 216, 239–40
fisheries: globalization and, 86; sustainable, 510–11
Flanders, 315
flexible accumulation, 493
floods, 179, 196, 198
food: animals as, 150, 248–9, 345, 355–6; changing preferences and, 380–1; consumption of, 378–82; cooking of, 99, 379; "ethnic," 378–80; fast, 379; global inequalities and, 183–90; global production of, 184, 185–8; global trade in, 377–8; lost and wasted, 381–2; population and, 150, 155–7; price increases and, 188; shortages of, 184–8; specialty, 380; supply system for, 373–4, *374*; *see also* agriculture; diet
food aid, 188–90, 194
Food and Agricultural Organization (FAO), 110–11, 381, 520
forces of production, 358
Ford, Larry R., 426
Fordism, 88, 492, 500
foreign direct investment (FDI), 72, 76
forests: density of, 110; monitoring of, 107; temperate, *109*, 127–8; tropical rain, 110–11; *see also* deforestation
formal (or uniform) region, 36–7
formal cultural regions, 214–18; Europe as, 217; North American, 216–17, *218*; world, 215–16, *216*
forum, 403
fossil fuels, 476–83; alternatives to, 99–104, 124; impacts of, 96, 115, 119, 478, 481, 482, 483; *see also specific fuels*
Foucault, Michel, 45, 260, 272
Foxconn, *90*
Francaviglia, R.V., 226
France, xxxi, 176, 304, 312, 494; Basque separatism in, 310, 332; as former empire, 306, 307, 309; geography in, 14; as nuclear power, 334; as state, 303, 309
Frank, Andre Gunder, 176
"freedom fighters," 332
Freedom House, 336, 520
free trade, 67–71, 80, 82, 85
French language, 235; in Belgium, *236*, 236, 237, 315; in Louisiana, 220–1, 240; in North America, *238*; in Quebec, 230, 236, 237, 239, 311; in Switzerland, *236*, 236
Friedman, Thomas L., 76, 100
Friends of the Earth, 363, 364
front-office activities, 448, 498
functional (or nodal) region, 37
Gaelic languages, 234–5; in Ireland, 236, 237, 239
Gaia hypothesis, 124
Gama, Vasco da, 10
gambling, 294
Gandhi, Mohandas K., 332
garden city, 450–1, 453, 454
Garreau, Joel, 219–20
gated communities, 440, 442
gateway cities, 414
gay pride parades, 278–9, 280
gays and lesbians, 33, 272, 276–9; *see also* homosexuality
Gaza Strip, 310
gazetteer of Ptolemy, 6
Geldof, Bob, 189
Gemeinschaft, 227, 474
gender, 31–2, 271–6; agriculture and, 273–4, 275, 374; health and, 275; human development/equity and, 275–6; labour and, 273–5, 459, 488, 500–1; landscape and, 272–4; of refugees, 193; religion and, 241; urban poverty and, 443; violence and, 274, 276, 283, 284, 285; work and, 274–5
Gender Inequality Index (GII), 276
General Agreement on Tariffs and Trade (GATT), 80, 81
genetic modification (GM), 362–4
genocide, 262–5
gentrification, 434–5, 442
geo-economic regions, 80, *81*, 338
"geographical imagination," 508, 515–16
geographical societies, 13, *13*
geographic information systems (GIS), 44, 48–9, 51, 512, 514–15; environment and, 96, 126
geographic literacy, 33, 34
geographic writings/descriptions, 4–6, 9, 11
géographie Vidalienne (or *la tradition Vidalienne*), 14
geography: agricultural, 342–84; applied, 19; careers in, xxxii, 126–7; classical, 4–6; critical, 258; culture and, 214–52; definition of, 12; as "discipline in distance," 61–2; economic, 344; family, 223–4; general vs special, 12, 13, 17; global issues and, 19; history of, 4–21; identity/difference and, 256–99; industrial, 483–501; institutionalization of, 13–15; literary, 4–6, 9, 11; mathematical, 6, 9; "new cultural," 258; "new economic," 205–6; of peace and war, 328–34; political, 300–40; population, 134–69; preclassical, 4; social, 430, 431; technology and, 20, 44–5, 49–51, 512, 514–15; as term, xxiv, 6; transport, 62–6; universal, 12–13; urban, 388–416; Varenius on, 11–12, 17; welfare, 281; world industrial, 483–92; *see also* human geography; physical geography; regional geography
geomancy, 401
geopolitics, 307–10, 333, 511; future of, 334–8; theories of, 307–8
geopolitik, 307
Georgia, 145, 317
geosophy, 5
Geospatial Web/Geoweb, 45
geothermal energy, 103
Germany, 176; creation of, 303; division/reunification of, 191, 310, 338; environment and, 103, 105, 122; geography in, 11–14, 307; industry in, 484, 494; under Nazi regime, 167, 265, 307, 309, 323; refugees and, 191, *192*; Ruhr coalfields of, 404, 476, 485
gerrymandering, 326–7
Gesellschaft, 474
Ghemawat, Pankaj, 76; and Steven A. Altman, 78
ghetto, 268; as historical space, 249, 435; poverty in, 442; stigma/segregation of, 268, 285, 436
Gilbert, Emily, 70
Githongo, John, 207
Glace Bay, NS, 499
global cities, 412–16; characteristics of, 412–14; as command centres, 412, 415, 456; geo-economic power of, 80, *81*; hierarchy/rankings of, 414–16, *415*; spheres of influence of, 416, *417*; transnationals and, 60, 413, 415
Global Cities Index, 414, *415*
global inequalities, 174–209; disasters and, 194, 196–8; disease and, 198–202; economic growth prospects and, 202–5; explanations for, 174–6; of hunger/famine, 183–90; identifying, 176–3; possible solutions to, 205–9; of refugees, 191–4, 195; *see also* less developed world
globalization, xxvi, xxxi, 34, 60–90; agriculture and, 81–2, 89, 370–8; assessment of, 85–90; cities and, 60, 80, 412–16, 456; as concept, 45–6; cultural, 60, 80–3, 89–90, 287–9, 379; distance and, 60, 61–2; economic, 78–80; history of, 79–80; industry and, 468–73, 483–501; interpretations of, 76; local experience and, 60–1, 83; measurement of, 76–8; opposition to, 85–7; political, 83–4; protests against, 76, 81; state and, 67–78, 80, 84; support for, 87–90; theses of, 76, 77, 80, 83, 86, 87; trade and, 66–71; transmission of information and, 61, 62, 72–5; transnationals and, 46, 71–2; transportation and, 61, 62–6
Globalization and World Cities (GaWC) Research Network, 414, 520
Global Malaria Action Plan, 198
global positioning system (GPS), 44, 51, *51*, 514
Global Slavery Index, 267
"global village," 60, 90
global warming: *see* climate change
"glocalization," 374
Gobineau, Joseph Arthur de, 264
Golden Temple, Amritsar, 250, *251*
gold standard, 76, 80, 492
"Gonneville land," 10
Gould, Stephen Jay, 262
government: agriculture and, 69, 345, 346, 374, 376–7; Canadian industry and, 487, 502–3; cities and, 390, 400, 404, 430, 432, 450, 453–4, 470–1, 480–6; devolved, 314, 315; food shortages and, 186–7, 188; forms of, 323; global cities and, 413–14, 416; homelessness and, 444–6; industry and, 469, 476, 484, 485–6, 491, 494; in less developed world, 186–91, 207, 208; map-making by, 13; municipal, 391, 422, 430, 434–5, 458–9; parks created by, 290–1; population and, 144–6; world, 323
Grameen Bank, 112, 204
Gramsci, Antonio, 260
"Great American Desert," 240

Great Bear Rainforest, 128
Great Lakes, 333, 485
Great Lakes–St Lawrence Lowlands, 487
Great Lakes Water Quality Agreement, 118
Great Wall of China, 309
Greece, 69, 240, 250, 487, 503; ancient, 4–6, 323, 402
green belt, 450
Greenberg, Charles, 291–2
Greene, Graham, 180
greenhouse effect: human-induced, 119–20, 123; natural, 119
greenhouse gases, 103–4, 111, 116, 119–20, 125; agriculture and, 359, 381; reduction of, 106, 120–3, 125
Greenpeace, 102, 128
"green revolution," 188, 205, 359–62, 366, 367
grid pattern, of city streets, 401, 403, 426, 449–50
Gritzner, Charles, xxvii
gross domestic product (GDP), 78, 89, 178; agriculture and, 373
gross national income (GNI), 71–2, 178, *181*
gross national product (GNP), 178
Group of 7/8 (G7/G8), 89
Group of 20 (G20), 89
groups: disadvantaged, 279–85; ethnic/minority, 268–71, 427, 430, 435–7; human, 174, 262, 263; more vs less developed, 176–83; size of, for study, 41–2, 279–80; of states, 321–3; *see also* ethnic groups, *and entries following*; minority populations (or groups)
growth: in less developed world, 202–5; limits to, 153, 154–5; population, 149–58; stages of, 182
growth poles, 502
guano, 359
Guinea-Bissau, 331

"Haber–Bosch process," 359
Hadrian's Wall, 309
Hägerstrand, Torsten, 42, 227
Haiti, 196; case study of, 180–1
Hallman, Bonnie C., 223–4
Hamas, 310
Hamilton, ON, 259, 501
Hansen, James, 124
happiness: geography of, 282–3, *285*
Hardin, Garrett, 106
Harris, Chauncey, and Edward Ullman, 424–5
Hartshorne, Richard, 17, 35, 308
Harvey, David, 90, 375, 516
Haushofer, Karl, 307
health: agriculture and, 361, 363–4, 375; gender and, 275; industry and, 488, 491, 499–500; MDGs and, 206–7; nutrition and, 183–4, 379, 380–1; poverty/homelessness and, 443–6; in slums, 460; well-being and, 281–2; *see also* disease
hearths: Indo-Gangetic, 241, 243, 244; Semitic, 241, 244, 246; urban, *399*, 400
heartland, 487; theory of, 307, *308*, 338
Heavily Indebted Poor Countries (HIDC) Initiative, 203
hegemony, 260, 261, 262, 272
Henry of Portugal, 9
Herbertson, A.J., 17
Herodotus, 4–5
heterosexuality, 276–9
Hettner, Alfred, 12, 14, 17, 63
hierarchical effect, 42, 227, 228–9
hierarchies, urban, 409–10; central place theory and, 406–8; of global cities, 414–16; primate cities and, 411–12; rank-size distribution and, 410–11
high-technology industries, 487, 488, 493, 494
highways, 403, 444, 449, 454, 455
high-yielding varieties (HYVs), 359–60
Hilton, James, 296
Hinduism, 88, 241, *242*, 243–4, 245, 246, 250
hinterland, 407–8, 487
Hipparchus, 6
Hippodamus of Miletus, 449
historical materialism, 29
historical sites, 290, 294
history: "end of," 334–5
HIV (human immunodeficiency virus), 199; *see also* AIDS
hockey associations/teams, 408, 409
Holocaust, 248, 262, 264, 265
Holocene epoch, 96
Holy Land (biblical), 222
holy land, concept of, 249–50
homeland, 218
homelessness, 285, 438, 442, 443–6, 520
home ownership, 430–2, 434
Homo erectus, 108
Homo sapiens, 263
Homo sapiens sapiens, 262, 263
homosexuality, 33, 156, 272, 276–9, 286; acceptance of, 278, *279*; landscape and, 278–9, 287; persecution of, 265, 278; secret language of, 280–1; terminology and, 277
Hong Kong, 84, 168, 304
Hong Kong–Shenzhen–Guangzhou region, 398, 490
Horn of Africa: drought in, *185*, 195; refugees in, 193, 194, 195–6
horses, 355, 357
houselessness, 443
housing, 430–7; of Aboriginal peoples, 284; as commodity vs right, 430–2; of immigrants, 435–7, 438; inadequate, 443; insecure, 443; ownership of, 430–2, 434; poverty and, 440–6; public, 431, 439; relocation and, 432–4; value of, 431–2, 434, 439
housing markets, 430–2
Hovorka, Alice J., 273–4
Howard, Ebenezer: garden city of, 450, *451*
Hoyt, Homer, 424
Human Development Index (HDI), 182–3, 276
Human Development Report, 182–3
Human Genome Project, 263
human geography, xxiv–xxxii; changing nature of, xxvi–xxvii, 508–16; concepts of, 33–46; data handling and, 514–15; definition of, xxiv, xxvii, 17; early, 13–14; goal of, xxvii–xxviii, 508; introduction to, 4–20; landscape and, xxiv, xxvi; modern, 18–19; philosophies of, 26–33; physical geography and, xxiv, xxvii, 15, 16, 17–18, 513–14; political world and, xxix–xxx; problem-solving by, 515; regional studies and, xxvi, 18; space/place and, 512, 515; spatial analysis and, xxvi, 18–19, 515; subdisciplines of, xxix, 17–19; as synthesizing, 514; techniques of, 46–53; values of, 515; work of, xxix– xxxi, 515–16
human impacts, 107–24, *125*, 510–11, 514; on air, 116, 149–50; on animals, 112–15; on climate, 119–24; on ecosystems, 96, 98, 104–7; energy needs and, 98–104; on land, 115; on plants, 107–12; on soil, 115–16; on water, 104, 111, 115, 116–19
humanism, 28–9, 35; phenomenology and, 27, 36; positivism and, 26, 27, 52
Human Rights Watch, 276
humans: early migration of, 5–6; evolution/origins of, 262, 263; land and, xxiv, xxvi, 15, 17; *see also* human impacts
human trafficking, 267
Humboldt, Alexander von, 12–13, 14, 17, 18, 42, 104
Hume, David, 264
hunting, 99, 113
Huntington, Ellsworth, 15, 16
Huntington, Samuel P., 316, 333, 336–8
Hurricane Katrina, 197–8, 444
Hutterites, 251
Hutu, 265
hydraulic fracturing ("fracking"), 481, 483
hydroelectricity, 100–1, 117, 477; tidal-/wave-powered, 101
hyperglobalist thesis, of globalization, 76, 77, 80, 83, 86, 87
hypothesis, 26

ibn-Battuta, 9
Ibn al-Haitham, 248
ibn-Khaldun, 9
ice, Arctic, 121, 122, 123–4
Iceland, 74
iconography, 259
idealism, 27
identity, 256–99; consumption and, 288; contesting of, 280–1; ethnicity and, 268–71, 437; food and, 378–80; future of, 511–12; gender and, 271–6; globalization and, 68, 80, 82–4, 90; group, 220–1; language and, 235–8; local, 90; microcultural, 279–81; music and, 287, 288–9; national, 302–3, 307, 308, 310, 312–13, 314–19, 321, 336, 337; neighbourhood, 434–5; regional, 221–2, 311; religion and, 246–8, 249, 250; *see also entry below*
identity/difference, 256–99, 511–12; "cultural turn" and, 258–61; folk/popular culture and, 285–9; landscape and, 258–9, 279–81; myth of race and, 262–8; power relations and, 271–4, 276–8, 280, 290, 293–4; theories of, 260–1; tourism and, 289–97; well-being and, 281–5; *see also* difference; ethnicity; gender; racism; sexuality
idiographic approach, to human geography, 35
al-Idrisi, 9; map of, *9*
IDRISI (software), 48
immigration: to Canada, 267, 268, 269–71; neighbourhoods and, 423–4, 435, 437, 438, 440; restricted, 264, 267–8; *see also* migration
imperialism: *see* empires
import substitution, 204–5
indentured labour, 266–7, 305, 321, 369, 456; *see also* slavery

India, 161, 182–3, 202, 208, 226, 240, 329, 334; agriculture in, 150, 360, 362, 363, 375, 380; cities in, 451, 457, *457–8*; ethnic tensions in, 318–19, 321, 380; industry in, 491–2; language in, 232, 235, 237; outsourcing in, 497–8; partition of, 191, 318–19; population in, 137, 143, 146, 147, 148, 152, 159, 160, 188; religion in, 243–5, 246; sacred spaces in, 249, 250, *251*
Indian Agricultural Research Foundation, 362
indigenous peoples, 281, 284, 288; as Others, 260, 279; as tourist attractions, 292–4; *see also* Aboriginal peoples
individual, 8; environment and, 106, 107, 119
Indo-European language family, 233–5, *235*
Indo-Gangetic hearth, 241, 243, 244
Indonesia, 198, 329, 332, 333, 477, 480, 487, 498, *500*
industrialization: cities and, 400, 404–5, 416, 429–30, 443, 446–7; countries new to, 182, 203, 400, 416, 487–8; diffusion of, 476; in less developed world, 178, 202–5, 456
industrial location, 404, 468–73, 482; behavioural approaches to, 472–3; early theories of, 468; factors related to, 469; least-cost theory of, 468–71, 472, 473, 475, 484; market-area analysis of, 471–2, 484
Industrial Revolution, xxxi, *30*, 32, 99, 223, 290, 473–6, 483; agriculture and, 358, 373, 377, 379, 473–4; cities and, 388, 391, 398, 400, 401, 404–5; coal and, 473, 475, 476; landscapes of, *30*, 404–5, 440–1, 473, 475, 476; origins of, 474–5; population and, 138, 150–1, 157–8, 160
industry: capitalism and, 468, 484, 498–501; energy and, 476–83; environmental impact of, 115–19, 476, 483; geography of, 483–501; location theory of, 404, 468–73, 482; service, 495–7; society and, 498–501; world, 483–501; *see also* labour, industrial
inequalities: in cities, 440–6; gender, 275–6; health and, 184, 189, 198–202, 206–7; public goods and, 324; well-being and, 282; *see also* global inequalities
infanticide, 147, 148
infant mortality rate (IMR), 140; in US, 142
informal sector, 459
informal settlements and slums, 455, 456, 458–60; employment in, 459; growth of, 456, 458; health/environmental issues in, 460; incidence of, 458, *459*; locations of, 426, 427, 455, 456, 458; poor governance in, 458–9
information, transmission of, 61, 72–5
information communication technologies (ICTs), 72–4
information technology (IT) industry, 493–5, 496; in India, 492, 498
infrastructure: Marxist, 513; urban, 424, 439, 454–5, 460
Inge, W.R., 302–3
inner city, 432, 446; as cultural capital, 454; gentrification of, 434–5; slums in, 455, 456, 458–60; vs suburbs, 422, 439, 442, 455
Innis, Harold, 62
innovations, 62; and S-shaped curve, 63–4, *64*, 263–4
intensity theory, 349–50, 352, 354
interaction, 38
Intergovernmental Panel on Climate Change (IPCC), 100, 121–2, 124
internally displaced persons (IDPs), 192
International Code of Ethics for Canadian Business, 87
International Court of Justice, 414
International Criminal Court, 414
international division of labour, 72
International Energy Authority (IEA), 99, 103, 591
internationalization: of cities, 431; economic, 78–9, 90
International Labour Organization, 267
International Monetary Fund (IMF), 80, 186, 203, 205, 316
International Trade Organization (ITO), 80
Internet, 62, 72–5, 82; cities and, 414, 416, 447, 448; maps on, 45, 49; social media on, 74–5
intra-urban transportation/communication, 405, 414, 447
Inuit, 123, 215, 241
Inuka Kenya Trust, 207
Iran, 246, 310, 312, 329, 334
Iraq, 310, 312, 331, 332, 334; US invasion of, xxvi, 76, 247, 329, 333, 336, 511
Ireland, 503; language in, 236, 237, 239; *see also* Northern Ireland
Irish Republican Army (IRA), 314–15, 332
iron and steel industry, 472, 482, 493; in Canada, 259, 499, 501; in Britain, 475–6, 494; in Europe, 485–6
irredentism, 310–11
irrigation, 115, 118, 344, 360, 361, 368
Irving, David, 248
Islam, 241, 243, 244, 246; Christianity and, 246, 247–8, 316; in early Africa, 179, 312; geography and, 8–9, 11, 19; Shia–Sunni split in, 246, 247–8, 511
Islamic State, 75, 247, 310, 312, 331, 511
isochrones, 39, *39*, *40*
isodapanes, 470, *471*
isolated state theory, 348–50, 352–3
isotims, 470, *471*
Israel, 244, 247, 261, 264, 309–10, 334
Italy, 306, 314, 323, 484, 485; creation of, 236, 303

Jainism, 241, 244
James, Preston, 214, 248, 252
Japan, 82, 85, 89, 122, 168, 176, 196, 306, 338, 478, 482, 490; industry in, 72, 476, 484, 485, 486–7, 488, 494; nuclear plant failures in, 102–3; population in, 145, 155, 160; religion in, 244, 250
al-Jazari, 248
Jenkins, Simon, 508
Jerusalem, 36, 50, 240, *252*, 309–10
jihadism, 247
Johnston, R.J., 511, 513
Jordan, T.G., 217, 240; and Matti Kaups, 225
Judaism, 88, 241, 244, 245, 246, 248; migration and, 167, 244, 309; *see also* Israel

Kadeer, Rebiya, *318*
Kant, Immanuel, 12, 14, 35, 334
Kaplan, Robert, 124, 208
Kasheshewan reserve, ON, 284
Kashmir, 319
Kazakhstan, 109, 317, 334, 478
Kensington Market, Toronto, *61*
Kenya, 112, 184, 185, 190, 191, 198, 228–9, 295
Kenyan Green Belt Movement, 112
Kerguelen Island, 10
Keynes, John Maynard, 88, 492
Kim Jong-un, 324
Kinshasa, DRC, 199, 397
Kjellén, Rudolf, 307
KOF Index of Globalization, 77–8, *78*
Kohr, Leopold, 321, *322*
Kolkata (Calcutta), 250
Korea, 308, 310; conflict in, 329; *see also* North Korea; South Korea
Kosovo, 306, 315–16
Krebheil, Edward, 325
Kropotkin, Peter, 323
Krugman, Paul, 205
Kurds, 265, 312, 511
Kyoto Protocol, 106, 120, 122
Kyrgyzstan, 317

labour: agricultural, 273–4, 358, 367, 369, 371, *373*, 375, 377; by children, *30*, 459; gender and, 273–5, 459, 488, 500–1; indentured, 266–7, 305, 321, 369, 377, 456; trade and, 67, 68; *see also entries below;* employment; workforce
labour, industrial: availability of, 468, 469, 485, 491, 495; as commodity, 89, 474, 499, 500; cost of, 72, 468, 469, 470–1, 484, 486, 487–8, 490, 493, 494, 497, 499, 500; by country, *496*; in early factories/mills, *30*, 404, 475, 476; exploitation of, 499, 500; Fordism/post-Fordism and, 492, 500, 501; gender and, 500–1; mobility/migration of, 82, 468, 469, 472, 476, 499; offshore, 72, 497–8
labour, international division of, 67, 72, 493, 500; colonialism and, 180, 266–7, 321, 369, 377, 456; in low-cost zones, 488, 489–90; outsourcing as, 497–8
Lake Chad, 118
Lake Erie, 118
land: human impacts on, 115; humans and, xxiv, xxvi, 15, 17; postwar recovery of, 329–30; value of, 423, *423*, 449
landrace, 366
Landsat, 49–50, 107
landscape, xxiv, xxvi; agricultural, 344–6, 357–66; as contested, 261–2, 280–1; cultural, 222–9; deforestation and, 107–11; elitist, 283–5, 426, 430; epidemiologic, 499–500; gender and, 272–4; industrial (early), *30*, 404–5, 440–1, 473, 475, 476; industrial (modern), 259, 491, 493; language and, 238–41; religion and, 225, 226, 247–50, 251; as represented, 259; sexuality and, 278–9, 287; of stigma, 285, 286; as symbolic, 258, 259; wind turbines and, 103
landscape geography, xxvi, 15; contemporary, 18
landscape school, 15, 225, 227; new cultural geography and, 18, 258, 261, 286, 288
Landschaftskunde, 15
land use, 347; agriculture and, 348–54; urban, 422–30, 439, 450–1; urban transportation and, 449; zoning and, 449, 454
language(s), 229–41; artificial, 238; classification/distribution of, 232–5, *233*, *234*; as cultural variable, 230; disappearing/endangered, 231–2, 250; early, 230; globalization and,

76, 83; identity and, 235–8; landscape and, 238–41; major, 231; minority, 230, 237–8, 239; in multilingual states, 236, 303; nationalism/separatism and, 230, 235–6, 311, 312, 314; number/speakers of, 83, 231, 232–3, 234; official, 235, 236–8, 240; secret, 280–1; state instability and, 308; universal, 238
Laos, 324
Las Vegas, 294
Latin America, 115, 198, 458; cities in, 422, 426, *426*, 427, 455
Latin American Integration Association, 323
latitude, 6, 10, 11, 35, 50
law: religion and, 246, 248; in scientific method, 26
League of Nations, 309, 331
least-cost theory, 468–71, 472, 484; agglomeration/deglomeration and, 470–1, 475; departures from, 472–73, 493–4, 498, 500; labour costs and, 470; tests of, 472; transport costs and, 469–70, *472*
least developed countries, 177–8, *178*, 183, 209
least effort, principle of, 61
lebensraum, 307
Le Corbusier, 451
Lee Kuan Yew, 177
Léros (Greek island), 286
lesbians and gays, 33, 272, 276–9; *see also* homosexuality
Lesotho, 311–12
less developed world, xxviii, xxx–xxxi, 174–209; agriculture in, 345, 347, 359–62, 363, 365, 366, 371, 376, 377; business in, 187; case studies of, 179–81; cities in, 392–4, 396–7, 455–61; colonialism and, 175–6, 185–6, 196, 204, 207, 305–7, 309, 312–14, 455–6; conflict in, 190, 328, 333, 337; countries in, 177–8, *178*; debt crisis in, 202–3; development in, 178–83; digital divide in, 73–4; disasters and, 194, 196–8; disease and, 198–202; economic growth prospects in, 202–5; environmental issues of, 188, 194, 196, 203, 207; explanatory theories of, 174–6; fertility in, 136; food aid to, 188–90; food loss in, 381; future of, 208–9; globalization and, 76, 82, 85–7; governance issues of, 186–91, 207, 208; hunger/famine in, 183–90; vs "least developed" world, 177–8; MDGS and, 206–7; migration and, 162, 164, 166, 167; population and, 134, 136–8, 153, 156, 158; population aging and, 148–9, 152; population policies in, 145–6; possible solutions for, 205–9; reasons for hope in, 190–1; refugees and, 162, 191–4, 195; security in, 207; socialism in, 323–4; terminology and, 176–8; tourism in, 285, 295–7; trade and, 67; transnationals and, 72; transportation in, 62–3
Lewis, Bernard, 247
Lewthwaite, G.R., 16
Liberia, 189, *314*
Libya, 330, 331, 334, 336, 338; oil in, 478
life cycle, 166, 433
life expectancy (LE), 140; by country, *141*; population aging and, 152
limits to growth, 153, 154–5
lingua franca, 237
liquid natural gas (LNG), 481–2
literacy, 158, 178, 208; geographic, 33, 34
lithosphere, 97
livestock, 350, 366, 368, 369, 374; crops and, 346, 368; as food, 380–1; *see also* animals
locale, 290
location, xxvi; of cities, 405–6, 447; as concept, 35; of informal settlements/slums, 456, 458; post-industrial, 431, 439–40; residential, 431–2, 437; retail, 447
locational figure, 470, 471
location theory, 347; agricultural, 344–50; industrial, 404, 468–73, 482; urban, 406–8
logging, 127–8
London, 403, 425, 454–5; as global city, 80, *81*, 412, 413–14, 416, *417*; as megacity, 396, 397, 398; riots in, 74–5
London agreement, on CFCs, 116
Long, Stephen, 240
longitude, 6, 10, 11, 35
Los Angeles, 116, 397, 447, 514; homelessness in, 445
Lösch, August, 471–2
Louisiana, 220–1, *222*
Love Canal area, Niagara Falls, NY, 50, *50*, 118
Lovelock, James, 124
Ludlow, Natalie C., 499–500
Lunenburg, NS, *19*

Maathai, Wangari, 112
Macao, 304
McDonald's, 60, *231*, 379
Macedonia, 240, 316
Mackinder, Halford J., 14; heartland theory of, 307, *308*, 338
McLean, Malcolm P., 66
McLuhan, Marshall, 90
MacMillan Bloedel, 128
Magellan, Ferdinand, 8, 10, 61
Maghreb Union, 71, 323
Make Poverty History campaign, 189
malapportionment, 327
malaria, 198, *199*, 201, 202, 206, 207, 460
Malawi, 189, 190, 199–200
Malaysia, 144–5, 167, 168
malnutrition, 184
Malte-Brun, Conrad, 12
Malthus, Thomas, 104, 153, 155–7, 158, 184, 190, 208
Malthusian theory, 155–6, 157
Manchester: cotton mills in, *30*
Mandarin, 232
Mandela, Nelson, 332
Mandeville, John, 11
Manitoba: urban centres in, *41*
manufacturing, 468; in Canada, 487; in cities, 422, 427, 429, 439–40, 443, 446; global, 483–92; location of, 468–73, 482; outsourcing and, 497–8; *see also* industry
Maoism, 324
Mao Zedong, 324
maps, 46–8, 512; Chinese, 8; choropleth, 46, *48*; digital, 48; dot, 46, *48*; early, 5–10; early science and, 10–11; GA.SUR or Nuzi, *4*; governments and, 13; Internet, 45, 49; isopleth, 46–7, *48*; mental, 42, 165, 268, *269*, 285; participatory creation of, 44–5; Portolano, 7, 11; power of, 49–50; printing and, 9, 11; "T-O," 7, 11, 50; scale of, 46, *47*; world population and, *141*, *143*, 159–60
maquiladoras, 488
market-area analysis, 471–2, 484
marriage, 136–7; delayed, 137, 146, 156; same-sex, 278, *279*
Marsh, George Perkins, 14, 104–5, 514
Marx, Karl, 29–30, 89, 223, 372, 468, 474; population theory and, 156
Marxism, 41–2, 44–5, 156, 174, 175; agriculture and, 344, 358, 372; on capitalism, 29–30, 31, 36, 44–5, 82, 88, 89; cultural geography and, 30, 31, 223, 227–9, 258, 271, 282, 513–14; feminism and, 30–1, 35, 271; on geographical values/imagination, 515, 516; humanism and, 26, 29–30, 258, 271, 282, 515; industrial location and, 468, 472, 474, 498–500; nation-states and, 303; positivism and, 26, 29–30, 35, 41; *see also* political economy
mass market, 490
material index, 470
matrilineal system, 271
Mauritius, *189*, 266–7
Mauro, Fra, 10
meat-eating, 117, 248–9, 345, 368, 380–1
Mecca, 246, 249, 250, *428*
mechanization, agricultural, 361, 369, 374
Médecins sans Frontières/Doctors Without Borders, 202
megacities, 396–8, *399*, 406, 416; vs global cities, 412
megaregions, 398
men: AIDS and, 199, 201; feminist geography and, 271–6; violence and, 274, 276, 283, 284, 285; *see also* gender
meningitis, 198
Mennonites, 249, 251
mental illness, 286; homelessness/poverty and, 285, 442
mental images, 42, *43*
mental map, 42, 165; ethnic identity and, 268, *269*
Menzies, Gavin, 8
mercantilism, 155, 403–4
Mercator, Gerardus, 11
Mercier, Michael, 34
MERCOSUR, 34, *69*, 71
Mesopotamia (urban hearth), 4, *399*, 400–2, 403
methane, 111, 119, 121, 125
metropolitan area, 391
Mexico, 4, 321, 478, 487, 497; debt crisis in, 203; drug wars in, 331; maquiladoras in, 488; steel industry in, 472; US and, *163*, 168, 309, *320*
Mexico City, 396, 415, 416, 472
microcultures, 279–81
microloans, 204
Middle East, 311, 477, 478, 483
migration, 160–8; behavioural approach to, 165; chain, 268; colonialism and, 162, 266–8, 303, 474; definition of, 161; as diffusion, 225; ethnic, 268–71; forced, 167, 244; free, 167; illegal, 166, 168, 309; impelled, 167; labour and, 266–8; language and, 232–5; mass, 167–8; mobility transition theory of, 163–5; moorings approach to, 165–6; population growth and, 134; primitive, 167; push-pull logic of, 161–3; racism and, 266–8; Ravenstein's laws of, 163, 164; rural–urban, 163–4, 167, 196, 206, 391, 392, 395, 404, 446, 456; selectivity of, 166; world (1500–1900), *166*; *see also* immigration; refugees

Migratory Birds Convention Act, 115
Mill, John Stuart, 468
Millennium Development Goals (MDGs), 203, 206–7, 276, 519
Mills, C. Wright, 515
mining industry, 482, 483, 484–5, 491, 492, 495, 499–500
minority populations (or groups), 435; cultural, 436; ethnicity of, 268–71, 427, 430, 435–7; genocide against, 262–5; languages of, 230, 237–8, 239; policies against, 267–8; segregation of, 265–6, 268, 285, 435–7; visible, 262, 268, 436; *see also* ethnicity; identity/difference
Mississippi: voting in, 327, *328*
moai (stone statues on Easter Island), 98
mobility, 448; residential, 432–4, 437–40, 448, 455; retailing and, 439, 446–7, 448
mobility transition theory, 163–5, 168, 182
model, use of, 52
models of cities, 422–30; Asian, 427–9, *428–9*; classic (North American), 423–5, *424*; classic (revised), 429–30, *430*; European, 425, *426*; Latin American, 426, *426*; sub-Saharan African, 426–7
mode of production, 29, 60, 88–9, 156
modernism, 32
modernization: population and, xxxi, 138
Mogae, Festus, 190
monarchy, 303, 323
Monbiot, George, 102
monetary union, 70
Mongolia, 368, 484–5
Monmonier, Mark, 50
Monsanto, 364
Monte Carlo simulation, 227
Montenegro, 316
Montesquieu, Charles-Louis de Secondat, baron de, 16
Montreal, 410, 415, 438, 487
Montreal Protocol, 105, 107
moorings, 165–6
more developed world, xxviii, xxx–xxxi; agriculture in, 345, 347, 359, 370–1, 374, 376–8; AIDS in, 199, 201, 202; cities in, 423–5, 446–9; countries in, 177, *178*; debt crisis and, 203; digital divide and, 73–4; energy use in, 477, 479–80, 482; fertility in, 139; food shortages and, 185–6; food waste in, 381–2; gendered employment in, 500–1; globalization and, 76, 82, 85–6; housing in, 430–2, 434–5, 437–40; industrial regions in, 485–7; industrial restructuring in, 446, 492–3, 494, 498; migration to, 164, 167, 192, 193; migration within, 168, 191, 194; natural disasters in, 197–8; population and, 134, 151, 153, 157; population aging and, 148–9, 152; terminology and, 176–8; trade and, 67; transnationals and, 72; uneven development in, 501–3
Mormons, 225, 226
Morris Motors, 473
mortality, 139–42; age/sex structure and, 146, 148, *149*, *151*; child, 140, 183, 195, 198, 206–7; factors affecting, 140; government policies to reduce, 144–5; infant, 140, 147, 157, 158, 180, 199, 206; levels of, 135; maternal, 207; measures of, 134, 139–40, *141*; population aging and, 148–9, 152; projections of, 153–5; variations in, 140, 142
mortality transition theory, 153
mortgages, 432, 434, 442
mosques, 249, 250, *252*, 427, *428*
Mozambique, 184, 187, 200, 329
Mugabe, Robert, 190–1
Muhammad, 8, 246, 248
multiculturalism, 84, 269–71, 437
Multilateral Debt Relief Initiative (MDRI), 203
multilingual states, 236, 303, 310, 311, 315
multinational states, 303, 310–19, 321–2; in Africa, 204, 207, 303, 311–14; in Europe, 303, 314–18; in South Asia, 318–21
multiple nuclei model, *424*, 424–5, 430; examples of, 427, 428
Mumbai, 457, *457–8*
municipalities: *see* cities
Münster, Sebastian, 11
music: identity and, 222, 280, 287, 288–9, 294
Myanmar, 240; *see also* Burma

Naess, Arne, 107
Napoleon, 490
National Islamic Front (Sudan), 86
nationalism, 302–3, 310, 311, 312, 321, 326; fascism as, 323; language and, 230, 235–6, 312
National Topographic System, 46
nations, 302; in Africa, 204, 207, 311–14; in Europe, 314–15; *see also* nation-states; states
nation-states, 302; European, 303, 314–18; globalization and, 76, 80, 84; nationalism and, 302–3; *see also* states
nativism, 267
NATO (North Atlantic Treaty Organization), 330, 338
natural gas, 477, 480–2
natural increase, 142–4, 152, 157; by country, *143*
natural resources: in Canada, 487; colonialism and, 176, 204, 305–6, 483; conflict and, 330, 333–4; corruption/poor governance and, 189, 207; human values and, 99, 104; stock vs renewable, 99; trade and, 68, 86
navigation, 8, 9, 20
Nazi Germany, 167, 265, 307, 309, 323
neighbourhoods, 430–7; decline vs revitalization of, 433–4; ethnic, 268, 285, 435–6, 442; gentrification in, 434–5; segregated, 283–5, 435–7, 440, 442; "welfare," 442–3
neo-colonialism, 361
neo-Malthusian theory, 156
Nepal, *197*, 206, 335
Nestlé, 72, 377
Netherlands, 120, 176, 184, 323, 414
New Haven, CT, 443–4
newly industrializing countries (NICs), 144–5, 182, 203, 400, 416, 487–8
New Urbanism, 452–3, *453*
New York, 391, 439, *434*; as global city, 80, *81*, 412, 414, 416, *417*; as "largest" city/megacity, 390, 396, 397, 398
New Zealand, 267, 278, 283, 444–5, 446
Niagara Falls, 36, *38*
Nigeria, 187, 202, 203, 312, 336; oil in, 477, 478, 480, 481
Nike, 60, 87–8, 500
Nile Valley (urban hearth), *399*, 402
Nisbet, Robert, 5
nitrous oxide, 104, 111
nomothetic approach, to human geography, 35
non-religious people, 243
Norberg, Johan, 87–8
normative theory, 348–9
"North," 177, *177*
North Africa, 103, 136, 478
North America: conurbations in, 439; cultural adaptation in, 225, 226; environmental issues in, 107, 109–10, 111, 113, 116, 127–8; global cities of, 414–16, *417*; industry in, 476, 485, 487, 488; monetary union in, 70; outsourcing from, 491, 497–8; regions of, 216–18; vernacular regions of, 219–22
North American Free Trade Agreement (NAFTA), 34, 67–8, *69*, 71, 321, 323, 338, 431
North Antelope Rochelle coal mine, WY, 483
Northern Ireland, 246–7, 314–15, 325, *332*
North Korea, 323, 324, 334
North Sea, 64, 86, 107, 483
Northwest Passage, 123, 333
Norway, 323, 333
no-till farming, 364–5
nuclear energy, 99, 100, 101–3, 104, 477; disasters associated with, 102–3, 105
Nuclear Non-Proliferation Treaty, 334
nuclear weapons, 324, 334
nuclei, multiple, model of, *424*, 424–5, 430; examples of, 427, 428
Nunavut, 284
nuptiality, 136
nutrition, 135, 183–4, 195, 206, 379, 380–1
Nuu-chah-nulth, of Clayoquot Sound, 127–8

objectivity, 26
"Occupy" protests, 280, 325
oceans, 116; human impact on, 96, 119, 123
office work, 430, 439–40, 446; outsourcing of, 498; by telecommuters, 448
Official Languages Act, 237
offshoring, 491–2, 496, 497–8, *498*
oil, 476–80, 483, 485–6; China and, 478, 490–1; conflict and, 333–4, 480; global movement of, 478, *479*; vs natural gas, 480–1; price of, 100, 477, 480, 490, 492; reserves/production of, 478–80; *see also entries below*
oil industry, 99; environmental impact of, 86–7, 104, 115, 119, 123, 480
oil sands, Alberta, 115, 480, 491
oil spills, 119
oligarchy, 323
Omo people, 179
Ontario Human Rights Commission, 501
Ordnance Survey (England), 13
Organization of African Unity (OAU), 313
Organization of Petroleum Exporting Countries (OPEC), 323, 477, 480, 492
Orientalism, 260–1, 279
Orinoco River, 66
Orontius, 10

Ortelius, Abraham, 11
Other, 260–1, 279
Ottoman Empire, 312, 315, 316
outsourcing, 491–2, 496, 497–8, *498*
overpopulation, 145, 156; food and, 184, 186, 208
Overseas Development Institute, 208
Oxfam, 189, 205
ozone layer, 116; Antarctic hole in, 105, 116

Pacific Asia: industry in, 484, 487–8
Pakistan, 137, 188, 196, 321, 329, 332, 334, 336; creation/division of, 191, 193, 318–19
Palestine, 247, 249–50, 309–10, 332
Panama Canal, 64, *65*, 66, 123
pandemics, 199, 228–9
Papua New Guinea, 233
Paris, 247, 403, 425; as global city, 397, 412, 414, 416
Paris Club, 203
parks and recreational areas: in apartheid era, *266*; environment and, *113*, 126, 128, 291; government, 290–1; as landscapes of fear, 285; theme, 223–4, 288, 294, 414
participant observation, 51
participatory mapping, 44–5
pastoral nomadism, 367–8
patriarchy, 31, 271, 278
Peace, Walter, 34; and John Eyles, 259
peace, "perpetual," 334–5
Pearl, Raymond, 153
Pearl River Delta, 485, 490
Peet, Richard, 351, 375
perception, of environment, 42
periphery, 176; *see also* core–periphery models
"persons of concern," 192–3
Peru: early civilization in, 150, 160, *215*; guano in, 359
phenomenology, 27, 36
phenotypes, 262
Philippines, 196, 329, 335, 336; industry in, 487, 488, *498*
phonemes, 230
photography: aerial, 49; family, 223–4
physical geography, xxiv; early civilization and, 174–5; human geography and, xxiv, xxvii, 15, 16, 17–18, 513–14; population and, 159–60; *see also* earth; environments, global
pidgin languages, 237–8
Pike, Zebulon, 240
pipelines: oil and gas, 86, 478, 481–2
place, 35–6, 512, 515; control of, 218, 239, 248; creation of, 218–22, 279–81, 292–7; naming/renaming, 238–40; as sacred, 36, 222, 249–50; sense of, 36, 218–19, 222; voting and, 326–8, *328*
placelessness, 36, 82–3
place utility, 165
plague, 153
planning, 449–55; of cities/suburbs, 450–3; origins of, 449–50; in practice, 453–4; transportation and, 454–55; zoning and, 454
plants: global distribution of, *7*; human impacts on, 107–12; lost species of, 111, 365–6
Plato, 16, 104, 323
Pleistocene epoch, 356
plough, use of, 357, 364–5, 368
pluralism: industrialization and, 204, 456
pneumonia, 198
Poland, 338
polar bears, 291–2
Polari (language), 280–1
policies: agricultural, 69, 345, 346, 374, 376–7; environmental, 104, 105–6, 107, 120, 122–3, 128, 485; immigration, 168, 264, 267–8; industrial, 487, 491, 492, 494, 502–3; population, 135, 136, 137–8, 139, 144–6, 155; structural adjustment, 186, 205
polio, 198
political economy: agriculture and, 370, 372, 374; culture and, 227–9
political units, xxix–xxx
politics, 302–38; agriculture and, 345, 346; cities and, 401, 411, 413–14, 416, 422, 427; electoral, 325–8; environment and, 100, 105–6, 122; food shortages and, 186–7, 188; globalization and, 83–4; homelessness and, 444–6; housing and, 431–2, 436; informal settlements and, 458–9; religion and, 246–8, 261, 314–16, 333, 336–8, 511; world, 307–10, 333, 334–8, 511
pollution, 100, 101; air, 105, 116; water, 117–19
Polo, Marco, 7, 8, 403
population, 134–69, 510; age/sex structure of, 146–8; aging of, 148–9, 152; agriculture and, 150–1, 156–8; density of, 5, 12, 159–60, 422, 425, 428, 451; distribution of, xxvii, 160, 161; of earth, all-time, 154; food and, 150, 155–7; future of, 510; government policies and, 144–6; Industrial Revolution and, 138, 150–1, 157–8, 160; modernization and, xxxi, 138; projections of, 153–5; surplus, 156; urban vs rural, 388, 391–3, 395, *396*, 439; US shift in, 405, 406
population growth: culture and, 149–51, 224–5; history of, 149–55; migration and, 134; natural increase and, 142–4; theories of, 155–8
population momentum, 142, 153, 158
population pyramids, 146, 147, 148, *148*, *149*, *151*, 152, 158
Population Reference Bureau, 154, 177, 519
Portolano charts, 7, 11
Portugal, 304, 306, 307, 487, 503
positionality, 261
positivism, 26–27, 28, 32, 35, 52, 63; diffusion and, 42; group scale focus of, 41–2; humanism and, 26, 27, 52; Marxism and, 26, 29–30
possibilism, 15, 17
post-colonial research, 260–1
post-Fordism, 88, 492–3, 500, 501
postmodernism, 32–3, 35, 258, 260–1; diversity of, 33; feminism and, 271; of suburbs, 439; tourism and, 288, 294
post-suburbanization, 439–40
poverty: AIDS and, 200, 202; in cities, 440–6; cycle of, 442; food shortages and, 184, 186; globalization and, 89; MDGs and, 206; poor governance and, 186–90; population growth and, 156; social groups susceptible to, 443; stipend program to remedy, 191
power, 260–1, 263; political, 302, 314; state, 106, 119, 280
power centres, retail, 439, 447
power relations, 260–1, 280; gender and, 271–4, 276, 278, 283; sexuality and, 276–7, 278; tourism and, 290, 293–4
pragmatism, 27
precipitation, 116, *117*, 123–4
primary activities, 468
primate cities, 411–12, 415, 456
principle of least effort, 61
producer services, 440, 457
projection, 47; as developed by Mercator, 11
pro-natalist policies, 135, 144–5, 155
protest, public, 325, 335; anti-Apple, *90*; anti-WTO, 76, 81; Arab Spring, 74, 330, 331, 334, 336, 338; environmental, 100, 102, 128, 478; by Indonesian workers, *500*; by Nigerian parents, *75*; "Occupy," 280, 325; Palestinian, *309*; social media and, 74–5
Protestant Reformation, 245
Ptolemy, 6, 9, 10, 20, 50; map of, *6*
public goods, 324
push–pull factors, of migration, 161–3, 165, 167–8; housing and, 432–3
Putin, Vladimir, 145

al-Qaeda, 247, 332–3, 336
qualitative methods, of data collection/analysis, 51–2; examples of, 52–3
quality of life, xxx–xxxi
quantitative methods, of data collection/analysis, 52–3
quaternary activities, 468
Quebec, 222, *269*, 374, 487; language in, 230, 236, 237, 239; secessionism in, 84, 303, 310, 311
queer theory, 277
questionnaire, use of, 52

rabbits, 113, *113*, 114
race, 174, 176, 262–8; vs species, 262, 263
racism: apartheid and, 265–6; definition/history of, 264; genocide and, 262–5; migration and, 266–8; multiculturalism as, 270
Radarsat, 51
Radburn, NJ, model of, 450–1, 452, 453
railways, 64–6, 68, 485; in Britain, 63, 473, 476; in Canada, 63, 487; cities and, 422, 446; in Russia, 63
rain forests, tropical: past/present location of, *110*; removal of, 110–11
Raitz, K.B., 268
ranching, 111, 350, 369, 370
random sampling, 52
range, 408
rank-size distribution, 410–11
rape, 283
raster method (GIS), 48
rate of natural increase (RNI), 142–4
rational choice theory, 349
Ratzel, Friedrich, 13–14, 15, 63, 307
Ravenstein, E.G., 163, 164, 165
Réclus, Elisée, 14, 323
recycling, 125
redlining, 432, 436
"reflexivity," 52
refugees, 162, 191–4, 195–6; asylum countries for, 191, 193, 194; definition of, 191; number of, 191, *192*; "persons of concern" and, 192–3; solutions to issue of, 194; source countries of, 192, *193*; from Syria, 191, *192*, 193
regional associations, political/economic, 34
regional geography, xxvi; contemporary, 18; history of, 12, 13, 14, 17; states and, 84; trade and, 67–71, 80, *81*
regionalism: in Europe, 314–18

regionalization, 36–7; culture and, 214–22
regions, xxvi; as concept, 36–7; cultural, 214–22; environmental; ethnic, 268, 312, 315–19, 321; formal (uniform), 36–7; functional (nodal), 37; literary, 222; mega-, 398; North American, 216–17, *218*, 219–20, *220–1*; vernacular, 37, 218–22; wine-producing, 222; world (cultural), 215–16, *216*; world (geo-economic), 80, *81*, 338
reindustrialization, 492, 493
religion, 241–51; abortion and, 138; agriculture and, 249, 345; AIDS and, 201; architecture and, 249, 250; civilization and, 215; display of, 249; ethnic, 243–4; gender and, 241; identity/conflict and, 246–8, 249, 250, 261, 314–16, 333, 336–8, 511; landscape and, 225, 226, 247–50, 251; law and, 246, 248; nature and, 248–9; origins/diffusion of, 241, *242*, 243; origins/distribution of, 241, *242*, 243–6; population policies and, 144, 145; sacred spaces of, 36, 222, 249–50; state and, 244–5, 246, 247–8; state instability and, 308, 310, 318; terrorism and, 247, 261, 332–3; universalizing, 243, 244–6
religious adherents, 241, 243, 519; pilgrimages by, 246, 250
remote sensing, 44, 49–51, 514; environment and, 96, 107, 110–11, 118, 119, 126
renewable energy, 99–104, 477
renewable resources, 99
rent: ceiling, 347; economic, 347–50, 353
rent gap hypothesis, 434, 462
repatriation, voluntary, 194
representation, 261
reserves, Native, 284
resettlement, of refugees, 194
restructuring: agricultural, 370–8; cities and, 431, 440, 442, 444, 446, 457; industrial, 444, 492–3
retailing, 446–7; in Asian cities, 427; decentralization of, 425, 430, 439, 440, 447; online, 447
returnees, 192
revolution: of earth, *see also* agrarian revolution; agricultural revolution; "green revolution"; Industrial Revolution
Rhine River, 118
Rhine–Main–Danube Canal, 64
Ricardo, David, 347, 468
Richthofen, Ferdinand von, 13, 14, 17
rimland theory, 307–8
riots: social media and, 74–5
Ritter, Carl, 12–13, 14, 17, 18
rivers: as boundaries, 309; cities and, 400–1, 405, 437
road transportation, 63, 66
Roché, Anthony de la, 10
Rokkan, Stein, 314
Roman Catholicism, 245, 250, 314–15, 316
Roman Empire, 6, 63; cities of, 401, 402–3, 449
Romania, 69, 137, 144
rooflessness, 443
Rostow, W.W., 182
Rotestein, Frank, 153
Royal Geographical Society (RGS), *13*, 14
Ruhr Valley, Germany, 404, 476, 485
rural settlements, 388, 391, 401
rural–urban migration, 163–4, 167, 206, 391, 392, 395, 404, 446, 456
Russia (imperial), 109, 168, 306, 307, 315, 317; industry in, 476; *see also* Union of Soviet Socialist Republics
Russia (modern), 78, 86, 89, 196, 317, 325, 338, 498; AIDS in, 202; Arctic and, 123, 333; democracy in, 336; industry in, 484, 485–6; as nuclear power, 334; oil/natural gas in, 477, 478, 482; population in, 137, 139, 145, 160; transportation in, 63
Rwanda, 98, 206–7, 265

sacred spaces, 36, 222, 249–50
Sahel region, Africa, 105, 111, 122
Said, Edward, 260–1
St Helena (island), 304–5
same-sex marriage, 278, *279*
sampling, 52
São Paulo, *396*, 411, 415, 416
São Tomé and Principe, 480
Saro-Wiwa, Ken, 478
Saskatchewan, 370
satellites, 49–51
satisficing behaviour, 349, 469, 472–3
Saudi Arabia, 32, 247–8, 278; women in, *32*, 276
Sauer, Carl, 18, 268, 356; landscape school of, 15, 225, 227, 258, 261, 286, 288
Saul, John Ralston, 76
Save the Children, 189
scale: as concept, 40–2; group, 41–2, 279–81; local, 327–8; on maps, 46, *47*; social, 41, 279–80; spatial, 40–1, *41*; temporal, 41
Scanlink, 65, 68
Schaefer, F.W., 17
Schlüter, Otto, 15, 18
scientific method, 26, *27*
Scotland, 315
sea levels, rising, 120
secession, 310, 313, 316, 329; *see also* separatism
secondary activities, 468
sector model, 424, *424*; examples of, 425, 426
sectors, employment, *495*, 495–6; agricultural, 371, 373, *373*, 495; industrial, 483–92, 495, *496*; service, 468, 493, 495–7
seed banks, 366
segregation: barriers/walls for, 309, 338; ethnic, 268, 285, 435–7; racial, 265–6; residential, 284–5, 435–7
Selassie, Haile, 179
selection: artificial, 355–6, 364; natural, 355–6
semi-periphery, 176; *see also* core–periphery models
Semitic hearth, 241, 244, 246
Semple, Ellen Churchill, 14, 15, 16
Sen, Amartya, 186
sense of place, 36
Seoul, 390, 403, 416
separatism: in Asia, 317–18, 319, 321; in Europe, 310, 314–15; Kurdish, 312; Quebec, 310, 311
September 11, 2001 terrorist attacks, xxvi, 203, 247, 511
Serbia, 315–16, 335
serfdom, 266
service industries, 468, 493, 495–7; deindustrialization and, 494; location of, 496; *see also* outsourcing
settlements: first effective, 252, 274–5; informal/squatter, 426, 427, 456, 458–60; permanent, 388; for refugees, 194; rural, 388, 391, 401; *see also* cities; urbanization
Seversky, Alexander de, 308
sexism, 273
sex ratio, 148
sex structure, 146–8, *149*, *151*
sexuality, 33, 272, 276–9; landscape and, 278–9, 287; *see also* homosexuality
shale: gas from, 481, 483; oil from, 480
Shanghai, 390, 396, *397*, 403, 411
Shangri-La, 296–7
shariah (Islamic law), 246
Shia Islam, 246, 511
Shinto, 244, 250
shopping malls, 274, 275, 287–8, 290, 294, 447
Shrubsole, Dan, 126–7
Sierra Club, 127
Sikhism, 241, 244, 249, 250, *251*, 321
Silicon Valley, 488
simulation, 227
Singapore, 168, 177; industry in, 487, 488; population of, 144–5
Sinhalese, 321, 511
site, 35, *35*
situatedness, 261
situation, 35, *36*
skeptic thesis, of globalization, 76, 77, 80, 83, 86, 87, 90
slavery, 29, 167, 203, 266, 369; *see also* indentured labour
Slovenia, 316
slums, 441–2, 455, 456, 458–60, 461; *favelas*, *460*, 461; as gateways to prosperity, 461; of industrial era, 404, 441, 456; *see also* informal settlements and slums
Small, Ernie, 292
Smil, Vaclav, 96, 115, 124, 129, 190
Smith, Adam, 348, 468
Smith, George, 226
smog, 116
socialism, 29, 323–4
social media, 74–5
social movements/protests, 280, 325, 335; *see also* Arab Spring; protest, public
social theory, 63, 498, 500
society, xxviii; constructionist views of, 260–1; developmental stages of, 182; feminism and, 30–2, 271–6; governance and, 207; human geography as serving, 4, 11, 18, 19, 512–15; humanism and, 27, 28–9; industry and, 498–501; Marxism and, 29–30; nation and, 302–3; post-industrial, 493, 495, 510; postmodernism and, 32–3; segregation/assimilation and, 435, 436–7, 443; spatial aspects of, 435, 501–3; state and, 323
sociolinguistics, 241
"sociological imagination," 515
sociology, 36, 51, 288
soil: agriculture and, 344–5, 358–9, 364–5; human impacts on, 115–16
solar power, 103
Somalia, 184, 190, 331, 332, 335; famine in, *185*, 189, 195–6
"South," 177, *177*
South Africa, 240, 267, 311; apartheid in, 80, 265–6, 332, 334, 435
South American Common Market (MERCOSUR), 34, *69*, 71

South Asia: conflict in, 318–19, *320*, 321
South Carolina, 326, *328*
South China Sea, 333, 335
Southern African Development Coordination Conference (SADCC), 71
"southern continent," 10
South Korea, 184, 482, 484, 487, 488, 489
South Ossetia, 317
South Sudan, 66, 478, 491; creation of, xxx, 84, 86, 303, 313–14, 511
sovereignty: Aboriginal, 311; Arctic, 123, 333–4; state, 302, 308, 315, 321–2, 331
space, 35, 512, 515; absolute vs relative, 35
Spain, 306, 310, 332, 487, 503
spatial analysis, xxvi, 27, 35, 37–8, 41, 52, 515; contemporary, 18–19; culture and, 227, 230, 239; history of, 12, 15, 17
spatial monopoly, 471–2; in Cape Breton, 499
spatial preferences, 165
special economic zones, in China, 489–90, *490*
species, 262; extinction of, 111, 113–15; introduction of, 113, 114; vs race, 262, 263
spectacles, 294
sport: cities and, 409, 414; identity and, 289
Spykman, Nicholas, 307–8
Squamish language, *233*
squatter settlements: *see* informal settlements and slums
Srebenica massacre, 316
Sri Lanka, 318–19, 336, 511; case study of, 179–80; civil war in, 192, 321; tourism in, 296
S-shaped curve, 42, *42*, 155; innovation and, 63–4, *64*, 227
staples theory, 502
state apparatus, 324
state creation, 302–7, 311, 312–14; exploration/colonialism and, 304–7; nationalism and, 302–3
state power, 324–5
states: absolutist, 303; in Africa, 204, 207, 303, 311–14; boundaries of, 308–9; conflicts of, 328–34; creation of, 302–7; divided, 310; elections and, 325–8; ethnicity and, 312–22; in Europe, 314–15; failed, 195, 330–1; geopolitics and, 307–10; globalization and, 67–78, 80, 84; groupings of, 321–3; ideal, 303; micro-, 311–12; modern, 331; multilingual, 236, 303, 310, 311, 315; multinational, 303, 310–19, 321–2; narco, 331; as organisms, 307; peace/democracy in, 334–6; pre-/postmodern, 331; role of, 323–5; socialist, 323–4; social movements/protests against, 280, 325, 335; sovereignty of, 302, 308; unstable, 308, 310–21; *see also* nation-states
Statistics Canada, 390, 519, 520
steel industry, 472, 499, 501; *see also* iron and steel industry
Stein, Charles, 450
Stelco, 501
sterilization, 146, 147
stigma: landscapes of, 285, 286
Stirling, Ian, 291
stock resources, 99
Strabo, 6, 20
structural adjustment policies, 186, 205
structuration theory, 258
subaltern, 260
subcultures, of youth, 280
subdisciplines, of human geography, xxix, 17–19
subspecies, 262, 263, 264
suburbanization, 437–40; sprawl and, 392, 439, 455
suburbs, 391, 432–3, 437–40; affluence of, 424, 426, 430, 442, 444; density of, 422, 450, 454; in Europe, 425, 439; future of, 439–40; gender and, 274–5; land use in, 424, 425, 427; redlining and, 432
Sudan, 140, 153, 190, 193, 195, 332; oil in, 86–7, 330, 333, 478, 491; *see also* South Sudan
Suez Canal, 64, *65*, 305
sun, 97
Sunni Islam, 246, 248, 511
supply and demand, 346–7, *347*
supranational organizations, 413
Suriname, 238
Süssmilch, Johann Peter, 208
sustainable development, 127–9
sustainability, 126–9, 207
Sweden, 104, 138, 139
Switzerland, 69, 303
Sydney Coalfield, 499–500
symbolic approach, to culture, 33, 258, 259; genocide and, 263–5; sexism and, 273–4; social movements and, 280; tourism/ecotourism and, 290, 292, 296–7
Syncrude, 115
syphilis, 198
Syria, 309, 312; civil war in, 191, 310, 330; refugees from, 191, *192*, 193
systems, open/closed, 96–7

Tablighi Jamaat, 249
Taliban, xxvi, 247
Talisman Oil, 86–7
Tamils, 192, 321, 511
Tansley, A.G., 97
Tanzania, 324
Taoism, 244, 245
tariffs, 67, 68, 71; in Canada, 487, 502, 503
Taylor, Griffith, 15, 16
tea, 321, 369, 377; Darjeeling, 380
technologies: agriculture and, 357–66; energy and, 99–100, 115, 475, 476, 480, 481; environmental, 96, 100, 103–4, 111–12, 122–3, 124–5; geography and, 20, 44–5, 49–51, 512, 514–15; population growth and, 149–53, 154–5; *see also* biotechnology; high-technology industries
Teixeira, Carlos, 438
telecommuting, 448
teleology, 6, 104
temperature: agriculture and, 120, 159, 344–5; CFCs and, 116; earth's, 103, 119; population and, 159–60; *see also* climate change
territory, 328, 333; nation-state and, 302, 310
terroir, 222
terrorism, xxvi–xxvii, 329, 331–3; definition of, 331–2; global, 332–3; religious conflict and, 247, 261, 332–3
tertiary activities, 468
tetanus, 198
text, 33
textile industry, 485, 492, 493; early factories of, *30*, 475; globalization of, 85
theme parks, 223–4, 288, 294, 414
theory, 26
Third World, 176–7
Three Gorges dam, 100–1, *101*
threshold, 408
Tibet, 65, 329
tidal/wave power, 101
time-space convergence, 61–2
Tito, Josip Broz, 315
tobacco, 198
Tobler, Waldo R., 38
Tokyo, 404, 441; as global city, 80, *81*, 412, 414, 416, *417*; as "largest" city/megacity, 390, 396–7, 398, 412
Tomlinson, Roger, 48
Tönnies, Ferdinand, 227, 474
topography, 65
toponyms, 238–40
topophilia, 36, 218–19
topophobia, 36, 218–19
Toronto, 259, 395, 450, 451, 454, 487; as global city, 409–10, 415; greater area of, 390, 439; office space in, 446; physical space/time space in, *40*
total fertility rate (TFR), 134–5
tourism, 289–97; alternative, 291–2, 293; artificiality and, 294; attractions for, *30*, 290, 292–4; as consumption, 290, 294; ecological impact of, 291; mass, 290–1, 293; place construction and, 292–7, *294*; religious, 250
tourist gaze, 290, 293
Toynbee, A.J., 215
trade, 66–71; barriers to, 67, 68, 79, 80, 85–6; factors affecting, 67; fair, 361; free, 67–71, 80, 82, 85; regional integration of, 67–71, *69*, 323
trade flows, 67, 68, 81
trading blocs, 67–71, *69*, 323; predicted future of, 80, *81*, 338
"tragedy of the commons," 106, 119, 325
transgendered people, 278
transformationalist thesis, of globalization, 76, 77, 80, 83, 86, 87
transition theories, of population: demographic, 135, 157–8, 168, 180, 182, 391; fertility, 135–6, 153, 510; mobility, 163–5, 168, 182; mortality, 153
transnational corporations, 46, 71–2, 79, 80, 82, 83, 85–8; African aid money and, 190; agriculture and, 377; decentralization by, 484, 495; environment and, 86, 500; Fordism and, 492, 500; global cities and 60, 413, 415; history of, 492; low-wage labour and, 72, 85, 488, 490, 493, 500; political/ethical issues of, 86–7; territorial interpenetration by, *79*
transportation, 61, 62–6; agriculture and, 62–3, 349–50, 351, 352, 357, 368; evolution of, 63–4; of food, 184, 377, 381; industrial location and, 469–70; modes of, 64–6; resource extraction and, 481–2, 487, 499; urban, 425, 448, 449, 453, 454–5
tribalism, 84
Trudeau, Justin, *328*
Trudeau, Pierre Elliott, 270

Tshiluba language, 232
tsunami, Asian (December 2004), 196–7, *197*, 291, 296
Tuan, Yi-Fu, 27, 28–9, 36
tuberculosis, 180, 198, 201, 202, 207, 443, 499
Tunisia, 336
tunnels, railway, 65, 68
Turkey, 69, 332; Kurds in, 310, 312; refugees in, 191, 193; *see also* Ottoman Empire
Turkmenistan, 145, 317, 325
turnpike roads, in Britain, 63
Tutsi, 265
Twitter, 74–5
Tyson's Corner, VA, 440

Uganda, 191, 201, 208, 312
Uighurs, 317–18
Ukraine, 315, 317, 334, 335; industry in, 476, 485–6
Ullman, Edward, 68; and Chauncey Harris, 424–5
undernutrition, 184
Union of Soviet Socialist Republics (USSR), 311, 315–18, 321, 334; bloc controlled by, 85, 323, 338; collapse of, 80, 84, 315, 317, 338; ethnic groups in, 317–18; industry in, 485–6; *see also* Russia (imperial)
United Kingdom, 310, 314; *see also* Britain
United Nations, 60, 112, 177, 203, 282, 413, 443; Africa and, 265; Charter of, 276; conflict and, 323, 328–9, 338; development measurement by, 182–3, *183*; disasters and, 194, 195–6; in globalization era, 83; MDGs of, 206–7; membership of, 302, 306; place names and, 240; population and, 134, 153; Security Council of, 334; on slum dwellers, 458, *459*; terrorism and, 331–2; trafficking and, 267; websites of, 519, 520
United Nations, conferences/summits of: on environment/development, 106, 128; on human environment, 105; on racism, 203, 264; on sustainable development, 106
United Nations, conventions/declarations of: on biodiversity, 364; on genocide, 262–3; on refugees, 191; on rights of indigenous peoples, 266
United Nations, organizations of: FAO, 110–11, 381, 520; UNESCO, 231, 290, 295–6, 519; WHO, 140, 286
United Nations AIDS Agency, 201
United Nations Biosphere Reserve, 128
United Nations Department of Economic and Social Affairs, 392
United Nations Environment Programme, 188
United Nations Habitat, 458, 520
United Nations Health Agency, 202
United Nations High Commissioner for Refugees (UNHCR), 191, 192–3, 194, 196, 519
United Nations Plan of Action, 111–12
United States, 34, 176, 338, 492; agriculture and, 345, 346, 351–2, 365, 366, 376, 378; Arctic and, 123, 333; cultural hegemony of, 60, 82, *231*, 379; elections in, 326–7, *326–8*; energy and, 477–8, 480, 481–2, 483; environment and, 100, 101, 105, 118, 122; geography in, 14; geopolitics and, 307–8, 333, 336; globalization and, 67–8, 71, 72, 76, 79–80, 82, 89, 492; GM crops in, 363; homelessness in, 443–5; immigration to, 167–8, 309; industry in, 476, 483–4, 485, 492; Islam and, 247–8, 333; regionalization of, 216–17, *218*; suburbs/urban sprawl in, 391, 439–40; territorial expansion of, 305, *305*; terrorism and, xxvi, 203, 247, 511; urban planning in, 450–1, 453
United States Geological Survey, 334
Universal Declaration of Human Rights, 90
Unruh, Jon D., 329–30
Ur, Babylon (modern-day Iraq), *401*, 402
urban areas, definition of, 389–90, 391
urban hearths, *399*, 400
urbanism, 400; "new," 452–3
urbanization, 388–416, 421–61; central place theory of, 406–8, 409; future of, 441, 510; hierarchies of, 409–12, 414–16; Industrial Revolution and, 404–5; in less/more developed worlds, 391–2, *392*, 393–4, 405, 416; *see also* cities
urban land values, 423, *423*
urban sprawl, 392, 439, 455
urban structure, 422–3
Uruguay, 324, 352–3, *354*

vacations, 290
vaccination, 199
Vancouver, 74, 410, 415, 435, 438, 440, 449
Varenius, Bernhardus, 11–12, 17, 20
Vatican, 312
vector method (GIS), 48
Vedism, 243; *see also* Hinduism
vegetation: *see* plants
Venezuela, 324, 336; oil in, 333, 477, 478, 481
Venice, 120, 425
vernacular regions, 37, 218–22; French Louisiana as, 220–1, *222*; North American (Garreau), 219–20, *221*; North American (Zelinsky), 219, *220*; regional identity and, 221–2
verstehen, 27
Vidal, Paul de la Blache, 14, 16, 18, 83
Vietnam, 87–8, 191, 308, 324, 329, *429*
Vikings, 235, 240
violence: gendered, 274, 276, 283, 284, 285; location and, 283, 285; political, 302, 316, 318, 321, 323, 328, 332; riots and, 74–5
volcanic eruptions, 183, 194, 196–7
Voltaire, 264
volunteered geographic information (VGI), 45
von Thünen, Johann Heinrich, 347, 348
von Thünen theory, 348–50, 373–4, 382, 406–7; testing of, 350–4
voting, 325–8

Wahhabism, 247
Wales, 230, 236, 239, 315
Walk Free Foundation, 267
Wallerstein, Immanuel, 175–6
Wallonia, 315
Walmart, 72, 76, 447
war, 328–34; civil, 329–30; conflicts and, 302–3, 328–9; failed states and, 330–1; nuclear, 334; recovery of land/property after, 329–30; resource competition and, 330, 333–4; terrorism and, 331–3; UN and, 328–9; *see also* conflicts
Warsaw Pact countries, 85, 323, 338
waste: electronic, *87*; as energy source, 99; food, 381–2; as nitrogen source, 358–9; as pollutant, 115, 117–8, 482, 503
water, 64, 97, 206; Aboriginal peoples and, 284; Arctic, 123–4; availability of, 116–17; conflict and, 179, 333; disease and, 197, 228; energy from, 100–1, 117, 477; global cycle of, 116–17, *117*; human impacts on, 104, 111, 115, 116–19; industry and, 475, 476, 485, 491; pollution of, 117–19; population and, 157, 159–60; resource extraction and, 115, 481, 482, 485; urban supply/quality of, 449, 450, 455, 459, 460
Watts, Michael, 375
wave/tidal power, 101
Weber, Alfred, 468–71, 472, 475, 482, 493–4, 500
Weber, Max, 468
websites, 518–21
well-being, 281–5; fear of crime and, 282, 283; happiness and, 282–3, *285*; health/health care and, 282; landscapes and, 283–5; measurement of, 281–2; service industries of, 496
West Bank, 309–10
West Edmonton Mall, 287, *289*, 294
Weyerhaeuser, 128
Whittlesey, Derwent, 366
whooping cough, 198
Wildwood Park, MB, 452
Wilson, E.O., 113–15
wind power, 103, *103*
wine, 222, 249, 345, 380
Winnipeg, *41*, 452; site and situation of, 35, *35*, *36*
women: AIDS and, 199, 201; education of, 136, 146, 169; feminist geography and, 271–6; fertility/family planning and, 134–9, 144–6; human geography and, 30–2; MDGs and, 276; in urban environment, 283, 443, 445, 461; violence and, 274, 276, 283, 284, 285; work of, 274–5, 459; *see also* gender
work: labour-intensive, 369, 475, 488, 489, 490, 499; paid/unpaid, 274–5
workforce: aging population and, *150*, 152; deindustrialization and, 494; Fordism and, 501; in outsourcing jobs, 497–8; "reserve," 156; *see also* employment; labour
World Bank, 80, 100, 110, 122, 177, 316, 329; development measurement by, 181–2, 187, 190, 205; on food shortages, 184; on governance, 187, 207; new economic geography and, 205–6; world debt crises and, 186, 203, 205, *314*; on world oil wealth, 478
World Commission on Environment and Development, 105, 127
World Conservation Union, 128
World Database of Happiness, 282
World Development Report, 123–4, 181, 187, 205
World Economic Forum, 87
World Health Organization (WHO), 140, 286
World Heritage sites, 290, *291*, 295–6
"world island" (Europe–Asia), 307
World Islands, Dubai, 441
World Social Forum, 87
world systems theory, 175–6, *177*, 185–6
World Trade Organization (WTO), 67, 76, 80, 81–2, 519; agriculture and, 81–2, 89, 363, 377; criticisms of, 81–2, 85

World Travel and Tourism Council, 289, 520
World Wildlife Fund (WWF), 114, 292
Wright, Henry, 450

youth, subcultures of, 280
Yugoslavia (former), 84, 194, 245, 265, 315–16, 335, 511
Yunus, Muhammad, 204

Zelinsky, Wilbur, 163–4, 216, 219
Zhongdian, China (Shangri-La), 296–7
ziggurat, 401; of Ur, *401*
Zimbabwe, 190–1, 198, 331
Zionism, 264
zoning, land-use, 449, 454; *see also* concentric zone models

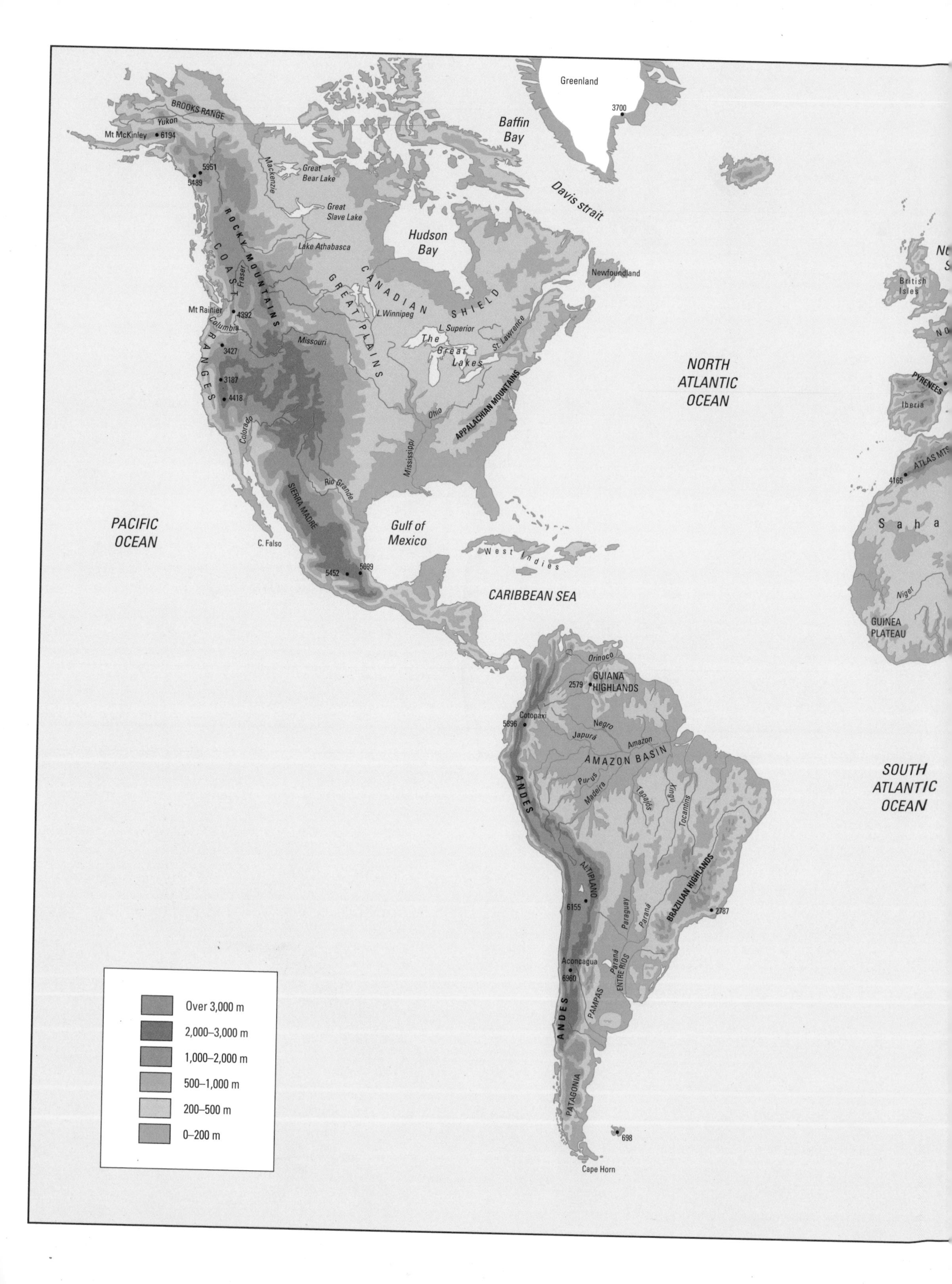

Greenland
3700
Baffin Bay
Davis strait
BROOKS RANGE
Yukon
Mt McKinley 6194
5951
5489
Mackenzie
Great Bear Lake
Great Slave Lake
Hudson Bay
Lake Athabasca
ROCKY MOUNTAINS
COAST RANGES
Fraser
CANADIAN SHIELD
GREAT PLAINS
Newfoundland
L.Winnipeg
Mt Rainier 4392
Columbia
L. Superior
St. Lawrence
The Great Lakes
Missouri
3427
3187
4418
NORTH ATLANTIC OCEAN
APPALACHIAN MOUNTAINS
Ohio
Colorado
Mississippi
Rio Grande
SIERRA MADRE
PACIFIC OCEAN
C. Falso
Gulf of Mexico
West Indies
5452
5699
CARIBBEAN SEA
British Isles
PYRENEES
Iberia
ATLAS MTS
4165
Sahara
Niger
GUINEA PLATEAU
Orinoco
GUIANA HIGHLANDS
2579
Cotopaxi
5896
Negro
Japurá
Amazon
AMAZON BASIN
ANDES
Purus
Madeira
Tapajós
Xingu
Tocantins
SOUTH ATLANTIC OCEAN
ALTIPLANO
6155
BRAZILIAN HIGHLANDS
2787
Paraguay
Paraná
Aconcagua
6960
Paraná
ENTRE RIOS
PAMPAS
PATAGONIA
698
Cape Horn
Over 3,000 m
2,000–3,000 m
1,000–2,000 m
500–1,000 m
200–500 m
0–200 m